G Proteins

G Proteins

EDITED BY

Ravi Iyengar
Department of Pharmacology
Mount Sinai School of Medicine
of the City University of New York
New York, New York

Lutz Birnbaumer
Department of Cell Biology
Baylor College of Medicine
Houston, Texas

ACADEMIC PRESS, INC.
Harcourt Brace Jovanovich, Publishers

San Diego   New York   Berkeley   Boston
London   Sydney   Tokyo   Toronto

This book is printed on acid-free paper. ∞

Copyright © 1990 by Academic Press, Inc.
All Rights Reserved.
No part of this publication may be reproduced or transmitted in any form or
by any means, electronic or mechanical, including photocopy, recording, or
any information storage and retrieval system, without permission in writing
from the publisher.

Academic Press, Inc.
San Diego, California 92101

United Kingdom Edition published by
Academic Press Limited
24–28 Oval Road, London NW1 7DX

Library of Congress Cataloging-in-Publication Data

G proteins / edited by Ravi Iyengar, Lutz Birnbaumer.
        p.   cm.
    Includes index.
    ISBN   0-12-377450-0   (alk. paper)
    1. G proteins.   I. Iyengar, Ravi.   II. Birnbaumer, Lutz.
    [DNLM: 1. Guanine Nucleotide Regulatory Protein—genetics.
    2. Guanine Nucleotide Regulatory Protein—physiology.   QU 55 G1101]
    QP552.G16G2   1990
    574.19'245—dc20
    DNLM/DLC
    for Library of Congress   89-6946
                                                                          CIP

Printed in the United States of America
89   90   91   92   9   8   7   6   5   4   3   2   1

CONTENTS

v

Part II Coupling

Part III Systems Regulated by G Proteins

23. G Proteins in Growth Factor Action
Jacques Pouysségur

24. G Proteins in Yeast *Saccharomyces cerevisiae*
Janet Kurjan

25. GTP-Binding Proteins and Exocytotic Secretion
Bastien D. Gomperts

CONTRIBUTORS

Numbers in parentheses indicate the pages on which the authors' contributions begin.

Jeffrey L. Benovic (295), Departments of Cell Biology, Biochemistry, and Medicine, Howard Hughes Medical Institute, Duke University Medical Center, Durham, North Carolina 27710

Lutz Birnbaumer (1, 267, 317), Department of Cell Biology, Baylor College of Medicine, Houston, Texas 77030

J. Bockaert (81), Centre CNRS-INSERM de Pharmacologie-Endocrinologie, Rue de la Cardonille, 34094 Montpellier, France

Henry R. Bourne (17), Departments of Pharmacology and Medicine, and the Cardiovascular Research Institute, University of California, San Francisco, California 94143-0450

Michel Bouvier (295), Departments of Cell Biology, Biochemistry, and Medicine, Howard Hughes Medical Institute, Duke University Medical Center, Durham, North Carolina 27710

José Luis Boyer (317), Department of Pharmacology, University of North Carolina, Chapel Hill, North Carolina 27514

P. Brabet (81), Centre CNRS-INSERM de Pharmacologie-Endocrinologie, Rue de la Cardonille, 34094 Montpellier, France

Arthur M. Brown (241, 267), Department of Molecular Physiology and Biophysics, Baylor College of Medicine, Houston, Texas 77030

Richard C. Bruch (411), Monell Chemical Senses Center, Philadelphia, Pennsylvania 19104

Marc G. Caron (295), Departments of Cell Biology, Biochemistry, and Medicine, Howard Hughes Medical Institute, Duke University Medical Center, Durham, North Carolina 27710

Donna Carty (267), Department of Pharmacology, Mount Sinai School of Medicine of the City University of New York, New York, New York 10029

Marc Chabre (215), Laboratoire de Biophysique Moléculaire et Cellulaire, (Unité 520 du CNRS), LBIO DRF, Centre d'Etudes Nucléaires, BP 85, F38041, Grenoble, France

David E. Clapham (41), Department of Pharmacology, Physiology and Biophysics, Mayo Foundation, Rochester, Minnesota 55905

Juan Codina (241, 267), Department of Cell Biology, Baylor College of Medicine, Houston, Texas 77030

Hans Deckmyn[1] (429), Division of Hematology–Oncology, Depart-

[1] Current address: Leuvin University, Leuvin, Belgium

ments of Internal Medicine and Biological Chemistry, Washington University School of Medicine, St. Louis, Missouri 63110

Philippe Deterre (215), Laboratoire de Biophysique Moléculaire et Cellulaire, (Unité 520 du CNRS), LBIO DRF, Centre d'Etudes Nucléaires, BP 85, F38041, Grenoble, France

Henrik Dohlman (295), Departments of Cell Biology, Biochemistry, and Medicine, Howard Hughes Medical Institute, Duke University Medical Center, Durham, North Carolina 27710

Kathleen Dunlap (357), Department of Physiology, Tufts University School of Medicine, Boston, Massachusetts 02111

J. Gabrion (81), Laboratoire de Neurobiologie-Endocrinologie Université des Sciences et Techniques du Languedoc, Place Eugène Bataillon, 34000 Montpellier, France

Bastien D. Gomperts (601), Department of Physiology, University College London, London WC1E 6JJ, United Kingdom

T. Kendall Harden (317), Department of Pharmacology, University of North Carolina, Chapel Hill, North Carolina 27514

William P. Hausdorff (295), Departments of Cell Biology, Biochemistry, and Medicine, Howard Hughes Medical Institute, Duke University Medical Center, Durham, North Carolina 27710

Warren Heideman[2] (17), Departments of Pharmacology and Medicine, and the Cardiovascular Research Institute, University of California, San Francisco, California 94143-0450

Jürgen Hescheler (383), Physiologisches Institut der Universität des Saarlandes, D-6650 Homburg/Saar, Federal Republic of Germany

Mark Hnatowich (295), Departments of Cell Biology, Biochemistry, and Medicine, Howard Hughes Medical Institute, Duke University Medical Center, Durham, North Carolina 27710

V. Homburger (81), Centre CNRS-INSERM de Pharmacologie-Endocrinologie, Rue de la Cardonille, 34094 Montpellier, France

Miles D. Houslay (521), Molecular Pharmacology Group, Institute of Biochemistry, University of Glasgow, Glasgow G12 8QQ, Scotland, United Kingdom

Hiroshi Itoh (63), Institute of Medical Science, University of Tokyo, 4-6-1, Shirokanedai, Minatoku, Tokyo 108, Japan

Ravi Iyengar (1, 147, 267), Department of Pharmacology, Mount Sinai School of Medicine of the City University of New York, New York, New York 10029

Richard A. Kahn (201), Building 37, Room 5D-02, Laboratory of Biolog-

---

[2] Current address: School of Pharmacy, University of Wisconsin, Madison, Wisconsin 53706

ical Chemistry, Division of Cancer Treatment, National Cancer Institute, National Institutes of Health, Bethesda, Maryland 20892

Yoshito Kaziro (63), Institute of Medical Science, University of Tokyo, 4-6-1, Shirokanedai, Minatoku, Tokyo 108, Japan

Brian K. Kobilka (295), Departments of Cell Biology, Biochemistry, and Medicine, Howard Hughes Medical Institute, Duke University Medical Center, Durham, North Carolina 27710

Itaru Kojima (503), Cell Biology Research Unit, Fourth Department of Internal Medicine, University of Tokyo School of Medicine, Tokyo 112, Japan

Janet Kurjan (571), Department of Biological Sciences, Columbia University, New York, New York 10027

Emmanuel M. Landau (497), Departments of Psychiatry and Pharmacology, Mount Sinai School of Medicine of the City University of New York, New York, New York 10029, and Department of Psychiatry, Veterans Administration Medical Center, Bronx, New York 10468

Robert J. Lefkowitz (295), Departments of Cell Biology, Biochemistry, and Medicine, Howard Hughes Medical Institute, Duke University Medical Center, Durham, North Carolina 27710

Irene Litosch (453), Department of Pharmacology, University of Miami School of Medicine, Miami, Florida 33101

Philip W. Majerus (429), Division of Hematology–Oncology, Departments of Internal Medicine and Biological Chemistry, Washington University School of Medicine, St. Louis, Missouri 63110

Michael W. Martin (317), Department of Pharmacology, University of North Carolina, Chapel Hill, North Carolina 27514

Rafael Mattera (267), Department of Cell Biology, Baylor College of Medicine, Houston, Texas 77030

John M. May (317), Department of Pharmacology, University of North Carolina, Chapel Hill, North Carolina 27514

Thomas M. Moriarty (479), Department of Psychiatry, Mount Sinai School of Medicine of the City University of New York, New York, New York 10029

Joel Moss (179), Laboratory of Cellular Metabolism, National Heart, Lung, and Blood Institute, National Institutes of Health, Bethesda, Maryland 20892

Masato Nakafuku (63), Institute of Medical Science, University of Tokyo, 4-6-1, Shirokanedai, Minatoku, Tokyo 108, Japan

Eva J. Neer (41), Department of Medicine, Brigham and Women's Hospital, Harvard Medical School, Boston, Massachusetts 02115

Brian F. O'Dowd (295), Departments of Cell Biology, Biochemistry, and Medicine, Howard Hughes Medical Institute, Duke University Medical Center, Durham, North Carolina 27710

Elena Padrell   (267), Department of Pharmacology, Mount Sinai School of Medicine of the City University of New York, New York, New York 10029

Jacques Pouysségur   (555), Centre de Biochimie, CNRS, Université de Nice, Parc Valrose, 06034 Nice, France

Richard T. Premont   (147), Department of Pharmacology, Mount Sinai School of Medicine of the City University of New York, New York, New York 10029

Stanley G. Rane[3]   (357), Department of Physiology, Tufts University School of Medicine, Boston, Massachusetts 02111

Walter Rosenthal   (383), Institut für Pharmakologie der Freie Universität Berlin, D-1000 Berlin 33, Federal Republic of Germany

B. Rouot   (81), Centre CNRS-INSERM de Pharmacologie-Endocrinologie, Rue de la Cardonille, 34094 Montpellier, France

Günter Schultz   (383), Physiologisches Institut der Universität des Saarlandes, D-6650 Homburg/Saar, Federal Republic of Germany

Allen M. Spiegel   (115), Molecular Pathophysiology Branch, National Institute of Diabetes, Digestive, and Kidney Disease, National Institutes of Health, Bethesda, Maryland 20892

M. Toutant   (81), Centre CNRS-INSERM de Pharmacologie-Endocrinologie, Rue de la Cardonille, 34094 Montpellier, France

Wolfgang Trautwein   (383), Physiologisches Institut der Universität des Saarlandes, D-6650 Homburg/Saar, Federal Republic of Germany

Antonius M. J. VanDongen   (267), Department of Molecular Physiology and Biophysics, Baylor College of Medicine, Houston, Texas 77030

Martha Vaughan   (179), Laboratory of Cellular Metabolism, National Heart, Lung, and Blood Institute, National Institutes of Health, Bethesda, Maryland 20892

Brian J. Whiteley   (429), Division of Hematology–Oncology, Departments of Internal Medicine and Biological Chemistry, Washington University School of Medicine, St. Louis, Missouri 63110

Atsuko Yatani   (241, 267), Department of Molecular Physiology and Biophysics, Baylor College of Medicine, Houston, Texas 77030

---

[3] Current address: Department of Biological Sciences, Purdue University, West Lafayette, Indiana 47907

PREFACE

The aim of this book is to provide an introduction to one class of systems used for signal transduction at the cell surface. The defining feature of these systems is the utilization of a heterotrimeric GTP-binding protein (G protein) to mediate the transfer of information across the plasma membrane, from receptor to effector. In the past decade, the application of biochemical and molecular biological techniques has substantially increased our understanding of the mechanisms of transduction. It has become apparent that G proteins play a central role in the transduction process and that the basic format used for transduction in the hormone-stimulated adenylyl cyclase system is widely employed by the cell in other hormone-sensitive systems. Hence we felt it would be useful to have a book focusing on G proteins and their interactions with receptors and effectors.

The book is divided into three sections. The first focuses on the structural aspects of G proteins with substantial emphasis on alpha subunits. The second deals with the mechanism of G protein coupling to effector systems, using the hormone-regulated adenylyl cyclase and light-regulated cGMP phosphodiesterase as models. Receptors and effector systems are also discussed. The third deals with cellular functions which may be regulated by heterotrimeric G proteins. Some of these systems are relatively well characterized, while others are not. However, it is certain that all will be the focus of further study.

As one would expect in a fast-moving area of research, not all of the systems regulated by G proteins are dealt with. Even as this book was in production, our understanding of some systems had moved forward. However, we hope that it will serve as a resource for those working on various aspects of G protein-mediated signal transduction and will provide for others a detailed introduction to this area of research.

Ravi Iyengar
Lutz Birnbaumer

CHAPTER 1

Overview

Ravi Iyengar

Department of Pharmacology, Mount Sinai School of Medicine of the City
University of New York, New York, New York 10029

Lutz Birnbaumer

Department of Cell Biology, Baylor College of Medicine, Houston, Texas 77030

Information transfer at the cell surface is very important in order for the cell to adjust its metabolism and responsiveness. The role of cell surface systems in converting extracellular signals into metabolic changes through intracellular second messengers was first described by Sutherland and co-workers in 1956 (Rall *et al.,* 1956). In 1971, Rodbell and co-workers showed that in conjunction with hormones, GTP regulates the signal transduction pathway that uses the enzyme adenylyl cyclase to convert ATP into the intracellular second messenger cAMP (Rodbell *et al.,* 1971). Nonhydrolyzable analogs of GTP such as Gpp(NH)p were shown to extensively and persistently stimulate adenylyl cyclase (Londos *et al.,* 1974), and hormone receptors were shown to increase this rate of stimulation (Salomon *et al.,* 1975). From these studies, Rodbell inferred two central concepts. The first is that the signal transduction system might be a three-component system consisting of a receptor that specifically binds hormone, a transducer, and an effector, as depicted in Fig. 1. The second concept is that the GTP-binding site might also be a GTPase, since GTP analogs that are enzymatically unhydrolyzable between the β and γ phosphates can extensively stimulate adenylyl cyclase even in the absence of hormones.

Cassel and Selinger, in experiments that were technically very demanding, demonstrated that a hormone-stimulated GTPase indeed exists (Cassel

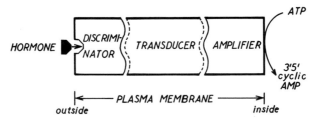

Fig. 1. An early version of the signal-transducing system as proposed by Rodbell and co-workers in 1970 (Rodbell *et al.,* 1971). The discriminator is the hormone receptor that specifically recognizes the appropriate hormone. The existence of several distinct receptors that could couple to adenylyl cyclase had been shown in fat cells (Birnbaumer and Rodbell, 1969). The transducer was at that time a hypothetical component. The amplifier is the catalyst of adenylyl cyclase, which would produce several thousand cAMP molecules for each adenylyl cyclase molecule activated by one receptor molecule.

and Selinger, 1976). They went on to show that GTP hydrolysis serves as the turn-off mechanism for the activation of adenylyl cyclase and that GDP remains bound to the system to hold it in the inactive state. Hormone receptors were thought to activate the system by promoting the release of GDP (Cassel and Selinger, 1978). These observations led to the formulation of the GTP regulatory cycle model (Fig. 2).

Independent of this functional analysis of the hormone-stimulated

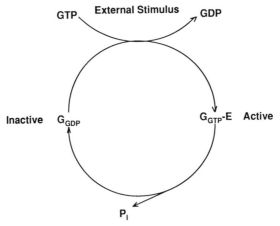

Fig. 2. The GTP/GDP regulatory cycle for a G protein. This scheme is adapted from the original version proposed by Cassel and Selinger (1978) for G_s regulation of adenylyl cyclase. In the basal state, G proteins exist in GDP-liganded, inactive form. In the presence of external stimulus, the GDP is released. The nucleotide-binding site is occupied by GTP, which allows the G protein to be in its active state and thus modulate the activity of the effector (E). The GTP is hydrolyzed at the binding site, resulting in release of P_i, and the GDP-liganded G protein then dissociates from the effector and returns to the basal state.

adenylyl cyclase system, Ross and Gilman (1977) proved Rodbell's three-component proposal by showing that communication between hormone receptors and adenylyl cyclase is disrupted in the cyc^- mutant of the murine lymphoma cell line S49 due to the lack of a GTP-binding protein. They initially named this component G/F, since it is responsible for both guanine nucleotide and fluoride stimulation of adenylyl cyclase. It is now known as G_s, the guanine nucleotide-binding regulatory protein that stimulates adenylyl cyclase. Using reconstitution of the cyc^- adenylyl cyclase as the assay, Gilman and co-workers purified this transducing protein to homogeneity (Northup et al., 1980). G_s was found to be a multimeric protein composed of α, β, and γ subunits (Northup et al., 1980; Hildebrandt et al., 1984).

Studies on the retinal phototransduction system also revealed the involvement of GTP in a regulatory step (Wheeler and Bitensky, 1977). The GTP-binding site was localized to a soluble protein named transducin, which was later determined to be a multimeric protein composed of α, β, and γ subunits (Godchaux and Zimmerman, 1979; Kuhn, 1980; Fung et al., 1981). Studies of transduction in the visual system showed a remarkable similarity to the erythrocyte adenylyl cyclase system (Fung et al., 1981; Cassel and Selinger, 1978). The idea of a close relationship between the two systems was greatly strengthened when Manning and Gilman (1983) showed that the β subunits of transducin and of G_s are very similar. Fairly detailed information about the components of both systems as well as their mechanisms of action is now available, and it appears that the hormone-stimulated adenylyl cyclase and the light-activated cGMP phosphodiesterase systems are very similar in overall organization.

It is now clearly established that G proteins involved in signal transduction are $\alpha\beta\gamma$ heterotrimers. The α subunits have the guanine nucleotide-binding sites, as well as the sites for bacterial toxin-catalyzed ADP-ribosylation. The α subunit is also thought to be the specific effector arm of the G protein, since the GTP-liganded α subunit regulates the activity of the effectors adenylyl cyclase (Northup et al., 1983), cGMP phosphodiesterase (Fung et al., 1981), and K^+ channels (Codina et al., 1987). The β subunits occur in two forms, β_{36} and β_{35}, which are highly homologous although they are the products of separate genes (Gao et al., 1987). Transducin (G_t) is found to contain only the β_{36} subunit, while other G proteins contain a mixture of β_{36} and β_{35}. The γ subunits have not been very well characterized, although immunological and peptide mapping studies have shown that there are at least three distinct γ subunits.

Studies using the bacterial toxin from Bordetella pertussis (pertussis toxin, or islet-activating protein, IAP) have shown that there are several substrates for ADP-ribosylation by the toxin. These proteins in the 39–41 kDa range are the α subunits of G proteins (Katada and Ui, 1983; Bokoch *et al.,* 1983; Codina *et al.,* 1983). One of these G proteins has been implicated in the hormonal inhibition of adenylyl cyclase, and was called G_i. However, cloning and immunological studies have shown that there are three G_i proteins (Jones and Reed, 1987), and at the current time it is unclear whether one or perhaps all three are involved in hormonal inhibition of adenylyl cyclase.

All of the G proteins serve as interfaces between receptors and effectors. The list of these systems is growing, and includes the originally described effector system adenylyl cyclase, the cGMP phosphodiesterase in the retina, K^+ channels, voltage-gated Ca^{2+} channels, the membrane-bound phosphatidylinositol-specific phospholipase C, and possibly the membrane-bound phospholipase A_2. Further, sensory transduction involving vision, olfaction, and taste are all based on G-protein-mediated transduction pathways. As shown in Table I, about 80% of all known hormones and neurotransmitters, as well as neuromodulators and autocrine and paracrine factors, elicit their cellular responses by combining with receptors which are coupled to effector functions through G proteins. A close examination of Table I reveals that though the primary messengers are many, the number of distinct receptors that mediate their action is even larger. Each hormone or neurotransmitter may couple to several distinct subtypes of receptors, which in turn couple to often-distinct effector systems.

It is also now becoming apparent that a single G protein can regulate multiple effector functions. The best studied example of this is G_s, which has been shown to stimulate adenylyl cyclase and activate Ca^{2+} channels (Yatani *et al.,* 1987). These multiple effects have been observed using both the native purified G_s as well as the bacterially expressed α_s subunits, unequivocally demonstrating bifurcation of the signal at the level of the G protein (Mattera *et al.,* 1989).

The mechanism of transduction by G proteins has been best studied in the adenylyl cyclase and cGMP phosphodiesterase systems. This cycle is summarized in Fig. 3. The main reactions can be summarized as follows: (1) The hormone-activated receptor binds the GDP-liganded G protein and promotes dissociation of GDP. This is a Mg^{2+}-dependent reaction, and in the absence of hormone very high concentrations of Mg^{2+} (50–100 mM) are required to promote GDP dissociation. The dissociation of GDP appears to occur without the dissociation of the G protein subunits. (2) The hormone receptor promotes the binding of GTP to the G protein. (3) The GTP-liganded G protein α subunit then dissociates from the $\beta\gamma$ subunits, and the hormone-occupied receptor provides the appropriate stratum for this dissociation process. The GTP-liganded and hence active α subunit (α^*) and the $\beta\gamma$ subunits dissociate from the receptor. Current data do not allow us to conclude which subunit dissociates first; however, it

Table I Receptors Acting on Cells via G Proteins

Type of receptor	Membrane function/ system affected[a]	Effect	Coupling protein involved	Examples of target cells(s)/organs
Neurotransmitters				
Adrenergic				
β_1	AC	Stimulation	G_s	Heart, fat, sympathetic synapse
	Ca^{2+} channel	Stimulation	G_s	Heart, skeletal muscle
β_2	AC	Stimulation	G_s	Liver, lung
α_1	PLC	Stimulation	G_{PLC}	Smooth muscle, liver
	PLA_2	Stimulation	G_{PLA}	FRTL-1 cells
α_{2a}, α_{2b}	AC	Inhibition	G_i	Platelet, fat (human)
	Ca^{2+} channel	Closing	G_o (G_p)?	NG-108, sympathetic presynapse
Dopamine				
D-1	AC	Stimulation	G_s	Caudate nucleus
D-2	AC	Inhibition	G_i	Pituitary lactotrophs
Acetylcholine				
Muscarinic M_1	PLC	Stimulation	G_{PLC}	Pancreatic acinar cell
	K^+ channel (M)	Closing	?	CNS, sympathetic ganglia
Muscarinic M_2	AC	Inhibition	G_s	Heart
	K^+ channel	Opening	G_k (G_i?)	Heart, CNS
	PLC	Stimulation	G_p	Heart, transfected cells
$GABA_B$	Ca^{2+} channel	Closing	G_o (G_p?)	Neuroblastoma N1E
	K^+ channel	Opening	G_i (G_k?)	Sympathetic ganglia
Purinergic P1				
Adenosine A-1 or Ri	AC	Inhibition	G_i	Pituitary, CNS, heart
	K^+ channel	Opening	G_k (G_i?)	Heart
Adenosine A-2 or Ra	AC	Stimulation	G_s	Fat, kidney, CNS
Purinergic P_{2X} and P_{2Y}	PLC (PIP_2)	Stimulation	G_{PLC}	Turkey erythrocytes
	PLC (PC)	Stimulation	G°(?)	Liver

(Table continues)

Table I (*Continued*)

Type of receptor	Membrane function/system affected[a]	Effect	Coupling protein involved	Examples of target cells(s)/organs
Serotonin (5-HT)				
S-1a (5-HT-1a)	AC	Inhibition	G_i	Pyramidal cells
	K^+ channel	Opening	G_K (G_i?)	Pyramidal cells
S-1c (5-HT-1c)	PLC	Stimulation	G_{PLC}	Aplysia
S-2 (5-HT-2)	AC	Stimulation	G_s	Skeletal muscle
Histamine				
H-1	PLC	Stimulation	G_{PLC}	Smooth muscle, macrophages
	PLA_2 (?)			
H-2	AC	Stimulation	G_s	Heart
H-3	AC	Inhibition	G_i	Presynaptic CNS, lung; mast
Peptide hormones				
Pituitary				
Adrenocorticotropin (ACTH)	AC	Stimulation	G_s	Fasciculata, glomerulosa
Opioid (μ,κ,δ)	AC	Inhibition	G_i	NG-108
	Ca^{2+} channel	Closing	G_o (G_p?)	NG-108
Luteinizing hormone (LH)	AC	Stimulation	G_s	Granulosa, luteal, Leydig
Follicle-stimulating hormone (FSH)	AC	Stimulation	G_s	Granulosa
Thyrotropin (TSH)	AC	Stimulation	G_s	Thyroid, FRTL-5
	Phospholipase?	Stimulation	G_p(?)	Thyroid
Melanocyte-stimulating hormone (MSH)	AC	Stimulation	G_s	Melanocytes
Hypothalamic				
Corticotropin-releasing hormone (CRF)	AC	Stimulation	G_s	Corticotroph, hypothalamus
Growth hormone-releasing hormone (GRF)	AC	Stimulation	G_s	Somatotroph
Gonadotropin-releasing hormone (GnRH)	PLA_2	Stimulation	G_{PLA}	Gonadotroph
	PLC	Stimulation	G_{PLC}	Gonadotroph
	Ca^{2+} channel	Opening	G_i-type	GH_3

Thyrotropin-releasing hormone (TRH)	PLC	Stimulation	G_{PLC}	Lactotroph, thyrotroph
	AC	Inhibition	G_i	GH_4C_1
Somatostatin (SST or SRIF)	AC	Inhibition	G_i	Pituitary cells, endocrine, pancreatic
	K^+ channel	Opening	$G\ (G_i?)$	Pituitary cells, endocrine pancreatic
	Ca^{2+} channel	Closing	?	Pituitary cells
Other hormones				
Chorionic gonadotropin	AC	Stimulation	G_s	Granulosa, luteal, Leydig
Glucagon	AC	Stimulation	G_s	Liver, fat, heart
	Ca^{2+} pump	Inhibition	$G_s(?)$	Liver, heart (?)
	PLC	Stimulation	?	Liver
Cholecystokinin (CCK)	PLC	Stimulation	G_{PLC}	Pancreatic acini
Secretin	AC	Stimulation	G_s	Pancreatic duct, fat
Vasoactive intestinal peptide (VIP)	AC	Stimulation	G_s	Pancreatic duct, CNS
	PLC	Stimulation	G_{PLC}	Sensory ganglia, CNS
Vasopressin				
V-1a (vasopressor, glycogenolytic)	PLC	Stimulation	G_{PLC}	Smooth muscle, liver, CNS
	AC	Inhibition	G_i	Liver
V-1b (pituitary)	PLC	Stimulation	G_{PLC}	Pituitary
V-2 (antidiuretic)	AC	Stimulation	G_s	Distal and collecting tubule
Oxytocin	PLC	Stimulation	G_{PLC}	Uterus, CNS
Angiotensin II	PLC	Stimulation	G_{PLC}	Liver, glomerulosa cells
	AC	Inhibition	G_i	Liver, glomerulosa cells
	Ca^{2+} channel	Stimulation	G_i-type	Y1 adrenal cells
Other regulatory factors				
Chemoattractant (fMet-Leu-Phe or fMLP)	PLC	Stimulation	G_{PLC}	Neutrophils
Thrombin	PLC	Stimulation	G_{PLC}	Platelets, fibroblasts
	AC	Inhibition	G_i	Platelets
Bombesin	PLC	Stimulation	G_{PLC}	Fibroblasts
IgE	PLC	Stimulation	G_{PLC}	Mast cells
Bradykinin	PLC	Stimulation	G_{PLC}	Lung, fibroblasts, NG-108
	PLA_2	Stimulation	G_{PLA}	Fibroblasts, endothelial
	K^+ channel	Stimulation	$G_k\ (G_i?)$	NG-108
	AC	Inhibition	G_i	NG-108

(Table continues)

7

Table I (*Continued*)

Type of receptor	Membrane function/ system affected[a]	Effect	Coupling protein involved	Examples of target cells(s)/organs
Neurokinin/Tachykinin				
NK1 (Substance P)	PLC	Stimulation	G_{PLC}	CNS, salivary gland, endothelial
NK2 (Neurokinin A or Substance K)	PLC	Stimulation	G_{PLC}	CNS, sympathetic, smooth muscle
NK3 (Neurokinin B)	PLC (?)	Stimulation	G_{PLC}	CNS, smooth muscle
Tumor necrosis factor (TNF)	?	?	?	Monocytes
Colony-stimulating factor (CSF-1)	?	?	?	Monocytes
Serum calcium ion	PLC	Stimulation	G_{PLC} (PTX)	Parathyroid cell
Prostanoids				
Prostaglandin E_1, E_2	AC	Inhibition	G_i	Fat, kidney
Prostacyclin (PGI_2, PGE_1, PGE_2)	AC	Stimulation	G_s	Luteal cells, endothelial, kidney
Thromboxanes	PLC	Stimulation	G_{PLC}	Platelets
Platelet-activating factor (PAF)	PLC	Stimulation	G_{PLC}	Platelets
Leukotriene D_4, C_4	PLA_2	Stimulation	G_{PLA}	Endothelial cells
Sensory				
Light (Rhodopsins)	cGMP-PDE	Stimulation	$Tr(G_{t-r})$	Retinal rod cells (night)
	cGMP-PDE	Stimulation	$Tc(G_{t-c})$	Retinal cone cells (color)
Olfactory signals	AC	Stimulation	G_{OLF}	Olfactory cilia
	Phospholipases?		G_p?	
Taste signals	AC	Stimulation	G_s	Taste epithelium
	Phospholipases	Stimulation	G_p?	

[a] AC, Adenylyl cyclase; PLC, unless denoted otherwise, phospholipase C with specificity for phosphatidylinositol bisphosphate; PLA_2, phospholipase A_2 (substrate specificity unknown); PIP_2, phosphatidylinositol bisphosphate; PC, phosphatidylcholine.

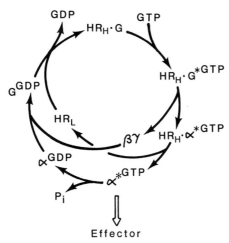

Fig. 3. The regulatory cycle of a G protein in terms of its subunit interactions and ligand binding characteristics through its interactions with a receptor and effector, as it is currently understood. The low-affinity form of the receptor (HR_L) interacts with the GDP-liganded heterotrimeric G protein (G–GDP) to form the high-affinity state of the receptor (HR_H) as a hormone receptor–G protein complex ($HR_H \cdot G$). Occupation of HR_H by hormone activates the receptor to stimulate the release of GDP. GTP then binds to the nucleotide site on the α subunit of the G protein, promoting the release of the G protein from the hormone receptor and the dissociation of the G protein $\beta\gamma$ complex from the α subunit. Though current data suggest the presence of $HR_H \cdot \alpha^*$–GTP, its presence is not certain. The GTP-liganded and active α subunit (α^*–GTP) then modulates the activity of the effector. The GTP is hydrolyzed by the intrinsic GTPase activity to form α–GDP, which has high affinity for the $\beta\gamma$ complex and associates with it to form the heterotrimeric resting state of the G protein (G–GDP). The hormone receptor, upon dissociation of α^*–GTP, reverts to the low affinity state (HR_L) until it encounters another G–GDP.

appears likely that the $\beta\gamma$ complex is released first. (4) The α^*-subunit is then released from the receptor and can activate the effector. The lifetime of the activated state of the effector depends on the lifetime of the activated α^* subunit. This is in turn determined by the K_{cat} of the GTPase activity of the α subunit. This terminal phosphatase activity is slow (4/min) and does not appear to be regulated by the association of the α^* subunit with hormone receptor, $\beta\gamma$, or effector. However, GDP dissociates at only one-tenth this rate (ca. 0.4/min), and the $\beta\gamma$ complex greatly stabilizes the GDP-liganded state of the α subunit, such that the heterotrimer exists as a stable, inactive protein–GDP complex until it encounters a hormone-occupied receptor complex.

In the signal-transducing G proteins, the GTPase serves as a clock that

Table II GTP-Binding Proteins

Proteins		Characteristics	
Type A	Type B[a]	Type A[a]	Type B[a]
ras gene products	G_s, 4 splice products of 1 gene	20–25 kDa	85–90 kDa
rho gene products	G_t, 2 distinct gene products	Monomeric	Heterotrimeric ($\alpha\beta\gamma$)
YPT gene products	G_{i1}, G_{i2}, G_{i3}, 3 distinct gene products	GTPase activity	GTPase activity
sec gene product	G_o, 2 proteins, 1 (?) gene	ADP-ribosylation by BTXc,d	ADP-ribosylation by CTX and PTX
smg-25 gene products	G_{43}, 43-kDa PTX substrate, G_{13}-like	PTX site absent or modified	All have CTX site (only G_s is substrate)
	G_z, 1 gene product	Limited homology with α subunits of heterotrimeric G proteins (and other GTP-binding proteins)	G_s and G_z lack PTX site

[a] PTX, pertussis toxin; CTX, cholera toxin; BTX, botulinum toxin.

allows the system to cycle at a fixed rate. It is tempting to speculate that there are cellular factors that can stimulate or inhibit the GTPase by altering its catalytic rate, either damping or amplifying the hormonal signal. As yet, no such factors have been identified for the heterotrimeric G proteins. However, a GTPase-activating protein (GAP) for the products of the *ras* gene family has been identified (Traney and McCormick, 1987), purified (Gibbs *et al.*, 1988), and cloned (Vogel *et al.*, 1988).

The G proteins that couple receptors to effectors are one class of GTP-binding regulatory proteins. Members of this class of GTP-binding protein have certain identifiable features: (1) They are heterotrimeric. (2) Their α subunits, which contain the GTP-binding site, are 38–52 kDa in size. (3) Functions regulated by these proteins are stimulated by nonhydrolyzable analogs of GTP. (4) Addition of excess $\beta\gamma$ subunits from these G proteins generally inhibits communication between the receptor and effector.

There is a second class of GTP-binding regulatory proteins which appear monomeric in nature. As with the heterotrimeric G proteins, there are several families within the class of monomeric GTP-binding proteins. These GTP-binding proteins include those encoded by the H-, K-, and N-*ras, rho, ral,* YPT, and SEC4 genes (Capon *et al.*, 1983; Madaule and Axel, 1985; Schmidtt *et al.*, 1986). Several of the proteins have been purified from mammalian sources, including the *rho* gene product (Yamamoto *et al.*, 1988). cDNAs have been isolated for a family of 25-kDa GTP-binding proteins (Matsui *et al.*, 1988). These 25-kDA proteins, called smg-25A, -B, and -C, share 77–85% homology. In contrast, there is only 28% homology between the smg-25 protein family and the *ras* protein family. The *ras* proteins are proto-oncogenes and are involved in an as yet unidentified manner with cell transformation. The yeast YPT1 gene product has been shown to be localized in the Golgi apparatus and to be involved in protein trafficking through the Golgi (Segev *et al.*, 1988). The monomeric class of GTP-binding proteins also has its identifying characteristics. These include (1) effector functions which are inhibited by nonhydrolyzable analogs of GTP, and (2) a lack of measurable interaction with the $\beta\gamma$ subunits of heterotrimeric G proteins. A reasonable hypothesis explaining the differences in the responsiveness of the two classes of GTP-binding proteins to nonhydrolyzable GTP analogs has recently been published (Bourne, 1988). Table II lists the GTP-binding proteins belonging to these two classes.

This book focuses on the heterotrimeric variety of GTP-binding regulatory proteins known as G proteins. The one exception is the GTP-binding protein ARF (ADP-ribosylating factor), which was identified as an essential cofactor for the cholera toxin-mediated ADP-ribosylation of G_s.

REFERENCES

Birnbaumer, L., and Rodbell, M. (1969). Adenyl cyclase in fat cells. II. Hormone receptors. *J. Biol. Chem.* **244,** 3477–3482.

Bokoch, G. M., Katada, T., Northup, J. K., Hewlett, E. L. and Gilman, A. G. (1983). Identification of the predominant substrate for ADP-ribosylation by islet activating protein. *J. Biol. Chem.* **258,** 2071–2075.

Bourne, H. R. (1988). Do GTPases direct membrane traffic in secretion? *Cell* **53,** 669–671.

Capon, D. J., Chen, E. Y., Levinson, A. D., Seeburg, P. H. and Goeddel, D. V. (1983). Complete nucleotide sequence of the T24 human bladder carcinoma oncogene and its normal homologue. *Nature* **302,** 33–37.

Cassel, D., and Selinger, Z. (1976). Catecholamine stimulated GTPase activity in turkey erythrocyte membranes. *Biochim. Biophys. Acta* **452,** 538–551.

Cassel, D., and Selinger, Z. (1978). Mechanism of adenylate cyclase activation through beta-adrenergic receptors: Catecholamine induced displacement of bound GDP. *Proc. Natl. Acad. Sci. U.S.A.* **75,** 2669–2673.

Codina, J., Hildebrandt, J. D., Iyengar, R., Birnbaumer, L., Sekura, R. D., and Manclark, C. R. (1983). Pertussis toxin substrate, the putative N_i of adenylyl cyclases, is an alpha/beta heterodimer regulated by guanine nucleotide and magnesium. *Proc. Natl. Acad. Sci. U.S.A.* **80,** 4276–4280.

Codina, J., Yatani, A., Grenet, D., Brown, A. M., and Birnbaumer, L. (1987). The alpha-subunit of the GTP-binding protein G_k opens atrial potassium channels. *Science* **236,** 442–445.

Fung, B. K. K., Hurley, J. B., and Stryer, L. (1981). Flow of information in light triggered cyclic nucleotide cascade of vision. *Proc. Natl. Acad. Sci. U.S.A.* **78,** 152–156.

Gao, B., Gilman, A. G., and Robishaw, J. D. (1987). A second form of the beta-subunit of signal transducing G-proteins. *Proc. Natl. Acad. Sci. U.S.A.* **84,** 6122–6125.

Gibbs, J. B., Schaber, M. D., Allard, W. J., Sigal, I. S., and Scolnick, E. M. (1988). Purification of *ras* GTPase activating protein from bovine brain. *Proc. Natl. Acad. Sci. U.S.A.* **85,** 5026–5030.

Godchaux, W., III, and Zimmerman, W. F. (1979). Membrane-dependent guanine nucleotide binding and GTPase activating of soluble protein from bovine rod cell outer segments. *J. Biol. Chem.* **254,** 7874–7884.

Hildebrandt, J. D., Codina, J., Risinger, R., and Birnbaumer, L. (1984). Identification of a gamma-subunit associated with the adenylyl cyclase regulatory proteins N_s and N_i. *J. Biol. Chem.* **259,** 2039–2042.

Jones, D. T., and Reed, R. R. (1987). Molecular cloning of five GTP binding protein cDNA species from rat olfactory neuroepithelium. *J. Biol. Chem.* **262,** 14241–14249.

Katada, T., and Ui, M. (1983). Direct modification of the membrane adenylate cyclase system by islet activating protein due to ADP-ribosylation of a membrane protein. *Proc. Natl. Acad. Sci. U.S.A.* **79,** 3129–3133.

Kuhn, H. (1980). Light and GTP-regulated interaction of GTPase and other proteins with bovine photoreceptor membranes. *Nature* **283**, 587–589.

Londos, C., Salomon, Y., Lin, M. C., Harwood, J. P., Schramm, M., Wolff, J., and Rodbell, M. (1974). 5'-Guanylylimidodiphosphate. A potent activator of adenylate cyclase in eukaryotic systems. *Proc. Natl. Acad. Sci. U.S.A.* **71**, 3087–3090.

Madaule, P., and Axel, R. (1985). A novel *ras*-related gene family. *Cell* **41**, 31–40.

Manning, D. R., and Gilman, A. G. (1983). The regulatory components of adenylate cyclase and transducin. A family of structurally homologous guanine nucleotide-binding proteins. *J. Biol. Chem.* **258**, 7059–7063.

Matsui, Y., Kikuchi, A., Kondo, J., Hishida, T., Teranishi, Y., and Takai, Y. (1988). Nucleotide and deduced amino acid sequences of a GTP-binding protein family with molecular weights of 25,000 from bovine brain. *J. Biol. Chem.* **263**, 11071–11074.

Mattera, R., Graziano, M., Yatani, A., Birnbaumer, L., Gilman, A. G., and Brown, A. M. (1989). The recombinant alpha$_s$ subunits open Ca^{2+} channels. *Science* **243**, 804–807.

Northup, J. K., Sternweis, P. C., Smigel, M. D., Schleifer, L. S., Ross, E. M., and Gilman, A. G. (1980). Purification of the regulatory component of adenylate cyclase. *Proc. Natl. Acad. Sci. U.S.A.* **77**, 6516–6520.

Northup, J. K., Smigel, M. D., Sternweis, P. C., and Gilman, A. G. (1983). The subunits of the stimulatory regulatory component of adenylate cyclase. Resolution of the activated 45,000 Dalton (alpha) subunit. *J. Biol. Chem.* **258**, 11369–11376.

Rall, T. W., Sutherland, E. W., and Wosilait, W. D. (1956). The relationship of epinephrine and glucagon to liver phosphorylase. III. Reactivation of liver phosphorylase in slices and in extracts. *J. Biol. Chem.* **218**, 483–495.

Rodbell, M., Birnbaumer, L., Pohl, S. L., and Krans, H. M. J. (1971). The glucagon-sensitive adenyl cyclase system in plasma membranes of rat liver. V. An obligatory role of guanyl nucleotides in glucagon action. *J. Biol. Chem.* **246**, 1877–1882.

Ross, E. M., and Gilman, A. G. (1977). Resolution of some components of adenylate cyclase necessary for catalytic activity. *J. Biol. Chem.* **252**, 6966–6969.

Salomon, Y., Lin, M. C., Londos, C., Rendell, B., and Rodbell, M. (1975). The hepatic adenylate cyclase. I. Evidence for transition states and structural requirements for guanine nucleotide activation. *J. Biol. Chem.* **250**, 4239–4245.

Schmidtt, H. D., Wagner, P., Pfaff, E., and Gallwitz, D. (1986). The *ras* related YPT1 gene product in yeast: a GTP-binding protein that might be involved in microtubule organization. *Cell* **47**, 401–412.

Segev, N., Mulholland, J., and Botstein, D. (1988). The yeast GTP binding YPT1 protein and a mammalian counterpart are associated with the secretion machinery. *Cell* **52**, 915–924.

Trahey, M., and McCormick, F. (1987). A cytoplasmic protein stimulated N-*ras* p21 GTPase but does not affect oncogenic mutants. *Science* **238**, 542–545.

Vogel, U. S., Dixon, R. A. F., Schaber, M. D., Diehl, R. E., Marshall, M. S.,

Scolnick, E. M., Sigal, I. S., and Gibbs, J. B. (1988). Cloning of bovine GAP and its interaction with oncogenic *ras* p21. *Nature* **335,** 90–93.

Wheeler, G. L., and Bitensky, M. W. (1977). A light activated GTPase in vertebrate photoreceptors: Regulation of light activated cyclic GMP phosphodiesterase. *Proc. Natl. Acad. Sci. U.S.A.* **74,** 4238–4242.

Yamamoto, K., Kondo, J., Hishida, T., Teranishi, Y., and Takai, Y. (1988). Purification and characterization of a GTP-binding protein with a molecular weight of 20,000 in bovine brain membranes. Identification of the *rho* gene product. *J. Biol. Chem.* **263,** 9926–9932.

Yatani, A., Codina, J., Imoto, Y., Reeves, J. P., Birnbaumer, L., and Brown, A. M. (1987). A G-protein directly regulates mammalian cardiac calcium channels. *Science* **238,** 1288–1292.

PART I

Structural Aspects

CHAPTER 2

Structure and Function of G-Protein α Chains

Warren Heideman*, Henry R. Bourne

Departments of Pharmacology and Medicine, and the Cardiovascular Research
Institute, University of California, San Francisco, California 94143-0450

I. INTRODUCTION

Knowledge of the structure and function of G proteins is growing rapidly
(for general reviews of G proteins, see Stryer and Bourne, 1986; Gilman,
1987). Molecular cloning and biochemical experiments have added new
members to the G-protein family and have established similarities between

* Current address: School of Pharmacy, University of Wisconsin, Madison, Wisconsin
53706

G proteins and other GTP-binding proteins. The evidence indicates that all G proteins are driven by a single kind of molecular machine, which is conserved within and beyond the G-protein family. The present review will summarize current understanding of the structure and function of this molecular machine.

Complete understanding of G-protein function will eventually identify the structural features of these proteins that allow them to transduce chemical signals by interacting with receptors, which detect sensory or hormonal stimuli outside the cell, and effectors, which regulate accumulation of intracellular second messengers. The G protein responds to the receptor by releasing a molecule of bound GDP, which is rapidly replaced by GTP. Thus loaded with GTP, the G protein alters its character; in the active state the G protein no longer interacts with the receptor, but proceeds to regulate activity of the effector. The G protein then returns to its original inactive state by virtue of an intrinsic GTPase activity, which converts bound GTP to GDP. G proteins exist as a complex of three subunits: α, β, and γ. Only two distinct β chains and a similar number of γ chains (perhaps three) have been identified in mammalian cells. Although the $\beta\gamma$ complex appears to play a role in anchoring G proteins to the plasma membrane, additional functions of the β and γ polypeptides are not yet clear. The number of different α chains is much larger — at least seven are known, and the number is still increasing. The α chains contain the sites for binding and hydrolysis of GTP, for ADP-ribosylation by cholera and pertussis toxins, and for specific recognition of receptors and effector molecules.

This review will begin with a brief overview of our current understanding of the biochemical mechanisms by which G proteins transduce signals from receptor to effector elements. This overview will be followed by a comparison of the function of a well-studied G protein with that of another GTP-binding protein, the bacterial elongation factor Tu (EF-Tu). Functional differences between the two proteins emphasize several issues that are also addressed and focuses on our knowledge of the relation between structure and function of G-protein α chains. Finally, we conclude by posing questions for future research.

A. Function

What do G proteins offer that they should have become so pervasive in biological signaling systems? G proteins provide both directionality and amplification. Directionality is provided by the GTPase. The essentially irreversible hydrolysis of GTP prevents the signaling cycle of the G protein from running backward between effector and receptor. This contrasts with allosteric regulation, in which inactivation is simply the reverse of activation. Because the GTP that activates a G protein is hydrolyzed (rather than simply released from the binding site), reactivation of the system requires replacement of GDP by GTP, i.e., reactivation requires that the G protein return to the receptor. This means that, in contrast to allosteric regulation, the rate of inactivation of a G protein is set by the GTPase rather than by dissociation of the activating ligand. Constraining the reaction to a single direction has other important biochemical consequences, as described below.

Amplification is provided by two mechanisms: First, activation of a single receptor molecule can lead to the activation of a much larger number of G proteins. This mechanism of amplification is especially prominent in retinal phototransduction, where a single photoactivated rhodopsin can activate 500 molecules of transducin (Fung and Stryer, 1980). The second amplification mechanism can operate even if the receptor activates only a single G protein: The effector protein regulated by a G protein may produce a great many second-messenger molecules before the G protein is inactivated by GTP hydrolysis. This period of activation may extend well beyond the duration of the ligand–receptor interaction that initiated it. This mechanism is likely to mediate amplification in the transduction of hormonal signals across membranes, where the hormone may occupy the receptor for a few milliseconds, whereas the resulting activation of a G protein and its effector enzyme may last several seconds.

Figure 1 illustrates the prevailing view of the guanine nucleotide-regulated protein–protein interactions that take place in transmembrane signaling mediated by G proteins. In order to understand current ideas that relate G-protein structure to function, it will be necessary to review this scheme in some detail. The G protein itself cycles between two principal conformations: an inactive $\alpha\beta\gamma$ form, in which GDP is bound to the α subunit, and an active form, in which the activated α subunit binds GTP and is dissociated from $\beta\gamma$. As shown in the lower part of Fig. 1, the active α–GTP complex binds with high affinity to the effector, converting it from an inactive form (E) to an active form (E*, complexed to α–GTP) that catalyzes generation of the second messenger (X → Y). The intrinsic GTPase activity of α terminates activation of the effector, which dissociates from the inactive α–GDP complex. Reactivation of the α subunit is more complicated. As depicted in the upper part of Fig. 1, reactivation requires both $\beta\gamma$ and the activated receptor (signal detector, D*). According to this scheme, α–GDP binds with high affinity to $\beta\gamma$, and the $\alpha\beta\gamma$ heterotrimer in turn binds to D*. The interaction of $\alpha\beta\gamma$ with D* catalyzes replacement of GDP by GTP because it accelerates dissociation of GDP from the guanine nucleotide-binding site. D* and guanine nucleotides

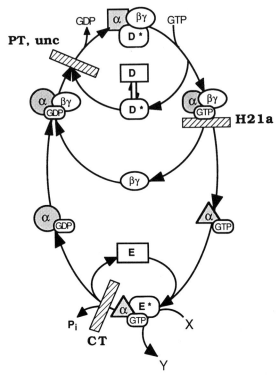

Fig. 1.   The GTP-driven cycle of G-protein subunits in transmembrane signaling. See text for details. α and $\beta\gamma$ represent the α subunit and the $\beta\gamma$ complex of the G-protein hetero-trimer; different conformational states of α are depicted by symbols of different shape (circle, square, triangle, etc.). D and D* represent the inactive and active (stimulated) form of the detector/receptor element; similarly, E and E* represent the inactive and active forms of the effector element (e.g., adenylyl cyclase, retinal cGMP phosphodiesterase, etc.), which produces the intracellular second messenger (Y; e.g., cAMP) from a precursor, X. Cross-hatched rectangles indicate sites at which the cycle is interrupted by mutations (*unc* and *H21a*) or bacterial toxins (CT, cholera toxin; PT, pertussis toxin).

heterotropically displace one another from binding to $\alpha\beta\gamma$; consequently, entry of a GTP molecule into the nucleotide-binding site causes $\alpha\beta\gamma$ to dissociate from D*. The resulting $\alpha\beta\gamma$–GTP complex is apparently tran-sient, presumably because the low affinity of α–GTP for $\beta\gamma$ causes its rapid dissociation.

At the heart of G-protein function, then, is a GTP-sensitive mechanism that cycles α subunits between an inactive conformation, which binds GDP and $\beta\gamma$, and an active conformation, which dissociates from $\beta\gamma$ and binds to the effector. In order to combine information about G-protein structure and function into a testable model of this GTP-driven mechanism, we will eventually need to answer many difficult questions: Which parts of the α subunits interact with $\beta\gamma$? With receptors? With effectors? Which regions make distinct G proteins able to transduce specific signals? Are these regions distinct, or do they overlap? Are these regions similarly located in α chains of different G proteins? How does the activated receptor open the guanine nucleotide-binding site and accelerate dissociation of GDP? Conversely, how do guanine nucleotides reduce affinity of $\alpha\beta\gamma$ for receptor? How do the GTP-and GDP-bound conformations of a G protein differ? How does the α subunit discriminate between GTP and GDP? How does binding of GTP decrease the affinity of the α subunit for binding $\beta\gamma$ and increase its affinity for effector?

Satisfying answers to these questions will come from a combination of experimental approaches, utilizing the tools of classical biochemistry, immunologic probes, and X-ray crystallography, as well as manipulations of protein structure and expression via alterations in nucleic acid sequence. As this chapter will show, these techniques are already beginning to provide strong hints regarding the relation of structure and function in G protein-mediated signaling.

B. Transducin versus EF-Tu

Before focusing on molecular structure it may be instructive to take a closer look at the different functions of the common GTP-dependent cycle shared by two members of the superfamily of GTP-binding proteins. We will compare the functions of transducin, the G protein that mediates retinal phototransduction, and EF-Tu, a bacterial protein involved in ribosomal protein synthesis. Both EF-Tu and transducin bind and hydrolyze GTP in the course of a regulatory cycle (for reviews, see Stryer, 1986; Kaziro, 1978). Both proteins require an accessory protein to catalyze guanine-nucleotide exchange — photoexcited rhodopsin in the case of transducin and EF-Ts for EF-Tu. Binding of GTP to each of these proteins causes dissociation from the protein that catalyzes guanine-nucleotide exchange and promotes binding to different molecular target proteins; these are the effector enzyme, cGMP phosphodiesterase, in the case of transducin, and aminoacyl-tRNA and the ribosome in the case of EF-Tu. The two proteins share key functional properties: (1) A guanine nucleotide-regulated change in conformation radically changes the binding specificity of the guanine nucleotide-binding protein; (2) receptor-catalyzed guanine-nucleotide exchange regulates the rate of activation; (3) GTP hydrolysis serves to turn the protein off in a fashion that cannot operate in reverse, with the result that the cycle must always begin with binding to the receptor or Ts element.

Despite these similarities, the regulatory cycles of the two GTP-binding proteins differ strikingly. Each emphasizes different aspects of the same basic mechanism.

1. Different Roles for GTP Hydrolysis

Transducin regulates a signaling system that is characterized by extraordinary sensitivity and speed, very high gain, and very low noise. A key feature is the strict requirement for an activated receptor to catalyze guanine-nucleotide exchange. The very tight binding of GDP to inactive transducin makes GTP-dependent activation of transducin a rare event in the dark. This property is essential for vision, in order to prevent the retina from "seeing" spurious photons. This point of view also makes clear one way in which GTP hydrolysis may be superior to simple dissociation of GTP as a means of turning off a transducin molecule that has been activated by photolyzed rhodopsin: Hydrolysis of GTP, combined with subsequent tight binding of GDP, makes it impossible for the protein to be reactivated by simple reversal of the inactivation process.

The chain elongation process in protein synthesis does not require so effective a barrier against noise. Instead, it is important that the process moves with maximum efficiency in a single direction. The principal function of EF-Tu is to assure that only aminoacylated tRNA binds to the ribosome and that the EF-Tu leaves the aminoacylated tRNA behind when it leaves. Consequently, precise regulation of guanine-nucleotide exchange is not so important. Indeed, although EF-Ts rapidly accelerates dissociation of GDP from EF-Tu, the dissociation rate in the absence of EF-Ts is much faster than the rate of GDP dissociation from transducin in the dark. If EF-Tu by itself were able to exchange nucleotides rapidly, protein synthesis might still be possible, although inefficient; in contrast, uncontrolled binding of GTP to transducin would make vision impossible.

With EF-Tu it is the direction of the cycle that counts, i.e., that the GTP form binds to the aminoacyl-tRNA and to the ribosome, and that the GDP form releases both. Note that the hydrolysis of GDP, triggered by ribosome binding, plays an indispensable role in making the cycle unidirectional. GTP-dependent transfer of aminoacyl-tRNA to the ribosome could be accomplished, although less efficiently, by a system in which binding to the ribosome simply triggers release of GTP from EF-Tu. Such a process would be readily reversible, however; if the GTP can be released while EF-Tu is bound to the ribosome, then it can go back on as well. This would allow the process to go equally well in either direction. GTP could bind to an empty EF-Tu on the ribosome and allow the EF-Tu–aminoacyl-tRNA complex to leave. Indeed, the presence of a hydrolysis-resistant GTP analog does make the system lose its direction—EF-Tu repeatedly binds to and unbinds from the ribosome while still carrying its aminoacyl-tRNA (Kaziro, 1978).

The GTPase reaction is thus the key element in the function of both proteins. This reaction requires the protein to shuttle constantly back and forth between the receptor (or EF-Ts) and the effector (or ribosome). With transducin, the cycle ensures tight receptor control of activation. With EF-Tu the cycle serves not so much to regulate the amount of active EF-Tu as to assure that EF-Tu leaves the ribosome before rebinding GTP, and therefore that it has a chance to bind aminoacyl-tRNA before heading back.

The energy barrier that prevents release of GDP (and the resulting activation that follows binding of GTP) also prevents dissociation of GTP; in other words, the same barrier that limits activation also blocks one mechanism of inactivation. The tight binding of guanine nucleotides to G-protein α chains makes necessary specific mechanisms for guanine-nucleotide exchange. Activation occurs when the receptor element catalyzes replacement of GDP by GTP. In the case of inactivation the energy barrier is breached by a "reverse" exchange reaction (GTP replaced by GDP) that is performed simply by hydrolyzing GTP. In this way the tight binding of guanine nucleotides, regulated by receptor-catalyzed exchange and an intrinsic GTPase activity, serves as an effective strategy for maintaining sensitivity and reducing signal noise.

2. Roles of $\beta\gamma$ and EF-Ts

The exchange of guanine nucleotide bound to EF-Tu requires only EF-Ts. To accomplish the same feat, however, a G-protein α chain requires at least three accessory polypeptides—receptor, β, and γ—in addition to the sensory or hormonal signal that regulates the receptor. Because receptor proteins are required to catalyze exchange of guanine nucleotides in the G proteins, it has been supposed that receptors are analogs of EF-Ts (or its dimeric counterpart in mammals, EF-1$\beta\gamma$). This analogy ignores the fact that the G-protein $\beta\gamma$ complex is also necessary for the exchange reaction. Indeed, neither receptor nor $\beta\gamma$ alone suffices to catalyze guanine-nucleotide exchange. It is therefore possible that $\beta\gamma$, rather than the receptor, is the analog of EF-Ts and functions to exchange the nucleotide. From this alternative point of view, the role of the receptor would be to regulate the function of $\beta\gamma$. (EF-Tu is apparently unregulated, so that EF-Ts has no need for a receptor.)

Although little experimental evidence is available to support or refute this alternative interpretation, it may serve to help in designing future experiments to probe the function of $\beta\gamma$. Addition of $\beta\gamma$ to the α subunit of G_i or G_o does accelerate guanine-nucleotide exchange in the absence of receptor (Higashijima *et al.*, 1987c). Because the effect is small compared to that observed with $\beta\gamma$ plus receptor, it is not clear whether it represents the basal (unstimulated) activity of $\beta\gamma$ or a nonspecific effect unrelated to normal guanine-nucleotide exchange. It may be worthwhile to compare primary structures of the β and γ chains of EF-1 to those of their potential counterparts in the G proteins.

II. STRUCTURE OF G-PROTEIN α CHAINS

In their GDP-bound forms, G proteins appear to exist as heterotrimers consisting of α, β, and γ subunits in a 1:1:1 ratio. How the three subunits are arranged in the complex is not known. Although still rudimentary, our present understanding of the structure and function of G-protein α chains greatly exceeds understanding of β and γ. Molecular cloning has revealed the primary structures of one γ and two β polypeptides; further experimentation will be necessary to relate these structures to the two known functions of $\beta\gamma$, attachment of the G protein to membrane (Sternweis, 1986), and presentation of the α chain to the receptor (Fung, 1983; Stryer and Bourne, 1986; Gilman, 1987). In the rest of this chapter we will focus on the structure and function of G-protein α chains.

A. Primary Structure

cDNA sequences predict the primary structures of seven G-protein α chains, including those of G_s, G_o, three proteins currently classified as G_i (or G_i-like), and the two distinct transducins of retinal rod and cone cells. Comparison of these amino acid sequences, aligned with a hypothetical "average" α chain, α_{avg} (Masters *et al.*, 1986), reveals a consistent pattern of conserved and divergent sequences (Fig. 2, *top*). Halliday (1984) identified four stretches of primary sequence that are conserved among the GTP-binding proteins of ribosomal protein synthesis and the 21,000-D products (p21^{ras}) of the *ras* oncogenes. The same stretches of sequence (designated A, C, E, and G, according to Halliday's nomenclature) are also conserved in the α chains of the G proteins (Fig. 2, *bottom*). Halliday's prediction that these conserved sequences would play important roles in the structure of the guanine nucleotide-binding site has been confirmed by the crystal structures of EF-Tu and p21^{ras}, as described below.

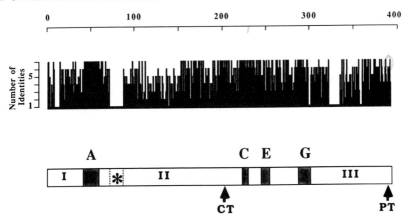

Fig. 2. Conservation of primary structure and domains of G-protein α chains. The graph at the top plots amino acid residue number in a composite α chain, α_{avg} (Masters *et al.*, 1986) versus the number of identical residues at each position among the seven α chains whose amino acid sequences are known (α_s, α_{i1}, α_{i2}, α_{i3}, α_o, and the α chains of transducin in rod and cone cells. The bar at the bottom shows, on the same axis, the sites ADP-ribosylated by cholera (CT) and pertussis (PT) toxins, the location of the peptide sequences inserted by alternative splicing of the α_s gene (*), the locations of four regions that contribute to the GDP-binding domain [designated A, C, E, and G, as first defined by Halliday (1984)], and three additional presumed domains (I, II, III) of the α chains.

The pattern of conservation and divergence is in itself conserved; sequences that are conserved between two α chains also tend to be conserved among the rest of the α chain family. These conserved regions probably perform similar or identical functions in the different G-protein α chains. Conversely, it is tempting to imagine that divergent sequences reflect specific and distinct functions of individual G proteins, such as specific interactions with different sets of receptors and effectors. These attractive generalizations are being tested. The functional significance of variation in primary structure is certainly not yet clear in one very prominent case: the four different α_s polypeptides that are produced by alternative splicing of the α_s gene (Bray *et al.*, 1986). Two of these polypeptides behave in polyacrylamide gels like proteins of about 45 kDa, and two like 52-kDa polypeptides. They differ over a stretch of 14 or 15 residues, located in the amino-terminal half of the polypeptide (indicated by an asterisk in the bottom part of Fig. 2). Whether differences in function result from these structural differences has not yet been tested.

B. Guanine Nucleotide Binding and GTPase Activity

Although no G protein has yet been crystallized, the crystal structure of EF-Tu (Jurnak, 1985; La Cour *et al.*, 1985) proved useful in constructing a model for a putative GDP-binding region of α_{avg} (Masters *et al.*, 1986); this model was based upon regions of conserved sequence and predicted secondary structure. As Halliday (1984) predicted, all four of the regions conserved among GTP-binding proteins turned out to form parts of the GDP binding site in EF-Tu. Similarly, the recently solved crystal structure of p21^{ras} (de Vos *et al.*, 1988) reveals a GDP-binding site made up of these conserved regions. Figure 3 depicts a schematic version of the predicted GDP-binding region of α_{avg}, based upon the EF-Tu structure; residue numbers refer to positions in α_{avg} (Masters *et al.*, 1986). Our knowledge and predictions regarding the structure and function of specific regions of the GDP-binding site come not only from the two crystal structures, but also from the phenotypes produced by a large number of natural and

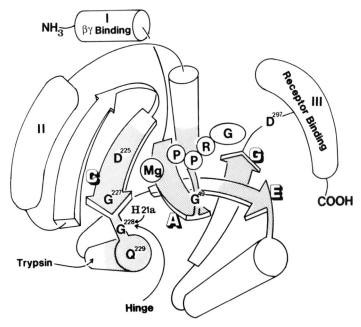

Fig. 3. Diagram of the predicted GDP-binding domain of α_{avg} (Masters *et al.*, 1986), based on the crystal structure of the corresponding domain of EF-Tu (Jurnak, 1985). The GDP molecule (G, guanine ring; R, ribose; P, phosphoryl) nestles into a binding site bounded by turns between β strands and α helices; these turns are located in the stippled regions, designated A, C, E, and G, which correspond to the four stretches of similar amino-acid sequence first noted by Halliday (1984), and which correspond to the same regions shown in Fig. 2. Key amino-acid residues, numbered in accord with α_{avg}, are indicated by single letters (D, Asp; G, Gly; Q, Gln). The predicted hinge region where flexion is thought to be blocked by the H21a mutation at Gly^{228} is indicated by an arrow, as is the site where tryptic cleavage is prevented by activation of the α chain. Domains I, II, and III correspond to Fig. 2.

man-made mutations in corresponding regions of p21^{ras} and a smaller number of mutations in the α chain of G$_s$.

Halliday region A constitutes a loop, located near the phosphoryl groups of GDP, that is thought to play a role in regulating (and perhaps in catalyzing) hydrolysis of bound GTP. In p21^{ras}, mutational replacement of Gly^{12} (corresponding to Gly^{49} in α_{avg}) by almost any other amino acid (Seeburg et al., 1984) produces a protein with enhanced capacity to cause malignant transformation of fibroblasts. The same mutations also decrease the ability of the ras protein to hydrolyze bound GTP (Barbacid, 1987); the inhibitory effect of the mutations is much more prominent when the GTPase is stimulated by a cytoplasmic GTPase-activating protein (GAP) (Trahey and McCormick, 1987). Will mutational replacement of the corresponding glycine residue by valine in α_s produce a G$_s$ protein with reduced GTPase activity and markedly enhanced capacity to stimulate adenylyl cyclase? If so, we will conclude that Gly^{49} of α_{avg} plays a role similar to that of Gly^{12} in p21^{ras}; by extension, such a result will strongly confirm the suspected structural and functional similarity between the GDP-binding regions of p21^{ras} and G-protein α chains.

Region C in EF-Tu is composed of a β strand connected by a loop to an amphipathic α helix. An aspartate residue in the β strand (Asp^{225} in α_{avg}) interacts with a Mg^{2+} ion that in turn is coordinated with the phosphoryl groups of GDP. Transforming mutations in region C of p21^{ras} have been identified at positions corresponding to residues 227, 229, and 231 of α_{avg} (for review, see Barbacid, 1987). Substitution of threonine for alanine at the position corresponding to α_{avg} residue 227 produces a transforming ras protein in which the threonine side chain is phosphorylated by GTP; this protein also has reduced GTPase activity (Gibbs et al., 1984; Temeles et al., 1985). This mutation implies close proximity of this residue in p21^{ras} to the γ-phosphoryl group of GTP.

Region E contains several conserved hydrophobic amino acids which, in conjunction with residues in region G, form a hydrophobic pocket for the guanine ring. Region G contains a conserved aspartate (Asp^{297} in α_{avg}) that interacts with the 2-amino group of the guanine ring. Substitution of alanine for this aspartate in p21^{ras} lowers the binding affinity for guanine nucleotides without a proportional change in affinity for inosine nucleotides; the mutant protein no longer discriminates between GDP and IDP, which differ only by the presence or absence of the 2-amino group (Sigal et al., 1986a). The strong conservation of amino acid sequence in this region among the G proteins, EF-Tu, and p21^{ras} led to the prediction (Masters et al., 1986) that the guanine ring of the bound nucleotide is located in a pocket defined by corresponding amino-acid residues in all three sets of proteins. The crystal structure of a p21^{ras} protein (de Vos et al., 1988) strikingly confirms this prediction, at least with respect to EF-Tu and ras.

C. GTP-Induced Conformational Change

The crystal structures of p21^{ras} and EF-Tu provide only a static picture of the GDP-bound conformation of these proteins. Key questions, both for these proteins and for the G proteins, center on the profound change in conformation that occurs when bound GDP is replaced by GTP. The G proteins are converted to the active conformation upon binding GTP, synthetic GTP analogs, or GDP plus AlF$_4^-$ ion; the latter combination is thought to resemble GTP, with AlF$_4^-$ replacing the γ-phosphoryl group of GTP (Bigay et al., 1987). G-protein activation is most easily measured by assessing ability of the protein to regulate activity of effectors, such as adenylyl cyclase. Activation is accompanied by changes in many other properties, including altered binding affinity for receptors and $\beta\gamma$, as well as effectors (Gilman, 1987), enhanced intrinsic tryptophan fluorescence of α chains (Higashijima et al., 1987a,b), altered reactivity of cysteine sulfhydryls (Winslow et al., 1987), susceptibility to ADP-ribosylation by cholera and pertussis toxins (Van Dop et al., 1984), and sensitivity to cleavage by trypsin (Fung and Nash, 1983).

The GTP-induced conformational change requires Mg^{2+} ion. In the absence of this ion, G proteins do not activate effector enzymes or hydrolyze GTP. Measurements of the intrinsic tryptophan fluorescence of α_i and α_o highlight the importance of Mg^{2+} for the GTP-induced conformational change: Binding of GTP in the absence of Mg^{2+} produces a small change in fluorescence, which is greatly augmented upon addition of Mg^{2+}; the fluorescence then decays in parallel with conversion of the bound GTP to GDP (Higashijima et al., 1987a,b). One attractive but untested interpretation of these observations is that the change in fluorescence reflects a change in conformation that results from entry of the Mg^{2+} ion into a location corresponding to that of Mg^{2+} in the crystal structure of EF-Tu, i.e., between the phosphoryl groups of GTP and an aspartate residue in region C (Asp^{225} of α_{avg}; see Fig. 3).

Useful clues regarding the GTP-dependent conformational change have come from studies of the phenotype of the S49 mutant cell line, H21a. The H21a phenotype (Bourne et al., 1981) results from replacement of a glycine by an alanine residue in α_s, at a position corresponding to Gly^{228} of α_{avg} (R. T. Miller et al., 1988). This residue is located at the $\beta \rightarrow \alpha$ turn in Halliday region C (Fig. 3). Although G$_s$ in H21a can bind GTP (as shown by the ability of GTP analogs to regulate the affinity of β-adrenoceptors for agonists in H21a membranes), the mutant G$_s$ cannot stimulate adenylyl cyclase in response to GTP analogs, AlF$_4^-$, hormones, or cholera toxin (Bourne et al., 1981). In addition, the GTP- and AlF$_4^-$-induced dissociation of α_s from $\beta\gamma$, detected by assessing altered migration of normal (S49 wild type) α_s in sucrose density gradients, does not occur with G_s from H21a (R. T. Miller *et al.,* 1988).

Studies of α_s cleavage by trypsin provide further evidence that G_s in H21a cannot undergo the conformational change that is normally induced by GTP. Activation of transducin and of normal G_s by GTP analogs protects the α chains of these proteins from cleavage by trypsin at a conserved site, corresponding to Arg^{233} of α_{avg}. In contrast to the α_s of wild type S49 cells, the mutant α_s of H21a is *not* protected from proteolysis by GTP analogs (R. T. Miller *et al.,* 1988). Thus the *H21a* mutation appears to have created a G protein that knows it has bound guanine nucleotide but that cannot tell GTP from GDP.

The functional defect of α_s in H21a can be precisely located in the GTP-driven activation cycle (see Fig. 1): The $\alpha\beta\gamma$ complex of H21a can apparently assume the "empty site" conformation, which binds tightly to the activated detector/receptor element and enhances the affinity of receptor for binding agonists. Furthermore, $\alpha\beta\gamma$ in H21a can be heterotropically displaced from association with the receptor upon binding guanine nucleotide, as shown by the ability of guanine nucleotides to lower receptor affinity for agonists in H21a membranes. The *H21a* mutation exerts its effect at the next step: Because the mutation prevents α_s from taking on its "active" conformation in the GTP-bound state, α_s does not dissociate from $\beta\gamma$ and cannot interact with the effector, adenylyl cyclase.

It is clear that the GTP-induced change from an inactive to an active conformation is common to all G proteins. We would like to know how the active GTP-bound conformation differs, in three dimensions, from the inactive GDP-bound conformation. If the guanine nucleotide-binding pocket of G-protein α chains resembles those of EF-Tu and p21^{ras}, it is obvious that the γ-phosphoryl group of GTP must occupy space that is not available in the GDP-bound protein. One attractive possibility is that replacement of GDP by GTP somehow displaces the β strand and induces flexion of the $\beta \rightarrow \alpha$ turn in Halliday region C (Fig. 3). In EF-Tu this turn begins at a conserved glycine residue that corresponds to the mutated residue in the H21a α chain; this glycine residue may serve as a hinge that allows relative movement of separate domains of the protein. The Gly $\rightarrow$ Ala substitution in H21a could decrease flexibility of the hinge in this region and thereby prevent a critical conformational change.

Several observations are in keeping with this speculative interpretation: (1) The precedent of adenylyl kinase. The *H21a* mutation is located in a region that is strongly conserved, not only among GTP-binding proteins but also with respect to enzymes, such as adenylyl kinase, that bind adenine nucleotides. Indeed, residues 109–131 of adenylyl kinase resemble Halliday region C both in amino acid sequence and in three-dimensional structure (Fry *et al.*, 1986). The glycine residue that corresponds to the site of mutation in H21a is thought to function in adenylyl kinase as a hinge that allows a conformational change induced by binding different adenine nucleotides (Sachsenheimer and Schulz, 1977). In the diagram in Fig. 3, the analogous conformational change would involve movement of the β strand containing Asp^{225} slightly down and to the reader's right; it seems reasonable to speculate that in H21a an added methyl group (resulting from the Gly → Ala substitution) in the side chain of residue 228 could limit flexion about this turn and thereby limit GTP-induced movement of the α helix situated immediately downstream. (2) An antibody that neutralizes $p21^{ras}$. The Y13-259 monoclonal antibody binds to a stretch of sequence in $p21^{ras}$ just distal to the glycine residue corresponding to Gly^{228} of α_{avg}, and thereby inhibits the transforming activity of the protein (Sigal *et al.*, 1986b; Mulcahy *et al.*, 1985). It seems possible that the inhibitory effect of this antibody results from increased rigidity of Halliday region C, which could prevent $p21^{ras}$ from taking on an active conformation. (3) Tryptic cleavage at a neighboring arginine residue. Binding of GTP prevents tryptic cleavage of α_s and the α chain of transducin at a site (Arg^{233} of α_{avg}) just five residues downstream from the residue that is altered by mutation in *H21a*. Because GTP can bind to both normal α_s and to α_s bearing the *H21a* mutation, it is likely that GTP protects against trypsin not by covering the tryptic cleavage site but rather by altering conformation of the presumptive α helix in which it is located (see Fig. 3). This result, like the effect of the anti-$p21^{ras}$ antibody, points to the turn in Halliday region C, which may transmit critical leverage in regulating relative positions of different parts of the guanine nucleotide-binding domain (and perhaps of other domains as well).

D. Site for Interaction with Receptors

Comparison of the primary structures of G-protein α chains reveals three major regions of divergent amino acid sequence, located in the domains designated I, II, and III in Fig. 2. Efforts to identify effector and receptor binding domains have focused on these regions.

Several lines of evidence point to a carboxy-terminal region of the α chains as one of the sites involved in receptor interaction. The pertussis toxin-catalyzed ADP-ribosylation of a cysteine located four residues from the carboxy terminus prevents G_i, G_o, and transducin from responding to stimulation by their respective receptors (Ui *et al.*, 1984). Indeed, pertussis toxin-catalyzed ADP-ribosylation prevents transducin from binding to rhodopsin (Van Dop *et al.*, 1984). The *unc* mutation in α_s causes a similar phenotype—uncoupling of G_s from interactions with the receptors that stimulate adenylyl cyclase (Haga *et al.*, 1977). The *unc* mutation causes replacement of an arginine by a proline residue near the extreme carboxy terminus of α_s (Sullivan *et al.*, 1987; Rall and Harris, 1987). Both the *unc* mutations of α_s and the cysteines modified by pertussis toxin in α_i, α_o, and α_t are located in a predicted α helix at the carboxy terminus of the composite G-protein α chain, α_{avg} (Masters *et al.*, 1986).

A retinal protein, arrestin, provides a third line of evidence implicating the predicted α helix at the carboxy terminus as a contact site of α chains with receptors. Arrestin competes with transducin for binding to the phosphorylated form of photorhodopsin. Because arrestin contains a stretch of internal sequence very similar to that of the carboxy terminus of α_t (Wistow *et al.*, 1986), it is reasonable to propose that these conserved regions are both involved in rhodopsin binding.

Further evidence has come from phenotypes produced by two different recombinant chimeric α chains. One such chimera is composed of the amino-terminal 60% of an α_i (mouse α_{i2}) and the carboxy-terminal 40% of α_s. Expression of this α_i/α_s chimera in S49 cyc^- cells (which are genetically deficient in normal α_s) produces a cell in which β-adrenoceptors stimulate adenylyl cyclase (Masters *et al.*, 1988). Other experiments showed that the α_i/α_s chimera did not interact with somatostatin receptors, which are coupled to G_i in S49 cells. These results imply that the carboxy-terminal portion of the chimera, which was contributed by α_s, confers on the α chain its specificity for interacting with receptors.

A second chimeric α chain, in which the carboxy-terminal 38 residues of α_s are replaced by the carboxy-terminal 36 residues of α_{i2}, has been expressed in a fibroblast cell line (G. L. Johnson, personal communication). In comparison to normal cells or cells expressing recombinant normal α_s, adenylyl cyclase in the cells expressing this α_s/α_{i2} chimera behaves as if it is constitutively activated by receptor, even in the absence of receptor agonists: i.e., cAMP synthesis is elevated in the presence of GTP alone, and hydrolysis-resistant GTP analogs activate cAMP synthesis with a very short lag. In adenylyl cyclase systems coupled by normal α_s, the temporal lag in activation by GTP analogs is due to slow dissociation of GDP from α_s; hormone–receptor complexes normally abolish the lag by accelerating the rate of GDP dissociation from α_s. Accordingly, the simplest and most exciting interpretation of the phenotype produced by the α_s/α_{i2} chimera is that it is due to accelerated release of GDP and a consequent increase in the rate at which GTP binds to the α chain. In short, the effect of replacing the normal α_s carboxy terminus with the cognate region of α_{i2} is to mimic the effect of hormone–receptor complex on normal G_s.

By combining the α_s/α_{i2} result with the other evidence cited above, we can construct a highly speculative but attractive and testable model of the mechanism by which hormone–receptor complexes are able to open the guanine nucleotide-binding site, promote release of GDP, and set in motion the transduction of hormonal signals. The model suggests that the carboxy-terminal α helix of a G protein α chain normally suppresses release of GDP when the G protein is in the inactive conformation. The hormone–receptor complex is imagined to bind to this α helix and to dislodge it from the position in which it suppresses release of GDP. This interpretation would account for the uncoupling effects of the Arg → Pro mutation in *unc* α_s and of the ADP–ribose attached to the helix by pertussis toxin. It would also account for the phenotype produced by the α_s/α_{i2} chimera, if we imagine that the substituted carboxy terminus derived from α_{i2} fits poorly into the site at which the cognate portion of α_s normally suppresses GDP release.

Although further experiments will be required to support or refute this interesting interpretation, it is in keeping with the crystal structure of p21^{ras} and the effect of a carboxy-terminal deletion in the RAS2 gene product of the budding yeast, *Saccharomyces cerevisiae*. In the GDP-bound form of p21^{ras}, residues 152–165 form an α helix that lies directly over the GDP-binding domain, near Halliday region G (de Vos *et al.*, 1988). In the yeast, GTP-dependent stimulation of the adenylyl cyclase system is mediated by the RAS2 protein, which has a primary structure (and therefore presumably a three-dimensional structure) almost identical to that of p21^{ras}, especially in its guanine nucleotide-binding domain. Like the mammalian G proteins, RAS2 normally requires the presence of a GDP–GTP exchange protein, called CDC25, in order to stimulate cAMP synthesis. The requirement for CDC25 is bypassed by deletion of a portion of RAS2 near the carboxy terminus (Marshall *et al.*, 1987); the deleted portion of RAS2 is located in the right place to serve as a homolog of the α helix (residues 152–165) in p21^{ras}. We propose, in turn, that this α-helical segment of p21^{ras} is the functional homolog of the carboxy-terminal portion of G-protein α chains, in that both are critical regulators of GDP release.

E. Effector Contact Region

The location of the effector binding region of G-protein α chains is not known. Masters *et al.* (1986) proposed that domain II (Fig. 2) would prove to mediate interactions with effectors. This proposal was based in part on the heterogeneity of amino-acid sequence in this domain among α chains that regulate different effectors, and in part on suggestive evidence that the much shorter cognate regions of EF-Tu (Laursen *et al.*, 1981) and p21^{ras} (Sigal *et al.*, 1986b) may be involved in interactions with proteins analogous to the effectors of G-protein-mediated signaling systems.

The 60:40 α_i/α_s chimera described above allowed a simple test of the notion that region II mediates effector interactions (Masters *et al.*, 1988). If this notion were correct, the chimera would not be expected to stimulate adenylyl cyclase, because domain II in the chimera is derived from α_{i2} rather than from α_s. Instead, the chimeric α chain behaves very much like α_s, in that it mediates *stimulation* of adenylyl cyclase activity by receptors that stimulate cAMP synthesis in normal cells (e.g., β-adrenoceptors and PGE$_1$ receptors). Thus the phenotype produced by expression of the α_i/α_s chimera makes it unlikely that domain II specifies stimulation of adenylyl cyclase.

It would be attractive to extend this conclusion to suggest that domain III, instead of domain II, contains the amino-acid sequences that are critical for a specific stimulatory interaction with catalytic adenylyl cyclase (and, by implication, the sequences in other α chains that specify regulation of their effectors). This more comprehensive conclusion is probably not justified, however, in part because it is possible that α_{i2}, the α chain of a G$_i$-like protein, may itself interact with catalytic adenylyl cyclase. If so, activity of the α_i/α_s chimera used in these experiments would not necessarily distinguish regions important for binding to effector enzymes from regions that determine how the effector will respond; parts of domain II might be important for binding to adenylyl cyclase in both α_s and α_i, while sequences in domain III could specify whether the enzyme is stimulated or inhibited.

In summary, the location of the effector contact region of G-protein α chains remains undefined. Although the α_i/α_s chimera suggests that such a region may be located in domain III, further experiments may show that this model is too simple, i.e., it is quite possible that sequences in domains II and III are spatially intertwined in the folded protein, and that sequences from either or both may be required for interaction with any other protein, including not only effectors but receptors and the $\beta\gamma$ complex as well.

F. Contact with $\beta\gamma$ Complex

Proteolytic removal of the amino-terminal 21 amino acids of the α chain of rod cell transducin destroys its ability to bind to $\beta\gamma$ (Navon and Fung,

1987). The inability to interact with $\beta\gamma$ prevents the truncated α_t chain from binding to rhodopsin and from undergoing photon-triggered exchange of guanine nucleotide. The simplest (but not the only) interpretation of these results is that $\beta\gamma$ binds directly to the extreme amino terminus of the α chain. Although observations that $\beta\gamma$ complexes can be functionally interchanged among several G proteins (Kanaho *et al.*, 1984) suggest that the parts of α chains that interact with $\beta\gamma$ should be conserved, the primary structure of the α chain amino terminus (domain I) is *not* highly conserved (see Fig. 2, *top*). Masters *et al.* (1986) predicted a conserved α helical secondary structure for domain I in all G protein α chains. Perhaps some feature of this conserved α helix, not dependent on strict conservation of a long stretch of amino acid sequence, specifies interaction with $\beta\gamma$. Alternatively, this region may be involved in regulating interactions with $\beta\gamma$ without contacting the complex directly.

III. SUMMARY AND PROSPECTS

This review has focused on the structure and function of a GTP-driven molecular machine that has found many uses in biology. One of the most adaptable versions of this machine, installed in each member of the G-protein family, is used to transduce hormonal and/or sensory signals across cell membranes. We have discussed the features that have made this machine so adaptable, including its sensitivity and ability to transduce signals without noise, its enormous capacity for amplification of the signal, and the built-in assurance that information can only flow through the machine in one direction. Each of these features depends on an intricate and finely tuned cycle that begins with receptor-triggered release of GDP, followed by binding of GTP and activation of the effector; the cycle terminates transduction of the signal when GTP is hydrolyzed.

In conclusion it may be useful to summarize what we know or surmise about each event in the G-protein signaling cycle (Fig. 1) and to pose a few of the most pressing questions that arise in relation to each. We will begin with the inactive GDP-bound $\alpha\beta\gamma$ heterotrimer (Fig. 1, *upper left*).

a. The Unstimulated (Basal) State Most of the evidence agrees in suggesting that G proteins in the resting state are $\alpha\beta\gamma$ heterotrimers bound to GDP. Here we should remind ourselves of the vast extent of our ignorance about this structure. First, our best structural and functional information relates to α chains, while the β and γ polypeptides remain shrouded in relative mystery. Further, this review has not even touched on the host of important unsolved questions: What is the topographic location of G proteins in the plasma membrane? Are they confined to two-dimensional domains, tethered to the cytoskeleton, etc.? Are the α chains bound to the membrane principally (only?) via $\beta\gamma$? What prevents α chains from wandering off into the cytoplasm when they are activated? Where are inactive G proteins located in relation to receptor and effector? Are our models correct in depicting the unstimulated heterotrimer as completely dissociated from other protein components of the signaling pathway?

b. Interaction with Receptor and Release of GDP   Abundant biochemical evidence and the phenotypes produced by mutant and chimeric α chains indicate that a presumed α helix at the carboxy terminus of the α chain represents an important site for contact with the receptor. Perhaps we can perceive, however dimly, a possible mechanism by which binding of the activated receptor to this region could open the guanine nucleotide-binding site and accelerate dissociation of GDP. Rapidly expanding structural and functional information about the family of receptors that talk to G proteins (Dohlman *et al.,* 1987), combined with site-directed mutations in both receptor and G protein components, will soon begin to sketch a picture of the complex structure symbolized by $D^*-\alpha\beta\gamma$, in which the guanine nucleotide site is apparently empty (Fig. 1, *top*). Such a picture will indicate how both the receptor and the G protein can alter the conformation and function of its partner in the complex.

We must probably await the solution of crystal structures of several G proteins to understand some of the most subtle differences among different G proteins. For example, what structural differences between G_s and transducin account for the large differences in rates of GDP release from these proteins in the unstimulated state?

c. The Transient "Empty" State and Guanine Nucleotide-Induced Dissociation of $\alpha\beta\gamma$ from the Receptor   Little is known of the conformational state in which G proteins with empty guanine nucleotide-binding sites bind tightly to activated receptors and enhance the receptors' affinity for agonists. We imagine that the high concentration of GTP in the intact cell makes this "empty" state short lived indeed. The apparently heterotropic interaction by which guanine nucleotides and receptor displace one another from binding to $\alpha\beta\gamma$ can be described in kinetic terms (De Lean *et al.,* 1980), but we know little or nothing about how entry of a guanine nucleotide into the α-chain binding site causes $\alpha\beta\gamma$ to dissociate from the receptor. The mystery will persist until sensitive biochemical probes are developed to dissect this receptor–G protein complex or until mutations that stabilize the $D^*-\alpha\beta\gamma$ complex are discovered in either the receptor or the G protein.

d. GTP-Induced Conformational Change   The *H21a* mutation prevents the GTP-induced conformational change in α_s (see Fig. 1, *top right*). The observation that this mutation replaces a conserved glycine residue in the C region of the guanine nucleotide-binding domain, combined with studies of analogous conformational changes in adenylyl kinase, provide a hint as to how the conformation of nearby structures may change on binding of GTP. The more widespread changes in conformation that cause α to dissociate from $\beta\gamma$ and to associate with the effector remain completely mysterious. As in so many other cases, understanding of these changes must await crystallographic or other studies of the physical states of G proteins and their signaling partners in different functional conformations.

e. Interaction of α with $\beta\gamma$ The part of α that interacts with $\beta\gamma$ remains unknown, although domain I, at the α chain amino terminus, may play a role. Our ignorance of the structures involved in this interaction is matched by our ignorance of the critical functional role of the $\beta\gamma$ complex. Although $\beta\gamma$ is thought to attach G proteins to the membrane and to aid in presenting α chains to the receptor, it is not clear why two additional polypeptides are required for these rather mundane functions—especially in view of the fact that related proteins like EF-Tu and (probably) p21^{ras} seem to function without $\beta\gamma$ homologs.

f. Interaction of α with Effector   The α-chain contact site for effectors has also not been identified. From the phenotype produced by an α_i/α_s chimeric protein, we know only that the effector contact region of G_s does not require the amino-acid sequences in the amino terminal 60% of α_s that differ from corresponding sequences in α_{i2}. We can hope that the combination of biochemical studies and site-directed mutations will soon identify the effector contact region.

g. Hydrolysis of GTP   The delayed hydrolysis of bound GTP remains one of the most fascinating and mysterious functions of the G protein machine. Although several mutations in p21^{ras} appear to reduce the rate of GTP hydrolysis, structural and functional determinants of GTPase activity in G proteins remain largely unexplored. The time delay before GTP is hydrolyzed plays a key role in mediating amplification of hormonal signals. What features of the G proteins determine the duration of this delay? ADP-ribosylation by cholera toxin of a residue in domain II, outside the presumptive guanine nucleotide-binding domain, stabilizes the active, GTP-bound form of α_s; what structural features of α chains account for this effect? Does any component of G-protein-mediated signaling systems play a role similar to that of GAP, the recently discovered protein that stimulates GTPase activity of p21^{ras}? This question may be particularly relevant to retinal phototransduction, where GTP bound to the active form of transducin appears to be hydrolyzed after a delay of several seconds (Navon and Fung, 1984); vision would clearly be impossible if this delay, measured *in vitro,* operated to transmit a similarly long-lasting signal to the intact retinal rod cell.

The list of unanswered questions is long. On the other hand, investigators have made extraordinary progress in the years that have passed since the first G protein was discovered (Ross and Gilman, 1977). At that time, none of the questions we ask now could have been dreamed of. From that perspective it seems possible — perhaps even likely — that our present questions will be solved in the near future.

REFERENCES

Barbacid, M. (1987). *ras* Genes. *Annu. Rev. Biochem.* **56,** 779–827.

Bigay, J., Deterre, P., Pfister, C., and Chabre, M. (1987). Fluoride complexes of aluminum or beryllium act on G-proteins as reversibly bound analogues of the γ phosphate of GTP. *EMBO J.* **6,** 2907–2913.

Bourne, H. R., Kaslow, D., Kaslow, H. R., Salomon, M., and Licko, V. (1981). Hormone sensitive adenylate cyclase: Mutant phenotype with normally regulated beta-adrenergic receptors uncoupled from catalytic adenylate cyclase. *Mol. Pharmacol.* **20,** 435–441.

Bray, P., Carter, A., Simons, C., Guo, V., Puckett, C., Kamholz, J., Spiegel, A., and Nirenberg, M. (1986). Human cDNA clones for four species of G α_s signal transduction protein. *Proc. Natl. Acad. Sci. U.S.A.* **83,** 8893–8897.

De Lean, A., Stadel, M., and Lefkowitz, R. J. (1980). A ternary complex model explains the agonist-specific binding properties of the adenylate cyclase-coupled β-adrenergic receptor. *J. Biol. Chem.* **255,** 7108–7117.

de Vos, A. M., Tong, L., Milburn, M. V., Matias, P. M., Jancarik, J., Miura, K., Ohtsuka, E., Noguchi, S., Nishimura, S., and Kim, S.-H. (1988). Three-dimensional structure of an oncogene protein: Catalytic domain of human cH-*ras* p21. *Science* **239,** 888–893.

Dohlman, H. G., Bouvier, M., Benovic, J. L., Caron, M. G., and Lefkowitz, R. J. (1987). The multiple membrane spanning topography of the β_2-adrenergic receptor. *J. Biol. Chem.* **262,** 14282–14288.

Fry, D. C., Kuby, S. A., and Mildvan, A. S. (1986). ATP-binding of adenylate kinase: Mechanistic implications of its homology with *ras*-encoded p21, F_1-ATPase, and other nucleotide-binding proteins. *Proc. Natl. Acad. Sci. U.S.A.* **83,** 907–911.

Fung, B. K.-K. (1983). Characterization of transducin from bovine retinal rod outer segments. *J. Biol. Chem.* **258,** 10495–10502.

Fung, B. K.-K., and Nash, C. R. (1983). Characterization of transducin from bovine retinal rod outer segments. *J. Biol. Chem.* **258,** 10503–10510.

Fung, B. K.-K., and Stryer, L. (1980). Photolyzed rhodopsin catalyzes the exchange of GTP for bound GDP in retinal rod outer segments. *Proc. Natl. Acad. Sci. U.S.A.* **77,** 2500–2504.

Gibbs, J. B., Sigal, I. S., Poe, M., and Scolnick, E. M. (1984) Intrinsic GTPase activity distinguishes normal and oncogenic *ras* p21 molecules. *Proc. Natl. Acad. Sci. U.S.A.* **81,** 5704–5708.

Gilman, A. G., (1987). G Proteins: Transducers of receptor-generated signals. *Annu. Rev. Biochem.* **56,** 615–649.

Haga, T., Ross, E. M., Anderson, H. J., and Gilman, A. G. (1977). Adenylate cyclase permanently uncoupled from hormone receptors in a novel variant of S49 mouse lymphoma cells. *Proc. Natl. Acad. Sci. U.S.A.* **74,** 2016–2020.

Halliday, K. (1984). Regional homology in GTP-binding proto-oncogene products and elongation factors. *J. Cyclic Nucleotide Res.* **9,** 435–448.

Higashijima, T., Ferguson, K. M., Sternweis, P. C., Ross, E. M., Smigel, M. D., and Gilman, A. G. (1987a). The effect of activating ligands on the intrinsic fluorescence of guanine nucleotide-binding regulatory proteins. *J. Biol. Chem.* **262,** 752–756.

Higashijima, T., Ferguson, K. M., Smigel, M. D., and Gilman, A. G. (1987b). The effect of GTP and Mg^{2+} on the GTPase activity and the fluorescent properties of G_o. *J. Biol. Chem.* **262,** 757–761.

Higashijima, T., Ferguson, K. M., Sternweis, P. C., Smigel, M. D., and Gilman, A. G. (1987c). Effects of Mg^{2+} and the $\beta\gamma$-subunit complex on the interactions of guanine nucleotides with G proteins. *J. Biol. Chem.* **262,** 762–766.

Jurnak, F. (1985). Structure of the GDP domain of EF-Tu and location of the amino acids homologous to *ras* oncogene proteins. *Science* **230,** 32–36.

Kanaho, Y., Tsai, S. C., Adamik, R., Hewlett, E. L., Moss, J., and Vaughan, M. (1984). Rhodopsin-enhanced GTPase activity of the inhibitory GTP-binding protein of adenylate cyclase. *J. Biol. Chem.* **259,** 7378–7381.

Kaziro, Y. (1978). The role of guanosine 5'-triphosphate in polypeptide chain elongation. *Biochim. Biophys. Acta* **505,** 95–127.

La Cour, T. F. M., Nyborg, J., Thirup, S., and Clark, B. F. C. (1985). Structural details of the binding of guanosine diphosphate to elongation factor Tu from *E. coli* as studied by X-ray crystallography. *EMBO J.* **4,** 2385–2388.

Laursen, R. A., L'Italien, J. J., Nagarkatti, S., and Miller, D. L. (1981). The amino acid sequence of elongation factor Tu of *Escherichia coli*. The complete sequence. *J. Biol. Chem.* **256,** 8102–8109.

Marshall, M. S., Gibbs, J. B., Scolnick, E. M., and Sigal, I. S. (1987). Regulatory function of the yeast *RAS* c-terminus. *Mol. Cell. Biol.* **7,** 2309–2315.

Masters, S. B., Stroud, R. M., and Bourne, H. R. (1986). Family of G protein α chains: Amphipathic analysis and predicted structure of functional domains. *Protein Engineering* **1,** 47–54.

Masters, S. B., Sullivan, K. A., Beiderman, B., Lopez, N. G., Ramechandron, J., and Bourne, H. R. (1988). The carboxy-terminal domain of $G_s\alpha$ specifies coupling of receptor to stimulation of adenylyl cyclase. *Science* **241,** 448–451.

Miller, R. T., Sullivan, K. A., Masters, S. B., Beiderman, B., and Bourne, H. R. (1988). A mutation that prevents GTP-dependent activation of the α chain of G_s. *Nature* **334**, 712–715.

Mulcahy, L. S., Smith, M. R., and Stacey, D. W. (1985). Requirement for *ras* proto-oncogene function during serum-stimulated growth of NIH 3T3 cells. *Nature* **313**, 241–243.

Navon, S. E., and Fung, B. K.-K. (1984). Characterization of transducin from bovine retinal rod outer segments—Mechanism and effects of cholera toxin-catalyzed ADP-ribosylation. *J. Biol. Chem.* **259**, 6686–6693.

Navon, S. E., and Fung, B. K.-K. (1987). Characterization of transducin from bovine retinal rod outer segments. Participation of the amino-terminal region of T_α in subunit interaction. *J. Biol. Chem.* **262**, 15746–15751.

Rall, T., and Harris, B. A. (1987). Identification of the lesion in the stimulatory GTP-binding protein of the uncoupled S49 lymphoma. *FEBS Lett.* **224**, 365–371.

Ross, E. M., and Gilman, A. G. (1977). Resolution of some components of adenylate cyclase necessary for catalytic activity. *J. Biol. Chem.* **252**, 6966–6969.

Sachsenheimer, W., and Schulz, G. E. (1977). Two conformations of crystalline adenylate kinase. *J. Mol. Biol.* **114**, 23–36.

Seeburg, P. H., Colby, W. W., Capon, D. J., Goeddel, D. V., and Levinson, A. D. (1984). Biological properties of human c-HA-*ras*1 genes mutated at codon 12. *Nature* **312**, 71–75.

Sigal, I. S., Gibbs, J. B., D'Alonzo, J. S., Temeles, G. L., Wolanski, B. S., Socher, S. H., and Scolnick, E. M. (1986a). Mutant *ras*-encoded proteins with altered nucleotide binding exert dominant biological effects. *Proc. Natl. Acad. Sci. U.S.A.* **83**, 952–956.

Sigal, I. S., Gibbs, J. B., D'Alonzo, J. S., and Scolnick, E. M. (1986b). Identification of effector residues and a neutralizing epitope of Ha *ras* p21. *Proc. Natl. Acad. Sci. U.S.A.* **83**, 4725–4729.

Sternweis, P. C. (1986). The purified α subunits of G_o and G_i from bovine brain require $\beta\gamma$ for association with phospholipid vesicles. *J. Biol. Chem.* **261**, 631–637.

Stryer, L. (1986) Cyclic GMP cascade of vision. *Annu. Rev. Neurosci.* **9**, 87–119.

Stryer, L., and Bourne, H. R. (1986). G proteins: A family of signal transducers. *Annu. Rev. Cell Biol.* **2**, 391–419.

Sullivan, K. A., Miller, R. T., Masters, S. B., Beiderman, B., Heideman, W., and Bourne, H. R. (1987). Identification of receptor contact site involved in receptor-G protein coupling. *Nature* **330**, 758–760.

Temeles, G. L., Gibbs, J. B., D'Alonzo, J. S., Sigal, I. S., and Scolnick, E. M. (1985). Yeast and mammalian *ras* proteins have conserved biochemical properties. *Nature* **313**, 700–703.

Trahey, M., and McCormick, F. (1987). A cytoplasmic protein stimulates normal N-*ras* p21 GTPase, but does not affect oncogenic mutants. *Science* **238**, 542–545.

Ui, M., Katada, T., Murayama, T., Kurose, H., Yajima, M., Tamura, M., Nakamura, T., and Nogimori, K. (1984). Islet-activating protein, pertussis toxin: A specific uncoupler of receptor-mediated inhibition of adenylate cyclase. *Adv. Cyclic Nucleotide Protein Phosphorylation Res.* **17**, 145–151.

Van Dop, C., Yamanaka, G., Steinberg, F., Sekura, R. D., Manclark, C. R., Stryer, L., and Bourne, H. R. (1984). ADP-ribosylation of transducin by pertussis toxin blocks the light-stimulated hydrolysis of GTP and cGMP in retinal photoreceptors. *J. Biol. Chem.* **259**, 23–26.

Winslow, J. W., Bradley, J. D., Smith, J. A., and Neer, E. J. (1987). Reactive sulfhydryl groups of α_{39}, a guanine nucleotide-binding protein from brain. Location and function. *J. Biol. Chem.* **262**, 4501–4507.

Wistow, G. J., Katial, A., Craft, C., and Shinohara, T. (1986). Sequence analysis of bovine retinal S-antigen. Relationships with α transducin and G proteins. *FEBS Lett.* **196**, 23–28.

CHAPTER 3

Structure and Function of G-Protein $\beta\gamma$ Subunit

Eva J. Neer
Department of Medicine, Brigham and Women's Hospital, Harvard Medical
School, Boston, Massachusetts 02115

David E. Clapham
Department of Pharmacology, Physiology and Biophysics, Mayo Foundation,
Rochester, Minnesota 55905

The guanine nucleotide-binding proteins which transmit signals from a
variety of hormone and neurotransmitter receptors to cellular enzymes
and ion channels are heterotrimers composed of α, β, and γ subunits.

Similar proteins in retinal cones and rod outer segments couple light activation of rhodopsin to activation of cGMP phosphodiesterase and to other enzymes (for reviews see Stryer and Bourne, 1986; Gilman, 1987, and other chapters in this volume). The α subunits of G proteins, which can be classified by their ability to serve as substrates for modification by toxins from *Vibrio cholera* or *Bordetella pertussis,* all bind and hydrolyze GTP[1]. In solution, binding of a nonhydrolyzable analog of GTP [Gpp (NH)p or GTPγS] to the α subunit persistently activates it and causes the heterotrimer to dissociate into α and $\beta\gamma$ subunits (Northup et al., 1983a,b; Sternweis et al., 1981; Codina et al., 1983; Huff et al., 1985). Since the monomeric form of the α subunit is active and the heterotrimer is inactive, it is clear that at least one role of $\beta\gamma$ is to modulate the function of the α subunit. In addition, recent work from this and other laboratories suggests that $\beta\gamma$ itself r..ay play an active role in regulating the function of membrane enzymes and ion channels. In this review, we will discuss the structure and function of the $\beta\gamma$ subunit and speculate on the regulatory significance of systems which may simultaneously generate two signaling molecules: activated α and $\beta\gamma$ subunits.

I. STRUCTURE OF $\beta\gamma$ SUBUNIT

Although the β and γ subunits are not covalently linked to each other, the native proteins cannot be dissociated without denaturation and therefore form a single functional unit. Thus, although the G proteins contain three polypeptides, they are functionally dimers. We have tried to dissociate native β from γ by varying pH, salt concentration, type of detergent, and temperature with no success (E. J. Neer, unpublished). To date the only report of isolation of native β without γ is by Yamazaki et al. (1987) who observed a small amount of transducin β, which spontaneously appeared to be free of γ, in the course of purifying transducin $\beta\gamma$ subunits. It is not known whether functions attributed to the $\beta\gamma$ subunit are due to β, to γ, or both proteins.

Much more is known about the structure of β than of γ subunits. There are at least two very similar forms of the β subunit, a 36-kDa and a 35-kDa protein (Sternweis and Robishaw, 1984; Winslow et al., 1986; Evans et al.,

---

[1] The α subunits which are the substrates for cholera toxin are called α_s. They have several functions including stimulation of adenylyl cyclase. The substrates for modification by pertussis toxin are called α_i and α_o. The term G_i or G_o refers to the $\alpha\beta\gamma$ heterotrimer which includes the particular α subunit. These proteins are involved in mediating hormonal inhibition of adenylyl cyclase, hormonal regulation of phosphoinositol turnover, and other cell enzymes.

1987). In most tissues the predominant form of β is the 36-kDa one, while the 35-kDa β protein is a minor component (Woolkalis and Manning, 1987). However, in placenta, the 35-kDa protein is the major form of β (Evans *et al.,* 1987). Rod outer segments appear to contain only the 36-kDa form of the β subunit. The structural similarity of transducin and receptor-linked β subunits was originally demonstrated by Manning and Gilman (1983) and confirmed when cDNAs corresponding to the β subunit were cloned (Fong *et al.,* 1986; Fong *et al.,* 1987; Gao *et al.,* 1987). The deduced amino-acid sequence of the β subunit from transducin is identical to that cloned from a human hepatocyte library (Codina *et al.,* 1987). cDNA for a form of β subunit which corresponds to the 35-kDa form has been identified by Fong *et al.* (1987) and Gao *et al.* (1987). The amino-acid sequence deduced from this second form of β, denoted β_2, is 90% identical to the 36-kDa β_1 form. Nevertheless, some antibodies to β_1 do not recognize β_2 (Roof *et al.,* 1985). The mRNA that encodes the β_2 subtype is expressed at lower levels than β_1 mRNA in a number of human tissues (Fong *et al.,* 1987).

The amino-acid sequences of β_1 and β_2 are highly conserved in different mammalian species. The sequences of bovine and human β_1 are identical, as are bovine and human β_2. This strict conservation supports the idea that the β_1 and β_2 subunits have distinct functions. Neither sequence predicts membrane spanning or especially hydrophobic regions nor is there any signal sequence. Analysis of the deduced amino acid sequence for the β subunits by Fong *et al.* (1986) revealed that the β polypeptide contains four similar repeats of approximately 86 residues. Both forms of the β subunit are similar to the carboxyl-terminal region of the yeast *CDC4* gene, a cell division cycle gene. Although the two may share a common evolutionary origin it is not yet clear whether there is any similarity in function between $\beta\gamma$ and *CDC4*.

The two β proteins are not the result of alternative splicing but appear to be products of separate genes which are located on different human chromosomes (β_1 on chromosome 1 and β_2 on chromosome 7; Lochrie and Simon, 1988). The 5' and 3' nontranslated regions of the two β cDNAs show no significant similarity.

Much less is known about the γ subunits (Hildebrandt *et al.,* 1984a) than about the β subunits. There is very good evidence that the γ subunit associated with transducin is different from the γ subunit associated with the hormone receptor-linked G proteins. Hildebrandt *et al.* (1985) showed that the pattern of tryptic peptides from iodinated γ of transducin is different from that seen with γ of red cells or of brain proteins. However, there was no detectable difference between $\beta\gamma$ associated with α_s and $\beta\gamma$ associated with α_i. The γ subunit associated with bovine brain $\beta\gamma$ can be resolved into two bands on sodium dodecyl sulfate-polyacrylamide gel electrophoresis (SDS-PAGE) (Fig. 1). It is not yet known whether these two bands represent different γ proteins or result from covalent modification of the γ subunit.

The transducin γ subunit is antigenically different from γ associated with nonretinal β (Roof et al., 1985; Audigier et al., 1987; Gierschik et al., 1985). Antibodies to transducin γ, which is quite antigenic, do not cross-react with the γ subunit from hormone receptor-linked G proteins. In contrast, the nontransducin γ subunit is a poor antigen. Many laboratories have immunized rabbits with $\beta\gamma$ and have usually obtained antibodies only to the β subunit (Huff et al., 1985). Evans et al. (1987) immunized rabbits with the 35-kDa form of β and its γ and did obtain antibodies to the placental γ subunit. These antibodies did not cross react with the γ subunits of other G proteins. On immunologic grounds, therefore, it appears that there are several different kinds of γ subunits, one in rod outer segments and two or three in nonretinal tissues. The cDNA corresponding to the γ subunit of transducin has been cloned by Van Dop et al. (1984) and by

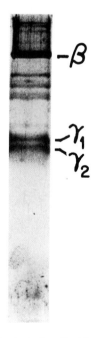

Fig. 1. $\beta\gamma$ purified from bovine brain as described by Neer et al. (1984) was analyzed by SDS-PAGE as described by Giulian and Graham (1988). The gel was silver stained as described by Merril et al. (1981).

Yatsunami *et al.* (1985). The cDNA predicts a hydrophilic protein of 8.7 kDa. Estimates of the molecular weights of other γ subunits range from 5,000–10,000.

A major difference between the $\beta\gamma$ subunits of transducin and of non-retinal G proteins is the manner of their association with the plasma membrane. Both α and $\beta\gamma$ subunits of transducin can be released from the rod outer-segment membrane simply by treating the membrane with a nonhydrolyzable guanine nucleotide and a moderately high concentration of salt (Fung, 1983a). In contrast, the $\beta\gamma$ subunit of the G proteins can only be solubilized by detergents. Once solubilized, the $\beta\gamma$ subunits vary considerably in hydrophobicity. Table I summarizes the hydrodynamic properties of solubilized $\beta\gamma$ subunits. The amount of detergent bound by a solubilized membrane protein may reflect either the extent of hydrophobic surface available for binding the detergent or the capacity of the protein to insert into a detergent micelle. The $\beta\gamma$ subunit purified from brain is very hydrophobic and binds considerably more detergent than the $\beta\gamma$ subunit from red blood cells (Huff *et al.*, 1985; Codina *et al.*, 1983). The $\beta\gamma$ subunit of transducin is the most hydrophilic since it is soluble without detergents (Fung, 1983a; Wessling-Resnick and Johnson, 1987).

Since the amino acid sequence of the β_1 subunit is identical in transducin and non-retinal G proteins, the differences in hydrophobicity may be

Table I Physical Properties of Solubilized $\beta\gamma$ Subunits

Parameter	Source		
	Bovine brain[a]	Human erythrocyte[b]	Rabbit liver[c]
Sedimentation coefficient $s_{20,w}$(S)	2.9	2.0	3.52
Stokes radius, a(Å)	36	51.0	34.6
Partial specific volume, $\bar{v}$ (ml/g)[d]	0.83	0.751	[0.74]
Detergent bound (%)	69	8	[14][e]
Molecular weight of protein	41,000	43,600	[51,000][e]

[a] Huff *et al.* (1985).

[b] Codina *et al.* (1984).

[c] Northup *et al.* (1983b).

[d] The values for $\bar{v}$ of $\beta\gamma$ from bovine brain and human erythrocytes were calculated by sedimentation of Lubrol 12A9-solubilized proteins through sucrose density gradients made up in H_2O or D_2O. The value for $\bar{v}$ of $\beta\gamma$ from rabbit liver was taken to be that of an average protein since the hydrodynamic studies were carried out in cholate, a detergent with a $\bar{v}$ value very similar to that of a protein.

[e] Because the $\bar{v}$ of cholate is similar to that of a protein, the amount of detergent bound could not be estimated and corrected. The weight of 51,000 therefore includes protein and any bound detergent. Since it is known that the molecular weight of β is 36,000 and of γ about 8,000, the molecular weight of protein should be 44,000. Thus, 7,000 out of the 51,000 combined detergent and protein molecular weight may represent detergent.

explained either by differences in posttranslational modification of β or by differences in the hydrophobicity of the γ subunit. Although the α subunits of some G proteins have been shown to be myristylated at the amino terminus, a modification which would render the subunits more hydrophobic, the $\beta\gamma$ subunits are apparently not modified (Buss et al., 1987). Thus, differences in the γ subunit are the likeliest reason for the different properties of the $\beta\gamma$ complex.

II. DISTRIBUTION OF β SUBUNITS

Since the predominant 36-kDa β subunits from many tissues are apparently identical, polyclonal antibodies cross-react over many tissues and species and can be used to quantitate the amount of β in membranes as well as to assess its distribution. Huff et al. (1985) estimated that there was at least as much $\beta\gamma$ in brain as the sum of all α subunits. Brain has the highest concentration of G proteins of any tissue; G proteins make up about 1% of brain membranes (Huff et al., 1985; Gierschik et al., 1985). β immunoreactivity was found at lower levels in all tissues examined (Huff et al., 1985; Audigier et al., 1987). Tissues from a variety of vertebrates including fish and reptiles contained an immunoreactive protein with the same molecular weight as β but tissues from invertebrates such as molluscs or insects did not (Homburger et al., 1987).

III. MEMBRANE ASSOCIATION OF $\beta\gamma$

The hydrophobicity of the $\beta\gamma$ subunits, together with the hydrophilic character of the α proteins, has led Sternweis (1986) to propose that the α subunits are held into the plasma membrane through their association with $\beta\gamma$. Evidence for this proposal comes from the observation that pure α subunits do not associate with phospholipid vesicles unless $\beta\gamma$ is also present. While the hydrophobic regions of $\beta\gamma$ may interact with hydrophobic membrane components, the subunit may also interact with cytoskeletal elements (Carlson et al., 1986). Whereas the $\beta\gamma$ subunits may play an important role in anchoring the α subunits, they are not likely to be the only attachment of the α subunit to the membrane. If they were, one would expect that activation of the G protein with a nonhydrolyzable analog of GTP, which causes dissociation of α from $\beta\gamma$, would release α subunits into the supernatant. Activation of G proteins in membranes with GTPγS does not in itself release α subunits (E. J. Neer, unpublished). Myristylation of α subunits may help to fix some α subunits to the plasma membrane, or α subunits may interact with other membrane proteins in addition to $\beta\gamma$.

IV. FUNCTIONAL REGIONS OF $\beta\gamma$ SUBUNITS

In contrast to the α subunit, very little is known about functional regions of the $\beta\gamma$ subunit. $\beta\gamma$ binds no known small molecules or metal ions. The amino acid sequence of β_2 includes 14 cysteine residues (Fong *et al.,* 1986). Ho and Fung (1984) reported that two were accessible to 5,5'-dithiobis(2-nitrobenzoic) acid in native transducin. The rate of reaction of the sulfhydryls in brain β with *N*-ethylmaleimide is not affected by association of $\beta\gamma$ with a (E. J. Neer, in preparation).

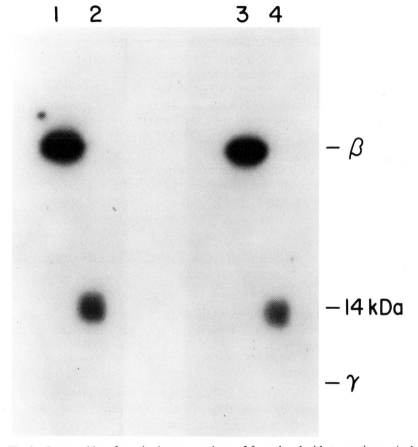

Fig. 2. Immunoblot of tryptic cleavage products of $\beta\gamma$ analyzed with two antisera raised against β. Protein was cleaved with trypsin for 30 min as described by Winslow *et al.,* 1986. After SDS-PAGE, the duplicate samples of β were transferred to nitrocellulose and probed with two different rabbit antisera to β (Huff and Neer, 1986), followed by incubation with ^{125}I-labeled Protein A. An autoradiogram is shown, exposed for 91 hr. Lanes 1 and 3 show uncleaved β; lanes 2 and 4 show β cleaved for 30 min.

Tryptic cleavage of the native $\beta\gamma$ subunit results in cleavage of the β subunit into two stable fragments with molecular weights of 26,000 and 14,000 (Fung and Nash, 1983; Halliday et al., 1984; Winslow et al., 1986). Although the two fragments are not covalently linked by disulfide bridges, they cannot readily be dissociated. In fact, the cleaved molecule is fully functional in supporting ADP-ribosylation of the α subunit (E. J. Neer, manuscript in preparation). The cleavage of the molecule into two stable large fragments suggests that $\beta\gamma$ may have two functional domains. However, the difficulty of dissociating the two pieces suggests, instead, that the two fragments may interdigitate to form a compact globular molecule. Since trypsin cleaves at polar amino acids, it is likely that the tryptic cleavage site would be in a hydrophilic region of the molecule and not buried in the lipid bilayer of the plasma membrane. This interpretation is consistent with the observation that the site of tryptic cleavage is accessible when the $\beta\gamma$ subunit is in the membrane (Winslow et al., 1986).

Polyclonal antibodies raised in rabbits against native $\beta\gamma$ and against SDS-denatured β which has been separated from the γ subunit all react virtually exclusively with epitopes on the 14-kDa tryptic fragment (Winslow et al., 1986; Fig. 2). Tainer et al. (1984) have suggested that mobile regions of proteins are the most antigenic. Correze et al. (1987) have shown that antibodies raised against transducin $\beta\gamma$ can affect adenylyl cyclase activity in brain membranes. However, the epitopes recognized by their antibody were not defined.

V. EFFECT OF $\beta\gamma$ SUBUNIT ON FUNCTION OF α SUBUNIT

A. Interaction of $\beta\gamma$ with α_s

In solution, activation of the G proteins by nonhydrolyzable analogs of GTP such as GTPγS or Gpp(NH)p is accompanied by dissociation of the α from the $\beta\gamma$ subunits.

$$\alpha\beta\gamma + \text{GTP}\gamma\text{S} \rightleftharpoons \alpha \cdot \text{GTP}\gamma\text{S} + \beta\gamma$$

By mass action, increasing the concentration of $\beta\gamma$ subunits should favor association of the subunits and hence deactivation of the G protein. This prediction was initially made and tested by Northup et al. (1983a,b) who showed that increasing the $\beta\gamma$ concentration reversed the activation of adenylyl cyclase by GTPγS-liganded α_s. The finding was subsequently confirmed by other laboratories (Neer et al., 1987). These observations led to the proposal that elevation of free $\beta\gamma$ concentration could be a general mechanism for hormone receptor-mediated inhibition of adenylyl cyclase.

Excess free $\beta\gamma$ in the membrane would be expected to blunt adenylyl cyclase activity (Bokoch, 1987). Since the $\beta\gamma$ subunits which associate with different α subunits are very similar to each other, it might follow that $\beta\gamma$ released from any abundant source in the plasma membrane would promote deactivation of G_s and hence inhibition of adenylyl cyclase. In this hypothesis, which will be discussed further below, inhibition of adenylyl cyclase is entirely due to reversal of stimulation by α_s.

The ability of $\beta\gamma$ to promote deactivation of α_s led to the first sensitive and specific assay for $\beta\gamma$ subunit function (Northup et al., 1983b). An important feature of this assay is that it appears to detect only the free $\beta\gamma$ subunits, and not to measure the $\beta\gamma$ already complexed with α subunits. The availability of an assay for $\beta\gamma$ function allows a comparison of the functional interchangeability of $\beta\gamma$ subunits derived from different sources. $\beta\gamma$ subunits derived from various hormone-responsive tissues and from the photoreceptor were all able to deactivate α_s, although there were subtle differences in concentration dependence (Cerione et al., 1987).

B. Interaction of $\beta\gamma$ Units with α Subunits Which Are Pertussis Toxin Substrates

A different kind of assay of $\beta\gamma$ function is based on the interaction of $\beta\gamma$ with the pertussis toxin substrate. Although pertussis toxin modifies the α subunit of G_i and G_o, the true substrate is not the resolved α subunit but the $\alpha\beta\gamma$ heterotrimer (Neer et al., 1984; Tsai et al., 1984; Katada et al., 1986b; Mattera et al., 1987). Using the ability of $\beta\gamma$ to allow pertussis toxin-catalyzed ADP-ribosylation of α subunits, Huff and Neer (1986) showed that the 39-kDa (α_o) and 41-kDa (α_i) α subunits from bovine brain differed in their apparent affinity for $\beta\gamma$. The apparent affinity of α_o for $\beta\gamma$ is about threefold lower than that of α_i. A similar observation was made by Katada et al. (1986). Since the $\beta\gamma$ subunits may be shared among many different kinds of α subunits, the observation that α subunits vary in their affinity for $\beta\gamma$ suggests that α subunits may be differentially capable of releasing $\beta\gamma$ when they are activated.

Treatment of cells or membranes with pertussis toxin prevents hormonal inhibition of adenylyl cyclase. A potential mechanism might be that ADP-ribosylation of the α subunit locks the heterotrimer in the associated state. Hormone receptors would then fail to inhibit adenylyl cyclase because the heterotrimer could not dissociate to release $\beta\gamma$. In solution, Huff and Neer (1986) showed that GTPγS is still able to dissociate ADP-ribosylated α_o and α_i subunits from $\beta\gamma$. GTP was not able to dissociate the subunits in solution whether or not they were ADP-ribosylated. Although ADP-ribosylation blocks the GTPase activity of transducin, some cleavage of GTP to GDP may have occurred at the active site of α_o or α_i. It is possible that failure to observe dissociation by GTP was due to its hydrolysis to GDP, a nucleotide which does not cause subunit dissociation. Another line of evidence supports the conclusion that ADP-ribosylated $\beta\gamma$ heterotrimers are able to dissociate. Casey et al. (1988) showed that $\beta\gamma$ acted catalytically to support ADP-ribosylation of purified subunits. The moles of [^{32}P]ADP-ribose incorporated into α subunits were substantially greater than the moles of $\beta\gamma$ present. These observations suggest that $\beta\gamma$ can dissociate from ADP-ribosylated α subunits and interact with other α subunits in solution. While covalent modification of α subunits by pertussis toxin may lead to subtle modifications of α and $\beta\gamma$ interactions, the primary consequence of pertussis toxin-catalyzed ADP-ribosylation is probably not stabilization of the inactive heterotrimer but uncoupling of the hormone receptor from the α subunit.

C. Effect of $\beta\gamma$ on Guanine Nucleotide Binding by α Subunit

Association of $\beta\gamma$ with α_s promotes the dissociation of GTPγS but increases the affinity of the α subunit for GDP. Brandt and Ross (1985) reported the differential effects of $\beta\gamma$ on nucleotide dissociation from α_s. The kinetics of this effect have been studied further with α_o (Higashijima et al., 1987) and transducin (Yamazaki et al., 1987). The affinity of α_o for GDP is increased by association with $\beta\gamma$ from a K_d of 40 nM to a K_d of 0.1 nM (Higashijima et al., 1987). Mg^{2+} diminishes the effect of $\beta\gamma$ on the affinity of α_o for GDP. The studies of Higashijima et al. (1987) suggest that the effect of Mg^{2+} is primarily on the rate of association of GDP with α_o and not on the rate of dissociation. Since $\beta\gamma$ inhibits the dissociation of GDP from α_o, it would be expected to inhibit the GTPase activity of the subunit and, at low concentration of Mg^{2+} this has indeed been observed (Higashijima et al., 1987). In the studies of Higashijima et al. (1987), α and $\beta\gamma$ subunits were reconstituted without hormone receptors. As described below, $\beta\gamma$ has the opposite effect when the reconstitution includes a receptor for hormone or light.

D. Effect of $\beta\gamma$ on α Subunit – Receptor Interactions

The $\beta\gamma$ subunit has been thought to enhance receptor association with α. The enhancement may be the consequence of interaction of $\beta\gamma$ with α or of $\beta\gamma$ with the receptor. Im et al. (1988) have obtained evidence to suggest that $\beta\gamma$ may interact with the β-adrenergic receptor.

   Fung (Fung, 1983; Fung and Nash, 1983) reconstituted transducin α and $\beta\gamma$ subunits into phospholipid membranes containing rhodopsin and found that the $\beta\gamma$ subunits enhanced the GTPase activity of the α subunit. The effect saturated at $\beta\gamma$ concentrations substantially lower than α subunit concentrations. Fung suggested that the $\beta\gamma$ subunits act catalytically to activate many α subunits and that they do so by enhancing the ability of the α subunit to interact with the receptor, rhodopsin. Interaction of α with a receptor promotes dissociation of GDP and binding of GTP; both effects increase GTPase activity. Florio and Sternweis (1985) reached a similar conclusion on the basis of different experiments. They found that the $\beta\gamma$ subunit alone had no effect on agonist binding to reconstituted muscarinic receptors from brain but did increase the effect of α subunits about two-fold.

VI. INTERACTION OF $\beta\gamma$ WITH EFFECTORS

A. Adenylyl Cyclase

There is general agreement that the dissociated guanine nucleotide triphosphate-liganded α_s subunit directly activates the adenylyl cyclase catalytic unit (Northup et al., 1983b). It is also clear that an excess of dissociated $\beta\gamma$ can reverse the activation (Northup et al., 1983a; Neer et al., 1987). Two extensions of these observations are, however, subjects of debate. Although there is suggestive evidence that $\alpha\beta\gamma$ actually does dissociate in the membrane, this has not been explicitly proven. While the release of a shared $\beta\gamma$ subunit from a variety of activated G proteins provides a mechanism for hormonal inhibition of adenylyl cyclase, it seems that this is not the only mechanism. If reversal of adenylyl cyclase activation by α_s were the only mechanism of hormonal inhibition of adenylyl cyclase, then inhibition ought never to be observed in the absence of stimulation by α_s. In fact, somatostatin is able to inhibit the adenylyl cyclase of mutant S49 cells (cyc^-) which are entirely lacking α_s (Hildebrandt et al., 1982). This important observation suggests that there must be an alternative pathway for hormonal inhibition of adenylyl cyclase. One possibility is that the α subunit of G_i directly inhibits adenylyl cyclase. Hildebrandt et al. (1984b) have argued for this mechanism on the basis of kinetic studies which showed that G_i and G_s act noncompetitively. Direct demonstration of inhibition of the isolated, solubilized adenylyl cyclase catalytic unit by purified α_i has been much more difficult than the analogous experiment showing stimulation of the catalytic unit with α_s. In fact, inhibition of catalytic activity by GTPγS-liganded α_i has only been shown when adenylyl cyclase is first stimulated by G_s (Katada et al., 1986a).

Paradoxically, purified α_i stimulates the activity of wild-type S49 cells

(Katada *et al.*, 1984a,b). The stimulatory effect of pure α_i in wild-type S49 cell membranes can be explained by sequestration of $\beta\gamma$ and facilitated activation of α_s. In *cyc⁻* cells, pure α_i inhibits adenylyl cyclase activity (Katada *et al.*, 1984a,b). S49 cell wild-type and mutant membranes provide different backgrounds in which to test the function of individual subunits. Both α_i and $\beta\gamma$ can act as inhibitors of adenylyl cyclase depending on the balance of G proteins in the particular membrane. Furthermore, the same α_i subunit can have opposite effects on adenylyl cyclase depending on the stoichiometry of the subunits.

Katada *et al.* (1986a) have proposed that the $\beta\gamma$ subunit may inhibit the catalytic unit by a mechanism independent of the α subunits. They have shown that preparations of adenylyl cyclase catalytic unit from bovine brain which are, as far as can be determined, free of any contaminating α_s, can still be inhibited by excess $\beta\gamma$. These studies were done using the resolved adenylyl cyclase catalytic unit from brain, which is a form of adenylyl cyclase particularly sensitive to activation by Ca^{2+}-calmodulin (Salter *et al.*, 1981). Katada *et al.* (1987) demonstrated that calmodulin-activated adenylyl cyclase can be inhibited by relatively low concentrations of the $\beta\gamma$ subunits. This inhibition was attributed to association of the $\beta\gamma$ subunits with calmodulin and removal of calmodulin activation. Association of $\beta\gamma$ with calmodulin was first proposed by Asano *et al.* (1986) although they did not directly demonstrate binding of calmodulin to $\beta\gamma$ itself. It remains to be shown whether the ability to bind calmodulin is a function of the particularly hydrophobic brain $\beta\gamma$ or whether it is a general property of $\beta\gamma$ subunits from other sources.

B. Phospholipase A₂ and Phospholipase C

Illumination of the rod outer segments of the bovine retina activates phospholipase A_2. This process depends on rhodopsin activation of transducin. Jelsema and Axelrod (1987) showed that addition of purified $\beta\gamma$ to dark-adapted, transducin-poor rod outer segments stimulated phospholipase A_2 activity. The α subunit of transducin slightly stimulated phospholipase A_2 but the predominant effect of transducin α was to inhibit the $\beta\gamma$-induced stimulation of phospholipase A_2. The blockade of $\beta\gamma$-induced stimulation of phospholipase A_2 appeared to be the result of subunit reassociation, since the GTPγS-liganded α subunit (which cannot associate with $\beta\gamma$) was not able to inhibit the activation of phospholipase A_2 induced by $\beta\gamma$. The concentration of $\beta\gamma$ required to activate phospholipase A_2 was relatively high (0.6 μM); however, this concentration is well within the range achievable in the rod outer segment where the molar concentration of transducin has been calculated to be about 500 mM (Chabre, 1985).

G protein regulation of phospholipase C activity is, as yet, poorly understood, but may follow a scheme similar to activation of adenylyl cyclase. Moriarty *et al.* (1988) have shown that in the *Xenopus* oocyte, muscarinic acetylcholine receptor activated Cl⁻ current by way of inositol 1,4,5-trisphosphate. Injection of human erythrocyte or bovine brain $\beta\gamma$ (but not injection of α subunits) inhibits 95% of the evoked Cl⁻ current.

C. Muscarinic-Gated K⁺ Channels

In the heart, muscarinic cholinergic receptors stimulate a potassium-selective ionic current ($i_{K \cdot ACh}$) through activation of a G protein. Pfaffinger *et al.* (1985) demonstrated that this activation apparently involves a pertussis toxin-sensitive G protein. Since the activity of the muscarinic-gated K⁺ channel can be studied in isolated patches of membrane, it is possible to reconstitute the ion channels with purified α or $\beta\gamma$ subunits of G proteins. Logothetis *et al.* (1987) reported that nanomolar concentrations of purified brain $\beta\gamma$ subunits activated $i_{K \cdot ACh}$ in chick embryonic atrial patches. This result was surprising since it had been expected that α subunits would be the only active agents. At about the same time, Codina *et al.* (1987) independently reported that picomolar concentrations of activated erythrocyte α subunits, but not $\beta\gamma$ subunits, open the K⁺ channel in guinea pig atrial patches. Further studies from Clapham and Neer's laboratories have extended the initial observations. The $\beta\gamma$ subunits originally used were from bovine brain and were a mixture of the 35-kDa and 36-kDa forms of the $\beta\gamma$ subunit, predominantly the 36-kDa form. $\beta\gamma$ subunits from human placenta, which are entirely the 35-kDa form, were subsequently shown to be equally effective in activating the K⁺ channel whereas transducin $\beta\gamma$ subunit were not (Logothetis *et al.,* 1988). Since the β component of transducin is the same as the 36-kDa β component of brain $\beta\gamma$, the discrepancy between $\beta\gamma$ from bovine brain and from bovine retina again emphasizes the potential importance of the role of the γ subunit.

The $\beta\gamma$ subunits open K⁺ channels at concentrations above 200 pM and maximally activate the channels at concentrations of 10 nM or greater. The purified brain $\beta\gamma$ preparations were contaminated with less than 0.01% α subunit, all inactive, and therefore the opening of the channel could not be attributed to contamination of the $\beta\gamma$ with activated α components. The detergent 3-[(3-cholamidopropyl)dimethylammonio]-1-propanesulfonate (CHAPS), which was used to suspend the $\beta\gamma$ subunits for these experiments, did not activate the channel by itself or in the presence of heat-inactivated $\beta\gamma$. An extremely important control was the demonstration that activation by $\beta\gamma$ was prevented by incubating $\beta\gamma$ with inactivated α subunits to form $\alpha\beta\gamma$ heterotrimers. Heat-inactivated α subunits were not able to form heterotrimers and did not affect $\beta\gamma$ activation. It is important to note that the detergent concentration was constant in these experiments and $\beta\gamma$ was present throughout. These experiments argue strongly for the specificity of the effect of $\beta\gamma$.

The initial experiments of Logothetis *et al.* (1987) were performed using membrane patches from chick atria. Chick atria appear to be less sensitive to stimulation by the α subunit than are the atria of mammalian hearts. Clapham's group was able to confirm the findings of Codina *et al.* (1987) that the 40-kDa α subunit from erythrocytes [called α_k by Codina *et al.* (1987)] was able to activate the K^+ channel in both chick and rat atrial patches. Purified α_o subunit (α_{39}) from bovine brain was able to activate the rat atrial patches at approximately the same concentration as the erythrocyte protein (Logothetis *et al.*, 1988). Yatani *et al.* (1988) have shown that all three α_i subtypes are equally active in stimulating this channel. The observation that different α subunits appear to be of approximately equal effectiveness in activating the K^+ channel is not surprising since the α_i and α_o subunits are extremely similar to each other. In many other cases, α subunits have been shown to be interchangeable (for example, see Florio and Sternweis, 1985).

By electrophysiological criteria it is clear that α and $\beta\gamma$ subunits activate the same type of channel in chick and mammal. The α subunit acts at a lower concentration than does the $\beta\gamma$ subunit although the concentration differences are hard to interpret. Since $\beta\gamma$ is hydrophobic, it is likely that it must first be incorporated into the membrane before it can diffuse within the membrane to its site of action. Recent studies by Kim *et al.* (1989) have shown that the action of $\beta\gamma$ on the K^+ channel may be mediated by phospholipase A_2. By comparison, the hydrophilic α protein may interact with the channel directly from the aqueous phase.

Release of two signal transduction elements by activation of the muscarinic receptor might explain some complex kinetics which have previously been observed. For example, desensitization of muscarinic-induced current decays in two exponential phases (Carmeliet and Mubagawa, 1986; Kurachi *et al.*, 1987) suggests two separate processes controlling the net current. This could reflect two pathways, one through αGTP and one through $\beta\gamma$, involving phospholipase A_2.

D. $\beta\gamma$ Subunits in the Yeast, *Saccharomyces cerevisiae*

Yeasts respond to mating pheromones through a transmembrane signalling system with many similarities to that which transduces mammalian responses to hormones. The deduced amino acid sequence of mating factor receptor predicts a protein with seven membrane spanning regions similar to a variety of mammalian hormone receptors (Nakayuma *et al.,* 1985). The yeast also contains a protein similar to the α subunit of G proteins which is part of the mating pheromone response pathway of haploid cells (Dietzel and Kurjan, 1987; Nakafuku *et al.,* 1987). Whiteway *et al.* (1989) have shown that the products of two other genes which are required for haploid cell mating (STE 4 and STE 18) encode proteins which are structurally similar to β and γ.

Analysis of the consequences of null mutations in the α,β and γ subunit genes suggests that the putative β and γ subunits play a positive role in initiating the mating response since null mutations in either β or γ are defective, both in mating and in the response to mating pheromone (Whiteway *et al.,* 1989). On the other hand, loss of α subunits produces a constitutive mating response leading to cell cycle arrest (Dietzel and Kurjan, 1987). This result suggests the α subunit is a negative modulator of the response. By analogy to the mammalian systems, activation of the pheromone receptor would lead to dissociation of α from $\beta\gamma$ subunits and subsequent activation of an as yet unknown effector enzyme by $\beta\gamma$. Reassociation with α would turn the reaction off.

VII. CONCLUSION

There is an appealing logic to the idea that after dissociation of $\alpha\beta\gamma$ heterotrimers, one molecule would activate a specific effector and another would turn off a competing pathway. For instance, activation of a G protein might simultaneously activate an enzyme such as phospholipase C through the α subunit and inhibit adenylyl cyclase by tying up its stimulator, α_s, in an inactive $\alpha\beta\gamma$ heterotrimer. Reality seems to be more complicated. First, $\beta\gamma$ can only be an effective deactivator when it is released from an abundant type of G protein. G_s is generally less abundant in membranes than the G_i group of proteins. Therefore, activation of G_s may release $\beta\gamma$ but probably not in sufficient quantity to affect G_i function. Thus, the subunit stoichiometry will, in part, determine whether dissociated, activated α and $\beta\gamma$ can oppose each other. An example of differing effects of α_i and $\beta\gamma$ depending on the background levels of G proteins was described above in the studies of Katada *et al.* (1986a,b) in S49 *cyc⁻* and wild-type membranes. Both α_i and $\beta\gamma$ could inhibit adenylyl cyclase.

The cardiac K^+ channel is an example of an effector which can be activated by both α and $\beta\gamma$ subunits. There is no strict specificity for the α-subunit subtype since several kinds of α subunits (α_{i-1}, α_{i-2}, α_{i-3}, α_o) can activate the same type of channel approximately equally (Logothetis *et al.,* 1987, 1988; Yatani *et al.,* 1988). The ability of both subunits of a G protein to have an active role in regulating the function of membrane effectors may give the cell greater flexibility in finely adjusting critical functions. However, we will not know the precise molecular details until all the components have been purified and reconstituted or until genetic manipulations allow us to better define the functions of the subunits.

ACKNOWLEDGMENTS

This work was supported by NIH Grants GM36259 to EJN and HL34873 to DEC. The authors thank Susan McHale for expertly typing the manuscript.

REFERENCES

Asano, T. Ogasawara, N., Kitajima, S., and Sano, M. (1986). Interaction of GTP-binding proteins with calmodulin. *FEBS Lett.* **203**, 135–138.

Audigier, Y., Pantaloni, C., Bigay, J., Deterre, P., Bockaert, J., and Homburger, V. (1987). Tissue expression and phylogenetic appearance of the β and γ subunits of GTP binding proteins. *FEBS Lett.* **189**, 107.

Bokoch, G. M. (1987). The presence of free G protein β/γ subunits in human neutrophils results in suppression of adenylate cyclase activity. *J. Biol. Chem.* **262**, 589–594.

Brandt, D. R., and Ross, E. M. (1985). GTPase activity of the stimulatory GTP-binding regulatory protein of adenylate cyclase, G_s. *J. Biol. Chem.* **260**, 266–272.

Buss, J. E., Mumby, S. M., Casey, P. J., Gilman, A. G., and Sefton, B. M. (1987). Myristylated α subunits of guanine nucleotide-binding regulatory proteins. *Proc. Natl. Acad. Sci. U.S.A.* **84**, 7493–7497.

Carlson, K. E., Woolkalis, M. J., Newhouse, M. G., and Manning, D. R. (1986). Fractionation of the β subunit common to guanine nucleotide-binding regulatory proteins with the cytoskeleton. *Mol. Pharmacol.* **30**, 463–468.

Carmeliet, E., and Mubagawa, K. (1986). Desensitization of the acetylcholine-induced increase of potassium conductance in rabbit cardiac Purkinje fibers. *J. Physiol.* **371**, 239–255.

Casey, P. J., Graziano, M. P., and Gilman, A. G. (1989). G protein beta/gamma subunits from bovine brain and retina: Equivalent catalytic support of ADP-ribosylation of alpha subunits by pertussis toxin but differential interactions with G_{salpha}. *Biochemistry* **28**, 611–616.

Cerione, R. A. Gierschik, P., Staniszewski, C., Benovic, J. L., Codina, J., Somers, R. Birnbaumer, L., Spiegel, A. M., Lefkowitz, R. J., and Caron M. G. (1987). Functional differences in the $\beta\gamma$ complexes of transducin and the inhibitory guanine nucleotide regulatory protein. *Biochemistry* **26**, 1485–1491.

Chabre, M. (1985). Trigger and amplification mechanisms in visual phototransduction. *Annu. Rev. Biophys. Chem.* **14**, 331–360.

Codina, J., Hildebrandt, J., Iyengar, R., Birnbaumer, L., Sekura, R. D., and Manclark, C. R. (1983). Pertussis toxin substrate, the putative N_i component of adenylyl cyclases is an $\alpha\beta$ heterodimer regulated by guanine nucleotide and magnesium. *Proc. Natl. Acad. Sci. U.S.A.* **80**, 4276–4280.

Codina, J., Hildebrandt, J. D., Sekura, R. D., Birnbaumer, M., Bryan, J., Manclark, C. R., Iyengar, R., and Birnbaumer, L. (1984). N_s and N_i, the stimulatory and inhibitory regulatory components of adenylyl cyclases. *J. Biol. Chem.* **259**, 5871–5886.

Codina, J., Stengel, D., Woo, S. L. C., and Birnbaumer, L. (1986). Beta-subunits of the human liver G_s/G_i signal-transducing proteins and those of bovine retinal rod cell transducin are identical. *FEBS Lett.* **207**, 187–192.

Codina, J., Yatani, A., Grenet, D., Brown, A. M., and Birnbaumer, L. (1987). The *alpha* subunit of G_k opens atrial potassium channels. *Science* **236**, 442–445.

Correze, C., d'Aalyer, J., Coussen, F., Berthillier, G., Deterre, P., and Monneron, A. (1987). Antibodies directed against transducin β subunits interfere with the regulation of adenylate cyclase activity in brain membranes. *J. Biol. Chem.* **262**, 15182–15187.

Dietzel, C., and Kurjan, J. (1987). The yeast SCG1 gene: A G_{alpha}-like protein implicated in the a- and alpha-factor response pathway. *Cell* **50**, 1001–1010.

Evans, T., Fawzi, A., Fraser, E. D., Brown, M. L., and Northup, J. K. (1987). Purification of a β_{35} form of the $\beta\gamma$ complex common to G-proteins from human placental membranes. *J. Biol. Chem.* **262**, 176–181.

Florio, V. A., and Sternweis, P. C. (1985). Reconstitution of resolved muscarinic cholinergic receptors with purified GTP-binding proteins. *J. Biol. Chem.* **260**, 3477–3483.

Florio, V. A., and Sternweis, P. C. (1989). Mechanisms of muscarinic receptor action on G_o in reconstituted phospholipid vesicles. *J. Biol. Chem.* **264**, 3909–3915.

Fong, H. K. W., Hurley, J. B., Hopkins, R. S., Miake-Lye, R., Johnson, M. S., Doolittle, R. F., and Simon, M. I. (1986). Repetitive segmental structure of the transducin β subunit: Homology with the CDC4 gene and identification of related mRNAs. *Proc. Natl. Acad. Sci. U.S.A.* **83**, 2162–2166.

Fong, H. K. W., Amatruda, T. T., III, Birren, B. W., and Simon, M. I. (1987). Distinct forms of the β subunit of GTP-binding regulatory proteins, identified by molecular cloning. *Proc. Natl. Acad. Sci. U.S.A.* **84**, 3792–3796.

Fung, B. K. K. (1983). Characterization of transducin from bovine retinal rod outer segments. I. Separation and reconstitution of the subunits. *J. Biol. Chem.* **258**, 10495–10502.

Fung, B. K. K., and Nash C. R. (1983). Characterization of transducin from bovine retinal rod outer segments. II. Evidence for distinct binding sites and conformational changes revealed by limited proteolysis with trypsin. *J. Biol. Chem.* **258**, 10503–10510.

Gao, B., Gilman, A. G., and Robishaw J. D. (1987). A second form of the β subunit of signal-transducing G proteins. *Proc. Natl. Acad. Sci. U.S.A.* **84**, 6122–6125.

Gierschik, P., Codina, J., Simons, C., Birnbaumer, L., and Spiegel, A. (1985). Antisera against a guanine nucleotide binding protein from retina cross-react with the β subunit of adenylyl cyclase-associated guanine nucleotide binding proteins, N_s and N_i. *Proc. Natl. Acad. Sci. U.S.A.* **82,** 727–731.

Gilman, A. G. (1987). G Proteins: Transducers of receptor generated signals. *Annu. Rev. Biochem.* **56,** 615–649.

Giulian, G., and Graham, J. (1988). The electrophoretic separation of low molecular weight polypeptides in polyacrylamide gels. *Hoefer Scientific Instruments Catalogue,* pp. 134–135.

Halliday, K. R., Stein, P. J., Chernoff, N., Wheeler, G. L., and Bitensky, M. W. (1984). Limited trypsin proteolysis of photoreceptor GTP-binding protein. *J. Biol. Chem.* **259,** 516–525.

Hekman, M., Holzhofer, A., Gierschik, P., Im, M. J., Jakobs, K. H., Pfeuffer, T., and Helmreich, E. J. M. (1987). Regulation of signal transfer from beta$_1$-adrenoceptor to adenylate cyclase by beta/gamma subunits in a reconstituted system. *Eur. J. Biochem.* **169,** 431–439.

Higashijima, T., Ferguson, K. M., Sternweis, P. C., Smigel, M. D., and Gilman A. G. (1987). Effects of Mg^{2+} and the $\beta\gamma$-subunit complex on the interactions of guanine nucleotides with G proteins. J. Biol. Chem. **262,** 762–766.

Hildebrandt, J. D., Hanoune, J., and Birnbaumer, L. (1982). Guanine nucleotide inhibition of cyc⁻ S49 mouse lymphoma cell membrane adenylyl cyclase. *J. Biol. Chem.* **257,** 14723–14725.

Hildebrandt, J. D., Codina, J., Risinger, R., and Birnbaumer, L. (1984a). Identification of a γ subunit associated with adenylyl cyclase regulatory proteins N_s and N_i. *J. Biol. Chem.* **259,** 2039–2042.

Hildebrandt, J. D., Codina, J., and Birnbaumer, L. (1984b). Interaction of the stimulatory and inhibitory regulatory proteins of the adenylyl cyclase system with the catalytic component of cyc⁻-S49 cell membranes. *J. Biol. Chem.* **259,** 13178–13185.

Hildebrandt, J. D., Codina, J., Rosenthal, W., Birnbaumer, L., Neer, E. J., Yamazaki A., and Bitensky M. W. (1985). Characterization by two-dimensional peptide mapping of the gamma subunits of N_s and N_i, the regulatory proteins of adenylyl cyclase, and of transducin, the guanine nucleotide-binding protein of rod outer segments of the eye. *J. Biol. Chem.* **260,** 14867–14872.

Ho, Y.K., and Fung, B. K. (1984). Characterization of transducin from bovine retinal rod outer segments. *J. Biol. Chem.* **259,** 6694–6699.

Homburger, V., Brabet, P., Audigier, Y., Pantaloni, C., Bockaert, J., and Ruout, B. (1987). Immunological localization of the GTP-binding protein G_o in different tissues of vertebrates and invertebrates. *Mol. Pharmacol.* **31,** 313–319.

Huff, R. M., and Neer, E. J. (1986). Subunit interactions of native and ADP-ribosylated α_{41} and α_{39}, two guanine nucleotide-binding proteins from bovine cerebral cortex. *J. Biol. Chem.* **261,** 1105–1110.

Huff, R. M., Axton, J. M., and Neer, E. J. (1985). Physical and immunological characterization of a guanine nucleotide finding protein purified from bovine cerebral cortex. *J. Biol. Chem.* **260,** 10864–10871.

Im, M. J., Holzhofer, A., Bottinger, H., Pfeuffer, T., and Helmreich, E. J. M. (1988). Interactions of pure $\beta\gamma$-subunits of G-proteins with purified β_1-adrenoceptor. *FEBS Lett.* **227,** 225–229.

Jelsema, C. L., and Axelrod, J. (1987). Stimulation of phospholipase A_2 activity in bovine rod outer segments by the $\beta\gamma$ subunits of transducin and its inhibition by the alpha subunit. *Proc. Natl. Acad. Sci. U.S.A.* **84,** 3623–3627.

Katada, T., Northup, J. K., Bokoch, G. M., Ui, M., and Gilman, A. (1984a). The inhibitory guanine nucleotide-regulatory component of adenylate cyclase. *J. Biol. Chem.* **259,** 3578–3585.

Katada, T., Bokoch, G. M., Smigel, M. D., Ui, M., and Gilman, A. G. (1984b). The inhibitory guanine nucleotide-binding regulatory component of adenylate cyclase. *J. Biol. Chem.* **259,** 3586–3595.

Katada, T., Oinuma, M., and Ui, M. (1986a). Mechanisms for inhibition of the catalytic activity of adenylate cyclase by the guanine nucleotide-binding proteins serving as the substrate of islet-activating protein, pertussis toxin. *J. Biol. Chem.* **261,** 5215–5221.

Katada, T., Oinuma, M., and Ui, M. (1986b). Two guanine nucleotide-binding proteins in rat brain serving as the specific substrate of islet-activating protein, pertussis toxin. *J. Biol. Chem.* **261,** 8182–8191.

Katada, T., Kusakabe, K., Oinuma, M., and Ui, M. (1987). A novel mechanism for the inhibition of adenylate cyclase via inhibitory GTP-binding proteins. Calmodulin-dependent inhibition of the cyclase catalyst by the $\beta\gamma$ subunits of GTP-binding proteins. *J. Biol. Chem.* **262,** 11897–11900.

Kim, D., Lewis, D. L., Graziadei, L., Neer, E. J., Bar-Sagi, D., and Clapham, D. E. (1989). G-protein beta/gamma subunits activate the cardiac muscarinic K^+-channel via phospholipase A_2. *Nature* **337,** 557–560.

Kurachi, Y., Nakajima, T., and Sugimoto, T. (1987). Short term desensitization of muscarinic K^+ channel current in isolated atrial myocytes and possible role of GTP binding proteins. *Pfluegers Arch.* **410,** 227–233.

Lochrie, M., and Simon, M. I. (1988). G protein multiplicity in eukaryotic signal transduction systems. *Biochemistry* **27,** 4957–4961.

Logothetis, D. E., Kurachi, Y., Galper, J., Neer, E. J., and Clapham, D. E. (1987). The $\beta\gamma$ subunits of GTP-binding proteins activate the muscarinic K^+ channel in heart. *Nature* **325,** 321–326.

Logothetis, D. E., Kim, D., Northup, J. K., Neer, E. J., and Clapham, D. E. (1988). Specificity of action of G protein subunits on the cardiac muscarinic K^+ channel. *Proc. Natl. Acad. Sci. U.S.A.* **85,** 5814–5818.

Manning, D. R., and Gilman, A. G. (1983). The regulatory components of adenylate cyclase and transducin. A family of structurally homologous guanine nucleotide-binding proteins. *J. Biol. Chem.* **258,** 7059–7063.

Mattera, R., Codina, J., Sekura, R. D., and Birnbaumer, L. (1987). Guanosine 5'-0-(3-thiotriphosphate) reduces ADP-ribosylation of the inhibitory guanine nucleotide-binding regulatory protein of adenylyl cyclase (N_i) by pertussis toxin without causing dissociation of the subunits of N_i. *J. Biol. Chem.* **262,** 11247–11251.

Merril, C. R., Goldman, D., Sedman, S. A., and Ebert, M. H. (1981). *Science* **211**, 1437–1438.

Nakayama, N., Miyajima, A., Arai, K. (1985). Nucleotide sequences of STE2 and STE3, cell type-specific sterile genes from *Saccharomyces cerevisiae*. *EMBO J.* **4**, 2643–2648.

Neer, E. J., Lok, J. M., and Wolf, L. G. (1984). Purification and properties of the inhibitory guanine nucleotide regulation unit of brain adenylate cyclase. *J. Biol. Chem.* **259**, 14222–14229.

Neer, E. J., Wolf, L. G., and Gill, D. M. (1987). The stimulatory guanine-nucleotide regulatory unit of adenylate cyclase from bovine cerebral cortex. *Biochem. J.* **241**, 325–336.

Northup, J. K., Sternweis, P. C., and Gilman, A. G. (1983a). The subunits of the stimulatory regulatory component of adenylate cyclase. *J. Biol. Chem.* **258**, 11361–11368.

Northup, J. K., Smigel, H., Sternweis, P. C., and Gilman, A. G. (1983b). The subunits of the stimulatory regulatory component of adenylate cyclase. *J. Biol. Chem.* **258**, 11369–11376.

Pfaffinger, P. J., Martin, J. M., Hunber, D. D., Nathanson, N. M., and Hille, B. (1985). GTP-binding proteins couple cardiac muscarinic receptors to a K^+ channel. *Nature* **317**, 536–538.

Ransnas, L. A., and Insel, P. A. (1988). Subunit dissociation is the mechanism for hormonal activation of the G_s protein in native membranes. *J. Biol. Chem.* **263**, 17239–17242.

Roof, D. J., Applebury, M. L., and Sternweis, P. C. (1985). Relationships within the family of GTP-binding proteins isolated from bovine central nervous system. *J. Biol. Chem.* **260**, 16242–16249.

Salter, R. S., Krinks, M. H., Klee, C. B., and Neer, E. J. (1981). Calmodulin activates the isolated catalytic unit of brain adenylate cyclase. *J. Biol. Chem.* **256**, 9830–9833.

Sternweis, P. C. (1986). The purified α subunits of G_o and G_i from bovine brain require $\beta\gamma$ for association with phospholipid vesicles. *J. Biol. Chem.* **261**, 631–637.

Sternweis, P. C., and Robishaw, J. D. (1984). Isolation of two proteins with high affinity for guanine nucleotides from membranes of bovine brain. *J. Biol. Chem.* **259**, 13806–13813.

Sternweis, P. C., Northup, J. K., Smigel, M. D., and Gilman, A. G. (1981). The regulatory component of adenylate cyclase. *J. Biol. Chem.* **256**, 11517–11526.

Stryer, L., and Bourne, H. R. (1986). G Proteins: A family of signal transducers. *Annu. Rev. Cell Biol.* **2**, 391–419.

Tainer, J. A., Getzoff, E. D., Alexander, H., Houghten, R. A., Olson, A. J., Lerner, R. A., and Hendrickson, W. A. (1984). The reactivity of anti-peptide antibodies is a function of atomic mobility of sites in a protein. *Nature* **312**, 127–134.

Tsai, S. C., Adamik, R., Kanaho, Y., Hewlett, E. L., and Moss, J. (1984). Effects of guanyl nucleotides and rhodopsin on ADP-ribosylation of the inhibitory GTP-binding component of adenylate cyclase by pertussis toxin. *J. Biol. Chem.* **259**, 15320–15323.

Van Dop, C., Medynski D., Sullivan, K., Wu, A. M., Fung, B. K.-K., and Bourne, H. R. (1984). Partial cDNA sequence of the gamma subunit of transducin. *Biochem. Biophys. Res. Commun.* **124,** 250–255.

Wessling-Resnick, M., and Johnson, G. L. (1987). Kinetic and hydrodynamic properties of transducin: Comparison of physical and structural parameters for GTP-binding regulatory proteins. *Biochemistry* **26,** 4316–4323.

Whiteway, M., Hougan, L., Dignard, D., Thomas, D. Y., Bell, L., Saari, G. C., Grant, F. J., O'Hara, P., and MacKay, V. L. (1989). The STE4 and STE18 genes of yeast encode potential beta and gamma subunits of the mating factor receptor-coupled G protein. *Cell* **56,** 467–477.

Winslow, J. W., Van Amsterdam, J. R., and Neer, E. J. (1986). Conformations of the α_{39}, α_{41}, and $\beta\gamma$ components of brain guanine nucleotide-binding proteins. *J. Biol. Chem.* **261,** 7571–7579.

Woolkalis, M. J., and Manning, D. R. (1987). Structural characteristics of the 35- and 36-kDa forms of the β subunit common to GTP-binding regulatory proteins. *Mol. Pharmacol.* **32,** 1–6.

Yamazaki, A., Tatsumi, M., Torney, D. C., and Bitensky, M. W. (1987). The GTP-binding protein of rod outer segments. I. Role of each subunit in the GTP hydrolytic cycle. *J. Biol. Chem.* **262,** 9316–9323.

Yatani, A., Mattera, R., Codina, J., Graf, R., Okabe, K., Padrell, E., Iyengar, R., Brown, A. M., and Birnbaumer, L. (1988). The G protein-gated atrial K^+ channel is stimulated by three distinct G_i alpha-subunits. *Nature* **336,** 680–682.

Yatsunami, K., Pandya, B. V., Oprian, D. D., and Khorana, H. G. (1985). cDNA-derived amino acid sequence of the gamma subunit of GTPase from bovine rod outer segments. *Proc. Natl. Acad. Sci. U.S.A.* **82,** 1936–1940.

CHAPTER 4

Organization of Genes Coding for G-Protein α Subunits in Higher and Lower Eukaryotes

Yoshito Kaziro, Hiroshi Itoh, Masato Nakafuku
Institute of Medical Science, University of Tokyo, 4-6-1, Shirokanedai, Minatoku, Tokyo 108, Japan

G proteins are involved in a variety of transmembrane signaling systems as transducers (for reviews see Gilman, 1987; Stryer and Bourne, 1986; and other chapters of this volume). Two G proteins, G$_s$ and G$_i$, are involved in hormonal stimulation and inhibition, respectively, of adenylyl cyclase, whereas G$_o$ (another G protein), which is present predominantly in brain tissues, may be involved in neuronal responses. Two transducins, G$_{t1}$ and G$_{t2}$, which are present in retinal rods and cones, respectively, regulate cGMP phosphodiesterase activity and mediate visual signal transduction.

There is evidence suggesting the presence of additional G proteins, which may be involved in the activation of phospholipase C and phospholipase A_2, as well as the gating of K^+ and Ca^{2+} channels.

In this article, we give a brief review of the structure of cDNAs for various G-protein α subunits (Gαs), the organization of human genes for Gαs and the occurrence of two Gα genes in *Saccharomyces cerevisiae.*

I. ISOLATION OF cDNA CLONES FOR G-PROTEIN α SUBUNITS FROM MAMMALIAN CELLS

The cDNA sequences for α subunits of numerous G proteins have been reported (Table I). These studies have revealed that there are at least three subtypes of $G_i\alpha$, which have closely related but distinct structures. The presence of multiple $G_i\alpha$ subspecies had been suggested from the molecular heterogeneity, immunological distinctness, and functional differences of the pertussis-toxin substrates in mammalian cells (reviewed in Itoh *et al.,* 1988a and 1988b). Transducins (G_t) also have two subtypes, $G_{t1}\alpha$ and $G_{t2}\alpha$, which are expressed in rods and cones, respectively (Lerea *et al.,* 1986). On the other hand, different $G_s\alpha$ cDNAs are generated by alternative splicing as described below.

Studies of G-protein α subunits from rat C6 glioma cells using cloned cDNAs revealed the structure of $G_s\alpha$, $G_{i2}\alpha$, $G_{i3}\alpha$, and $G_o\alpha$, consisting of 394, 355, 354, and 354 amino acid residues, respectively, with molecular weights of 45,663; 40,499; 40,522; and 40,068. $G_{i1}\alpha$, which is the most abundant among $G_i\alpha$ subtypes in brain tissues, was not expressed in C6 glioma cells. However, $G_{i1}\alpha$ cDNA clones were obtained from rat olfactory epithelium (Jones and Reed, 1987), and brain (H. Itoh *et al.* unpublished observations). We have isolated a new G_α clone (designated as $G_x\alpha$) which is apparently insensitive to pertussis toxin (Matsuoka *et al.,* 1988). Human $G_z\alpha$ cDNA, isolated independently by Fong *et al.* (1988) from retina, may be the counterpart of $G_x\alpha$. Rat $G_{i1}\alpha$ and $G_x\alpha$ code for 354 and 355 amino acid residues, respectively, giving molecular weights of 40,345 and 40,879. The amino acid sequences of rat $G_s\alpha$, $G_{i2}\alpha$, $G_{i3}\alpha$, $G_{i1}\alpha$, $G_o\alpha$, and $G_x\alpha$ deduced from the nucleotide sequences are shown in Fig. 1.

Figure 2 shows a schematic representation of the structure of *E. coli* EF-Tu, G-protein α subunits (Gα), yeast RAS2 protein, and mammalian H-ras p21 protein. A remarkable homology was found in two regions, designated as P and G sites, of all GTP-binding proteins. In EF-Tu, earlier biochemical studies indicated that the region around Cys^{137} of EF-Tu (G site) is responsible for interaction with guanine nucleotides (Kaziro, 1978),

Table I Molecular Cloning of G_s, G_i, G_o, G_x, and $G_t\alpha$ Subunit Genes and cDNAs

DNA library	$G_s\alpha$	$G_{i1}\alpha$	$G_{i2}\alpha$	$G_{i3}\alpha$	$G_o\alpha$	$G_x\alpha$	$G_{t1}\alpha$	$G_{t2}\alpha$	Reference
Genomic library									
Human	$G_s\alpha$								Kozasa et al. (1988)
Human		$G_{i1}\alpha$	$G_{i2}\alpha$	$G_{i3}\alpha$					Itoh et al. (1988b)
Human			$G_{i2}\alpha$						Weinstein et al. (1988)
Human						$G_x\alpha$			Matsuoka et al. (1988)
cDNA library									
Human brain	$G_s\alpha$	$\alpha_{i\text{-}1}$							Bray et al. (1986, 1987)
Human brain					$G_o\alpha$				Lavu et al. (1988)
Human T-cells			α_{i2}	α_{i3}					Beals et al. (1987)
Human monocytes (U-937)			$G_i\alpha$						Didsbury et al. (1987)
Human granulocytes (HL-60)				$G_x\alpha$					Didsbury and Snyderman (1987)
Human liver	$G_s\alpha$								Mattera et al. (1986)
Human liver				$\alpha_{i\text{-}3}$					Suki et al. (1987)
Human liver				$\alpha_{i\text{-}3}$					Codina et al. (1988)
Human retina					$G_z\alpha$				Fong et al. (1988)
Bovine brain	$G_s\alpha$								Harris et al. (1985)
Bovine adrenal gland	$G_s\alpha$								Robishaw et al. (1986b)
Bovine cerebral cortex	$G_s\alpha$	$G_i\alpha$							Nukada et al. (1986a,b)
Bovine pituitary gland		α_i	α_h						Michel et al. (1986)
Bovine retina					$G_o\alpha$				Van Meurs et al. (1987)
Bovine cerebellum					G_{39}				Ovchinnikov et al. (1987)
Bovine retina							$T\alpha$		Tanabe et al. (1985)
Bovine retina							$T\alpha$		Medynsky et al. (1985)
Bovine retina							$G_t\alpha$		Yatsunami and Khorana (1985)
Bovine retina								$T\alpha$	Lochrie et al. (1985)
Rat brain					$G_x\alpha$				Matsuoka et al. (1988)
Rat glioma cells (C6)	$G_s\alpha$		$G_{i2}\alpha$	$G_{i3}\alpha$	$G_o\alpha$				Itoh et al. (1986, 1988b)
Rat olfactory epithelium	$G\alpha_s$	$G\alpha_{i1}$	$G\alpha_{i2}$	$G\alpha_{i3}$	$G\alpha_o$				Jones and Reed (1987)
Mouse macrophages (PU-5)	α_s		α_i						Sullivan et al. (1986)
Mouse lymphoma cells (S49)	$G_s\alpha$								Sullivan et al. (1987)
Mouse lymphoma cells (S49)	$G_s\alpha$								Rall and Harris (1987)

```
Gsα : MGCLGNSKTEDQRNEEKAQREANKKIEKQLQKDKQVYRATHRLLLLGAGESGKSTIVKQM  60
Gi2α: MGCTVSA-------EDKAAAERSKMIDKNLREDGEKAAREVKLLLLGAGESGKSTIVKQM  53
Gi3α: MGCTLSA-------EDKAAVERSKMIDRNLREDGEKAAKEVKLLLLGAGESGKSTIVKQM  53
Gi1α: MGCTLSA-------EDKAAVERSKMIDRNLREDGEKAAREVKLLLLGAGESGKSTIVKQM  53
Goα : MGCTLSA-------EERAALERSKAIEKNLKEDGISAAKDVKLLLLGAGESGKSTIVKQM  53
Gxα : MGCRQSS-------EEKEAARRSRRIDRHLRSESQRQRREIKLLLLGTSNSGKSTIVKQM  53

Gsα : RILHVNGFNGEGGEEDPQAARSNSDGEKATKVQDIKNNLKEAIETIVAAMSNLVPPVELA 120
Gi2α: KIIHEDGYSEE---------------ECRQYRAVVYSNTIQSIMAIVKAMGNLQIDFADP  98
Gi3α: KIIHEDGYSED---------------ECKQYKVVVYSNTIQSIIAIIRAMGRLKIDFGEA  98
Gi1α: KIIHEAGYSEE---------------ECKQYKAVVYSNTIQSIIAIIRAMGRLKIDFGDA  98
Goα : KIIHEDGFSGE---------------DVKQYKPVVYSNTIQSLAAIVRAMDTLGVEYGDK  98
Gxα : KIIHSGGFNLE---------------ACKEYKPLIIYNAIDSLTRIIRALAALKIDFHNP  98

Gsα : NPENQFRVDYILSVMNVPNFDFPPEFYEHAKALWEDEGVRACYERSNEYQLIDCAQYFLD 180
Gi2α: QRADDARQLFALSCAAEEQGMLPEDLSGVIRRLWADHGVQACFGRSREYQLNDSAAYYLN 158
Gi3α: ARADDARQLFVLAGSAEE-GVMTSELAGVIKRLWRDGGVQACFSRSREYQLNDSASYYLN 157
Gi1α: ARADDARQLFVLAGAAEE-GFMTAELAGVIKRLWKDSGVQACFNRSREYQLNDSAAYYLN 157
Goα : ERKADSKMVCDVVSRMEDTEPFSAELLSAMMRLWGDSGIQECFNRSREYQLNDSAKYYLD 158
Gxα : DRAYDAVQLFALTGPAESKGEITPELLGVMRRLWADPGAQACFGRSSEYHLEDNAAYYLN 158

Gsα : KIDVIKQADYVPSDQDLLRCRVLTSGIFETKFQVDKVNFHMFDVGGQRDERRKWIQCFND 240
Gi2α: DLERIAQSDYIPTQQDVLRTRVKTTGIVETHFTFKDLHFKMFDVGGQRSERKKWIHCFEG 218
Gi3α: DLDRISQTNYIPTQQDVLRTRVKTTGIVETHFTFKELYFKMFDVGGQRSERKKWIHCFEG 217
Gi1α: DLDRIAQPNYIPTQQDVLRTRVKTTGIVETHFTFKDLHFKMFDVGGQRSERKKWIHCFEG 217
Goα : SLDRIGAADYQPTEQDILRTRVKTTGIVETHFTFKNLHFRLFDVGGQRSERKKWIHCFED 218
Gxα : DLERIAAPDYIPTVEDILRSRDMTTGIVENKFTFKELTFKMVDVGGQRSERKKWIHCFEG 218

Gsα : VTAIIFVVASSSYNMVIREDNQTNRLQEALNLFKSIWNNRWLRTISVILFLNKQDLLAEK 300
Gi2α: VTAIIFCVALSAYDLVLAEDEEMNRMHESMKLFDSICNNKWFTDTSIILFLNKKDLFEEK 278
Gi3α: VTAIIFCVALSDYDLVLAEDEEMNRMHESMKLFDSICNNKWFTDTSIILFLNKKDLFEEK 277
Gi1α: VTAIIFCVALSDYDLVLAEDEEMNRMHESMKLFDSICNNKWFTDTSIILFLNKKDLFEEK 277
Goα : VTAIIFCVALSGYDQVLHEDETTNRMHESLMLFDSICNNKFFIDTSIILFLNKKDLFGEK 278
Gxα : VTAIIFCVELSGYDLKLYEDNQTSRMAESLRLFDSICNNNWFINTSLILFLNKKDLLSEK 278

Gsα : VLAGKSKIEDYFPEFARYTTPEDATPEPGEDPRVTRAKYFIRDEFLRISTASGDGRHYCY 360
Gi2α: IT--QSPLTICFPEYTGANKYDE-------------AASYIQSKFEDLNKRKDTKEIY-- 321
Gi3α: IK--RSPLTICYPEYTGSNTYEE-------------AAAYIQCQFEDLNRRKDTKEVY-- 320
Gi1α: IK--KSPLTICYPEYAGSNTYEE-------------AAAYIQCQFEDLNKRKDTKEIY-- 320
Goα : IK--KSPLTICFPEYPGSNTYED-------------AAAYIQTQF-ESKNRSPNKEIY-- 320
Gxα : IR--RIPLSVCFPEYKGQNTYEE-------------AAVYIQRQFEDLNRNKETKEIY-- 321

Gsα : PHFTCAVDTENIRRVFNDCRDIIQRMHLRQYELL 394
Gi2α: THFTCATDTKNVQFVFDAVTDVIIKNNLKDCGLF 355
Gi3α: THFTCATDTKNVQFVFDAVTDVIIKNNLKECGLY 354
Gi1α: THFTCATDTKNVQFVFDAVTDVIIKNNLKDCGLF 354
Goα : CHMTCATDTNNIQVVFDAVTDIIIANNLRGCGLY 354
Gxα : SHFTCATDTSNIQFVFDAVTDVIIQNNLKYIGLC 355
```

Fig. 1. Deduced amino acid sequences of α subunits of rat $G_s\alpha$, $G_{i2}\alpha$, $G_{i3}\alpha$, $G_{i1}\alpha$, $G_o\alpha$, and $G_x\alpha$. The sequences of $G_s\alpha$ and $G_{i2}\alpha$ are from Itoh *et al.* (1986); $G_{i3}\alpha$ from Itoh *et al.* (1988b); $G_{i1}\alpha$ from Jones and Reed (1987); $G_o\alpha$ from Itoh *et al.* (1986) and Jones and Reed (1987); and $G_x\alpha$ from Matsuoka *et al.* (1988).

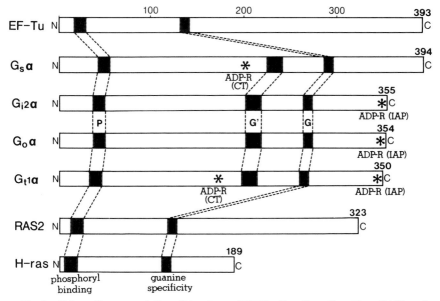

Fig. 2. Schematic representation of structures of EF-Tu, $G_s\alpha$, $G_{i2}\alpha$, $G_o\alpha$, $G_{t1}\alpha$, *RAS2,* and H-*ras* p21. ADP-R (CT) and ADP-R (IAP) are the ADP-ribosylation sites by cholera toxin and pertussis toxin, respectively.

and later the four residues Asn-Lys-Cys-Asp were found to be situated close to the guanine ring by X-ray analysis (Jurnak, 1985). On the other hand, it has been shown that the mutation of amino acid residue 12 of p21 from Gly to Val decreases GTPase activity and increases transforming activity. The sequence homologous to this region (P site) was found in all

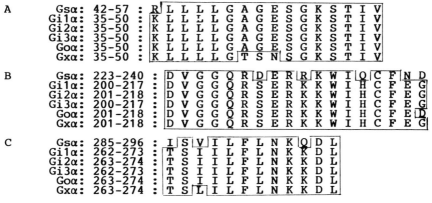

Fig. 3. Conserved sequences of G proteins. Sequences of the P site (A), G' site (B), and G site (C) are shown.

GTP-binding proteins. G' site is a unique sequence which is highly conserved in all G proteins but is absent in other GTP-binding proteins except for ARF proteins (see Sewell and Kahn, 1988). The deduced amino acid sequences of $G\alpha s$ in the P, G', and G sites are shown in Fig. 3A, 3B, and 3C, respectively.

It must be noted that the predicted amino acid sequence of $G_x\alpha$ is different from other $G\alpha$ proteins at the P site. As shown in Fig. 3A, three amino acid residues in the consensus GTP-hydrolysis site of $G_x\alpha$ (Thr-Ser-Asn, at positions 41–43) are different from the corresponding residues in other $G\alpha$ proteins (Ala-Gly-Glu). It remains to be seen whether the kinetics of the $G_x\alpha$-mediated signal transduction may be different from the other systems due to the replacement of Gly (which corresponds to Gly^{12} of p21) to Ser.

II. ISOLATION OF HUMAN $G\alpha$ GENES

We have screened human genomic libraries with the above rat cDNA clones and thus isolated human genes coding for $G_s\alpha$, $G_{i1}\alpha$, $G_{i2}\alpha$, $G_{i3}\alpha$, and $G_o\alpha$. So far, we have determined the gene organization and nucleotide sequences of total exons of $G_s\alpha$, $G_{i2}\alpha$, and $G_{i3}\alpha$, and obtained the partial sequences (exons 1, 2, and 3) of $G_{i1}\alpha$ (Kozasa et al., 1988; Itoh et al., 1988b) (see Fig. 4). The human $G_o\alpha$ gene is a huge gene spanning at least 90 kilobases (kb); however, the organization of exons is completely identical to those of the $G_i\alpha$ subfamily (T. Tsukamoto, H. Itoh, and Y. Kaziro, unpublished observations).

III. STRUCTURE OF HUMAN $G_s\alpha$ GENE AND GENERATION OF FOUR $G_s\alpha$ cDNAs BY ALTERNATIVE SPLICING

We have screened the human genomic libraries with rat $G_s\alpha$ cDNA and have isolated the human $G_s\alpha$ gene (Kozasa et al., 1988). Three overlapping clones were isolated which cover the entire human $G_s\alpha$ gene. The gene contains 13 exons and 12 introns and spans about 20 kb of genomic DNA (Fig. 4A).

The presence of two species of $G_s\alpha$ protein with different molecular masses (45 and 52 kDa) was known (Northup et al., 1980). Bray et al. (1986) isolated four different $G_s\alpha$ cDNAs ($G_s\alpha$-1 to -4) from human brain and characterized their partial structure. $G_s\alpha$-1 and $G_s\alpha$-3 are identical except that $G_s\alpha$-3 lacks a single stretch of 45 nucleotides. $G_s\alpha$-2 and $G_s\alpha$-4

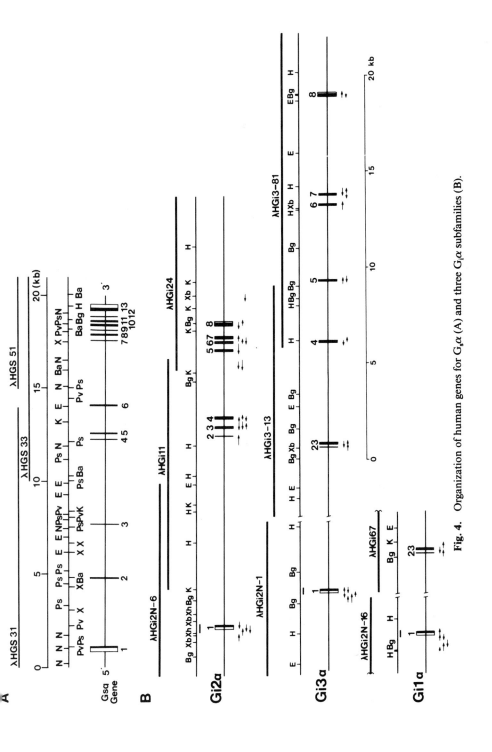

Fig. 4. Organization of human genes for $G_s\alpha$ (A) and three $G_i\alpha$ subfamilies (B).

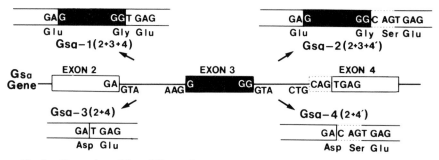

Fig. 5. Generation of four different $G_s\alpha$ mRNAs by alternative splicing. The $G_s\alpha$ gene is shown in the center. $G_s\alpha$ mRNAs are indicated by $G_s\alpha$-1, -2, -3, and -4. For details see Kozasa *et al.* (1988).

have 3 additional nucleotides (CAG) compared to $G_s\alpha$-1 and $G_s\alpha$-3 located 3′ to the above 45 nucleotides. Robishaw *et al.* (1986a) also isolated two $G_s\alpha$ cDNAs from bovine adrenal gland that correspond to $G_s\alpha$-1 and $G_s\alpha$-4, and showed that these two cDNAs generated a 52- and a 45-kDa protein when expressed in COS-m6 cells. Mattera *et al.* (1986) also isolated two $G_s\alpha$ cDNAs from human liver that correspond to $G_s\alpha$-1 and $G_s\alpha$-4.

Comparison of the four types of human $G_s\alpha$ cDNAs reported by Bray *et al.* (1986) with the sequence of the human $G_s\alpha$ gene of Kozasa *et al.* (1988) suggests that four types of $G_s\alpha$ mRNAs may be generated from a single $G_s\alpha$ gene by alternative splicing as shown in Fig. 5. $G_s\alpha$-1 has a sequence identical to exons 2, 3, and 4, whereas $G_s\alpha$-3 lacks exon 3. $G_s\alpha$-2 and $G_s\alpha$-4 have 3 additional nucleotides (CAG) to $G_s\alpha$-1 and $G_s\alpha$-3, respectively, at the 5′ end of exon 4. In the genomic sequence of the 3′ splice site of intron 3, this CAG sequence is found. Although the 5′ adjacent nucleotides to the CAG are TG and do not match with the 3′ splice consensus sequence AG, this 3′ splice site may be used for the production of $G_s\alpha$-2 and $G_s\alpha$-4. One additional serine residue in $G_s\alpha$-2 and $G_s\alpha$-4 may be the potential site for phosphorylation by protein kinase C, and the alternative use of these splice sites may provide $G_s\alpha$ proteins with differential regulatory properties.

IV. HUMAN GENES FOR $G_i\alpha$ SUBTYPES

The coding region of the human $G_{i2}\alpha$ and $G_{i3}\alpha$ genes is split into eight exons and seven introns (Itoh *et al.*, 1988b) (Fig. 4B). There is an additional exon (exon 9) in the 3′ noncoding region of $G_{i2}\alpha$ and $G_{i3}\alpha$, but this is not included in the figure. For human $G_{i1}\alpha$, we only have the sequences of exons 1 to 3 at present. All introns begin with the sequence GT at the 5′ end and end with the sequence AG at the 3′ end. Remarkably, the positions of the splice junctions on the sequence of cDNA for $G_{i2}\alpha$ and $G_{i3}\alpha$ are identical, although the length of the introns is different (Itoh *et al.*, 1988b). The same splice sites are also conserved in the partial sequence (exons 1, 2, and 3) of the human $G_{i1}\alpha$ gene as well as in the human $G_o\alpha$ gene (T. Tsukamoto, unpublished). From the Southern blot analysis, it appears that each of the three $G_i\alpha$ genes occurs as a single copy per haploid human genome.

V. ORGANIZATION OF HUMAN Gα GENES

The exon–intron organization of the $G_s\alpha$, $G_{i2}\alpha$, $G_{i3}\alpha$, and $G_o\alpha$ genes was compared with the predicted functional domain structure of proteins (Fig. 6). The NH_2-terminal domain encoded by exon 1 is hydrophilic and contains the site for limited tryptic digestions. Although this region may be involved in interaction with $\beta\gamma$ subunits, its precise function has not yet been shown. Exon 2 encodes a short region (24 and 14 amino-acid residues, respectively, for $G_s\alpha$ and $G_i\alpha$s), which is the most conserved among all $G\alpha$ proteins and is responsible for GTP hydrolysis. Exon 3 of $G_s\alpha$ is the one which is unique to $G_s\alpha$. The most structurally divergent domain is encoded by exons 4 to 6 of $G_s\alpha$ and 3 to 4 of $G_i\alpha$. The amino acid sequences of residues 110–140 of $G_s\alpha$ and 90–130 of $G_i\alpha$ are remarkably diverse. Exon 8 of $G_s\alpha$ contains Arg^{201}, which is ADP-ribosylated in the presence of cholera toxin (Van Dop *et al.*, 1984). ADP-ribosylation of $G_s\alpha$

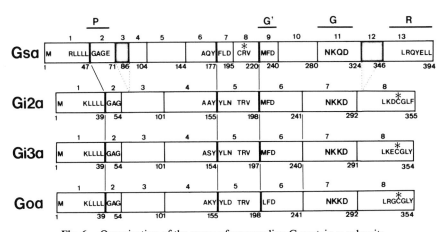

Fig. 6. Organization of the exons of mammalian G-protein α subunits.

by cholera toxin causes a decrease in affinity for $\beta\gamma$ subunits (Kahn and Gilman, 1984). Arg^{179} in exon 5 of $G_{i2}\alpha$ corresponds to this arginine residue. The domain encoded by exons 9 to 11 of $G_s\alpha$, and 6 to 7 of $G_i\alpha$ is strongly conserved among all $G\alpha$ proteins. This domain is involved in formation of a core structure for GTP binding, along with the region encoded by exon 2 (Masters *et al.,* 1986). The sequence Asn-Lys-Xaa-Asp, common to all guanine nucleotide-binding proteins, occurs in exon 11 of $G_s\alpha$ and exon 7 of $G_i\alpha$. The conserved Asp^{223} in exon 9 of $G_s\alpha$, and Asp^{201} in exon 6 of $G_{i2}\alpha$, may form a salt bridge to Mg^{2+}, which is linked to the β-phosphoryl group of GDP (Jurnak, 1985). The exchange of GDP to GTP may result in displacement of the surrounding region residues 230–238 in exon 9 of $G_s\alpha$. A nonhydrolyzable GTP analog, but not GDP, prevents tryptic cleavage at Lys^{210} in $G_o\alpha$ or Lys^{205} in $G_{t1}\alpha$ (Hurley *et al.,* 1984).

Exon 12 of $G_s\alpha$ is unique to $G_s\alpha$. Exon 13 of $G_s\alpha$ and exon 8 of $G_i\alpha$ encode the COOH terminus region. This domain may be involved in interaction with a receptor, since the Cys residue that is ADP-ribosylated by pertussis toxin is present in this region of $G_i\alpha$; the structure of this region is heterogeneous. In $G_x\alpha$, the Cys residue is replaced by Ile indicating that $G_x\alpha$ is probably refractory to modification by pertussis toxin. $G_s\alpha$, which is also resistant to pertussis toxin, contains Tyr instead of Cys in this position. It has been shown that the replacement of Arg with Pro at position −6 of $G_s\alpha$ gives rise to a mutant protein which does not couple with β-adrenergic receptor in S49 cells (Sullivan *et al.,* 1987). Further studies including site-directed mutagenesis and construction of chimeric genes may throw more light on the structure–function relationships of $G\alpha$ proteins.

Comparison of the exon organization of $G_o\alpha$ and the $G_i\alpha$ subfamily with that of $G_s\alpha$ indicates that some of the exon junctions are conserved between the $G_i\alpha$ subfamily and $G_s\alpha$. Thus, 3 out of 12 splice sites of the human $G_s\alpha$ gene are shared with the human $G_i\alpha$ genes, and exon 1 and exons 7 and 8 of $G_s\alpha$ correspond to exon 1 and exon 5 of $G_i\alpha$, respectively.

VI. CONSERVATION OF PRIMARY STRUCTURE OF EACH $G\alpha$ AMONG MAMMALIAN SPECIES

Table II shows that, in addition to the remarkable homologies of the overall structure, there is a strong conservation of the amino acid sequence in each subtype of G-protein α subunit. The amino acid sequence of $G_s\alpha$ is strongly conserved between human and rat; only 1 out of 394 amino acids are different. The sequence of $G_{11}\alpha$ is identical between bovine and human.

Table II **Conservation of G-Protein α Subunit Sequences Among Different Mammalian Species**[a]

Species	Amino acid sequences	Nucleotide sequences
rG$_s$α vs. hG$_s$α	393/394 (99.7%)	1128/1182 (95.4%)
bG$_{i1}$α vs. hG$_{i1}$α	354/354 (100%)	998/1062 (94.0%)
rG$_{i2}$α vs. hG$_{i2}$α	350/355 (98.6%)	985/1065 (92.4%)
rG$_{i3}$α vs. hG$_{i3}$α	349/354 (98.6%)	981/1062 (92.4%)
rG$_o$α vs. bG$_o$α	348/354 (98.3%)	992/1062 (93.4%)
rG$_x$α vs. hG$_x$α	349/355 (98.3%)	977/1065 (91.7%)

[a] h, Human; r, rat; b, bovine

For G$_{i2}$α, G$_{i3}$α, G$_x$α, and G$_o$α, over 98% identity of amino acid sequences is maintained among different mammalian species.

The strong conservation of the amino acid sequence of each G-protein α subunit among distant mammalian species may reflect the presence of evolutionary pressure to maintain the specific physiological function of each G-protein gene product. Each Gα protein may be linked to a specific receptor and thereby involved in a specific signal transducing pathway.

An evolutionary tree of G-protein α subunits can be drawn based on the homologies of the predicted amino acid sequences obtained from various mammalian sources (Fig. 7). It is remarkable that the homologies among three G$_i$α species are higher than that between rod (G$_{t1}$α) and cone (G$_{t2}$α) transducin α subunits.

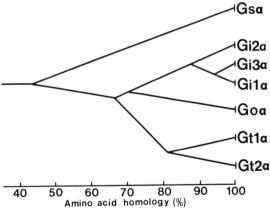

Fig. 7. Relationships among the different mammalian Gα species. For details see Itoh *et al.*, 1988b.

VII. G PROTEINS FROM *SACCHAROMYCES CEREVISIAE*

A family of GTP-binding proteins, the RAS family, is widely distributed among eukaryotes (see Barbacid, 1987, for review) including the yeasts *Saccharomyces cerevisiae* (DeFeo-Jones *et al.,* 1983; Powers *et al.,* 1984) and *Schizosaccharomyces pombe* (Fukui and Kaziro, 1985). It has been suggested that the *RAS2* gene in *S. cerevisiae* is involved in the activation of adenylyl cyclase (Toda *et al.,* 1985; Broek *et al.,* 1985) and mimics the role of mammalian G_s. However, in view of the strong conservation of the amino acid sequences of each G-protein species among different organisms (see Table II), we speculated that G proteins may also occur in yeast. We have searched for a G protein-homologous gene in yeast and isolated two genes, *GPA1* (Nakafuku *et al.,* 1987) and *GPA2* (Nakafuku *et al.,* 1988) from *S. cerevisiae,* which are homologous to cDNAs for mammalian G-protein α subunits.

GPA1 and *GPA2* code for sequences of 472 and 449 amino acid residues, respectively, with calculated M_r 54,075 and 50,516. When aligned with the α subunit of mammalian G proteins to obtain maximal homology, GP1α (*GPA1*-encoded protein) and GP2α (*GPA2*-encoded protein) were found to contain additional stretches of 110 and 83 amino acid residues, respectively, located near the NH_2 terminus (Fig. 8).

VIII. COMPARISON OF AMINO ACID SEQUENCES OF YEAST GP1α AND GP2α WITH THOSE OF RAT BRAIN $G_i α$ AND $G_o α$

The deduced amino acid sequences of yeast GP1α and GP2α are highly homologous with those of rat brain $G_i α$ and $G_o α$. As shown in Fig. 9, the

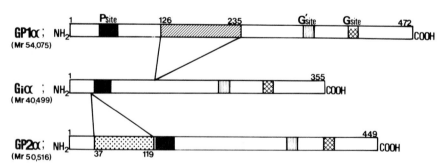

Fig. 8. Schematic representation of the structure of yeast GP1α, yeast GP2α, and mammalian $G_{i2} α$.

```
 Gi2α:   1- 29: MG-CTVSA-----------EDKAAAERSKMIDKNLREDGEK-------------------
 Gsα:    1- 36: MG-CLGNSKTEDQRN----EEKAQREANKKIEKQLQKDKQV-------------------

 GP1α:          ------------------------------------------------------------
 GP2α:  56-115: RQQQPSPHNVKDRKEQNGSINNAISPTATANTSGSQQINIDSALRDRSSNVAAQPSLSDA
 Gi2α:          ------------------------------------------------------------
 Gsα:           ------------------------------------------------------------

 GP1α:  38- 78: ----DKNEIKLLLLGAGESGKSTVLKQLKLLHQGGFSHQ---------------ERLQYA
 GP2α: 116-160: SSGSNDKELKVLLLGAGESGKSTVLQQLKILHQNGFSEQ---------------EIKEYI
 Gi2α:  30- 70: ----AAREVKLLLLGAGESGKSTIVKQMKIIHEDGYSEE---------------ECRQYR
 Gsα:   37- 92: ----YRATHRLLLLGAGESGKSTIVKQMRILHVNGFNGEGGEEDPQAARSNSDGEKATKV

 GP1α:  79-136: QVIWADAIQSMKILIIQAR-KLGIQLDCDDPINN-KDLFACKRILLKAKALDYINASVAG
 GP2α: 161-210: PLIYQNLLEIGRNLI-QARTRFNVNLEPECELTQ-QDLSRTMSYEMPNNYTG--------
 Gi2α:  71-118: AVVYSNTIQSIMAIVKAMG-NLQIDFADPQRADDARQLFALSCAAEEQG-----------
 Gsα:   93-140: QDIKNNLKEAIETIVAAMS-NLVPPVELANPENQFRVDYILSVMNVPNF-----------

 GP1α: 137-197: GSDFLNDYVLKYSERYETRRRVQSTGRAKAAFDEDGNISNVKSDTDRDAETVTQNEDADR
 GP2α:          ------------------------------------------------------------
 Gi2α:          ------------------------------------------------------------
 Gsα:           ------------------------------------------------------------

 GP1α: 197-255: NNSSRINLQDICKDLNQEGDDQMFVRKTSREIQGQNRRNLIHEDIAKAIKQLWNND-KGI
 GP2α: 211-231: ---------------------------------------QFPEDIAGVISTLWALPSTQD
 Gi2α: 119-137: ---------------------------------------MLPEDLSGVIRRLWA-D-HGV
 Gsα:  141-159: ---------------------------------------DFPPEFYEHAKALWE-D-EGV

 GP1α: 256-314: KQCFARSNEFQLEGSAAYYFDNIEKFASPNYVCTDEDILKGRIKTTGITETEFNIGS-SK
 GP2α: 232-291: LVNGPNASKFYLMDSTPYFMENFTRITSPNYRPTQQDILRSRQMTSGIFDTVIDMGSDIK
 Gi2α: 138-196: QACFGRSREYQLNDSAAYYLNDLERIAQSDYIPTQQDVLRTRVKTTGIVETHFTFKD-LH
 Gsα:  160-218: RACYERSNEYQLIDCAQYFLDKIDVIKQADYVPSDQDLLRCRVLTSGIFETKFQVDK-VN

 GP1α: 315-374: FKVLDAGGQRSERKKWIHCFEGITAVLFVLAMSEYDQMLFEDERVNRMHESIMLFDTLLN
 GP2α: 292-351: MHIYDVGGQRSERKKWIHCFDNVTLVIFCVSLSEYDQTLMEDKNQNRFQESLVLFDNIVN
 Gi2α: 197-256: FKMFDVGGQRSERKKWIHCFEGVTAIIFCVALSAYDLVLAEDEEMNRMHESMKLFDSICN
 Gsα:  219-278: FHMFDVGGQRDERRKWIQCFNDVTAIIFVVASSSYNMVIREDNQTNRLQEALNLFKSIWN

 GP1α: 375-421: SKWFKDTPFILFLNKIDLFEEKVK--SMPIRKYFPDYQGRVGDAEAGL-----------K
 GP2α: 352-417: SRWFARTSVVLFLNKIDLFAEKLR--KVPMENYFPDYTGG-SDINKAA-----------K
 Gi2α: 257-302: NKWFTDTSIILFLNKKDLFEEKIT--QSPLTICFPEYTGANKYDEAAS------------
 Gsα:  279-338: NRWLRTISVILFLNKQDLLAEKVLAGKSKIEDYFPEFARYTTPEDATPEPGEDPRVTRAK

 GP1α: 422-472: YFEKI-FLSLNK--TNKP--IYVKRTCATDTQTMKFVLSAVTDLIIQQNLKKIGII-
 GP2α: 418-449: YILWR-FVQLNR--ANLS--IYFHVTQATDTSNIRLVFAAIKETILENTLKDSGVLQ
 Gi2α: 303-355: -YIQSKFEDLNKRKDTKE--IYTHFTCATDTKNVQFVFDAVTDVIIKNNLKDCGLF-
 Gsα:  339-394: YFIRDEFLRISTASGDGRHYCYPHFTCAVDTENIRRVFNDCRDIIQRMHLRQYELL-
```

Fig. 9. Alignment of the predicted amino acid sequences of yeast GP1α and GP2α with those of rat brain $G_{i2}\alpha$ and $G_s\alpha$. Identical or conservative amino acid residues are enclosed within solid lines. From Nakafuku *et al.*, 1988. In GP2α, *F* at amino acid residue 435 should read P.

homology is most remarkable in the region of GTP hydrolysis (P site) (amino acid residues 43–58 of GP1α, 125–140 of GP2α, and 35–50 of G$_{i2}\alpha$). The region responsible for GTP binding (G site) containing amino acid residues 381–397 of GP1α, 358–374 of GP2α, and 263–279 of G$_{i2}\alpha$, is also highly homologous. Another region of homology, the G' site, is found in amino acid residues 321–334 of GP1α, 298–311 of GP2α, and 203–216 of G$_{i2}\alpha$, where a sequence of 14 contiguous amino acids is completely identical in yeast GP1α, GP2α, and rat G$_{i2}\alpha$.

The overall homology in nucleotide and amino acid sequences of yeast GP1α, yeast GP2α, rat G$_{i2}\alpha$, and rat G$_s\alpha$ is remarkable. Disregarding the unique sequences present in GP1α (residues 126–235) and GP2α (residues 37–119), the proteins are 60% homologous if conservative amino acid substitutions are considered to be homologous. The homology is smaller than that between rat G$_{i2}\alpha$ and G$_o\alpha$ (85%) but is comparable to that between rat G$_{i2}\alpha$ and G$_s\alpha$ (60%).

As is described elsewhere in detail (Miyajima *et al.,* 1987; Dietzel and Kurjan, 1987), *GPA1* is a haploid-specific gene involved in the mating factor signal transduction. On the other hand, *GPA2* is expressed both in haploid and diploid cells, and may be involved in the regulation of cAMP levels in *S. cerevisiae* (Nakafuku *et al.,* 1988). The demonstration of the occurrence of G proteins in yeast may open the way for a detailed genetic analysis of the function of G proteins in eukaryotic cells.

REFERENCES

Barbacid, M. (1987). *ras* Genes. *Annu. Rev. Biochem.* **56,** 779–827.

Beals, C. R., Wilson, C. B., and Perlmutter, R. M. (1987). A small multigene family encodes Gi signal-transduction proteins. *Proc. Natl. Acad. Sci. U.S.A.* **84,** 7886–7890.

Bray, P., Carter, A., Simons, C., Guo, V., Puckett, C., Kamholz, J., Spiegel, A., and Nirenberg, M. (1986). Human cDNA clones for four species of Gαs signal transduction protein. *Proc. Natl. Acad. Sci. U.S.A.* **83,** 8893–8897.

Bray, P., Carter, A., Guo, V., Puckett, C., Kamholz, J., Spiegel, A., and Nirenberg, M. (1987). Human cDNA clones for an α subunit of Gi signal-transducing protein. *Proc. Natl. Acad. Sci. U.S.A.* **84,** 5115–5119.

Broek, D., Samiy, N., Fasano, O., Fujiyama, A., Tamanoi, F., Northup, J., and Wigler, M. (1985). Differential activation of yeast adenylate cyclase by wild-type and mutant *RAS* proteins. *Cell* **41,** 763–769.

Codina, J., Olate, J., Abramowitz, J., Mattera, R., Cook, R. G., and Birnbaumer, L. (1988). αi-3 cDNA encodes the α subunit of Gk, the stimulatory G protein of receptor-regulated K$^+$ channels. *J. Biol. Chem.* **263,** 6746–6750.

DeFeo-Jones, D., Scolnick, E. M., Koller, R., and Dhar, R. (1983). *ras*-Related gene sequences identified and isolated from *S. cerevisiae. Nature (London)* **306**, 707–709.

Didsbury, J. R., and Snyderman, R. (1987). Molecular cloning of a new human G protein. Evidence for two Giα-like protein families. *FEBS Lett.* **219**, 259–263.

Didsbury, J. R., Ho, Y., and Snyderman, R. (1987). Human Gi protein α-subunit: Deduction of amino acid structure from a cloned cDNA. *FEBS Lett.* **211**, 160–164.

Dietzel, D., and Kurjan, J. (1987). The yeast *SCG1* gene: A Gα-like protein implicated in the **a**- and α-factor response pathway. *Cell* **50**, 1001–1010.

Fong, H. K. W., Yoshimoto, K. K., Eversole-Cire, P., and Simon, M. I. (1988). Identification of a GTP-binding protein α subunit that lacks an apparent ADP-ribosylation site for pertussis toxin. *Proc. Natl. Acad. Sci. U.S.A.* **85**, 3066–3070.

Fukui, Y., and Kaziro, Y. (1985). Molecular cloning and sequence analysis of a *ras* gene from *Schizosaccharomyces pombe. EMBO J.* **4**, 687–691.

Gilman, A. G. (1987). G proteins: Transducers of receptor-generated signals. *Annu. Rev. Biochem.* **56**, 615–649.

Harris, B. A., Robishaw, J. D., Mumby, S. M., and Gilman, A. G. (1985). Molecular cloning of complementary DNA for the alpha subunit of the G protein that stimulates adenylate cyclase. *Science* **229**, 1274–1277.

Hurley, J. B., Simon, M. I., Teplow, D. B., Robishaw, J. D., and Gilman, A. G. (1984). Homologies between signal transducing G proteins and *ras* gene products. *Science* **226**, 860–862.

Itoh, H., Kozasa, T., Nagata, S., Nakamura, S., Katada, T., Ui, M., Iwai, S., Ohtsuka, E., Kawasaki, H., Suzuki, K., and Kaziro, Y. (1986). Molecular cloning and sequence determination of cDNAs for α subunit of the guanine nucleotide-binding proteins Gs, Gi, and Go from rat brain. *Proc. Natl. Acad. Sci. U.S.A.* **83**, 3776–3780.

Itoh, H., Katada, T., Ui, M., Kawasaki, H., Suzuki, K., and Kaziro, Y. (1988a). Identification of three pertussis toxin substrates (41, 40 and 39 kDa proteins) in mammalian brain. *FEBS Lett.* **230**, 85–89.

Itoh, H., Toyama, R., Kozasa, T., Tsukamoto, T., Matsuoka, M., and Kaziro, Y. (1988b). Presence of three distinct molecular species of Gi protein α subunit. *J. Biol. Chem.* **263**, 6656–6664.

Jones, D. T., and Reed, R. R. (1987). Molecular cloning of five GTP-binding protein cDNA species from rat olfactory neuroepithelium. *J. Biol. Chem.* **262**, 14241–14249.

Jurnak, F. (1985). Structure of the GDP domain of EF-Tu and location of the amino acids homologous to *ras* oncogene proteins. *Science* **230**, 32–36.

Kahn, R. A., and Gilman, A. G. (1984). ADP-ribosylation of Gs promoters the dissociation of its α and β subunits. *J. Biol. Chem.* **259**, 6235–6240.

Kaziro, Y. (1978). The role of guanosine 5′-triphosphate in polypeptide chain elongation. *Biochim. Biophys. Acta* **505**, 95–127.

Kozasa, T., Itoh, H., Tsukamoto, T., and Kaziro, Y. (1988). Isolation and characterization of human Gsα gene. *Proc. Natl. Acad. Sci. U.S.A.* **85**, 2081–2085.

Lavu, S., Clark, J., Swarup, R., Matsushima, K., Paturu, K., Moss, J., and Kung, H.-F. (1988). Molecular cloning and DNA sequence analysis of the human guanine nucleotide-binding protein Goα. *Biochem. Biophys. Res. Commun.* **150**, 811–815.

Lerea, C. L., Somers, D. E., Hurley, J. B., Klock, I. B., and Bunt-Milan, A. H. (1986). Identification of specific transducin α subunits in retinal rod and cone photoreceptors. *Science* **324**, 77–80.

Lochrie, M. A., Hurley, J. B., and Simon, M. I. (1985). Sequence of the alpha subunit of photoreceptor G protein: Homologies between transducin, *ras,* and elongation factors. *Science* **228**, 96–99.

Masters, S. B., Stroud, R. M., and Bourne, H. R. (1986). Family of G protein α chains: Amphipathic analysis and predicted structure of functional domains. *Protein Eng.* **1**, 47–54.

Matsuoka, M., Itoh, H., Kozasa, T., and Kaziro, Y. (1988). Sequence analysis of cDNA and genomic DNA for a putative pertussis toxin-insensitive guanine nucleotide-binding regulatory protein α subunit. *Proc. Natl. Acad. Sci. U.S.A.* **85**, 5384–5388.

Mattera, R., Codina, J., Crozat, A., Kidd, V., Woo, S. L. C., and Birnbaumer, L. (1986). Identification by molecular cloning of two forms of the α-subunit of the human liver stimulatory (Gs) regulatory component of adenylate cyclase. *FEBS Lett.* **206**, 36–41.

Medynsky, D. C., Sullivan, K., Smith, D., Van Dop, C., Chang, F.-H., Fung, B. K.-K., Seeburg, P. H., and Bourne, H. R. (1985). Amino acid sequence of the α subunit of transducin deduced from the cDNA sequence. *Proc. Natl. Acad. Sci. U.S.A.* **82**, 4311–4315.

Michel, T., Winslow, J. W., Smith, J. A., Seidman, J. G., and Neer, E. J. (1986). Molecular cloning and characterization of cDNA encoding the GTP-binding protein αi and identification of a related protein, αh. *Proc. Natl. Acad. Sci. U.S.A.* **83**, 7663–7667.

Miyajima, I., Nakafuku, M., Nakayama, N., Brenner, C., Miyajima, A., Kaibuchi, K., Arai, K., Kaziro, Y., and Matsumoto, K. (1987). *GPA1,* a haploid-specific essential gene, encodes a yeast homolog of mammalian G-protein which may be involved in mating factor signal transduction. *Cell* **50**, 1011–1019.

Nakafuku, M., Itoh, H., Nakamura, S., and Kaziro, Y. (1987). Occurrence in *Saccharomyces cerevisiae* of a gene homologous to the cDNA coding for the α subunit of mammalian G proteins. *Proc. Natl. Acad. Sci. U.S.A.* **84**, 2140–2144.

Nakafuku, M., Obara, T., Kaibuchi, K., Miyajima, I., Miyajima, A., Itoh, H., Nakamura, S., Arai, K., Matsumoto, K., and Kaziro, Y. (1988). Isolation of a second yeast *Saccharomyces cerevisiae* gene *(GPA2)* coding for guanine nucleotide-binding regulatory protein: Studies on its structure and possible functions. *Proc. Natl. Acad. Sci. U.S.A.* **85**, 1374–1378.

Northup, J. K., Sternweise, P. C., Smigel, M. D., Shleifer, L. S., Ross, E. M., and Gilman, A. G. (1980). Purification of the regulatory component of adenylate cyclase. *Proc. Natl. Acad. Sci. U.S.A.* **77**, 6516–6520.

Nukada, T., Tanabe, T., Takahashi, H., Noda, M., Haga, K., Haga, T., Ichiyama,

A., Kangawa, K., Hiranaga, M., Matsuo, H., and Numa, S. (1986a). Primary structure of the α subunit of bovine adenylate cyclase-inhibiting protein deduced from the cDNA sequence. *FEBS Lett.* **197**, 305.

Nukada, T., Tanabe, T., Takahashi, H., Noda, M., Hirose, T., Inayama, S., and Numa, S. (1986b). Primary structure of the α-subunit of bovine adenylate cyclase-stimulating G-protein deduced from the cDNA sequence. *FEBS Lett.* **195**, 220–224.

Ovchinnikov, Y. A., Slepak, V. Z., Pronin, A. N., Shlensky, A. B., Levina, N. B., Voeikov, V. L., and Lipkin, V. M. (1987). Primary structure of bovine cerebellum GTP-binding protein G_{39} and its effect on the adenylate cyclase system. *FEBS Lett.* **226**, 91–95.

Powers, S., Kataoka, T., Fasano, O., Goldfarb, M., Strathern, J., Broach, J., and Wigler, M. (1984). Genes in *S. cerevisiae* encoding proteins with domains homologous to the mammalian *ras* proteins. *Cell* **36**, 607–612.

Rall, T., and Harris, B. A. (1987). Identification of the lesion in the stimulatory GTP-binding protein of the uncoupled S49 lymphoma. *FEBS Lett.* **224**, 365–371.

Robishaw, J. D., Russell, D. W., Harris, B. A., Smigel, M. D., and Gilman, A. G. (1986a). Deduced primary structure of the α subunit of the GTP-binding stimulatory protein of adenylate cyclase. *Proc. Natl. Acad. Sci. U.S.A.* **83**, 1251–1255.

Robishaw, J. D., Smigel, M. D., and Gilman, A. G. (1986b). Molecular basis for two forms of the G protein that stimulates adenylate cyclase. *J. Biol. Chem.* **261**, 9587–9590.

Sewell, J. L., and Kahn, R. A. (1988). Sequences of the bovine and yeast ADP-ribosylation factor and comparison to other GTP-binding proteins. *Proc. Natl. Acad. Sci. U.S.A.* **85**, 4620–4624.

Stryer, L., and Bourne, H. R. (1986). G proteins: A family of signal transducers. *Annu. Rev. Cell Biol.* **2**, 391.

Suki, W. N., Abramowitz, J., Mattera, R., Codina, J., and Birnbaumer, L. (1987). The human genome encodes at least three non-allellic G proteins with αi-type subunits. *FEBS Lett.* **220**, 187–192.

Sullivan, K. A., Liao, Y.-C., Alborzi, A., Beiderman, B., Chang, F.-H., Masters, S. B., Levinson, A. D., and Bourne, H. R. (1986). Inhibitory and stimulatory G proteins of adenylate cyclase: cDNA and amino acid sequences of the α chains. *Proc. Natl. Acad. Sci. U.S.A.* **83**, 6687–6691.

Sullivan, K. A., Miller, R. T., Masters, S. B., Beiderman, B., Heideman, W., and Bourne, H. R. (1987). Identification of receptor contact site involved in receptor—G protein coupling. *Nature (London)* **330**, 758–760.

Tanabe, T., Nukada, T., Nishikawa, Y., Sugimoto, K., Suzuki, H., Takahashi, H., Noda, M., Haga, T., Ichiyama, A., Kangawa, K., Minamino, N., Matsuo, H., and Numa, S. (1985). Primary structure of the α-subunit of transducin and its relationship to *ras* proteins. *Nature (London)* **315**, 242–245.

Toda, T., Uno, I., Ishikawa, T., Powers, S., Kataoka, T., Broek, D., Cameron, S., Broach, J., Matsumoto K., and Wigler, M. (1985). In yeast, *RAS* proteins are controlling elements of adenylate cyclase. *Cell* **40**, 27–36.

Van Dop, C., Tsubokawa, M., Bourne, H. R., and Ramachandran, J. (1984). Amino acid sequences of retinal transducin at the site ADP-ribosylated by cholera toxin. *J. Biol. Chem.* **259,** 696–698.

Van Meurs, K. P., Angus, C. W., Lavu, S., Kung, H.-F., Czarnecki, S. K., Moss, J., and Vaughan, M. (1987). Deduced amino acid sequence of bovine retinal Goα: Similarities to other guanine nucleotide-binding proteins. *Proc. Natl. Acad. Sci. U.S.A.* **84,** 3107–3111.

Weinstein, L. S., Spiegel, A. M., and Carter, A. D. (1988). Cloning and characterization of the human gene for the α-subunit of Gi2, a GTP-binding signal transduction protein. *FEBS Lett.* **232,** 333–340.

West, R. E., Moss, J., Vaughan, M., Liu, T., and Liu, T.-Y. (1985). Pertussis toxin-catalysed ADP-ribosylation for transducin. *J. Biol. Chem.* **260,** 14428–14430.

Yatsunami, K., and Khorana, H. G. (1985). GTPase of bovine rod outer segments: The amino acid sequence of the α subunits as derived from the cDNA sequence. *Proc. Natl. Acad. Sci. U.S.A.* **82,** 4316–4320.

CHAPTER 5

Structural, Immunobiological, and Functional Characterization of Guanine Nucleotide-Binding Protein G$_o$

J. Bockaert, P. Brabet
Centre CNRS-INSERM de Pharmacologie-Endocrinologie, Rue de la Cardonille
34094 Montpellier, France

J. Gabrion
Laboratoire de Neurobiologie-Endocrinologie, Université des Sciences et
Techniques du Languedoc, Place Eugène Bataillon, 34000 Montpellier, France

V. Homburger, B. Rouot, M. Toutant
Centre CNRS-INSERM de Pharmacologie-Endocrinologie Rue de la Cardonille,
34094 Montpellier, France

I. INTRODUCTION

One of the intriguing questions concerning cellular communication is how a great diversity of messages can be generated by a limited number of signaling components. This seems to be particularly true for the communication between neurons. Most of our ideas on how diversity is achieved are based on the concept of multiple receptors. However, biochemistry and molecular biology of intracellular pathway components indicate a variety of elements which probably exceed those known for surface receptors. The best example of this new concept is provided by the GTP-binding proteins (G proteins).

G proteins are a family of proteins which transduce extracellular signals (detected by receptors for light, odors, neurotransmitters, and hormones) into cellular responses (Birnbaumer *et al.*, 1987; Bockaert *et al.*, 1987; Gilman, 1987). G proteins are thought to be attached to the inner (cytoplasmic) face of plasma membranes and consist of three subunits: α (39–52 kDa), β (35–36 kDa), and γ (8–10 kDa). It is believed that $\beta\gamma$ subunits are interchangeable among different G proteins. In contrast, the distinctive α chains, which contain the guanine nucleotide-binding site, allow each G protein to interact with its own effectors and receptors. The G_s proteins $G_{s1a}\alpha$, $G_{s1b}\alpha$, $G_{s2}\alpha$; 42–52 kDa) (Birnbaumer *et al.*, 1987; Gilman, 1987) activate adenylyl cyclase. It is likely, but not conclusively demonstrated, that at least one of the G_i proteins ($G_{i1}\alpha$, $G_{i2}\alpha$, $G_{i3}\alpha$; 40–41 kDa) (Suki *et al.*, 1987; Van Meurs *et al.*, 1987) inhibits adenylyl cyclase (Gilman, 1987). In the visual system, photorhodopsin of retinal rods stimulates a cGMP phosphodiesterase through transducin 1 (T_1) (Stryer and Bourne, 1986) while transducin 2 (T_2) is thought to mediate the stimulation of a similar enzyme by color-discriminating rhodopsins in retinal cones (Lerea *et al.*, 1986). In addition to these three main G proteins, at least one other has been purified. A G protein named G_o, the subscript standing for other (other than G_s and G_i), was obtained from brain tissues (Sternweis and Robishaw, 1984; Neer *et al.*, 1984; Waldo *et al.*, 1987). This protein has several interesting features: (1) G_o is highly concentrated in brain (0.5 to 1% of membrane-bound proteins) (Asano *et al.*, 1987; Gierschik *et al.*, 1986b; Homburger *et al.*, 1987). (2) G_o is poorly expressed in most peripheral tissues (Asano *et al.*, 1987; Gierschik *et al.*, 1986b; Homburger *et al.*, 1987). (3) Epitope(s) of G_o are highly conserved within a class of G proteins almost exclusively represented in nervous tissues from vertebrates and invertebrates (Homburger *et al.*, 1987). It is likely that G_o triggers functions which may be, if not exclusively represented in the nervous system, at least highly expressed in this tissue.

This chapter focuses on the studies devoted to determining the tissue, cellular, and subcellular distributions of G_o as well as its structural, immunological, and functional properties.

II. PURIFICATION AND BIOPHYSICAL AND BIOCHEMICAL CHARACTERIZATIONS OF BRAIN G_o PROTEIN

A. Purification

In an attempt to purify G_i from bovine brain, two groups almost simultaneously discovered $G_o\alpha$ (Sternweis and Robishaw, 1984; Neer et al., 1984). In fact, bovine brain demonstrates about 10 times more GTPγS binding than other tissues. This GTPγS binding comigrates with G_i when the different steps of the G_i purification procedure are carried out (cholate solubilization, anion-exchange chromatography, gel filtration on AcA 34 and hydrophobic chromatography using heptylamine-Sepharose). The mixture obtained from the final step of this purification contains two substrates ADP-ribosylated by pertussis toxin (PTX). One is $G_i\alpha$ (41 kDa), the other is $G_o\alpha$ (39 kDa).

Sternweis and Robishaw (1984) indicated that $G_o\alpha$ and $G_i\alpha$ have different sensitivities to trypsin digestion. The $\beta\gamma$ subunits associated with $G_o\alpha$ were apparently identical to those associated with $G_i\alpha$ and $G_s\alpha$. The G_o protein was later purified from rat (Katada et al., 1986a) and porcine (Katada et al., 1987) brain. In both studies, the G_o protein was further purified from the G_o/G_i mixture by anion-exchange chromatography either on Fractogel HW 65 (Toyopearl) or on Mono Q (Pharmacia). The resolution of $G_o\alpha$ from G_o holoprotein was achieved using NaF, $AlCl_3$ and $MgCl_2$, a mixture known to dissociate the α subunit from $\beta\gamma$ subunits. Thus, $G_o\alpha$ was obtained after preincubation and elution in the presence of the ionic mixture either by heptylamine-Sepharose or DEAE (Neer et al., 1984).

Although not always reproducible, it should be noted that pure $G_o\alpha$ sometimes elutes from the first heptylamine chromatography, even in the absence of preincubation with dissociating agents.

B. Purity of Isolated G_o Protein

In the first report by Neer et al. (1984), most of the work was devoted to substrates, α_{39} ($G_o\alpha$) and α_{41} ($G_i\alpha$). However, in this study, the ADP-ribosylation of the G_i/G_o mixture revealed the presence of a third labeled band between 41 kDa and 39 kDa, estimated at 40 kDa. Not much attention was paid to this substrate during the two years that followed this initial publication. Subsequently, a PTX substrate at 40 kDa, different from the brain $G_i\alpha$ and $G_o\alpha$, was described in C_6 glioma cells (Milligan *et al.*, 1986) and neutrophils (Gierschik *et al.*, 1986a). We also demonstrated that under appropriate sodium dodecyl sulfate-polyacrylamide gel electrophoresis (SDS-PAGE) conditions, such a 40-kDa PTX substrate can be observed in most of the tissues studied, including chromaffin cells, human platelets, adipocytes, red blood cells, neurons, glial cells, and adenohypophysis (Toutant *et al.*, 1987a; Journot *et al.*, 1987; Brabet *et al.*, 1988a,b; Rouot *et al.*, 1989).

Figure 1 shows that under SDS-PAGE conditions, three brain polypeptides at 41, 40, and 39 kDa can be resolved from a mixture of G proteins purified from bovine brain (lanes 2 and 4). However, it is obvious from Fig. 1 that the substrate at 40 kDa ($G\alpha_{40}$) is only present in small amounts compared to $G_i\alpha$ and $G_o\alpha$.

A comparison with the mobility of the untreated samples (lanes 1 and 5) with those treated with N-ethylmaleimide (NEM) (lanes 2 and 4) reveals that under standard conditions, the small 40-kDa protein band is probably overwhelmed by the wide staining of $G_o\alpha$. This raises the question of the

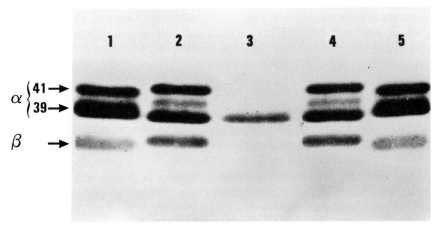

Fig. 1. Resolution by SDS-PAGE of $G_o\alpha$ from two other brain PTX substrates. The brain PTX substrate mixture (lanes 1, 2, 4, 5) and enriched $G_o\alpha$ fraction (lane 3) were untreated (lanes 1 and 5) or treated (lanes 2, 3, 4) with excess N-ethylmaleimide over dithiothreitol concentration, according to Evans *et al.*, (1986). Samples were loaded on modified Laemmli gels (10% polyacrylamide, 0.13% bisacrylamide) (as previously described by Toutant *et al.*, 1987a,b; Brabet *et al.*, 1988) and after electrophoresis the gel was stained with silver.

purity of $G_o\alpha$ preparations obtained and used for reconstitution experiments (Section VI). Katada *et al.* (1987) have been able to purify on a Mono Q column a PTX substrate from brain which seems to have an intermediate mobility between $G_i\alpha$ and $G_o\alpha$ on SDS-PAGE. A 40-kDa $G\alpha$ protein has also been purified from neutrophils and human leukemic (HL-60) cells (Dickey *et al.*, 1987; Oinuma *et al.*, 1987; Gierschik *et al.*, 1987).

C. Biophysical and Biochemical Characterizations

Comparison of the hydrodynamic properties of $G_o\alpha$ and $G_i\alpha$ shows the similarity between the two proteins (Huff *et al.*, 1985; Rouot *et al.*, 1987). An interesting clue is given by the amount of detergent bound to the different G subunits. It appears that the $\beta\gamma$ component is the more hydrophobic of the subunits whereas $G_o\alpha$ binds little detergent (10%) (Huff *et al.*, 1985). This raises the problem of the interaction of $G_o\alpha$ with the plasma membrane.

The hydrophilic nature of $G_o\alpha$ is confirmed by the deduced amino acid sequence of the protein (Itoh *et al.*, 1986; Van Meurs *et al.*, 1987; Jones and Reed, 1987). However, the myristylation of $G_o\alpha$, $G_i\alpha$ and $G\alpha_{40}$ has been demonstrated (Buss *et al.*, 1987; Schultz *et al.*, 1987). Myristate may play an important role in stabilizing interactions of G proteins with phospholipid or with membrane-bound proteins. However, myristylated $G_o\alpha$ is poorly incorporated into phospholipid vesicles *in vitro,* while $\beta\gamma$ totally associates (Sternweis, 1986). Nevertheless, $G_o\alpha$ can bind to phospholipid vesicles which have already incorporated $\beta\gamma$ subunits. Thus, the $\beta\gamma$ subunit may act as an anchor which provides a way of associating the $G_o\alpha$ subunit to the membranes. In any case the exact mode of interaction of the $G_o\alpha-\beta\gamma$ complex with plasma membranes remains to be determined.

The properties of guanine nucleotide binding on $G_o\alpha$ and the properties of the GTPase activity of this protein have been reviewed (Rouot *et al.*, 1987). Two tryptophan residues easily detected by fluorescence are present in $G_o\alpha$. The intensity of fluorescence is increased by adding GTPγS, especially when Mg^{2+} is present, indicating that at least one of the tryptophan residues is sensitive to conformational changes in the protein (Higashijima *et al.*, 1987a). This property has been used to study the kinetics of binding of GTP (or GTP analogs), the GTPase activity, and their regulation by $\beta\gamma$ subunits, Mg^{2+} or Cl^- (Higashijima *et al.*, 1987b,c,d).

PTX-catalyzed ADP-ribosylation of G proteins requires the integrity of the heterotrimer $\alpha\beta\gamma$. ADP-ribosylation of the pure $G_o\alpha$ subunit requires the addition of three times more $\beta\gamma$ subunits than for $G_i\alpha$ (Katada *et al.*, 1986a; Huff and Neer, 1986), indicating that the $\beta\gamma$ component has more affinity for $G_i\alpha$ than for $G_o\alpha$. This relatively weak association of $G_o\alpha$ with $\beta\gamma$ is probably related to the possibility, already discussed in Section I.A, of purifying $G_o\alpha$ even in the absence of dissociating agents. This also raises the interesting question of the possibility of having free $G_o\alpha$ in addition to $G_o\alpha - \beta\gamma$ in cells (see discussion in Section VI).

III. STRUCTURE OF $G_o\alpha$

A partial cDNA sequence of $G_o\alpha$ has been reported by Itoh *et al.* (1986). These authors purified $G_o\alpha$ from rat brain, carried out tryptic digestion and obtained several peptides which were sequenced. A probe corresponding to the amino acid sequence of one of the peptides was used to screen a C_6 glioma cell cDNA library. They obtained a cDNA clone identified as a specific $G_o\alpha$ cDNA clone, since the homology between the amino acid sequences of six tryptic peptides and the deduced amino acid sequence from the nucleotide sequence was perfect. This strategy was necessary because the $G_o\alpha$ subunit was refractory to Edman degradation, suggesting that the NH_2-terminal amino acid was blocked (Hurley *et al.*, 1984). The myristylation of the Gly-terminal residue was probably the reason for the blockade (Buss *et al.*, 1987). The complete coding sequence of $G_o\alpha$ was recently reported by two groups (Van Meurs *et al.*, 1987; Jones and Reed, 1987). Both groups used oligonucleotide probes to screen either a bovine retinal cDNA library (Van Meurs *et al.*, 1987) or a rat olfactory neuro-epithelium cDNA library (Jones and Reed, 1987). The $G_o\alpha$ subunit has 73% identity (82% homology) with $G_{i1}\alpha$, 60–61% identity (76–78% homology) with $T_1\alpha$ and $T_2\alpha$, whereas it has only 34% identity (50% homology) with $G_s\alpha$ (Van Meurs *et al.*, 1987).

Two regions in GTP-binding proteins are believed to be involved in guanine nucleotide binding and hydrolysis (Lochrie *et al.*, 1985; Tanabe *et al.*, 1985; Medynski *et al.*, 1985; Yatsunami and Khorana 1985; Robishaw *et al.*, 1986; Itoh *et al.*, 1986; Sullivan *et al.*, 1986; Van Meurs *et al.*, 1987; Dever *et al.*, 1987). In $G_o\alpha$, these two sequences are localized between residues 36 and 53 ("GTP signature 1") and 264 to 279 ("GTP signature 2") respectively (Fig. 2).

In "GTP signature 1" (18 amino acids), the sequence $Gly - Xaa_1 - Xaa_2 - Xaa_3 - Xaa_4 - Gly - Lys$ is found in all members of GTP-binding families including $G\alpha$ subunits, *ras* proteins, EF-Tu (elongation factor Tu), EF-G (elongation factor G) and PEPCK (phosphoenolpyruvate carboxykinase) families (Dever *et al.*, 1987). Lys^{24} in EF-Tu (Lys^{46} in $G_o\alpha$) is thought to contribute to the neutralization of the β-phosphate of GDP (based on the X-ray structure of the GTP domain of EF-Tu) (La Cour *et al.*, 1985;

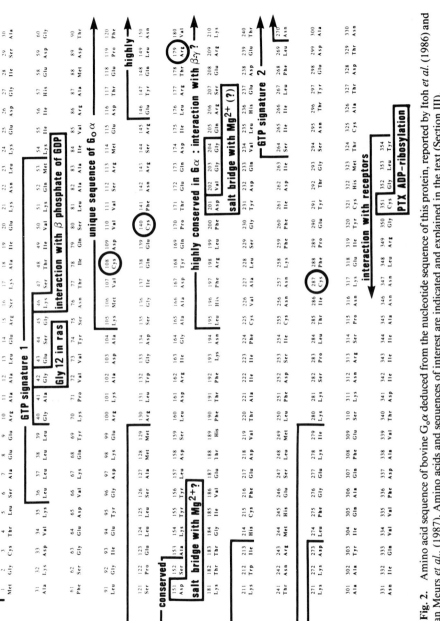

Fig. 2. Amino acid sequence of bovine $G_o\alpha$ deduced from the nucleotide sequence of this protein, reported by Itoh *et al.* (1986) and Van Meurs *et al.*, (1987). Amino acids and sequences of interest are indicated and explained in the text (Section III).

Jurnak, 1985). In all Gα subunits, the Xaa$_1$ – Xaa$_2$ – Xaa$_3$ – Xaa$_4$ sequence is Ala – Gly – Glu – Ser. "GTP signature 1" is also implicated in GTPase activity. This assumption is based on the observation that mutation of Gly^{12} of the mammalian *ras* protein (Gly^{42} in G$_o\alpha$) results in a decrease in GTPase activity and an increase in transforming activity (Tabin *et al.*, 1982; Mc Grath *et al.*, 1984). The "GTP signature 1" is flanked on both sides by either Arg (G$_s\alpha$) or Lys (G$_{i1,2,3}\alpha$, G$_o\alpha$) (Jones and Reed, 1987; Suki *et al.*, 1987). The meaning of this curiosity is unknown. "GTP signature 2" (Fig. 2) includes the consensus sequence Asn – Lys – Xaa – Asp which is found in all GTP-binding proteins and is believed to interact with the guanine ring of bound nucleotides (La Cour *et al.*, 1985; Jurnak, 1985).

Van Meurs *et al.* (1987) proposed that Asp^{151} is equivalent to Asp^{80} of EF-Tu. In EF-Tu, Asp^{80} forms a salt bridge with Mg^{2+} (Jurnak, 1985). Alternatively, since the Asp – Xaa – Xaa – Gly sequence (80 – 84 in EF-Tu) seems to be a consensus sequence in all GTP-binding proteins and corresponds to the highly conserved sequence Asp – Val – Gly – Gly in Gα subunits (Asp^{201} to Gly^{204} in G$_o\alpha$), Dever *et al.* (1987) suggested that it is in fact Asp^{201} in G$_o\alpha$ that forms a salt bridge with Mg^{2+} (Fig. 2).

The domain of Gα which interacts with $\beta\gamma$ has not yet been clearly defined. The first possibility is an interaction between the NH$_2$-terminal region of Gα subunits. Indeed, an amino-terminal tryptic fragment is required for the binding of α to $\beta\gamma$ subunits in the case of transducin (Medynski *et al.*, 1985). The differences between the Gα sequences in this region could account for the different affinities of Gα subunits for $\beta\gamma$. The second possibility is an interaction with a sequence (Leu^{195} to Ile^{213} in G$_o\alpha$) well conserved in Gα proteins and not found in other GTP-binding protein families (Robishaw *et al.*, 1986; Jones and Reed, 1987).

Other regions of homology are the sites for ADP-ribosylation by PTX. The cysteine residue located in the fourth position from the COOH-terminal position (Cys^{351} in G$_o\alpha$) is present in all Gα proteins except G$_s\alpha$, and is ADP-ribosylated by PTX. Since ADP-ribosylation by PTX results in functional uncoupling of the G protein from the receptor (West *et al.*, 1985), this COOH-terminal region is thought to be involved in receptor binding. Both G$_i\alpha$ and G$_o\alpha$ can replace Tα and display light-stimulated GTPase activity when reconstituted with rhodopsin and $\beta\gamma$ subunits (Kanaho *et al.*, 1984), while G$_s\alpha$ is much less effective (Cerione *et al.*, 1986). This is in agreement with the homologies observed in the COOH-terminal region between Tα, G$_{i1,2,3}\alpha$ and G$_o\alpha$ whereas G$_s$ appears to be different. It is also important to note that in the COOH-terminal region of G$_i\alpha$-G$_o\alpha$ proteins, the differences between G$_o\alpha$ on one hand and G$_{i1,2,3}\alpha$ on the other are higher than those (almost negligible) within the G$_{i1,2,3}\alpha$ family. ADP-ribosylation by cholera toxin (CT) of Arg^{174} in Tα, and Arg^{201} in G$_s\alpha$, results in decreased GTPase activity (Abood *et al.*, 1982; Angus *et al.*, 1986). In the vicinity of the Arg^{179} in $G_o\alpha$ which could be modified by CT, Thr–Arg–Val–Lys sequence (178–181) is somehow different from the Cys–Arg–Val–Leu (200–203) sequence in $G_s\alpha$. This could explain why, unlike G_s, G_o and G_i are not easily ADP-ribosylated by CT.

When the sequences of all known $G\alpha$ subunits are compared to those of *ras* proteins and elongation factors, one domain appears highly specific to $G_o\alpha$ (Lys^{105} to Met^{129}). This region may contain a domain involved in the interaction of $G_o\alpha$ with its specific effector(s).

As in the three G_i sequences, 10 cysteine residues are present in $G_o\alpha$. Among them, only three are reactive to alkylation by NEM; these are located at positions 108, 140, and 287 (Winslow *et al.*, 1987).

When the GTP site is occupied by GTPγS, alkylation by NEM is unaffected on Cys^{108}, reduced on Cys^{140} and suppressed on Cys^{287}. Cys^{140} and Cys^{287} are highly conserved, while residue Cys^{108} is only found in $G_o\alpha$. Alkylation of Cys^{108} blocks ADP-ribosylation by PTX (Winslow *et al.*, 1987), suggesting that this modification alters the three-dimensional structure of $G_o\alpha$. However, the same alkylation does not affect other functional parameters which are likely to be sensitive to protein conformation, such as the ability of $G_o\alpha$ to interact with $\beta\gamma$, and the GTPase activity of $G_o\alpha$ (Winslow *et al.*, 1987).

IV. PRODUCTION AND SPECIFICITY OF ANTI-$G_o\alpha$ ANTIBODIES

To produce antibodies against the $G_o\alpha$ protein, several routes were taken. Concerning the antigens, three forms were chosen to raise antibodies against: the holoprotein, i.e., the ternary complex composed of the $\alpha\beta\gamma$ subunits (Gierschik *et al.*, 1986b), the α subunit (Huff *et al.*, 1985; Katada *et al.*, 1987; Mumby *et al.*, 1986; Rapiejko *et al.*, 1986; Tsai *et al.*, 1987; Homburger *et al.*, 1987), or some peptides synthesized from the determined nucleotide sequence of the α subunit of the G_o protein (Mumby *et al.*, 1986). All the antibodies obtained were polyclonal antibodies from rabbits injected with either nondenatured (Huff *et al.*, 1986; Katada *et al.*, 1987; Mumby *et al.*, 1986; Rapiejko *et al.*, 1987; Homburger *et al.*, 1987) or denatured antigens (Gierschik *et al.*, 1986b). Characterization of the specificity of the antibodies was accomplished by immunoblotting after SDS-PAGE of a mixture of G proteins or alternatively by immunospotting using a native or denatured α subunit. Since $G_{i1,2,3}\alpha$ and $G_o\alpha$ are ADP-ribosylated by PTX, comparison of the mobilities of ADP-ribosylated $G\alpha$

proteins after SDS-PAGE, and those of Gα proteins recognized by the antibodies, completed the characterization of the antibodies.

In order to avoid cross-reactivity with proteins unrelated to G proteins, antisera were often affinity purified on G$_o\alpha$, either bound to nitrocellulose or coupled to activated sepharose gels.

A. Characterization of G$_o\alpha$ Antibodies

1. Antibodies against Trimeric Structure of G$_o$

Antibodies were raised in rabbits by Gierschik *et al.* (1986b) with purified G proteins from bovine brain, using as antigens a mixture of G$_o$ and G$_i$ which had been treated with 6 M urea. On immunoblots, the antiserum termed RV$_3$ recognized a protein of 39 kDa in brain membranes for several species corresponding to the α subunit of G$_o$ and a band at 35–36 kDa related to the β subunits of G proteins. No cross-reactivity was obtained with the α subunits of bovine transducin, or with G$_s$ and G$_i$ proteins of human erythrocytes.

2. Antibodies against G$_o\alpha$

Several groups used G$_o\alpha$ as antigen to obtain polyclonal antibodies against the α subunit of G$_o$. G$_o\alpha$ antigens were either devoid of G$_i\alpha$ or contaminated with G$_i\alpha$ (Mumby *et al.*, 1986; Asano *et al.*, 1987). Polyclonal antisera obtained did not react with the β subunit but sometimes displayed a slight cross-reactivity with G$_i\alpha$ (Huff *et al.*, 1985; Katada *et al.*, 1987; Mumby *et al.*, 1986). This phenomenon can be due either to the presence of G$_i\alpha$ in the preparations or to a cross-reactivity of G$_o\alpha$ antibodies with G$_i\alpha$. This is not surprising in view of the homologies between these proteins.

3. Antibodies against Synthetic Peptides Derived from G$_o\alpha$ Subunit

Antibodies specific for particular α subunits were prepared using synthetic peptides corresponding to distinct sequences. Mumby *et al.* (1986) used the amino acid sequence, Asn–Leu–Lys–Glu–Asp–Gly–Ile–Ser–Ala–Ala–Lys–Asp–Val–Lys of G$_o\alpha$ to raise polyclonal antibodies which recognized G$_o\alpha$ but no other G proteins.

B. Cross-Reactivity between G$_o\alpha$ and 40–41 kDa PTX Substrates

As already discussed in Section I.B, several lines of evidence have been provided for a new GTP-binding protein, serving as a specific substrate for

PTX, with a molecular mass of 40 kDa in SDS-PAGE and differing from $G_i\alpha$ (41 kDa) and $G_o\alpha$ (39 kDa) (Oinuma *et al.*, 1987; Katada *et al.*, 1987). Fig. 3B shows that under SDS-PAGE conditions already described for Fig. 1, three PTX substrates can be separated in neurons (39, 40, and 41 kDa), whereas only two are present in C_6 glioma cell membranes (40 and 41 kDa). In the same samples (Fig. 3A), the anti-$G_o\alpha$ antibody that we prepared (Homburger *et al.*, 1987) only recognizes the protein with the highest electrophoretic mobility (39 kDa) in neurons, but does not recognize any proteins in C_6 glioma cells. Therefore, in nervous tissues the anti-$G_o\alpha$ antibody recognizes a single 39-kDa band in preparations in which three PTX substrates of 39, 40, and 41 kDa are present. The confirmation that in brain cells, the anti-$G_o\alpha$ antibody recognizes neither the 40 kDa nor the 41 kDa PTX substrates, is given in Fig. 3C showing that a PTX substrate (at 39 kDa) is immunoprecipitated in neurons and not in C_6 glioma cells.

Previous studies comparing the mobilities of proteins $[^{32}P]$ADP-ribosylated by PTX and immunodetected by anti-$G_o\alpha$ antibodies also suggested that the G_o protein is expressed in rat adipocytes (Rapiejko *et al.*, 1986) and preadipocytes (NIH 3T3 L1 cells) (Gierschik *et al.*, 1986c; Watkins *et al.*, 1987). This assumption is supported by the fact that a positive immunoreactivity was detected at 39–41 kDa with anti-$G_o\alpha$ antibodies and that

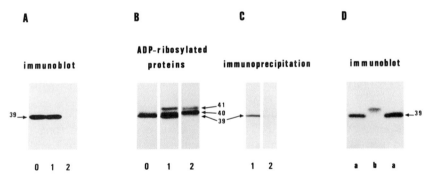

Fig. 3. Specificity of affinity-purified anti-$G_o\alpha$ antibodies. (A) and (D) Immunoblot with affinity-purified anti-$G_o\alpha$ antibodies. Enriched $G_o\alpha$ preparation (0 or a), particulate fractions (about 100 μg) of mouse cortical neurons in primary culture (1), C_6 glioma cells (2) and purified membranes from human adipocytes (b) were subjected to electrophoresis, immunoblotted, and labeled with affinity-purified anti-$G_o\alpha$ antibodies. (B) ADP-ribosylation with PTX. Enriched $G_o\alpha$ preparation (0) and particulate fractions from cortical neurons (1) and C_6 glioma cells (2) were $[^{32}P]$ADP-ribosylated with PTX and analyzed by electrophoresis on 10% SDS-polyacrylamide gels as previously described (Toutant *et al.*, 1987a,b; Brabet *et al.*, 1988). (C) ADP-ribosylated proteins of cortical neurons (1) and C_6 glioma cells (2). Particulate fractions were solubilized in cholate and subjected to immunoprecipitation with anti-$G_o\alpha$ IgG and formalin-fixed *Staphylococcus aureus* cells. After washing, the precipitates were boiled with Laemmli SDS buffer and analyzed by SDS-PAGE as described in (B).

two proteins (39–41 kDa) could be ADP-ribosylated by PTX toxin (Rapiejko et al., 1986; Owens et al., 1985).

However, a detailed analysis of this question using anti-$G_o\alpha$ antibodies and ADP-ribosylation shows that specific affinity-purified antibodies against $G_o\alpha$ subunits reacts with a protein migrating at 40–41 kDa and not at 39 kDa on immunoblot after SDS-PAGE of rat, human, and dog adipocyte membranes (Rouot et al., 1989). Figure 3D shows the detection of this 40–41 kDa protein in human adipocytes with our affinity-purified anti-$G_o\alpha$ antibodies. The SDS-PAGE conditions described in Fig. 1 provide a way of clearly demonstrating that in adipocytes the protein recognized by our anti-$G_o\alpha$ antibody does not have a molecular mass of 39 kDa. Similarly, it should be mentioned that ADP-ribosylation by PTX clearly shows the absence of a [^{32}P]ADP-ribosylated protein migrating at 39 kDa. (Rouot et al., 1989). This result suggests that human, adipocytes express a G_o-like protein of 40–41 kDa that is not present in brain cells. It also indicates that at least the following criteria should be retained to identify the $G_o\alpha$ protein in a tissue:

1. SDS-PAGE should be performed under experimental conditions allowing a clear separation between at least three PTX substrates (39, 40, and 41 kDa). A preparation of nervous tissues must be used as control.
2. When studied under these conditions the tissue tested should contain a PTX substrate having the electrophoretic mobility of the 39-kDa protein.
3. Immunoblotting with affinity-purified anti-$G_o\alpha$ antibodies should recognize one band at 39 kDa. The experimental conditions described in (1) should be used and preparations of nervous tissue used as control.

It is likely that in addition to transducin, at least four PTX substrates have already been identified: one of 39 kDa, two of 40 kDa (one from brain tissues, the other from adipocytes and possibly other tissues), and one of 41 kDa. More sophisticated experimental techniques will probably provide a way of separating more PTX substrates of 39–41 kDa. We think that the $G_o\alpha$ protein described here can be identified as the protein derived from the $G_o\alpha$ cDNA clone isolated by Van Meurs et al. (1987), Jones and Reed (1987), and Itoh et al. (1986). In contrast, it is still not possible to identify the other PTX substrates (40–41 kDa) as any of the proteins encoded by the three isolated $G_i\alpha$ cDNA clones ($G_{i1,2,3}\alpha$) (Suki et al. 1987; Didsbury and Snyderman, 1987; Bray et al., 1987; Jones and Reed, 1987). One can also notice that the separation by SDS-PAGE of three PTX substrates (Fig. 3B) is probably not based on a difference in the molecular masses of $G_o\alpha$ and $G_{i1,2,3}\alpha$, which are expected to be almost identical according to their nucleotide sequences. This raises the problem of possible posttranscriptional modifications of these proteins which could modify their electrophoretic behavior.

V. TISSUE DISTRIBUTIONS OF $G_o\alpha$ PROTEIN AND mRNA CODING FOR $G_o\alpha$

A. Tissue Distribution of $G_o\alpha$ Protein

The use of selective antibodies has provided a way of detecting and quantifying the G_o protein in neuronal tissues and in nonneuronal tissues. These studies were carried out on immunoblots after electrotransfer of the proteins from the SDS-PAGE gels.

1. Neuronal Tissues

In the central nervous system, it has been shown that the $G_o\alpha$ subunit is a major membrane protein which accounts for approximately $0.5-1\%$ of total membrane protein (Huff et al., 1985; Gierschik et al., 1986b; Asano et al., 1987; Homburger et al., 1987). However, the concentration of the G_o protein is not homogenous in the brain. Regional distribution in bovine brain revealed that G_o is more abundant in cerebral cortex, thalamus, hypothalamus and cerebellum than in medulla and pons (Gierschik et al., 1986b; Asano et al., 1987). $G_o\alpha$ is also present in retina (Terashima et al., 1987a).

In addition to bovine, porcine, and rat brain [from which G_o has been purified (Sternweis and Robishaw, 1984; Neer et al., 1984; Katada et al., 1986a, 1987)], its presence was also detected immunologically in nervous tissues of other vertebrate species, such as human, chicken, frog, and *Torpedo marmorata* (Gierschik et al., 1986b; Homburger et al., 1987; Toutant et al., 1987b). Furthermore, this immunolocalization has been extended to invertebrate nervous tissues of snail and locust (Homburger et al., 1987). Such data are illustrated in Fig. 4. In all brain tissues of vertebrates and invertebrates, we found an immunoreactivity with our anti-$G_o\alpha$ antibodies. This result suggests a high conservation of at least some $G_o\alpha$ epitopes during evolution. A detailed analysis of the protein recognized in snail neurons indicates that its molecular mass is 40 kDa rather than 39 kDa. It is also the unique PTX-toxin substrate in this invertebrate nervous tissue (Harris-Warrick et al., 1988).

Several peripheral neuronal or related tissues also exhibited G_o immunoreactivity. They include peripheral neuronal tissues, like cervical spinal

Immunoblot

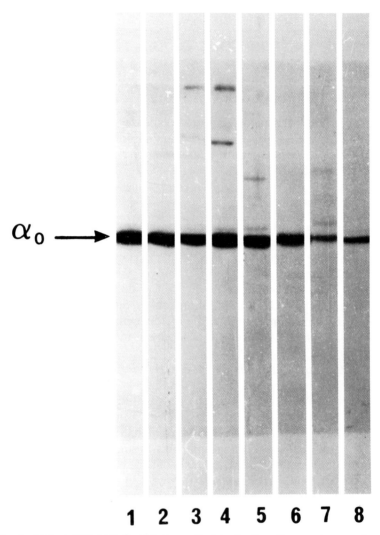

$\alpha_0 \longrightarrow$

1   2   3   4   5   6   7   8

Fig. 4. Immunoblot analysis of brain particulate fractions from various species with anti-$G_o\alpha$ antiserum. 100 μg of membrane protein was subjected to electrophoresis through 10% polyacrylamide gels, blotted onto nitrocellulose and analyzed with anti-$G_o\alpha$ antiserum. Proteins loaded were from: lane 1, calf; lane 2, rat; lane 3, turtle; lane 4, pigeon; lane 5, frog; lane 6, trout; lane 7, snail; lane 8, locust.

cord (Gierschik *et al.*, 1986b), sciatic nerve (Homburger *et al.*, 1987), and adrenal medulla (Mumby *et al.*, 1986; Toutant *et al.*, 1987a).

2. Nonneuronal Tissues

Although less abundant than in neuronal tissue, $G_o\alpha$ immunoreactivity (which has not always been characterized with the criteria proposed in Section III.B) was also detected in various nonneuronal tissues such as sea urchin eggs (Oinuma *et al.*, 1986), adenohypophysis (Homburger *et al.*, 1987; Journot *et al.*, 1987), and heart (Mumby *et al.*, 1986; Luetje *et al.*, 1987; Huff *et al.*, 1985; Liang *et al.*, 1986; Martin *et al.*, 1987). Some $G_o\alpha$ immunoreactivity was observed in trachea (Mumby *et al.*, 1986), kidney medulla (Huff *et al.*, 1985) and olfactory epithelium (Anholt *et al.*, 1987). Since $G_o\alpha$-immunoreactive material was present in the peripheral nervous system (Homburger *et al.*, 1987), part of the immunoreactivity found in some peripheral tissues may be due to their innervation. This assumption is supported by the observations by Luetje *et al.* (1987) and Liang *et al.* (1986) that an increase in $G_o\alpha$ levels occurs during the onset of functional parasympathetic innervation of the chick heart.

$G_o\alpha$ immunoreactivity has also been detected in endocrine cells from pancreatic islets of Langerhans (Terashima *et al.*, 1987b). This finding and our observations that $G_o\alpha$ is present in adenohypophysis (Homburger *et al.*, 1987; Journot *et al.*, 1987) suggest that G_o is not only present in neuronal tissues but is also expressed in some endocrine tissues.

Furthermore, Ui's group has suggested that G_o may be a marker from cells which derive from the neural tube and crest (Terashima *et al.*, 1987b). This assumption may explain $G_o\alpha$ detection in a variety of tissues and cells such as the enteric nervous system, chromaffin cells, and possibly endocrine cells from pancreas (Terashima *et al.*, 1987b).

B. Tissue Distribution of mRNA Coding for $G_o\alpha$

Two reports have used oligodeoxynucleotide probes for measurement of $G_o\alpha$ mRNA in different tissues (Brann *et al.*, 1987; Jones and Reed, 1987). Brann *et al.* (1987) found that the $G_o\alpha$ probe labeled three mRNA species; the most abundant had a size of 3500 bases. As expected, this mRNA was highly expressed in brain and weakly in peripheral tissues (Brann *et al.*, 1987). In contrast, Jones and Reed (1987) found that their $G_o\alpha$ probe hybridized to two messengers of 4100 and 4500 bases. One of them (4100 bases) was only expressed in olfactory and brain tissues whereas the other (4500 bases) was expressed in all tissues. The significance of such a result, as well as the clear differences observed by Brann *et al.*

(1987) in the distribution of the expression of mRNA encoding Gα sub-units on one hand and the expression of the Gα subunit proteins on the other hand, remain to be further studied.

VI. CELLULAR AND SUBCELLULAR IMMUNOLOCALIZATION OF G$_o\alpha$

Although the tissue distribution of G$_o\alpha$ has been relatively well studied, the cellular and subcellular localizations of G$_o$ are still poorly known.

Immunohistochemical studies in rat and bovine brain indicate that G$_o\alpha$ is widely distributed in the molecular layers of both cerebral cortex and cerebellum, the neuropil of the hippocampus, and the substantia nigra, pars reticulata, and neurohypophysis (Worley *et al.*, 1986; Asano *et al.*, 1987). In the spinal cord, the gray matter is more stained than the white matter, especially in the dorsal band corresponding to the substantia gelatinosa. In the rat retina, the most intense immunostaining is observed in the inner and outer plexiform layers which are considered synaptic regions (Terashima *et al.*, 1987a; Lad *et al.*, 1987). Together with the biochemical evidence that the synaptosomes from *Aplysia* neurons are enriched in G$_o$ (Chin *et al.*, 1987), immunohistochemical data clearly demonstrate higher G$_o$ concentrations in synapse-rich neuropil than in regions rich in neuronal cell bodies.

Using primary cultures of differentiated striatal neurons, cerebellar granules, and glial or ependymal cells that we have previously character-ized (Weiss *et al.*, 1986; Bockaert *et al.*, 1986; Gabrion *et al.*, 1988), we studied the cellular and subcellular localizations of G$_o\alpha$. The antibody used was previously characterized by Homburger *et al.*, (1987) (see also Figs. 3 and 4). In different types of cultured brain cells, the antibody only recog-nizes a 39-kDa protein both in immunoblot and by immunoprecipitation (Fig. 3B, C). Indeed, a 39-kDa G$_o\alpha$ subunit was identified by immunobio-chemical evidence and by PTX-dependent ADP-ribosylation in primary cultures of striatal or cortical neurons, in cerebellar granule cells, as well as in primary cultures of astrocytes from striatum, colliculi, and cerebral cortex but not in astrocytes from cerebellum (Brabet *et al.*, 1988a). C$_6$ glioma cells do not express the G$_o\alpha$ protein (Milligan *et al.*, 1986; Brabet *et al.*, 1988a; and Fig. 3A, B). In nondifferentiated NIE 115 neuroblastoma cells, G$_o\alpha$ is barely detectable (Brabet *et al.*, 1988). In contrast, differentia-tion of NIE 115 cells induces an increased G$_o\alpha$ expression. Interestingly, differentiation increased the expression of a G$_o\alpha$ protein having a pHi of 5.5, whereas in nondifferentiated forms the G$_o\alpha$ isoform had a pHi of 5.8 (Brabet *et al.*, 1988b; Brabet *et al.*, submitted). In choroidal and ciliated

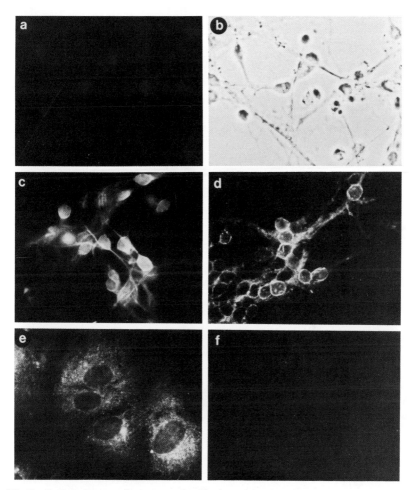

Fig. 5. Indirect immunofluorescence localization of $G_o\alpha$ in cerebellar granule cells, striatal neurons, and glial cells. Cells cultured on glass coverslips were fixed and permeabilized with Triton X-100. Adsorption of affinity-purified anti-$G_o\alpha$ antibodies was carried out with a 10-fold molar excess of $G_o\alpha$. Secondary antibodies were rhodamine-conjugated goat anti-rabbit IgG. Magnification: ×620. (a) Prior incubation of affinity-purified anti-$G_o\alpha$ antibodies with a 10-fold molar excess of $G_o\alpha$ prevents specific staining of striatal neurons. (b) The same field as shown in a, viewed by phase-contrast optics. Striatal neurons (c), cerebellar granule cells (d), and striatal glial cells (e) stained with unadsorbed affinity-purified anti-$G_o\alpha$ antibodies. Note the marked immunolabeling at neuronal cell–cell contact areas. (f) Absence of staining of meningeal cells with affinity-purified anti-$G_o\alpha$ antibodies.

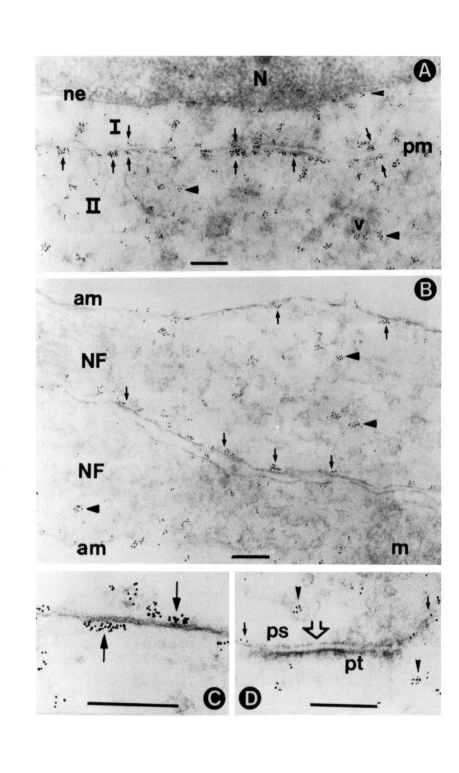

ependymocytes, $G_o\alpha$ is recognized both by immunoblot and PTX-dependent ADP-ribosylation (Peraldi *et al.*, 1989).

An intense immunofluorescent reaction is observed in perikarya of cultured neurons labeled with anti-$G_o\alpha$ antibody. (Brabet *et al.*, 1988a) (Fig. 5). This positive reaction is especially pronounced at the periphery of cerebellar granule cells in culture, whereas cytoplasm and neurite arborizations are more intensively stained in striatal neurons (Fig. 5). In both neuron types, a reinforcement of the immunoreaction is observed in cell–cell or cell–neurite contacts.

Surprisingly, very few studies have been devoted to the ultrastructural localization of G proteins, especially of G_o. Preliminary results have only concerned the ultrastructural immunolocalization of the β subunit of G proteins (Lombardo *et al.*, 1985). Electron immunocytochemistry using anti-IgG antibodies coupled with 5-nm gold particles subsequent to affinity-purified anti-$G_o\alpha$ antibodies (Gabrion *et al.*, 1989) has confirmed the immunofluorescence studies discussed above (Fig. 6). In synapses no positive reaction was detected with pre- or postsynaptic membranes. In contrast, clusters of gold particles have been noted near the synaptic structures on the cytoplasmic face of the plasma membrane, mainly in areas where the plasma membranes of neighboring cells are in close association. Gold particles are seen at the level of the synaptic endings as well as at the level of cell–cell, cell–neurite or neurite–neurite contacts (Fig. 6). Along these contacting membranes, the positive reaction is sometimes asymmetrically distributed. In addition, noncontacting membranes can display an intense, although less frequent labeling. Gold particles have also been seen dispersed in the cytoplasmic matrix, especially between rough endoplasmic reticulum and Golgi cisternae and at the cytoplasmic face of coated vesicles

Fig. 6. Ultrastructural localization of G_o in cultured striatal neurons. Affinity purified anti-$G_o\alpha$ antibodies and 5-nm immunogold probes were used after 0.05% Triton X-100 permeabilization. Epon-embedded cultures were ultrathin sectioned and observed under the electron microscope without lead citrate contrast. (A) The gold labeling is observed on the inner face of plasma membranes (pm) of two neighboring cells (I and II) in symmetrical (→ ←) or asymmetrical (→) distributions. The cytoplasmic matrix (◄) displays a diffuse labeling of gold particles which are evenly associated with cytoplasmic vesicles (v) or nuclear envelope (ne). The nuclear matrix (N) is unlabeled. (B) In two varicose nerve fibers (NF), running parallel to culture substrates, gold labeling is observed with a distribution similar to that described in perikarya (in A): the inner face (↓) of axolemmal membranes (am) and the cytoplasmic matrix (◄) display gold labeling. m, mitochondria. (C) Close contacts between perikarya and neurite membranes are frequently lined with clusters of gold particles (↓). (D) In synaptic contacts between nerve endings, gold particles are never seen in association with the presynaptic (ps) or postsynaptic (pt) membranes, while they are found on the inner face of membranes, near the negative synaptic cleft (⇓), or in the pre- or postsynaptic cytoplasm (▼). Bar = 0.2 μm.

and endosomal membranes (Gabrion *et al.*, 1989). In neurons, (Fig. 6) as well as in the electric organ of *Torpedo marmorata* (Toutant *et al.*, 1987b), we never observe $G_o\alpha$ on synaptic vesicles (Gabrion *et al.*, 1989). In contrast, $G_o\alpha$ is present on granules of chromaffin cells (Toutant *et al.*, 1987a).

In astrocytes and ependymal cells, the distribution of the immunostaining appears quite different, according to the cell type considered. While astrocyte cell cultures derived from cerebellum are poorly reactive to anti-$G_o\alpha$ antibodies, those derived from striatum display specific immunostaining which appears relatively diffuse in the cytoplasmic matrix; G_o is never observed on the plasma membrane or at the cell–cell contacts of 3-week-old glial cell cultures (Brabet *et al.*, 1988a) (Fig. 5). In ependymal cells, purified anti-$G_o\alpha$ antibodies recognize a protein with a highly asymmetrical distribution (Peraldi *et al.*, 1987, 1989). In choroidal ependymocytes *in vivo*, as well as in culture, $G_o\alpha$ is abundant at the apical pole of the cell, in the cytoplasmic matrix, and also under the apical plasma membrane where it is occasionally associated with coated vesicles. In ciliated ependymocytes, $G_o\alpha$ is only recognized at the surface of the axonema in apical kinocilia (Peraldi *et al.*, 1987, 1989). It is interesting to associate these results with two previously reported observations.

First, $G_o\alpha$ is also found in a partially purified preparation of isolated chemosensory cilia from olfactory epithelium, whereas the respiratory cilia, which line the tracheal cavity, do not contain $G_o\alpha$ (Anholt *et al.*, 1987). Second, ciliated ependymocytes and specifically ependymal kinocilia are a preferential target in the infectious pathogenesis induced by *Bordetella pertussis* (Berenbaum *et al.*, 1960; Hopewell *et al.*, 1972).

In nonnervous tissues, where the presence of $G_o\alpha$ has been biochemically or immunochemically demonstrated, the subcellular distribution of $G_o\alpha$ frequently remains to be specified. In pancreatic islets of Langerhans, antibodies against $G_o\alpha$ have been used in the immunohistochemical localization of this protein at the periphery of the cells (Terashima *et al.*, 1987b). It would be interesting to establish whether the subcellular distribution of $G_o\alpha$ is exclusively focused on the inner face of the plasma membrane in endocrine cells and in other cells derived from neural crest, or if $G_o\alpha$ is only present in the subplasmalemmal cytoplasmic matrix as noted in astrocytes.

VII. ROLES OF G_o

Both the cellular and subcellular immunolocalization of G_o discussed above give some indications about its possible functions. We are now sure that $G_o\alpha$ is expressed as a 39-kDa detectable and PTX-sensitive protein in cells of the central and peripheral nervous system (Worley *et al.*, 1986; Homburger *et al.*, 1987; Asano *et al.*, 1987; Brabet *et al.*, 1988a), adenohypophysis (Journot *et al.*, 1987), chromaffin cells (Toutant *et al.*, 1987a), choroidal and ciliated ependymal cells (Peraldi *et al.*, 1987), and perhaps in cardiac cells (Sternweis *et al.*, 1984; Mumby *et al.*, 1986; Malbon *et al.*, 1985). Additional experiments are required to clearly establish its presence in other tissues for which the criteria proposed in Section IV.B to identify $G_o\alpha$ have not always been met. The cells in which $G_o\alpha$ is expressed are generally excitable and secretory cells containing ionic channels.

Subcellular localizations of $G_o\alpha$ indicate that it could mediate different functions in neurons and astrocyte cells. As already pointed out, electron microscopic studies show that in neurons, $G_o\alpha$ is present both at the inner face of the plasma membrane and in the cytoplasmic matrix (Fig. 6), whereas in astrocyte cells the majority of $G_o\alpha$ is in the cytoplasmic matrix. Additional experiments are required to determine the possible functions associated with such an unexpected localization. It is important to point out that we do not know whether or not our antibody against $G_o\alpha$ recognizes both $G_o\alpha$ and $G_o\alpha\beta\gamma$ by the immunolocalization experiments reported in Section VI. It would be particularly interesting to determine if, for example, $G_o\alpha$ and $G_o\alpha\beta\gamma$ are localized on the cytoplasmic matrix and on plasma membrane respectively. In neurons in culture, $G_o\alpha$ is not localized at the synaptic cleft, but rather at nonsynaptic cell–cell contacts including those present in synaptic areas (Fig. 6). This is not necessarily in contradiction with immunohistochemical localizations of $G_o\alpha$ in rat and bovine brain reported by others in which they found that $G_o\alpha$ had a higher density in neuropil rather than in cell bodies (Worley *et al.*, 1986; Asano *et al.*, 1987). Indeed, there is more and more evidence showing that typical synapses are not the only area of neuronal communication (Descarries *et al.*, 1977). Furthermore, receptors which are susceptible to interact with $G_o\alpha$ [muscarinic, γ-aminobutyric acid (GABA B), opiates, α_2-adrenergic (Florio and Sternweis, 1985; Haga *et al.*, 1986; Kurose *et al.*, 1986; Asano *et al.*, 1985; Hescheler *et al.*, 1987; Cerione *et al.*, 1986)] are receptors known to be modulators of transmitter release rather than classical synaptic transmitters (North *et al.*, 1987). The possibility of an overexpression of $G_o\alpha$ in cell bodies of neurons in culture compared to neurons *in vivo* cannot be excluded. $G_o\alpha$ is not present in synaptic vesicles of neurons in culture (Fig. 6) or in those of the electric organ of *Torpedo marmorata* (Toutant *et al.*, 1987b). However, $G_o\alpha$ is present in chromaffin granules (Toutant *et al.*, 1987a). This may be related to differences in the regulation of transmitter release in these two secretory systems.

Reconstitution experiments have been undertaken to clarify the functions of $G_o\alpha$. However, the purity of the proteins used has not always been carefully controlled.

It has been observed that both G_o and G_i (generally not entirely pure) have similar potencies to interact with partially purified or highly purified muscarinic receptors (Florio and Sternweis, 1985; Haga *et al.*, 1986; Kurose *et al.*, 1986) and GABA B receptors (Asano *et al.*, 1985). In all cases, G_o and G_i could increase the proportion of receptors having a high affinity for agonists which was previously reduced either by eliminating G proteins during purification or after treatment with PTX. Functional reconstitution of pure muscarinic (Kurose *et al.*, 1986) and α_2-adrenergic receptors (Cerione *et al.*, 1986) with pure G proteins was performed in phospholipid vesicles. Muscarinic receptors stimulate the GTPase of both G_i and G_o but with slight differences. The time courses for carbachol stimulation of GTPγS binding and Mg^{2+} were different when G_i or G_o were used, suggesting that coupling was not identical (Kurose *et al.*, 1986). On the other hand, α_2-adrenergic receptors stimulated equally the GTPase activity of G_i and G_o, whereas they were less potent in stimulating the GTPase of T or G_s (Cerione *et al.*, 1986). After treatment with PTX, a mixture of G_o and G_i reconstituted the stimulation of GTPase, or adenylyl cyclase stimulation by opiates and bradykinin in membranes of neuroblastoma × glioma hybrid cells (N × G cells) (Higashida *et al.*, 1986; Milligan and Klee, 1985).

In leukemia cells HL-60, G_o and G_i were able to reconstitute the coupling between f-met-leu-phe (FMLP) receptors and phospholipase C after a preliminary blockade of this receptor function with PTX (Kikuchi *et al.*, 1986). This experiment suggests that $G_o\alpha$ can activate phospholipase C at least under the conditions tested. However, in HL-60, another G protein is likely to be involved in phospholipase C activation, since $G_o\alpha$ is not present in these cells. A role for $G_o\alpha$ in phospholipase C regulation has also been suggested by the correlation between the histochemical localizations of $G_o\alpha$ and those of 4β-[^{3}H] phorbol 12, 13-dibutyrate binding (Worley *et al.*, 1986). Although these results are interesting, one should note that the antibody used to detect $G_o\alpha$ also recognized some $G_i\alpha$ (10% crossreactivity) (Huff *et al.*, 1985) and that 4β-[^{3}H] phorbol 12, 13-dibutyrate binding has not been proved to reliably measure the density of receptor–G protein–phospholipase C systems. Furthermore, the coupling between neurotransmitter receptors and phospholipase C in brain has not been reported to be particularly affected by PTX as could be expected if G_o was the major G protein involved in this system.

The experiments described so far indicate that, in reconstitution experiments, $G_o\alpha$ and $G_i\alpha$ behave similarly. This is not really surprising in view of the high degree of homology between these proteins. However, in three reconstitution models, G_i and G_o were reported to have different efficacies. Two of them concern the interaction of G_i and G_o with adenylyl cyclase: (1) In membranes of *cyc⁻* ($G_s\alpha$- deficient) mutants of S49 lymphoma cells,

$G_i\alpha$ activated with GTPγS was found to inhibit adenylyl cyclase, whereas the $G_o\alpha$-activated protein was inactive (Roof et al., 1985). (2) Similarly, activated $G_i\alpha$ inhibited the partially purified catalyst of adenylyl cyclase stimulated by activated $G_s\alpha$, whereas activated $G_o\alpha$ had no effect (Katada et al., 1986b). These experiments indicate that $G_o\alpha$ is not able to interact with the components of the adenylyl cyclase system, at least under the experimental conditions used. Additional experiments are required to definitely conclude this point.

The other system in which G_o has been reported to behave differently than G_i is the regulation of voltage-sensitive Ca^{2+} channels (VSCC) in N $\times$ G cells (Hescheler et al., 1987). There is increasing evidence that neurotransmitter receptors can be coupled to ionic channels directly through G proteins. This is well documented for the activation of K^+ channels by muscarinic receptors in atrial cells (Pfaffinger et al., 1985; Yatani et al., 1986), and less for the inhibition of VSCC by α_2-adrenergic and GABA-B receptors in dorsal root ganglia (Holz et al., 1986; Dolphin and Scott, 1987). In N $\times$ G cells, Hescheler et al. (1987) found that opiate receptors inhibit VSCC. Pretreatment with PTX almost completely abolishes the opiate effect. It can be restored by intracellular injection of G_i and G_o. As the α subunit of G_o (with or without $\beta\gamma$) was 10 times more potent than G_i, Hescheler et al. (1987) proposed that G_o is involved in the functional coupling of opiate receptors to VSCC. We carried out similar experiments in snail (Helix aspersa) neurons (Harris-Warrick et al., 1988). In these neurons, dopamine (DA) inhibits VSCC. Intracellularly injected nonhydrolyzable GTP analogs mimics and subsequently blocks the DA-induced response. Injection of activated PTX also blocks the DA response (Fig. 7), whereas injection of activated $G_o\alpha$ purified from bovine brain mimics the DA effect. Since it was difficult to control the amount of $G_o\alpha$ injected with the technique used, we did not compare the effects of $G_o\alpha$ and $G_i\alpha$. Instead, we used specific affinity-purified anti-$G_o\alpha$ antibodies (Homburger et al., 1987; Brabet et al., 1988a; Toutant et al., 1987a,b) to determine the nature of the G protein involved in the coupling between DA receptors and VSCC in snail neurons. We found that this antibody markedly reduces the DA-induced decrease in the Ca^{2+} current (Fig. 7). This indicates that in snail neurons, a G protein having homologies with $G_o\alpha$ is somehow involved in the coupling between the DA receptor and VSCC. As already discussed in Section V.A, we have shown that snail neurons contain a unique ADP-ribosylated PTX substrate (in experimental conditions under which we were able to separate three PTX substrates in brain tissues (see Brabet et al., 1988a; Toutant et al., 1987 a,b; and Fig. 3) and that on immunoblots, the antibody directed against $G_o\alpha$ recognized this protein. Its molecular weight is 40,000 (Harris-Warrick et al., 1988).

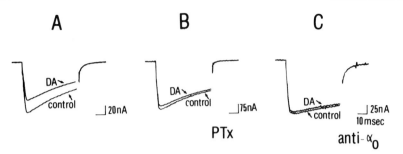

Fig. 7. Role of a G_o-like protein in the coupling between dopamine receptors and voltage-sensitive Ca^{2+} channels (VSCC in snail neurons). Snail neurons were voltage-clamped. They were bathed in a saline solution in which Ca^{2+} was replaced by Ba^{2+} and containing tetraethylammonium (TEA) (30 mM). The neurons were held at -50 mV and depolarized to -5 mV for 60 msec. (A) Effect of dopamine (DA) (500 nM) on the Ca^{2+} current. DA induced a decrease in the voltage-dependent Ca^{2+} current. (B) Effect of PTX on DA-induced decrease in the voltage-dependent Ca^{2+} current. PTX was pressure-injected for 30 min and the cell was held in Ba^{2+}/TEA saline solution for 90 min before recording the inward current. The traces were recorded before and during the first application of DA. Before PTX, DA evoked a 30–35% decrease in the voltage-dependent Ca^{2+} current (left panel). After PTX, DA evoked a $5 \pm 3\%$ decrease of the current. Heat inactivated PTX had no effect (data not shown). (C) Effect of intracellular injections of purified anti-$G_o\alpha$ antibodies on DA-induced decrease in the voltage-dependent Ca^{2+} current. Anti-$G_o\alpha$ antibodies (Homburger *et al.*, 1987; Toutant *et al.*, 1987a,b; Brabet *et al.*, 1988) were injected for 18 min and a 15 min period was allowed for antibody diffusion. Subsequent to this injection, application of DA only causes a slight decrease in Ca^{2+} current ($7 \pm 3\%$ in eight experiments). Heat-inactivated anti-$G_o\alpha$ antibodies had no effect. Results presented in this figure are from H. M. Gerschenfeld's Laboratory (Harris-Warrick *et al.*, 1988)

As already discussed, it is remarkable that one (or several) epitopes of the vertebrate $G_o\alpha$ has been conserved during evolution in a family of G proteins specifically associated with neurons. It would be highly interesting to determine whether these epitopes are those involved in the coupling of $G\alpha$ to VSCC in vertebrates and invertebrates.

In conclusion, the localization of the $G_o\alpha$ protein (associated or not with $\beta\gamma$) both under the surface of plasma membranes and on the cytoplasmic matrix suggests the possibility of at least two distinct roles for this protein. One is likely to be a classical role in the coupling of neurotransmitter receptors to effectors, the other has yet to be discovered. Concerning the coupling receptor–effector, the adenylyl cyclase system seems to be excluded. It is possible that $G_o\alpha$ is involved in the coupling between receptor–phospholipase C but this remains to be proven under normal physiological conditions. The best candidate for a role of G_o is probably the coupling between receptors and ionic channels. It is remarkable that all the possible receptor candidates for an interaction with G$_o$ [muscarinic, α_2-adrenergic and GABA B receptors, and probably opiates, dopaminergic (D$_2$) serotonin (5-HT$_{1A}$) and somatostatin] are receptors which are involved in the inhibition of transmitter or hormonal release (North et al., 1987). Such an inhibition is generally mediated either by an activation of K$^+$ channels or by an inhibition of VSCC channels. A role for G$_o\alpha$ in the activation of K$^+$ channels has not yet been investigated, but we do know that activating effects of receptors on these channels are generally PTX-sensitive. Although several other models have to be tested, the inhibition of VSCC by G$_o\alpha$ reported in N × G cells and snail neurons is likely to be one of the important functions of this protein.

ACKNOWLEDGMENTS

We would like to thank A. L. Turner-Madeuf, P. Costagliola, and M. Passama for their expert secretarial assistance in the preparation of this manuscript. We would also like to thank J. Carrette, C. Pantaloni, S. Peraldi and M. Sebben for their major contributions in experiments reported in this review and Y. Audigier and L. Journot for helpful and stimulating discussions.

REFERENCES

Abood, M. E., Hurley, J. B., Pappoue, M.-C., Bourne, H. R., and Stryer, L. (1982). Functional homology between signal coupling proteins. Cholera toxin inactivates the GTPase activity of transducin. J. Biol. Chem. **257**, 10540–10543.

Angus, C. W., Van Meurs, K. P., Tsai, S. C., Adamik, R., Miedel, M. C., Pan, Y. C., Fung, H. F., Moss, J., and Vaughan, M. (1986). Identification of the probable site of choleragen-catalysed ADP-ribosylation in a G$_o\alpha$-like protein based on cDNA sequence. Proc. Natl. Acad. Sci. U.S.A. **83**, 5813–5846.

Anholt, R. R. H., Mumby, S. M., Stoffers, D. A., Griard, P. R., Kuo, J. F., and Snyder, S. H. (1987). Transduction proteins of olfactory receptor cells: Identification of guanine nucleotide binding proteins and protein kinase C. Biochemistry **26**, 788–795.

Asano, T., Ui, M., and Ogasawara, N. (1985). Prevention of the agonist binding to γ-aminobutyric acid β receptors by guanine nucleotides and islet activating protein, pertussis toxin, in bovine cerebral cortex. Possible coupling of the toxin sensitive GTP binding proteins to receptors. J. Biol. Chem. **260**, 12653–12658.

Asano, T., Semba, R., Ogasawara, N., and Kato, K. (1987). Highly sensitive immunoassay for the α subunit of the GTP binding protein G$_o$ and its regional distribution in bovine brain. J. Neurochem **48**, 1617–1623.

Berenbaum, M. C., Ungar, J., and Stevens, M. K. (1960). Intracranial infection of mice with *Bordetella pertussis. J. Gen. Microbiol.* **22,** 313–322.

Birnbaumer, L., Codina, J., Mattera, R., Yatani, A., Scherer, N., Toro, M. J., and Brown A. M. (1987). Signal transduction by G proteins. *Kidney Int., Suppl.* **23,** S14–S42.

Bockaert, J., Gabrion, J., Sladeczek, F., Pin, J. P., Récasens, M., Sebben, M., Kemp, D., Dumuis, A., and Weiss, S. (1986). Primary culture of striatal neurons: A model of choice for pharmacological and biochemical studies of neurotransmitter receptors. *J. Physiol.* **81,** 219–227.

Bockaert, J., Homburger, V., and Rouot, B. (1987). GTP binding proteins: A key role in cellular communication. *Biochimie* **69,** 329–338.

Brabet, P., Dumuis, A., Sebben, M., Pantaloni, C., Bockaert, J., and Homburger, V. (1988a). Immunocytochemical localization of the guanine nucleotide-binding protein G_o in primary culture of neuronal and glial cells. *J. Neurosci.* **8,** 701–708.

Brabet, P., Pantaloni, C., Rouot, B., Toutant, M., Garcia-Sainz, A., Bockaert, J., and Homburger, V. (1988b). Multiple species and isoforms of Bordetella Pertussis substrates. *Biochem. Biophys. Res. Commun.* **152,** 1185–1192.

Brann, M. R., Collins, R. M., and Spiegel, A. (1987). Localization of mRNAs encoding the α subunit of signal-transducing G-proteins within rat brain and among peripheral tissues. *FEBS Lett.* **222,** 191–198.

Bray, P., Carter, A., Guo, V., Puckett, C., Kamholz, J., Spiegel, A., and Nirenberg, M. (1987). Human cDNA clones for an α subunit of G_i signal transduction protein. *Proc. Natl. Acad. Sci. U.S.A.* **84,** 5115–5119.

Buss, J. E., Mumby, S., Casey, P. J., Gilman, A. G., and Sefton, B. M. (1987). Myristylated α subunits of guanine nucleotide-binding regulatory proteins. *Proc. Natl. Acad. Sci. U.S.A.* **84,** 7493–7497.

Cerione, R. A., Regan, J. W., Nakata, H., Codina, J., Benovic, J. L., Gierschik, P., Somers, R. L., Spiegel, A. M., Birnbaumer, L., Lefkowitz, R. J., and Caron, M. G. (1986). Functional reconstitution of the α_2-adrenergic receptor with guanine nucleotide regulatory proteins in phospholipid vesicles. *J. Biol. Chem.* **261,** 3901–3909.

Chin, G. J., Vogel, S. S., Mumby, S. M., and Schwartz, J. H. (1987). Characterization and localization of G_s and G_o protein molecules in *Aplysia* neurons. *Soc. Neurosci. Abs.* **12,** 315.

Descarries, L., Watkins, K. C., and Lapierre, Y. (1977). Noradrenergic axon terminals in the cerebral cortex of the rat. III Topometric ultrastructural analysis. *Brain Res.* **133,** 197–222.

Dever, T. E., Glynias, M. J., and Merrick, W. C. (1987). GTP binding domain: Three consensus sequence elements with distinct spacing. *Proc. Natl. Acad. Sci. U.S.A.* **84,** 1814–1818.

Dickey, B. F., Pyun, H. Y., Williamson, K. C., and Navarro, J. (1987). Identification and purification of a novel G protein from neutrophils. *FEBS Lett.* **219,** 289–292.

Didsbury, J. A., and Snyderman, R., (1987). Molecular cloning of a new human G protein — Evidence for two $G_i\alpha$-like protein families. *FEBS Lett.* **219**, 259–263.

Dolphin, A. C., and Scott, R. H. (1987). Calcium channel currents and their inhibition by (-) baclofen in rat sensory neurons: Modulation by guanine nucleotides. *J. Physiol.* **386**, 1–7.

Evans, T., Brown, M. L., Fraser, E. D., and Northup, J. K. (1986). Purification of the major GTP-binding proteins from human placental membranes. *J. Biol. Chem.* **261**, 7052–7059.

Florio, V. A., and Sternweis, P. C. (1985). Reconstitution of resolved muscarinic cholinergic receptors with GTP-binding proteins. *J. Biol. Chem.* **260**, 3477–3483.

Gabrion, J., Peraldi, S., Faivre-Bauman, A., Klotz, C., Ghandour, M. S., Paulin, D., Assenmacher, I., and Tixier-Vidal, A. (1988). Characterization of ependymal cells in hypothalamic and choroidal primary cultures. *Neuroscience* **24**, 993–1007.

Gabrion, J., Brabet, P., Nguyen-Than Dao, B., Homburger, V., Dumuis, A., Sebben, M., Rouot, B., and Bockaert, J. (1989). Ultrastructural localization of the GTP-binding protein G_o in neurons. *Cellular Signalling* **1**, 107–123.

Gierschik, P., Falloon, J., Milligan, G., Pines, M., Gallin, J. I., and Spiegel, A. (1986a). Immunochemical evidence for a novel pertussis toxin substrate in human neutrophils. *J. Biol. Chem.* **261**, 8058–8062.

Gierschik, P., Milligan, G., Pines, M., Goldsmith, P., Codina, J., Klee, N., and Spiegel, A. (1986b). Use of specific antibodies to quantitate the guanine nucleotide-binding protein G_o in brain. *Proc. Natl. Acad. Sci. U.S.A.* **83**, 2258–2262.

Gierschik, P., Morrow, B., Milligan, G., Rubin, C., and Spiegel, A. (1986c). Changes in the guanine nucleotide-binding proteins, G_i and G_o, during differentiation of 3T3-L1 cells. *FEBS Lett.* **199**, 103–106.

Gierschik, P., Sidiropoulos, D., Spiegel, A., and Jakobs, K. H. (1987). Purification and immunological characterization of the major pertussus toxin-sensitive guanine nucleotide binding protein of bovine neutrophil membranes. *Eur. J. Biochem.* **165**, 185–194.

Gilman, A. G. (1987). G proteins: Transducers of receptor-generated signals. *Annu. Rev. Biochem.* **56**, 615–649.

Haga, K., Haga, T., and Ichiyama, A. (1986). Reconstitution of the muscarinic acetylcholine receptor-guanine nucleotide-sensitive high affinity binding of agonists to purified muscarinic receptors reconstituted with GTP binding proteins (G_i and G_o). *J. Biol. Chem.* **261**, 10133–10140.

Harris-Warrick, R., Hammond, C., Paupardin-Tritsch, D., Homburger, V., Rouot, B., Bockaert, J., and Gerschenfeld, H. M. (1988). An $\alpha 40$ subunit of a GTP-binding protein immunologically related to G_o mediates a dopamine-induced decrease in snail neurons. *Neuron* **1**, 27–32.

Hescheler, J., Rosenthal, W., Trautwein, W., and Schultz, G. (1987). The GTP-binding protein, G_o, regulates neuronal calcium channels. *Nature* **325**, 445–447.

J. Bockaert et al.

Higashida, H., Streatly, R. A., Klee, W., and Nirenberg, M. (1986). Bradykinin-activated transmembrane signals are coupled via N_o or N_i to production of inositol 1,4,5-trisphosphate, a second messenger in NG 108-15 neuroblastomaglioma hybrid cells. *Proc. Natl. Acad. Sci. U.S.A.* **83**, 942–946.

Higashijima, T., Kenneth, M., Ferguson, G., Sternweis, P. C., Ross, E., M., Smigel, M. D., and Gilman, A. G. (1987a). The effect of activating ligands on the intrinsic fluorescence of guanine nucleotide-binding regulatory proteins. *J. Biol. Chem.* **262**, 752–756.

Higashijima, T., Kenneth, M., Ferguson, G., Smigel, M. D., and Gilman, A. G. (1987b). The effect of GTP and Mg^{2+} on the GTPase activity and the fluorescent properties of G_o. *J. Biol. Chem* **262**, 757–761.

Higashijima, T., Kenneth, M., Ferguson, G., Sternweis, P. C., Smigel, M. D., and Gilman, A. G. (1987c). Effects of Mg^{2+} and the $\beta\gamma$-subunit complex on the interactions of guanine nucleotides with G proteins. *J. Biol. Chem.* **262**, 762–766.

Higashijima, T., Kenneth, M., Ferguson, G., and Sternweis, P. C. (1987d). Regulation of hormone-sensitive GTP-dependent. Regulatory proteins by chloride. *J. Biol. Chem.* **262**, 3597–3602.

Holz, G., Rane, S. G., and Dunlap, K. (1986). GTP binding proteins mediate transmitter inhibition of voltage-dependent calcium channels. *Nature* **319**, 670–672.

Homburger, V., Brabet, P., Audigier, Y., Pantaloni, C., Bockaert, J., and Rouot, B. (1987). Immunological localization of the GTP binding proteins G_o in different tissues of vertebrates and invertebrates. *Mol. Pharmacol.* **31**, 313–319.

Hopewell, J. W., Holt, L. B., and Desombre, T. R. (1972). An electron microscope study of intracerebral infection of mice with low-virulence *Bordetella pertussis. J. Med. Microbiol.* **5**, 154–157.

Huff, R. M., and Neer, E. J. (1986). Subunit interactions of native and ADP-ribosylated α_{39} and α_{41}, two guanine nucleotide binding proteins from bovine cerebral cortex. *J. Biol. Chem.* **261**, 1105–1110.

Huff, R. M., Axton, J. M., and Neer, E. (1985). Physical and immunological characterization of a guanine nucleotide-binding protein purified from bovine cerebral cortex. *J. Biol. Chem.* **260**, 10864–10871.

Hurley, J. B., Simon M. I., Teplow, D. B., Robishaw, J. D., and Gilman, A. G. (1984). Homologies between signal transducing G proteins and *ras* gene products. *Science* **226**, 860–862.

Itoh, H., Kozasa, T., Nagata, S., Nakamura, S. I., Katada, T., Ui, M., Iwai, S., Ohtsuka, E., Kawasaki, H., Suzuki, K., and Kaziro, Y. (1986). Molecular cloning and sequence determination of cDNA s for α subunits of the guanine nucleotide-binding proteins G_s, G_i and G_o from brain. *Proc. Natl. Acad. Sci. U.S.A.* **83**, 3776–3780.

Jones, D. T., and Reed, R. R. (1987). Molecular cloning of five GTP-binding protein cDNA species from rat olfactory neuroepithelium. *J. Biol. Chem.* **262**, 14241–14249.

Journot, L., Homburger, V., Pantaloni, C., Priam, M., Bockaert, J., and Enjalbert,

A. (1987). An Islet activator protein-sensitive G protein is involved in dopamine inhibition of angiotensin and thyrotropin-releasing hormone stimulated inositol phosphate production in anterior pituitary cells. *J. Biol. Chem.* **262**, 15106–15110.

Jurnak, F. (1985). Structure of the GDP domain of EF-Tu and location of the amino acids homologous to ras oncogene proteins. *Science* **230**, 32–36.

Kanaho, Y., Tsai, S-C., Adamik, R., Hewlett, E. L., Moss, J., and Vaughan, M. (1984). Rhodopsin-enhanced GTPase activity of the inhibitory GTP binding protein of adenylate cyclase. *J. Biol. Chem.* **259**, 7378–7381.

Katada, T., Oinuma, M., and Ui, M. (1986a). Two guanine nucleotide-binding proteins in rat brain serving as the specific substrate of islet-activating protein, pertussis toxin. *J. Biol. Chem.* **261**, 8182–8191.

Katada, T., Oinuma, M., and Ui, M. (1986b). Mechanisms for inhibition of the catalytic activity of adenylate cyclase by the guanine nucleotide-binding proteins serving as the substrate of islet-activating protein, pertussis-toxin. *J. Biol. Chem.* **261**, 5215–5221.

Katada, T., Oinuma, M., Kusakabe, K., and Ui, M. (1987). A new GTP-binding protein in brain tissues serving as the specific substrate of islet-activating protein, pertussis toxin. *FEBS Lett.* **321**, 353–358.

Kikuchi, A., Kozawa, O., Kaibuchi, K. Katada, T., Ui, M., and Takai, Y. (1986). Direct evidence of involvement of a guanine nucleotide-binding protein in chemotactic peptide stimulated formation of inositol bisphosphate and trisphosphate in differentiated human leukemic (HL-60) cells. Reconstitution with G$_i$ or G$_o$ of the plasma membranes ADP-ribosylated by pertussis toxin. *J. Biol. Chem.* **261**, 11558–11562.

Kurose, H., Katada, T., Haga, T., Haga, K., Ichiyama, A., and Ui, M. (1986). Functional interaction of purified muscarinic receptors with purified inhibition guanine regulatory proteins reconstituted in phospholipid vesicles. *J. Biol. Chem.* **261**, 6423–6428.

La Cour, T. F. M., Nyborg, J., Thirup, S., and Clark, B. F. C. (1985). Structural details of the binding of guanosine diphosphate to elongation factors Tu from E. coli as studied by X-ray crystallography. *EMBO J.* **4**, 2385–2388.

Lad, R. P., Simons, C., Gierschik, P., Milligan, G., Woodard, C., Griffo, M., Goldsmith, P., Ornberg, R., Gerfen, C. R., and Spiegel, A. (1987). Differential distribution of signal transducing G proteins in retina. *Brain Res.* **423**, 237–246.

Lerea, C. L., Somers, D. E., Hurley, J. B., Klock, I. B., and Bunt-Milam, A. H. (1986). Identification of specific transducin α subunits in retinal rod and cone photoreceptors *Science* **234**, 77–80.

Liang, B. T., Helmreich, M. R., Neer, E. J., and Galper, J. B. (1986). Development of muscarinic cholinergic inhibition of adenylate cyclase in embryonic chick heart. *J. Biol. Chem.* **261**, 9011–9021.

Lochrie, M. A., Hurley, J. B., and Simon, M. I. (1985). Sequence of the α subunit of photoreceptor G protein: Homologies between transducin, *ras*, and elongation factors. *Science* **228**, 96–99.

Lombardo, B. P., Anglade, G., Deterre, P., and Monneron, A. (1985). Immunocytochemistry of the α and β subunit of transducin and related proteins in retina and brain sections. *J. Cell. Biol.* **101**, 466a.

Luetje, C. W., Gierschik, P., Milligan, G., Unson, C., Spiegel, A., and Nathanson, N. M. (1987). Tissue specific regulation of GTP-binding protein and muscarinic acetylcholine receptor levels during cardiac development. *Biochemistry* **26**, 4876–4884.

Malbon, C. C., Mangano, T. J., and Watkins, D. C. (1985). Heart contains two substrates ($M = 40.000$ and 41.000) for pertussis toxin-catalyzed ADP-ribosylation that co-purify with N_s. *Biochem. Biophys. Res. Commun.* **128**, 809–815.

Martin, J. M., Subers, E. M., Halvorsen, S. W., and Nathanson, N. M. (1987). Functional and physical properties of chick atrial and ventricular GTP binding proteins: Relationship to muscarinic acetylcholine receptor mediated responses. *Mol. Pharmacol.* **240**, 683–688.

McGrath, J. P., Capon, D. J., Goeddel, D. V., and Levinson, A. D. (1984). Comparative biochemical properties of normal and activated human *ras* p 21 protein. *Nature* **310**, 644–649.

Medynski, D. C., Sullivan, K., Smith, K., Van Dop, C., Chang, F. H., Fung, B. K. K., Seeburg, P. H., and Bourne, H. R. (1985). Amino acid sequence of the α subunit of transducin deduced from cDNA sequence. *Proc Natl. Acad. Sci. U.S.A.* **82**, 4311–4315.

Milligan, G., and Klee, W. A. (1985). The inhibitory guanine nucleotide-binding protein (N_i) purified from bovine brain is a high affinity GTPase *J. Biol. Chem.* **260**, 2057–2063.

Milligan, G., Gierschik, P., Spiegel, A. M., and Klee, W. (1986). The GTP-binding regulatory proteins of neuroblastoma x glioma, NG 108-15 and glioma, C_6, cells. Immunochemical evidence of a pertussis toxin substrate that is neither N_i or N_o. *FEBS Lett.* **195**, 225–230.

Mumby, S. M., Kahn, R. A., Manning, D. R., and Gilman, A. G. (1986). Antisera of designed specificity for subunits of guanine nucleotide-binding regulatory proteins. *Proc. Natl. Acad. Sci. U.S.A.* **83**, 265–269.

Neer, E. J., Lok, J. M., and Wolf, L. G. (1984). Purification and properties of the inhibitory guanine nucleotide regulatory unit of brain adenylate cyclase. *J. Biol. Chem.* **259**, 14222–14229.

North, R. A., Williams, J. T., Surprenant, A., and Cristie, M. J. (1987). μ and δ receptors belong to a family of receptors that are coupled to potassium channels. *Proc. Natl. Acad. Sci. U.S.A.* **84**, 5487–5491.

Oinuma, M., Katada, T., Yokosawa, H., and Ui, M. (1986). Guanine nucleotide-binding protein in sea urchin eggs serving as the specific substrate of islet-activating protein, pertussis toxin. *FEBS Lett.* **207**, 28–34.

Oinuma, M., Katada, T., and Ui, M. (1987). A new GTP-binding protein in differentiated human leukemic (HL 60) cells serving as the specific substrate of islet-activating protein, pertussis toxin. *J. Biol. Chem.* **262**, 8347–8353.

Owens, J. R., Frame, L. T., Ui, M., and Cooper, D. M. F (1985). Cholera toxin ADP-ribosylates in the islet-activating protein substrate in adipocyte membranes and alters its function. *J. Biol. Chem.* **260**, 15946–15952.

Peraldi, S., Nguyen Than Dao, B., Brabet, P., Homburger, V., Toutant, M., Rouot, B., Assenmacher, I., Bockaert, J., and Gabrion, J. (1987). Immunodetection of α subunits of G$_o$ proteins in choroidal and ciliated ependymocytes. *Ann. Endocrinol.* **48**, 37.

Peraldi, S., Nguyen Than Dao, B., Brabet, P., Homburger, V., Rouot, B., Toutant, M., Bouillé, C., Assenmacher, I., Bockaert, J., and Gabrion, J. (1989). Apical localization of the α subunit of the GTP binding protein G$_o$ in choroidal and ciliated ependymocytes. *J. Neurosci.,* **9**, 806–814.

Pfaffinger, P. J., Martin, J. M., Hunter, D. D., Nathanson, N. M., and Hille, B. (1985). GTP binding proteins couple cardiac muscarinic receptors to a K$^+$ channel. *Nature* **317**, 536–538.

Rapiejko, P. J., Northrup, J. K., Evans, T., Brown, J. E., and Malbon, G. C. (1986). G-proteins of fat cells. Role in hormonal regulation of intracellular inositol 1,4,5-trisphosphate. *Biochem. J.* **240**, 35–40.

Robishaw, J. D., Russell, D. W., Harris, B. A., Smigel, M. D., and Gilman, A. G. (1986). Deduced primary structure of the α subunit of the GTP-binding stimulatory protein of adenylate cyclase. *Proc. Natl. Acad. Sci. U.S.A.* **83**, 1251–1255.

Roof, D., Applebury, M. L., and Sternweis, P. C. (1985). Relationships within the family of GTP-binding proteins isolated from bovine central nervous system. *J. Biol. Chem.* **260**, 16242–16249.

Rouot, B., Brabet, P., Homburger, V., Toutant, M., and Bockaert, J. (1987). G$_o$, a major brain GTP binding protein in search of a function: Purification, immunological and biochemical characteristics. *Biochimie* **69**, 339–349.

Rouot, B., Carrette, J., Lafontan, M., Tran, P. L., Fehrentz, J. A., Bockaert, J., and Toutant, M. (1989). The adipocyte G$_o$α-immunoreactive polypeptide is different from the α subunit of the brain G$_o$ protein. *Biochem. J.,* **260**, 307–310.

Schultz, A. M., Tsai, S.-C., Kung, H.-F., Oroszlan, S., Moss, J., and Vaughan, M. (1987). Hydroxylamine-stable covalent linkage of myristic acid in G$_o$α, a guanine nucleotide binding protein in bovine brain. *Biochem. Biophys. Res. Commun.* **146**, 1234–1239.

Sternweis, P. C. (1986). The purified α subunits of G$_o$ and G$_i$ from bovine brain require βγ for association with phospholipid vesicles. *J. Biol. Chem.* **261**, 631–637.

Sternweis, P. C., and Robishaw, J. D. (1984). Isolation of two proteins with high affinity for guanine nucleotides from membranes of bovine brain. *J. Biol. Chem.* **259**, 13806–13813.

Stryer, L., and Bourne, H. (1986). G proteins: A family of signal transducers. *Annu. Rev. Cell. Biol.* **2**, 391–419.

Suki, W. N., Abramowitz, J., Mattera, R., Codina, J., and Birnbaumer, L. (1987). The human genome encodes at least three non allelec G proteins with α$_i$-type subunits. *FEBS Lett.* **220**, 187–192.

Sullivan, K. A., Lia, Y.-C., Alborzi, A., Beiderman, B., Chang, F. H., Masters, S. B., Levinson, A., and Bourne, H. R. (1986). Inhibition and stimulatory G proteins of adenylate cyclase: cDNA and amino acid sequences of the α chains. *Proc. Natl. Acad. Sci. U.S.A.* **83**, 6687–6691.

Tabin, C. J., Bradley, S. M., Bargmann, C. J., Weinberg, R. A., Papageorge, A. G., Scolnick, E. M., Dhar, R., Lowy, D. R., and Chang, E. H. (1982). Mechanism of activation of a human oncogene. *Nature* **300**, 143–149.

Tanabe, T., Nukada, T., Nishikawa, Y., Sugimoto, K., Suzuki, H., Takagashi, H., Noda, M., Haga, T., Ichiyama, A., Kangawa, K., Minamoto, N., Matsuo, H., and Numa, S. (1985). Primary structure of the α subunit of transducin and its relationship to *ras* protein. *Nature* **315**, 242–245.

Terashima, T., Katada, T., Oinuma, M., Inoue, Y., and Ui, M. (1987a). Immunohistochemical localization of guanine nucleotide binding proteins in rat retina. *Brain Res.* **410**, 97–100.

Terashima, T., Katada, T., Oinuma, M., Inoue, Y., and Ui, M. (1987b). Endocrine cells in pancreatic islets of Langerhans are immunoreactive to antibody against guanine nucleotide-binding protein (G_o) purified from rat brain. *Brain Res.* **417**, 190–194.

Toutant, M., Aunis, D., Bockaert, J., Homburger, V., and Rouot, B. (1987a). Presence of three pertussis toxin substrates and $G_o\alpha$ immunoreactivity in both plasma and granule membranes of chromaffin cells. *FEBS Lett.* **215**, 339–344.

Toutant, M., Bockaert, J., Homburger, V., and Rouot, B. (1987b). G proteins in *Torpedo marmorata* electric organ. Differential distribution in pre- and postsynaptic membranes and synaptic vesicles. FEBS Lett. **222**, 51–55.

Tsaï, S. C., Adamik, R., Kanaho, Y., Halpern, H. L., and Moss, J. (1987). Immunological and biochemical differentiation of guanyl nucleotide binding proteins: Interaction of $G_o\alpha$ with rhodopsin, anti-$G_o\alpha$ polyclonal antibodies and a monoclonal antibody against transducin α subunit of $G_i\alpha$. *Biochemistry* **26**, 4728–4733.

Van Meurs, K. P., Angus, W., Lavu, S., Kung, H. F., Czarnecki, S. K., Moss, J., and Vaughan, M. (1987). Deduced amino acid sequence of bovine retinal: Similarities to other guanine nucleotide binding proteins. *Proc. Natl. Acad. Sci. U.S.A.* **84**, 3107–3111.

Waldo, G. L., Evans, T., Fraser, E. D., Northup, J. K., Martin, M. W., and Harden, T. K. (1987). Identification and purification from bovine brain of a guaninenucleotide binding protein distinct from G_s, G_i and G_o. *Biochem J.* **246**, 431–439.

Watkins, D. C., Northup, J. K., and Malbon, C. C. (1987). Regulation of G-protein in differentiation. *J. Biol. Chem.* **262**, 10651–10657.

Weiss, S., Pin, J. P., Sebben, M., Kemp, D. E., Sladeczek, F., Gabrion, J., and Bockaert, J. (1986). Synaptogenesis of cultured striatal neurons in serum-free medium: A morphological and biochemical study. *Proc. Natl. Acad. Sci. U.S.A.* **83**, 2238–2242.

West, R. E., Moss, J. Vaughan, M., Liu, T., and Liu, T. Y. (1985). Pertussis toxin catalyzed ADP-ribosylation of transducin. Cysteine 347 in the ADP-ribose acceptor site. *J. Biol. Chem.* **260**, 14428–14430.

Winslow, J. W., Bradley, J. D., Smith, J. A., and Neer, E. J. (1987). Reactive sulfhydryl groups of α 39, a guanine nucleotide-binding protein from brain-location and function. *J. Biol. Chem.* **262**, 4501–4507.

Worley, P. F., Baraban, J. M., Van Dop, C., Neer, E., and Snyder, S. H. (1986). G_o, a guanine nucleotide binding protein: Immunohistochemical localization in rat brain resembles distribution of second messenger systems. *Proc. Natl. Acad. Sci. U.S.A.* **83,** 4561–4565.

Yatani, A., Codina, J., Brown, A. M., and Birnbaumer, L. (1986). Direct activation of mammalian atrial muscarinic K^+ channels by human erythrocyte pertussis toxin sensitive G protein (Gk). *Science* **235,** 207–211.

Yatsunami, K., and Khorana, H. G. (1985). GTPase of bovine rod outer segments: The amino acid sequence of the α subunit as derived from the cDNA sequence. *Proc. Natl. Acad. Sci. U.S.A.* **82,** 4316–4320.

CHAPTER 6

Immunologic Probes for Heterotrimeric GTP-Binding Proteins

Allen M. Spiegel

Molecular Pathophysiology Branch, National Institute of Diabetes, Digestive, and Kidney Disease, National Institutes of Health, Bethesda, Maryland 20892

I. INTRODUCTION

G proteins convey extracellular signals from cell membrane receptors to effectors such as adenylyl cyclase, retinal cGMP phosphodiesterase, phospholipase C, and several types of ion channel. Receptor–effector coupling G proteins are composed of three subunits, α, β, and γ, each a separate gene product. At least seven different G protein α subunits have been identified by purification and cDNA cloning; β and γ subunits also show heterogeneity. Specific G-protein functions (receptor and effector coupling) have been correlated with a structurally defined G protein in only a minority of cases; indeed, for several cDNA clones, the predicted protein product has yet to be identified definitively. The ability of bacterial toxins to catalyze ADP-ribosylation of G-protein α subunits was, for some time, the only method for specific identification of G-protein α subunits in crude membranes. This method has several limitations, not the least of which is the relative lack of specificity of pertussis toxin which may well be capable of modifying six of the seven known α subunits.

Antibodies specific for G-protein subunits, prepared in several laboratories, have proved to be powerful tools for studies of G proteins. Antisera raised against purified G holoproteins and subunits provided initial evidence for similarities (β subunits) and differences (α and γ subunits) between members of the G-protein family (Fig. 1 and Table I). Immunochemical evidence for diversity in pertussis toxin substrates (Gierschik *et al.*, 1986a) corresponded to cloning of cDNAs encoding multiple types of putative pertussis toxin substrate. Antisera raised against synthetic peptides corresponding to amino acid sequences predicted by cloned cDNAs have

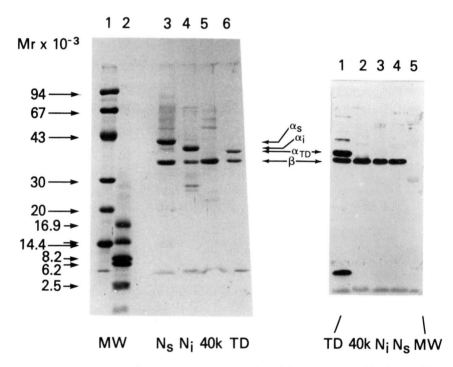

Fig. 1. Analysis of purified G proteins by protein staining and immunoblotting. Purified G proteins [G_s and G_i (N_s and N_i) from human erythrocytes, and bovine transducin (TD) from retina] were separated by urea-SDS-PAGE and either stained with Coomassie blue (A) or immunoblotted (B) with an antiserum raised against holotransducin. Lanes 1 and 2 in (A) contain molecular weight markers. "40K" denotes the βγ complex purified from human erythrocytes. The positions of the individual subunits are marked by arrows. The immunoblot shows reactivity of the antiserum with all three transducin subunits, but only with the β subunits of G_s and G_i. (From Gierschik *et al.*, 1985.)

Table I   Antisera Directed against G Proteins

Name	Antigen[a]	Specificity
Antisera raised against purified G proteins		
AS/1	Holotransducin	Transducin α,β,γ crossreacts G.β
CW/6	Holotransducin	Transducin α,β,γ crossreacts Gβ, $G_{i1}\alpha$
GI/2	Transducin α	Transducin α (rod only)
SP/3	Transducin β,γ	Transducin β,γ crossreacts Gβ
RV/3	Brain G_i/G_o	G_o-alpha, Gβ crossreacts transducin β
Antisera raised against synthetic peptides		
AS/6, AS/7	Transducin α (aa 341–350)	Transducin α $G_{i1}\alpha$, $G_{i2}\alpha$, (slight)$G_{i3}\alpha$
GO/1	$G_o\alpha$ (aa 345–354)	$G_o\alpha$
IM/1	$G_o\alpha$ (aa 22–35)	$G_o\alpha$
GC/1	$G_o\alpha$ (aa 2–16)	$G_o\alpha$; slightly crossreacts with $G_{i1,i2,i3}\alpha$s
LE/2, LE/3	$G_{i2}\alpha$ (aa 160–169)	$G_{i2}\alpha$
LD/1, LD/2	$G_{i1}\alpha$ (aa 159–168)	$G_{i1}\alpha$
SQ/2, SQ/3	$G_{i3}\alpha$ (aa 159–168)	$G_{i3}\alpha$

[a] aa, Amino acid.

allowed identification of the protein products of the cDNAs in several cases. G-protein antibodies have also proven invaluable in studies of G-protein function, tissue and subcellular distribution, and posttranslational modification.

II. IMMUNOCHEMICAL STUDIES OF G-PROTEIN SUBUNIT STRUCTURE

A. γ Subunits

γ Subunits are low molecular weight ($\sim 8,000–11,000$) proteins tightly, but noncovalently, associated with β subunits to form the $\beta\gamma$ complex of G proteins. The $\beta\gamma$ complex reversibly dissociates from the α subunit upon activation of solubilized G proteins by fluoride or nonhydrolyzable GTP analogs. The primary structure of transducin γ, a very hydrophilic 8.4 kDa protein, has been defined by cDNA cloning (Hurley *et al.*, 1984). In two instances (Van Dop *et al.*, 1984; Yatsunami *et al.*, 1985), transducin γ cDNA clones were obtained by screening λgt11 expression libraries with transducin antisera.

Antisera raised against holotransducin or its $\beta\gamma$ complex readily react with transducin γ on immunoblots but fail to recognize the γ subunits of other G proteins (Gierschik et al., 1985; Fig. 1). Immunochemical distinction between transducin γ and Gγ subunits has also been shown by Roof et al. (1985), Audigier et al. (1985), and Evans et al. (1987). This result is also consistent with the results of Northern blot analysis using a transducin γ cDNA probe; a hybridizing band is observed only for retinal mRNA (Van Dop et al., 1984).

As yet, the structures of other Gγ subunits have not been defined. Silver staining of purified G proteins suggests possible diversity among G-protein γ subunits other than transducin (Sternweis and Robishaw, 1984). Antisera raised against a $\beta\gamma$ complex purified from human placenta recognized placental γ subunits on immunoblot, but failed to react with either transducin γ or the γ subunits of other G proteins purified from platelet, liver, and brain (Evans et al., 1987). Species-specific differences could not account for these findings since placental and platelet proteins were both human. In summary, immunochemical studies suggest substantial diversity among G-protein γ subunits. Since γ subunits bind to similar if not identical β subunits, some common structural features may yet be found, but apparently these are not apparent as antigenic epitopes.

B. β Subunits

Two forms of G-protein β subunit, β_{36} and β_{35}, have been purified (Sternweis and Robishaw, 1984). Transducin β contains only the 36-kDa form; in brain, β_{36} predominates over the 35-kDa form, and in placenta, β_{35} predominates (Evans et al., 1987). Antisera raised against transducin β recognize the corresponding subunit in all G proteins (Fig. 1; Gierschik et al., 1985; Roof et al., 1985), and antisera raised against β subunits of other G proteins (e.g. brain, Fig. 2) recognize transducin β (Gierschik et al., 1986b; Evans et al., 1987). cDNA clones encoding the 36-kDa form of β predict identical 340-amino acid sequences for transducin- and Gβ (Sugimoto et al., 1985; Fong et al., 1986; Codina et al., 1986), thus accounting for immunologic crossreactivity.

Immunochemical differences have been noted between β_{36} and β_{35}. Some transducin β antisera react preferentially with β_{36} (Roof et al., 1985; Evans et al., 1987); an antiserum raised against a synthetic peptide representing residues 130–145 of β_{36} reportedly recognized β_{36} but not β_{35} (Mumby et al., 1986). cDNAs clearly related to but distinct from β_{36} have recently been cloned (Fong et al., 1987; Gao et al., 1987a). The purified β_{35} protein has not yet been sequenced, but antisera to unique peptide se-

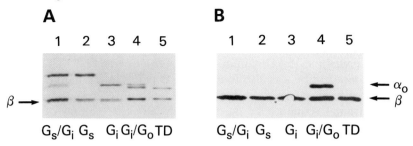

Fig. 2. Coomassie blue staining and immunoblotting of purified G proteins. Purified G proteins (G_s and G_i from human erythrocytes, G_i/G_o from brain, and transducin (TD) from bovine retina) were separated by SDS-PAGE, and either stained with Coomassie blue (A) or immunoblotted with antiserum RV/3 directed against $G_o\alpha$ and $G_o\beta$ (B). Only α and β subunit regions are shown. The antiserum reacts only with the β subunits and with the $G_o\alpha$ subunit. Reactivity with the upper band (G_i) of the doublet shown in panel A, lane 4 has been excluded by experiments not shown here. (From Gierschik *et al.*, 1986b.)

quences predicted by the novel cDNA suggest that the latter corresponds to β_{35} (Gao *et al.*, 1987b).

Trypsin cleaves soluble native $\beta\gamma$ complexes preferentially at arg[129] of the β subunit (Sugimoto *et al.*, 1985; Mumby *et al.*, 1986). This results in two fragments of β, a 15-kDa amino-terminal and a 26-kDa carboxy-terminal, as well as the intact γ subunit. An identical pattern is observed on immunoblots after tryptic cleavage of retinal rod outer segment membranes and brain plasma membranes (Pines *et al.*, 1985b). This implies that the tryptic cleavage site between residues 129–130 of β is exposed in the native membrane-bound protein.

The majority of antisera raised against β subunits recognize epitope(s) within the 15-kDa amino-terminal fragment (Zaremba *et al.*, 1988). Screening a retinal cDNA expression vector library with a β antiserum led to cloning of a cDNA (designated clone A) distinct from the β subunit but sharing with the latter a short region of sequence homology (Table II). Six

Table II Comparison of Predicted Amino Acid Sequences of Homologous Domains of Proteins Encoded by Clone A[a] and β Subunit cDNAs

Unit	Residues	Sequence[b]	Ref.
Clone A	17–26	K S E L D E L Q E E	—
β (36-kDa form)	1–10	M S E L D Q L R Q E	Sugimoto *et al.* (1985)
β	1–10	M S E L E Q L R Q E	Gao *et al.* (1987)

[a] See text.
[b] Single letter amino acid code.

of nine residues are identical with the amino terminus of β_{36}. Studies with cDNA clone A indicate that the amino-terminal decapeptide contains a major epitope of β_{36} and may account for the predominance of antibodies directed against the 15-kDa amino-terminal tryptic fragment. The amino-terminal decapeptide was also found to be exposed in the native $\beta\gamma$ complex (Zaremba *et al.*, 1988). A synthetic decapeptide corresponding to the amino-terminal sequence of β_{36} effectively blocked binding of most β antisera to the 15-kDa but not 26-kDa tryptic fragment of β (Fig. 3). The matrix-bound peptide could also be used to affinity-purify from crude β antisera antibodies capable of binding both the 36- and 35-kDa forms of β (Fig. 4). This suggests that the amino acid sequence of the amino-terminal decapeptide of β_{35} is sufficiently similar to that of β_{36} that the peptide-puri-

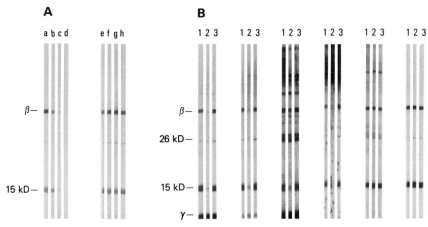

Fig. 3. Competition by synthetic peptides for binding to β subunit antibodies. Trypsin-cleaved and uncleaved transducin $\beta\gamma$ were separated on 15% polyacrylamide gels and transferred to nitrocellulose. Individual strips were cut and incubated with sera and peptides (final volume of first antibody solution, 5 ml) as follows. (A): Antibodies affinity-purified from AS/1 by adsorption to filters containing clone-A fusion protein (dilution approximately 1/400); lanes a, e: no peptide; b, f: 1 μg; c, g: 10 μg; d, h: 100 μg peptide; lanes b, c, d: peptide MSELDQLRQE-amide; lanes f, g, h: peptide RLKDIFGESA. (B): Antisera, in order from left to right, and each at 1/100 dilution: SP/3, CW/1, AS/1, AS/6 (note the latter is from immunization with holotransducin, and is distinct from peptide antiserum AS/6), PG/4, and MP/6 (latter two are nonimmune sera). Lane 1: no peptide; lane 2: 100 μg MSELDQLRQE-amide; lane 3: 100 μg RLKDIFGESA. The positions of the intact β and γ subunits and of the 26- and 15-kDa fragments of β are indicated. The faintly stained, thin band seen in several lanes near the position of the 26-kDa tryptic fragment is distinct from the latter and may represent a minor tryptic fragment that includes the amino terminus. (From Zaremba *et al.*, 1988.)

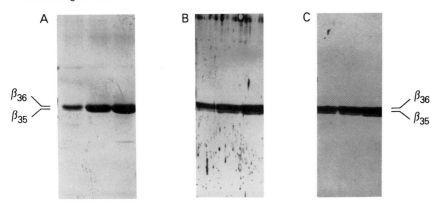

Fig. 4. Immunoblot of purified β_{35} and β_{36}-kDa proteins with crude and affinity-purified β antibodies. Fractions from DEAE-Toyopearl chromatography containing purified bovine brain $\beta\gamma$ subunits (approximately 5, 10, and 15 μg/lane from left to right) were separated by SDS-PAGE (10% gel), and either stained with Coomassie blue (A) or transferred to nitrocellulose paper for immunoblotting (B, C). In (B), crude AS/1 antiserum (1/100 dilution) was used as first antibody. In (C), the first antibody solution contained antibodies (1/20 dilution) affinity-purified from AS/1 by passage over a matrix with covalently bound amino-terminal β decapeptide. Only the portion of the gel and blots containing the β subunits is shown. The positions of the 36- and 35-kDa forms are indicated. (From Zaremba et al., 1988.)

fied antibodies crossreact. The sequence of the novel β cDNA differs in this region from that of β_{36} by a single conserved residue (Table II), consistent with the idea that this cDNA corresponds to β_{35}.

β subunits are highly conserved among vertebrate but not invertebrate species (Gierschik et al., 1986b; Audigier et al., 1985; Codina et al., 1986). On two-dimensional gel electrophoresis both β_{36} and β_{35} display charge heterogeneity (Heydorn et al., 1986; Rosenthal et al., 1986; Woolkalis and Manning, 1987). The significance of this heterogeneity is not yet clear.

The function of the $\beta\gamma$ complex is controversial. Possibilities (not mutually exclusive) include receptor interaction, direct interaction with effectors, indirect modulation of effectors by inhibition of α subunit activation, and anchoring α subunits to the cell membrane. Immunochemical differences between γ subunits, and between β_{36} and β_{35}, clearly correspond to structural differences, but the functional significance of this structural heterogeneity is unknown. Transducin $\beta\gamma$ complexes are interchangeable with those of other G proteins with respect to qualitative interactions with α subunits, although quantitative differences have been observed (Cerione et al., 1987).

C. α Subunits

1. Transducin

Transducin is the G protein that couples opsins to cGMP phosphodiesterase in retinal photoreceptors. Transducin, readily purified from retinal rod outer segment membranes, consists of an α_{39}, β_{36}, and 8.4-kDa γ. Polyclonal antisera (Table I) raised against the holoprotein (Fig. 1) or against the resolved α_{39} (Fig. 5), in general are specific for transducin α (Gierschik *et al.,* 1985; Mumby *et al.,* 1986). Rarely (2/18 times in our experience), a transducin α antiserum (e.g. CW/6, Table I) was obtained that showed clear crossreactivity with another Gα. Monoclonal antibodies raised against transducin α also generally failed to crossreact with other Gαs, but exceptions have been observed (Halpern *et al.,* 1986; Tsai *et al.,* 1987). Failure of antisera raised against individual purified Gα subunits to crossreact with other Gαs is somewhat surprising given the relatively high degree of sequence homology (see below) noted among members of the G-protein family. In particular, several regions corresponding to GTP-binding domains (Jurnak, 1985) show almost total sequence conservation. It would appear that these regions are relatively poorly antigenic.

Two distinct cDNA clones corresponding to transducin α were obtained upon screening bovine retinal libraries (Lochrie *et al.,* 1985; Tanabe *et al.,* 1985). The latter corresponded perfectly to the amino acid sequence derived from the purified α_{39} protein, whereas the former showed several

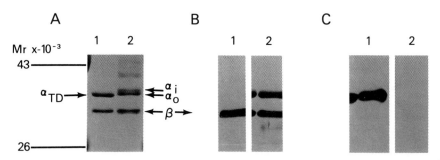

Fig. 5. Specificity of antisera RV/3 and GI/2 versus G proteins. Transducin (A, B, C; lane 1), purified from bovine rod outer segments, and G_i/G_o (A, B, C; lane 2), purified from bovine brain, were separated by SDS-PAGE (10% gel). Approximately 6 μg of each protein were loaded per lane. (A): Proteins stained with Coomassie blue. (B and C): Proteins were transferred to nitrocellulose and immunoblotted with either RV/3 (B; affinity-purified antibodies at a final dilution of 1/20) or with crude antiserum GI/2 (C; final dilution 1/100). The positions of α and β subunits and of molecular weight markers are indicated by arrows. Only the region of the gel and blots between approximately 26- and 43-kDa markers is shown; no other bands were seen on immunoblots. (From Lad *et al.,* 1987.)

discrepancies. This suggested that the cDNA cloned by Lochrie *et al.* encodes a protein closely related to, but not identical with, α_{39} purified from rod outer segment membranes. Antisera raised against α_{39} purified from bovine rod outer segment membranes showed specific staining of rod outer segments but failed to stain cone outer segments in chick retina (Grunwald *et al.*, 1986). Antisera raised against synthetic peptides corresponding to sequences predicted by the two forms of transducin cDNA (Table III) proved capable of distinguishing between rod α_{39} and an α_{41} present in retinal homogenates (Lerea *et al.*, 1986). The peptide antiserum corresponding to the rod-type cDNA stained only rod outer segments on immunohistochemical analysis, consistent with the results of Grunwald *et al.* (1986). The peptide antiserum based on the alternative transducin cDNA specifically stained cone outer segments. This cDNA may correspond to a cone-specific form of transducin; however, this protein has not yet been purified and sequenced.

2. G_s

G_s is the G protein that mediates agonist stimulation of adenylyl cyclase. Cholera toxin-catalyzed ADP-ribosylation of $G_s\alpha$, and purification of the protein on the basis of a complementation assay (restoration of adenylyl cyclase activation to the G_s-deficient *cyc*$^-$ cell line), reveal α subunits variously referred to as 42- and 48-, or 45- and 52 kDa (I will use the latter hereafter). Antisera raised against the purified protein have not been reported. A cDNA cloned by Harris *et al.* (1985) upon screening a brain library with an oligonucleotide based on the sequence of a GTP-binding domain common to all $G\alpha s$ predicted a protein sequence unrelated to those sequenced to date. Nukada *et al.* (1986b) cloned an identical cDNA.

Table III   Comparison of Amino Acid
Sequences of Residues 159–168 in G_{i1} and
Corresponding Residues in Other $G\alpha s$

α Subunit	Sequence[a]
Transducin (rod)[b]	L D R I T A P D Y L
Transducin (cone)[b]	L E R L V T P G Y V
G_{i1}	L D R I A Q P N Y I
G_{i2}	L E R I A Q S D Y I
G_{i3}[c]	L D R I S Q S N Y I
G_o	L D R I G A A D Y Q

[a] Single letter amino acid code.
[b] Bovine.
[c] Human.

A protein-derived amino acid sequence for G_s, however, has not been reported (undoubtedly due to its relatively low abundance).

Synthetic peptide antisera identified the protein encoded by the cDNA as $G_s\alpha$ (Harris *et al.*, 1985). An antiserum against a peptide predicted by the cDNA but also found in other $G\alpha s$ reacted with all purified $G\alpha s$ on immunoblot. In contrast, an antiserum against a peptide sequence unique to this cDNA reacted exclusively with purified $G_s\alpha$ (both 45- and 52-kDa forms). Subsequently, additional cDNAs related to $G_s\alpha$ were obtained (Robishaw *et al.*, 1986; Bray *et al.*, 1986). Bray *et al.* (1986) showed that four forms of $G_s\alpha$ mRNA could be derived by alternative splicing from a single gene. The major difference between these forms is the presence or absence of a fifteen amino-acid peptide (residues 72–86). Robishaw *et al.* (1986), using a synthetic peptide antiserum corresponding to the fifteen-amino-acid peptide, showed that the "long" forms of mRNA encode 52-kDa α subunits and the "short" mRNAs encode 45-kDa α subunits. As yet the functional significance, if any, of this heterogeneity is unclear.

3. G_o

Upon purification of pertussis toxin substrates from bovine brain, Sternweis and Robishaw (1984) and Neer *et al.* (1984) discovered an abundant and novel G protein (α_{39}), termed G_o by Sternweis and Robishaw. cDNAs cloned from rat (Itoh *et al.*, 1986; Jones and Reed, 1987) and bovine (Van Meurs *et al.*, 1987) libraries predict a protein sequence that corresponds precisely to purified α_{39}-derived amino acid sequence. Antisera prepared by immunization with purified brain G holoprotein (Gierschik *et al.*, 1986b) or with resolved α_{39} subunit (Huff *et al.*, 1985; Tsai *et al.*, 1987) distinguish $G_o\alpha$ from other $G\alpha$ subunits (Figs. 2 and 5). Antisera prepared against synthetic peptides with sequences predicted by $G_o\alpha$ cDNAs (Table I; Fig. 6) also show unique reactivity with $G_o\alpha$ (Mumby *et al.*, 1986; Eide *et al.*, 1987). Full crossreactivity of $G_o\alpha$ antibodies raised against the bovine protein with the protein found in brains of other mammals, chicks, and toads implies high conservation of primary structure for $G_o\alpha$ in vertebrates (Gierschik *et al.*, 1986b).

4. G_i

G_i was initially defined functionally as the G protein coupling receptors to inhibition of adenylyl cyclase. G_i is uncoupled from receptors by pertussis toxin-catalyzed ADP-ribosylation (Ui, 1984). In some cell types, e.g. neutrophils, a pertussis toxin-sensitive G protein couples receptors to phospholipase C. Since inhibition of agonist-stimulated phospholipase C activity by pertussis toxin correlated with ADP-ribosylation of a 40–41-kDa

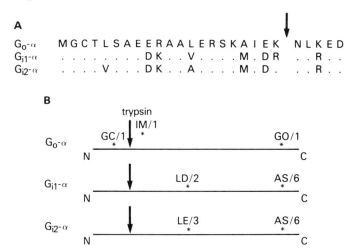

Fig. 6. (A): Predicted amino acid sequence (single letter code) of the amino terminus of the indicated cDNAs. Dots indicate identical residues. For the residues shown, the sequence of G_{i3} is identical to that of G_{i1}. The large arrow indicates a tryptic cleavage site. (B): Schematic diagram of α subunits of three indicated G proteins. N and C refer to amino and carboxy termini, respectively. The epitope locations of synthetic peptide antibodies are indicated with asterisks. GO/1 and AS/6 were raised against the carboxy-terminal peptides indicated in Table IV. LD/2 and LE/3 were raised against the decapeptides indicated in Table III. (Modified from Eide *et al.*, 1987.)

protein in neutrophil membranes, it was assumed that this protein is G_i (Ui, 1984).

Neutrophil membranes contain an abundant (> 20 pmol/mg membrane protein) pertussis toxin substrate(s). Immunoblots with transducin- and $G_o\alpha$-specific antisera show that these pertussis toxin substrates are lacking (or present in very low concentrations) in neutrophil membranes (Gierschik *et al.*, 1986a). A transducin antiserum, CW/6 (Table I), was found to crossreact with the 41-kDa form of $G_i\alpha$ in brain, but not with $G_o\alpha$ (Pines *et al.*, 1985a). Immunoblots with CW/6 showed that the amount of "G_i" in neutrophil membranes also could not account for the abundant pertussis-toxin substrate (Gierschik *et al.*, 1986a). These studies provided immuno-chemical evidence for a novel pertussis-toxin substrate in neutrophils.

cDNA cloning likewise suggested heterogeneity of pertussis toxin sub-strates. In addition to $G_o\alpha$, three closely related forms of "$G_i\alpha$" were cloned (Itoh *et al.*, 1986; Nukada *et al.*, 1986a; Sullivan *et al.*, 1986; Didsbury and Snyderman, 1987; Didsbury *et al.*, 1987; Suki *et al.*, 1987). Differences in nucleotide sequence of untranslated regions of mRNA per-mitted distinction between the subtypes within a given species (Bray *et al.*,

1987). Cloning of distinct cDNAs for the three forms within a single species, rat (Jones and Reed, 1987), also indicated that each is a distinct gene product. The three types, arbitrarily designated G_{i1}, G_{i2}, and G_{i3} in order of cloning, all contain a cysteine, the putative pertussis-toxin ADP-ribosylation site (West et al., 1985), as the fourth residue from the carboxy terminus. The predicted amino acid sequence of $G_{i1}\alpha$ cDNA matches the amino acid sequence obtained for the α_{41} purified from bovine and rat brain (Nukada et al., 1986a; Itoh et al., 1986). The G_{i2}- and $G_{i3}\alpha$ cDNAs, however, were not correlated with the sequence of a purified protein.

Peptide antisera (Fig. 6, Table I) proved useful in identifying proteins corresponding to G_i cDNAs. Antisera AS/6 and AS/7, directed against the carboxy-terminal decapeptide of transducin (Table IV) recognize the abundant pertussis-toxin substrate of neutrophils, as well as brain G_i (Falloon et al., 1986). In contrast, antisera LE/2 and 3, directed against residues 160–169 predicted by G_{i2} cDNA (Table III), recognize an abundant 40-kDa neutrophil membrane protein and only a faint 40-kDa band in brain membranes (Goldsmith et al., 1987). LE antisera react specifically with a 40-kDa pertussis toxin substrate purified from bovine neutrophils (Gierschik et al., 1987), and do not react with transducin, brain α_{41} (G_{i1}), or G_o (Fig. 7). On this basis the neutrophil α_{40} is likely to be $G_{i2}\alpha$. What is likely an identical protein has also been purified from HL-60 cells (Oinuma et al., 1987; Uhing et al., 1987), and from brain (Katada et al., 1987; Goldsmith et al., 1988a). An antiserum raised against a synthetic peptide (residues 3–17) of $G_{i2}\alpha$ also reacts specifically with an α_{40} (Lang and Costa, 1987).

The identity of the protein corresponding to G_{i3} cDNA has not been definitively established, but is likely to be an α_{41} present in many tissues and cells such as HL-60 (Murphy et al., 1987; Goldsmith et al., 1988b).

Table IV   Comparison of Amino Acid
Sequences of Carboxy-Terminal
Decapeptide of G-Protein
α Subunits

α Subunits	Sequence[a]
Transducin[b]	K E N L K D C G L F
G_{i1} and G_{i2}	K N N L K D C G L F
G_{i3}	K N N L K E C G L Y
G_o	A N N L R G C G L Y
G_s	R M H L R Q Y E L L

[a] Single letter amino acid code.
[b] Both the rod and cone forms.

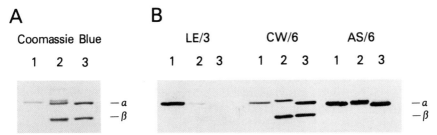

Fig. 7. Coomassie blue staining and immunoblots of purified G proteins. Purified G proteins (the major pertussis-toxin substrate purified as an α_{40} from bovine neutrophils in lane 1, bovine brain G_i/G_o in lane 2, and bovine retinal transducin in lane 3) were separated by SDS-PAGE and either stained with Coomassie blue (A) or immunoblotted with LE/3 (a G_{i2}-specific peptide antiserum), CW/6 (raised against holotransducin and partially crossreactive with other $G_i\alpha$ subunits), and AS/6 (a G_i and transducin α-specific peptide antiserum). Positions of α and β subunits are indicated. (From Goldsmith et al., 1987.)

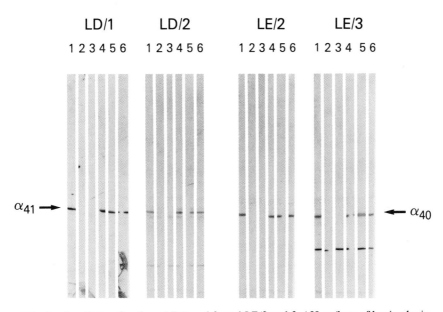

Fig. 8. Specificity of antisera LD/1 and 2, and LE/2 and 3. 150 µg/lane of bovine brain membrane cholate extract was separated by SDS-PAGE (10% gel) and the proteins transferred to nitrocellulose for immunoblotting. Each strip was incubated with 5 ml of a 1/100 dilution of the indicated antiserum. Peptides at indicated concentrations were added to first antibody solutions as follows: lane 1, no peptide; lanes 2 and 3, 1 and 10 µg, respectively, of $G_{i1}\alpha$ (amino acids 159–168) peptide for LD/1 and LD/2, and $G_{i2}\alpha$ (amino acids 160–169) peptide for LE/2 and LE/3; lanes 4, 5, 6, 1, 10, and 100 µg, respectively, of $G_{i3}\alpha$ (amino acids 159–168) peptide. Arrows mark the positions of $G_{i1}\alpha$ (α_{41}) and $G_{i2}\alpha$ (α_{40}).

Antisera raised against peptides corresponding to the region shown in Table III are highly specific for individual $G\alpha$s. Thus, LD and LE antisera (Table I and III) are specific for G_{i1} and G_{i2}, respectively. The synthetic peptide antigen blocks reactivity of the corresponding antiserum on immunoblots, but the homologous G_{i3} synthetic peptide (Table III) does not (Fig. 8). SQ antisera, raised against the corresponding G_{i3} peptide (Tables I and III), are specific for a 41-kDa protein purified from HL-60 cells, and

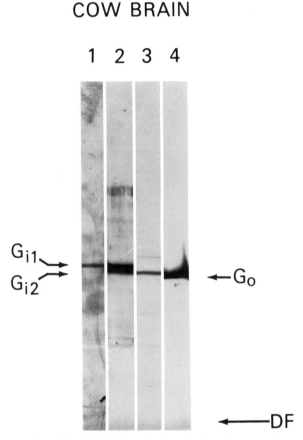

Fig. 9. Immunoblot of bovine brain cerebral cortical membranes with peptide antisera specific for G-protein α subunits serving as pertussis toxin substrates. Cow brain membrane cholate extract (150 μg/lane) was separated by SDS-PAGE (10% gel; 0.13% bis acrylamide). The separated proteins were transferred to nitrocellulose and immunoblotting performed with: lane 1, LD/1 (1/100 dilution); lane 2, AS/7 (1/250 dilution); lane 3, LE/2 (1/100 dilution); lane 4, GO/1 (1/250 dilution). Arrows denote the positions of specific immunoreactive bands corresponding to $G_i\alpha$ and $G_o\alpha$ subunits. DF indicates the dye front.

fail to react with either brain 41- (G_{i1}) or 40-kDa (G_{i2}) proteins (Goldsmith *et al.*, 1988b).

Antisera specific for G_o and G_s may be raised by immunization with the corresponding carboxy-terminal sequences (Table IV); antisera against the carboxy-terminal sequence of transducin α, in contrast, crossreact with G_{i1}, G_{i2}, and weakly with G_{i3}, but not with G_o. Specific peptide antisera identify at least three distinct pertussis toxin substrates in brain (Fig. 9). Two-dimensional gel electrophoresis combined with immunoblotting is a powerful method for specific identification of $G\alpha s$, since α subunits poorly separated by sodium dodecyl sulfate-polyacrylamide gel electrophoresis (SDS-PAGE) are readily separated by pI (Backlund *et al.*, 1988).

Some caution in defining the specificity of peptide antisera is warranted, since there are undoubtedly additional G proteins with unknown degrees of homology to defined sequences. Thus a 43-kDa pertussis-toxin substrate of unknown primary structure has been purified (Iyengar *et al.*, 1987), and a pertussis toxin-insensitive G protein coupling to phospholipase C remains to be identified.

III. IMMUNOCHEMICAL STUDIES OF G-PROTEIN FUNCTION

G-protein subunits interact with each other, with guanine nucleotides, receptors, effectors, membranes, and perhaps with other as yet unidentified entities. Immunologic probes have provided new insights into some of these functional interactions.

Reconstitution studies suggest relative specificity in receptor–G-protein coupling. Under certain conditions, particularly agonist pretreatment, receptors copurify with G protein(s). Antibodies to G proteins should be useful in identifying the endogenous G protein coupled to a given receptor, and might also be applied to affinity purification. The anterior pituitary D_2-dopamine receptor copurifies with a pertussis-toxin substrate. Immunochemical studies suggest the latter may include both G_o as well as a form of G_i (Senogles *et al.*, 1987). In liver, the vasopressin receptor couples to a pertussis toxin-insensitive G protein (G_p) that stimulates phospholipase C. The receptor copurifies with a G protein shown to contain immunoreactive β subunit (Fitzgerald *et al.*, 1986). This suggests that the pertussis toxin-insensitive G_p is a heterotrimeric G protein.

G-protein–receptor interaction requires the holoprotein rather than dissociated α subunits. Immunochemical detection of β subunits copurifying with receptors (Senogles *et al.*, 1987; Fitzgerald *et al.*, 1986) is consistent with a requirement for the G holoprotein in receptor interaction, but does not clarify whether or not the $\beta\gamma$ complex interacts directly with receptor domains or is required to induce the appropriate α subunit conformation. Immunization with transducin $\beta\gamma$ subunits reportedly led to formation of antibodies directed against rhodopsin (Halpern *et al.*, 1987). These were presumed to be antiidiotypic antibodies, and were interpreted as evidence for physical association of $\beta\gamma$ with receptor (rhodopsin). Further studies are needed to verify this hypothesis.

Several lines of evidence suggest that the carboxy terminus of Gα subunits is involved in receptor interaction. These include the ability of pertussis toxin-catalyzed ADP-ribosylation of cysteine (four residues from the C-terminus) to uncouple G proteins from receptors, and limited homology of the retinal 48K protein, known to bind to rhodopsin, with a carboxy-terminal sequence of transducin α (Wistow *et al.*, 1986). Antibodies against the carboxy-terminal decapeptide of transducin α (AS/6 and 7, Table I) block rhodopsin – transducin interaction without interfering with G-protein subunit interaction or with effector interaction (Fig. 10, Cerione *et al.*, 1988). Such antibodies also block opiate-stimulated GTPase in NG-108 cells. This effect presumably reflects antibody-induced uncoupling of opiate receptors from G$_i$ rather than G$_o$, since AS/6 and 7 antisera crossreact with G$_i$ but do not react with G$_o$ (McKenzie *et al.*, 1988). The comparable G$_o$ peptide antiserum, e.g. GO/1 (Table I), may be expected to similarly uncouple G$_o$ from receptors. Studies with such antisera may be helpful in defining specificity of receptor–G-protein coupling in endogenous rather than reconstituted systems.

Antibodies have proved useful in studying α and $\beta\gamma$ subunit interactions. G$_o$ or G$_i$, radioactively labeled by pertussis toxin-catalyzed ADP-ribosylation, can be immunoprecipitated with anti-β subunit antibodies (Huff and Neer, 1986). Similarly, a monoclonal antibody directed against an internal sequence of transducin α immunoprecipitates the holoprotein (Navon and Fung, 1987). If, however, transducin α is activated by guanine nucleotide and dissociated from $\beta\gamma$, antibody fails to coprecipitate $\beta\gamma$. Proteolytic cleavage of the 1–2-kDa amino-terminal portion of transducin α also prevented coimmunoprecipitation of $\beta\gamma$ subunits, even when transducin was not activated. This suggests that the amino terminus of transducin α is required for $\beta\gamma$ interaction (Navon and Fung, 1987). Antibodies have been prepared against the comparable domain of G$_o\alpha$ (GC/1, see Table I and Fig. 6) and should prove useful in defining the role of this region in $\beta\gamma$ interaction.

$\beta\gamma$ subunits may serve to anchor α subunits to the cytoplasmic side of the plasma membrane. Under activating conditions (0.1 mM GTPαS), how-

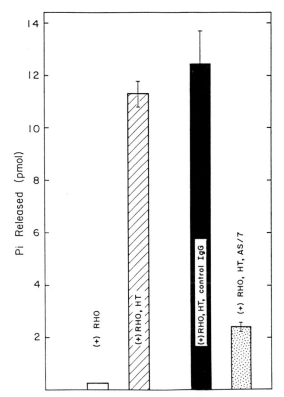

Fig. 10. The effects of affinity-purified AS/7 antibodies on the rhodopsin-stimulated GTPase activity of holotransducin. Five μl of purified holotransducin (13 pmol) were incubated with 5 μl of either H_2O, nonimmune rabbit IgG (500 $\mu g/ml$) or affinity-purified AS/7 antibodies (500 $\mu g/ml$) for 1.5 hr at 0°C. Following these incubations, 40 μl of phosphatidylcholine vesicles containing pure rhodopsin (about 2 pmol) was added to each mixture. Aliquots of the resultant mixtures were then assayed for GTPase activity for 20 min at 23°C. (+)RHO designates the basal activity for the vesicles containing rhodopsin only. The data are the mean of duplicate determinations and the bars show the range.

ever, α subunits remain associated with plasma membranes in brain and neutrophil (Eide *et al.,* 1987). This may suggest that subunit dissociation does not occur for membrane-bound G proteins or that α subunits can bind to membranes independent of $\beta\gamma$. Tryptic cleavage of the amino-terminus of $G_o\alpha$, $G_{i1}\alpha$, or $G_{i2}\alpha$ releases these proteins from the membrane as determined by immunoblot (Figs. 6 and 11). This suggests a role for the amino terminus of α subunits in membrane attachment, and may reflect myristoylation of amino-terminal glycine residues in $G\alpha$ subunits. Immu-

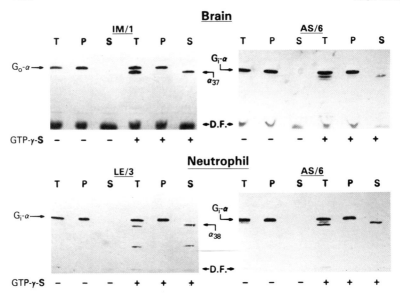

Fig. 11. Immunoblots of particulate and soluble Gα subunits. Brain and neutrophil plasma membranes were subjected to tryptic digestion, with (+) or without (−) pretreatment with GTPγS. Total (T) membrane digest, or digest separated into particulate (P) and soluble (S) fractions were subject to SDS-PAGE and then immunoblotted with the indicated antisera (refer to Fig. 6 for sites of antibody reactivity and for schematic diagram of tryptic cleavage site). The positions of Gα subunits are indicated, as are the positions of the large tryptic fragments of the α subunits. Note that these fragments are found in the soluble fraction whereas the undigested Gα remains membrane-bound even after GTPγS treatment. See text for further discussion of these results. (From Eide *et al.*, 1987.)

noprecipitation of cells labeled with [³H]myristate reveals that several forms of $G_i\alpha$ are myristoylated (Buss *et al.*, 1987). This mode of membrane attachment contrasts with palmitoylation of a carboxy-terminal cysteine in the *ras* gene products (Willumsen *et al.*, 1984). Neither palmitate nor myristate was found to be associated with β subunits (Buss *et al.*, 1987). On immunoblots of membranes extracted with various detergents, β subunits were found to be associated with cytoskeletal elements (Carlson *et al.*, 1986).

Polyclonal antibodies (of unspecified epitope) directed against β subunits were found to impair brain G protein – guanine-nucleotide exchange, and thus interfere with regulation of adenylyl cyclase (Correze *et al.*, 1987). This is consistent with a role for β subunits in modulating α subunit – guanine-nucleotide exchange. Monoclonal antibodies against β subunits can block G_s deactivation by added β subunits (Lingham *et al.*, 1986).

Monoclonal antibodies have also been raised against transducin α and used in functional studies. Several monoclonals directed against a unique sequence within the 23-kDa tryptic fragment of transducin α were found to block rhodopsin-stimulated GTPase. Surprisingly, another monoclonal directed against an epitope conserved in several Gαs failed to block GTPase activity (Halpern *et al.*, 1986). A monoclonal antibody raised against frog rod outer segments binds to transducin α and blocks GTP binding (Hamm and Bownds, 1984). Proteolysis was used to map the epitopes of this and several other transducin α monoclonals (Deretic and Hamm, 1987). These data suggest a binding site near the carboxy terminus. These antibodies were found to block rhodopsin–transducin interaction, consistent with a role for the carboxy terminus in receptor interaction (see above), but also released holotransducin from rod outer segment membranes in the dark under isosmotic conditions (Hamm *et al.*, 1987).

IV. QUANTITATION AND DISTRIBUTION OF G PROTEINS

G proteins vary in tissue distribution. G$_s$ is essentially ubiquitous, whereas transducin is found only in photoreceptors; some, like G$_o$ and G$_{i2}$, vary widely in abundance. Within a given tissue or cell type, the amount of G protein may vary as a function of differentiation, development, and other physiologic or pathologic changes. While bacterial toxin-catalyzed ADP-ribosylation may be used to identify and quantitate G proteins, this has several drawbacks. It applies only to α subunits, may be influenced by secondary factors such as β-subunit concentration, and, for pertussis toxin, is not specific for a single G protein. Antibodies, used in immunoblotting, radioimmunoassay, and immunohistochemistry, offer advantages in quantitating G proteins and studying their distribution.

Monospecific transducin α antibodies localize this protein to photoreceptor rod outer segments (Grunwald *et al.*, 1986). Specific peptide antibodies, moreover, permit distinction between rod and cone forms of transducin α (Lerea *et al.*, 1986). Photoreceptor cells in avian and amphibian pineal organs also contain transducin (Van Veen *et al.*, 1986). Immunoelectronmicroscopy with transducin α antibodies precisely localizes the protein to outer segment membranes (Rodriguez *et al.*, 1987). There may, however, be a diurnal translocation of transducin from inner segment (day) to outer segment (night) (Brann and Cohen, 1987). Antisera specific for either transducin α or G$_o\alpha$ (Fig. 5) show differential distributions of these proteins on immunohistochemistry of rat retina (Fig. 12). G$_o$, unlike transducin, is found in inner and outer plexiform layers rather than outer segments (Lad *et al.*, 1987; Terashima *et al.*, 1987a). This suggests a role

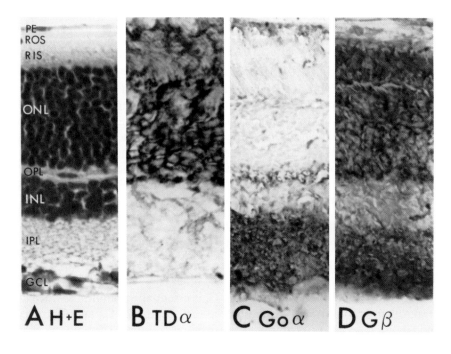

Fig. 12. (A): Hematoxylin and eosin (H + E) stained section of the rat retina shows its laminar organization. Laminae distinguished, from top (outer surface) to bottom (inner surface), are (1) pigment epithelium (PE); (2) receptor cell layers, which include the receptor outer segment (ROS, location of photoreceptor segments), receptor inner segments (RIS) and outer nuclear layer (ONL, location of the cell bodies of photoreceptors); (3) outer plexiform layer (OPL, location of synaptic contacts between photoreceptor cells and bipolar cells); (4) bipolar cell layers, which include the inner nuclear layer (INL, location of bipolar cell bodies) and inner plexiform layer (IPL, location of synaptic contacts between the bipolar cells and ganglion cells); and (5) the ganglion cell layer (GCL). (B): Section stained for transducin α immunoreactivity. Labeling is evident throughout the photoreceptor layers, including the outer and inner receptor layers and outer nuclear layer. (C): Section stained for $G_o\alpha$ immunoreactivity. The pattern of labeling is complementary to that for transducin α. Labeling is distributed in the pigment epithelium, the inner and outer plexiform layers and the ganglion cell layer. Notably there is little labeling in the inner nuclear layer. (D): Section stained for $G\beta$ immunoreactivity. The pattern of labeling is the sum of that for $G_o\alpha$ and transducin α immunoreactivity. (From Lad *et al.*, 1987.)

for G_o in synaptic transmission, but not in photon transduction. Another sensory tissue, olfactory epithelium, contains G_o as well, but the olfactory cilia (comparable to photoreceptor outer segments) appear selectively enriched in G_s immunoreactivity, consistent with a role for the latter in olfactory transduction (Anholt *et al.*, 1987).

G_o is the most abundant G protein in brain and is highest in concentration in forebrain regions (Gierschik *et al.*, 1986b; Huff *et al.*, 1985; Homburger *et al.*, 1987; Asano *et al.*, 1987). Immunohistochemistry localizes G_o to neuropil rather than cell bodies in brain; some regional variation is apparent. Colocalization with protein kinase C was interpreted to suggest a role for G_o in coupling to phospholipase C (Worley *et al.*, 1986). Further data are needed on this point. In addition to brain, G_o is found in some, e.g. heart, anterior pituitary, and adrenal medulla, but not all peripheral tissues (Huff *et al.*, 1985; Homburger *et al.*, 1987; Toutant *et al.*, 1987a). In pancreas, G_o was localized immunohistochemically to islet cells (Terashima *et al.*, 1987b). G_o was found in many neuroendocrine tumors, and was even present in plasma of patients with neuroblastomas (Kato *et al.*, 1987).

G proteins are highly conserved during evolution. Antibodies raised against mammalian G proteins crossreact with those from other vertebrate species (Gierschik *et al.*, 1986b; Van Veen *et al.*, 1986). Invertebrates also contain G proteins identified by toxin-catalyzed ADP-ribosylation. Pertussin toxin, for example, labels a 39-kDa protein in sea urchin eggs and sperm, and in *Torpedo* electric organ (Kopf *et al.*, 1986; Oinuma *et al.*, 1986; Toutant *et al.*, 1987b). In some cases immunologic crossreactivity with $G_o\alpha$ antibodies was found (Toutant *et al.*, 1987b).

Different forms of "G_i" vary in their tissue distribution (Brann *et al.*, 1987). G_{i2} is particularly abundant in phagocytic cells including neutrophils, macrophages, and glia (Goldsmith *et al.*, 1987; Backlund *et al.*, 1988). Increases in this protein occur with differentiation of U-937 monocytes (Falloon *et al.*, 1986) and HL-60 cells (Murphy *et al.*, 1987). In contrast, G_o is particularly abundant in neural type cells (Klinz *et al.*, 1987). Changes in both G_i and G_o occur during ontogeny in brain (Milligan *et al.*, 1987b) and in heart (Leutje *et al.*, 1987). Some of these changes parallel acquisition of receptors. This suggests the possibility of coordinate regulation of receptor and G-protein synthesis. In 3T3-L1 cells, G_s, G_i, and G_o concentrations change during adipocyte differentiation (Gierschik *et al.*, 1986c; Watkins *et al.*, 1987); this may underlie changes in adenylyl cyclase responsiveness.

Immunochemical studies have revealed several examples of physiologic and pathologic changes in G proteins leading to altered signal transduction. Increases in $G_i\alpha$ and β subunits have been observed in adipocytes from hypothyroid rats (Milligan *et al.*, 1987a). This may explain reduced β-adrenergic response in hypothyroidism. Reduction in liver G_i in experimental diabetes mellitus (Gawler *et al.*, 1987) could explain some of the effects of insulin deficiency. Rous sarcoma virus-transformed fibroblasts show alterations in adenylyl cyclase response and in β subunits. In dopamine-resistant pituitary tumors, a selective deficiency of G_o was found

(Collu *et al.*, 1988). In animals with inherited forms of retinitis pigmentosa, transducin subunits are immunochemically detectable in retina prior to degeneration, and then disappear during degeneration of photoreceptors (Navon *et al.*, 1987). Given the critical role of G proteins in signal transduction, future immunochemical studies are likely to reveal additional examples of alterations in G proteins as the basis for altered response to extracellular signals.

ACKNOWLEDGMENT

I gratefully acknowledge the collaboration of B. Eide, P. Gierschik, P. Goldsmith, G. Milligan, M. Pines, K. Rossiter, W. Simonds, C. Unson, and R. Vinitsky in the immunochemical studies from our own laboratory.

REFERENCES

Anholt, R. R. H., Mumby, S. M., Stoffers, D. A., Girard, P. R., Kuo, J. F., and Snyder, S. H. (1987). Transduction proteins of olfactory receptor cells: Identification of guanine nucleotide binding proteins and protein kinase C. *Biochemistry* **26**, 788–795.

Asano, T., Semba, R., Ogasawara, N., and Kato, K. (1987). Highly sensitive immunoassay for the alpha subunit of the GTP-binding protein G_o and its regional distribution in bovine brain. *J. Neurochem.* **48**, 1617–1623.

Audigier, Y., Pantaloni, C., Bigay, J., Deterre, P., Bockaert, J., and Homburger, V. (1985). Tissue expression and phylogenetic appearance of the beta and gamma subunits of GTP binding proteins. *FEBS Lett.* **189**, 1–7.

Backlund, P. S., Jr., Aksamit, R. R., Unson, C. G., Goldsmith, P., Spiegel, A. M., and Milligan, G. (1988). Immunochemical and electrophoretic characterization of the major pertussis toxin substrate of the RAW264 macrophage cell line. *Biochemistry* **27**, 2040–2046.

Brann, M. R., and Cohen, L. V. (1987). Diurnal expression of transducin mRNA and translocation of transducin in rods of rat retina. *Science* **235**, 585–587.

Brann, M. R., Collins, R. M., and Spiegel, A. (1987). Localization of the mRNAs encoding the alpha subunits of signal-transducing G-proteins within rat brain and among peripheral tissues. *FEBS Lett.* **222**, 191–198.

Bray, P., Carter, A., Simons, C., Guo, V., Puckett, C., Kamholtz, J., Spiegel, A., and Nirenberg, M. (1986). Human cDNA clones for four species of G-alpha$_s$ signal transduction protein. *Proc. Natl. Acad. Sci. U.S.A.* **83**, 8893–8897.

Bray, P., Carter, A., Guo, V., Puckett, C., Kamholz, J., Spiegel, A., and Nirenberg, M. (1987). Human cDNA clones for an alpha subunit of G_i signal transduction protein. *Proc. Natl. Acad. Sci. U.S.A.* **84**, 5115–5119.

Buss, J. E., Mumby, S. M., Casey, P. J., Gilman, A. G., and Sefton, B. M. (1987). Myristoylated alpha subunits of guanine nucleotide-binding regulatory proteins. *Proc. Natl. Acad. Sci. U.S.A.* **84,** 7493–7497.

Carlson, K. E., Woolkalis, M. J., Newhouse, M. G., and Manning, D. R. (1986). Fractionation of the beta subunit common to guanine nucleotide-binding regulatory proteins with the cytoskeleton. *Mol. Pharmacol.* **30,** 463–468.

Cerione, R. A., Gierschik, P., Stanizsewski, C., Benovic, J. L., Codina, J., Somers, R., Birnbaumer, L., Spiegel, A. M., Lefkowitz, R. J., and Caron, M. G. (1987). Functional differences in the beta gamma complexes of transducin and the inhibitory guanine nucleotide regulatory protein. *Biochemistry* **26,** 1485–1491.

Cerione, R. A., Kroll, S., Rajaram, R., Unson, C., Goldsmith, P., and Spiegel, A. M. (1988). An antibody directed against the carboxy-terminal decapeptide of the retinal GTP-binding protein, transducin: Effects on transducin function. *J. Biol. Chem.* **263,** 9345–9352.

Codina, J., Stengel, D., Woo, S., and Birnbaumer, L. (1986). Beta-subunits of the human liver G_s/G_i signal-transducing proteins and those of bovine retinal rod cell transducin are identical. *FEBS Lett.* **207,** 187–193.

Collu, R., Bouvier, C., Lagace, G., Unson, C., Milligan, G., Goldsmith, P., and Spiegel, A. (1988). Selective deficiency of guanine nucleotide binding protein, G_o, in two dopamine-resistant pituitary tumors. *Endocrinology,* **122,** 1176–1178.

Correze, C., d'Alayer, J., Coussen, F., Berthillier, G., Deterre, P., and Monneron, A. (1987). Antibodies directed against transducin beta subunits interfere with the regulation of adenylate cyclase activity in brain membranes. *J. Biol. Chem.* **261,** 15182–15187.

Deretic, D., and Hamm, H. E. (1987). Topographic analysis of antigenic determinants recognized by monoclonal antibodies to the photoreceptor guanyl nucleotide-binding protein, transducin. *J. Biol. Chem.* **262,** 10839–10847.

Didsbury, J. R., and Snyderman, R. (1987). Molecular cloning of a new human G protein: evidence for two G_{ialpha}-like protein families. *FEBS Lett.* **219,** 259–263.

Didsbury, J. R., Ho, Ye-Shih, and Snyderman, R. (1987). Human G_i protein alpha subunit: Deduction of amino acid structure from a cloned cDNA. *FEBS Lett.* **211,** 160–164.

Eide, B., Gierschik, P., Milligan, G., Mullaney, I., Unson, C., Goldsmith, P., and Spiegel, A. (1987). GTP-binding proteins in brain and neutrophil are tethered to the plasma membrane via their amino termini. *Biochem. Biophys. Res. Commun.* **148,** 1398–1405.

Evans, T., Fawzi, A., Fraser, E. D., Brown, M. L., and Northup, J. K. (1987). Purification of a $beta_{35}$ form of the beta/gamma complex common to G-proteins from human placental membranes. *J. Biol. Chem.* **262,** 176–181.

Falloon, J., Malech, H., Milligan, G., Unson, C., Kahn, R., Goldsmith, P., and Spiegel, A. (1986). Detection of the major pertussis toxin substrate of human leukocytes with antisera raised against synthetic peptides. *FEBS Lett.* **209,** 352–356.

Fitzgerald, T. J., Uhing, R. J., and Exton, J. H. (1986). Solubilization of the vasopressin receptor from rat liver membranes. *J. Biol. Chem.* **261**, 16871–16877.

Fong, H. K. W., Hurley, J. B., Hopkins, R. S., Miake-Lye, R., Johnson, M. S., Doolittle, R. F., and Simon, M. I. (1986). Repetitive segmental structure of the transducin beta subunit: Homology with the CDC4 gene and identification of related mRNAs. *Proc. Natl. Acad. Sci. U.S.A.* **83**, 2162–2166.

Fong, H. K. W., Amatruda, T. T., III, Birren, B. W., and Simon, M. I. (1987). Distinct forms of the beta subunit of GTP-binding regulatory proteins identified by molecular cloning. *Proc. Natl. Acad. Sci. U.S.A.* **84**, 3792–3796.

Gao, B., Gilman, A. G., and Robishaw, J. D. (1987a). A second form of the beta subunit of signal-transducing G proteins. *Proc. Natl. Acad. Sci. U.S.A.* **84**, 6122–6125.

Gao, B., Mumby, S., and Gilman, A. G. (1987b). The G protein beta$_2$ complementary DNA encodes the beta$_{35}$ subunit. *J. Biol. Chem.* **262**, 17254–17257.

Gawler, D., Milligan, G., Spiegel, A. M., Unson, C. G., and Houslay, M. D. (1987). Abolition of the expression of inhibitory guanine nucleotide regulatory protein G$_i$ activity in diabetes. *Nature* **327**, 229–232.

Gierschik, P., Codina, J., Simons, C., Birnbaumer, L., and Spiegel, A. (1985). Antisera against a guanine nucleotide binding protein from retina cross-react with the beta subunit of the adenylyl cyclase associated guanine nucleotide binding proteins, N$_s$ and N$_i$. *Proc. Natl. Acad. Sci. U.S.A.* **82**, 727–731.

Gierschik, P., Falloon, J., Milligan, G., Pines, M., Gallin, J. I., and Spiegel, A. (1986a). Immunochemical evidence for a novel pertussis toxin substrate in human neutrophils. *J. Biol. Chem.* **261**, 8058–8062.

Gierschik, P., Milligan, G., Pines, M., Goldsmith, P., Codina, J., Klee, W., and Spiegel, A. (1986b). Use of specific antibodies to quantitate the guanine nucleotide-binding protein G$_o$ in brain. *Proc. Natl. Acad. Sci. U.S.A.* **83**, 2258–2262.

Gierschik, P., Morrow, B., Milligan, G., Rubin, C., and Spiegel, A. (1986c). Changes in the guanine nucleotide binding proteins, G$_i$ and G$_o$, during differentiation of 3T3-L1 cells. *FEBS Lett.* **199**, 103–106.

Gierschik, P., Sidiropoulos, D., Spiegel, A., and Jakobs, K. H. (1987). Purification and immunochemical characterization of the major pertussis-toxin sensitive guanine-nucleotide-binding protein of bovine-neutrophil membranes. *Eur. J. Biochem.* **165**, 185–194.

Goldsmith, P., Gierschik, P., Milligan, G., Unson, C., Vinitsky, R., Malech, H. L., and Spiegel, A. M. (1987). Antibodies directed against synthetic peptides distinguish between GTP-binding proteins in neutrophil and brain. *J. Biol. Chem.* **262**, 1–6.

Goldsmith, P., Backlund, P. S., Jr., Rossiter, K., Carter, A., Milligan, G., Unson, C. G., and Spiegel, A. (1988a). Purification of heterotrimeric GTP-binding proteins from brain: Identification of a novel form of G$_o$. *Biochem.* **27**, 7085–7090.

Goldsmith, P., Rossiter, K., Carter, A., Simonds, W., Unson, C. G., Vinitsky, R., and Spiegel, A. M. (1988b). Identification of the GTP-binding protein encoded by G_{i3} complementary DNA. *J. Biol. Chem.*, **263**, 6476–6479.

Grunwald, G. B., Gierschik, P., Nirenberg, M., and Spiegel, A. (1986). Detection of alpha-transducin in retinal rods but not cones. *Science* **231**, 856–859.

Halpern, J. L., Tsai, S-U., Adamik, R., Kanaho, Y., Bekesi, E., Kung, H-F., Moss, J., and Vaughan, M. (1986). Structural and functional characterization of guanyl nucleotide-binding proteins using monoclonal antibodies to the alpha-subunit of transducin. *Mol. Pharmacol.* **29**, 515–519.

Halpern, J. L., Chang, P. P., Tsai, S-C., Adamik, R., Kanaho, Y., Sohn, R., Moss, J., and Vaughan, M. (1987). Production of antibodies against rhodopsin after immunization with beta/gamma-subunits of transducin: Evidence for interaction of beta/gamma-subunits of guanosine 5′-triphosphate binding proteins with receptor. *Biochemistry* **26**, 1655–1658.

Hamm, H. E., and Bownds, D. (1984). A monoclonal antibody to guanine nucleotide binding protein inhibits the light-activated cyclic GMP pathway in frog rod outer segments. *J. Gen. Physiol.* **84**, 265–280.

Hamm, H. E., Deretic, D., Hofman, K. P., Schleicher, A., and Kohl, B. (1987). Mechanism of action of monoclonal antibodies that block the light activation of the guanyl nucleotide-binding protein, transducin. *J. Biol. Chem.* **262**, 10831–10838.

Harris, B. A., Robishaw, J. D., Mumby, S. M., and Gilman, A. G. (1985). Molecular cloning of complementary DNA of the alpha subunit of the G protein that stimulates adenylate cyclase. *Science* **229**, 1274–1277.

Heydorn, W. E., Gierschik, P., Creed, G. J., Milligan, G., Spiegel, A., and Jacobowitz, D. M. (1986). The beta subunit of the guanine nucleotide regulatory proteins: Identification of its location on two-dimensional gels of brain tissue and its regional and subcellular distribution in brain. *J. Neuro. Res.* **16**, 541–552.

Homburger, V., Brabet, P., Audigier, Y., Pantaloni, C., Bockaert, J., and Rouot, B. (1987). Immunological localization of the GTP-binding protein G_o in different tissues of vertebrates and invertebrates. *Mol. Pharmacol.* **31**, 313–319.

Huff, R. M., and Neer, E. J. (1986). Subunit interactions of native and ADP-ribosylated $alpha_{39}$ and $alpha_{41}$, two guanine nucleotide-binding proteins from bovine cerebral cortex. *J. Biol. Chem.* **261**, 1105–1110.

Huff, R. M., Axton, J. M., and Neer, E. J. (1985). Physical and immunological characterization of a guanine nucleotide-binding protein purified from bovine cerebral cortex. *J. Biol. Chem.* **260**, 10864–10871.

Hurley, J. B., Fong, H. K. W., Teplow, D. B., Dreyer, W. J., and Simon, M. I. (1984). Isolation and characterization of a cDNA clone for the gamma subunit of bovine retinal transducin. *Proc. Natl. Acad. Sci. U.S.A.* **81**, 6948–6952.

Itoh, H., Kozasa, T., Nagata, S., Nakamura, S., Katada, T., Ui, M., Iwai, S., Ohtsuka, E., Kawasaki, H., Suzuki, K., and Kaziro, Y. (1986). Molecular cloning and sequence determination of cDNAs for alpha subunits of the guanine nucleotide-binding proteins G_s, G_i, and G_o from rat brain. *Proc. Natl. Acad. Sci. U.S.A.* **83**, 3776–3780.

Iyengar, R., Rich, K. A., Herberg, J. T., Grenet, D., Mumby, S., and Codina, J. (1987). Identification of a new GTP-binding protein. *J. Biol. Chem.* **262,** 9239–9245.

Jones, D. T., and Reed, R. R. (1987). Molecular cloning of five GTP-binding protein cDNA species from rat olfactory neuroepithelium. *J. Biol. Chem.* **262,** 14241–14249.

Jurnak, F. (1985). Structure of the GDP domain of EF-Tu and location of the amino acids homologous to *ras* oncogene proteins. *Science* **230,** 32–36.

Katada, T., Oinuma, M., Kusakabe, K., and Ui, M. (1987). A new GTP-binding protein in brain tissues serving as the specific substrate of islet-activating protein, pertussis toxin. *FEBS Lett.* **213,** 353–358.

Kato, K., Asano, T., Kamiya, N., Haimoto, H., Hosoda, S., Nagasaka, A., Ariyoshi, Y., and Ishiguro, Y. (1987). Production of the alpha subunit of guanine nucleotide-binding protein G_o by neuroendocrine tumors. *Cancer Res.* **47,** 5800–5805.

Klinz, F-J., Yu, V. C., Sadee, W., and Costa, T. (1987). Differential expression of alpha-subunits of G-proteins in human neuroblastoma-derived cell clones. *FEBS Lett.* **224,** 43–48.

Kopf, G., Woolkalis, M. J., and Gerton, G. L. (1986). Evidence for a guanine nucleotide-binding regulatory protein in invertebrate and mammalian sperm. *J. Biol. Chem.* **261,** 7327–7331.

Lad, R. P., Simons, C., Gierschik, P., Milligan, G., Woodard, G., Griffo, M., Goldsmith, P., Ornberg, R., Gerfen, C. R., and Spiegel, A. (1987). Differential distribution of signal-transducing G-proteins in retina. *Brain Res.* **423,** 237–246.

Lang, J., and Costa, T. (1987). Antisera against the 3-17 sequence of rat $G_{alpha-i}$ recognize only a 40 kDa G-protein in brain. *Biochem. Biophys. Res. Commun.* **148,** 838–848.

Lerea, C. L., Somers, D. E., Hurley, J. B., Klock, I. B., and Bunt-Milan, A. H. (1986). Identification of specific transducin alpha subunits in retinal rod and cone photoreceptors. *Science* **234,** 77–80.

Lingham, R. B., Brown, P. J., Holcombe, V., and Schreiber, C. L. (1986). Monoclonal antibodies to the guanine-nucleotide binding proteins of adenylate cyclase. *Biochem. J.* **236,** 267–271.

Lochrie, M. A., Hurley, J. B., and Simon, M. I. (1985). Sequence of the alpha subunit of photoreceptor G protein: Homologies between transducin, *ras,* and elongation factors. *Science* **228,** 96–99.

Luetje, C. W., Gierschik, P., Milligan, G., Unson, C., Spiegel, A., and Nathanson, N. M. (1987). Tissue-specific regulation of GTP-binding protein and muscarinic acetylcholine receptor levels during cardiac development. *Biochemistry* **26,** 4876–4884.

McKenzie, F. R., Kelly, E. C. H., Unson, C. G., Spiegel, A. M., and Milligan, G. (1988). Antibodies which recognize the C-terminus of the inhibitory guanine nucleotide binding protein (G_i) demonstrate that opioid peptides and foetal calf serum stimulate the high affinity GTPase activity of two separate pertussis toxin substrates. *Biochem. J.* **249,** 653–659.

Milligan, G., Spiegel, A. M., Unson, C. G., and Saggerson, E. D. (1987a). Chemically induced hypothyroidism produces elevated levels of the alpha subunit of the inhibitory guanine nucleotide binding protein (G_i) and the beta subunit common to all G-proteins. *Biochem. J.* **247**, 223–227.

Milligan, G., Streaty, R. A., Gierschik, P., Spiegel, A. M., and Klee, W. A. (1987b). Development of opiate receptors and GTP-binding regulatory proteins in neonatal rat brain. *J. Biol. Chem.* **262**, 8626–8630.

Mumby, S. M., Kahn, R. A., Manning, D. R., and Gilman, A. G. (1986). Antisera of designed specificity for subunits of guanine nucleotide-binding regulatory proteins. *Proc. Natl. Acad. Sci. U.S.A.* **83**, 265–269.

Murphy, P. M., Eide, B., Goldsmith, P., Brann, M., Gierschik, P., Spiegel, A., and Malech, H. L. (1987). Detection of multiple forms of G_{ialpha} in HL60 cells. *FEBS Lett.* **221**, 81–86.

Navon, S. E., and Fung, B. K.-K. (1987). Characterization of transducin from bovine retinal rod outer segments. *J. Biol. Chem.* **262**, 15746–15751.

Navon, S. E., Lee, R. H., Lolley, R. N., and Fung, B. K.-K. (1987). Immunological determination of transducin content in retinas exhibiting inherited degeneration. *Exp. Eye Res.* **44**, 115–125.

Neer, E. J., Lok, J. M., and Wolf, L. G. (1984). Purification and properties of the inhibitory guanine nucleotide regulatory unit of brain adenylate cyclase. *J. Biol. Chem.* **259**, 14222–14229.

Nukada, T., Tanabe, T., Takahashi, H., Noda, M., Haga, K., Haga, T., Ichiyama, A., Kangawa, K., Hiranaga, M., Matsuo, H., and Numa, S. (1986a). Primary structure of the alpha-subunit of bovine adenylate cyclase-inhibiting G-protein deduced from the cDNA sequence. *FEBS Lett.* **197**, 305–310.

Nukada, T., Tanabe, T., Takahashi, H., Noda, M., Hirose, T., Inayama, S., and Numa S. (1986b). Primary structure of the alpha-subunit of bovine adenylate cyclase-stimulating G-protein deduced from the cDNA sequence. *FEBS Lett.* **195**, 220–224.

Oinuma, M., Katada, T., Yokosawa, H., and Ui, M. (1986). Guanine nucleotide-binding protein in sea urchin eggs serving as the specific substrate of islet-activating protein, pertussis toxin. *FEBS Lett.* **207**, 28–34.

Oinuma, M., Katada, T., and Ui, M. (1987). A new GTP-binding protein in differentiated human leukemic (HL-60) cells serving as the specific substrate of islet-activating protein, pertussis toxin. *J. Biol. Chem.* **262**, 8347–8353.

Pines, M., Gierschik, P., Milligan, G., Klee, W., and Spiegel, A. (1985a). Antibodies against the C-terminal 5kDa peptide of the alpha subunit of transducin crossreact with the 40 but not the 39kDa guanine nucleotide binding protein from brain. *Proc. Natl. Acad. Sci. U.S.A.* **82**, 4095–4099.

Pines, M., Gierschik, P., and Spiegel, A. (1985b). The tryptic and chymotryptic fragments of the beta-subunit of guanine nucleotide binding proteins in brain are identical to those of retinal transducin. *FEBS Lett.* **182**, 355–359.

Robishaw, J. D., Smigel, M. D., and Gilman, A. G. (1986). Molecular basis for two forms of the G protein that stimulates adenylate cyclase. *J. Biol. Chem.* **261**, 9587–9590.

Rodriguez, M., Hackett, J., Wiggert, B., Gery, I., Speigel, A., Krishna, G., Stein, P., and Chader, G. (1987). Immunoelectron microscopic localization of photoreceptor-specific markers in monkey retina. *Curr. Eye Res.* **6**, 369–380.

Roof, D. J., Applebury, M. L., and Sternweis, P. C. (1985). Relationships within the family of GTP-binding proteins isolated from bovine central nervous system. *J. Biol. Chem.* **260**, 16242–16249.

Rosenthal, W., Koesling, D., Rudolph, U., Kleuss, C., Pallast, M., Yajima, M., and Schultz, G. (1986). Identification and characterization of the 35-kDa beta subunit of guanine-nucleotide-binding proteins by an antiserum raised against transducin. *Eur. J. Biochem.* **158**, 255–263.

Senogles, S. E., Benovic, J. L., Amlaiky, N., Unson, C., Milligan, G., Vinitsky, R., Spiegel, A. M., and Caron, M. G. (1987). The D_2-dopamine receptor of anterior pituitary is functionally associated with a pertussis toxin-sensitive guanine nucleotide binding protein. *J. Biol. Chem.* **262**, 4860–4867.

Sternweis, P. C., and Robishaw, J. D. (1984). Isolation of two proteins with high affinity for guanine nucleotides from membranes or bovine brain. *J. Biol. Chem.* **259**, 13806–13813.

Sugimoto, K., Nukada, T., Tanabe, T., Takahashi, H., Noda, M., Minamino, N., Kangawa, K., Matsuo, H., Hirose, T., Inayama, S., and Numa, S. (1985). Primary structure of the beta-subunit of bovine transducin deduced from the cDNA sequence. *FEBS Lett.* **191**, 235–240.

Suki, W. N., Abramowitz, J., Mattera, R., Codina, J., and Birnbaumer, L. (1987). The human genome encodes at least three non-allellic G proteins with alpha-i-type subunits. *FEBS Lett.* **220**, 187–192.

Sullivan, K. A., Liao, Y-C., Alborzi, A., Beiderman, B., Chang, F-H., Masters, S. B., Levinson, A. D., and Bourne, H. R. (1986). Inhibitory and stimulatory G proteins of adenylate cyclase: cDNA and amino acid sequences of the alpha chains. *Proc. Natl. Acad. Sci. U.S.A.* **83**, 6687–6691.

Tanabe, T., Nukada, T., Nishikawa, Y., Sugimoto, K., Suzuki, H., Takahashi, H., Noda, M., Haga, T., Ichiyama, A., Kangawa, K., Minamino, N., Matsuo, H., and Numa, S. (1985). Primary structure of the alpha subunit of transducin and its relationship to ras proteins. *Nature* **315**, 242–245.

Terashima, T., Katada, T., Oinuma, M., Inoue, Y., and Ui, M. (1987a). Immunohistochemical localization of guanine nucleotide-binding protein in rat retina. *Brain Res.* **410**, 97–100.

Terashima, T., Katada, T., Oinuma, M., Inoue, Y., and Ui, M. (1987b). Endocrine cells in pancreatic islets of langerhans are immunoreactive to antibody against guanine nucleotide-binding protein (G_o) purified from rat brain. *Brain Res.* **417**, 190–194.

Toutant, M., Aunis, D., Bockaert, J., Homburger, V., and Rouot, B. (1987a). Presence of three pertussis toxin substrates and $G_o$alpha immunoreactivity in both plasma and granule membranes of chromaffin cells. *FEBS Lett.* **215**, 339–344.

Toutant, M., Bockaert, J., Homburger, V., and Rouot, B. (1987b). G-proteins in torpedo marmorata electric organ. *FEBS Lett.* **222**, 51–55.

Tsai, S-C., Adamik, R., Kanaho, Y., Halpern, J. L., and Moss, J. (1987). Immunological and biochemical differentiation of guanyl nucleotide binding proteins: Interaction of G_{oalpha} with rhodopsin, anti-G_{oalpha} polyclonal antibodies, and a monoclonal antibody against transducin alpha subunit and G_{ialpha}. *Biochemistry* **26**, 4728–4733.

Uhing, R. J., Polakis, P. G., and Snyderman, R. (1987). Isolation of GTP-binding proteins from myeloid HL-60 cells. *J. Biol. Chem.* **262**, 15575–15579.

Ui, M. (1984). Islet-activating protein, pertussis toxin: A probe for functions of the inhibitory guanine nucleotide regulatory component of adenylate cyclase. *Trends Pharmacol. Sci.* **5**, 277–279.

Van Dop, C., Medynski, D., Sullivan, K., Wu, A. M., Fung, BK-K., and Bourne, H. R. (1984). Partial cDNA sequence of the gamma subunit of transducin. *Biochem. Biophys. Res. Commun.* **124**, 250–255.

Van Meurs, K. P., Angus, C. W., Lavu, S., Kung, H-F., Czarnecki, S. K., Moss, J., and Vaughan, M. (1987). Deduced amino acid sequence of bovine retinal G_{oalpha}: Similarities to other guanine nucleotide-binding proteins. *Proc. Natl. Acad. Sci. U.S.A.* **84**, 3107–3111.

Van Veen, T., Ostholm, T., Gierschik, P., Spiegel, A., Somers, D., and Klein, D. (1986). Alpha-transducin immunoreactivity in retinae and sensory pineal organs of adult vertebrates. *Proc. Natl. Acad. Sci. U.S.A.* **83**, 912–916.

Watkins, D. C., Northup, J. K., and Malbon, C. C. (1987). Regulation of G-proteins in differentiation. *J. Biol. Chem.* **262**, 10651–10657.

West, R. E., Jr., Moss, J., Vaughan, M., Liu, T., and Liu, T-Y. (1985). Pertussis toxin-catalyzed ADP-ribosylation of transducin. Cysteine 347 is the ADP-ribose acceptor site. *J. Biol. Chem.* **260**, 14428–14430.

Willumsen, B. M., Norris, K., Papageorge, A. G., Hubbert, N. L., and Lowy, D. R. (1984). Harvey murine sarcoma virus p21 *ras* protein: Biological and biochemical significance of the cysteine nearest the carboxy terminus. *EMBO J.* **3**, 2581–2585.

Wistow, G. J., Katial, A., Craft, C., and Shinohara, T. (1986). Sequence analysis of bovine retinal S-antigen: Relationships with alpha-transducin and G-proteins. *FEBS Lett.* **196**, 23–28.

Woolkalis, M. J., and Manning, D. R. (1987). Structural characteristics of the 35- and 36-kDa forms of the beta subunit common to GTP-binding regulatory proteins. *Mol. Pharmacol.* **32**, 1–6.

Woolkalis, M., Nakada, M. T., and Manning D. R. (1986). Alterations in components of adenylate cyclase associated with transformation of chicken embryo fibroblasts by rous sarcoma virus. *J. Biol. Chem.* **261**, 3408–3413.

Worley, P. F., Baraban, J. M., Van Dop, C., Neer, E. J., Snyder, S. H. (1986). G_o, a guanine nucleotide-binding protein: Immunohistochemical localization in rat brain resembles distribution of second messenger systems. *Proc. Natl. Acad. Sci. U.S.A.* **83**, 4561–4565.

Yatsunami, K., Pandya, B. V., Oprian, D. D., and Khorana, H. G. (1985). cDNA-derived amino acid sequence of the gamma subunit of GTPase from bovine rod outer segments. *Proc. Natl. Acad. Sci. U.S.A.* **82**, 1936–1940.

Zaremba, T., Gierschik, P., Pines, M., Bray, P., Carter, A., Kahn, R., Simons, C., Vinitsky, R., Goldsmith, P., and Spiegel, A. (1988). Immunochemical studies of the 36kDa common beta subunit of guanine nucleotide binding proteins: Identification of a major epitope. *Mol. Pharmacol.,* **33**, 257–264.

PART II

Coupling

CHAPTER 7

Adenylyl Cyclase and Its Regulation by G$_s$

Richard T. Premont, Ravi Iyengar
Department of Pharmacology, Mount Sinai School of Medicine of the City
University of New York, New York, New York 10029

I. INTRODUCTION

The importance of adenylyl cyclase in the homeostasis of the cell is widely recognized. Perhaps of equal importance, the adenylyl cyclase system serves as a model for other, less well-characterized G protein-regulated systems, both as a basic analogy and as a point of departure for understanding how G proteins function in the cell. Regulation of adenylyl cyclase by receptors coupled to G proteins has been extensively studied in the last 15 years. Our understanding of the general features of the system and many aspects of regulation is substantial, yet several details remain to be worked out. Some of these issues will be addressed here.

In recent years, enzymatically nonhydrolyzable guanine nucleotide analogs and certain bacterial toxins/mono-ADP-ribosyltransferases have been used to assess the involvement of G proteins and other GTP-binding proteins in diverse cellular regulatory events. Many such systems have been identified, including phospholipase C, phospholipase A_2, K^+ channels, Ca^{2+} channels, and controls of exocytosis. The regulation of these other effectors, both enzymes and ion channels, by G proteins is discussed in other chapters of this volume.

II. REGULATION OF ADENYLYL CYCLASE BY G_s

A. Guanine Nucleotide and Mg^{2+} Activation of Adenylyl Cyclase

The requirement of GTP to obtain glucagon stimulation of liver adenylyl cyclase (Rodbell et al., 1971) was the first indication of the regulatory role of that nucleotide in signal transduction systems. It was subsequently found that nonhydrolyzable analogs of GTP, such as Gpp(NH)p, could extensively and persistently activate adenylyl cyclase (Londos et al., 1974; Schramm and Rodbell, 1975). This stimulation is a time-dependent phenomenon accompanied by a characteristic lag that cannot be abolished by high concentrations of guanine nucleotides. A typical experiment showing the time-dependent activation of liver plasma membrane adenylyl cyclase is shown in Fig. 1.

In the liver plasma membrane adenylyl cyclase system, three distinct features are observed: (1) Various nonhydrolyzable analogs of GTP give differing lag times for the activation of adenylyl cyclase (Birnbaumer et al., 1980). (2) In the presence of saturating concentrations of differing nonhydrolyzable analogs of GTP, the primary effect of the activated hormone–receptor complex is to abolish the lag and thus increase the rate of the guanine nucleotide-mediated activation process (Salomon et al., 1975; Fig. 1). (3) Increasing concentrations of Mg^{2+} lead to progressive decrease in lag

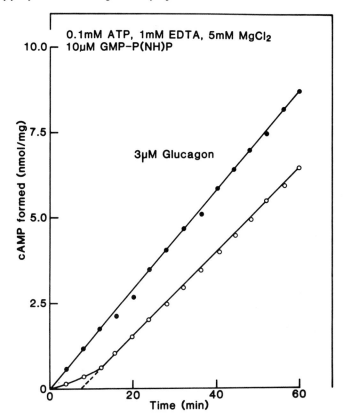

Fig. 1. Effect of hormones on the activation of liver adenylyl cyclase by Gpp(NH)p. Note the characteristic lag in the Gpp(NH)p activation curve (O), and the effect of the hormone (●) in abolishing this lag. The final velocity of adenylyl cyclase in the presence and absence of hormone is very similar. This experiment is like one first published by Salomon *et al.*, 1975.

times such that by 25 mM Mg^{2+}, no lag is observed for the Gpp(NH)p activation of adenylyl cyclase (Iyengar and Birnbaumer, 1981). From these observations the concept arose that the hormone–receptor complex activated adenylyl cyclase by modulating the guanine nucleotide activation process.

In contrast, Schramm and Rodbell (1975) found that the frog erythrocyte adenylyl cyclase system required the presence of a hormone–receptor complex for the Gpp(NH)p activation of the enzyme. Cassel and Selinger first described the presence of a hormone-stimulated GTPase activity (Cassel and Selinger, 1976), and demonstrated that cholera toxin stimulation of adenylyl cyclase and its promotion of hormonal stimulation was accompanied by an inhibition of this hormone-stimulated GTPase activity (Cassel and Selinger, 1977). In the turkey erythrocyte system, the hormone–

receptor complex promotes the dissociation of bound GDP. This facilitation of the release of GDP was postulated to be the mechanism by which hormone receptors acted (Cassel and Selinger, 1978).

B. Identification and Purification of G_s

Ross and Gilman (1977a,b) analyzed the defect in the cyc^- mutant of the S49 murine lymphoma cell line by reconstitution analysis, and found that the missing component was a guanine nucleotide-binding entity that promoted the coupling between the hormone-receptor complex and the catalyst of adenylyl cyclase. They named this component G/F for its ability to restore guanine nucleotide and fluoride stimulation to the cyc^- adenylyl cyclase. It is now called G_s; the name N_s has also been used previously.

Studies on G_s and its reconstitution of stimulation of cyc^- S49 cell membrane adenylyl cyclase have shown the following: (1) The persistent activation of adenylyl cyclase by nonhydrolyzable analogs of GTP is due to the persistent activation of G_s (Ross et al., 1978). (2) Cholera toxin has its effect on G_s (Ross et al., 1978; Johnson et al., 1978; Howlett and Gilman, 1980). (3) Activation of G_s is a slow, Mg^{2+}-dependent process that is the rate-limiting step in the activation of the adenylyl cyclase system (Iyengar, 1981; Fig. 2). (4) Association of activated G_s with the catalyst is a fast event (Iyengar, 1981).

G_s was first purified from rabbit liver by its ability to reconstitute cyc^- S49 cell membrane adenylyl cyclase (Northup et al., 1980). G_s has subsequently been purified from turkey (Hanski et al., 1981) and human (Codina et al., 1984a) erythrocytes. The liver and human erythrocyte proteins have been extensively characterized, and shown to be $\alpha\beta\gamma$ heterotrimers.

The α subunit carries the GTP-binding site, as well as the site for cholera toxin-mediated modification. The α subunit is responsible for activation of the catalyst of adenylyl cyclase (Northup et al., 1983a). The α subunit of G_s appears as a 43- or 52-kDa protein following sodium dodecylsulfate-polyacrylamide gel electrophoresis (SDS-PAGE) of purified proteins or following labeling with cholera toxin and $[^{32}P]NAD^+$. These α_s forms are distributed differently in different tissues: liver and brain contain both 43- and 52-kDa species, while erythrocytes contain only the 43-kDa form. These different species all stimulate adenylyl cyclase. Cloning has revealed that α_s exists in four distinct forms, two small and two large. The structural basis for these multiple forms will be described in a later section.

The β and γ subunits are tightly associated in a dimeric complex under native conditions. The $\beta\gamma$ complex, by association with the α subunit, inhibits the capability of the α subunit to activate adenylyl cyclase (Northup et al., 1983b). The $\beta\gamma$ complex is very hydrophobic, and would appear to anchor the G protein to the plasma membrane (Sternweis, 1986). There are two forms of the G protein β subunit, of 36 and 35 kDa. The γ

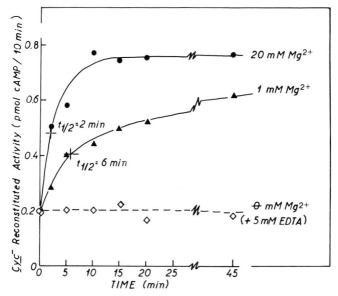

Fig. 2. Effect of increasing concentrations of Mg^{2+} on the rate of activation of liver G_s by saturating concentrations of Gpp(NH)p. Note that both the rate and the extent of activation increase with increasing Mg^{2+} concentration. (From Iyengar, 1981.)

subunits appear to occur in several forms, of 7–10 kDa. The role of $\beta\gamma$ complexes in G protein function is further discussed by Neer and Clapham in Chapter 3 of this volume.

C. Activation and Reconstitution of G_s

It was found that on activation by guanine nucleotides, G proteins undergo a decrease in sedimentation coefficient (Howlett and Gilman, 1980; Sternweis, 1986). From these observations it was suggested that activation of G_s is associated with the dissociation of the G_s α and $\beta\gamma$ subunits. While native (inactive) G_s sediments at 4S, the persistently activated α_s subunit resolved from $\beta\gamma$ complex sediments at 2S in linear sucrose gradients, as does activated G_s (Iyengar *et al.*, 1988).

Studies on [^{35}S]GTPγS binding to purified G_s in detergent solution have shown that the binding reaction is slow and concomitant with subunit dissociation. The rate of [^{35}S]GTPγS binding is similar over a wide range (10^{-4}–$10^{-8}M$) of GTPγS concentrations. However, increasing concentrations of Mg^{2+} greatly stimulate the rate of [^{35}S]GTPγS binding (Northup *et al.*, 1982). Sedimentation analysis has shown that the nondissociated form of G_s can be liganded with nonhydrolyzable analogs of GTP (Codina *et al.*,

1984b). While high concentrations of Mg^{2+} do not promote subunit dissociation (Codina et al., 1984b), they do promote GDP dissociation from the G protein complex (Hagashijima et al., 1987). These data, combined with observations that various nonhydrolyzable analogs of GTP give differing activation lag times (Birnbaumer et al., 1980), would favor the following sequence of reactions (1) for the activation of G_s:

$$\alpha\text{-GDP-}\beta\gamma \overset{\text{GDP}}{\underset{}{\rightleftharpoons}} \alpha\beta\gamma \overset{\text{GTP}}{\underset{}{\rightleftharpoons}} \alpha\text{-GTP-}\beta\gamma \rightleftharpoons \alpha^*\text{-GTP} + \beta\gamma \qquad (1)$$

The GTPase activity of G_s has been extensively studied by Ross and co-workers (Brandt et al., 1983; Brandt and Ross, 1985). They found that the GTPase activity has a maximum observed turnover number of 1.5 min^{-1}. The K_m for GTP is 0.3 μM. During steady state hydrolysis, 20–40% of G_s is found to be in the GDP-liganded state, while only 1–2% is in the GTP-liganded state. The rate of GDP release is observed to be biphasic. $\beta\gamma$ subunits greatly stabilize GDP binding to the α_s subunit and thus inhibit GDP release. This sequence of reactions (2) may be summarized as follows:

$$\alpha^*\text{-GTP} \overset{P_i}{\underset{}{\rightleftharpoons}} \alpha\text{-GDP} \overset{\beta\gamma}{\underset{}{\rightleftharpoons}} \alpha\text{-GDP-}\beta\gamma \qquad (2)$$

Reaction sequences (1) and (2) describe the basic cycle of G_s through its activation and deactivation phases, and account for most of the observed behavior of G_s during the guanine nucleotide and Mg^{2+} activation of adenylyl cyclase.

D. Hormone–Receptor-Mediated Activation of G_s

The earliest indication that hormone–receptor activation of adenylyl cyclase resulted from the modulation of guanine nucleotide-mediated activation came from the observations of Salomon et al. (1975), who showed that hormones abolished the lag in Gpp(NH)p-activation of the liver adenylyl cyclase (see Fig. 1). Detailed kinetic analysis of this activation process by Tolkovsky and Levitzki (1978) indicated that a single hormone-occupied receptor molecule would be able to activate several adenylyl cyclase molecules, and thus function in a catalytic manner. This mode of activation was termed "collision coupling".

When it became apparent that activation of adenylyl cyclase was mediated through G_s, receptor-dependent activation of G_s became the focus of attention. Citri and Schramm (1980) were able to utilize resolved β-adrenergic receptors and G_s to demonstrate receptor-dependent activation of adenylyl cyclase. The mechanism of this activation was addressed by Iyengar and Birnbaumer (1981), who showed that in the presence of saturating

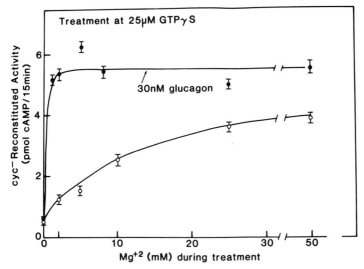

Fig. 3. Effect of hormones on the Mg^{2+} requirement of the G_s activation by guanine nucleotides. Treated (C^-) liver membrane cholate suspension plus cyc^- S49 cell membranes. Note that in the presence of a hormone–receptor complex (●) at the lowest concentration of free Mg^{2+} tested (0.5 mM), full activation of G_s is observed. (From Iyengar and Birnbaumer, 1982.)

concentrations of guanine nucleotide, hormone receptors greatly increase the rate of activation of G_s at low, physiologically relevant concentrations of Mg^{2+} (Fig. 3).

Detailed analysis of hormone–receptor complex interactions with G_s, using purified proteins reconstituted in phospholipid vesicles, has been performed by Ross and co-workers. They found that the hormone–receptor complex is able to enhance the rate of GTPγS binding to G_s at low Mg^{2+} concentrations. Binding of GTPγS to G_s in these situations occurs concomitant with activation of G_s, as assessed by its ability to stimulate the cyc^- S49 cell membrane adenylyl cyclase. A single hormone–receptor complex is able to activate several G_s molecules in a sequential manner, confirming the validity of the "collision coupling" model (Asano *et al.,* 1984).

Kinetic analysis of the binding of GTPγS to G_s indicates apparent first-order rate constants when guanine nucleotide, hormone, and Mg^{2+} concentrations are varied (Asano and Ross, 1984). On this basis, it was suggested that the hormone receptors stabilize the active state of G_s. Detailed kinetic analysis of the GTPase regulatory cycle displayed by purified and reconstituted β-adrenergic receptors and G_s led Brandt and Ross (1986) to conclude that both dissociation of tightly bound GDP and high-affinity binding of GTP are promoted by the hormone–receptor complex. Comparison of the rates of dissociation of bound GDP and of association of GTPγS shows that hormone-stimulated GDP dissociation is twice as fast as GTPγS binding, indicating that GDP dissociation is not the rate-limiting step in the activation of G_s by hormone. These observations of Ross and co-workers were obtained using purified liver G_s, and are in general agreement with previous results from the membrane-bound liver adenylyl cyclase system (Birnbaumer *et al.*, 1980). However, the general applicability of these conclusions remains to be determined. For instance, if turkey erythrocyte G_s were used, the slow step regulated by the hormone–receptor complex might be GDP dissociation rather than GTP binding.

An interesting extension of these complex interactions between the agonist-bound receptor and G_s is the observation by May and Ross (1988) of burst kinetics in GTPγS binding when vesicles containing β-adrenergic receptors are preincubated with agonist in the absence of GTPγS. These observations allow for the first time the separation of the GDP dissociation reaction from the GTP binding reaction, and indicate that a hormone–receptor–G_s complex with a vacant GTP binding site can be formed and that this entity is capable of very rapid association with GTP.

Since activation of G_s in detergent solution is accompanied by subunit dissociation, it has been presumed that subunit dissociation occurs in the membrane environment as well. This has not as yet been demonstrated in the reconstituted system. In liver membranes, glucagon-dependent activation of G_s is accompanied by subunit dissociation as assessed by determination of the sedimentation coefficient of the cyc^- S49 cell membrane adenylyl cyclase-reconstituting activity (Iyengar *et al.*, 1988; Fig. 4). Since the activation treatment is done in the membrane environment and the activating agents (hormone, guanine nucleotide, and Mg^{2+}) are removed prior to extraction of G_s, these observations provide reasonable verification that receptor-mediated activation of G_s is accompanied by subunit dissociation. The following sequence of reactions (3) can be written for receptor-mediated activation of G_s:

$$\underset{\alpha-GDP-\beta\gamma}{\overset{GDP}{}} \overset{[HR]}{\underset{}{\rightleftharpoons}} \alpha\beta\gamma \overset{GTP}{\underset{[HR]}{\rightleftharpoons}} \alpha-GTP-\beta\gamma \rightleftharpoons \alpha^*-GTP + \beta\gamma \qquad (3)$$

It is important to note that the hormone–receptor complex does not directly alter the catalytic rate of the GTPase activity of G_s. Hormonal stimulation of GTPase activity is the result of the hormone–receptor-mediated increase in the enzyme–substrate complex, i.e., the $\alpha-GTP-\beta\gamma$ complex.

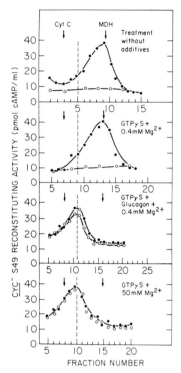

Fig. 4. Effect of activation of G_s in the native membrane environment on the size of the activated α_s-subunit, as assessed by reconstitution of the cyc^- adenylyl cyclase. Liver membranes were treated without any additives *(top panel)*, with GTPγS and 0.4 mM Mg^{2+} in the absence and presence of hormone *(second and third panels)*, and with GTPγS and 50 mM Mg^{2+} *(bottom panel)*. After removal of the activating ligands, the size of G_s was assessed by sucrose density gradient ultracentrifugation. Gradient fractions were assayed without (O) or with (●) GTPγS to assess the extent of persistent activation of G_s during the membrane treatment. The positions of the markers used, malate dehydrogenase (MDH, 4.3S) and cytochrome c (Cyt c, 1.71S), is indicated with arrows. It can be readily seen that upon activation, the cyc^--reconstituting activity shifts to a lower S-value (4S to 2S), indicating subunit dissociation. (From Iyengar *et al.*, 1988.)

III. CLONING AND CHARACTERIZATION OF PRODUCTS OF α_s GENE

The recent and continued cloning of G-protein subunits has greatly advanced our understanding of the structure of the G proteins as a family, and will facilitate the elucidation in coming years of how G proteins function. Both the α and β subunits of G_s have been cloned and characterized. Only the α_s subunit will be discussed here, as the common $\beta\gamma$ subunits are discussed by Neer and Clapham in Chapter 3 of this volume. The relationship of structure to function of the α_s subunit, as modeled by comparison to other G proteins and GTP-binding proteins, is discussed by Heideman and Bourne in Chapter 2 of this volume.

A. Cloning of α_s cDNAs

Harris *et al.* (1985) were the first to report the isolation of a partial cDNA clone for the α subunit of G_s. The clone was identified by screening a bovine brain cDNA library with a degenerate mixture of synthetic oligonucleotides whose sequences were derived from the known partial amino acid sequences common to purified α subunits of G_t and G_o (Hurley *et al.*, 1984). The authenticity of the clone was verified by preparing an anti-peptide antibody to a region of the deduced amino acid sequence, which reacted with purified G_s but not G_t, G_o, or G_i, and by demonstrating the specific lack of mRNA hybridizing to this clone in the cyc^-, but not the wild-type, S49 lymphoma cell line, which lacks the G_s function.

Screening a bovine adrenal cDNA library with the 5' end of this partial clone, Robishaw *et al.* (1986a) obtained a full-length α_s cDNA. The deduced amino acid sequence indicated that this form of α_s was a 394-amino acid protein with a calculated molecular weight of 45,000. Transient expression of this full-length cDNA in COS cells led to the expression of the 52K form of α_s, as measured by immunoblotting of membranes from transfected cells. Nukada *et al.* (1986), Itoh *et al.* (1986), Jones and Reed (1987), and Harris (1988) also reported the cloning of 394-amino acid α_s cDNAs from bovine cerebral cortex, rat C_6 glioma cell, rat olfactory epithelial, and human fetal adrenal cDNA libraries, respectively.

A second form of α_s cDNA was cloned by Robishaw *et al.* (1986b) after rescreening a bovine adrenal cDNA library with a 5' fragment of their initial full-length clone. The two coding regions were identical except for a deletion of 42 nucleotides in the second cDNA leading to a loss of 14 amino acids, with two flanking amino acid changes. In contrast to the previous 394-amino acid, 52K form of α_s, the new cDNA encoded a 380-amino acid protein with a predicted molecular weight of 44K. This smaller form of α_s cDNA directed the expression of the 43K form of α_s in transiently transfected COS cells. S_1 nuclease-protection analysis of bovine adrenal RNA indicated the presence of two distinct mRNA species in this tissue. Mattera *et al.* (1986) isolated the equivalent long and short clones from a human liver cDNA library. Both groups suggested alternative splicing of a single gene transcript as a mechanism for generating α_s diversity.

Sullivan *et al.* (1986) isolated yet a third α_s cDNA upon screening a cDNA library prepared from RNA from the PU-5 murine macrophage cell line. This cDNA encoded a protein similar to the shorter form of α_s

described above, but lacking a single serine residue at the point where a 14-amino acid insertion distinguished the long and short proteins. Soon thereafter, Bray *et al.* (1986) isolated cDNA clones encoding four distinct forms of α_s from a human brain cDNA library. These four differed in the presence or absence of a 15-amino acid block with or without an additional serine residue. Alternative splicing of a single exon (encoding the 45-nucleotide insertion) and the use of a secondary splice-acceptor site shifted by three nucleotides (to encode the additional serine) was offered as an explanation for the generation of these four mRNAs.

The four α_s protein sequences have been designated α_{s1} through α_{s4} (Bray *et al.*, 1986). The initially described 394-amino acid, 52-K form, which has the 15-amino acid insertion following amino acid 71 but lacks the additional serine, is α_{s1}. The 395-amino acid form, with the additional serine, is α_{s2}. The 379-amino acid form, lacking both the insertion and the serine, is α_{s3}. The 380-amino acid, 43K form, with only the serine insertion, is α_{s4}.

Cloning of the complete α_s gene from a human genomic library by Kozasa *et al.* (1988) has established that the four forms of α_s mRNA arise from alternative splicing of a single gene transcript. The gene does not appear to have the potential to generate any further alternatively spliced products (see Kaziro *et al.*, Chapter 4 in this volume), so that all forms of α_s now appear to have been described. The gene for α_s has been localized to human chromosome 20 using hybridization of the human cDNA to a panel of human–mouse somatic cell hybrids (Blatt *et al.*, 1988).

B. Expression of α_s

The regulation of α_s expression may be controlled on several levels. The α_s gene appears to be a housekeeping gene, and possibly lacks acute regulation. Chronic regulation has been demonstrated by reports that glucocorticoids increase the levels of α_s mRNA in GH$_3$ cells (Chang and Bourne, 1987) and that chronic ethanol decreases the α_s mRNA in NG108-15 cells (Mochly-Rosen *et al.*, 1988). Clearly, there is regulation of alternative RNA splicing to produce tissue-specific distributions of the four α_s forms, although this has not yet been studied. While the factors controlling the relative expression of these various spliced mRNAs (and their respective protein products) remain to be defined, some generalizations concerning their relative expression can be made. Considering the number of reports of clones isolated for particular forms of α_s, and that these various clones were obtained from cDNA libraries from diverse tissues, α_{s1} appears to be the most highly expressed form of α_s. Five groups have reported cloning α_{s1}, two groups each α_{s3} and α_{s4}, and one group α_{s2}. Indeed, of five α_s clones detected in a bovine adrenal cDNA library, four represented α_{s1} (Robishaw *et al.*, 1986b).

Functional expression of the α_{s1} cDNA in cyc^- S49 lymphoma cells was reported by Nukada *et al.* (1987). Cells were transformed to electroporation, and transformants were selected for G418 (neomycin) resistance. α_{s1} expression was driven using the SV40 early promoter, and the 3' untranslated region was extended to distinguish this transcript from any endogenous α_s mRNA. G418-resistant clones regained the ability of β-adrenergic ligands to inhibit cell growth and induce cytolysis. Northern blot analysis of RNA from transformed cells indicated expression of the extended α_s mRNA, and membranes from transformed cells regained the 52K cholera toxin substrate. Analysis of cAMP accumulation in the presence of isoproterenol directly indicated that transformants had regained functional G_s activity.

Since the α_s gene occurs in only a single copy in the human haploid genome, all of these α_s proteins are derived from this single gene (Kozasa *et al.*, 1988). Within a species, the four alternative products of this gene transcript are therefore identical in nucleotide and amino acid sequence, except for the variable region following amino acid 70. A relevant question is whether the four different forms of G_s exhibit different functional properties. Graziano *et al.* (1987, 1989) have begun to address this question through the use of bacterially expressed recombinant α_s subunits.

Two forms of α_s have been expressed in *Escherichia coli* by Graziano *et al.* (1987). α_{s1} and α_{s4} cDNAs were placed under the control of the phage T7 promoter in a plasmid carrying an ampicillin resistance marker. A heat shock-inducible T7 RNA polymerase was used to initiate α_s accumulation to 0.1% of the total bacterial protein. α_s was partially purified from induced bacteria using DEAE-Sephacel and Mono Q FPLC. The α_{s1} clone directed the expression of a 52K protein and the α_{s4} clone a 43K protein as detected by α_s-specific immunoblotting and by cholera-toxin labeling. Both recombinant α_{s1} and α_{s4} were capable of stimulating the cyc^- S49 cell adenylyl cyclase, although both proteins appeared only 1% as active as purified liver G_s. Addition of purified $\beta\gamma$ increased this stimulatory activity significantly. The reduced activity of the bacterially produced α_s was attributed to the lack of $\beta\gamma$ with which to associate as the α_s is folded. Mattera *et al.* (1989) have similarly expressed recombinant α_{s1} and α_{s4} as fusion proteins with nine amino acid extensions at the amino terminus. These $r\alpha_s$ subunits were also only 1–2% as active as native purified G_s in stimulating cyc^- S49 cell adenylyl cyclase.

Graziano *et al.* (1989) have purified their bacterially expressed α_{s1} and α_{s4} to homogeneity from *E. coli* lysates. The purified α_s subunits bind GTPγS stoichiometrically, contain intrinsic GTPase activity which can be inhibited by addition of purified $\beta\gamma$ complex, and can reconstitute cyc^- S49 cell adenylyl cyclase stimulation by isoproterenol, GTPγS, and NaF. Taken together, this indicates that the bacterially expressed α_s subunits are able to functionally interact with β-adrenergic receptors, $\beta\gamma$ complex, and adenylyl cyclase (although with reduced potency). Interestingly, the bacterially expressed α_{s1} and α_{s4} exhibit similar k_{cat} values for the hydrolysis of GTP, but differ in their rates of GDP dissociation. This difference results in the more rapid activation of α_{s1} than α_{s4} by GTPγS and is reflected by more rapid stimulation of adenylyl cyclase by α_{s1}.

Olate *et al.* (1988) have prepared α_{s1} and α_{s4} *in vitro* using T7 polymerase to make mRNA, and reticulocyte lysates to express the protein. Both α_{s1} and α_{s4} cDNAs were able to express protein (of lower than expected sizes) and reconstitute *cyc*⁻ S49 cell adenylyl cyclase activity. α_{s1} was shown to activate adenylyl cyclase activity in the presence of GTPγS with a lag which was abolished by addition of isoproterenol, indicating that the expressed protein was capable of properly coupling to the S49 cell β-adrenergic receptor as well as adenylyl cyclase. *In vitro*-expressed α_{s1} was estimated to be 0.75 times as active as purified human erythrocyte G$_s$ in stimulating the *cyc*⁻ S49 cell adenylyl cyclase, indicating that α_s can be synthesized in an essentially native state in the absence of $\beta\gamma$ and of the machinery of co- and posttranslational modification of the protein.

C. Generation of Antipeptide Antisera to G$_s$

One of the practical benefits arising thus far from the molecular cloning of α_s cDNAs has been the use of the deduced amino acid sequences to prepare synthetic peptides corresponding to defined regions of the α_s protein for use as antigens in the generation of α_s-specific antibodies in rabbits. The high degree of similarity among G proteins makes the prospect of producing antisera specific for only one G protein using purified protein as the immunogen seem remote, although such an antiserum has been produced for G$_t$ (Pines *et al.*, 1985). Previous attempts to use purified protein as an antigen indicated that G$_s$ was not very immunogenic, and such attempts were basically unsuccessful.

Harris *et al.* (1985) prepared antiserum to a synthetic peptide with a sequence found in their initial partial cDNA clone which differed from the known sequence for α_t. Using this antiserum, they determined in immunoblots that it reacted with purified α_s, but not α_t, α_o, or α_i. Additionally, this antiserum reacted with α_s in membranes from S49 lymphoma cells, but detected no such protein in the *cyc*⁻ variant of S49 cells, which lacks G$_s$ function. Mumby *et al.* (1986) used the same approach to design antisera specific for particular G-protein subunits. They prepared and characterized antisera specific for α_s, α_o, and β_{36}. In addition, noting the presence of a region of strict conservation among known α-subunit sequences, they prepared an antiserum crossreacting with all α subunits.

Ransnas and Insel (1988) have prepared antipeptide antisera specific for α_s, and developed an ELISA assay allowing direct quantitation of α_s.

Membranes from S49 murine lymphoma cells were found to contain 18.9 ± 2.3 pmol α_s per mg protein. Such a direct measure of protein concentration has advantages over the previous 'functional' methods for determining G_s concentrations, cholera toxin labeling with [^{32}P]NAD$^+$ or reconstitution of cyc^- S49 cell adenylyl cyclase activity. α_s-specific antisera have several possible uses: (1) The levels of G_s can be determined by ELISA or immunoblotting independent of function. This approach has been used by Mochly-Rosen *et al.* (1988) to measure changes in the levels of G_s during chronic ethanol treatment of NG108 cells. Another use could be to calibrate the levels of $r\alpha_s$ in stably reconstituted cyc^- S49 cell membranes in order to accurately assess the true specific activity of bacterially ex- pressed $r\alpha_s$ in stimulating adenylyl cyclase. (2) The distribution of the α_s proteins can be determined immunohistochemically in various tissues, such as brain. (3) Rapid purification of α_s by immunoprecipitation has been employed by Buss *et al.* (1987) to demonstrate the apparent lack of palmitylation or myristylation of α_s from ^{3}H-labeled 1321N1 human as- trocytoma cells. The possibility of other covalent modifications of the α_s proteins can be similarly tested.

IV. ADENYLYL CYCLASE: MOLECULAR CHARACTERIZATION

Adenylyl cyclase itself remains the least characterized component of the hormone-stimulated adenylyl cyclase pathway. Despite successful solubili- zation from membranes of mammalian tissues, early attempts to purify this catalytic activity could produce resolved enzyme, but were generally unsuccessful at extensive purification due to both the lability of the protein and its low abundance in the membrane. Success in the purification and reconstitution of the other components of the hormone-stimulated ade- nylyl cyclase pathway has renewed interest in the purification and molecu- lar characterization of adenylyl cyclase. Affinity chromatography tech- niques for the purification of adenylyl cyclase have been developed, and the enzyme has recently been purified to apparent homogeneity.

It has become clear that there exist distinct forms of adenylyl cyclase activity in mammalian tissues. Most, if not all, tissues contain the hor- mone-stimulated plasma membrane adenylyl cyclase. This adenylyl cy- clase is activated by G_s as described above. Adenylyl cyclase in brain (and a few other tissues, particularly heart, pancreatic islets, and adrenal medulla) possesses the ability to be activated by Ca^{2+}-occupied calmodulin (Bros- trom *et al.*, 1975; Cheung *et al.*, 1975). Early studies of brain adenylyl cyclase activity indicated that it was composed of a mixture of calmodulin-sensitive and -insensitive forms (Brostrom *et al.*, 1977). The regulatory and structural similarity between these two adenylyl cyclase forms remains to be defined, but much progress has been made.

Genetic evidence for distinct forms of adenylyl cyclase has been obtained from studies of behavioral mutants of the fly, *Drosophila melanogaster*. Thus *rutabaga* is an X-linked recessive mutation which appears to reduce the ability to solidify associative memory (Tempel *et al.*, 1983). Biochemical analysis of *rutabaga* mutants revealed that, in contrast to wild-type flies, *rutabaga* flies lacked the ability of Ca^{2+}–calmodulin to stimulate adenylyl cyclase (Livingstone *et al.*, 1984; Livingstone, 1985). However, adenylyl cyclase from *rutabaga* flies retained normal stimulation by neurotransmitters, guanine nucleotides, and forskolin. The *rutabaga* phenotype therefore appears to be due to the loss of a calmodulin-stimulated adenylyl cyclase catalytic subunit distinct from adenylyl cyclase stimulated by G_s.

A. Purification of Adenylyl Cyclase Using Forskolin Affinity Matrix

Recognition of the ability of the diterpine forskolin to stimulate the activity of adenylyl cyclase (Seamon *et al.*, 1981), even in the absence of G_s in cyc^- S49 cell membranes (Seamon and Daly, 1981), made it the first known pharmacological agent to interact directly with the catalytic subunit of adenylyl cyclase. Following this lead, Pfeuffer and Metzger (1982) prepared 7-*O*-hemisuccinimidyl-7-deacetylforskolin, which was coupled to aminoethyl-Sepharose as an adenylyl cyclase affinity matrix. Using this affinity support, specific adsorption and elution with forskolin of Lubrol-solubilized rabbit myocardial adenylyl cyclase was demonstrated. This eluted activity retained stimulation by Mn^{2+} and forskolin, but appeared to have been resolved from G_s. Solubilized rat brain membrane adenylyl cyclase eluted from the deacetylforskolin matrix also lost the ability to be stimulated by added Ca^{2+} by having been separated from calmodulin (Pfeuffer and Metzger, 1982).

Myocardial membranes were treated with isoproterenol and Gpp(NH)p to persistently activate G_s prior to solubilization, and adenylyl cyclase eluted from the forskolin affinity matrix remained in a highly active state indicating the copurification of G_s (Pfeuffer and Metzger, 1982). Sedimentation in sucrose gradients of adenylyl cyclase activity eluted from the forskolin matrix demonstrated that the Gpp(NH)p-activated form of adenylyl cyclase appeared larger (8.7S versus 7.0S) than the unactivated form, and that [^{35}S]GTPγS-bound α_s copurified with this larger, activated form of the enzyme (Pfeuffer *et al.,* 1983). Further, unactivated adenylyl cyclase while bound to the forskolin matrix was able to specifically retain purified [^{3}H]G$_s$ added to the column, and was shown to have restored guanine-nucleotide stimulation to the eluted adenylate cyclase–G$_s$ complex (Pfeuffer *et al.,* 1983).

Using forskolin–Sepharose, Gpp(NH)p-activated adenylyl cyclase was purified 2000-fold from Lubrol extracts of rabbit myocardial membranes (Pfeuffer *et al.,* 1985a). In combination with HPLC gel filtration, a 60,000-fold purification was achieved over the starting membranes, achieving 16 μmol cAMP/min/mg protein. Purified preparations consisted of a broad band of material migrating on SDS-PAGE near 150K, as well as a band at 43K presumed to be α_s. The 150K band was identified as adenylyl cyclase by measuring a gel shift following crosslinking of reconstituted adenylyl cyclase and [^{32}P]ADP-ribosylated α_s with disuccinimidyl suberate.

Unactivated adenylyl cyclase was purified 10,000-fold from bovine cerebral cortical membranes by Pfeuffer *et al.* (1985b), utilizing two cycles of passage over forskolin–Affi-Gel 10 to obtain purification to 7 μmol cAMP/min/mg protein in the presence of forskolin. This preparation was stimulated 2-fold by calmodulin. They demonstrated binding of purified adenylyl cyclase to wheat germ lectin (WGL)–Sepharose and its elution by *N*-acetylglucosamine. A single protein band of 115K was evident in purified preparations. Photoactivated cross-linking of 4-azidobenzoyl [^{32}P]ADP-ribosyl-α_s (the product of reacting *N*-hydroxysuccinimidyl 4-azidobenzoate with purified, cholera toxin-catalyzed [^{32}P]ADP-ribosylated G$_s$) with purified adenylyl cyclase led to the appearance of a 160K radiolabeled band, suggesting a size near 115K for the catalytic subunit of adenylyl cyclase (Pfeuffer *et al.,* 1985b).

Monoclonal antibodies raised against forskolin-affinity-purified unactivated adenylyl cyclase from bovine brain were able to immunoprecipitate adenylyl cyclase activity from partially purified preparations (Mollner and Pfeuffer, 1988). When ^{125}I-iodinated crude bovine brain adenylyl cyclase preparations were immunoprecipitated, SDS-PAGE and autoradiography of proteins eluted from the immune complex revealed the specific retention of a 115K protein.

Smigel (1986) used forskolin–Affi-Gel 102 followed by WGL–Sepharose chromatography to purify unactivated adenylyl cyclase 15,000-fold from bovine brain membranes, to 10 μmol cAMP/min/mg in the presence of forskolin. This preparation was stimulated twofold by calmodulin and fourfold by G$_s$. Purified preparations consisted of a 120K band on SDS-PAGE, and it was reported that this band could be covalently labeled by radiolabeled 2′,3′-dialdehyde of ATP, demonstrating that it is an ATP-binding protein as would be expected for adenylyl cyclase.

B. Purification of Adenylyl Cyclase Using Calmodulin Affinity Matrix

The ability of calmodulin to stimulate brain adenylyl cyclase activity has also been used as a tool for the purification of adenylyl cyclase. Wescott *et al.* (1979) demonstrated the Ca^{2+}-dependent binding of adenylyl cyclase solubilized from bovine brain to calmodulin–Sepharose, and eluted adenylyl cyclase activity by removing free Ca^{2+}. They also noted the resolution of a calmodulin-insensitive adenylyl cyclase activity in the flow-through of their calmodulin affinity column. Adenylyl cyclase preparations partially purified from bovine brain using this calmodulin affinity matrix were labeled by crosslinking to azido[^{125}I]iodocalmodulin to tentatively identify the calmodulin-stimulated adenylyl cyclase as a protein with a size near 150K (Andreason *et al.*, 1983). Adenylyl cyclase purified 700-fold from bovine brain using calmodulin–Sepharose and heptanediamine–Sepharose retained stimulation by guanine nucleotides and forskolin (to 2 μmol cAMP/min/mg protein), as well as calmodulin (Yeager *et al.*, 1985). ADP-ribose labeling of the purified fractions using [^{32}P]NAD$^+$ catalyzed by cholera and pertussis toxins demonstrated the presence of contaminating G-protein α subunits (Yeager *et al.*, 1985).

Minocherhomjee *et al.* (1987) purified the enzyme 3000-fold from Gpp(NH)p-activated membranes by sequential chromatography on both calmodulin and forskolin affinity columns. The purified protein, with a specific activity of 2 μmol cAMP/min/mg protein in the presence of Ca^{2+}–calmodulin, migrated as a single 135K band, and bound to both [^{125}I]iodocalmodulin and [^{125}I]iodoWGL in gel overlays following separation by SDS-PAGE. This adenylyl cyclase retained stimulation by Ca^{2+}–calmodulin following the forskolin affinity purification, indicating that both forskolin and calmodulin can bind to and stimulate the same molecular form of adenylyl cyclase.

Polyclonal antisera were raised against this purified Ca^{2+}–calmodulin stimulated adenylyl cyclase preparation from bovine brain (Rosenberg and Storm, 1987). Adenylyl cyclase activity in purified fractions could be immunoprecipitated by this antiserum, and immunoprecipitation of ^{125}I-iodinated purified adenylyl cyclase led to the recovery in the pellet of a 135K protein. While essentially all calmodulin-stimulated adenylyl cyclase activity eluted from the calmodulin affinity column could be immunoprecipitated, in the flow-through fraction containing calmodulin-insensitive activity there was no precipitation. This antiserum was also unable to immunoprecipitate adenylyl cyclase activity from tissues in which activity is insensitive to calmodulin.

Partial purification (500-fold) of adenylyl cyclase from rat brain synaptosomes using gel filtration, calmodulin affinity chromatography, and ion exchange FPLC was described by Coussen *et al.* (1985). Purified adenylyl cyclase had a specific activity of only 76 nmol cAMP/min/mg protein in the presence of Ca^{2+}–calmodulin. Adenylyl cyclase was identified as a 135K protein by its ability to bind [^{125}I]iodocalmodulin in an SDS-PAGE gel overlay assay.

Coussen *et al.* (1986) further purified rat synaptosomal adenylyl cyclase using gel filtration, forskolin affinity chromatography, and finally calmodulin affinity chromatography. They noted that in their initial gel filtration step, forskolin-stimulated adenylyl cyclase activity gave two apparent activity peaks rather than the single peak stimulated by Ca^{2+}–calmodulin. These two activity peaks remained distinct through rechromatography and sucrose density sedimentation. Purification of the calmodulin-stimulated adenylyl cyclase gel filtration peak from the forskolin matrix yielded a 155K major protein with a specific activity of 20 μmol cAMP/min/mg protein in the presence of forskolin. This activity could also be stimulated by calmodulin, and further purification of this fraction on calmodulin–Sepharose led to the reproducible elution of a 135K protein (Monneron *et al.*, 1987). After applying the non-calmodulin–sensitive adenylyl cyclase gel filtration peak to forskolin gel, fractions with 6 μmol cAMP/min/mg protein were eluted, which appeared devoid of both 135K and 155K proteins, but were enriched in a 105K protein (Monneron *et al.*, 1987). The relationships among these various putative forms of adenylyl cyclase remain to be determined.

Monneron *et al.* (1988) have prepared polyclonal antisera against both purified calmodulin-stimulated rat brain adenylyl cyclase and the calmodulin-stimulated adenylyl cyclase toxin of *Bordetella pertussis*. Rat brain adenylyl cyclase eluted from a forskolin affinity column was separated by SDS-PAGE and transferred to nitrocellulose. The 155K major band was cut from the paper and used as the immunogen. The resulting antisera reacted with several bands in immunoblots of purified rat brain adenylyl cyclase preparations, as well as crossreacting with the purified *Bordetella pertussis* adenylyl cyclase. The sera were unable to immunoprecipitate the *Bordetella pertussis* adenylyl cyclase, and their ability to immunoprecipitate rat brain adenylyl cyclase activity was not examined.

C. Purification of Adenylyl Cyclase from Liver

Our laboratory has undertaken the purification of adenylyl cyclase from porcine liver, utilizing the forskolin affinity- and WGL-affinity-purification procedures of Smigel (1986), with the addition of an intervening Mono Q FPLC ion-exchange step. In contrast to brain and myocardium, liver has only the calmodulin-insensitive form of adenylyl cyclase.

Liver membrane adenylyl cyclase was extracted with 1% Lubrol and passed over a forskolin-Affi-Gel 102 column. The column was washed and adenylyl cyclase activity eluted with 50 μM forskolin. The forskolin eluate was injected onto a Mono Q FPLC column and eluted with a 100–400 mM NaCl gradient. The peak of adenylyl cyclase activity was released at 300 mM NaCl. Eluted adenylyl cyclase was adsorbed onto WGL–Sepharose, and eluted with 300 mM N-acetylglucosamine. Liver adenylyl cyclase preparations are purified 10,000-fold from the starting membrane extract, and consist of protein bands at 130 and 90K.

Purified adenylyl cyclase is not significantly stimulated by NaF or guanine nucleotides, indicating that little G$_s$ has copurified. Addition of purified human erythrocyte G$_s$ reconstitutes stimulation by GTPγS as shown in Fig. 5, with a half-maximal concentration of 25 pM G$_s$ as determined using Eadie-Hofstee plots of data as in Fig. 5. Purified liver adenylyl cyclase is not stimulated by 10 μM calmodulin.

We have identified the molecular form of adenylyl cyclase by crosslinking of [α-^{32}P]ATP to the purified protein by treatment with UV light. ATP is specifically linked to the 130K protein band, since a 20-fold excess of unlabeled ATP can abolish the labeling, while a 20-fold excess of unlabeled GTP does not (Fig. 6, *left panel*). We have used this [α-^{32}P]ATP-labeled protein to determine the size of the enzymatically deglycosylated protein

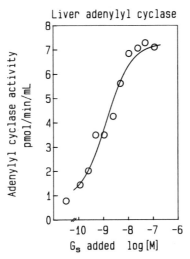

Fig. 5.   Stimulation of purified porcine liver adenylyl cyclase by purified human erythrocyte G$_s$. Adenylyl cyclase was extracted from purified porcine liver membranes in 1% Lubrol, and purified using forskolin–agarose, Mono Q FPLC, and WGL–Sepharose chromatography. G$_s$ was diluted and mixed with an aliquot of purified adenylyl cyclase for 10 min on ice in the absence of other assay components. The G$_s$–cyclase mixture was then added to the assay substrate mixture and incubated at 32°C for 20 min. Activity was stimulated by 100 μM GTPγS with the indicated concentrations of G$_s$ in the assay.

backbone of adenylyl cyclase. Crosslinked adenylyl cyclase was digested for 16 hours with glycopeptidase F to reveal that the 130K protein is reduced to 110K (Fig. 6, *right panel*).

D. Characteristics of Purified Adenylyl Cyclases

Forskolin is known to stabilize the interaction of activated G_s with adenylyl cyclase (Seamon, 1987; Bouhelal *et al.,* 1985). Based on this, purification of adenylyl cyclase using forskolin affinity has been performed in two distinct manners. The basal or unactivated adenylyl cyclase has been purified from brain and liver, and appears essentially free of G_s (Pfeuffer *et al.,* 1985b; Smigel, 1986; Fig. 5). However, preactivation of G_s with non-hydrolyzable GTP analogs allows the copurification of the stabilized G_s–adenylyl cyclase complex, as has been purified from myocardium (Pfeuffer *et al.,* 1985a). Liver adenylyl cyclase preactivated with Gpp(NH)p is able to bind much more extensively to the forskolin–affinity matrix than is the unactivated adenylyl cyclase, indicating that the G_s–adenylyl cyclase com-

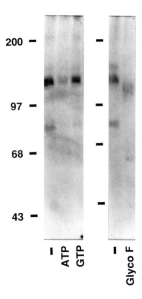

Fig. 6. Crosslinking of [α-^{32}P]ATP to adenylyl cyclase. Partially purified porcine liver adenylyl cyclase (Mono Q eluate) was incubated with 2 μM [α-^{32}P]ATP (50 Ci/mmol) in the presence of 25 mM HEPES (pH 7.5), 1 mM EDTA, 5 mM Mn^{2+} and 50 μM forskolin for 2 min at room temperature. The reaction was then moved under a UV lamp (15 cm) on ice for 10 min. The proteins were precipitated with 10% TCA, washed with diethyl ether, and separated by SDS-PAGE on a 7.5% gel. Left panel: assays contained no addition (−), 20 μM ATP (ATP), or 20 μM GTP (GTP). Right panel: assays without competitor were performed as described. The ether-washed pellet was resuspended in 25 mM HEPES (pH 6.5), 1 mM EDTA, and incubated overnight at 32°C wih no addition (−), or with 100 mU of glycopeptidase F (Glyco F). Molecular weight markers ($\times10^{-3}$).

plex has tighter binding to forskolin than the catalyst alone (unpublished observations).

Both the calmodulin-sensitive and -insensitive types of adenylyl cyclase activity from brain bind directly to forskolin affinity matrix, indicating that both catalytic units interact directly with this stimulator. Calmodulin-affinity-purified adenylyl cyclase can be stimulated by forskolin, and the brain forskolin-affinity-purified adenylyl cyclase can be stimulated by calmodulin, indicating that a single molecular form of adenylyl cyclase can be regulated by both molecules. Smigel (1986) has shown that exogenous G_s can stimulate forskolin affinity-purified calmodulin-sensitive adenylyl cyclase from bovine brain; however, it is not yet clear whether a single molecular species responds to both G_s and calmodulin. Both purified brain (Smigel, 1986) and liver (Fig. 5) adenylyl cyclase exhibit surprisingly similar dose-dependent stimulation by added G_s. The activity of purified brain adenylyl cyclase was not significantly affected by either GTPγS-activated G_o or G_i, but was slightly inhibited by βγ complex (Smigel, 1986). Purified brain adenylyl cyclase reconstituted into phospholipid vesicles with both G_s and β-adrenergic receptors was shown to exhibit catecholamine stimulation of adenylyl cyclase activity, indicating that these three components are sufficient for hormonal stimulation in this signal transduction system (May *et al.*, 1985). Hormonal inhibition of adenylyl cyclase is addressed by Codina *et al.* in Chapter 12 of this volume.

The calmodulin-stimulated adenylyl cyclase in brain and the calmodulin-insensitive adenylyl cyclase of liver appear to be proteins migrating near 120–130K in SDS-PAGE. Some degree of variance among groups purifying the protein can be explained by the use of differing SDS-PAGE gel systems (gradient versus slab gels). Further differences may be due to misidentification of purified proteins as adenylyl cyclase, since both major purification schemes rely on affinity chromatography using forskolin or calmodulin. Calmodulin is known to interact with a wide variety of cellular components; however, both groups utilizing calmodulin-affinity purification have relied at least in part on interactions of their purified preparations with [125I]iodocalmodulin to identify the molecular species carrying adenylyl cyclase activity (Yeager *et al.*, 1985; Coussen *et al.*, 1985; Minocherhomjee *et al.*, 1987). Similarly, in addition to stimulation of adenylyl cyclase, forskolin is known to directly inhibit the glucose transporter (Sergeant and Kim, 1985; Joost and Steinfelder, 1987). Recently, non-cAMP-mediated effects of forskolin have been reported for desensitization of the muscle nicotinic acetylcholine receptor (Wagoner and Pallotta, 1988; White, 1988), as well as inhibition of PC-12 cell K_z^+ channels (Hoshi *et al.*, 1988). Thus, many cellular components may bind to forskolin specifically and possibly copurify with adenylyl cyclase. This caveat also makes suspect the common association of [3H]forskolin binding sites with adenylyl cyclase (Gehlert, 1986; Worley *et al.*, 1986).

The binding of adenylyl cyclase activity to WGL–Sepharose and its elution by *N*-acetylglucosamine indicates that the enzyme is a glycoprotein, and as such presumably spans the plasma membrane with the sugar chains oriented extracellularly. The liver adenylyl cyclase has been enzymatically deglycosylated, and appears to contain N-linked sugars on a protein backbone of 110K (Fig. 6, *right panel*).

V. DESENSITIZATION OF ADENYLYL CYCLASE

In addition to instantaneous regulation by hormones through the activity state of G_s, the adenylyl cyclase system displays time-dependent modulation of its responsiveness. Prolonged stimulation of adenylyl cyclase leads to desensitization, the waning of the ability of the system to be further stimulated. Desensitization responses have been functionally classified as homologous or heterologous. Homologous desensitization is loss of response to only the inducing hormone, and has been shown to be due to regulation at the level of the hormone receptor (Sibley and Lefkowitz, 1985). Heterologous desensitization is the loss of responsiveness to other hormones as well, and often to nonhormonal stimulators such as NaF. Although cAMP-dependent phosphorylation of hormone receptors has been demonstrated as a cause of heterologous desensitization (Sibley *et al.,* 1984), direct regulation of adenylyl cyclase has only recently been implicated in heterologous desensitization. Only this latter aspect will be discussed here.

A. Direct Modulation of Adenylyl Cyclase

In addition to regulation of receptors and G_s, adenylyl cyclase itself has recently been implicated as an important locus in the regulation of adenylyl cyclase responsiveness. We had previously found that treatment of chick hepatocytes with glucagon induces a rapid and rapidly reversible heterologous desensitization, while treatment with 8-Br-cAMP is capable of inducing only partial heterologous desensitization (Premont and Iyengar, 1988). Further characterization of this heterologous desensitization revealed the involvement of cAMP-dependent modulation of adenylyl cyclase (unpublished observations).

Addition of purified G_s to untreated and glucagon-treated cell membranes increased the NaF-stimulated adenylyl cyclase activity, but the activity of the glucagon-treated cell membranes remained significantly lower than that of the untreated membranes. This inability of even an enormous excess of exogenous G_s to overcome the heterologous desensitization indicates that adenylyl cyclase in the desensitized membranes is unable to express stimulation by G_s to the same extent as that in untreated cell membranes. Addition of G_s to 8-Br-cAMP-treated hepatocyte membranes was unable to abolish the difference between the treated and untreated cell membrane adenylyl cyclase activities, indicating that this alteration in adenylyl cyclase is cAMP-dependent. Since this heterologous desensitization does not appear to reduce forskolin-stimulated adenylyl cyclase activity, sequestration of adenylyl cyclase in treated hepatocytes does not appear to be responsible. The rapid timecourse of induction of and recovery from heterologous desensitization (Premont and Iyengar, 1988) favors the possibility of reversible modification of the enzyme, such as by phosphorylation.

Phosphorylation of adenylyl cyclase has been demonstrated in frog erythrocytes in response to treatment with tetradecanoyl phorbol acetate by Yoshimasa *et al.* (1987). In many cells and tissues, phorbol esters have been observed to markedly increase (Bell *et al.*, 1985; Johnson *et al.*, 1986) or decrease (Mallorga *et al.*, 1980; Heyworth *et al.*, 1984; Rebois and Patel, 1985) adenylyl cyclase activity in a heterologous manner, presumably due to action of protein kinase C. In frog erythrocytes, phorbol esters increase adenylyl cyclase activity (Sibley *et al.*, 1986). Yoshimasa *et al.* (1987) prelabeled frog erythrocytes with ortho[^{32}P]phosphate, incubated the cells with phorbol ester, and purified the Gpp(NH)p-activated adenylyl cyclase from membranes using the forskolin affinity method of Smigel (1986). SDS-PAGE and autoradiography of the forskolin column eluate demonstrated phorbol ester-induced phosphorylation of a 130K protein presumed to be adenylyl cyclase on the basis of specific elution by forskolin, but not analogs of forskolin unable to activate adenylyl cyclase. Yoshimasa *et al.* (1987) also purified adenylyl cyclase activity from bovine cerebellum by forskolin affinity chromatography, and demonstrated the *in vitro* phosphorylation of the major 150K protein by purified preparations of protein kinase C.

The regulation of adenylyl cyclase by direct modification is a new, longer-term level of control which must now be considered to be operating on top of the instantaneous regulation by G proteins. Modulation of the activity state of adenylyl cyclase by protein phosphorylation, either by protein kinase C or by the cAMP-dependent protein kinase, is an appealing mechanism for the cell to use to control its level of responsiveness, and is an area of active investigation in our laboratory.

VI. EDITOR'S COMMENTS (BY RAVI IYENGAR)

A. G_s Action

Studies by Johnson and co-workers (Woon *et al.*, 1989) have shown that two site-specific mutations in the α subunit of G_s result in persistently activated forms of G_s. Both mutations are analogous to those occurring in oncogenic forms of the *ras* proteins. The mutations that have been made in α_s by *in vitro* mutagenesis are $Gly^{49} \rightarrow Val$ and $Gly^{225} \rightarrow Thr$. Both mutant forms of α_s, when expressed following transfection COS-1 cells, persistently activate adenylyl cyclase. Stable expression in CHO cells results in constitutive elevation of intracellular cAMP levels and increased cAMP-dependent kinase activity.

In pituitary adenocarcinomas, it has been shown that an apparently altered form of G_s raises cAMP levels and causes a high rate of growth hormone secretion (Vallar *et al.*, 1987). Since growth in pituitary somatotrophs is regulated by cAMP, a persistently activated G_s could function as an oncogene (Bourne, 1987). Studies with mutant forms of G_s may prove to be useful in understanding the pathophysiology of G_s-regulated processes.

An olfactory neuron-specific isoform of G_s has recently been identified by cDNA cloning, and named G_{olf} (see Editor's Comments to Chapter 17 for further details).

B. Structure of Adenylyl Cyclase

A cDNA encoding an adenylyl cyclase from bovine brain has recently been cloned (Krupinski *et al.*, 1989). A bovine brain cDNA library was screened using oligonucleotides derived from partial amino acid sequences of bovine brain adenylyl cyclase purified by the procedure of Smigel (1986). The cDNA predicts an 1134 amino acid protein with an internally repeated structure and multiple transmembrane spans. Each half contains an initial hydrophobic region which may contain six transmembrane spans, followed by a large hydrophilic domain predicted to be intracellular. The large intracellular domains may be involved in nucleotide binding and catalysis, since they both contain similarities to yeast adenylyl cyclase and several guanylyl cyclases.

Transfection of COS cells with an expression vector containing the full length cDNA indicates that it does encode an adenylyl cyclase which can be stimulated by forskolin, although the stimulation was very low. It also appears likely that this adenylyl cyclase will be regulated by calmodulin, since a homologous genomic clone isolated by cross hybridization from a *Drosophila* genomic library was mapped by *in situ* hybridization to the region of the X chromosome which contains the *rutabaga* locus. Stimulation of brain adenylyl cyclase by calmodulin is blocked by synthetic peptides corresponding to the calmodulin binding sequence of myosin light chain kinase, indicating that the calmodulin binding site on adenylyl cyclase may be similar (Blumenthal *et al.*, 1989). However, no such related sequence was found in the cloned adenylyl cyclase, so the site of calmodulin interaction remains to be defined. The mammalian adenylyl cyclase

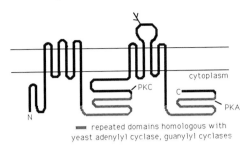

Fig. 7.

sequence contains a single consensus protein phosphorylation site for protein kinase C and another for protein kinase A. Currently published data
do not allow for any conclusions on other regulatory features of this cloned
adenylyl cyclase.

Since many tissues contain G_s-regulated but calmodulin-insensitive
adenylyl cyclase, there is at least one more, and in all likelihood, several
additional forms of adenylyl cyclase. It remains to be seen whether most
mammalian adenylyl cyclases are regulated by G_s, or whether like the testis
adenylyl cyclase there are G_s-insensitive adenylyl cyclases in other tissues
as well. Figure 7 is adapted from Krupinski *et al.* (1989), and depicts some
of the features of mammalian brain adenylyl cyclase.

REFERENCES

Blumenthal, D. K., Charbonneau, H., Edelman, A. M., Hinds, T. R., Rosenberg,
    G. B., Storm, D. R., Vincenzi, F. F., Beavo, J. A., and Krebs, E. G. (1988).
    Synthetic peptides based on the calmodulin-binding domain of myosin light
    chain kinase inhibit activation of other calmodulin-dependent enzymes. *Bio-
    chem. Biophys. Res. Comm.* **156**, 860–865.
Bourne, H. R. (1987). Discovery of a new oncogene in pituitary tumours? *Nature*
    330, 517–518.
Krupinski, J., Coussen, F., Bakalyar, H. A., Tang, W. J., Feinstein, P. G., Orth, K.,
    Slaughter, C., Reed, R. R., and Gilman, A. G. (1989). Adenylyl cyclase amino
    acid sequence: Possible channel or transporter like structure. *Science* **244**,
    1558–1564.
Vallar, L., Spada, A., and Giannattasio, G. (1987). Altered G_s and adenylate
    cyclase activity in human GH-secreting pituitary adenomas. *Nature* **330**, 566–
    568.
Woon, C. W., Heasley, L., Osawa, S., and Johnson, G. L. (1989). Mutation of
    glycine-49 to valine in the alpha-subunit of G_s results in the constitutive
    elevation of cAMP synthesis. *Biochemistry* **28**, 4547–4551.

ACKNOWLEDGMENTS

Supported by NIH grants CA 44948 and DK 38761. RI was an Established Investigator of the American Heart Association during the course of the work presented here. RTP is the recipient of a National Science Foundation Predoctoral Fellowship.

REFERENCES

Andreason, T. J., Heideman, W., Rosenberg, G. B., and Storm, D. R. (1983). Photoaffinity labeling of brain adenylate cyclase preparations with azido[^{125}I] iodocalmodulin. *Biochemistry* **22**, 2757–2562.

Asano, T., and Ross, E. M. (1984). Catecholamine-stimulated guanosine 5'-O-(3-thiotriphosphate) binding to the stimulatory GTP-binding protein of adenylate cyclase: kinetic analysis in reconstituted phospholipid vesicles. *Biochemistry* **23**, 5467–5471.

Asano, T., Pedersen, S. E., Scott, C. W., and Ross, E. M. (1984). Reconstitution of catecholamine-stimulated guanosine 5'-O-(3-thiotriphosphate) to the stimulatory GTP-binding protein of adenylate cyclase. *Biochemistry* **23**, 5460–5467.

Bell, J. D., Buxton, I. L. O., and Brunton, L. L. (1985). Enhancement of adenylate cyclase activity in S49 lymphoma cells by phorbol esters. Putative effect of C kinase on alpha$_s$-GTP-catalytic subunit interaction. *J. Biol. Chem.* **260**, 2625–2628.

Birnbaumer, L., Swartz, T.L., Abramowitz, J., Mintz, P. W., and Iyengar, R. (1980). Transient and steady state kinetics of the interaction of guanyl nucleotides with the adenylyl cyclase system from rat liver plasma membranes. Interpretation in terms of a simple two-state model. *J. Biol. Chem.* **255**, 3542–3551.

Blatt, C., Eversole-Cire, P., Cohn, V. H., Zollman, S., Fournier, R. E. K., Mohandas, L. T., Nesbitt, M., Lugo, T., Jones, D. T., Reed, R. R., Weiner, L. P., Sparkes, R. S., and Simon, M. I. (1988). Chromosomal localization of genes encoding guanine nucleotide-binding protein subunits in mouse and human. *Proc. Natl. Acad. Sci. U.S.A.* **85**, 7642–7646.

Bouhelal, R., Guillon, G., Homburger, V., and Bockaert, J. (1985). Forskolin-induced change in the size of adenylate cyclase. *J. Biol. Chem.* **260**, 10901–10904.

Brandt, D. R., and Ross, E. M. (1985). GTPase activity of the stimulatory GTP-binding regulatory protein of adenylate cyclase, G$_s$. *J. Biol. Chem.* **260**, 266–272.

Brandt, D. R., and Ross, E. M. (1986). Catecholamine-stimulated GTPase cycle. Multiple sites of regulation by beta-adrenergic receptor and Mg^{2+} studied in reconstituted phospholipid receptor-G$_s$ vesicles. *J. Biol. Chem.* **261**, 1656–1664.

Brandt, D. R., Asano, T., Pedersen, S. E., and Ross, E. M. (1983). Reconstitution of catecholamine-stimulated guanosine-triphosphatase activity. *Biochemistry* **22**, 4357–4362.

Bray, P., Carter, A., Simons, C., Guo, V., Puckett, C., Kamholz, J., Spiegel, A., and

Nirenberg, M. (1986). Human cDNA clones for four species of $G_{alpha-s}$ signal transduction protein. *Proc. Natl. Acad. Sci. U.S.A.* **83**, 8893–8897.

Brostrom, C. O., Huang, Y., Breckenridge, B. McL., and Wolff, D. J. (1975). Identification of a calcium-binding protein as a calcium-dependent regulator of brain adenylate cyclase. *Proc. Natl. Acad. Sci. U.S.A.* **72**, 64–68.

Brostrom, C. O., Brostrom, M. A., and Wolff, D. J. (1977). Calcium-dependent adenylate cyclase from rat cerebral cortex. Reversible activation by sodium fluoride. *J. Biol. Chem.* **252**, 5677–5685.

Buss, J. E., Mumby, S. M., Casey, P. J., Gilman, A. G., and Sefton, B. M. (1987). Myristoylated alpha subunits of guanine nucleotide-binding regulatory proteins. *Proc. Natl. Acad. Sci. U.S.A.* **84**, 7493–7497.

Cassel, D., and Selinger, Z. (1976). Catecholamine-stimulated GTPase activity in turkey erythrocyte membranes. *Biochem. Biophys. Acta.* **452**, 538–551.

Cassel, D., and Selinger, Z. (1977). Mechanism of adenylate cyclase activation by cholera toxin: Inhibition of GTP hydrolysis at the regulatory site. *Proc. Natl. Acad. Sci. U.S.A.* **74**, 3307–3311.

Cassel, D., and Selinger, Z. (1978). Mechanism of adenylate cyclase activation through the beta-adrenergic receptor: Catecholamine-induced displacement of bound GDP by GTP. *Proc. Natl. Acad. Sci. U.S.A.* **75**, 4155–4159.

Chang, F.-H., and Bourne, H. R. (1987). Dexamethasone increases adenylyl cyclase activity and expression of the alpha-subunit of G_s in GH_3 cells. *Endocrinology* **121**, 1711–1715.

Cheung, W. Y., Bradham, L. D., Lynch, T. J., Lin, Y. M., and Tallant, E. A. (1975). Protein activator of cyclic $3':5'$-nucleotide phosphodiesterase of bovine or rat brain also activates its adenylate cyclase. *Biochem. Biophys. Res. Commun.* **66**, 1055–1062.

Citri, Y., and Schramm, M. (1980). Resolution, reconstitution and kinetics of the primary action of a hormone receptor. *Nature* **287**, 297–300.

Codina, J., Hildebrandt, J. D., Sekura, R. D., Birnbaumer, M., Bryan, J., Manclark, C. R., Iyengar, R., and Birnbaumer, L. (1984a). N_s and N_i, the stimulatory and inhibitory regulatory components of adenylyl cyclase. Purification of the human erythrocyte proteins without the use of activating ligands. *J. Biol. Chem.* **259**, 5871–5886.

Codina, J., Hildebrandt, J. D., Birnbaumer, L., and Sejura, R. D. (1984b). Effects of guanine nucleotides and Mg on human erythrocyte N_i and N_s, the regulatory components of adenylyl cyclase. *J. Biol. Chem.* **259**, 11408–11418.

Coussen, F., Haiech, J., D'Alayer, J., Monneron, A. (1985). Identification of the catalytic subunit of brain adenylate cyclase: A calmodulin binding protein of 135 kDa. *Proc. Natl. Acad. Sci. U.S.A.* **82**, 6736–6740.

Coussen, F., Guermah, M., d'Alayer, J., Monneron, A., Haiech, J., and Cavadore, J-C. (1986). Evidence for two distinct adenylate cyclase catalysts in rat brain. *FEBS Lett.* **206**, 213–217.

Gehlert, D. R. (1986). Regional modulation of [^{3}H]forskolin binding in the rat brain by guanylyl-5'-imidodiphosphate and sodium fluoride: comparison with the distribution of guanine nucleotide binding sites. *J. Pharmacol. Exp. Ther.* **239**, 952–958.

Graziano, M. P., Casey, P. J., and Gilman, A. G. (1987). Expression of cDNAs for G proteins in *Escherichia coli. J. Biol. Chem.* **262**, 11375–11381.

Graziano, M. P., Freissmuth, M., and Gilman, A. G. (1989). Expression of $G_{s\text{-alpha}}$ in *Escherichia coli:* Purification and properties of two forms of the protein. *J. Biol. Chem.* **264**, 409–418.

Hagashijima, T., Ferguson, K. M., Sternweis, P. C., Smigel, M. D., and Gilman, A. G. (1987). Effects of Mg^{2+} and the beta-gamma-subunit complex on the interactions of guanine nucleotides with G proteins. *J. Biol. Chem.* **262**, 762–766.

Hanski, E., Sternweis, P. C., Northup, J. K., Dromeric, A. W., and Gilman, A. G. (1981). The regulatory component of adenylate cyclase: Purification and properties of the turkey erythrocyte protein. *J. Biol. Chem.* **256**, 12911–12919.

Harris, B. A. (1988). Complete cDNA sequence of a human stimulatory GTP-binding protein alpha subunit. *Nucleic Acids Res.* **16**, 3585.

Harris, B. A., Robishaw, J. D., Mumby, S. M., and Gilman, A. G. (1985). Molecular cloning of complementary DNA for the alpha subunit of the G protein that stimulates adenylate cyclase. *Science* **229**, 1274–1277.

Heyworth, C. M., Whetton, A. D., Kinsella, A. R., and Houslay, M. D. (1984). The phorbol ester, TPA inhibits glucagon-stimulated adenylate cyclase activity. *FEBS Lett.* **170**, 38–42.

Hoshi, T., Garber, S. S., and Aldrich, R. W. (1988). Effect of forskolin on voltage-gated K^+ channels is independent of adenylate cyclase activation. *Science* **240**, 1652–1655.

Howlett, A. C., and Gilman, A. G. (1980). Hydrodynamic properties of the regulatory component of adenylate cyclase. *J. Biol. Chem.* **255**, 2861–2866.

Hurley, J. B., Simon, M. I., Teplow, D. B., Robishaw, J. D., and Gilman, A. G. (1984). Homologies between signal transducing G proteins and *ras* gene products. *Science* **226**, 860–862.

Itoh, H., Kozasa, T., Nagata, S., Nakamura, S., Katada, T., Ui, M., Iwai, S., Ohtsuka, E., Kawasaki, H., Suzuki, K., and Kaziro, Y. (1986). Molecular cloning and sequence determination of cDNAs for alpha subunits of the guanine nucleotide-binding proteins G_s, G_i, and G_o from rat brain. *Proc. Natl. Acad. Sci. U.S.A.* **83**, 3776–3780.

Iyengar, R. (1981). Hysteretic activation of adenylyl cyclases II. Mg ion regulation of the activation of the regulatory component as analyzed by reconstitution. *J. Biol. Chem.* **256**, 11042–11050.

Iyengar, R., and Birnbaumer, L. (1981). Hysteretic activation of adenylyl cyclases I. Effect of Mg ion on the rate of activation by guanine nucleotides and fluoride. *J. Biol. Chem.* **256**, 11036–11041.

Iyengar, R., and Birnbaumer, L. (1982). Hormone receptor modulates the regulatory component of adenylyl cyclase by reducing its requirement for Mg^{2+} and enhancing its extent of activation by guanine nucleotides. *Proc. Natl. Acad. Sci. U.S.A.* **79**, 5179–5183.

Iyengar, R., Rich, K. A., Herberg, J. T., Premont, R. T., and Codina, J. (1988). Glucagon receptor-mediated activation of G_s is accompanied by subunit dissociation. *J. Biol. Chem.* **263**, 15348–15353.

Johnson, G. L., Kaslow, H. R., and Bourne, H. R. (1978). Reconstitution of cholera toxin activated adenylate cyclase. *Proc. Natl. Acad. Sci. U.S.A.* **75**, 3113–3117.

Johnson, J. A., Goka, T. J., and Clark, R. B. (1986). Phorbol ester-induced augmentation and inhibition of epinephrine-stimulated adenylate cyclase in S49 lymphoma cells. *J. Cyclic Nucleotide Protein Phosphorylation Res.* **11**, 199–215.

Jones, D. T., and Reed, R. R. (1987). Molecular cloning of five GTP-binding protein cDNA species from rat olfactory neuroepithelium. *J. Biol. Chem.* **262**, 14241–14249.

Joost, H. G., and Steinfelder, H. J. (1987). Forskolin inhibits insulin-stimulated glucose transport in rat adipose cells by direct interaction with the glucose transporter. *Mol. Pharmacol.* **31**, 279–283.

Kozasa, T., Itoh, H., Tsukamoto, T., and Kaziro, Y. (1988). Isolation and characterization of the human G$_s$-alpha gene. *Proc. Natl. Acad. Sci. U.S.A.* **85**, 2081–2085.

Livingstone, M. S. (1985). Genetic dissection of *Drosophila* adenylate cyclase. *Proc. Natl. Acad. Sci. U.S.A.* **82**, 5992–5996.

Livingstone, M. S., Sziber, P. P., and Quinn, W. G. (1984). Loss of calcium/calmodulin responsiveness in adenylate cyclase of *rutabaga,* a *Drosophila* learning mutant. *Cell* **37**, 205–215.

Londos, C., Salomon, Y., Lin, M. C., Harwood, J. P., Schramm, M., Wolff, J., and Rodbell, M. (1974). 5'-Guanylyl-imidodiphosphate, a potent activator of adenylate cyclase systems in eukaryotic cells. *Proc. Natl. Acad. Sci. U.S.A.* **71**, 3087–3090.

Mallorga, P., Tallman, J. F., Henneberry, R. C., Hirata, F., Strittmatter, W. T., and Axelrod, J. (1980). Mepacrine blocks beta-adrenergic agonist-induced desensitization in astrocytoma cells. *Proc. Natl. Acad. Sci. U.S.A.* **77**, 1341–1345.

Mattera, R., Codina, J., Crozat, A., Kidd, V., Woo, S. L. C., and Birnbaumer, L. (1986). Identification by molecular cloning of two forms of the human liver stimulatory (G$_s$) regulatory component of adenylyl cyclase. *FEBS Lett.* **206**, 36–41.

Mattera, R., Yatani, A., Kirsch, G. E., Graf, R., Okabe, K., Olate, J., Codina, J., Brown, A. M., and Birnbaumer, L. (1989). Recombinant alpha$_i$-3 subunit of G protein activates G$_k$-gated K$^+$ channels. *J. Biol. Chem.* **264**, 465–471.

May, D. C., and Ross, E. M. (1988). Rapid binding of guanosine 5'-O-(3-thiotriphosphate) to an apparent complex of beta-adrenergic receptor and the GTP-binding regulatory protein G$_s$. *Biochemistry* **27**, 4888–4893.

May, D. C., Ross, E. M., Gilman, A. G., and Smigel, M. D. (1985). Reconstitution of catecholamine-stimulated adenylate cyclase activity using three purified proteins. *J. Biol. Chem.* **260**, 15829–15833.

Minocherhomjee, A. M., Selfe, S., Flowers, N. J., and Storm, D. R. (1987). Direct interaction between the catalytic subunit of the calmodulin-sensitive adenylate cyclase from bovine brain with ^{125}I-labeled wheat germ agglutinin and ^{125}I-labeled calmodulin. *Biochemistry* **26**, 4444–4448.

Mochly-Rosen, D., Chang, F.-H., Cheever, L., Kim, M., Diamond, I., and Gordon, A. S. (1988). Chronic ethanol causes heterologous desensitization of receptors by reducing alpha$_s$ messenger RNA. *Nature* **333**, 848–850.

Mollner, S., and Pfeuffer, T. (1988). Two different adenylyl cyclases in brain distinguished by monoclonal antibodies. *Eur. J. Biochem.* **171**, 265–271.

Monneron, A., d'Alayer, J., and Coussen, F. (1987). Purification of the catalytic subunit of adenylate cyclase in vertebrates: State of the art in 1987. *Biochimie*

69, 263–269.

Monneron, A., Ladant, D., d'Alayer, J., Bellalou, J., Bârzu, O., and Ullmann, A. (1988). Immunological relatedness between *Bordetella pertussis* and rat brain adenylyl cyclases. *Biochemistry* **27,** 536–539.

Mumby, S. M., Kahn, R. A., Manning, D. R., and Gilman, A. G. (1986). Antisera of designed specificity for subunits of guanine nucleotide-binding regulatory proteins. *Proc. Natl. Acad. Sci. U.S.A.* **83,** 265–269.

Northup, J. K., Sternweis, P. C., Smigel, M. D., Schleifer, L. S., Ross, E. M., and Gilman, A. G. (1980). Purification of the regulatory component of adenylate cyclase. *Proc. Natl. Acad. Sci. U.S.A.* **77,** 6516–6520.

Northup, J. K., Smigel, M. D., and Gilman, A. G. (1982). The guanine nucleotide activating site of the regulatory component of adenylate cyclase: Identification by ligand binding. *J. Biol. Chem.* **257,** 11416–11423.

Northup, J. K., Smigel, M. D., Sternweis, P. C., and Gilman, A. G. (1983a). The subunits of the stimulatory regulatory component of adenylyl cyclase. Resolution of the activated 45,000-dalton (alpha) subunit. *J. Biol. Chem.* **258,** 11369–11376.

Northup, J. K., Sternweis, P. C., and Gilman, A. G. (1983b). The subunits of the stimulatory regulatory component of adenylyl cyclase. Resolution, activity, and properties of the 35,000-dalton (beta) subunit. *J. Biol. Chem.* **258,** 11361–11368.

Nukada, T., Tanabe, T., Takahashi, H., Noda, M., Hirose, T., Inayama, S., and Numa, S. (1986). Primary structure of the alpha-subunit of bovine adenylate cyclase-stimulating G-protein deduced from the cDNA sequence. *FEBS Lett.* **195,** 220–224.

Nukada, T., Mishina, M., and Numa, S. (1987). Functional expression of cloned cDNA encoding the alpha-subunit of adenylate cyclase-stimulating G-protein, *FEBS Lett.* **211,** 5–9.

Olate, J., Mattera, R., Codina, J., and Birnbaumer, L. (1988). Reticulocyte lysates synthesize an active alpha subunit of the stimulatory G protein G_s. *J. Biol. Chem.* **263,** 10394–10400.

Pfeuffer, E., Dreher, R.-M., Metzger, H., and Pfeuffer, T. (1985a). Catalytic unit of adenylate cyclase: Purification and identification by affinity crosslinking. *Proc. Natl. Acad. Sci. U.S.A.* **82,** 3086–3090.

Pfeuffer, E., Mollner, S., and Pfeuffer, T. (1985b). Adenylate cyclase from bovine brain cortex: Purification and characterization of the catalytic subunit. *EMBO J.* **4,** 3675–3679.

Pfeuffer, T., and Metzger, H. (1982). 7-O-Hemisuccinyl-deacetyl forskolin-sepharose: A novel affinity support for purification of adenylate cyclase. *FEBS Lett.* **146,** 369–375.

Pfeuffer, T., Gaugler, B., and Metzger, H. (1983). Isolation of homologous and heterologous complexes between catalytic and regulatory components of adenylate cyclase by forskolin-sepharose. *FEBS Lett.* **164,** 154–160.

Pines, M., Gierschik, P., Milligan, G., Klee, W., and Spiegel, A. (1985). Antibodies against the carboxyl-terminal 5-kDa peptide of the alpha subunit of transducin crossreact with the 40-kDa but not the 39-kDa guanine nucleotide binding protein from brain. *Proc. Natl. Acad. Sci. U.S.A.* **82,** 4095–4099.

Premont, R. T., and Iyengar, R. (1988). Glucagon-induced desensitization of adenylyl cyclase in primary cultures of chick hepatocytes: Evidence for multiple pathways. *J. Biol. Chem.* **263,** 16087–16095.

Ransnas, L. A., and Insel, P. A. (1988). Quantitation of the guanine nucleotide binding regulatory protein G_s in S49 cell membranes using antipeptide antibodies to alpha$_s$. *J. Biol. Chem.* **263,** 9482–9485.

Rebois, R. V., and Patel, J. (1985). Phorbol ester causes desensitization of gonadotropin-responsive adenylate cyclase in a murine leydig tumor cell line. *J. Biol. Chem.* **260,** 8026–8031.

Robishaw, J. D., Russell, D. W., Harris, B. A., Smigel, M. D., and Gilman, A. G. (1986a). Deduced primary structure of the alpha subunit of the GTP-binding stimulatory protein of adenylate cyclase. *Proc. Natl. Acad. Sci. U.S.A.* **83,** 1251–1255.

Robishaw, J. D., Smigel, M. D., and Gilman, A. G. (1986b). Molecular basis for two forms of the G protein that stimulates adenylate cyclase. *J. Biol. Chem.* **261,** 9587–9590.

Rodbell, M., Birnbaumer, L., and Pohl, S. L., and Krans, M. J. (1971). The glucagon-sensitive adenyl cyclase system in plasma membranes of rat liver. V. An obligatory role of guanyl nucleotides in glucagon action. *J. Biol. Chem.* **246,** 1877–1882.

Rosenberg, G. B., and Storm, D. R. (1987). Immunological distinction between calmodulin-sensitive and calmodulin-insensitive adenylate cyclases. *J. Biol. Chem.* **262,** 7623–7628.

Ross, E. M., and Gilman, A. G. (1977a). Resolution of some components of adenylate cyclase necessary for catalytic activity. *J. Biol. Chem.* **252,** 6966–6969.

Ross, E. M., and Gilman, A. G. (1977b). Reconstitution of catecholamine-sensitive adenylate cyclase activity: Interaction of solubilized components with receptor-replete membranes. *Proc. Natl. Acad. Sci. U.S.A.* **74,** 3715–3719.

Ross, E. M., Howlett, A. C., Ferguson, K. M., and Gilman, A. G. (1978). Reconstitution of hormone-sensitive adenylate cyclase activity with resolved components of the enzyme. *J. Biol. Chem.* **253,** 6401–6412.

Salomon, Y., Lin, M. C., Londos, C., Rendall, M., and Rodbell, M. (1975). The hepatic adenylate cyclase system I. Evidence for transition states and structural requirements for guanine nucleotide activation. *J. Biol. Chem.* **250,** 4239–4245.

Schramm, M., and Rodbell, M. (1975). A persistent active state of the adenylate cyclase system produced by the combined actions of isoproterenol and guanylyl imidodiphosphate in frog erythrocyte membranes. *J. Biol. Chem.* **250,** 2232–2237.

Seamon, K. B. (1987). Forskolin and adenylate cyclase. *ISI Atlas Sci.: Pharmacol.* **1,** 250–253.

Seamon, K. B., and Daly, J. W. (1981). Activation of adenylate cyclase by the diterpine forskolin does not require the guanine nucleotide regulatory protein. *J. Biol. Chem.* **256,** 9799–9801.

Seamon, K. B., Padgett, W., and Daly, J. W. (1981). Forskolin: Unique diterpine activator of adenylate cyclase in membranes and intact cells. *Proc. Natl. Acad. Sci. U.S.A.* **78,** 3363–3367.

Sergeant, S., and Kim, H. D. (1985). Inhibition of 3-O-methylglucose transport in human erythrocytes by forskolin. *J. Biol. Chem.* **260**, 14677–14682.

Sibley, D. R., and Lefkowitz, R. J. (1985). Molecular mechanisms of receptor desensitization using the beta-adrenergic receptor-coupled adenylate cyclase as a model. *Nature* **317**, 124–129.

Sibley, D. R., Peters, J. R., Nambi, P., Caron, M. G., and Lefkowitz, R. J. (1984). Desensitization of turkey erythrocyte adenylate cyclase. Beta-adrenergic receptor phosphorylation is correlated with attenuation of adenylate cyclase activity. *J. Biol. Chem.* **259**, 9742–9749.

Sibley, D. R., Jeffs, R. A., Daniel, K., Nambi, P., and Lefkowitz, R. J. (1986). Phorbol diesters promote beta-adrenergic receptor phosphorylation and adenylate cyclase desensitization in duck erythrocytes. *Arch. Biochem. Biophys.* **244**, 373–381.

Smigel, M. (1986). Purification of the catalyst of adenylate cyclase. *J. Biol. Chem.* **261**, 1976–1982.

Sternweis, P. C. (1986). The purified alpha subunits of G_o and G_i from bovine brain require beta-gamma for association with phospholipid vesicles. *J. Biol. Chem.* **261**, 631–637.

Sullivan, K. A., Liao, Y. C., Alborzi, A., Beiderman, B., Chang, F. H., Masters, S. B., Levinson, A. D., and Bourne, H. R. (1986). Inhibitory and stimulatory G proteins of adenylate cyclase: cDNA and amino acid sequences of the alpha chains. *Proc. Natl. Acad. Sci. U.S.A.* **83**, 6687–6691.

Tempel, B. L., Bonini, N., Dawson, D. R., and Quinn, W. G. (1983). Reward learning in normal and mutant *Drosophila. Proc. Natl. Acad. Sci. U.S.A.* **80**, 1482–1486.

Tolkovsky, A. M., and Levitski, A. (1978). Mode of coupling between the beta-adrenergic receptor and adenylate cyclase in turkey erythrocytes. *Biochemistry* **17**, 3795–3810.

Wagoner, P. K., and Pallotta, B. S. (1988). Modulation of acetylcholine receptor desensitization by forskolin is independent of cAMP. *Science* **240**, 1655–1657.

Wescott, K. R., LaPorte, D. C., and Storm, D. R. (1979). Resolution of adenylate cyclase sensitive and insensitive to Ca^{2+} and calcium-dependent regulatory protein (CDR) by CDR-sepharose affinity chromatography. *Proc. Natl. Acad. Sci. U.S.A.* **76**, 204–208.

White, M. W. (1988). Forskolin alters acetylcholine receptor gating by a mechanism independent of adenylate cyclase activation. *Mol. Pharmacol.* **34**, 427–430.

Worley, P. F., Baraban, J. M., De Souza, E. B., and Snyder, S. H. (1986). Mapping second messenger systems in the brain: Differential localizations of adenylate cyclase and protein kinase C. *Proc. Natl. Acad. Sci. U.S.A.* **83**, 4053–4057.

Yeager, R. E., Heideman, W., Rosenberg, G. B., and Storm, D. R. (1985). Purification of the calmodulin-sensitive adenylate cyclase from bovine cerebral cortex. *Biochemistry* **24**, 3776–3783.

Yoshimasa, T., Sibley, D. R., Bouvier, M., Lefkowitz, R. J., and Caron, M. G. (1987). Cross-talk between cellular signalling pathways suggested by phorbol ester-induced adenylate cyclase phosphorylation. *Nature* **327**, 67–70.

CHAPTER 8

Participation of Guanine Nucleotide-Binding Protein Cascade in Activation of Adenylyl Cyclase by Cholera Toxin (Choleragen)

Joel Moss, Martha Vaughan
Laboratory of Cellular Metabolism, National Heart, Lung, and Blood Institute, National Institutes of Health, Bethesda, Maryland 20892

I. Introduction
II. ADP-Ribosyltransferase and NAD$^+$ Glycohydrolase (NADase) Activities of Choleragen
III. Effect of ADP-Ribosylation Factor on Enzymatic Activities of Choleragen
IV. Similarities between Choleragen and *Escherichia Coli* Heat-Labile Enterotoxin
V. Evidence for ADP-Ribosylation Cycle Endogenous to Animal Cells
References

I. INTRODUCTION

Choleragen (cholera toxin) is a secretory product of *Vibrio cholerae* responsible in part for the pathogenesis of cholera (Finkelstein, 1973; Kelly, 1986). Its effects on the intestinal mucosa result from toxin-catalyzed activation of the hormone-sensitive adenylyl cyclase, leading to an increase in cAMP (Kelly, 1986).[1] This "second messenger" induces the abnormalities in fluid and electrolyte flux that are manifest in the diarrheal syndrome of clinical cholera (Kelly, 1986).

Choleragen is an oligomeric protein of 84 kDa, composed of one A subunit associated noncovalently with five (~ 11 kDa) B subunits (Gill, 1976a, 1977). The A subunit, synthesized as a single polypeptide chain (Gill and Rappaport, 1979), is nicked by proteases following secretion to yield two polypeptides, termed A_1 and A_2 (~ 21 and 6 kDa, respectively) linked through a single disulfide bond. The initial event in enterotoxin action is its binding to intestinal cells through specific interaction of the B subunits with the oligosaccharide moiety of the monosialoganglioside G_{M1} (van Heyningen et al., 1971; Cuatrecasas, 1973a,b,c; Holmgren et al., 1973, 1974). Modifications in structure of the G_{M1} oligosaccharide alter its capacity to bind to toxin. Gangliosides G_{M2} (lacking the terminal galactose) and G_{D1a} (with N-acetylneuraminic acid attached to the terminal galactose) bind much less well. Addition of G_{M1} to ganglioside-deficient, toxin-unresponsive cells resulted in uptake of the ganglioside, enhanced ^{125}I-labeled toxin binding and toxin responsiveness, the latter evidenced by increased intracellular cAMP and activation of adenylyl cyclase (Fishman et al., 1976; Moss et al., 1976a; unpublished data). Gangliosides that differ from G_{M1} in oligosaccharide side-chain structure were much less effective in increasing toxin sensitivity (Moss et al., 1976a).

It was initially observed by Gill (1975) that NAD is necessary for activation of cyclase by toxin. The subsequent demonstration that purified toxin in the absence of cellular components catalyzed the hydrolysis of NAD to ADP-ribose and nicotinamide (Moss et al., 1976b), as well as the ADP-ribosylation of arginine and other simple guanidino compounds (Moss and

---

[1] ARF, ADP-ribosylation factor; sARF, soluble form of ARF; mARF, membrane-associated form of ARF; GDPβS, guanosine 5'-O-[β-thio]diphosphate; Gpp(NH)p, guanylyl imidodiphosphate; GTPγS, guanosine 5'-O-[γ-thio]triphosphate; App(NH)p, adenylyl imidodiphosphate; G_S, stimulatory guanine nucleotide-binding protein of adenylate cyclase; G_i, inhibitory guanine nucleotide-binding protein of adenylate cyclase; G protein, guanine nucleotide-binding protein; G_α, α subunit of guanine nucleotide-binding protein; $G_{\beta\gamma}$, β and γ subunits of guanine nucleotide-binding protein; G_t, transducin (guanine nucleotide-binding protein involved in visual excitation); G_O, guanine nucleotide-binding protein present in brain; DMPC, dimyristoylphosphatidylcholine; CTA$_1$, choleragen A$_1$ protein; CTA, choleragen A subunit; SDS, sodium dodecyl sulfate; PAGE, polyacrylamide gel electrophoresis; G_{M1}, galactosyl-N-acetylgalactosaminyl-(N-acetylneuraminyl)galactosylglucosylceramide; G_{M2}, N-acetylgalactosaminyl-(N-acetylneuraminyl)galactosylglucosylceramide; G_{D1a}, N-acetylneuraminylgalactosyl-N-acetylgalactosaminyl-(N-acetylneuraminyl)galactosylglucosylceramide; CHAPS, 3-[(3-cholamidopropyl)dimethylammonio]-1-propane sulfonate; 12-APS-GlcN, 12-(4-azido-2-nitrophenyl)stearoyl[1-^{14}C]glucosamine; LT, Escherichia coli heat-labile enterotoxin; ST, Escherichia coli heat-stable enterotoxin.

Vaughan, 1977b), identified choleragen as an ADP-ribosyltransferase (Moss *et al.*, 1979c). After searching for an intracellular toxin substrate, several groups reported in 1978 the specific ADP-ribosylation of a 42-kDa membrane protein (Johnson *et al.*, 1978; Gill and Meren, 1978; Cassel and Pfeuffer, 1978). Association between the ADP-ribosylation of the 42-kDa protein and activation of adenylyl cyclase was established when Northup *et al.* (1980) demonstrated that $G_S\alpha$, a 42-kDa protein responsible for stimulation of the adenylyl cyclase catalytic unit, is ADP-ribosylated by choleragen. The ADP-ribosylation of $G_S\alpha$ and expression of the activated cyclase required GTP or a nonhydrolyzable GTP analog; other nucleoside triphosphates and related derivatives were inactive (Moss and Vaughan, 1977a; Lin *et al.*, 1978; Enomoto and Gill, 1980; Nakaya *et al.*, 1980).

The A subunit of choleragen possesses latent ADP-ribosyltransferase activity that is expressed optimally following proteolytic cleavage and reduction of the single disulfide bond linking the A_1 and A_2 proteins (Mekalanos *et al.*, 1979b). The isolated A_1 protein expresses ADP-ribosyltransferase activity and is capable of activating adenylyl cyclase in membrane preparations independent of A_2 or B proteins (Gill and King, 1975; Moss *et al.*, 1979c). Although the functions of both A and B components can be expressed independent of each other, intoxication of cells requires the intact holotoxin. Activation of adenylyl cyclase in intact cells occurs only after a significant delay (Gill and King, 1975; Kimberg *et al.*, 1971; Bennett and Cuatrecasas, 1975; Fishman, 1980; Kassis *et al.*, 1982), although *in vitro*, enzymatic activity of the toxin A_1 protein is constant from zero time. The delay in toxin action following toxin binding has been attributed to the several steps, such as the entry of the A subunit, release of the A_1 protein, and ADP-ribosylation of $G_S\alpha$ which is located on the inner membrane surface (Farfel *et al.*, 1979), that precede activation of the cyclase.

Each of the five B subunits in the holotoxin is capable of binding to the oligosaccharide moiety of G_{M1} (Fishman *et al.*, 1978; Sattler *et al.*, 1978; Schwarzmann *et al.*, 1978). Multivalent binding of the B subunits may lead to aggregation of the toxin on the cell surface (Craig and Cuatrecasas, 1975; Sedlacek *et al.*, 1976; Revesz and Greaves, 1975), a process that may facilitate entry of the A subunit and may also contribute to the delay in onset of toxin action (Fishman and Atikkan, 1980; Osborne *et al.*, 1982). Internalization of toxin was monitored in mouse thymocytes, using anti-choleragen antibodies coupled to peroxidase; the process, as expected, was temperature-dependent and not observed at $< 18°C$ (Hansson *et al.*, 1977). In mouse thymocytes, entry of both A and B subunits was observed (Tsuru *et al.*, 1982). Similar data were obtained with cultured chick sympathetic neurons (Joseph *et al.*, 1978) and neuroblastoma N2a cells (Joseph *et al.*, 1979). Toxin was detected in the Golgi – endoplasmic reticulum –

lysosomal system (GERL); entry of the toxin was not via coated pits (Montesano *et al.*, 1982). The toxin is enzymatically active only after proteolysis and reduction to release A_1. Thiol:protein-disulfide oxidoreductase can accelerate release of A_1 from purified toxin in the presence of a thiol (e.g., glutathione, cysteine, dithiothreitol), presumably by accelerating disulfide exchange (Moss *et al.*, 1980a).

Studies of the rate of formation of A_1 and activation of adenylyl cyclase in cultured cells yielded data consistent with the conclusion that generation of A_1 may, under certain circumstances, be a rate-limiting step in cyclase activation. In mouse neuroblastoma NB cells, activation of adenylyl cyclase by choleragen and increased cell cAMP content, as well as release of $[^{125}I]A_1$ from ^{125}I-labeled holotoxin all showed a similar ~ 15 min delay (Kassis *et al.*, 1982). Parallel delays in generation of $[^{125}I]A_1$, activation of adenylyl cyclase, and accumulation of intracellular cAMP were also found in Friend erythroleukemic cells (Kassis *et al.*, 1982). Toxin activation of adenylyl cyclase is temperature-dependent. In NB cells, cAMP elevation was favored at $37°C$ ($37°C > 28°C > 22°C > 4°C$) as was generation of the $[^{125}I]A_1$ protein from the precursor ^{125}I-labeled choleragen. Both processes were inhibited by anticholeragen antibodies (Kassis *et al.*, 1982). Response of rat glioma C_6 cells incubated first with G_{M1} and subsequently with toxin at $37°C$ was significantly higher than that of cells incubated in the absence of G_{M1} or exposed to toxin at $4°C$ after a preliminary incubation with G_{M1} (Kassis *et al.*, 1982). Generation of $[^{125}I]A_1$ was likewise enhanced by exposure of cells to G_{M1} and incubation at $37°C$ after toxin binding.

Independent data supporting the insertion of the A_1 protein into membranes were obtained using a photoreactive glycolipid probe, 12-(4-azido-2-nitrophenyl)stearoyl[1-^{14}C]glucosamine (12-APS-G1cN), and the G_{M1}-rich Newcastle disease virus (Wisnieski and Bramhall, 1981). 12-APS-G1cN inserted into the outer membrane monolayer readily labels proteins (Wisnieski and Bramhall, 1981). Following toxin binding, rapid labeling of the A_1 occurred at $37°C$, but not at $0°C$ (Wisnieski and Bramhall, 1981). The data were not compatible with channel formation by the B oligomer, as had been proposed (Gill, 1976a).

Activation of adenylyl cyclase by toxin is persistent (O'Keefe and Cuatrecasas, 1974; Chang *et al.*, 1983), perhaps a result of the continued presence of toxin in the cell (Chang *et al.*, 1983). ^{125}I-Labeled choleragen bound to human fibroblasts was degraded slowly with a $t_{1/2}$ of 2–3 days; radiolabel was lost more rapidly from the $[^{125}I]A_1$ than from the $[^{125}I]B$ proteins. Persistence of toxin on the cell surface was demonstrated by reaction with antitoxin, antisubunit A or antisubunit B antibodies (Chang *et al.*, 1983). After exposure of fibroblasts to toxin, ^{125}I-labeled toxin binding was decreased; recovery was slow with a $t_{1/2}$ of 7 days (Chang *et al.*,

1983). Thus, the data support a model wherein toxin binding is G_{M1}-dependent and poorly reversible. Activation of adenylyl cyclase is delayed in part due to the time necessary for generation of the A_1 protein; slow degradation of the toxin may explain the persistent activation of the cyclase.

The apparently stable nature of adenylyl cyclase activation by the toxin could also result if the ADP-ribosylation of $G_S\alpha$ is irreversible. Although animal cells possess an ADP-ribosylarginine hydrolase that can cleave [ADP-ribose]arginine bonds (presumably the linkage of ADP-ribose to $G_S\alpha$), whether this enzyme plays a role in reversing intoxication by choleragen remains to be established (Moss *et al.*, 1985).

II. ADP-RIBOSYLTRANSFERASE AND NAD⁺ GLYCOHYDROLASE (NADase) ACTIVITIES OF CHOLERAGEN

Choleragen catalyzes several different reactions, all based on its ability to activate the ribosyl–nicotinamide bond of NAD and to use a variety of compounds as ADP-ribose acceptors. The diarrheal syndrome characteristic of cholera is caused by adenylyl cyclase activation in cells of the intestinal mucosa resulting from toxin-catalyzed transfer of the ADP-ribose moiety of NAD to $G_S\alpha$, the stimulatory guanine nucleotide-binding protein of the cyclase system [reaction (1)] (Kelly, 1986; Northup *et al.*, 1980). $G_S\alpha$ is active when GTP is bound; ADP-ribosylation increases its sensitivity to the activator GTP (Lin *et al.*, 1978; Gill, 1976b; Nakaya *et al.*, 1980).

Although the mechanism of the choleragen effect is not entirely clear, it appears that the ADP-ribosylation of $G_S\alpha$ decreases its GTPase activity (Cassel and Selinger, 1977). $G_S\alpha$ is active when GTP is bound; hydrolysis of GTP to GDP by the intrinsic GTPase activity of $G_S\alpha$ reverses the activation (Gilman, 1987). A decrease in this GTPase activity would result in preservation of the active species, $G_S\alpha \cdot$ GTP. It has been found that ADP-ribosylation also facilitates release of GDP, thus opening the site to further binding of GTP and reactivation (Burns *et al.*, 1982, 1983).

In addition to ADP-ribosylation of $G_S\alpha$, several other reactions are catalyzed by choleragen. The toxin ADP-ribosylates $G_t\alpha$, the α subunit of transducin, a G protein found in retina (Van Dop *et al.*, 1984; Abood *et al.*, 1982). The ADP-ribosylated amino acid in $G_t\alpha$ was identified as arginine following isolation and analysis of an ADP-ribosylated tetrapeptide from a tryptic digest of ADP-ribosyl $G_t\alpha$ (Van Dop *et al.*, 1984).

$$NAD + G_s\alpha \rightarrow \text{ADP-ribosyl-}G_s\alpha + \text{nicotinamide} + H^+ \qquad (1)$$

Free arginine and other simple guanidino compounds, such as agmatine, can also serve as ADP-ribose acceptors in the toxin-catalyzed reaction [reaction (2)] (Mekalanos *et al.*, 1979a; Moss and Vaughan, 1977b; Moss *et al.*, 1979c). In contrast, citrulline, in which the guanidino moiety of arginine is replaced by a ureido group, is inactive. Guanidine itself can be ADP-ribosylated; the linkage appears to be between the [ADP-ribosyl]C-1′ and the guanidino N. The ADP-ribosylation reaction is stereospecific and proceeds with inversion of configuration; the substrate is β-NAD and the product is the α-anomer of ADP-ribosylarginine (Oppenheimer, 1978).

$$NAD + \text{arginine} \rightarrow \text{ADP-ribosylarginine} + \text{nicotinamide} + H^+$$

$$\text{(guanidino-R) (ADP-ribosylguanidino-R)}$$

$$(2)$$

Proteins other than $G_s\alpha$, including a variety of membrane and soluble proteins, can be ADP-ribosylated, presumably reflecting the presence of a suitably accessible arginine [reaction (3)] (Van Dop *et al.*, 1984; Cooper *et al.*, 1981; Moss and Vaughan, 1978).

$$NAD + \text{(arginine)protein} \rightarrow \text{ADP-ribosyl(arginine)protein} + \text{nicotinamide} + H^+$$

$$(3)$$

The toxin also catalyzes auto-ADP-ribosylation of the A_1 peptide [reaction (4)] (Moss *et al.*, 1980c; Trepel *et al.*, 1977). There appear to be at least three modification sites, one of which has been identified (Xia *et al.*, 1984; Moss *et al.*, 1980c; Lai, 1986). The automodified toxin retains catalytic activity (Moss *et al.*, 1980c).

$$NAD + \text{CT-}A_1 \rightarrow \text{ADP-ribosyl-CT-}A_1 + \text{nicotinamide} + H^+ \qquad (4)$$

The toxin can also use water as an ADP-ribose acceptor (Moss *et al.*, 1976b). This NAD^+ glycohydrolase (NADase) activity [reaction (5)] is considerably slower than the ADP-ribosylation of guanidino compounds by the toxin (Moss and Vaughan, 1977b).

$$NAD + HOH \rightarrow \text{ADP-ribose} + \text{nicotinamide} + H^+ \qquad (5)$$

Both the A subunit and the holotoxin are relatively inactive as ADP-ribosyltransferases when compared to the A_1 protein (Mekalanos *et al.*, 1979b). Expression of the latent activity of A_1 requires reduction of the single disulfide bond linking A_1 and A_2 (Moss *et al.*, 1976b; Mekalanos *et al.*, 1979b); this reaction is enhanced by thiol:protein oxidoreductase, which promotes disulfide exchange (Moss *et al.*, 1980a). Based on the amino acid sequence deduced from cDNA analysis, only two cysteines are present in the A chain (positions 187 and 199) (Mekalanos *et al.*, 1983).

Presumably their reduction is necessary for expression of catalytic activity. In addition to reduction, proteolytic nicking is required for activation of the toxin A chain (Mekalanos et al., 1979b). Proteolytic cleavage occurs at an internal region between the two cysteines (Makalanos et al., 1983; Lai, 1986). Thus, proteolysis and reduction results in the generation of the A_1 and A_2 fragments. The A_1 fragment, isolated following alkylation of the cysteine with iodoacetamide, exhibited both ADP-ribosyltransferase and NAD^+ glycohydrolase (NADase) activities, consistent with the conclusion that the cysteine residue is not critical for activity (Moss et al., 1979c). Proteolytic fragments of other bacterial toxins (e.g., diphtheria toxin) also have catalytic activity. These toxins, as well as pertussis toxin, require reduction for expression of activity (Lai, 1986; Moss et al., 1983; Vasil et al., 1977; Pappenheimer, 1977).

III. EFFECT OF ADP-RIBOSYLATION FACTOR ON ENZYMATIC ACTIVITIES OF CHOLERAGEN

Membrane and soluble factors from animal tissues promote the activation of adenylyl cyclase by choleragen (Gill, 1976b; Enomoto and Gill, 1980; Le Vine and Cuatrecasas, 1981; Pinkett and Anderson, 1982; Schleifer et al., 1982; Kahn and Gilman, 1984a,b, 1986; Gill and Meren, 1983). Kahn and Gilman (1984a,b, 1986), and Kahn (this volume) reported purification of a 21-kDa membrane protein from bovine brain that enhanced the toxin-catalyzed ADP-ribosylation of $G_S\alpha$; the protein, termed ARF for ADP-ribosylation factor, was later shown to bind guanine nucleotides (Kahn and

Table I   Effect of Nucleotides on ARF Stimulation of
NAD:Agmatine ADP-ribosyltransferase Activity of
Choleragen A Subunit[a]

Addition	NAD:agmatine ADP-ribosyltransferase activity ($nmol \cdot \mu g^{-1} \cdot h^{-1}$)
None	1.09
GTP	3.16
Gpp(NH)p	2.88
GTPγS	2.67
GDP	1.28
GDPβS	1.16
ATP	1.07
App(NH)p	1.20

[a] Data are derived from Tsai et al. (1987).

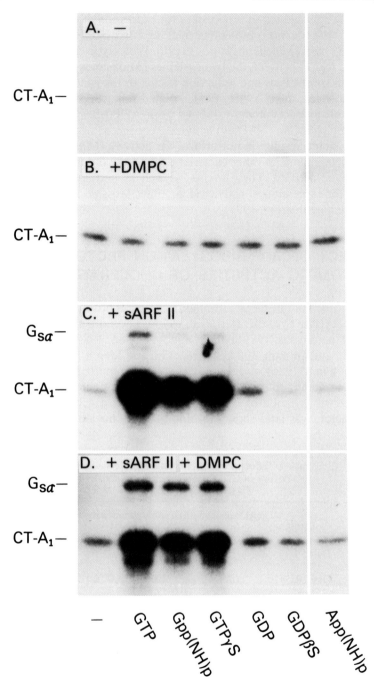

Gilman, 1986). It was proposed that ARF facilitates ADP-ribosylation by complexing with $G_S\alpha$ (Kahn and Gilman, 1984a).

To determine the mechanism of action of ARF its effects on $G_S\alpha$-dependent and $G_S\alpha$-independent ADP-ribosylation reactions were investigated (Tsai et al., 1987) (see Table I). ADP-ribosylation of $G_S\alpha$ in the presence of ARF was enhanced by dimyristoylphosphatidylcholine (DMPC) and GTP or its nonhydrolyzable analogs GTPγS and Gpp(NH)p; GDP, GDPβS and adenine nucleotides were inactive (Fig. 1) (Tsai et al., 1987). ARF also enhanced auto-ADP-ribosylation of the toxin A_1 protein; although the guanine nucleotide dependence was similar to that for ADP-ribosylation of $G_S\alpha$, DMPC was inhibitory (Fig. 1) (Tsai et al., 1987). The fact that the $G_S\alpha$-independent auto-ADP-ribosylation of the CTA_1 protein was stimulated by ARF was consistent with the hypothesis that effects of ARF were directly on the toxin rather than due to its interaction with $G_S\alpha$, as had been proposed. To address this question further, the effects of membrane (mARF) and soluble ARFs (sARF) on the NAD:agmatine ADP-ribosyltransferase activity of toxin were determined (Tsai et al., 1987, 1988). This activity was stimulated ~ three-fold by each of the purified ARF preparations (Tsai et al., 1987, 1988; unpublished data). The mARF-dependent enhancement of reaction rate required GTP or its analogs GTPγS and Gpp(NH)p; GDPβS and GDP were inactive as were the adenine nucleotides ATP and App(NH)p (Tsai et al., 1987). Further support for the direct interaction of toxin and ARF was the finding that ARF enhanced as well the toxin-catalyzed ADP-ribosylation of proteins unrelated to the cyclase system (Tsai et al., 1988) (Fig. 2).

To determine the site of action of ARF on toxin, the guanine nucleotide-dependent ARF activation of NAD:agmatine ADP-ribosyltransferase activity was assayed using holotoxin [(CTA_1-s-s-CTA_2)$B_{(5)}$], CTA, or alkylated CTA_1 (Noda et al., 1988). The last was prepared by reduction of CTA and alkylation with iodoacetamide; the CTA_1 and CTA_2 components were then separated by gel permeation chromatography under denaturing conditions (Noda et al., 1988). This purification scheme was used since reduction of toxin alone does not necessarily result in the dissociation of the CTA_1 protein. Holotoxin, CTA, and CTA_1 (reduced and alkylated) were all activated by sARF, consistent with the conclusion that sARF interacts directly with the catalytic unit of toxin (Noda et al., 1988). One difference in the reaction requirements with the different species of toxin was the lack

---

Fig. 1. Effects of DMPC, nucleotides, and sARF-II on ADP-ribosylation of $G_S\alpha$ and auto-ADP-ribosylation of choleragen A_1 protein. Standard assays contained [^{32}P]NAD, choleragen, and the indicated additions. After incubation, ADP-ribosylated proteins were separated by SDS-PAGE and identified by autoradiography. Data are from Tsai et al. (1988).

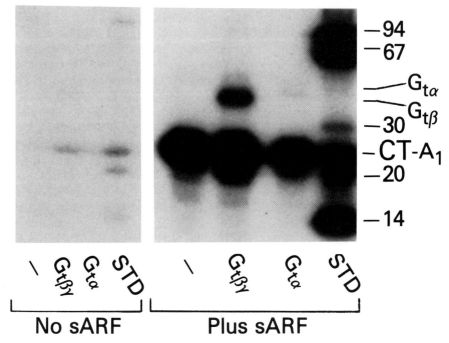

Fig. 2. Effects of ARF on ADP-ribosylation of proteins and auto-ADP-ribosylation of toxin A_1 protein. Assays with or without ARF contained choleragen, [^{32}P]NAD, and the indicated proteins, $G_t\alpha$, tranducin α; $G_t\beta\gamma$, transducin $\beta\gamma$; STD, standard proteins. After incubation, ADP-ribosylated proteins were separated by SDS-PAGE and identified by autoradiography. Data are from Tsai *et al.* (1988).

of a requirement for a high concentration of thiol with the reduced and alkylated CTA_1 for both basal and ARF-stimulated activity (Noda *et al.,* 1988).

Activation of toxin by ARF was enhanced selectively by certain detergents. SDS was highly effective in augmenting the ability of sARF to increase both NAD:CTA_1 auto-ADP-ribosylation and the NAD:agmatine ADP-ribosyltransferase activity of CTA (Noda *et al.,* 1988). Stimulation by SDS of the sARF-dependent NAD:protein (non-$G_S\alpha$) ADP-ribosyltransferase activity of CTA was dependent on the protein substrate; SDS stimulated ADP-ribosylation in some cases and inhibited in others (Noda *et al.,* 1988). On the NAD:agmatine ADP-ribosyltransferase activity, the effect of SDS was biphasic. Optimal stimulation was obtained at 0.003% SDS; higher concentrations inhibited both ARF-stimulated and basal activities (Noda *et al.,* 1988). Effects of cholate were similar. In contrast, Triton X-100, over a wide range of concentrations, inhibited sARF activation of toxin with little effect on the basal toxin activity (Noda *et al.,* 1988).

Choleragen-catalyzed ADP-ribosylation of simple guanidino compounds appears to proceed via a rapid-equilibrium, random-sequential mechanism (Osborne et al., 1985). Either NAD or the ADP-ribose acceptor may bind first to its site on the toxin. Analysis of reaction kinetics is consistent with the conclusion that binding of the first substrate results in an increase in apparent K_m for the second substrate (Osborne et al., 1985). Activation of toxin by ARF is associated with a decrease in K_m for both substrates, with no significant alteration in V_{max} (Noda et al., 1988). Thus, ARF to some extent reverses the negative cooperativity thereby facilitating the reaction between substrates (Noda et al., 1988).

Since ARF is a guanine nucleotide-binding protein as is $G_s\alpha$ (Kahn and Gilman, 1986), it would appear that a guanine nucleotide-binding protein cascade participates in the activation of adenylyl cyclase by choleragen (Tsai et al., 1988) (Fig. 3). ARF, in the presence of GTP or its analogues but not GDP or adenine nucleotides, interacts with the A_1 protein of toxin

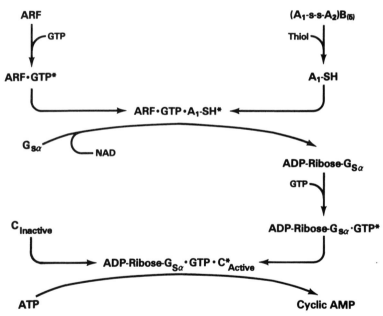

Fig. 3. A guanine nucleotide-binding protein cascade may participate in activation of adenylyl cyclase by choleragen. Choleragen, represented as its components, $(A_1\text{-s-s-}A_2)B_{(5)}$, is activated by thiol. Reduction of the disulfide yields the A_1-SH component, which may remain associated with the holotoxin, but exhibits increased enzymatic activity. ARF is activated by GTP; association of ARF·GTP with A_1-SH yields an enzymatically active complex that ADP-ribosylates $G_s\alpha$. In the presence of GTP, ADP-ribosyl $G_s\alpha$ activates the catalytic unit of adenylyl cyclase (C). The proposed pathway is further discussed in the text and in Tsai et al. (1988).

to yield an active complex that catalyzes the ADP-ribosylation of $G_s\alpha$ (Fig. 3). ADP-ribosyl-$G_s\alpha$, in the presence of GTP, activates the catalytic unit of adenylyl cyclase, resulting in increased formation of cyclic AMP from ATP (Fig. 3).

IV. SIMILARITIES BETWEEN CHOLERAGEN AND *ESCHERICHIA COLI* HEAT-LABILE ENTEROTOXIN

Certain strains of *E. coli* responsible for "traveler's diarrhea" secrete an enterotoxin known as *E. coli* heat-labile enterotoxin or LT (for labile toxin, to distinguish it from an *E. coli* heat-stable toxin). (For review, see Holmgren and Lönnroth, 1980; Carpenter, 1980; Moss and Vaughan, 1988a,b.) LT is an oligomeric protein similar in molecular weight to choleragen; it, too, is composed of A and B subunits, with properties and functions similar to those of the choleragen subunits (Clements and Finkelstein, 1979; Dallas *et al.,* 1979; Clements *et al.,* 1980; Robertson *et al.,* 1980). Genes for the A and B subunits of LT have been cloned. In contrast to choleragen (Vasil *et al.,* 1975), LT is not encoded in the bacterial chromosome, but rather on plasmids (So *et al.,* 1978). The deduced amino acid sequences of both A and B subunits exhibit considerable similarity to their choleragen counterparts (Spicer *et al.,* 1981; Spicer and Noble, 1982; Dallas and Falkow, 1980). LT, however, in addition to binding to G_{M1}, may use glycoproteins which are not receptors for choleragen (Moss *et al.,* 1979a, 1981a; Holmgren, 1973; Pierce, 1973; Holmgran *et al.,* 1982, 1985). LT possesses NAD: arginine ADP-ribosyltransferase activity, with the same stereospecificity as choleragen (Gill and Richardson, 1980; Moss and Richardson, 1978; Moss *et al.,* 1979b, 1981a). $G_s\alpha$ is presumably the cellular substrate; however, arginine, other low molecular weight guanidino compounds, and proteins unrelated to the cyclase system may also serve as ADP-ribose acceptors (Moss *et al.,* 1979b, 1981a). Activation of the intact A subunit of LT required dithiothreitol and tryptic digestion, similar to the findings with choleragen (Moss *et al.,* 1981a). Preliminary data support the conclusion that LT, like choleragen, is activated by the purified ADP-ribosylation factors from bovine brain (Chang *et al.,* 1988).

V. EVIDENCE FOR ADP-RIBOSYLATION CYCLE ENDOGENOUS TO ANIMAL CELLS

Demonstration of the ADP-ribosyltransferase activity of choleragen led to investigation of the possibility that the toxin might be mimicking an activity normally present in animal cells. Indeed, animal cells possess a family of enzymes that catalyze the ADP-ribosylation of arginine residues (Moss and Vaughan, 1978; Moss *et al.*, 1980b; Yost and Moss, 1983; West and Moss, 1986; Tanigawa *et al.*, 1984; Soman *et al.*, 1984) (Fig. 4). Like the toxin, these NAD:arginine ADP-ribosyltransferases can utilize guanidine, other simple guanidino compounds, and proteins as substrates. The reaction is stereospecific, resulting in the formation of α-ADP-ribosylarginine from β-NAD (Moss *et al.*, 1979d). NAD:arginine ADP-ribosyltransferases have been found in different species and tissues, and purified to homogeneity from turkey eythrocytes and chicken liver nuclei (Moss *et al.*, 1980b; Yost and Moss, 1983; Tanigawa *et al.*, 1984). Transferases from soluble, membrane, and nuclear fractions of turkey erythrocytes have been extensively purified and exhibit similar protomeric molecular weights of $\sim 30,000$ (Moss *et al.*, 1980b; Yost and Moss, 1983; West and Moss, 1986). Their kinetic and regulatory properties, however, differ considerably. Transferase "A", a soluble enzyme, exists in active protomeric and inactive oligomeric forms (Moss *et al.*, 1981b, 1982). Activation and conversion from oligomer to protomer is promoted by chaotropic salts and histones (Moss and Stanley, 1981; Moss *et al.*, 1981b, 1982); certain detergents (e.g., CHAPS) and phospholipids are also capable of activation (Moss *et al.*, 1984). The specific activity of the activated, purified transferase using agmatine as ADP-ribose acceptor far exceeds the activity of choleragen under its optimal assay conditions (Moss *et al.*, 1980b). As with toxin, the reaction is stereospecific with β-NAD serving as substrate and α-ADP-ribosylarginine being the reaction product (Oppenheimer, 1978; Moss *et al.*, 1979d). A second soluble NAD:arginine ADP-ribosyltransferase ("B") is readily distinguishable from transferase "A" by the facts that it is inhibited by chaotropic salts and not activated by histone; the specificity for model

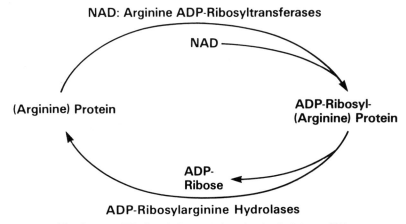

Fig. 4.   An ADP-ribosylation cycle in animal cells (Moss, 1987).

substrates (e.g., low molecular weight guanidino compounds, proteins) is different as well (Yost and Moss, 1983; unpublished data).

The NAD:arginine ADP-ribosyltransferases presumably catalyze the forward arm of an ADP-ribosylation cycle (Fig. 4). The opposing limb may be catalyzed by an enzyme, ADP-ribosylarginine hydrolase, found in many animal tissues and purified to >90% homogeneity from turkey erythrocytes (Smith *et al.*, 1985; Moss *et al.*, 1985, 1986; unpublished data). The hydrolase cleaves the [ADP-ribose]arginine linkage, with production of ADP-ribose and arginine (Moss *et al.*, 1985). The generation of arginine with an intact guanidino group was confirmed by its ability to serve as substrate for either choleragen or the erythrocyte NAD:arginine ADP-ribosyltransferase (Moss *et al.*, 1985; unpublished data). Hydrolase activity is dependent on Mg^{2+} and thiol (e.g., dithiothreitol); with optimal concentrations of both activity was increased more than 20-fold (Moss *et al.*, 1985). In addition to stimulation, Mg^{2+} also stabilized the hydrolase to thermal inactivation; in contrast, dithiothreitol promoted inactivation (Moss *et al.*, 1986).

Hydrolase activity was dependent on the presence of an intact ADP-ribose moiety; ADP-ribosylarginine was an excellent substrate, whereas phosphoribosylarginine and ribosylarginine were not readily cleaved (Moss *et al.*, 1986). In contrast to the apparent necessity for an intact ADP-ribose moiety, only the guanidino moiety of arginine was required for cleavage (Moss *et al.*, 1986). ADP-ribosylguanidine and ADP-ribosylarginine were approximately equally effective as substrates (Moss *et al.*, 1986). In agreement with these data was the finding that ADP-ribose, but not ADP, AMP or arginine, was a potent competitive inhibitor of the hydrolase reaction (Moss *et al.*, 1986). If the hydrolase were to participate in a cycle of reversible ADP-ribosylation, the product of the transferase reaction should be the hydrolase substrate. As noted, the transferase synthesizes exclusively α-anomeric ADP-ribosylarginine. Since the hydrolase preferentially cleaves α- rather than β-ADP-ribosylarginine (Moss *et al.*, 1986), all of the data are consistent with the view that an ADP-ribosylation cycle does exist in animal tissues, with NAD:arginine ADP-ribosyltransferases catalyzing the forward arm, and ADP-ribosylarginine hydrolases responsible for reversing the modification (Fig. 4).

ACKNOWLEDGMENTS

We thank Catherine Magruder for expert secretarial assistance.

REFERENCES

Abood, M. E., Hurley, J. B., Pappone, M.-C., Bourne, H. R., and Stryer, L. (1982). Functional homology between signal-coupling proteins. Cholera toxin inactivates the GTPase activity of transducin. *J. Biol. Chem.* **257**, 10540–10543.

Bennett, V., and Cuatrecasas, P. (1975). Mechanism of activation of adenylate cyclase by *Vibrio cholerae* enterotoxin. *J. Membr. Biol.* **22**, 29–52.

Burns, D. L., Moss, J., and Vaughan, M. (1982). Choleragen-stimulated release of guanyl nucleotides from turkey erythrocyte membranes. *J. Biol. Chem.* **257**, 32–34.

Burns, D. L., Moss, J., and Vaughan, M. (1983). Release of guanyl nucleotides from the regulatory subunit of adenylate cyclase. *J. Biol. Chem.* **258**, 1116–1120.

Carpenter, C. C. J. (1980). Clinical and pathophysiologic features of diarrhea caused by *Vibrio cholerae* and *Escherichia coli*. *In* "Secretory Diarrhea" (M. Field, J. S. Fordtran, and S. G. Schultz, eds.), pp. 67–83. American Physiological Society, Bethesda, Maryland.

Cassel, D., and Pfeuffer, T. (1978). Mechanism of cholera toxin action: Covalent modification of the guanyl nucleotide-binding protein of the adenylate cyclase system. *Proc. Natl. Acad. Sci. U.S.A.* **75**, 2669–2673.

Cassel, D., and Selinger, Z. (1977). Mechanism of adenylate cyclase activation by cholera toxin: Inhibition of GTP hydrolysis at the regulatory site. *Proc. Natl. Acad. Sci. U.S.A.* **74**, 3307–3311.

Chang, P. P., Fishman, P. H., Ohtomo, N., and Moss, J. (1983). Degradation of choleragen bound to cultured human fibroblasts and mouse neuroblastoma cells. *J. Biol. Chem.* **258**, 426–430.

Chang, P. P., Tsai, S.-C., Adamik, R., Twiddy, E. M., and Holmes, R. K. (1988). Activation of the ADP-ribosyltransferase activity of *Escherichia coli* heat-labile enterotoxins by 19kDa guanine nucleotide-binding proteins. *Clin. Res.* **36**, 578A.

Clements, J. D., and Finkelstein, R. A. (1979). Isolation and characterization of homogenous heat-labile enterotoxins with high specific activity from *Escherichia coli* cultures. *Infect. Immun.* **24**, 760–769.

Clements, J. D., Yancey, R. J., and Finkelstein, R. A. (1980). Properties of homogeneous heat-labile enterotoxin from *Escherichia coli*. *Infect. Immun.* **29**, 91–97.

Cooper, D. M. F., Jagus, R., Somers, R. L., and Rodbell, M. (1981). Cholera toxin modifies diverse GTP-modulated regulatory proteins. *Biochem. Biophys. Res. Commun.* **101**, 1179–1185.

Craig, S. W., and Cuatrecasas, P. (1975). Mobility of cholera toxin receptors on rat lymphocyte membranes. *Proc. Natl. Acad. Sci. U.S.A.* **72**, 3844–3848.

Cuatrecasas, P. (1973a). Interaction of *Vibrio cholerae* enterotoxin with cell membranes. *Biochemistry* **12**, 3547–3558.

Cuatrecasas, P. (1973b). Gangliosides and membrane receptors for cholera toxin. *Biochemistry* **12**, 3558–3566.

Cuatrecasas, P. (1973c). *Vibrio cholerae* choleragenoid. Mechanism of inhibition of cholera toxin action. *Biochemistry* **12**, 3577–3581.

Dallas, W. S., and Falkow, S. (1980). Amino acid sequence homology between toxin and *Escherichia coli* heat-labile toxin. *Nature* **288**, 499–501.

Dallas, W. S., Gill, D. M., and Falkow, S. (1979). Cistrons encoding *Escherichia coli* heat-labile toxin. *J. Bacteriol.* **139**, 850–858.

Enomoto, K., and Gill, D. M. (1980). Cholera toxin activation of adenylate cyclase. Roles of nucleoside triphosphates and a macromolecular factor in the ADP ribosylation of the GTP-dependent regulatory component. *J. Biol. Chem.* **255**, 1252–1258.

Farfel, Z., Kaslow, H. R., and Bourne, H. R. (1979). A regulatory component of adenylate cyclase is located on the inner surface of human erythrocyte membranes. *Biochem. Biophys. Res. Commun.* **90**, 1237–1241.

Finkelstein, R. A. (1973). Cholera. *CRC Crit. Rev. Microbiol.* **2**, 553–623.

Fishman, P. H. (1980). Mechanism of action of cholera toxin: Studies on the lag period. *J. Membr. Biol.* **54**, 61–72.

Fishman, P. H., and Atikkan, E. E. (1980). Mechanism of action of cholera toxin: Effect of receptor density and multivalent binding on activation of adenylate cyclase. *J. Membr. Biol.* **54**, 51–60.

Fishman, P. H., Moss, J., and Vaughan, M. (1976). Uptake and metabolism of gangliosides in transformed mouse fibroblasts. Relationship of ganglioside structure to choleragen response. *J. Biol. Chem.* **251**, 4490–4494.

Fishman, P. H., Moss, J., and Osborne, J. C., Jr. (1978). Interaction of choleragen with the oligosaccharide of ganglioside G_{M1}: Evidence for multiple oligosaccharide binding sites. *Biochemistry* **17**, 711–716.

Gill, D. M. (1975). Involvement of nicotinamide adenine dinucleotide in the action of cholera toxin *in vitro*. *Proc. Natl. Acad. Sci. U.S.A.* **72**, 2064–2068.

Gill, D. M. (1976a). The arrangement of subunits in cholera toxin. *Biochemistry* **15**, 1242–1248.

Gill, D. M. (1976b). Multiple roles of erythrocyte supernatant in the activation of adenylate cyclase by *Vibrio cholerae* toxin *in vitro*. *J. Infect. Dis.* **133**(suppl.), S55–S63.

Gill, D. M. (1977). Mechanism of action of cholera toxin. *In* "Advances in Cyclic Nucleotide Research", Vol. *8* (P. Greengard and G. A. Robison, eds.), pp. 85–118. Raven Press, New York.

Gill, D. M., and King, C. A. (1975). The mechanism of action of cholera toxin in pigeon erythrocyte lysates. *J. Biol. Chem.* **250**, 6424–6432.

Gill, D. M., and Meren, R. (1978). ADP-ribosylation of membrane proteins catalyzed by cholera toxin: Basis of the activation of adenylate cyclase. *Proc. Natl. Acad. Sci. U.S.A.* **75**, 3050–3054.

Gill, D. M., and Meren, R. (1983). A second guanyl nucleotide-binding site associated with adenylate cyclase. Distinct nucleotides activate adenylate cyclase and permit ADP-ribosylation by cholera toxin. *J. Biol. Chem.* **258**, 11908–11914.

Gill, D. M., and Rappaport, R. S. (1979). Origin of the enzymatically active A_1 fragment of cholera toxin. *J. Infect. Dis.* **139**, 674–680.

Gill, D. M., and Richardson, S. H. (1980). Adenosine diphosphate-ribosylation of adenylate cyclase catalyzed by heat-labile enterotoxin of *Escherichia coli:* Comparison with cholera toxin. *J. Infect. Dis.* **141**, 64–70.

Gilman, A. G. (1987). G proteins: Transducers of receptor-generated signals. *Annu. Rev. Biochem.* **56**, 615–649.

Hansson, H.-A., Holmgren, J., and Svennerholm, L. (1977). Ultrastructural localization of cell membrane G_{M1} ganglioside by cholera toxin. *Proc. Natl. Acad. Sci. U.S.A.* **74**, 3782–3786.

Holmgren, J. (1973). Comparison of the tissue receptors for *Vibrio cholerae* and *Escherichia coli* enterotoxins by means of gangliosides and natural cholera toxoid. *Infect. Immun.* **8**, 851–859.

Holmgren, J., and Lönnroth, I. (1980). Structure and function of enterotoxins and their receptors. *In* "Cholera and Related Diarrheas" (O. Ouchterlony and J. Holmgren, eds.), pp. 88–103. Karger, Basel.

Holmgren, J., Lönnroth, I., and Svennerholm, L. (1973). Tissue receptor for cholera exotoxin: Postulated structure from studies with G_{M1} ganglioside and related glycolipids. *Infect. Immun.* **8**, 208–214.

Holmgren, J., Månsson, J.-E., and Svennerholm, L. (1974). Tissue receptor for cholera exotoxin: Structural requirements of G_{M1} ganglioside in toxin binding and inactivation. *Med. Biol.* **52**, 229–233.

Holmgren, J., Fredman, P., Lindblad, M., Svennerholm, A.-M., and Svennerholm, L. (1982). Rabbit intestinal glycoprotein receptor for *Escherichia coli* heat-labile enterotoxin lacking affinity for cholera toxin. *Infect. Immun.* **38**, 424–433.

Holmgren, J., Lindblad, M., Fredman, P., Svennerholm, L., and Myrvold, H. (1985). Comparison of receptors for cholera and *Escherichia coli* enterotoxins in human intestine. *Gastroenterology* **89**, 27–35.

Johnson, G. L., Kaslow, H. R., and Bourne, H. R. (1978). Genetic evidence that cholera toxin substrates are regulatory components of adenylate cyclase. *J. Biol. Chem.* **253**, 7120–7123.

Joseph, K. C., Kim, S. U., Stieber, A., and Gonatas, N. K. (1978). Endocytosis of cholera toxin into neuronal GERL. *Proc. Natl. Acad. Sci. U.S.A.* **75**, 2815–2819.

Joseph, K. C., Stieber, A., and Gonatas, N. K. (1979). Endocytosis of cholera toxin in GERL-like structures of murine neuroblastoma cells pretreated with G_{M1} ganglioside-cholera toxin internalization into neuroblastoma GERL. *J. Cell Biol.* **81**, 543–554.

Kahn, R. A., and Gilman, A. G. (1984a). Purification of a protein cofactor required for ADP-ribosylation of the stimulatory regulatory component of adenylate cyclase by cholera toxin. *J. Biol. Chem.* **259**, 6228–6234.

Kahn, R. A., and Gilman, A. G. (1984b). ADP-ribosylation of G_S promotes the dissociation of its α and β subunits. *J. Biol. Chem.* **259**, 6235–6240.

Kahn, R. A., and Gilman, A. G. (1986). The protein cofactor necessary for ADP-ribosylation of G_S by cholera toxin is itself a GTP binding protein. *J. Biol. Chem.* **261**, 7906–7911.

Kassis, S., Hagmann, J., Fishman, P. H., Chang, P. P., and Moss, J. (1982).

Mechanism of action of cholera toxin on intact cells: Generation of A_1 peptide and activation of adenylate cyclase. *J. Biol. Chem.* **257**, 12148–12152.

Kelly, M. T. (1986). Cholera: A worldwide perspective. *Ped. Infect. Dis.* **5**, 5101–5105.

Kimberg, D. V., Field, M., Johnson, J., Henderson, A., and Gershon, E. (1971). Stimulation of intestinal mucosal adenyl cyclase by cholera enterotoxin and prostaglandins. *J. Clin. Invest.* **50**, 1218–1230.

Lai, C.-Y. (1986). Bacterial protein toxins with latent ADP-ribosyl transferases activities. *Adv. Enzymol.* **58**, 99–140.

Le Vine, H., III, and Cuatrecasas, P. (1981). Activation of pigeon erythrocyte adenylate cyclase by cholera toxin. Partial purification of an essential macromolecular factor by horse erythrocyte cytosol. *Biochim. Biophys. Acta* **672**, 248–261.

Lin, M. C., Welton, A. F., and Berman, M. F. (1978). Essential role of GTP in the expression of adenylate cyclase activity after cholera toxin treatment. *J. Cyclic Nucleotide Res.* **4**, 159–168.

Mekalanos, J. J., Collier, R. J., and Romig, W. R. (1979a). Enzymic activity of cholera toxin. I. New method of assay and the mechanism of ADP-ribosyl transfer. *J. Biol. Chem.* **254**, 5849–5854.

Mekalanos, J. J., Collier, R. J., and Romig, W. R. (1979b). Enzyme activity of cholera toxin. II. Relationships to proteolytic processing, disulfide bond reduction, and subunit composition. *J. Biol. Chem.* **254**, 5855–5861.

Mekalanos, J. J., Swartz, D. J., Pearson, G. D. N., Harford, N., Groyne, F., and de Wilde, M. (1983). Cholera toxin genes: Nucleotide sequence, deletion analysis and vaccine development. *Nature* **306**, 551–557.

Motesano, R., Roth, J., Robert, A., and Orci, L. (1982). Non-coated membrane invaginations are involved in binding and internalization of cholera and tetanus toxins. *Nature (London)* **296**, 651–653.

Moss, J. (1987). Signal transduction by receptor-responsive guanyl nucleotide-binding proteins: Modulation by bacterial toxin-catalyzed ADP-ribosylation. *Clin. Res.* **35**, 451–458.

Moss, J., and Richardson, S. H. (1978). Activation of adenylate cyclase by heat-labile *Escherichia coli* enterotoxin. Evidence for ADP-ribosyltransferase activity similar to that of choleragen. *J. Clin. Invest.* **62**, 281–285.

Moss, J., and Stanley, S. J. (1981). Histone-dependent and histone-independent forms of an ADP-ribosyltransferase from human and turkey erythrocytes. *Proc. Natl. Acad. Sci. U.S.A.* **78**, 4809–4812.

Moss, J., and Vaughan, M. (1977a). Choleragen activation of solubilized adenylate cyclase: Requirement for GTP and protein activator for demonstration of enzymatic activity. *Proc. Natl. Acad. Sci. U.S.A.* **74**, 4396–4400.

Moss, J., and Vaughan, M. (1977b). Mechanism of action of choleragen. Evidence for ADP-ribosyltransferase activity with arginine as an acceptor. *J. Biol. Chem.* **252**, 2455–2457.

Moss, J., and Vaughan, M. (1978). Isolation of an avian erythrocyte protein possessing ADP-ribosyltransferase activity and capable of activating adenylate cyclase. *Proc. Natl. Acad. Sci. U.S.A.* **75**, 3621–3624.

Moss, J., and Vaughan, M. (1988a). ADP-ribosylation of guanyl nucleotide-binding regulatory proteins by bacterial toxins. *Adv. Enzymol.* **61**, 303–379.

Moss, J., and Vaughan, M. (1988b). *In* "Handbook of Natural Toxins, Vol. 4, Bacterial Toxins" (A. T. Tu and M. C. Hardegree, eds.), pp 39–88. Dekker, New York.

Moss, J., Fishman, P. H., Manganiello, V. C., Vaughan, M., and Brady, R. O. (1976a). Functional incorporation of ganglioside into intact cells: Induction of choleragen responsiveness. *Proc. Natl. Acad. Sci. U.S.A.* **73**, 1034–1037.

Moss, J., Manganiello, V. C., and Vaughan, M. (1976b). Hydrolysis of nicotinamide adenine dinucleotide by choleragen and its A promoter: Possible role in the activation of adenylate cyclase. *Proc. Natl. Acad. Sci. U.S.A.* **73**, 4424–4427.

Moss, J., Garrison, S., Fishman, P. H., and Richardson, S. H. (1979a). Gangliosides sensitize unresponsive fibroblasts to *Escherichia coli* heat-labile enterotoxin. *J. Clin. Invest.* **64**, 381–384.

Moss, J., Garrison, S., Oppenheimer, N. J., and Richardson, S. H. (1979b). NAD-dependent ADP-ribosylation of arginine and proteins by *E. coli* heat-labile enterotoxin. *J. Biol. Chem.* **254**, 6270–6272.

Moss, J., Stanley, S. J., and Lin, M. C. (1979c). NAD glycohydrolase and ADP-ribosyltransferase activities are intrinsic to the A_1 peptide of choleragen. *J. Biol. Chem.* **254**, 11993–11996.

Moss, J., Stanley, S. J., and Oppenheimer, N. J. (1979d). Substrate specificity and partial purification of a stereospecific NAD- and guanidine-dependent ADP-ribosyltransferase from avian erythrocytes. *J. Biol. Chem.* **254**, 8891–8894.

Moss, J., Stanley, S. J., Morin, J. E., and Dixon, J. E. (1980a). Activation of choleragen by thiol: Protein disulfide oxidoreductase. *J. Biol. Chem.* **255**, 11085–11087.

Moss, J., Stanley, S. J., and Watkins, P. A. (1980b). Isolation and properties of an NAD- and guanidine-dependent ADP-ribosyltransferase from turkey erythrocytes. *J. Biol. Chem.* **255**, 5838–5840.

Moss, J., Stanley, S. J., Watkins, P. A., and Vaughan, M. (1980c). ADP-ribosyltransferase activity of mono- and multi-(ADP-ribosylated) choleragen. *J. Biol. Chem.* **255**, 7835–7837.

Moss, J., Osborne, J. C., Jr., Fishman, P. H., Nakaya, S., and Robertson, D. C. (1981a). *Escherichia coli* heat-labile enterotoxin: Ganglioside specificity and ADP-ribosyltransferase activity. *J. Biol. Chem.* **256**, 12861–12865.

Moss, J., Stanley, S. J., and Osborne, J. C., Jr. (1981b). Effect of self-association on activity of an ADP-ribosyltransferase from turkey erythrocytes: Conversion of inactive oligomers to active promoters by chaotropic salts. *J. Biol. Chem.* **256**, 11452–11456.

Moss, J., Stanley, S. J., and Osborne, J. C., Jr. (1982). Activation of an NAD:arginine ADP-ribosyltransferase by histone. *J. Biol. Chem.* **257**, 1660–1663.

Moss, J., Stanley, S. J., Burns, D. L., Hsia, J. A., Yost, D. A., Myers, G. A., and Hewlett, E. L. (1983). Activation by thiol of the latent NAD glycohydrolase and ADP-ribosyltransferase activities of *Bordetella pertussis* toxin (Islet Activating Protein). *J. Biol. Chem.* **258**, 11879–11882.

Moss, J., Osborne, J. C., Jr., and Stanley, S. J. (1984). Activation of an erythrocyte NAD : arginine ADP-ribosyltransferase by lysolecithin and nonionic and zwitterionic detergents. *Biochemistry* **23**, 1353–1357.

Moss, J., Jacobson, M. K., and Stanley, S. J. (1985). Reversibility of arginine-specific mono(ADP-ribosyl)ation: Identification in erythrocytes of an ADP-ribose-L-arginine cleavage enzyme. *Proc. Natl. Acad. Sci. U.S.A.* **82**, 5603–5607.

Moss, J., Oppenheimer, N. J., West, R. E., Jr., and Stanley, S. J. (1986). Amino acid specific ADP-ribosylation: Substrate specificity of an ADP-ribosylarginine hydrolase from turkey erythrocytes. *Biochemistry* **25**, 5408–5414.

Nakaya, S., Moss, J., and Vaughan, M. (1980). Effects of nucleoside triphosphates on choleragen-activated brain adenylate cyclase. *Biochemistry* **19**, 4871–4874.

Noda, M., Tsai, S.-C., Adamik, R., Moss, J., and Vaughan, M. (1988). Effects of detergents on activation of cholera toxin by 19kDa membrane and soluble brain proteins (ADP-ribosylation factors). *FASEB J.* **2**, 6091.

Northup, J. K., Sternweis, P. C., Smigel, M. D., Schleifer, L. S., Ross, E. M., and Gilman, A. G. (1980). Purification of the regulatory component of adenylate cyclase. *Proc. Natl. Acad. Sci. U.S.A.* **77**, 6516–6520.

O'Keefe, E., and Cuatrecasas, P. (1974). Cholera toxin mimics melanocyte stimulating hormone in inducing differentiation in melanoma cells. *Proc. Natl. Acad. Sci. U.S.A.* **71**, 2500–2504.

Oppenheimer, N. J. (1978). Structural determination and stereospecificity of the choleragen-catalyzed reaction of NAD^+ with guanidines. *J. Biol. Chem.* **253**, 4907–4910.

Osborne, J. C., Jr., Chang, P. P., and Moss, J. (1982). Kinetic analysis of agonist-receptor interactions: Model for the "irreversible" binding of choleragen to human fibroblasts. *J. Biol. Chem.* **257**, 10210–10214.

Osborne, J. C., Jr., Stanley, S. J., and Moss, J. (1985). Kinetic mechanisms of two NAD : arginine ADP-ribosyltransferases: The soluble, salt-stimulated transferase from turkey erythrocytes and choleragen, a toxin from *Vibrio cholerae*. *Biochemistry* **24**, 5235–5240.

Pappenheimer, A. M., Jr. (1977). Diphtheria toxin. *Annu. Rev. Biochem.* **46**, 69–94.

Pierce, N. F. (1973). Differential inhibitory effects of cholera toxoids and ganglioside on the enterotoxins of *Vibrio cholerae* and *Escherichia coli*. *J. Exp. Med.* **137**, 1009–1023.

Pinkett, M. O., and Anderson, W. B. (1982). Plasma membrane-associated component(s) that confer(s) cholera toxin sensitivity to adenylate cyclase. *Biochim. Biophys. Acta* **714**, 337–343.

Revesz, T., and Greaves, M. (1975). Ligand-induced redistribution of lymphocyte membrane ganglioside G_{M1}. *Nature (London)* **257**, 103–106.

Robertson, D. C., Kunkel, S. L., and Gilligan, P. H. (1980). Structure and function of *E. coli* heat-labile enterotoxin. *In* "Proceedings of the Fifteenth Joint Conference on Cholera," pp. 389–400, (DHEW Publ. No. (NIH) 80-2003).

Sattler, J., Schwarzmann, G., Knack, I., Röhm, K.-H., and Wiegandt, H. (1978).

Studies of ligand binding to cholera toxin. III. Cooperativity of oligosaccharide binding. *Hoppe-Seyler's Z. Physiol. Chem.* **359**, 719–723.

Schleifer, L. S., Kahn, R. A., Hanski, E., Northup, J. K., Sternweis, P. C., and Gilman, A. G. (1982). Requirements for cholera toxin-dependent ADP-ribosylation of the purified regulatory component of adenylate cyclase. *J. Biol. Chem.* **257**, 20–23.

Schwarzmann, G., Mraz, W., Sattler, J., Schindler, R., and Wiegandt, H. (1978). Comparison of the interaction of mono- and oligovalent ligands with cholera toxin–Demonstration of aggregate formation at low ligand concentrations. *Hoppe-Seyler's Z. Physiol. Chem.* **359**, 1277–1286.

Sedlacek, H. H., Stärk, J., Seiler, F. R., Ziegler, W., and Wiegandt, H. (1976). Cholera toxin induced redistribution of sialoglycolipid receptor at the lymphocyte membrane. *FEBS Lett.* **61**, 272–276.

Smith, K. P., Benjamin, R. C., Moss, J., and Jacobson, M. K. (1985). Identification of enzymatic activities which process protein bound mono(ADP-ribose). *Biochem. Biophys. Res. Commun.* **126**, 136–142.

So, M., Dallas, W. S., and Falkow, S. (1978). Characterization of an *Escherichia coli* plasmid encoding for synthesis of heat-labile toxin: Molecular cloning of the toxin determinant. *Infect. Immun.* **21**, 405–411.

Soman, G., Mickelson, J. R., Louis, C. F., and Graves, D. J. (1984). NAD:guanidino group specific mono ADP-ribosyltransferase activity in skeletal muscle. *Biochem. Biophys. Res. Commun.* **120**, 973–980.

Spicer, E. K., and Noble, J. A. (1982). *Escherichia coli* heat-labile enterotoxin. Nucleotide sequence of the A subunit gene. *J. Biol. Chem.* **257**, 5716–5721.

Spicer, E. K., Kavanaugh, W. M., Dallas, W. S., Falkow, S., Konigsberg, W. H., and Schafer, D. E. (1981). Sequence homologies between A subunits of *Escherichia coli* and *Vibrio cholerae* enterotoxins. *Proc. Natl. Acad. Sci. U.S.A.* **78**, 50–54.

Tanigawa, Y., Tsuchiya, M., Imai, Y., and Shimoyama, M. (1984). ADP-ribosyltransferase from hen liver nuclei. *J. Biol. Chem.* **259**, 2022–2029.

Trepel, J. B., Chuang, D.-M., and Neff, N. H. (1977). Transfer of ADP-ribose from NAD to choleragen: A subunit acts as catalyst and acceptor protein. *Proc. Natl. Acad. Sci. U.S.A.* **74**, 5440–5442.

Tsai, S.-C., Noda, M., Adamik, R., Moss, J., and Vaughan, M. (1987). Enhancement of choleragen ADP-ribosyltransferase activities by guanyl nucleotides and a 19-kDa membrane protein. *Proc. Natl. Acad. Sci. U.S.A.* **84**, 5139–5142.

Tsai, S.-C., Noda, M., Adamik, R., Chang, P. P., Chen, H.-C., Moss, J., and Vaughan, M. (1988). Stimulation of choleragen enzymatic activity by GTP and two soluble proteins purified from bovine brain. *J. Biol. Chem.* **263**, 1768–1772.

Tsuru, S., Matsuguchi, M., Ohtomo, N., Zinnaka, Y., and Takeya, K. (1982). Entrance of cholera enterotoxin subunits into cells. *J. Gen. Microbiol.* **128**, 497–502.

Van Dop, C., Tsubokawa, M., Bourne, H. R., and Ramachandran, J. (1984).

Amino acid sequence of retinal transducin at the site ADP-ribosylated by cholera toxin. *J. Biol. Chem.* **259**, 696–698.

van Heyningen, W. E., Carpenter, C. C. J., Pierce, N. F., and Greenough, W. B., III (1971). Deactivation of cholera toxin by ganglioside. *J. Infect. Dis.* **124**, 415–418.

Vasil, M. L., Holmes, R. K., and Finkelstein, R. A. (1975). Conjugal transfer of a chromosomal gene determining production of enterotoxin in *Vibrio cholerae*. *Science* **187**, 849–850.

Vasil, M. L., Kabat, D., and Iglewski, B. H. (1977). Structure-activity relationships of an exotoxin of *Pseudomonas aeruginosa*. *Infect. Immun.* **16**, 353–361.

West, R. E., Jr., and Moss, J. (1986). Amino acid specific ADP-ribosylation: Specific NAD:arginine mono-ADP-ribosyltransferases associated with turkey erythrocyte nuclei and plasma membranes. *Biochemistry* **25**, 8057–8062.

Wisnieski, B. J., and Bramhall, J. S. (1981). Photolabelling of cholera toxin subunits during membrane penetration. *Nature (London)* **289**, 319–321.

Xia, Q.-C., Chang, D., Blacher, R., and Lai, C.-Y. (1984). The primary structure of the COOH-terminal half of cholera toxin subunit A_1 containing the ADP-ribosylation site. *Arch. Biochem. Biophys.* **234**, 363–370.

Yost, D. A., and Moss, J. (1983). Amino acid-specific ADP-ribosylation: Evidence for two distinct NAD:arginine ADP-ribosyltransferases in turkey erythrocytes. *J. Biol. Chem.* **258**, 4926–4929.

CHAPTER 9

ADP-Ribosylation Factor of Adenylyl Cyclase: A 21-kDa GTP-Binding Protein

Richard A. Kahn

Laboratory of Biological Chemistry, Division of Cancer Treatment, National
Cancer Institute, National Institutes of Health, Bethesda, Maryland 20892

I. INTRODUCTION

ADP-ribosylation factor (ARF) is the 21-kDa GTP-binding protein cofactor in the cholera toxin-catalyzed ADP-ribosylation of the purified stimulatory regulatory subunit of adenylyl cyclase. The ARF activity in the toxin reaction is dependent on the binding of GTP. The lack of any measurable intrinsic GTPase activity suggests a novel mechanism of regulation of ARF activity. Although more abundant in the cytosol in some tissues, ARF appears to function in or at the cytosolic surface of the membranes and has been purified from membrane sources. Covalently attached myristic acid at the N-terminal glycine of ARF may explain the partial membrane localization and play a critical role in its function. ARF has been cloned and sequenced from bovine and yeast. Immunoblotting of cell or tissue lysates has revealed the presence of ARF in a wide variety of eukaryotic cells and tissues. Comparison of the primary structure of ARF to other GTP-binding proteins reveals that it is unique in having significant structural relationship with both the ras p21 and trimeric G proteins. The physiological role of ARF is likely to extend well beyond cholera or adenylyl cyclase and is currently under investigation.

A number of important factors, proteins, and detailed molecular mechanisms of cellular signal transduction systems have emerged from studies of the regulation of adenylate cyclase. Included in this list are the key role of guanine nucleotides as regulatory ligands, the identification of the family of regulatory GTP-binding proteins, and the subunit dissociation mechanism of activation of the G proteins. This volume is ample testimony to the current thinking in regard to the central role of G proteins as regulators of cell signaling and function. Among the more arcane factors first identified via interaction with adenylyl cyclase are aluminum (Sternweis and Gilman, 1982), or more accurately AlF^{4-}, and ARF (Schleifer et al., 1982). First identified as a cofactor in the cholera toxin-catalyzed activation of adenylyl cyclase, ARF appeared capable of much more when it was found to be a GTP-binding protein (Kahn and Gilman, 1986). The binding of GTP to ARF appears to regulate the interaction of ARF with another GTP-binding protein, $G_s\alpha$. This review will summarize what is known about the function and structure of ARF and why the author feels that ARF may prove to be an important regulatory protein. Whether its cellular role is limited to adenylyl cyclase is doubtful but unproven.

II. ROLE OF ADP-RIBOSYLATION FACTOR IN CHOLERA TOXIN ACTION

All of the pathophysiological effects of cholera toxin are a direct result of the irreversible activation of the stimulatory regulatory component of adenylyl cyclase, G_s.[1] ARF was first identified as the factor required in the cholera toxin-catalyzed ADP-ribosylation of purified G_s (Schleifer et al., 1982). The requirement for a cytosolic factor (CF) had first been described by Enomoto and Gill (1979) using pigeon erythrocyte membranes as the source of G_s and adenylyl cyclase. Gill and Meren (1983) have also described a membrane-associated protein with features similar to ARF. It now appears likely that all three of these factors are one and the same (J. Coburn, D. M. Gill, and R. Kahn, unpublished observation) and will be

---

[1] The actual substrate is the α subunit of G_s.

referred to herein as ARF. With the purification of G_s it became possible to study the relevant cholera toxin reaction and the consequences of this modification in greater detail than was previously possible. The demonstration that ADP-ribosylation of $G_s\alpha$ in membranes promotes subunit dissociation (Kahn and Gilman, 1984b) remains perhaps the clearest support for the subunit dissociation model as the mechanism of activation of G_s in membranes.

Using purified G_s as substrate it was possible to develop a quantitative assay and subsequently to purify ARF (Kahn and Gilman, 1984a). ARF has now been purified from three different membrane sources, rabbit liver, turkey erythrocyte, and bovine brain. The purified protein runs as a closely spaced doublet in sodium dodecyl sulfate (SDS) gels, with an apparent molecular weight of 21,000. More accurate hydrodynamic determination and later sequencing of the gene confirm this estimate of the molecular weight. No larger forms of ARF are detected either by immunoblotting or in the cDNA sequence of ARF. The two bands seen on SDS gels can be partially resolved and shown to have very similar specific activities (Kahn and Gilman, 1984a).

The basic features of the cholera toxin reaction were first described in 1978 by Cassel and Pfeuffer (1978) and Gill and Meren (1978). The enzymatic reaction is:

$$G_s + NAD \xrightarrow[\text{toxin}]{\text{cholera}} G_s - ADP\text{-ribose} + nicotinamide + H^+$$

Not shown here are the roles of the cofactors, GTP and ARF, in the reaction. The requirement for GTP was first shown by Enomoto and Gill (1979) and later amplified by Nakaya et al. (1980). It is now clear that GTP must bind to and activate ARF, allowing the formation of the $G_s \cdot ARF \cdot GTP$ complex, the substrate for cholera toxin (see below). Neither the GTP nor the ARF are altered in the reaction. Each cofactor is catalytic in that several molecules of G_s can be modified by a single $ARF \cdot GTP$. Thus, a more detailed sequence of events required for cholera toxin to activate G_s is shown below, broken down into four reactions:

$$ARF \cdot GDP + GTP \rightarrow ARF \cdot GTP + GDP \tag{1}$$

$$ARF \cdot GTP + G_s \rightarrow G_s \cdot ARF \cdot GTP \tag{2}$$

$$G_s \cdot ARF \cdot GTP + NAD \xrightarrow[\text{toxin}]{\text{cholera}} ADP\text{-ribose} - G_s \cdot ARF \cdot GTP \tag{3}$$

$$ADP\text{-ribose} - G_s \cdot ARF \cdot GTP \rightarrow ADP\text{-ribose} - G_s + ARF \cdot GTP \tag{4}$$

ARF was found to copurify with GDP in a 1 : 1 molar ratio. The binding of an activating ligand, either GTP or GTPγS, can be monitored by either radioligand binding or the enhancement of intrinsic fluorescence (Kahn and Gilman, 1986). In each case the ARF activity correlates very well with the extent of activation by guanine nucleotide triphosphate. Thus, it is likely that ARF exists in the cell in the inactive, GDP bound, form. The stimulus for activation is not known but should provide important clues to the role of ARF in normal cellular physiology.

Step (2), shown above, is currently the most controversial point in the cholera toxin/ARF field. While I indicate a direct interaction of ARF with $G_s\alpha$, others believe that ARF interacts with cholera toxin to enhance the enzymatic activity of the toxin, independent of substrate. In support of this latter view is the observation that artificial substrates (e.g. agmatine) of cholera toxin are modified to a greater extent with the inclusion of purified ARF (Tsai et al., 1987). Gill and Coburn (1987) reported the enhancement of incorporation of ADP-ribose equally into all of the protein substrates of cholera toxin after exposure of erythrocyte membranes to ARF. A point to be kept in mind is that cholera toxin is a very promiscuous enzyme capable of ADP-ribosylating virtually any protein with an arginine to a stoichiometry of at least 0.01–0.03. My own studies have concentrated on the effect of ARF on the physiological substrate, G_s, and are not directly comparable to the other results. When a number of purified proteins were examined for the ability to serve as substrates for cholera toxin, either alone or with added ARF or lipid we observed no consistent pattern (see Fig. 1). Because the stoichiometry of ADP-ribosylation was low it was necessary to use about 50-fold more protein than when G_s was the acceptor. The pattern of labeling seen in Fig. 1 is incompatible with a general increase in cholera toxin activity as a result of its interaction with ARF.

In separate experiments, when we varied the concentrations of the three proteins (G_s, ARF, and toxin) and examined the kinetics of the ADP-ribosylation reaction we observed an apparent K_m for ARF very near the concentration of G_s and more than tenfold below the concentration of toxin (Kahn and Gilman, 1984a). Finally, a lag was observed in the initial rate of ADP-ribosylation unless ARF, G_s, GTP, and lipid were preincubated prior to addition of the toxin. If the toxin was preincubated with ARF, GTP, and lipid the lag was present (Kahn and Gilman, 1984a). These three separate results are good support for the conclusion that ARF interacts with G_s first to form the complex. Unfortunately, no one has been able to demonstrate a direct association of ARF with another protein, presumably due to the presence of lipid which complicates such studies. A possible way to reconcile both sets of results is to conclude that ARF can bind both G_s and cholera toxin. In the absence of G_s the presence of ARF on the toxin would then stimulate the toxin to catalyze the modification of any other protein present.

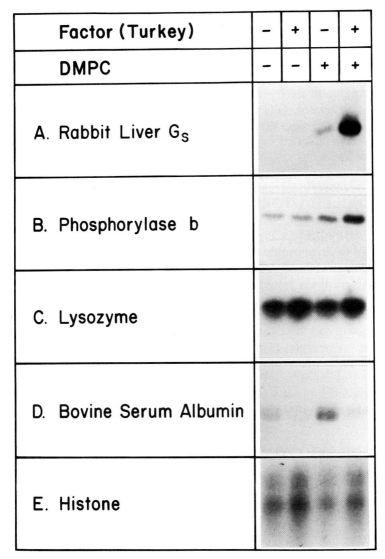

Factor (Turkey)	–	+	–	+
DMPC	–	–	+	+
A. Rabbit Liver G$_S$				
B. Phosphorylase b				
C. Lysozyme				
D. Bovine Serum Albumin				
E. Histone				

Fig. 1. Effects of ARF and lipid on cholera toxin-catalyzed ADP-ribosylation of purified proteins. Purified rabbit liver G$_s$ (10 μg/ml) or other purified proteins (500 μg/ml) were labeled with [^{32}P]NAD (20,000 CPM/pmol) and cholera toxin (40 μg/ml) for 1 hr at 30°C. The top panel indicates if purified turkey erythrocyte ARF (2 μg/ml) or lipid (dimyristoylphosphatidylcholine (DMPC), 3 mM) was present in the reaction. An autoradiogram of the region containing each protein from an 11% SDS-polyacrylamide gel is shown.

While the requirement for GTP to activate ARF is proven, the effect of GTP binding to the substrate G_s is less clear. ADP-ribosylation of G_s can occur at an apparently maximal rate when the substrate is inactive, i.e. in the GDP-bound form (Kahn and Gilman, 1986). It has not been possible to assess the effect, if any, of GTP binding to G_s on the ADP-ribosylation reaction due to the hydrolysis of GTP by G_s. However, using the nonhydrolyzable analog, GTPγS, showed that activated ($G_s\alpha \cdot GTP\gamma S$) is not a substrate for cholera toxin (Kahn and Gilman, 1984b). It will be interesting to learn if this effect results from a decreased affinity for either ARF or cholera toxin.

III. CHOLERA TOXIN AS PROBE FOR G_s

Using [^{32}P]NAD as the substrate in the cholera toxin reaction it was possible to demonstrate the presence of $G_s\alpha$ well before purification of G_s was achieved (Cassel and Pfeuffer, 1978). Many people have attempted to use cholera toxin labeling as a quantitative assay for G_s. This should be done with great caution. Even under optimal conditions cholera toxin is a slow and relatively indiscriminate enzyme. The requirements for metal, salt, lipid, and ARF make quantitation of G_s in crude mixtures very difficult to interpret. The lability of G_s and its inactivity as substrate for cholera toxin when nonhydrolyzable GTP analogs are bound further complicate true estimates of G_s content based on toxin labeling. The availability of specific antibodies directed against each subunit of the G proteins (Mumby et al., 1986) has hopefully made such methods unnecessary.

IV. ADP-RIBOSYLATION FACTOR AS GTP-BINDING PROTEIN

ARF has a single high-affinity binding site for guanine nucleotides. The affinity for GDP ($K_D = 40$ nM) is greater than that for GTP ($K_D = 90$ nM, Kahn and Gilman, 1986). Other nucleotides, including GMP, ATP, ADP, CTP, do not compete for GDP or GTP binding. The conditions necessary to detect nucleotide binding to ARF are unique. Binding of radiolabeled nucleotides to purified ARF requires high salt (e.g. 0.8 M NaCl), Mg^{2+} (μM), lipid (3 mM dimyristoylphosphatidylcholine), and detergent (2 mM sodium cholate). The lipid requirement can be met by a diverse array of lipids suggesting the need for a hydrophobic environment rather than a specific lipid. No binding was observed when ARF was assayed using conditions optimal for the $G\alpha$ or ras proteins. The combination of high ionic strength and a hydrophobic environment is probably best duplicated in the cell at a membrane/cytosol interface.

It appears that all of the GTP-binding proteins share a common turn-off mechanism; that is, hydrolysis of bound GTP. Inhibition of GTPase activity has been proposed as the mechanism of activation of $G_s\alpha$ by cholera toxin (Cassel and Selinger, 1977) and of a number of transforming mutations of *ras* p21 (e.g. see Gibbs *et al.*, 1985). However, unlike the G proteins or *ras*, purified ARF has no measurable intrinsic GTPase activity (Kahn and Gilman, 1986). When tested in a number of different buffers and conditions, no GTP hydrolysis over background was detected. The sensitivity of this assay allows us to put an upper estimate of 0.0015 min^{-1} for the GTPase activity of purified bovine brain ARF. Nevertheless, our working hypothesis is that ARF does have latent GTPase activity which is expressed only at the appropriate time or as a result of some unspecified protein interaction. Interaction with an effector is the leading candidate for a protein capable of activating the GTPase activity of ARF as is the case with EF-Tu (Kaziro, 1980). However, no ARF-dependent hydrolysis of GTP was detected when G_s was included in the assay (R. Kahn, unpublished observation). The presence in cytosol of a factor that can stimulate the GTPase activity of nontransforming *ras* proteins, without affecting the GTPase activity of the transforming forms of *ras* (Trahey and McCormick, 1987), suggests that there may exist a number of important and undiscovered regulatory elements in the action of the various GTP-binding proteins.

Like the $G\alpha$ and ras proteins, purified ARF contains tightly associated GDP. It is believed that all of these proteins exist in a basal state in the cell with tightly bound GDP. The cellular mechanisms for activation of ARF (binding GTP) and deactivation (hydrolysis of GTP) may both be distinct from the models proposed for the regulation of other important regulatory GTP-binding proteins.

V. CELLULAR LOCALIZATION OF ADP-RIBOSYLATION FACTOR

Although originally purified from membrane sources it is now clear that the majority of ARF is present in the cytosol. Both ARF activity and immunoreactivity are present in the cytosol and in very pure membrane preparations from a number of tissues (Kahn *et al.*, 1988). Based on the lipid and high-salt requirement for nucleotide binding we proposed that ARF is present primarily in the cytosol but is active at the membrane. The presence of myristic acid as the amino-terminal blocking group of ARF lends further support to the idea that ARF action occurs at the membrane.

Using a method for the detection of amino-terminal myristylglycine, we have demonstrated the presence of myristic acid as an N-terminal blocking group of purified bovine ARF. Thus, ARF joins such regulatory proteins as calcineurin B, pp60^{src}, and the catalytic subunit of cAMP-dependent protein kinase, as proteins which are myristylated and thought to act at the membrane surface (Levinson *et al.*, 1981; Krueger *et al.*, 1984; Politino and King, 1987).

The final step in the purification of bovine ARF is a molecular sieve column which partially resolves the ARF doublet. Testing fractions from this column for covalently attached myristic acid revealed that only the upper band of the doublet appears to be myristylated (C. Goddard and R. Kahn, unpublished observation). This is the first demonstration of any difference between the two bands, as they each have been shown to bind guanine nucleotides and have similar specific activities in the ARF assay. We cannot yet determine if myristylation is the only difference between the two bands seen in SDS gels. We know of no other protein found in cells which is partially myristylated. As myristylation is believed to be a co-translational event (Wilcox *et al.*, 1987), it is not clear why ARF is only partially myristylated. One possible explanation is based on studies of the substrate requirements for the enzyme *N*-myristyltransferase by Towler *et al.* (1987). Using synthetic peptides as substrates they observed that in addition to the absolute requirement for glycine at position 2 there is a strong preference for serine, threonine, or alanine at position 6 of the protein sequence, with alanine being about 500-fold less active. The sequence of bovine ARF (see below) includes an alanine in position 6. If the native proteins behave like the synthetic peptide substrates it is possible that the K_m for ARF is high enough that incomplete myristylation is observed. More likely, the lower band may represent a distinct gene product with conserved ARF activity.

VI. STRUCTURE OF ADP-RIBOSYLATION FACTOR

Partial amino acid sequencing of ARF was performed on purified bovine brain ARF. Oligonucleotide probes were then synthesized and used to clone the full-length cDNA from bovine adrenal chromaffin cells (Sewell and Kahn, 1988). The sequence of bovine ARF was obtained from translation of the adrenal ARF cDNA. The coding region of the bovine ARF gene was then used as a probe of genomic yeast DNA to obtain the yeast ARF gene. Both the bovine and yeast genes code for proteins of 181 amino acids

```
bARF:  MGNIFANLFKGLFGKKEMRILMVGLDAAGKTTILYKLKLGEIVTTIPTIG  50
yARF:  MGLFASKLFSNLFGNKEMRILMVGLDGAGKTTVLYKLKLGEVITTIPTIG  50
CPS1:  MGKVLSKIF   GNKEMRILMLGLDAAGKTTILYKLKLGQSVTTIPTVG  46


bARF:  FNVETVEYKNISFTVWDVGGQDKIRPLWRHYFQNTQGLIFVVDSNDRERV  100
yARF:  FNVETVQYKNISFTVWDVGGQDRIRSLWRHYYRNTEGVIFVVDSNDRSRI  100
CPS1:  FNVETVTYKNVKFNVWDVGGQDKIRPLWRHYYTGTQGLIFVVDCADRDRI  96


bARF:  NEAREELMRMLAEDELRDAVLLVFANKQDLPNAMNAAEITDKLGLHSLRH  150
yARF:  GEAREVMQRMLNEDELRNAAWLVFANKQDLPEAMSAAEITEKLGLHSIRN  150
CPS1:  DEARQELHRIINDREMRDAIILTFANKQDLPDAMKPHEIQEKLGLTRIRD  146


bARF:  RNWYIQATCATSGDGLYEGLDWLSNQLRNQK  181
yARF:  RPWFIQATCATSGEGLYEGLEWLSNSLKNST  181
CPS1:  RNWYVQPSCATTGDGLYEGLTWLTSNYKS   175
```

Fig. 2. Amino acid sequences of bovine ARF (bARF) and yeast ARF (yARF) and the processed chicken pseudogene CPS-1. Identities are shown in boxes.

(see Fig. 2). The bovine and yeast proteins share 76.8% identities, with many conservative substitutions. Nucleic acid sequences of the coding regions are 65% identical.

The amino acid sequence fully supports all of the biochemical findings in that consensus sequences for GTP binding and myristylation are present. Further, the first GTP-binding domain of ARF includes an aspartic acid residue at the position corresponding to position 12 of ras. This residue has been shown to be a critical determinant of intrinsic GTPase activity. Reduced levels of GTPase activity are observed in *ras* proteins with any residue other than glycine at this position.

In agreement with immunoblotting results the sequencing of ARF proves the unique identity of ARF. None of the other 21-kDa GTP-binding proteins identified, including ras, rho, ral, ypt, G_P, or the substrate for botulinum C3 toxin, have ARF activity, immunological cross-reactivity, or extensive sequence homology.

VII. COMPARISON TO OTHER GTP-BINDING PROTEINS

Scanning several computer databases for ARF-related sequences failed to produce any previously published ARF genes or highly homologous sequences. However, a review of the literature located a processed chicken pseudogene, termed CPS-1 (Alsip and Konkel, 1986), which shares 74.1%

sequence identity with bovine ARF and is probably derived from a chicken ARF gene. As the CPS-1 sequence retains a similar degree of identity, compared to bovine ARF, throughout the length of the sequence (see Fig. 2) it is likely that CPS-1 represents a close homolog of the expressed chicken ARF.

Alignment of bovine ARF to other GTP-binding proteins was performed both by eye and with the aid of the ALIGN computer program. No significant relationship was observed between ARF and any of the GTP-binding proteins that function as initiation or elongation factors in pro-karyotic or eukaryotic protein synthesis (e.g. EF-Tu, EF-G, etc.), or any of the published tubulins. Comparison of ARF to other GTP-binding proteins reveals a significant relationship with both the α subunits of the G proteins and ras families. These relationships can be seen either by alignment of short regions of high homology, as in Fig. 3, or more complex algorithms such as the ALIGN program, shown in Table I. Group I in Fig. 3 shows the alignment of the different proteins around the first GTP-binding domain. This is the region containing the residue (Asp^{26} of ARF and Gly^{12} of *ras* proteins) critical for determination of GTPase activity. Regions II and IV show alignments of ARF sequence with those in p21's or Gα's, respectively. In these cases there is no region in the G-protein α subunits which corresponds to region II and no region in the p21's which corresponds to region IV. Region III shows the longest stretch of conservation without any gaps. Included in this region is the sequence DVGGQ, which is absolutely conserved in all of the G proteins and ARF. The sequence DTAGQ is found in all forms of *ras* and the other 21-kDa proteins examined.

Table I contains the results of the ALIGN program in which two sequences are aligned and scored using a scoring matrix. The two sequences being compared are then randomized and scored again (in this case 100 times) to obtain a mean and standard deviation for random alignments. The numbers shown are the number of standard deviations away from the mean of the random scores that the optimally aligned sequences produce. It has been empirically determined that scores below 3 are not significant, scores between 3–6 are possibly significant, and scores above 6 are probably significant. Within this framework it is clear that only ARF has a significant relationship to both the Gα and *ras* families of GTP-binding proteins.

The significance of the described relationship between ARF and other GTP-binding proteins is of interest to two types of investigators; those attempting to describe the evolutionary origins of the GTP-binding proteins and those interested in functional crossreactivities between different GTP-binding proteins. The results shown in Table I are of more help to the former than the latter. Before becoming too enamored with the structural

```
                  I
     Gtα    33  LLLGAGESGKSTIVKQMK   50
     Giα    37  LLLGAGESGKSTIVKQMK   54
     Goα    37  LLLGAGESGKSTIVKQMK   54
     Gsα    44  LLLGAGESGKSTIVKQMR   61
I    ARF    21  LMVGLDAAGKTTILYKLK   38
     H-RAS   6  VVVGAGGVGKSALTIQLI   23
     RHO    14  VIVGDGACGKTCLLIVFS   31
     YPT    12  LLIGNSGVGKSCLLLRFS   29
     RAL    18  IMVGSGGVGKSALTLQFM   35


     ARF    46  IPTIG   50
     H-RAS  33  DPTIE   37
II   RHO    40  VPTIV   44
     YPT    38  ISTIG   42
     RAL    44  EPTIK   48


     Gtα   192  FRMFDVGGQRSERKKWIHCFEGVTCIIF   219
     Giα   196  FKMFDVGGQRSERKKWIHCFEGVTAIIF   223
     Goα   197  FRLFDVGGQRSERKKWIHCFEDVTATIF   224
     Gsα   219  FHMFDVGGQRDERRKWIQCFNDVTAIIF   246
III  ARF    63  FTVWDVGGQDKIRPLWRHYFQNTQGLIF    90
     H-RAS  53  LDILDTAGQEEYSAMRDQYMRTGEGFLC    80
     RHO    61  LALWDTAGQEDYDRLRPLSYPDSNVVLI    88
     YPT    59  LQIWDTAGQERFRTITSSYYRGSHGIII    86
     RAL    64  IDILDTAGQEDYAAIRDNYFRSGEGFLC    91


     ARF   158  TCAT   161
     Gtα   320  TCAT   323
IV   Giα   324  TCAT   327
     Goα   324  TCAT   327
     Gsα   364  TCAV   367


     Gtα   262  LFLNKKDV   269
     Giα   266  LFLNKKDL   273
     Goα   267  LFLNKKDL   274
     Gsα   289  LFLNKQDL   296
V    ARF   123  VFANKQDL   130
     H-RAS 113  LVGNKCDL   120
     RHO   120  LVGCKVDL   127
     YPT   118  LVGNKCDL   125
     RAL   124  LVGNKSDL   131
```

Fig. 3. Alignments of regions of bovine ARF with similar regions in members of the Gα family (G_s, G_i, G_o, G_t) or p21 GTP-binding proteins (H-ras, rho, ypt, ral). Amino acid identities are boxed. Alignments were made with the aid of the ALIGN program (see text).

Table I   Comparison of Amino Acid Sequences of 13 Different GTP-Binding Proteins Using the Program ALIGN[a]

	bARF	yARF	Gs	Gi	Go	Gt	RHO	YPT	RAL	R-RAS	yRAS	H-RAS	K-RAS
bARF		84.50	9.26	14.15	10.34	10.83	6.80	8.99	6.42	8.31	6.06	7.14	8.27
yARF	84.50		10.98	11.84	11.92	9.15	6.45	11.02	8.23	5.58	7.16	6.51	5.82
Gs	9.26	10.98		55.86	62.92	72.55	0.91	1.72	0.21	1.90	0.77	2.24	2.27
Gi	14.15	11.84	55.86		137.47	125.84	0.27	0.18	-0.06	3.30	2.42	2.03	2.09
Go	10.34	11.92	62.92	137.47		114.74	3.24	-0.32	1.87	2.79	1.59	3.26	3.28
Gt	10.83	9.15	72.55	125.84	114.74		0.81	1.07	2.04	3.22	1.19	1.96	1.62
RHO	6.80	6.45	0.91	0.27	3.24	0.81		21.74	19.14	24.84	22.61	20.15	21.03
YPT	8.99	11.02	1.72	0.18	-0.32	1.07	21.74		29.98	28.11	29.91	24.19	37.58
RAL	6.42	8.23	0.21	-0.06	1.87	2.04	19.14	29.98		53.41	47.28	66.98	48.50
R-RAS	8.31	5.58	1.90	3.30	2.79	3.22	24.84	28.11	53.41		63.80	58.39	61.05
yRAS	6.06	7.16	0.77	2.42	1.59	1.19	22.61	29.91	47.28	63.80		61.95	57.67
H-RAS	7.14	6.51	2.24	2.03	3.26	1.96	20.15	24.19	66.98	58.39	61.95		98.69
K-RAS	8.27	5.82	2.27	2.09	3.28	1.62	21.03	37.58	48.50	61.05	57.67	98.69	

[a] See text for an explanation of the ALIGN program. A gap penalty of 25 was used with a matrix bias parameter of 0. Boxes are drawn around clusters of similar scores to facilitate viewing only. bARF, bovine ARF; yARF, yeast ARF.

relationships between ARF and other GTP-binding proteins it is important to keep in mind that no functional crossreactivities have ever been observed between ARF and any of the other proteins mentioned above.

VIII. FUTURE DIRECTIONS

Most of the current work on ARF is directed at identifying the cellular effector. The presence of ARF in every eukaryotic cell tested and the high degree of conservation noted between divergent species argues strongly for a role more basic to cell survival than the role of cofactor in cholera intoxication. We are currently trying to exploit the unique nucleotide-binding properties and lack of GTP hydrolysis by ARF to identify cellular components capable of affecting either of these characteristics of ARF. Until a role for endogenous ADP-ribosyltransferases can be defined it will be difficult to test the notion that ARF may function as a regulator of this class of enzymes. Finally we are beginning to utilize yeast genetics to aid in the study of ARF expression and function. These studies should also provide the means to study the role of myristylation in ARF function and cellular localization.

REFERENCES

Alsip, G. R., and Konkel, D. A. (1986). A processed chicken pseudogene (CPS1) related to the *ras* oncogene superfamily. *Nucleic Acids Res.* **14**, 2123–2138.

Cassel, D., and Pfeuffer, T. (1978). Mechanism of cholera toxin action: Covalent modification of the guanyl nucleotide-binding protein of the adenylate cyclase system. *Proc. Natl. Acad. Sci. U.S.A.* **75**, 2669–2673.

Cassel, D., and Selinger, Z. (1977). Mechanism of adenylate cyclase activation by cholera toxin: Inhibition of GTP hydrolysis at the regulatory site. *Proc. Natl. Acad. Sci. U.S.A.* **74**, 3307–3311.

Enomoto, K. and Gill, D. M. (1979). Requirement for guanosine triphosphate in the activation of adenylate cyclase by cholera toxin. *J. Supramol. Struct.* **10**, 51–60.

Ferguson, K. M., Higashijima, T., Smigel, M. D., and Gilman, A. G. (1986). The influence of bound GDP on the kinetics of guanine nucleotide binding to G proteins. *J. Biol. Chem.* **261**, 7393–7399.

Gibbs, J. B., Sigal, I. S., and Scolnick, E. M. (1985). Biochemical properties of normal and oncogenic *ras* p21. *Trends Biochem. Sci.* **10**, 350–353.

Gill, D. M. and Coburn, J. (1987). ADP-ribosylation by cholera toxin: Functional analysis of a cellular system that stimulates the enzymic activity of cholera toxin fragment A$_1$. *Biochemistry* **26**, 6364–6371.

Gill, D. M. and Meren, R. (1978). ADP-ribosylation of membrane proteins catalyzed by cholera toxin: Basis of the activation of adenylate cyclase. *Proc. Natl. Acad. Sci. U.S.A.* **75**, 3050–3054.

Gill, D. M. and Meren, R. (1983). A second guanyl nucleotide-binding site associated with adenylate cyclase. *J. Biol. Chem.* **258**, 11908–11914.

Kahn, R. A., and Gilman, A. G. (1984a). Purification of a protein cofactor required for ADP-ribosylation of the stimulatory regulatory component of adenylate cyclase by cholera toxin. *J. Biol. Chem.* **259**, 6228–6234.

Kahn, R. A., and Gilman, A. G. (1984b). ADP-ribosylation of G$_s$ promotes the dissociation of its α and β subunits. *J. Biol. Chem.* **259**, 6235–6240.

Kahn, R. A., and Gilman, A. G. (1986). The protein cofactor necessary for ADP-ribosylation of G$_s$ by cholera toxin is itself a GTP binding protein. *J. Biol. Chem.* **261**, 7906–7911.

Kahn, R. A., Goddard, C., and Newkirk, M. (1988). Chemical and immunological characterization of the 21-kDa ADP-ribosylation factor of adenylate cyclase. *J. Biol. Chem.* **263**, 8282–8287.

Kaziro, Y. (1980). The role of guanosine 5'-triphosphate in polypeptide chain elongation. *Biochim. Biophys. Acta* **505**, 95–127.

Krueger, J. G., Garber, E. A., Chin, S. S.-M., Hanafusa, H., and Goldberg, A. R. (1984). Size-variant pp60src proteins of recovered avian sarcoma viruses interact with adhesion plaques as peripheral membrane proteins: Effects on cell transformation. *Mol. Cell. Biol.* **4**, 454–467.

Levinson, A. D., Courtneidge, S. A., and Bishop, M. J. (1981). Structural and functional domains of the Rous sarcoma virus transforming protein (pp60src). *Proc. Natl. Acad. Sci. U.S.A.* **78**, 1624–1628.

Mumby, S. M., Kahn, R. A., Manning, D. R., and Gilman, A. G. (1986). Antisera of designed specificity for subunits of guanine nucleotide-binding regulatory proteins. *Proc. Natl. Acad. Sci. U.S.A.* **83**, 265–269.

Nakaya, S., Moss, J., and Vaughan, M. (1980). Effects of nucleotide triphosphates on choleragen-activated brain adenylate cyclase. *Biochemistry* **19**, 4871–4874.

Politino, M., and King, M. M. (1987). Calcium- and calmodulin-sensitive interactions of calcineurin with phospholipids. *J. Biol. Chem.* **262**, 10109–10113.

Schleifer, L. S., Kahn, R. A., Hanski, E., Northup, J. K., Sternweis, P. C., and Gilman, A. G. (1982). Requirements for cholera toxin-dependent ADP-ribosylation of the purified regulatory component of adenylate cyclase. *J. Biol. Chem.* **257**, 20–23.

Sewell, J. L., and Kahn, R. A. (1988). Sequences of the bovine and yeast ADP-ribosylation factor and comparison to other GTP-binding proteins. *Proc. Natl. Acad. Sci. U.S.A.* **85**, 4620–4624.

Sternweis, P. C. and Gilman, A. G. (1982). Aluminum: A requirement for activation of the regulatory component of adenylate cyclase by fluoride. *Proc. Natl. Acad. Sci. U.S.A.* **79**, 4888–4891.

Towler, D. A., Eubanks, S. R., Towery, D. S., Adams, S. P., and Glaser, L. (1987). Amino-terminal processing of proteins by N-myristylation. *J. Biol. Chem.* **262**, 1030–1036.

Trahey, M. and McCormick, F. (1987). A cytoplasmic protein stimulates normal N-*ras* p21 GTPase but does not affect oncogenic mutants. *Science* **238**, 542–545.

Tsai, S.-C., Noda, M., Adamik, R., Moss, J., and Vaughan, M. (1987). Enhancement of choleragen ADP-ribosyltransferase activities by guanyl nucleotides and a 19 kDa membrane protein. *Proc. Natl. Acad. Sci. U.S.A.* **84**, 5139–5142.

Wilcox, C., Hu, J.-S., and Olson, E. N. (1987). Acylation of proteins with myristic acid occurs cotranslationally. *Science* **238**, 1275–1278.

CHAPTER 10

Transducin, Rhodopsin, and 3',5'-Cyclic GMP Phosphodiesterase: Typical G Protein-Mediated Transduction System

Marc Chabre, Philippe Deterre
Laboratoire de Biophysique Moléculaire et Cellulaire (Unité 520 du CNRS),
LBIO DRF, Centre d'Etudes Nucléaires, BP 85, F 38041 Grenoble, France

215

I. INTRODUCTION: RHODOPSIN – TRANSDUCIN – 3′,5′-CYCLIC GMP PHOSPHODIESTERASE CASCADE AS MODEL FOR G PROTEIN-MEDIATED PROCESSES

Among the various G proteins, transducin (T), the G protein that conveys the visual transduction message in vertebrate retinal rod outer segments (ROS), might appear as a special case. It is enormously abundant in the native organelles (10 to 20% of total protein content and about 50% of the soluble proteins), and has the convenient peculiarity of being soluble. Its extraction from ROS membranes and its manipulation in solution do not require the help of any detergent. Indeed, in the physiological transduction cycle, the guanine nucleotide-binding subunit Tα dissociates spontaneously from the membrane surface after the rhodopsin-catalyzed exchange of GDP for GTP or a nonhydrolyzable GTP analog. Tα–GTP may then take a cytoplasmic route to reach its effector protein, the cyclic GMP phosphodiesterase (PDE). This high solubility might be uniquely related to the very special morphology of the retinal rod outer segments, in which the cytoplasmic volume is confined to extremely thin layers about 150 Å thick between rod disc membrane surfaces. An activated Tα cannot drift very far away from the membrane that bears both the receptor, rhodopsin, and the effector enzyme, PDE, between which transducin shuttles in its amplifying cascade (Chabre, 1985; Stryer, 1986; Hurley, 1987).

However, transducin may not differ much from other G proteins: the Tβ subunit is strictly identical to the 36-kDa β subunit of G_s and G_i (Codina *et al.*, 1986). The small Tγ subunit, which is not dissociable from Tβ without denaturation of the complex, is indeed notably different from the other Gγ. Its sequence is quite hydrophilic (Hurley *et al.*, 1984) and this must be related to the fact that the undissociable T$\beta\gamma$ complex is only weakly membrane-bound and is solubilized from the membrane in low ionic-strength media. The unknown sequences of the γ subunits of other G proteins must then account for the much tighter membrane attachments of the corresponding G$\beta\gamma$ complexes. As for the guanine nucleotide-binding subunit Tα, its sequence does not differ from that of other Gα in a way that would suggest a deep membrane penetration of the latter. All Gα are fundamentally cytoplasmic proteins; all of them are terminated on their amino end by a glycine which is the binding site for a myristyl in $G_o\alpha$ and $G_i\alpha$, but which remains free in Tα and in $G_s\alpha$ (Muss *et al.*, 1987). This posttranslational modification is probably solely responsible for the different degrees of membrane attachment of the various Gα. Many observations, previously made on transducin because of its comparatively easy biochemistry, can be generalized for other G proteins, e.g., the existence of a γ subunit was first shown on transducin (Kühn, 1980). The dissociation of the GTP-binding form of the α subunit from the $\beta\gamma$ complex was first reported for transducin (Kühn, 1981; Fung et al., 1981), and the fact that an activated, i.e., GTP- or GTPγS-bound, α subunit may be soluble in the absence of detergent was recently observed also for $G_s\alpha$ (Lynch et al., 1986). Therefore, despite its apparent peculiarities, the high-gain rhodopsin–transducin–PDE cascade appears as an instructive model system for the coupling by G proteins of membrane receptors to effectors of cytoplasmic messengers.

On the receptor side, rhodopsin now stands as the archetype of a large class of 7α helix receptors which already includes the M1 and M2 muscarinic, the α_2-, β_1-, and β_2-adrenergic, and the substance K receptors (Dohlman et al., 1987), and is growing very fast. Rhodopsin is exceptionally abundant in the rod cells: as compared to the $10^4 - 10^5$ specific receptors usually found in the plasma membrane of an hormone-responsive cell, there are nearly 10^8 rhodopsin molecules in a mammalian retinal rod, and more than 10^9 in the large rods of batrachians. These enormous numbers are necessary because of the special physical character of the signal to be received. Unlike chemical transmitter molecules photons do not diffuse in the extracellular medium until they are captured by a receptor. Whereas a limited number of receptors, covering less than 1% of the plasma membrane surface, are sufficient to confer on a cell an efficiency close to the theoretical maximum for the capture of a diffusing molecule (Berg, 1983), a much larger number of pigmented receptors is needed to get a significant probability of absorbing photons that cross the cell only once on a straight line. Visual receptors are probably the only cells in which the number of receptors of a given type is in large excess over that of the G proteins to which these receptors couple. This anomalous stoichiometry has two major implications: (1) Conceptually, it excludes a priori any "precoupling" between the inactivated receptors and the G proteins. As there are ten times more receptors than transducins, if the latter are bound to unactivated receptors, a maximum of only 10% of the receptors could be precoupled and the remaining 90% would then be totally ineffective! (2) Technically, the excess of rhodopsin allows a very simple and efficient purification of transducin (Kühn, 1980). By photoexciting a population of rhodopsin that is larger than that of transducin, one induces the "postcoupling" of all transducins to the activated receptors, R*. This R*–T interaction allows the release of GDP bound to transducin and, if guanine nucleotides have been suppressed from the medium, transducin molecules with empty sites will remain tightly, and permanently, bound to R* and therefore to the membrane fraction. All other soluble proteins can then be washed out, before an addition of GTP or GTPγS selectively releases transducin from R* and from the membrane. This is essentially a one-step purification, to 90% homogeneity, employing affinity binding of the G protein to its activating receptor. The excess of rhodopsin and the easy quantitation of its activation by calibrated light flashes also offers unrivaled potential for studying the R*–T complex, an instructive model for the capital step of catalytic interaction between a receptor and the G protein which transmit its signal.

On the effector side, transducin activates a peripherally membrane-bound cGMP phosphodiesterase. Two other major classes of enzyme known to be activated by G proteins are adenylyl cyclases, which seem to be intrinsic transmembrane proteins, and phosphatidylinositol-specific phospholipases C (PLC) which are soluble or peripherally attached to the membrane. These three types of enzyme share the common feature of acting on a phosphodiester link, either to synthesize it (cyclase) or to cleave it (PDE and PLC). In all cases the G protein α subunit, in its GTP-binding form, seems to be solely responsible for the activation. In the case of transducin, the activation of PDE is the release of an inhibition, due, as we shall discuss, to the physical removal of inhibitory subunits by Tα. Whether the other types of Gα–GTP activate their respective effectors by an analogous process remains to be seen.

Rhodopsin and transducin (the three subunits) from cattle were the first receptor and G protein to be cloned and sequenced [see review by Hurley (1987) for data and references]. The sequence of the PDE inhibitor and that of the catalytic subunit PDEα have also recently been obtained (Ovchinnikov, 1987). To complete the protein inventory of the visual transduction cascade, only one other important protein remains to be cloned: rhodopsin kinase, which is not abundant and appears difficult to purify to homogeneity. By contrast, arrestin (also called 48K protein or S-antigen), the protein that blocks the activity of rhodopsin after its multiphosphorylation, is very abundant in the rod cell, and has been recently sequenced (Yamaki *et al.,* 1987).

II. "INACTIVE" Tα–GDP–T$\beta\gamma$ HOLOENZYME: MEMBRANE ATTACHMENT, REQUIREMENT FOR MAGNESIUM, ABSENCE OF PRECOUPLING

When no receptor has been activated by illumination, transducin binds to the ROS disc membrane as an heterotrimeric complex, T$\alpha\beta\gamma$. The stability of the complex and its attachment to the membrane depend on the type of guanine nucleotide, GDP or GTP, present in the binding site of Tα, and are also very sensitive to the ionic strength and composition of the medium. In a medium of ionic composition close to that of the cytoplasm (KCl or NaCl 100 mM, MgCl$_2$ 1–5 mM), in the dark (without any activated receptor), the binding of the holoenzyme T$\alpha\beta\gamma$ to ROS membranes is not very tight. An equilibrium exists between membrane-bound and solubilized transducin, and the membrane-bound pool can be significantly depleted if the membrane suspension is diluted to below 100 μg total proteins/ml. Like other Gα subunits, Tα binds its guanine nucleotide very strongly. When extracted quantitatively from dark-adapted ROS membrane under nondenaturing conditions, simply by washing with a medium of low ionic strength (~ 1 mM) but in the absence of divalent ion-chelating agent such as EDTA, isolated transducin always contains GDP. The holoenzyme Tα–GDP–T$\beta\gamma$ appears then as tightly associated but monomeric (Deterre *et al.*, 1984).

Magnesium ions are important for the binding of the nucleotide to Tα, for the association of Tα–GDP to T$\beta\gamma$, and the association of the heterotrimeric complex with the membrane. This multiplicity of roles sometimes leads to some confusion and misunderstanding as to a role for magnesium in regulating transducin or other G protein-mediated processes. One must always keep in mind that in the rod cell, as in any cell, the level of magnesium always remains stable at the millimolar range, thus excluding, *a priori*, a significant regulatory role for magnesium. Demonstrations of effects of lowering the magnesium level to concentrations clearly below the physiological limits only tell us that magnesium ions are necessary, and do not imply that they have a regulatory role. A magnesium ion participates in the binding of GDP to Tα and interacts at the level of the β-phosphate of the bound nucleotide (Yamanaka *et al.*, 1985). Its affinity is below 10^{-4} M; *in vitro,* in the absence of divalent ion chelating agents, sufficient residual magnesium is usually present in all solutions to ensure its persistence in the nucleotide site of Tα. Its removal by EDTA allows the fast release of GDP from transducin and impairs the R*-catalyzed GDP/GTP exchange reaction, hence also the GTPase rate. The effects of varying the magnesium concentration in the submillimolar range on the function of transducin and other G proteins must be related to this interaction with the guanine nucleotide bound to Gα. These concentrations of magnesium are, however, orders of magnitude below the physiological level, and are never reached *in vivo.* In the millimolar range, that is around the physiological concentration, the level of magnesium influences the binding to ROS membranes of the inactive holoenzyme Tα–GDP–T$\beta\gamma$ (Deterre *et al.,* 1984). This cannot be related to an action on the guanine nucleotide site, which is already saturated in magnesium at these concentrations, but must rather result from an effect on the electrostatic interactions between the lipid polar heads and the peripherally bound transducin. Concentrations of magnesium above 10 mM which is above the physiological range have other effects. These high levels of magnesium induce the dissociation of the holoenzyme, and the release of Tα–GDP from ROS disc membranes. This is not a pure ionic strength effect as it is not observed with solutions of equivalent ionic strength of monovalent ions, but can be reproduced with similar concentrations of calcium ions (Deterre *et al.*, 1984). It looks therefore like an unspecific divalent ion effect on protein–protein and/or protein–lipid contacts, as is often observed for membrane-bound peripheral proteins, rather than a specific action of magnesium at a single site on a component of the system.

At physiological concentrations of magnesium, Tα–GDP is unable to release, and therefore to exchange, its nucleotide over long periods (Bennett and Dupont, 1985). The concept of affinity of the nucleotide for Tα is therefore not very meaningful as an equilibrium is never reached for the dissociation of GDP from Tα, and the state devoid of nucleotide is never observable for isolated transducin under physiological conditions. Lines of evidence for a permanent binding of GDP to the inactive state of G protein were already given by Cassel and Selinger (1978) and by Pfeuffer (1979), and form the basis of the GDP/GTP exchange scheme. This scheme was clearly demonstrated for transducin in 1980 by Fung and Stryer (1980). In this context, it is quite surprising to see that it took until 1986 for many groups working on G$_s$ to realize that the isolated protein still contained stoichiometric amounts of GDP (Ferguson *et al.*, 1986), unless a harsh treatment [1 M (NH$_4$)$_2$SO$_4$ + 20% glycerol + 5 mM EDTA] had been imposed on the protein to remove the nucleotide.

The quantitative extractability of transducin from dark-adapted ROS membranes, simply by lowering the ionic strength in the absence of detergent and of divalent ion chelating agents, demonstrates the lack of significant interaction of this G protein with its unactivated receptor, which is present in large excess on the membrane.

III. TRANSDUCIN–PHOTOEXCITED RHODOPSIN INTERACTION

A. Catalytic Action of R* on Transducin: Opening of Nucleotide Site

The first evidence of a strong interaction of transducin with photoexcited rhodopsin (Fig. 1, Stages 1 and 2) was the observation by Kühn (1980) that full illumination of ROS membrane induces the quantitative binding of

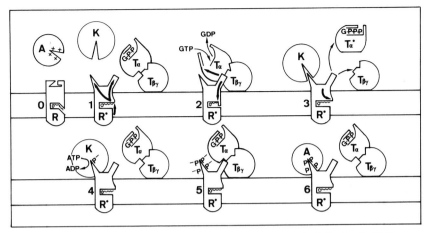

Fig. 1. Scheme representing the interaction between photoexcited rhodopsin (R*), transducin (T), rhodopsin kinase (K) and arrestin (A), as discussed in Section III. The thick arrows symbolize interactions between different sites within R* and Tα. For example, in stage 2 the binding of R* to Tα–GDP–Tβγ induces the release of GDP from its binding site, and conversely the presence of Tα-empty–Tβγ onto the cytoplasmic binding site of R* hinders the release of the chromophore from its binding site within the hydrophobic core of R*.

transducin to the membrane. The transducin is unextractable even at low ionic strength and in the presence of EDTA. Near-infrared light-scattering measurements demonstrated that the binding is stoichiometric — one R* binds one transducin — and is very rapid, i.e. in the millisecond time range (Kühn et al., 1981). But this binding is observable and permanent only in the total absence of GTP. Addition of GTP or analog immediately dissociates the complex and releases transducin into solution as two separated subunits, Tα–GTP and Tβγ, thus indicating that the binding to R* allows nucleotide exchange in Tα. Furthermore, when GTP was added before illumination, total release of hundreds of transducins was induced upon illumination of one rhodopsin, thus demonstrating the catalytic amplification of the nucleotide exchange. Although the catalytic action of R* is on Tα only, the presence of the complete holoenzyme Tα–GDP–Tβγ on the membrane is necessary for transducin to properly recognize the active receptor and bind to it; the βγ subunit appears necessary for Tα–GDP to rebind to the membrane after the hydrolysis of the bound GTP (Fung, 1983; Kühn, 1984). It is therefore not clear whether this requirement for Tβγ reflects the need of a direct interaction of the βγ subunit with the receptor for binding of the complex, or is due to a necessary modification of the α subunit by its interaction with βγ before it can bind to R*, or even simply reflects a kinetic enhancement of the reaction resulting from the binding and proper orientation of Tα–GDP on the membrane through its interaction with membrane bound Tβγ.

The main effect of the interaction between R* and transducin is to modify the conformation of Tα so that the nucleotide site opens and the bound GDP becomes rapidly exchangeable (Fig. 1). The K_{off} for GDP release is reduced from hours to milliseconds or less. It must be stressed that at this stage the replacement of the released GDP by a GTP is not particularly favored over a simple GDP/GDP exchange which can be observed by illuminating in the presence of radioactive GDP: very rapidly all the previously bound cold GDP is replaced by radioactive GDP; the affinity of R*-bound transducin for GDP is measurable and is in the range of 10^{-5} M (Bennett and Dupont, 1985). GDPβS can also enter into the open site and, if added in large excess, will kinetically compete with GTP (T. M. Vuong and M. Chabre, unpublished result). What drives the reaction toward the GTP-bound state is therefore not a significant higher affinity of the site for GTP, but the conformational change of Tα induced by the binding of GTP and the ensuing dissociation of the rhodopsin–transducin complex in three independent components: R*, Tα–GTP and Tβγ. Therefore, if GTP is present, R* is liberated and able to catalyze in sequence the GDP/GTP exchange on many transducins until eventually the whole pool of transducin is processed. But before detailing that step, further analysis of the intermediate state of the interaction of transducin with photoexcited rhodopsin is needed.

B. Action of Bound T on R* in R*–Tα-Empty–Tβγ Complex: GTP-Dependence of Chromophore Release

If all guanine nucleotides have been suppressed or if none is added, and the transducin concentration is micromolar or less, upon illumination and formation of the R*–T complex *in vitro* the open nucleotide site of Tα remains empty after the release of GDP; the normally transient R*-Tα-empty–Tβγ complex persists and transducin remains permanently bound to the membrane through its attachment to R* (see Fig. 1, Stage 2). Studies of the sensitivity of this binding to the presence of micromolar amounts of GDP (F. Bornancin *et al.*, 1989) indicate that the lifetime of the R*–T complex depends on the occupancy by GDP of the open nucleotide site, the binding being enhanced when the site is empty. Another effect of the interaction between R* and T is observable in the stabilized R*–T complex, in which the bound Tα-empty–Tβγ acts on the conformation of R* and modifies the chromophore site. The active form, R*, is the so-called Meta-II state in which the retinal has photoisomerized to the all-*trans* conformation but is still in the protein site. At physiological temperature in frog rods, this state is normally transient and free Meta-II rhodopsin evolves within seconds, and then the chromophore is released from the protein part, the opsin. It has, however, been observed that when rhodopsin is blocked in a $R^* - T\alpha$-empty - $T\beta\gamma$ complex, the Meta-II state is stabilized and the retinal decays much slower (Pfister *et al.*, 1983). The stabilization of Meta-II by its interaction with transducin was also observed on mammalian rods at low temperature, where it is seen as an enhancement of Meta-II over the earlier state Meta-I. This observation is the basis for the identification of Meta-II as the active R^* state (Emeis *et al.*, 1982). The retinal-binding site is located in the middle of the transmembrane core of rhodopsin, and the transducin-binding site must be on the cytoplasmic surface since transducin is peripherally attached and does not penetrate the membrane. These two distant sites therefore remain allosterically coupled after the photoexcitation. It was the isomerization of the chromophore in its hydrophobic site that triggered the formation of the transducin binding site on the surface, but conversely, the binding of transducin on the cytoplasmic surface of rhodopsin blocks the release of the isomerized chromophore.

These observations may seem anecdotal, yet they have important implications when one draws on the analogies between the rhodopsin - transducin model and other receptor - G protein systems. The observation can be reworded in terms of receptor - G protein coupling: the binding of transducin, a G protein, to R^*, an activated receptor, hinders the release of the bound chromophore, the equivalent of a liganded agonist. In the case of rhodopsin one cannot talk in terms of ligand affinity, as the chromophore is already bound to the site in an inactive form before its activation by light; the phenomenological observation is, however, that the release of the ligand from the activated receptor is slower in the absence of GTP than in its presence. But for a hormone receptor, a slower ligand release means a higher affinity, since the affinity is the ratio $K = k_{on}/k_{off}$ of the kinetic binding and release constants of the ligand. The observed GTP-dependence of chromophore decay in photoexcited rhodopsin is therefore equivalent to a GTP dependence of ligand affinity for a hormone receptor. In the case of rhodopsin, this GTP dependence clearly results from the coupling of transducin to photoactivated rhodopsin, rather than from a putative precoupling of transducin to the inactive dark-adapted rhodopsin. The same process can very simply account for the well-known GTP dependence of agonist affinity for other receptors (Lefkowitz *et al.*, 1983), without requiring any precoupling of the unliganded receptor to the G protein, contrary to what is often assumed (Chabre, 1987b). First one must remember that a measured affinity of a receptor for an agonist does not define a conformational state of the receptor, it only describes the equilibrium between two states, the free inactive receptor state R and the ligand bound state R*–L. As by definition an agonist is a ligand that changes the conformation of the receptor to make it active, the states R and R* must be conformationally different, since otherwise no signal would be generated by the liganded receptor. Then, for a hormonal or neurotransmitter receptor the causal sequence is as follows: (1) The ligand's binding to the receptor changes the receptor conformation, and (2) This induces the later coupling of the G protein to the receptor. Statement (2) is a consequence of (1), and not the opposite. The so-called "high-affinity state" observed when GTP is suppressed does not preexist before the binding of the ligand; it is an artificial consequence of the permanent binding of G protein to the liganded receptor that occurs when the ligand is added to assess the affinity.

C. Dissociation of R*–T Complex on Loading of GTP into Tα, Release of Tα–GTP, Kinetics, and Amplification

The dissociation of the R*–T complex on the binding of GTP (Fig. 1, stage 3) was first observed by Kühn (1980). The addition of GTP or a nonhydrolyzable analog to illuminated ROS membranes induced the release of transducin in a low ionic-strength aqueous medium, with the two subunits, Tα–GTP and Tβγ being dissociated as two independent entities (Fung et al., 1981; Kühn, 1981). However, in a medium of ionic composition close to that of the cytoplasm, the Tβγ subunit remains preferentially membrane-bound and only Tα–GTP goes into solution. Light-scattering studies have demonstrated that this release is nearly instantaneous, and that under conditions close to that in vivo, in the presence of several hundred millimolar GTP and GDP, the total duration of an exchange reaction is about 1 msec (Vuong et al., 1984). Even with a very low number of rhodopsin molecules activated, the exchange reaction is completed on the whole pool of transducin in a very short time, thus confirming the high amplification (several hundred transducins activated by one R*) gained at this step of the cascade. The kinetic limit might be imposed by the rate of encounter between R* and membrane-bound transducin in the Tα–GDP–Tβγ state. It is likely that in this process the mobility of transducin predominates over that of rhodopsin: while the lateral mobility of rhodopsin is high for a transmembrane protein that moves in the viscous lipid bilayer, the transducin molecules are only bound to the cytoplasmic surface of the membrane and move essentially in water that is two orders of magnitude more fluid than the lipid milieu. One can thus expect the lateral mobility of transducin to be even higher than that of rhodopsin. No direct measurement of transducin lateral diffusion has yet been published, but Phillips and Cone (1986) reported that the kinetics of the photoresponse in the intact isolated rods was slowed down by a factor of 1.7 when the cytoplasm viscosity was increased by exactly the same factor, thus suggesting that diffusion in the cytoplasm is the rate-limiting process. Furthermore, in our laboratory recent kinetic measurements using light-scattering techniques and photoexcitation by interference fringe laser flash indicate that the lateral mobility of the photoexcited rhodopsin is not rate-limiting for GDP/GTP exchange reaction, as measured through the release of $T\alpha$-GTP from the membrane (F. Bruckert and M. Chabre, in preparation). For other G-protein systems, when discussing collision coupling models, the emphasis is often put on the receptor mobility. It is, however, likely that, as for the visual transduction cascade, the most mobile components, which are responsible for the kinetics and amplification of the processes, are always the G proteins. It has been argued (Chabre, 1987a) that in all cases the G protein is outside the membrane and interacts with cytoplasmic domains of the receptor as well as of the effector; a G protein that is only attached to the membrane by a myristyl chain can have a lateral diffusion coefficient much higher than that of a necessarily transmembranous hormone receptor.

D. Competition between Kinase and Transducin for Interaction with R*, Phosphorylation of R*, and Blocking, by Arrestin

The spontaneous decay of the R* state, correlated with the release of the chromophore, is very slow (half-life on the order of $10-10^2$ sec), but a faster, ATP-dependent inactivation process of R* was first observed by Liebman and Pugh (1980), and elucidated mainly by Kühn (Kühn et al., 1984; Wilden et al., 1986) (Fig. 1, stages 4-6). It is based on the phosphorylation of R* by an R*-specific kinase (R*K), followed by the binding of a soluble protein, arrestin, that specifically recognizes phosphorylated R*. The kinase was detected long ago (Kühn et al., 1981) and its specific affinity for photoactivated rhodopsin was the basis for its isolation (Kühn, 1978). Its activity does not depend on any activator, but only on the presence of ATP and of its specific substrate. In the rods it will bind to and phosphorylate the serine- and threonine-rich C-terminal end of those rhodopsin molecules that have been photoactivated. Photoactivation unfolds the C-terminal end, as shown by the fact that it becomes more susceptible to proteolytic cleavage (Kühn et al., 1982). The kinase and transducin are therefore in competition to interact with the cytoplasmic pole of R*. Given the sizes of these two proteins, and the limited extension of rhodopsin into the cytoplasm, one may expect that steric hindrance would prevent the simultaneous binding of Rh*K and T on R*. If one photoexcites a number of rhodopsins smaller than that of the pool of transducin, in the absence of GTP, the transducins in excess bind quasi-permanently to all the R*. Under such conditions, in the presence of 15 μM ATP, the kinase activity is very low, but it is considerably enhanced on addition of 15 μM GTP (Pfister *et al.,* 1983). This does not result from a sensitivity of the kinase to GTP, but rather from the release of the transducin "cap" from R* after the GDP/GTP exchange. The multiphosphorylation (up to nine phosphates per rhodopsin) impedes but does not completely stop the catalytic action of R* on T. The final step is that arrestin, present in very large quantity in the rods, binds specifically to phosphorylated R*, thus definitely blocking further access to transducin. It must be pointed out that the binding sites for kinase and transducin seem different on R*: the kinase needs the presence of the C-terminal end to bind, whereas transducin still binds to R* after the C terminal polypeptide has been removed (Kühn and Hargrave, 1981). This regulation mechanism is based on the simultaneous formation, on photoexcitation, of a site triggering excitatory interaction with transducin and another site allowing the inhibitory action of kinase and arrestin (Fig. 1, stage 1).

This process is of broad interest as, once more, it seems generalizable to the regulation of the activity of other G protein-coupled receptors. Rh*K is probably the first of a new class of activated receptor-specific kinases, and arrestin the first phosphorylated receptor-specific blocker. The β-adrenergic and muscarinic receptors have phosphorylatable C-terminal ends, and it has already been shown that Rh*K recognizes and phosphorylates the β-adrenergic receptor exclusively in its agonist-liganded form, and that arrestin may block also these phosphorylated receptors (see contribution by Caron and Lefkowitz, this volume).

IV. ACTIVATED FORMS OF Tα: Tα-GTP, Tα-GTPγS AND Tα-GDP-A1F

The catalytic action of one R* results in the loading of GTP onto a large number of transducin molecules and in the subsequent release from the membrane of Tα-GTP. One usually thinks of Tα-GTP, or other Gα-GTP, as the activated form of the G protein, since it is in this GTP-bound state that G proteins couple to their effector enzymes, cyclase or PDE, and activate them. But this shouldn't be misunderstood as implying that the

Tα–GTP conformation is of higher energy than Tα–GDP. Until its hydrolysis, the bound GTP has not transferred any energy to the protein. This is clear from the fact that nonhydrolyzable analogs such as Gpp(NH)p or GTPγS, which cannot transfer any energy to their host protein, are perfectly able to activate Tα, and indeed block the protein indefinitely in this active conformation. As a catalyst, the receptor can only accelerate an energetically favored reaction. Recent time-resolved microcalorimetric measurements indicate that the exchange reaction:

$$T\alpha\text{–GDP} \xrightarrow[\text{GTP}\gamma\text{S} \quad \text{GDP}]{(R^*)} T\alpha\text{–GTP}\gamma\text{S}$$

is exothermic (T. M. Vuong and M. Chabre, to be published). One might correlate to this the fact that Tα is less vulnerable to proteolytic attack, hence presumably more compact, in its GTPγS-bound, active form, than in the GDP-bound form (Fung and Nash, 1983). It is only with the slow GTPase step that part of the free energy released from the hydrolysis of the γ-phosphate of the bound GTP will be transformed into conformational energy in the protein, to regenerate the high-energy state Tα–GDP, that will be able to undertake a new cycle. If the GTP has been replaced by an unhydrolyzable analog, the protein is trapped indefinitely in what is most probably its conformation of lowest internal energy.

Another artificial way of activating Tα–GDP has recently been elucidated, and now provides new insights into the role of GTP; this is the use of fluorometallic complexes that simulate the γ-phosphate of GTP. The long-known activating effect of fluoride had remained very puzzling until Sternweis and Gilman (1982) demonstrated that the activating effect of millimolar concentrations of F$^-$ anions depends on the presence of micromolar concentration of metallic ions such as aluminum or beryllium, which are generally etched from the glassware by the NaF solution. They suggested that the real activating species could be the ion complex AlF$_4^-$ or BeF$_3^-$. Bigay *et al.* (1985) then measured the stoichiometric binding of one aluminum per transducin. An absolute prerequisite for this activation is the binding of GDP on Tα. Analogs of this nucleoside diphosphate are also efficient, provided their β-phosphate remains unsubstituted. Noticing strong structural analogies between AlF$_4^-$ or BeF$_3^-$ and a phosphate group, we proposed that the metallofluoride complex binds in the nucleotide site next to the β-phosphate of GDP and simulates the presence of the γ-phosphate of GTP, hence conferring on transducin its active conformation. A set of experiments has now amply confirmed the value of this model, which furthermore seems to be valid for many other types of nucleotide- and phosphate-binding proteins (Bigay *et al.,* 1987). In the case of transducin, and other G proteins, AlF_4^--activation bypasses the requirement for a catalyzing receptor, since the previously bound GDP does not have to be exchanged. It is "complemented" in the site by an AlF_4^- that binds to form a "pseudo-nucleoside triphosphate": $GDP-AlF_3 \cdot T\alpha$-$GDP-AlF_3$ has the solubility, the proteolytic sensitivity and the PDE-activating capacity of $T\alpha$–GTP. This conformational change does not require the presence of $T\beta\gamma$. As the binding of the aluminofluoride complex is not covalent, activation can be reversed by elution of fluoride, or hindered by the addition of excess $T\beta\gamma$ which forces to the right the equilibrium:

$$T\alpha-GDP-AlF_3 + T\beta\gamma \rightleftharpoons T\alpha-GDP-T\beta\gamma + AlF_4^-$$

The activation of $T\alpha$–GDP can also be obtained with NaF and beryllium. The GDP is then complemented by the complex ion $BeF_3(OH_2)$, another structural analog of phosphate. By contrast with AlF_4^-, and with phosphorus compounds, $BeF_3(OH_2)$ is strictly tetrahedral; it cannot be penta- or hexacoordinated. The fact that the binding of this strictly tetrahedral beryllium also confers to $T\alpha$–GDP the activity observed with a bound GTP suggests that in the active $T\alpha$–GTP stage the γ phosphate is not constrained in the pentacoordinated conformation often found to be the intermediate state of interaction between the energy-rich nucleotide triphosphates and their host proteins. This stengthens our idea that in the active $T\alpha$–GTP state the trinucleotide has yet to confer its chemical potential energy to the protein: the γ phosphate of GTP is passively bound until the slow GTPase step. That the binding of these phosphate analogs to $T\alpha$–GDP is spontaneous and fast, and does not need receptor catalysis, also reinforces the conviction that the receptor acts only to facilitate the release of GDP which allows the spontaneous binding of GTP.

V. TRANSDUCIN AND 3',5'-CYCLIC GMP PHOSPHODIESTERASE

A. Activation of PDE by $T\alpha$–GTP: Release of PDE Inhibitor

Miki *et al.* (1975) determined that PDE can be rapidly activated by trypsin, thus suggesting it is maintained in its basal state by a trypsin-sensitive inhibitor. A heat-stable 11-kDa subunit (hereafter called I) was isolated by Hurley and Stryer (1982), and was shown to inhibit with high affinity the activity of the proteolysis-resistant but heat-sensitive catalytic subunits $PDE\alpha\beta$. These two large catalytic subunits are nearly identical, and the size of the enzyme indicates the presence of one unit of each in the complex (Baehr *et al.*, 1979). In the absence of solid biochemical data on the abundance of inhibitor, a 1:1:1 stoichiometry was usually assumed for the α, β, and I subunits in the inactive holoenzyme. The activation of PDE by activated Tα must correspond to a release of inhibition, but it was not clear whether this involved, as for activation by proteolysis, the physical removal of the inhibitory subunit from the catalytic complex. *A priori,* however, one can exclude the simple exchange reaction: Tα-GTP + I-PDE$\alpha\beta$ $\rightleftharpoons$ Tα-GTP-I + PDE$\alpha\beta$, that would rely on the spontaneous dissociation: I-PDE$\alpha\beta$ $\rightleftharpoons$ I + PDE$\alpha\beta$, to allow the binding of Tα-GTP to free I, without any interaction of Tα-GTP with the I-PDE$\alpha\beta$ complex. The affinity of I for PDE$\alpha\beta$ is too high ($K_d = 10^{-10}$ M) to allow the dissociation kinetics to be in the range of a second or faster, as would be required to account for the fast light-induced activation of PDE by Tα-GTP. It is therefore necessary that the first step of activation be the formation of a complex Tα-GTP-I-PDE$\alpha\beta$. In this complex Tα-GTP interacts with I and releases its inhibitory action on PDE$\alpha\beta$; alternatively Tα-GTP might interact directly with PDE$\alpha\beta$ and activate it, by substituting for I or by counteracting the action of I. The various possible schemes are (the stars on the PDE complexes denote active PDE):

Tα-GTP + I-PDE$\alpha\beta$ $\rightleftharpoons$ Tα-GTP-I-PDE$\alpha\beta$

Tα-GTP-I-PDE$\alpha\beta$ $\rightleftharpoons$ Tα-GTP-I* + PDE$\alpha\beta$ $\rightleftharpoons$ Tα-GTP-I* + PDE$\alpha\beta$ (1)

$\rightleftharpoons$ I-Tα-GTP-PDE$\alpha\beta$* (2)

$\rightleftharpoons$ Tα-GTP-I*-PDE$\alpha\beta$ $\rightleftharpoons$ I + Tα-GTP-PDE$\alpha\beta$* (3)

In reaction (1), I* denotes that the inhibitor is in a modified state, due to its interaction with Tα-GTP, so that the affinity of I for PDE$\alpha\beta$ is lowered to allow its rapid release with Tα-GTP. An analysis by ion-exchange chromatography of the products of interaction of purified Tα-GTPγS with native I-PDE$\alpha\beta$ provided strong evidence for this first scheme (Deterre *et al.,* 1986). As shown in Fig. 2, the elution profile of a mixture of Tα-GTPγS and I-PDE$\alpha\beta$ demonstrates the formation of a Tα-GTPγS-I complex, and not of the I-Tα-GTPγS-PDE$\alpha\beta$ complex expected in scheme (2), nor of the Tα-GTPγS-PDE$\alpha\beta$ complex that would be expected in scheme (3); these latter two schemes had been proposed by Sitaramayya *et al.* (1986). On a gel-filtration column the various protein complexes are not well resolved, but some inhibitor appears to be associated with a peak of Tα-GTPγS, at an elution volume corresponding to about 50 kDa, and no Tα-GTPγS is found associated with the main I-PDE$\alpha\beta$ peak that elutes at a volume corresponding to about 200 kDa. This indicates that the separation of Tα-GTPγS-I from PDE$\alpha\beta$ on the ion exchange column was not an artifact due to strong interaction of either

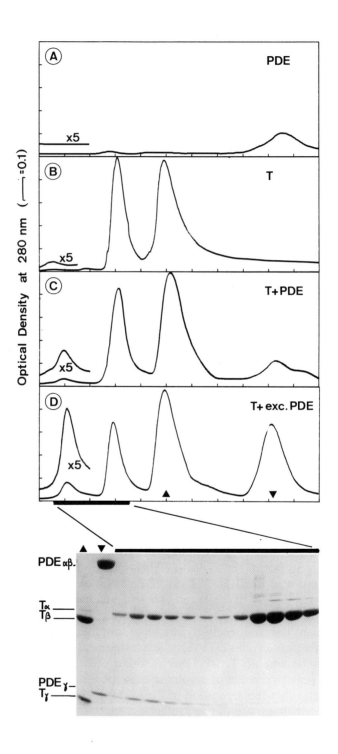

component with the resin, hence suggesting that the intermediate complex $T\alpha-GTP\gamma S-I^*-PDE\alpha\beta$ is transient or has a very weakened $I^*-PDE\alpha\beta$ bond. This, however, is only representative of the situation after the proteins have been extracted from the membrane by low ionic strength washing. At physiological ionic strength, we found, in contrast to previous results of Yamazaki et al. (1983) on frog PDE, that the soluble $T\alpha-GTP\gamma S$ that interacts with PDE becomes membrane-bound. There is no way to detect whether the membrane-bound $T\alpha-GTP\gamma S-I$ complex is physically dissociated from the activated PDE that is also membrane bound. The fact that low ionic strength extraction is sufficient to separate $T\alpha-GTP\gamma S-I$ from $PDE\alpha\beta$ suggests that, on the membrane, the interaction between I and $PDE\alpha\beta$ is very weak and no longer functional. The molecular weight of around 50K observed by gel filtration for the $T\alpha-GTP\gamma S-I$ complex suggests a 1:1 binding.

B. Two Inhibitory Subunits per PDE

Further analysis of the $T-PDE$ interaction by ion-exchange chromatography techniques demonstrated the existence of not just one but two inhibitory subunits per native PDE (Deterre et al., 1988). The formation of $T\alpha-GTP\gamma S-I$ complex, in reaction (1), should leave an equivalent amount of $PDE\alpha\beta$ stripped of inhibitor. Indeed, besides the excess native, inactive PDE peak, not one but two peaks of active PDE were observed (Fig. 3). Sodium dodecyl sulfate (SDS) gels of the peak fractions revealed that only the last peak was totally devoid of inhibitor, and that the first one still contained some inhibitor accompanying the catalytic subunits. Quantitative comparisons by scanning the stained gels of the peak fractions, indicated that the $I/PDE\alpha\beta$ ratio in this first active PDE peak was close to 50% of that found for native PDE. The simplest interpretation for the existence of three PDE peaks with 2:1:0 relative amounts of inhibitor for a normalized amount of $PDE\alpha\beta$ is that they correspond, respectively, to inactive PDE with two inhibitory subunits, $I_2-PDE\alpha\beta$, and to two differ-

---

Fig. 2. Identification of the $T\alpha-GTP\gamma S-I$ complex formed upon interaction of activated transducin with PDE, as analyzed by ion exchange chromatography on Pharmacia FPLC Polyanion SI column, and by SDS-polyacrylamide gels of the elution fractions. (A) Transducin extracted from 10 mg of ROS membrane. (B) PDE extracted from 10 mg of ROS membrane. (C) Transducin and PDE extracted from 10 mg of ROS membrane after inducing them to interact by illuminating in the presence of GTPγS before the extraction. (D) Same sample as in C but with an excess of PDE added. The first elution peak in C and D corresponds to the $T\alpha-I$ complex, as seen on the SDS gel. (Adapted from Deterre et al., 1986.)

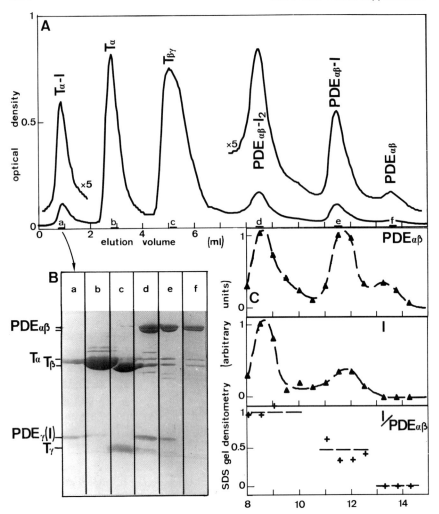

Fig. 3. Demonstration of the existence of two inhibitory subunits per PDE. (A) Complete elution profile (on Pharmacia Polyanion SI column) of a mixture of transducin and PDE extracted after inducing them to interact by illuminating in the presence of GTPγS before the extraction [as in Fig. 1 (C)]. Three peaks of PDE are present, the last two corresponding to activated PDE (activity profile not shown here). (B) Gels of elution fractions from the six different peaks: there is clearly some inhibitor present in the second PDE peak. (C) Densitometry of the stained gel fractions from the three PDE peaks, and quantitation of the I/PDEαβ ratio in these three peaks.

ent states of activated PDE with respectively one and zero inhibitory subunits. The same two species of activated PDE are observed upon activation of native PDE by very short tryptic proteolysis, which degrades and successively eliminates the two inhibitory subunits (Deterre *et al.*, 1988). The specific activities (V_{max}) are the same for the trypsin-activated species as for the corresponding transducin-activated PDE. For the I–PDE$\alpha\beta$ peak fractions this V_{max} is apparently about half of that measured on totally stripped PDE$\alpha\beta$. But by mixing purified PDE$\alpha\beta$ with native I_2–PDE, exchange of inhibitor is observed: PDE$\alpha\beta$ + I_2–PDE $\rightleftharpoons$ 2 I–PDE$\alpha\beta$. The possibility of this exchange precludes as yet an unambiguous estimate of the actual activity of the intermediate complex I–PDE$\alpha\beta$, since the elution fractions containing initially pure I–PDE$\alpha\beta$ had time to dissociate and reequilibrate before the PDE activity measurements. As the existence of the two inhibitory subunits was not suspected, their relative binding affinities for the catalytic complex had not yet been properly analysed. Quantitative inhibition of trypsin-activated PDE (that is, essentially fully stripped of inhibitor) by purified inhibitor (Wensel and Stryer, 1986) may give an estimate of the order of 10^{-9}–10^{-11} M for the binding constant of the first inhibitor to the fully stripped catalytic complex: I + PDE$\alpha\beta$ $\rightleftharpoons$ I–PDE$\alpha\beta$. On the other hand, we observed that activation by transducin *in vitro* produces mostly I–PDE$\alpha\beta$. This is probably due to the low concentrations of activated Tα reached *in vitro,* as compared to the very high concentration reached *in vivo* (see below). Sitaramayya *et al.* (1986) noticed that, in broken rods, light-activated PDE (that is, transducin-activated PDE) needed much higher inhibitor concentration to be fully inhibited than does the trypsin-activated enzyme. The low affinity (0.4 μM) evaluated from these inactivation curves may be taken as an estimate for the dissociation constant in the equilibrium: I + I–PDE$\alpha\beta$ $\rightleftharpoons$ I_2–PDE$\alpha\beta$.

C. Diffusion-Controlled Fast Inactivation of PDE?

The existence of two inhibitory subunits per PDE complex may allow for subtle regulation of PDE activity; the differential binding and exchangeability of inhibitor between native I_2–PDE$\alpha\beta$ and fully activated PDE$\alpha\beta$ raises the possibility of a fast, diffusion-controlled switch-off mechanism of the PDE activity after a flash, that could bypass the standard inactivation process (due to GTP hydrolysis in Tα) that will be discussed later. *In vivo,* within a few hundred milliseconds, all the transducins surrounding a photoexcited rhodopsin are activated and the local concentration of solubilized Tα–GTP in the cytoplasmic cleft between two discs may reach 500 μM. This high concentration may be necessary to overcome the high affinity inhibitor binding and fully strip PDE$\alpha\beta$. But very soon the photoactivated rhodopsin is blocked by phosphorylation and arrestin binding. As the pool of Tα–GTP dilutes in the cytoplasm, its concentration becomes too low to produce any more PDE$\alpha\beta$. The initially formed PDE$\alpha\beta$ complexes also diffuse away into membrane areas where they encounter an excess of native I$_2$–PDE$\alpha\beta$, from which they can regain one inhibitor: PDE$\alpha\beta$ + I$_2$–PDE$\alpha\beta$ $\rightleftharpoons$ 2 I–PDE$\alpha\beta$. It suffices then to assume that I–PDE$\alpha\beta$ has a lower activity or a higher K_m than PDE$\alpha\beta$, for this diffusion and exchange-controlled process to provide a rapid quenching of the PDE activity, which would precede the permanent inactivation of transducin resulting from GTP hydrolysis.

VI. HYDROLYSIS OF GTP IN Tα AND INACTIVATION OF TRANSDUCIN

The GTPase activity stimulated by agonist-liganded receptors was the first identified enzymatic characteristic of G proteins. For transducin on retinal rod membrane in total darkness, the basal GTPase is extremely low, due, as we have seen, to the very slow rate of release of GDP from Tα–GDP after the hydrolysis of GTP. Photoexcited rhodopsin, as an agonist-liganded receptor, accelerates the rate of GTP hydrolysis by transducin, not by increasing the intrinsic hydrolytic capacity of Tα–GTP on a bound GTP, but by catalytically accelerating the release of GDP from Tα–GDP, allowing the binding, hence the hydrolysis, of more GTP. But even upon strong illumination, when the number of photoexcited rhodopsins exceeds that of transducin, the turnover rate of GTP hydrolysis, measured *in vitro,* remains surprisingly low, on the order of 3 GTP/min/transducin. The hydrolytic step, Tα–GTP $\rightarrow$ Tα–GDP + P$_i$, which as we already pointed out, regenerates the high energy conformation Tα–GDP, is usually seen as an inactivation of Tα, since it corresponds to a loss of affinity of Tα for the inhibitor of PDE and a recovery of affinity for T$\beta\gamma$. Therefore GTP hydrolysis induces the release of I from Tα–I complex, hence the definitive inactivation of PDE, and allows the rebinding of Tα, which loses its solubility, to T$\beta\gamma$ on the membrane. The association to T$\beta\gamma$ is required for Tα–GDP to be able to bind to R* and undertake a new cycle of GDP/ GTP exchange. Fung (1983) observed, however, that the GTPase rate saturated already at the low level of 3/min/Tα, in reconstituted systems, when more than 0.1 T$\beta\gamma$ was added per Tα. The slow rate of GTPase poses a major problem, as it seems to indicate much too long a lifetime, about 20 sec for the active Tα–GTP state. This is to be compared to the shut-off time of 0.2 sec for the physiological response to a brief flash. A possible explanation could be that the GTPase turnover rate, as measured *in vitro*, is artificially limited, due for example to the dilution of $T\alpha$-GTP in the solution. Significant variations between the GTPase rates measured by different groups, from 1–5 per minute per transducin, are probably related to dilution and variations in ionic composition of the suspension. Indirect measurements by light-scattering techniques (Dratz *et al.*, 1987) indeed suggest that the higher the concentration, the faster the GTPase, down to the range of a few seconds.

A suggestion that could make the slow rate of GTPase compatible with fast turn-off kinetics would be that $T\alpha$-GTP becomes inactive, that is, takes on a conformation in which it loses its affinity for the PDE inhibitor, before the actual release of the γ phosphate of GTP. This occurs for other nucleotide-dependent systems, such as actin for example, where the fast hydrolysis ATP $\rightarrow$ ADP + bound P is followed by a much slower release of the bound phosphate (Korn *et al.*, 1987). For transducin, the intermediate state $T\alpha$-GDP-P would have lost its activating capacity for the PDE, but not yet regained the capacity to start a new cycle until the release of P_i. The 20-sec GTPase cycle time would then be divided in a short active time followed by a longer "deadtime" during which the inactive $T\alpha$-GDP-P would not be reactive. One way to look for possible intermediate states during the GTP hydrolysis process is to try to analyse the time course of the enthalpy release within the cycle: One GTP is hydrolyzed to GDP + free P_i and therefore 8 kcal are released for each complete cycle of GTPase. We have attempted to detect eventual early enthalpy changes by time-resolved microcalorimetric measurements of the heat released after a flash in the presence of GTP. Our preliminary results (T. M. Vuong and M. Chabre, to be published) suggest that the hydrolysis of GTP is energetically completed within a few seconds at 20°C in bovine rod suspensions of relatively high concentration (5 mg/ml of rhodopsin).

VII. CONCLUSION

More is known about the rhodopsin–transducin–PDE cascade than any other G-protein system, but it remains mostly restricted to the logistics of the process, i.e., which subunits are interacting. Understanding these interactions at the molecular level, however, will require the determination of the three-dimensional structure of transducin. The analysis of the two key protein–protein interactions, R^*–$T\alpha$ and $T\alpha$-GTP-I, would also require crystallographic studies of the corresponding complexes. The role of GTP in the system and the coupling of nucleotide binding and hydrolysis with functional changes of transducin, need to be further analyzed, perhaps by drawing on possible analogies not only with other GTP-binding proteins such as elongation factors and tubulin, but even with ATP-binding proteins like actin or other contractile proteins. The visual transduction cascade has proven to be a particularly suitable model system for G-protein transduction processes; biological phenomena are very diverse, but often derive from few common basic schemes.

REFERENCES

Baehr, W., Devlin, M. J., and Applebury, M. L. (1979). Isolation and characterization of cGMP phosphodiesterase from bovine rod outer segments. *J. Biol. Chem.* **254,** 11669–11677.

Bennett, N., and Dupont, Y. (1985). The G-protein of retinal rod outer segments (transducin). *J. Biol. Chem.* **260,** 4156–4168.

Berg, H. C. (1983). *In* "Random Walks in Biology," Chapter 2. Princeton University Press, Princeton, New Jersey.

Bigay, J., Deterre, P., Pfister, C., and Chabre, M. (1985). Fluoroaluminates activate transducin-GDP by mimicking the γ-phosphate of GTP in its binding site. *FEBS Lett.* **191,** 181–185.

Bigay, J., Deterre, P., Pfister, C., and Chabre, M. (1987). Fluoride complexes of aluminum and beryllium act on G-proteins as reversibly bound analogs of the γ-phosphate of GTP. *EMBO J.* **6,** 2907–2913.

Bornancin, F., Pfister, C., and Chabre, M. (1989). The transitory complex between photoexcited rhodopsin and transducin. *Eur. J. Biochem.* (in press).

Cassel, D., and Selinger, Z. (1978). Mechanism of adenylate cyclase activation through the β-adrenergic receptor: Catecholamine-induced displacement of bound GDP by GTP. *Proc. Natl. Acad. Sci. U.S.A.* **75,** 4155–4179.

Chabre, M. (1985). Molecular mechanism of visual phototransduction in retinal rod cells. *Ann. Rev. Biophys. Biophys. Chem.* **14,** 331–360.

Chabre, M. (1987a). The G protein connection: Is it in the membrane or the cytoplasm? *Trends Biochem. Sci.* **12,** 213–215.

Chabre, M. (1987b). Receptor-G protein precoupling: Neither proven or needed. *Trends Neurosci.* **10,** 355–356.

Codina, J., Stengel, D., Woo, S. L. C., and Birnbaumer, L. (1986). β-subunits of the human liver Gs/Gi signal-transducing proteins and those of bovine retinal rod cell transducin are identical. *FEBS Lett.* **207,** 187–192.

Deterre, P., Bigay, J., Pfister, C., and Chabre, M. (1984). Guanine nucleotides and magnesium dependence of the association states of the subunits of transducin. *FEBS Lett.* **178,** 228–232.

Deterre, P., Bigay, J., Robert, M., Kühn, H., and Chabre, M. (1986). Activation of retinal rod cyclic GMP-phosphodiesterase by transducin: Characterization of the complex formed by phosphodiesterase inhibitor and transducin α-subunit. *Proteins: Structure, Function and Genetics* **1,** 188–193.

Deterre, P., Bigay, J., Forquet, F., Robert, M., and Chabre, M. (1988). The cGMP phosphodiesterase of retinal rods is regulated by two inhibitory subunits. *Proc. Natl. Acad. Sci. U.S.A.* **85**, 2424–2428.

Dohlman, H. G., Caron, M. G., and Lefkowitz, R. J. (1987). A family of receptor coupled to guanine nucleotide regulatory proteins. *Biochemistry,* **26**, 2657–2664.

Dratz, E. A., Lewis, J. W., Schaechter, L. E., Parker, K. R., and Kliger, D. S. (1987). Retinal rod GTP-ase turn over rate increases with concentration: A key to the control of visual excitation. *Biochem. Biophys. Res. Commun.* **148**, 379–386.

Emeis, D., Kühn, H., Reichert, J., and Hofmann, K. P. (1982). Complex formation between metarhodopsin II and GTP-binding protein in bovine photoreceptor membranes leads to a shift of the photoproduct equilibrium. *FEBS Lett.* **143**, 29–34.

Ferguson, K. M., Higashijima, T., Smigel, M. D., and Gilman, A. G. (1986). The influence of bound GDP on the kinetics of guanine nucleotide binding to G proteins. *J. Biol. Chem.* **261**, 7393–7399.

Fung, B. K. K. (1983). Characterization of transducin from bovine retinal rod outer segments. I. Separation and reconstitution of the subunits. *J. Biol. Chem.* **251**, 10495–10502.

Fung, B. K. K., and Nash, C. R. (1983). Characterization of transducin from bovine retinal rod outer segments. II. Evidence for distinct binding sites and conformational changes revealed by limited proteolysis with trypsin. *J. Biol. Chem.* **258**, 10503–10510.

Fung, B. K. K., and Stryer, L. (1980). Photolyzed rhodopsin catalyses the exchange of GTP for bound GDP in retinal rod outer segments. *Proc. Natl. Acad. Sci. U.S.A.* **77**, 2500–2504.

Fung, B. K. K., Hurley, J. B., and Stryer, L. (1981). Flow of information in the light-triggered cyclic nucleotide cascade of vision. *Proc. Natl. Acad. Sci. U.S.A.* **78**, 152–156.

Hurley, J. B. (1987). Molecular properties of the cGMP cascade of vertebrate photoreceptors. *Annu. Rev. Physiol.* **49**, 793–812.

Hurley, J. B., and Stryer, L. (1982). Purification and characterization of the γ regulatory subunit of the cyclic GMP phosphodiesterase from retinal rod outer segments. *J. Biol. Chem.* **257**, 11094–11099.

Hurley, J. B., Fong, H. K. W., Teplow, D. B., Dreyer, W. J., and Simon, M. I. (1984). Isolation and characterization of a cDNA clone for the γ subunit of bovine retinal transducin. *Proc. Natl. Acad. Sci. U.S.A.* **81**, 6948–6952.

Korn, E. D., Carlier, M. F., and Pantoloni, D. (1987). Actin polymerization and ATP hydrolysis. *Science* **238**, 638–644.

Kühn, H. (1978). Light-regulated binding of rhodopsin kinase and other proteins to cattle photoreceptor membranes. *Biochemistry* **17**, 4389–4395.

Kühn, H. (1980). Light- and GTP-regulated interaction of GTP-ase and other proteins with bovine photoreceptor membranes. *Nature* **283**, 587–589.

Kühn, H. (1981). Interactions of rod cell proteins with the disk membrane: Influence of light, ionic strength and nucleotides. *In:* "Current Topics in Membranes and Transport" (W. H. Miller, ed.) Vol. 15, pp. 171–201. Academic Press, New York.

Kühn, H. (1984). Interactions between photoexcited rhodopsin and light-activated enzyme in rods. In: "Progress in Retinal Research" (N. Osborne and J. Chader, eds.) Vol. 1, pp. 123–156, Pergamon, New York.

Kühn, H., and Dreyer, W. J. (1972). Light-dependent phosphorylation of rhodopsin by ATP. *FEBS Lett.* **20**, 1–6.

Kühn, H., and Hargrave, P. A. (1981). Light-induced binding of GTP-ase to bovine photoreceptor membranes: Effects of limited proteolysis of the membranes. *Biochemistry* **20**, 2410–2417.

Kühn, H., Bennett, N., Michel-Villaz, M., and Chabre, M. (1981). Interactions between photoexcited rhodopsin and GTP-binding protein: Kinetic and stochiometric analyses from light-scattering changes. *Proc. Natl. Acad. Sci. U.S.A.* **78**, 6873–6877.

Kühn, H., Mommertz, O., and Hargrave, P. A. (1982). Light-dependent conformational change at rhodopsin's cytoplasmic surface detected by increased susceptibility to proteolysis. *Biochim. Biophys. Acta* **679**, 95–100.

Kühn, H., Hall, S. W., and Wilden, U. (1984). Light-induced binding of 48-kDa protein to photoreceptor membranes is highly enhanced by phosphorylation of rhodopsin. *FEBS Lett.* **176**, 473–478.

Lefkowitz, R. J., Stadel, J. M., and Caron, M. G. (1983). Adenylate cyclase-coupled β-adrenergic receptors: Structure and mechanisms of activation and desensitization. *Annu. Rev. Biochem.* **52**, 159–186.

Liebman, P. A., and Pugh, E. N. (1980). ATP mediates rapid reversal of cyclic GMP phosphodiesterase activation in visual receptor membranes. *Nature* **287**, 734–736.

Lynch, C. J., Morbach, L., Blackmore, P. F., and Exton, J. H. (1986). α-Subunits of N_s are released from the plasma membrane following cholera toxin activation. *FEBS Lett.* **200**, 333–336.

Miki, N., Baraban, J. M., Keirns, J. J., Boyce, J. J., and Bitensky, M. W. (1975). Purification and properties of the light-activated cyclic nucleotide phosphodiesterase of rod outer segments. *J. Biol. Chem.* **250**, 6320–6327.

Muss, J. E., Mumby, S. M., Casey, P. J., Gilman, A. G., and Sefton, B. M. (1987). Myristoylated α subunit of guanine nucleotide-binding regulatory proteins. *Proc. Natl. Acad. Sci. U.S.A.* **84**, 7493–7497.

Ovchinnikov, Y. A., Gubanov, V. V., Khramtsov, N. V., Ischenko, K. A., Zagranichny, V. E., Muradov, K. G., Shuvaeva, T. M., and Lipkin, V. M. (1987). Cyclic GMP phosphodiesterase from bovine retina. Amino acid sequence of the α-subunit and nucleotide sequence of the corresponding cDNA. *FEBS Lett.* **223**, 169–173.

Pfeuffer, T. (1979). Guanine nucleotide controlled interaction between components of adenylate cyclase. *FEBS Lett.* **101**, 85–89.

Pfister, C., Kühn, H., and Chabre, M. (1983). Interaction between photoexcited rhodopsin and peripheral enzymes in frog retinal rods. Influence of the post metarhodopsin II decay and phosphorylation rate of rhodopsin. *Eur. J. Biochem.* **136,** 489–499.

Phillips, E. S., and Cone, R. A. (1986). Do diffusion rates in the cytoplasm or the membrane limit the response speed of the vertebrate rod? *Biophys. J.* **49,** 277a.

Sitaramayya, A., Harkness, J., Parkes, J. H., Gonzalez-Ouva, C., and Liebman, P. A. (1986). Kinetic studies suggest that light-activated cyclic GMP phosphodiesterase is a complex with G-protein subunits. *Biochemistry* **25,** 651–655.

Sternweis, P. C., and Gilman, A. G. (1982). Aluminum: A requirement for activation of the regulatory component of adenylate cyclase by fluoride. *Proc. Natl. Acad. Sci. U.S.A.* **79,** 4888–4891.

Stryer, L. (1986). Cyclic GMP cascade in vision. *Annu. Rev. Neurosci.* **9,** 87–119.

Vuong, T. M., Chabre, M., and Stryer, L. (1984). Millisecond activation of transducin in the cyclic nucleotide cascade in vision. *Nature* **311,** 659–661.

Wensel, T. G., and Stryer, L. (1986). Reciprocal control of retinal rod cyclic GMP phosphodiesterase by its γ subunit and transducin. Proteins: Structure, Function and Genetics **1,** 90–99.

Wilden, U., Hall, S. W., and Kühn, H. (1986). Phosphodiesterase activation by photoexcited rhodopsin is quenched when rhodopsin is phosphorylated and binds the intrinsic 48-kDa protein of rod outer segments. *Proc. Natl. Acad. Sci. U.S.A.* **83,** 1174–1178.

Yamaki, K., Takahashi, Y., Sakuragi, S., and Matsubara, K. (1987). Molecular cloning of the S-antigen cDNA from bovine retina. *Biochem. Biophys. Res. Commun.* **142,** 904–910.

Yamanaka, G., Eckstein, F., and Stryer, L. (1985). Stereochemistry of the guanyl nucleotide binding site of transducin probed by phosphothioate analogues of GTP and GDP. *Biochemistry,* **24,** 8094–8101.

Yamazaki, A., Stein, P. J., Chernoff, N., and Bitensky, M. W. (1983). Activation mechanism of rod outer segment cyclic GMP phosphodiesterase. Release of inhibitor by the GTP/GDP-binding protein. *J. Biol. Chem.* **258,** 8188–8194.

CHAPTER 11

G Protein-Mediated Effects on Ionic Channels

Atsuko Yatani
Department of Molecular Physiology and Biophysics, Baylor College of Medicine,
Houston, Texas 77030

Juan Codina
Department of Cell Biology, Baylor College of Medicine, Houston, Texas 77030

Arthur M. Brown
Department of Molecular Physiology and Biophysics, Baylor College of Medicine,
Houston, Texas 77030

G proteins mediate effects between membrane receptors and ionic channels in two ways: indirectly after activating a membrane-associated enzyme which acts via cytoplasmic second messengers on ionic channels, and directly by activating the ionic channel itself (Fig. 1). We refer to the direct action as G-protein gating of ionic channels although the possibility that a membrane-delimited intermediary is activated cannot be excluded until the response is reconstituted with pure components. Direct G-protein gating (Fig. 2) takes two forms: obligatory and modulatory. In the obligatory case the opening probability, P_o, of the channel in the absence of agonist is low; examples are G-protein gating of muscarinic atrial K^+

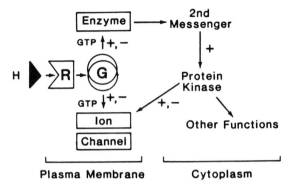

Fig. 1. Diagram of the role of G proteins in receptor-mediated regulation of ionic channels. The indirect pathway is via a membrane enzyme and cytoplasmic second messenger. G,
signal-transducing G protein with structure $\alpha\beta\gamma$, is shown as a double circle to denote the
possibility that a single receptor may interact with more than one type of G protein (such as
G_i and G_o) and/or that a single G protein may interact with more than one effector system
(such as G_s with adenylyl cyclase and the Ca^{2+} channel). +, stimulatory; −, inhibitory.

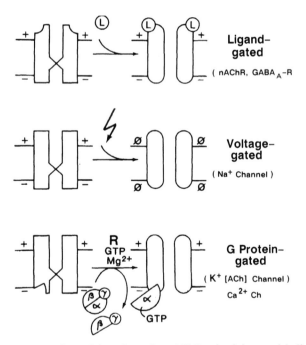

Fig. 2. Three types of gated ion channels. nAChR, nicotinic acetylcholine receptor,
GABA$_A$-R γ-aminobutyric acid receptor; K$^+$[ACh], atrial muscarinic potassium channel;
Ca^{2+} Ch, calcium channel.

channels and muscarinic and/or somatostatin K^+ channels in clonal anterior pituitary (GH_3) cells. In the modulatory case, P_o is determined primarily by membrane potential and not agonist; examples are G-protein modulation of voltage-gated dihydropyridine-sensitive Ca^{2+} channels in heart or skeletal muscle. How widespread the phenomenon is likely to be can be gauged from considerations of numbers of G proteins, numbers of receptor–G protein interactions, and numbers of ionic channels. G proteins are $\alpha\beta\gamma$ trimers; there are presently cDNAs for ten α subunits, two β subunits, and two γ subunits for a total of 40 permutations (Birnbaumer et al., 1988). G proteins may be linked to as many as 70 receptors (Birnbaumer et al., 1987) and could be linked to any one of numerous K^+ channels, several Ca^{2+} channels, and possibly Na^+ and Cl^- channels. With respect to ionic transport, membrane carriers may also be under the control of G proteins but the present chapter is restricted to ionic transport by membrane channels. Furthermore, we restrict ourselves to plasmalemmal ionic channels although it is likely that channels in intracellular membranes such as endoplasmic reticulum may also be regulated by G proteins. At present, direct G-protein gating has been proved for two distinct Ca^{2+} channels and two distinct K^+ channels but there is strong evidence for the involvement of many more channels. For example, we have recently shown that the major brain G protein, G_o, acts directly on hippocampal pyramidal cell K^+ channels (VanDongen et al., 1988). Consequently the repertoire of G-protein-gated ionic channels is far greater than the repertoire of ligand-gated channels such as nicotinic acetylcholine receptors (AChRs), $GABA_A$ (γ-aminobutyric acid), and glycine receptors in which the receptor and ionic channels are part of the same protein. The effectors for G proteins involved in the indirect category of G-protein gating of ion channels are a group of membrane enzymes including phosphodiesterases (PDE), adenylyl cyclase (AC), phospholipase C (PLC), phospholipase A_2 (PLA), and $3',5'$-cyclic GMP phosphodiesterase (PDE). These effectors change the cytoplasmic activity of molecules such as cyclic nucleotides, inositol polyphosphates, and Ca^{2+} that act as second messengers upon the ionic channels. This additional level of complexity increases still further the number of possible outcomes among receptors, G proteins, and ionic channels. Rather than enumerate all the possibilities in this chapter we shall examine the better understood cases beginning with direct G-protein gating in Section I and turning to indirect G-protein gating in Section IV.

I. DIRECT G-PROTEIN GATING OF K^+ CHANNELS

The history of direct G-protein gating is related to one of the earliest observations in physiology; namely, slowing of the heart. It is tied to the first demonstration of chemical transmission at synapses (Loewi and Navratil, 1926) and to some of the earliest measurements of intracellular cardiac membrane potentials (Hutter and Trautwein, 1956). The relationship is fundamental but, as we shall see, the mechanism by which the two events are associated remained elusive for many years. Loewi demonstrated chemical transmission in the classical studies on vagal slowing of the heart (Loewi and Navratil, 1926). The "Vagusstoff" was acetylcholine, the effect was blocked by atropine, and the receptor was defined as a muscarinic cholinergic receptor (mAChR). Modern electrophysiology made its next contribution when Trautwein and Dudel (1958) showed, using intracellular microelectrode recording, that ACh hyperpolarized the atrial membrane by increasing K^+ permeability. A subsequent result of significance was the latency of 50–150 msec which occurred following topical application of ACh or vagal nerve stimulation (Glitsch and Pott, 1978; Hill-Smith and Purves, 1978; Hartzell, 1980; Osterrieder *et al.*, 1981; Nargeot *et al.*, 1982, 1983). The latency at the neuromuscular junction where the nicotinic AChR (nAChR) and channel are one protein is about 1 msec and this comparison suggested a coupling process between the mAChR and the channel. The second messengers often considered in such processes are cAMP and cGMP, but it was found that neither hyperpolarizes the atrial membrane (Trautwein *et al.*, 1982; Nargeot *et al.*, 1983).

About 10 years earlier, biochemical studies showed that guanine nucleotides were important in the linkage between membrane receptors and cellular effectors (Rodbell *et al.*, 1971a; Rodbell *et al.*, 1971b). Binding studies showed that just as for other receptor–ligand interactions, guanine nucleotides modified the affinity with which muscarinic agonists displace the muscarinic antagonist QNB from mAChR's (Rosenberger *et al.*, 1979). Furthermore, pertussis toxin (PTX) substrates, sure evidence of the presence of G proteins, were identified in cardiac tissue (Rosenberger *et al.*, 1979; Hazeki and Ui, 1981; Kurose and Ui, 1983; Halvorsen and Nathanson, 1984). Nevertheless, a connection between guanine nucleotide binding and the latency of the ACh response recorded electrically was not made at this time nor for a considerable period thereafter, and the biochemical and electrophysiological approaches continued along their separate paths. The next big step in electrophysiology was the introduction of methods for dispersing single cardiac cells (Powell and Twist, 1976) and studying the isolated cells electrophysiologically with a suction pipette (Lee *et al.*, 1979), and, ultimately, with a patch pipette (Sakmann *et al.*, 1983). The pace now quickened significantly. Single-channel studies identified the inwardly rectifying single atrial K^+ currents that were activated by ACh (Sakmann *et al.*, 1983) and the bases for specifying these currents from other single K^+-channel currents in atria were established. Then, by perfusing the patch pipette, Soejima and Noma (1984) showed that ACh activated this K^+ channel independently of cytoplasmic mediators. This was an important observation and was followed by the report of Pfaffinger *et al.* (1985), which showed that the ACh response was blocked by pertussis toxin and that GTP was required for the ACh effect. At the same time, Breitwieser and Szabo (1985) showed that the ACh response became irreversible in the presence of the nonhydrolyzable GTP analog GMP-P(NH)P and they called the putative endogenous G protein G_k. These studies linked G proteins with the inwardly rectifying K^+ channel activated by ACh, a linkage that became stronger when GTPγS was shown to activate atrial K^+ channels in excised, inside-out membrane patches (Kurachi *et al.*, 1986a). The GTPγS had an absolute requirement for Mg^{2+} in producing this result (Kurachi *et al.*, 1986c). Direct involvement of a G protein was now evident, but the possibility that the G protein was acting indirectly through, for example, a membrane-associated enzyme such as protein kinase C (PKC) had not been specifically excluded although Kurachi *et al.* (1986a) noted that ATP was not required for the GTPγS effect. Nor did the experiments identify the G protein involved other than that it was a PTX substrate. A direct demonstration was required. Therefore, G proteins purified from human erythrocytes (Codina *et al.*, 1983, 1984a,b) were applied to ACh-sensitive K^+ channels in excised, inside-out membrane patches from mammalian atrial muscle (Yatani *et al.*, 1987a; Codina *et al.*, 1987a). The experimental approach is shown in Fig. 3. The basic strategy is to study single-channel currents which give less ambiguous results than whole-cell currents, and to demonstrate the independence of the G-protein effect from second messengers. To do this required using excised membrane patches or membrane vesicles incorporated into planar lipid bilayers (Fig. 3). Only the exogenous PTX-sensitive G protein called G_k reconstituted the mACh effect on K^+[ACh] channels. G_k was preactivated with GTPγS and is denoted as G_k^*; it was effective at picomolar concentrations. The molar ratio for binding is about 1 and GTPγS only has effects at 10 nM or greater. The cholera toxin (CTX)-sensitive G protein, G_s, which activates AC and is the other principal G protein purified from human red blood cells (RBC), had no effect after preactivation with GTPγS (Yatani *et al.*, 1987a; Codina *et al.*, 1987a). The G protein transducin was also ineffective even at nanomolar concentrations. Figure 4 shows single-channel currents which are identified by their ACh (in this case carbachol or carb) responsiveness, their slope conductance of about 40 pS, and their average open time of about 1.5 msec. The results were identical for single channel currents activated by ACh in cell-attached patches, ACh and GTP in the bath solution in excised inside-out patches, GTPγS, G_k^*, α_k^*, and unactivated G_k plus GTP in the presence of ACh. However, G_k^* and α_k^*

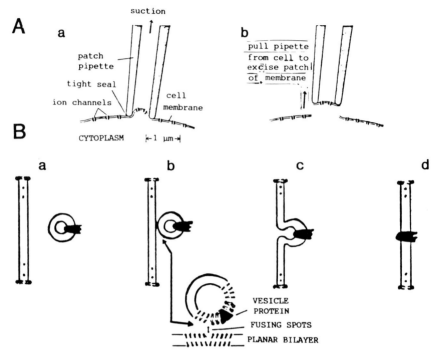

Fig. 3. (A) The patch clamp method. Current enters or leaves the pipette by passing through the channels in the patch of membrane. Recordings of the current through these channels can be made with the patch still attached to the rest of the cell, as in (a), or excised, as in (b). (b) was the method we usually employed. (B) Method of incorporating membrane vesicles into planar lipid bilayers. The vesicles added to the *cis* chamber are carried to the bilayer by an osmotic gradient (a) and fusion begins at the fusing spots (b–d). The orientation of the Ca^{2+} channels in the vesicle is usually right side out for cardiac sarcolemma and inside out for skeletal muscle T tubules. The conventions for current recording are the same as those used with patch clamp: positive current is outward.

effects persist even with washes as long as one hour, whereas the ACh effects cease after GTP is removed. P_o, in the absence of activation, is nearly zero and activation is an increase in P_o; neither conductance nor open times are affected and the frequency of simultaneous openings for the customary two to three channels in each atrial patch was fitted from binomial expectations. ADP-ribosylation with PTX blocked muscarinic activation and the response was reconstituted by the G protein G_k in the presence of GTP. Hence, endogenous G_k cannot be tightly coupled to either the ACh-sensitive K^+ or the muscarinic receptor since ADP-ribosylation with PTX would, in this case, have led to permanent loss of receptor–G protein–K^+ channel coupling. As noted, the α subunit of G_k preactivated with GTPγS, G_k^*, was added at picomolar concentrations and mimicked the muscarinic response while the $\beta\gamma$ dimer was ineffective at nanomolar concentrations. Moreover, GTP plus $\beta\gamma$ could not reconstitute the response whereas GTP plus G_k under identical conditions could. $\beta\gamma$ subunits on occasion inhibited muscarinic responses. Activated G_o from bovine brain had weak effects which may have been due to contamination with an activated G_k protein (Yatani et al., 1987a), but a direct effect of G_o could not be excluded.

The α subunit of the particular PTX-sensitive G protein we tested has not been shown to have an inhibitory effect on adenylyl cyclase. Hence, the G protein was not truly G_i. The ACh-sensitive K^+ channel was its first proved effector, and the active exogenous G protein was therefore called G_k, the "k" referring to K^+ channel (Yatani et al., 1987a). Henceforth we refer to ACh-induced K^+ currents as $I_{K[ACh]}$ or $I_{K[G_k]}$ and the ACh-sensitive K^+ channel as $K^+[ACh]$ or $K^+[G_k]$. To be effective, the G_k had to be activated either by ACh receptor plus GTP or by GTPγS and the first condition required Mg^{2+}. Activated G_k (G_k^*) was effective in the absence of Mg^{2+} and neither ATP nor App(NH)p had any effect. Phorbol esters also had no effect. These results excluded phosphorylation via kinases, including protein kinase C (PKC), and proved that a specific signal-transducing, GTP-activated G protein directly activated an ion channel. The activated α subunit of G_k, α_k^*, was as effective as G_k^*, and $\beta\gamma$ subunits had effects that were difficult to establish as we discuss in Section III. As noted, G_k^*, α_k^*, GTPγS, and ACh all produced activation in the same way (Table I); not by changing channel open time or conductance, but simply by increasing the frequency of channel openings (Yatani et al., 1987a; Codina et al., 1987a). In another laboratory, G proteins purified from bovine brain were applied to $K^+[G_k]$ channels in inside-out patches from chick atrial muscle (Logothetis et al., 1987). The authors concluded that the $\beta\gamma$ dimer at nanomolar concentrations, not the α subunit, mediates the G protein effect. The reasons for the different conclusions are discussed in Section III, but it has now been demonstrated in other laboratories, including the laboratory of Logothetis et al., (Neer and Clapham, 1988), that the α subunits are effective (Cerbai et al., 1988), so we may reasonably proceed with the idea that α_k mediates the G_k effect.

An analogous situation to that for muscarinic atrial K^+ channels exists for K^+ channels in GH_3 clonal rat anterior pituitary cells (Fig. 5; Table II; Yatani et al., 1987b; Codina et al., 1987b). Somatostatin (SST) inhibits secretion, reduces intracellular Ca^{2+} levels (Schlegel et al., 1985), and produces cAMP-independent membrane hyperpolarization (Yamashita et al., 1986). ACh has the same effects, possibly through the same type of muscarinic receptors present in heart. These results implicated K^+ channels, and ACh and SST were found to activate the same subset of GH_3 K^+ channels (Yatani et al., 1987a,b). Just as it did in heart, human RBC G_k^*

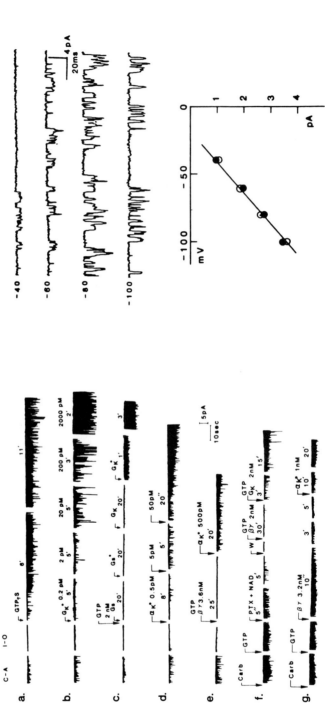

directly activated single K⁺-channel currents but, unlike the conductance of 40 pS for neonatal atrial $K^+[G_k]$ currents (Kirsch *et al.*, 1988), the conductance of these K^+ channels from this rat anterior pituitary cell line was 55 pS (Yatani *et al.*, 1987b). The rectifying properties of $K^+[G_k]$ channel were not fully determined because GH_3 cells, unlike atrial myocytes, have a variety of outward K^+ currents that complicate the record-

---

Fig. 4. Activation of an atrial muscarinic K^+ channel (A) and its single channel behavior (B). (A) Each line represents a separate experiment in which single-channel K^+ currents were recorded before (cell-attached or C-A) and after membrane patch excision to the inside out configuration (C-F or I-O). Records were obtained at holding potentials varied between -80 and -100 mV and using symmetrical 130 to 140 mM KCl or potassium-methanesulfonate solutions in 5 mM HEPES, pH 7.5, containing in addition either 1.8 mM $CaCl_2$ or 5 mM EGTA in the pipette, and 2 mM $MgCl_2$ in the 100 μl bathing chamber. Other additions are shown or described for each experiment. Numbers above the records denote time elapsed between the indicated additions and the beginning of the segment of record shown. Routinely, the first addition was made between 5 and 10 min after patch excision and subsequent additions were at 5 to 25 min intervals, depending on purpose of the experiments. Five min intervals were used when dose–response relationships were studied; 25 min intervals were used when substances added had no apparent effect. In some instances the bathing solutions were exchanged by perfusion at 1 to 2 ml/min. Experiments used atrial membrane patched from adult guinea pigs. Cells were obtained by collagenase digestion and used without further culturing. (a) Stimulation of single-channel K^+ currents by activation of the atrial G_k proteins with GTPγS (100 μM in the bath). (b) Stimulation of the atrial K^+ channels in a dose-dependent manner by exogenously added human erythrocyte pertussis-toxin substrate, referred to as G_k (formerly referred to as N_i or G_i), preactivated to G_k^* form by incubation with GTPγS and Mg^{2+} and dialyzed extensively to remove free GTPγS to ineffective levels. Threshold effects of G_k^* were obtained with 0.2–1 pM in separate membrane patches. (c) The bathing solution contained 100 μM GTP throughout, proteins were added at 2–nM each. Lack of stimulatory effects of 1 nM of either non-activated or GTPγS-activated G_s or of non-activated G_k. (d) The effect of GTPγS-activated holo-G_k is mimicked by equally low concentrations of the resolved GTPγS-activated α subunit of the protein (α_k^*). (e) Lack of effect of resolved human erythrocyte $\beta\gamma$ exposed to a G_k^*-sensitive membrane patch. (f) Stimulation of G_k-sensitive K^+ channels by the muscarinic receptor agonist carbachol (Carb) present in the pipette throughout, maintenance of receptor–G protein–effector coupling after patch excision into bathing solution with 100 μM GTP, and uncoupling of atrial G_k by treatment with pertussis toxin (PTX) and NAD added to the bathing solution. Lack of recoupling effect of resolved human erythrocyte $\beta\gamma$ and reconstitution of acetylcholine receptor K^+ channel stimulation by addition of native unactivated human erythrocyte G_k in the presence of GTP. Note that this contrasts with the result in (c) and demonstrates that the exogenously added G_k requires receptor participation for activation by GTP. (g) $\beta\gamma$ subunits at high concentrations may have an inhibitory effect on the muscarinic response. (B) Conductance of single muscarinic atrial K^+ channels. Representative single-channel currents from a C-A patch with 1 μM Carb in pipette at indicated holding potentials. Bathing solution contained isotonic K^+ and EGTA, which produced a resting potential of ~ 0 mV. There were at least two K^+[ACh] channels in this patch. Current–voltage relationships for the channel in A (●) and for another channel from another patch under the same experimental conditions (○). Each point is average value of at least 20 single openings. Slope conductance was 41 pS.

Table I   Comparison of Endogenous and Exogenous G_k-Activation of Atrial Muscarinic K$^+$ Channels[a]

Conditions (Concentration)		Amplitude (pA)	Mean open times (msec)	n
Agonist[b] + GTP	(100 μM)	2.2 ± 0.2	1.54 ± 0.6	14
GTPγS	(100 μM)	2.2 ± 0.3	1.70 ± 0.7	6
G_k^*	(20–50 pM)	2.1 ± 0.4	1.60 ± 0.4	8
α_k^*	(10–50 pM)	2.1 ± 0.3	1.52 ± 0.6	12
α_{i3}^*	(200–1000 pM)	2.2 ± 0.3	1.79 ± 0.3	4

[a] Single channel properties of agonist-activated K$^+$ channel, GTPγS, G_k^*, α_k^*, or recombinant α_{i3}^*-activated K$^+$ channels of inside-out membrane patches from adult guinea pig. Holding potential: -80 mV.
[b] Agonist: Carbachol or acetylcholine. Values are mean I.S.D.; n, number of experiments.

ings. Nevertheless, the data indicate that the atrial and clonal pituitary K$^+$ channels are different. Just as for heart, G_k^* had no effect on this K$^+$ channel. G_o^* effects were weak and, as for heart, possibly due to the presence of G_k. The GTP-binding α subunit of G_k, not the $\beta\gamma$ complex, was responsible for stimulating GH$_3$ K$^+$ channels just as it was responsible for stimulating muscarinic atrial K$^+$ channels.

While α_k^* appeared to be a single protein in Coomassie blue-stained sodium dodecyl sulfate-polyacrylamide gel electrophoresis (SDS-PAGE), when [^{32}P]NAD labeling by ADP-ribosylation with PTX was used it became clear that α_k was about 95% of the protein in the band and the remainder was another PTX substrate. Tryptic digest yielded amino acid sequences predicted by the α_{i3} cDNA cloned from human liver (Birnbaumer et al., 1988). Therefore, we attempted to express this clone in Escherichia coli cells using the pT7-7 expression plasmid (Mattera et al., 1989). We found that the protein encoded by α_{i3}, preactivated by GTPγS or AlF^{2-}, simulated α_k^* effects although at ~ 1–10% of α_k^*'s potency (Table I). Similar results have been obtained for proteins expressed by α_s cDNAs in the same constructs. In this case the α_s subunits also had lower potencies as AC stimulators than did native α_s, a result similar to that reported by Graziano et al. (1987). These results prove the α_{i3} can act as α_k. They do not prove that α_{i3} is α_k since the effects of the proteins encoded by α_{i2}, α_{i1}, and α_o remain to be tested.

The experiments to this point dealt mainly with reconstitution of the K$^+$[ACh] response. To probe the functional aspects we used a monoclonal antibody mAb 4A, that binds to α_T and α_k and had been shown to block rhodopsin–metarhodopsin changes (Hamm et al., 1988; Deretic and Hamm, 1988; Fig. 6). The results showed that mAb 4A blocked the mAChR effect and, furthermore, the block was irreversible only if endogenous G protein was activated. Only two conclusions were possible: (1) that

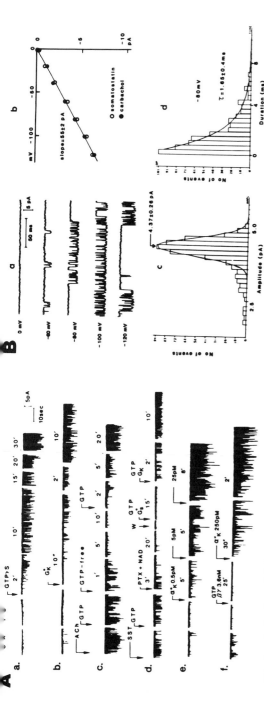

Fig. 5. Activation of GH₃ muscarinic and somatostatin K⁺ channels (A) and its single channel behavior (B). (a)–(f) are from different cells. C–A, cell-attached; I–O, inside-out. (A) Experiments with inside-out membrane patches from rat GH₃ pituitary tumor cells. Cells were grown as monolayers on cover slips and membrane patches excised from their upper membrane surface. (a) Effect of GTPγS added to the bathing medium. Gₖ (Gₖ). (c) Dependence of GTP and reversibility of receptor-mediated stimulation of Gₖ-sensitive K⁺ channel. Acetylcholine (ACh) was present in the pipette solution throughout. 100 μM GTP was present in the bathing medium, removed and readied as shown. (d) Stimulation of GH₃-cell Gₖ-sensitive K⁺ channels by somatostatin (SST): demonstration of PTX sensitivity of the GH₃ Gₖ protein and reconstitution of the signal transduction pathway by addition of unactivated native human erythrocyte Gₖ in the presence of GTP. (e) and (f) Mimicry of the effects of GTPγS-activated Gₖ by resolved GTPγS-activated α subunit of human erythrocyte Gₖ and lack of effect of resolved, α subunit-free, βγ dimers. (B) Somatostatin and acetylcholine activation of single-channel K⁺ currents in clonal anterior pituitary (GH₃) cells. (a) Representative records of activity in a patch with 10 μM carbachol in the pipette and 100 μM GTP in the bath obtained at the indicated holding potentials. (b) Current–voltage relationships for K⁺ channels activated by 10 μM carbachol (●) or 0.5 μM somatostatin (○). Each point is the mean value from four independent experiments with carbachol and five independent experiments with somatostatin; line is best least-squares fit. (c) Amplitude histogram of channel openings obtained in the presence of carbachol and GTP in the bath at −80 mV; line: best fit to a Gaussian distribution. (d) Open-time duration histogram of the 55 pS channel openings at −80 mV holding potential; line: best fit to single exponential decay function.

Table II Comparison of Endogenous and Exogenous G_k-Activation of GH$_3$ K$^+$ Channels[a]

Conditions (Concentration)		Amplitude (pA)	Mean open times (msec)	n
Agonist[b] + GTP	(100 μM)	4.2 ± 0.3	1.77 ± 0.3	4
GTPγS	(100 μM)	4.2 ± 0.2	1.54 ± 0.4	3
G_k^*	(200 pM)	4.0 ± 0.3	1.90 ± 0.3	3
α_k^*	(50 pM)	4.1 ± 0.1	1.70 ± 0.2	2

[a] Properties of somatostatin- and carbachol-activated single K$^+$-channel current in GH$_3$ cells recorded in symmetrical isotonic K$^+$ solution at -80 mV. Slope conductance between -120 and -40 mV was 55 pS.

[b] Agonist: somatostatin (0.5 μM) or carbachol (10 μM). Values are mean ± S.D.; n, number of experiments.

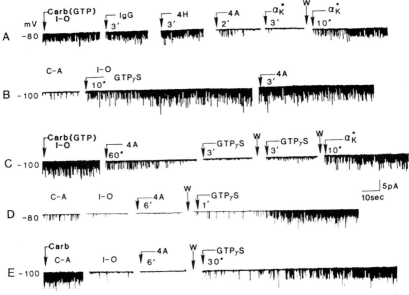

Fig. 6. Characteristics of mAb 4A-block of single muscarinic K$^+$ channel currents, i_k, in inside-out patches excised from atria of adult guinea pigs. The i_k was activated by carbachol at 0.1 μM in the patch pipette and GTP at 100 μM in the bathing solution. I-O refers to inside-out excised membrane patch configuration. The single channel currents were recorded with symmetrical isotonic K$^+$ (140 mM) solutions and the holding potentials are indicated. (A) Slow time-resolution recordings of test agent effects. Times on top of each trace are minutes (') or seconds (") elapsed between the addition of the test agent to the bathing solution and the illustrated recording. mAb 4H at 6.5 μM and control IgG (Sigma) at 6.5 μM had no effects, but mAb 4A at 6.5 μM blocked i_k. α_k^* at 10 pM added in the presence of mAb 4A was also ineffective, but after washing out (W) mAb 4A, α_k^* activated i_k. (B) mAb 4A at 6.5 μM does not block i_k previously activated by 100-μM GTPγS. The membrane patch

$\beta\gamma$ was liberated at a time when the mACh effect was blocked, or (2) that α-GTP$\beta\gamma$ and $\beta\gamma$ be equipotent. Neither result is consistent with a role for $\beta\gamma$ in mAChR activation of $K^+[ACh]$ (Yatani et al., 1988a).

II. DIRECT G-PROTEIN GATING OF Ca^{2+} CHANNELS

Atrial ACh-sensitive and GH_3 SST- and ACh-sensitive K^+ channels are not the only channels proven to be regulated directly by G proteins. G proteins also directly regulate Ca^{2+} channels from rabbit skeletal muscle T-tubules (Yatani et al., 1988b) and from guinea pig or bovine cardiac sarcolemma (Yatani et al., 1987c; Imoto et al., 1988). Unlike $K^+[G_k]$, however, where G-protein gating is obligatory, G-protein gating is modulatory in the case of Ca^{2+} channels; the G protein is not essential for channel opening but membrane depolarization is. Suggestions that G proteins might directly regulate Ca^{2+} channels have been given in the literature. Guanine nucleotides changed the binding of dihydropyridines to their receptors, thought to be Ca^{2+} channels, in cell-free membrane preparations of cardiac muscle (Triggle et al., 1986) and skeletal muscle T-tubules (Galizzi et al., 1984). Somatostatin inhibited Ca^{2+} currents in the anterior pituitary AtT-20 cell line by a PTX-sensitive mechanism and the effect was mimicked by GTPγS (Lewis et al., 1986). $GABA_B$ and α-adrenergic receptor ligands reduced Ca^{2+} currents in avian dorsal root ganglion cells (DRGs), and GDPγS blocked the effects (Holz et al., 1986). GTPγS inhibited the transient Ca^{2+} current in DRGs and potentiated the inhibition of transient and steady components of Ca^{2+} current caused by the $GABA_B$ agonist baclofen (Scott and Dolphin, 1986; Dolphin and Scott, 1987). PTX attenuated the effects of baclofen and GDPβS antagonized the effects of GTPγS (Dolphin and Scott, 1987). Opioid peptide agonists reduced Ca^{2+} currents in

---

shows little baseline activity in the cell-attached configuration in the absence of carbachol. After I-O excision GTPγS at 100 μM stimulated i_k. mAb 4A was ineffective in blocking this effect. (C) mAb 4A applied prior to GTPγS prevents activation. i_k in I-O patch were stimulated by carbachol at 0.1 μM in the patch pipette and GTP at 100 μM in the bath. Prior incubation with mAb 4A at 6.5 μM prevented the GTPγS response shown in (B). After washing out mAb 4A, GTPγS at 100 μM was ineffective, but α_k^* reactivated i_k. Hence, mAb 4A blocked endogenous atrial G_k irreversibly. (D) mAb 4A does not block when atrial G_k is not activated. Spontaneous i_k was reduced in mAb 4A after excision into bathing solution containing 100-μM GTP. Unlike in (C) however, GTPγS at 100 μM activated i_k currents after washing mAb 4A away. (E) mAb 4A does not block when the muscarinic receptor and atrial G_k are tightly coupled. Following activation by carbachol at 0.1 μM, the patch is excised into GTP-free solution. i_k disappears. As in (D), but again unlike in (C), GTPγS activated i_k after washing away mAb 4A.

neuroblastoma $\times$ glioma hybrid cells; the effect was PTX-sensitive and was restored by injecting the cells with G_i^* or G_o^* purified from porcine brain with G_o^* being at least 10 times more effective (Hescheler *et al.*, 1987). The α_o^* subunit was equipotent with G_o^*. Direct G-protein action might be responsible for these effects, especially since G_o has no known effectors, but Ca^{2+} currents are also reduced by activators of PKC such as OAG and phorbol esters (Rane and Dunlap, 1986; Holz IV, 1986; Lewis *et al.*, 1986). Thus, although a G protein, possibly G_o, mediates the effects, the G protein could be acting indirectly via PKC rather than directly.

To test for direct effects on cardiac Ca^{2+} channels (Yatani *et al.*, 1987c; Imoto *et al.*, 1988), GTPγS and G proteins were applied directly just as for the K^+ channels. However, the predominant dihydropyridine-sensitive Ca^{2+} channel, variously referred to as the high threshold, fast deactivating, long-lasting, or L channel, quickly inactivates or runs down following patch excision. Two approaches were used to deal with this problem. In one approach, ventricular Ca^{2+} channels were activated by the β-adrenoreceptor agonist isoproterenol (Iso), or the dihydropyridine agonist Bay K 8644 prior to patch excision. Iso phosphorylates Ca^{2+} channels via protein kinase A (PKA) (Kameyama *et al.*, 1985, 1986) and Bay K 8644 activates the channels directly. Both cause some prolongation of survival time after patch excision, but single channel currents still ceased quickly. In the other approach, Ca^{2+} channels in vesicles from skeletal muscle T-tubules and cardiac sarcolemma were incorporated into planar phospholipid bilayers. Although these channels are dihydropyridine-sensitive, the skeletal muscle channels have smaller conductances (Rosenberg *et al.*, 1986; Ma and Coronado, 1988; Yatani *et al.*, 1988b) and very different kinetics (Cota and Stefani, 1986; Caffrey *et al.*, 1987). Incorporated skeletal muscle T-tubule single Ca^{2+}-channel currents do not run down in the presence of Bay K 8644 (Affolter and Coronado, 1985; Rosenberg *et al.*, 1986) and are stable for relatively long periods when the unincorporated material has been removed (Yatani *et al.*, 1988b).

In inside-out membrane patches from guinea pig ventricles, GTPγS had striking effects on Iso-activated (phosphorylated) or Bay K 8644-stimulated Ca^{2+} channels. In both cases, Iso in the patch pipette was an absolute requirement, presumably to increase the off-rate of GDP from endogenous G protein to allow GTPγS activation to occur before rundown became irreversible. GTP was also effective in these situations. Protocols using combinations of ATP, cAMP, PKA, and forskolin were designed to mimic possible phosphorylating effects of GTPγS and they failed as did protocols designed to block AC activity (addition of 10 mM ADPβS) or PKA activity (addition of protein kinase inhibitor). Accessibility to the inside surface of the membrane patch was not a problem because the nucleotide GTPγS and the protein G_s^* had equivalent effects. Several lines of evidence indicate that the Ca^{2+}-channel regulatory protein is none other than G_s, the activator protein of AC. First, GTPγS, to be effective in excised patches, required an agonist-occupied β-adrenergic receptor and the latter couples to G_s. Second, only activated purified human erythrocyte G_s (G_s^*) and its α subunit (α_s^*) mimicked the effect of GTPγS. Third, the active G protein is a CTX substrate (Fig. 7). In the reconstitution experiments on skeletal muscle T-tubule Ca^{2+} channels (Fig. 8), G_s^* stimulation was more straightforward since the control single Ca^{2+}-channel currents were not running down; neither ATP nor Bay K 8644 were required and addition of exoge-

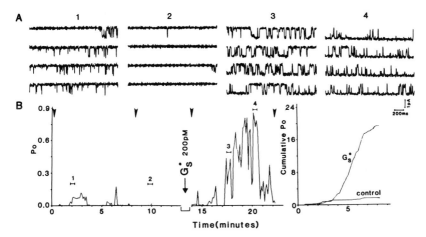

Fig. 7. Effects of G_s^* on single cardiac Ca^{2+} channels incorporated into a planar lipid bilayer. The number of functional Ca^{2+} channels incorporated into the bilayer was usually less than three and often one. As originally reported, cardiac sarcolemmal Ca^{2+} channels incorporated into this type of bilayer had properties similar to those of high threshold or L-type cardiac Ca^{2+} channels. In the presence of Bay K 8644 and with 100-mM Ba^{2+} as the charge carrier, the conductance between -50 and $+20$ mV was ~ 20 pS. The channel opened in bursts, and the probability of opening was voltage-dependent. Brief openings that dominate under control conditions could not be detected at the recording bandwidth, and the mean open times were fit to an exponential distribution ($\tau \sim 12 \pm 3$ msec, $n = 9$, at test potentials between 0 and $+20$ mV). The values are consistent with those obtained for Bay K 8644-stimulated channels in cell-attached patches. Current traces produced by depolarizing clamp steps to 0 mV from a holding potential of -40 mV are shown in (A) before [(1) and (2)] and after [(3) and (4)] addition of G_s^*. Pulses were applied every 30 seconds for 20 minutes. Leakage and capacitive currents were substrated. Traces (1) to (4) were taken at the times indicated in (B) and are 2-second segments from the 20-second pulses. (B) The entire experimental record. Note the decrease in activity with time in the control in (B) before addition of G_s^* (100 pM) to the trans chamber. The ordinate is given at P_o because n was 1 in this experiment. (C) Cumulative P_o's obtained between the arrows in (B). Cum P_o (G_s^*)/cum P_o (control) is 11.

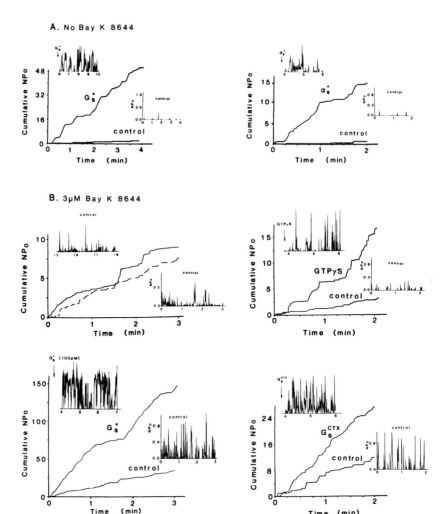

Fig. 8. Representative experiments showing direct regulation of the skeletal muscle T tubule dihydropyridine-sensitive Ca^{2+} channel by purified human erythrocyte G_s (PS/PE bilayer; charge carrier: Ba^{2+}). Experiments were carried out both in the absence (panel A) and the presence (panel B) of the dihydropyridine agonist Bay K 8644 in the *cis* chamber. Shown are the NPo (product of the number of channels and the opening probability of one channel) diaries prior and after addition of GTPγS-activated G_s, GTPγS-activated α_s, CTX-treated G_s (in the presence of GTP), or GTPγS (insets) and the corresponding cumulative NPo curves. The holding potentials were either 0 or +20 mV, and all experiments were done with $BaCl_2$ in the *cis* chamber. Average stimulations of activity were between 10- and 20-fold in the absence, and 2- to 3-fold in the presence, of 3-μM Bay K 8644. Note that Ca^{2+} channels incorporated into the bilayers can be stimulated either by exogenously activated G_s, or by activation of an exogenous T-tubule G protein, presumably also G_s, coincorporated with the Ca^{2+} channel. In other experiments it was established that to be effective GTPγS had to be added to the *cis* chamber, confirming the sidedness of the T tubule membranes.

nous G_s activated either with GTPγS or CTX was effective. Activation of coincorporated G_s by addition of either GTPγS or GTP plus Iso also increased single Ca^{2+}-channel currents. Moreover, activation by GTP plus Iso or G_s^* was asymmetrical. Iso was effective only from the extracellular side and GTP, GTPγS, and G_s^* were effective only from the intracellular side. GDPβS blocked the stimulation produced by activating coincorporated G_s with GTPγS. Human erythrocyte G_k^* had no effect and neither activated bovine brain G_{41} nor activated bovine brain G_{39} (forms of G_o) had any clear effect. Single Ca^{2+}-channel currents in lipid bilayers were also stimulated by phosphorylation catalyzed by PKA, and such phosphorylated channels were stimulated still further by addition of G_s^* (Imoto et al., 1988). Thus, G proteins directly modulate Ca^{2+}-channel activity in the case of muscle, producing stimulation. In the case of neurons and neurosecretory cells (Holz et al., 1986; Lewis et al., 1986; Scott and Dolphin, 1986; Dolphin and Scott, 1987; Hescheler et al., 1987), the anticipated direct G protein effect would be inhibitory. The direct G_s effect is at present of uncertain physiological significance. Both the indirect G_s effect via PKA-mediated phosphorylation and the direct effect are the consequences of occupancy of the β-adrenergic receptor (Fig. 9). Hence a single G protein appears to have two effectors and G proteins may therefore act as integrators in addition to their role as signal transducers. The $α_s$ subunits are four structures differing by a 15-amino acid insertion/deletion and a serine residue at the carboxyl end of the insertion/deletion. The $α_s$ of human erythrocytes is of the short type and it is not known if both forms are present. To determine if one $α_s$ alone could act on both effectors we expressed the $α_s$ short serine-minus form in vitro. We found that this $α$ subunit, when preactivated, stimulated both adenylyl cyclase and Ca^{2+} channels, proving that one $α_s$ can indeed activate two distinct effectors (Mattera et al., 1988). The nomenclature that has assumed a single effector such as G_s and G_i will have to be modified.

Before G_s can exert its stimulatory effect in heart, the Ca^{2+} channel may need first to be brought into a voltage-responsive state either by phosphorylation or by drug (Bay K 8644) as Fig. 7 shows. Alternatively, a direct effect independent of phosphorylation could occur as for skeletal muscle T-tubule Ca^{2+} channels. Also, in adrenal glomerulosa cells, adrenocorticotropic hormone (ACTH), acting via G_s, stimulates Ca^{2+} influx independently of changes in cAMP (Kojima and Ogata, 1986). In any event, the demonstration that G proteins can gate two distinct Ca^{2+} channels directly is important and it seems likely that G proteins may directly regulate a wide range of ion channels. In fact, a recent report by Horne et al. (1988) indicates that bovine brain G_o can change Ca^{2+} fluxes produced by solubilized Ca^{2+} channels from skeletal muscle T tubules that have been recon-

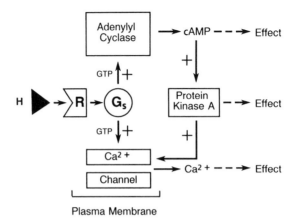

Fig. 9. Comparison of direct and indirect regulation of cardiac Ca^{2+} currents by G_s.

stituted in lipids. If this report is confirmed, it is strong evidence that G proteins directly regulate ionic channels.

III. G-PROTEIN SUBUNITS MEDIATING DIRECT IONIC CHANNEL GATING

Picomolar concentrations of human red blood cell (hRBC) G_k^* activated atrial (Yatani et al., 1987a) and GH_3 (Yatani et al., 1987b) $K^+[G_k]$ channels, whereas nanomolar concentrations of hRBC G_s^* were ineffective. G_k^* and G_s^* share a common $\beta\gamma$ dimer so the implicit conclusion was that, as for G_t and G_s, the α subunit (α_k) mediates the G_k effect on the K^+ channel (Yatani et al., 1987a,b). This was confirmed when the α_k^* at picomolar concentrations activated $K^+[G_k]$ channels (Codina et al., 1987a,b). By contrast, the $\beta\gamma$ dimers at nanomolar concentrations had no effect that could be attributed to them rather than to the accompanying detergent, in our case Lubrol, required to prevent $\beta\gamma$ dimer aggregation (Codina et al., 1987a,b). Furthermore, under identical conditions $\beta\gamma$ plus GTP could not reconstitute the PTX-blocked response whereas $\alpha\beta\gamma$ plus GTP could [Fig. 4A(f)]. In other experiments human erythrocyte α_k^* subunits activated $I_{K[G_k]}$ in membrane patches of neonatal rat atria and, importantly, 14-day-old embryonic chick atria (Kirsch et al., 1988). The activated α_s subunit (α_s^*) also mimicked the effect of G_s^* on single Ca^{2+}-channel currents in native and reconstituted systems. $\beta\gamma$ dimers were ineffective at concentrations where the effects of accompanying detergents could be ruled out (see below).

Against this is the report of stimulation of $K^+[G_k]$ channels by nanomolar concentrations of bovine brain $\beta\gamma$ dimers but not by bovine brain α subunits which led to the interpretation that $\beta\gamma$ subunits, not α subunits, mediate the G protein effect (Logothetis *et al.*, 1987). This conclusion has since been modified because Neer and Clapham (1988) have reported that the 39-kDa bovine brain α_{39} stimulates atrial muscarinic K^+ channels and confirmed our results with the 40-kDa hRBC α_{40}. Cerbai *et al.* (1988) have also reported that human erythrocyte α_{40} is effective at picomolar concentrations and that the $\beta\gamma$ dimer is effective at nanomolar concentrations. In fact, they used the same α_k^* that was ineffective initially when sent to Clapham and Neer. Cerbai *et al.* (1988) also found, as did Kirsch *et al.* (1988), that the detergent 3-[(3-cholamidopropyl)dimethylammonio]-1-propanesulfonate (CHAPS) by itself stimulated these channels. As noted in the preceding paragraph, the results with human red blood cell α_k^* have been extended from adult guinea pig atria to neonatal rat atria and chick embryo atria, the latter being the preparation used by Logothetis *et al.* (1987). Thus, the evidence shows that α subunits are the physiological mediators of the direct G-protein gating of ionic channels. A question that remains concerns the effectiveness and role of the $\beta\gamma$ dimer. The role as an activator would be modest at best given the 100 to 10,000-fold lower potency of the α subunit. Moreover, we are unconvinced that a stimulatory effect independent of detergent has been established. It is likely that $\beta\gamma$ has different and far more important roles, possibly to specify one of numerous similar α subunits and to arrange its proper alignment with receptor and/or effector.

IV. INDIRECT G-PROTEIN GATING OF IONIC CHANNELS

This category is distinguished from direct G-protein gating because in this case G proteins have as their effectors membrane enzymes, not ionic channels. The activated enzymes change cytoplasmic constituents and it is via this indirect pathway that ionic channel activity is modulated. Another distinction is that in all instances the effects are modulatory; an obligatory requirement for receptor activation has not yet been shown.

The sequence has been delineated most clearly for rhodopsin–transducin (G_t) and voltage-independent Na^+ channels in rods and for β-adrenoreceptors–G_s and dihydropyridine-sensitive Ca^{2+} channels in ventricular myoctyes. In this context rhodopsin may be viewed as a receptor and photons as the agonists. Activated rhodopsin promotes the exchange of GDP for GTP on transducin and the released α_t activates a membrane-associated cGMP phosphodiesterase. PDE hydrolyzes cGMP which was keeping the Na^+ channels open (Stryer, 1986). As a result the channels close and the membrane hyperpolarizes (Fesenko et al., 1985; Haynes and Yau, 1985). In rods the rhodopsin–transducin and cGMP-PDE are in the disc membranes and the fall in cGMP has to be translated to the plasma membrane where the Na^+ channels are located. An analogous sequence is thought to occur in cones, although rhodopsin–G_t and the channels are all in the plasma membrane (Mathews, 1987).

In heart, norepinephrine is the agonist for the β-adrenoreceptor and the activated G protein is G_s. G_s activates AC, cAMP is increased, and PKA is activated (Kameyama et al., 1985, 1986; Fig. 9). The Ca^{2+} channel is presumably phosphorylated just as the DHP receptor in skeletal muscle is (Flockerzi et al., 1987), and the opening probability is increased. That the effects can be prevented by protein kinase A inhibitor and muscle-specific phosphatase are additional evidence for a phosphorylating mechanism (Kameyama et al., 1986; Trautwein et al., 1982). This sequence is not involved in basal regulation, however (ibid.). The relative effects of direct and indirect regulation are compared in Figure 9. It will be seen that for the nonbasal situation direct regulation is minor. However, its role in basal conditions remains to be evaluated.

There are many other examples of indirect G-protein gating of ionic channels. The details are not as complete as in the preceding cases, but a similar pattern is present; the differences being the particular G protein involved and the intervening cytoplasmic steps. A significant number involve G-protein activation of phospholipase C. Diacylglycerol (DAG) and inositol trisphosphate (IP_3) are formed; DAG activates PKC which may phosphorylate K^+ or Ca^{2+} (Dunlap et al., 1987) channels, IP_3 releases Ca^{2+} from intracellular cells, and the Ca^{2+} can, in turn, alter the activity of a number of membrane channels, in particular Ca^{2+}-activated K^+- and nonselective cation channels (Berridge and Irvine, 1984).

V. CONCLUSIONS

The discovery of direct regulation of ion channels by G proteins clearly indicates that we must now reckon not only with ligand and voltage gating of ion channels, but also with direct G-protein gating of ion channels. Ligand-gated ion channels, such as the nicotinic acetylcholine, $GABA_A$, and glycine receptors are part of a superfamily of molecules having the receptor and channel incorporated into the same structure (Schofield et al., 1987; Grenningloh et al., 1987). Likewise, certain voltage-gated ion channels such as nerve Na^+ channel (Noda et al., 1987), muscle Ca^{2+} channel (Tanabe et al., 1987), and muscle K^+ channel (Tempel et al., 1987) appear to share structural features that explain their voltage sensitivity (Schofield *et al.*, 1987). In direct G-protein gating, the common design feature is the G proteins, not the receptors or the ionic channels. The direct G-protein link allows diverse receptors to be coupled to diverse ionic channels adding an interesting and undoubtedly important layer of complexity to membrane excitability. The frequency of indirect G-protein gating of ionic channels is even greater and has until recently received the lion's share of attention. Taken together with direct G-protein gating, the combination makes G proteins the preeminent players in coupling receptors to ionic channels.

REFERENCES

Affolter, H., and Coronado, R. (1985). Agonists Bay-K8644 and CGP-28392 open calcium channels reconstituted from skeletal muscle transverse tubules. *Biophys. J.* **48**, 341–347.

Berridge, M. J., and Irvine, R. F. (1984). Inositol triphosphate, a novel second messenger in cellular signal transduction. *Nature* **312**, 315–321.

Birnbaumer, L., Codina, J., Mattera, R., Yatani, A., Scherer, N., Toro, M. J., and Brown, A. M. (1987). Signal transduction by G proteins. *Kidney Int.* **32**, S-14–S-37.

Birnbaumer, L., Yatani, A., Codina, J., Mattera, R., Graf, R., Liao, C.-F., Themmen, A., Sanford, J., Hamm, H., Iyengar, R., Birnbaumer, M., and Brown, A. M. (1988). Signal transduction by G proteins, regulation of ion channels as seen with native and recombinant subunits and multiplicity of intramembrane transduction pathways. *In* "The Molecular and Cellular Endocrinology of the Testis" (B. A. Cook, ed.), Ares Serono Symposia, Milan (in press).

Bray, P., Carter, A., Simons, C., Guo, V., Puckett, C., Kamholz, J., Spiegel, C., and Nirenberg, M. (1986). Human cDNA clones for four species of G α_s signal transduction protein. *Proc. Natl. Acad. Sci. U.S.A.* **83**, 8893–8897.

Breitwieser, G. E., and Szabo, G. (1985). Uncoupling of cardiac muscarinic and β-adrenergic receptors from ion channels by a guanine nucleotide analogue. *Nature* **317**, 538–540.

Caffrey, J., Brown, A. M., and Schneider, M. D. (1987). Mitogens and oncogenes can block the induction of specific voltage-gated ion channels. *Science* **236**, 570–573.

Cerbai, E., Klöckner, U., and Isenberg, G. (1988). The α subunit of the GTP binding protein activates muscarinic potassium channels of the atrium. *Science* **240**, 1782–1783.

Codina, J., Hildebrandt, J. D., Iyengar, R., Birnbaumer, L., Sekura, R. D., and Manclark, C. R. (1983). Pertussis toxin substrate, the putative N_i of adenylyl cyclases, is an α/β heterodimer regulated by guanine nucleotide and magnesium. *Proc. Natl. Acad. Sci. U.S.A.* **80**, 4276–4280.

Codina, J., Hildebrandt, J. D., Sekura, R. D., Birnbaumer, M., Bryan, J., Manclark, C. R., Iyengar, R., and Birnbaumer, L. (1984a). N_s and N_i, the stimulatory and inhibitory regulatory components of adenylyl cyclases. Purification of the human erythrocyte proteins without the use of activating regulatory ligands. *J. Biol. Chem.* **259**, 5871–5886.

Codina, J., Hildebrandt, J. D., Birnbaumer, L., and Sekura, R. D. (1984b). Effects of guanine nucleotides and Mg on human erythrocyte N_i and N_s, the regulatory components of adenylyl cyclase. *J. Biol. Chem.* **259**, 11408–11418.

Codina, J., Yatani, A., Grenet, D., Brown, A. M., and Birnbaumer, L. (1987a). The α subunit of G_k opens atrial potassium channels. *Science* **236**, 442–445.

Codina, J., Grenet, D., Yatani, A., Birnbaumer, L., and Brown, A. M. (1987b). Hormonal regulation of pituitary GH_3 cell K^+ channels by G_k is mediated by its α-subunit. *FEBS Lett.* **216**, 104–106.

Cota, G., and Stefani, E. (1986). A fast-activated inward calcium current in twitch muscle fibers of the frog (Rana Montezume). *J. Physiol (London)* **370**, 151–163.

Deretic, D., and Hamm, H. E. (1988). Topographic analysis of antigenic determinations recognized by monoclonal antibodies to the photoreceptor guanyl nucleotide binding protein, transducin. *J. Biol. Chem.* **262**, 10839–10847.

Dolphin, A. C., and Scott, R. H. (1987). Calcium channel currents and their inhibition by (−)baclofen in rat sensory neurones: Modulation by guanine nucleotides. *J. Physiol.* **386**, 1–17.

Dunlap, K., Holz, G. G., and Rane, S. G. (1987). G proteins as regulators of ion channel function. *TINS* **10**, 241–244.

Fesenko, E. E., Kolesnikov, S. S., and Lyubarsky, A. L. (1985). Induction by cyclic GMP of cationic conductance in plasma membrane rod outer segment. *Nature* **313**, 310–313.

Flockerzi, V., Oeken, H. J., Hofmann, F., Pelzer, D., Cavalie, A., and Trautwein, W. (1987). Purified dihydropyridine-binding site from rabbit skeletal muscle t-tubules is a functional calcium channel. *Nature* **323**, 66–68.

Galizzi, J-P., Fossett, M., and Lazdunski, M. (1984). Properties of receptors for the Ca^{2+}-channel blocker verapamil in transverse-tubule membranes of skeletal muscle. *Eur. J. Biochem.* **144**, 211–215.

Glitsch, H. G., and Pott, L. (1978). Effects of acetylcholine and parasympathetic nerve stimulation on membrane potential in quiescent guinea pig atria. *J. Physiol. (London)* **279**, 655–668.

Graziano, M. P., Casey, P. J., and Gilman, A. G. (1987). Expression of cDNAs for G proteins in *Escherichia coli,* two forms of $G_{s\alpha}$ stimulate adenylate cyclase. *J. Biol. Chem.* **262**, 11375–11381.

Grenningloh, G., Rienitz, A., Schmitt, B., Methfessel, C., Zensen, M., Beyreuther, K., Gundelfinger, E. D., and Betz, H. (1987). The strychnine-binding subunit of the glycine receptor shows homology with anicotinic acetylcholine receptors. *Nature* **328**, 215–220.

Halvorsen, S. W., and Nathanson, N. M. (1984). Ontogenesis of physiological responsiveness and guanine nucleotide sensitivity of cardiac muscarinic receptors during chick embryonic development. *Biochemistry* **23**, 5813–5321.

Hamm, H. E., Deretic, D., Hofmann, K. P., Schleicher, A., and Kohn, B. (1987). Mechanism of action of monoclonal antibodies that block the light-activation of the guanyl nucleotide binding protein, transducin. *J. Biol. Chem.* **262**, 10831–10838.

Hartzell, H. (1980). Distribution of muscarinic acetylcholine receptors and presynaptic nerve terminals in amphibian heart. *J. Cell Biol.* **86**, 6–20.

Haynes, L., and Yau, K. W. (1985). Cyclic GMP-sensitive conductance in outer segment membrane of catfish cones. *Nature* **317**, 61–64.

Hazeki, O., and Ui, M. (1981). Modification by islet-activating protein of receptor-medicated regulation of cAMP accumulation in isolated rat heart cells. *J. Biol. Chem.* **256**, 2856–2862.

Hescheler, J., Rosenthal, W., Trautwein, W., and Schultz, G. (1987). The GTP-binding protein, G_o, regulates neuronal calcium channels. *Nature* **325**, 445–447.

Hill-Smith, I., and Purves, R. D. (1978). Synaptic delay in the heart: An ionophoretic study. *J. Physiol. (London)* **279**, 31–54.

Holz IV, G. G., and Dunlap, K. (1986). GTP-binding proteins mediate transmitter inhibition of voltage-dependent calcium channels. *Nature* **319**, 670–672.

Horne, W. A., Abdel-Ghany, M., Kacher, E., Weiland, G. A., Oswald, R. E., and Cerione, K. A. (1988). Functional reconstitution of skeletal muscle Ca^{2+} channels: Separation of regulatory and channel component. *Proc. Natl. Acad. Sci. U.S.A.* **85**, 3718–3722.

Hutter, O. F., and Trautwein, W. (1956). Vagal and sympathetic effects on the pacemaker fibers in sinus venosus of the heart. *J. Gen. Physiol.* **39**, 153–194.

Imoto, Y., Yatani, A., Reeves, J. P., Codina, J., Birnbaumer, L., and Brown, A. M. (1988). The α subunit of G_s directly activates cardiac calcium channels in lipid bilayers. *Am. J. Physiol.* **255**, H722–H728.

Kameyama, M., Hofmann, F., and Trautwein, W. (1985). On the mechanism of beta-adrenergic regulation of the Ca channel in the guinea-pig heart. *Pflüegers Arch.* **405**, 285–293.

Kameyama, M., Hescheler, J., Hofmann, F., and Trautwein, W. (1986). Modulation of Ca current during the phosphorylation cycle in the guinea pig heart. *Pflüegers Arch.* **407**, 123–128.

Kirsch, G., Yatani, A., Codina, J., Birnbaumer, L., and Brown, A. M. (1988). α-Subunit of G_k activates atrial K^+ channels of chick, rat and guinea pig. *Am. J. Physiol.* **23**, H1200–1205.

Kojima, I., and Ogata, E. (1986). Direct demonstration of adrenocorticotropin-induced changes in cytoplasmic free calcium with aequorin in adrenal glomerulosa cell. *J. Biol. Chem.* **261**, 9832–9838.

Kurachi, Y., Nakajima, T., and Sugimoto, T. (1986a). On the mechanism of activation of muscarinic K^+ channels by adenosine in isolated atrial cells: Involvement of GTP-binding proteins. *Pflüegers Arch.* **407**, 264–274.

Kurachi, Y., Nakajima, T., and Sugimoto, T. (1986b). Acetylcholine activation of K^+ channels in cell-free membrane of atrial cells. *Am. J. Physiol.* **251** Heart Circ. Physiol. **20**, H681–H684.

Kurachi, Y., Nakajima, T., and Sugimoto, T. (1986c). Role of intracellular Mg^{2+} in the activation of muscarinic K$^+$ channel in cardiac atrial cell membrane. *Pflüegers Arch.* **407**, 572–574.

Kurose, H., and Ui, M. (1983). Functional uncoupling of muscarinic receptors from adenylate cyclase in rat cardiac membranes by the active component of islet-activating protein, pertussis toxin. *J. Cyclic Nucleotide Protein Phosphorylation Res.* **9**, 305–318.

Lee, K. S., Weeks, T. A., Kao, R. L., Akaike, N., and Brown, A. M. (1979). Sodium current in single heart muscle cells. *Nature* **278**, 269–271.

Lewis, D. L., Weight, F. F., and Luini, A. (1986). A guanine nucleotide-binding protein mediates the inhibition of voltage-dependent calcium current by somatostatin in a pituitary cell line. *Proc. Natl. Acad. Sci. U.S.A.* **83**, 9035–9039.

Liuni, A., Lewis, D., Guild, S., Schofield, G., and Weight, F. (1986). Somatostatin, an inhibitor of ACh secretion decreases cytosolic free calcium and voltage-dependent calcium current in a pituitary cell line. *J. Neurosci.* **6**, 3128–3132.

Loewi, O., and Navratil, E. (1926). Über humorale Übertragbarkeit der herznervenwirkung. *Pflüegers Arch.* **214**, 678–688.

Logothetis, D. E., Kurachi, Y., Galper, J., Neer, E. J., and Clapham, D. E. (1987). The β subunits of GTP-binding proteins activate the muscarinic K$^+$ channel in heart. *Nature* **325**, (6102) 321–326.

Ma, J., and Coronado, R. (1988). Heterogeneity of conductance states in calcium channels of skeletal muscle. *Biophys. J.* **53**, 387–395.

Mathews, G. (1987). Single channel recordings demonstrate that cGMP opens the light sensitive ion channel of the rod photoreceptor. *Proc. Natl. Acad. Sci. U.S.A.* **84**, 299–302.

Mattera, R., Codina, J., Crozat, A., Kidd, V., Woo, S. L. C., and Birnbaumer, L. (1986). Identification by molecular cloning of two forms of the α subunit of the human liver stimulatory (G$_s$) regulatory component of adenylyl cyclase. *FEBS Lett.* **206**, 36–42.

Mattera, R., Yatani, A., Kirsh, G. E., Graf, R., Olate, J., Codina, J., Brown, A. M., and Birnbaumer, L. (1989). Recombinant α-3 subunit of CT protein activates T$_k$-gated K$^+$ channels. *J. Biol. Chem.* **264**, 465–471.

Nargeot, J., Lester, H. A., Birdsall, N. J. M., Stockton, J., Wassermann, N. H., and Erlanger, B. F. (1982). A photoisomerizable muscarinic antagonist: Studies of binding and of conductance relaxations in frog heart. *J. Gen. Physiol.* **79**, 657–678.

Nargeot, J., Nerbonne, J. M., Engels, J., and Lester, H. A. (1983). Time course of the increase in the myocardial slow inward current after a photochemically generated concentration jump of intracellular cAMP. *Proc. Natl. Acad. Sci. U.S.A.* **80**, 2395–2399.

Neer, E. J., and Clapham, D. C. (1988). Role of G protein subunits in transmembrane signalling. *Nature* **333**, 129–134.

Noda, M., Iheda, T., Kayano, T., Sujuki, H., Takeshima, H., Kuraschi, M., Takahashi, H., and Numa, S. (1987). Existence of distinct sodium channel messenger RNA's in rat brain. *Nature* **320**, 188–192.

Osterrieder, W., Yang, Q.-F., and Trautwein, W. (1981). The time course of the muscarinic response to ionophoretic acetylcholine application to the S-A node of the rabbit heart. *Pflüegers Arch.* **389**, 283–291.

Pfaffinger, P. J., Martin, J. M., Hunter, D. D., Nathanson, N. M., and Hille, B. (1985). GTP-binding proteins couple cardiac muscarinic receptors to a K channel. *Nature* **317**, 536–538.

Powell, T., and Twist, V. W. (1976). Isoprenaline stimulation of cyclic AMP production by isolated cells from adult rat myocardium. *Biochem. Biophys. Res. Commun. Chem. Pathol. Pharmacol.* **72**, 1218–1225.

Rane, S. O., and Dunlap, K. (1986). Kinase C activator 1,2-oleylacetylglycerol attenuates voltage-dependent calcium current in sensory neurones. *Proc. Natl. Acad. Sci. U.S.A.* **83**, 184–188.

Robishaw, J. D., Smigel, M. D., and Gilman, A. G. (1986). Molecular basis for two forms of the G protein that stimulates adenylate cyclase. *J. Biol. Chem.* **261**, 9587–9590.

Rodbell, M., Krans, H. M. J., Pohl, S. L., and Birnbaumer, L. (1971a). The glucagon-sensitive adenylyclase system in plasma membranes. IV. Binding of glucagon: Effects of guanylnucleotide. *J. Biol. Chem.* **246**, 1872–1876.

Rodbell, M., Birnbaumer, L., Pohl, S. L., and Krans, H. M. J. (1971b). The glucagon-sensitive adenyl cyclase system in plasma membranes of rat liver. V. An obligatory role guanyl nucleotides in glucagon action. *J. Biol. Chem.* **246**, 1877–1882.

Rosenberg, R. L., Hess, P., Reeves, J. P., Smilowitz, H., and Tsien, R. W. (1986). Calcium channels in planar lipid bilayers: Insights into mechanisms of ion permeation and gating. *Science* **231**, 1564–1566.

Rosenberger, L. B., Roeske, W. R., and Yamamura, H. I. (1979). The regulation of muscarinic cholinergic receptors by guanine nucleotides in cardiac tissue. *Eur. J. Pharmacol.* **56**, 179–180.

Sakmann, B., Noma, A., and Trautwein, W. (1983). Acetylcholine activation of single muscarinic K^+ channels in isolated pacemaker cells of the mammalian heart. *Nature* **303**, 250–253.

Scherer, N. M., Toro, M.-J., Mumby, S. M., Sekura, R. D., Gilman, A. G., Entman, M. L., and Birnbaumer, L. (1987). G protein distribution in canine cardiac sarcoplasmic reticulum and sarcolemma. Comparison to Lagomorph skeletal membranes, and brain and erythrocyte G proteins. *Arch. Biochem. Biophys.* **259**, 431–440.

Schlegel, W., Wuarin, F., Zbaren, C., Wolheim, C. B., and Zahnd, G. R. (1985). Pertussis toxin selectively abolishes hormone induced lowering of cytosolic calcium in GH_3 cells. *FEBS Lett.* **189**, 27–32.

Schofield, P. R., Darlison, M. G., Fujita, N., Burt, D. R., Stephenson, F. A., Rodriguez, H., Rhee, L. M., Ramachandran, J., Reale, V., Glencorse, T. A., Seeburg, P., and Barnard, E. A. (1987). Sequence and functional expression of the $GABA_A$ receptor shows a ligand-gated receptor super-family. *Nature* **328**, 221–227.

Scott, R. H., and Dolphin, A. C. (1986). Regulation of calcium currents by a GTP analogue: Potentiation of (-)-baclofen-mediated inhibition. *Neurosci. Lett.* **69**, 59–64.

Soejima, M., and Noma, A. (1984). Mode of regulation of the ACh-sensitive K-channel by the muscarinic receptor in rabbit atrial cells. *Pflüegers Arch.* **400**, 424–431.

Stryer, L. (1986). Cyclic GMP Cascade of Vision. *Annu. Rev. Neurosci.* **9**, 87–119.

Szabo, G., Pang, I. H., and Sternweis, P. C. (1988). Activation of a K^+ current by exogenous G protein in atrial myocytes. *Biophys. J.* **53**, 424a.

Tanabe, T., Takeshima, H., Mikami, A., Flockerzei, V., Takahashi, H., Kangawa, K., Kojima, M., Matsuo, H., Hirose, T., and Numa, S. (1987). Primary structure of the receptor for calcium channel blocker from skeletal muscle. *Nature* **328**, 313–318.

Tempel, B. L., Papajian, D. M., Schwartz, T. L., Jan, Y. N., and Jan, L. Y. (1987). Sequence of a probable potassium channel component encoded at a Shaker locus of Prosephila. *Science* **237**, 770–775.

Trautwein, W., and Dudel, J. (1958). Zum Mechanismus der Membranewirkung des Acetylcholins an der Herzmuskelfaser. *Pflüegers Arch.* **266**, 324–334.

Trautwein, W., Taniguchi, J., and Noma, A. (1982). The effect of intracellular cyclic nucleotides and calcium on the action potential and acetylcholine response of isolated cardiac cells. *Pflüegers Arch.* **392**, 307–314.

Triggle, D. J., Skattebol, A., Rampe, D., Joclyn, A., and Gengo, P. (1986). Chemical pharmacology of Ca^{2+} channel ligands. *In* "New Insights into Cell and Membrane Transport Processes" (G. Poste and S. T. Crooke, eds.), pp. 125–143. Plenum, New York.

VanDongen, A., Codina, J., Olate, J., Mattera, R., Joho, R., Birnbaumer, L., and Brown, A. M. (1988). Newly identified brain potassium channels gated by the guanine nucleotide binding protein G_o. *Science* **242**, 1433–1437.

Yamashita, N., Shibuya, N., and Ogata, E. (1986). Hyperpolarization of the membrane potential caused by somatostatin in dissociated human pituitary adenoma cells that secrete growth hormone. *Proc. Natl. Acad. Sci. U.S.A.* **83**, 6198–6202.

Yatani, A., Codina, J., Brown, A. M., and Birnbaumer, L. (1987a). Direct activation of mammalian atrial muscarinic K channels by a human erythrocyte pertussis toxin-sensitive G protein, G_k. *Science* **235**, 207–211.

Yatani, A., Codina, J., Sekura, R. D., Birnbaumer, L., and Brown, A. M. (1987b). Reconstitution of somatostatin and muscarinic receptor mediated stimulation of K^+ channels by isolated G_k protein in clonal rat anterior pituitary cell membranes. *Mol. Endocrinol.* **1**, 283–289.

Yatani, A., Codina, J., Imoto, Y., Reeves, J. P., Birnbaumer, L., and Brown, A. M. (1987c). Direct regulation of mammalian cardiac calcium channels by a G protein. *Science* **238**, 1288–1292.

Yatani, A., Hamm, H., Codina, J., Mazzoni, M. R., Birnbaumer, L., and Brown, A. M. (1988a). A monoclonal antibody to the alpha subunit of G_k blocks muscarinic activation of atrial K^+ channels. *Science* **241**, 828–831.

Yatani, A., Imoto, Y., Codina, J., Hamilton, S., Brown, A. M., and Birnbaumer, L. (1988b). The stimulatory G protein of adenylyl cyclase, G_s, also stimulates dihydropyridine-sensitive Ca^{2+} channels: Evidence for direct regulation independent of phosphorylation. *J. Biol. Chem.* **263** (20), 9887–9895.

CHAPTER 12

Receptor – Effector Coupling by Pertussis Toxin Substrates: Studies with Recombinant and Native G-Protein α Subunits

Juan Codina
Department of Cell Biology, Baylor College of Medicine, Houston, Texas 77030

Atsuko Yatani, Antonius M. J. VanDongen
Department of Molecular Physiology and Biophysics, Baylor College of Medicine, Houston, Texas 77030

Elena Padrell, Donna Carty
Department of Pharmacology, Mount Sinai School of Medicine, New York, New York 10029

Rafael Mattera
Department of Cell Biology, Baylor College of Medicine, Houston, Texas 77030

Arthur M. Brown
Department of Molecular Physiology and Biophysics, Baylor College of Medicine, Houston, Texas 77030

Ravi Iyengar
Department of Pharmacology, Mount Sinai School of Medicine, New York, New York 10029

Lutz Birnbaumer
Department of Cell Biology, Baylor College of Medicine, Houston, Texas 77030

I. Introduction
II. Pertussis Toxin: Structure and Conditions for Toxin-
   Catalyzed ADP-Ribosylation

I. INTRODUCTION

In recent years, pertussis toxin has been widely utilized to probe the role of G proteins in transmembrane signaling. The role of pertussis toxin in affecting signal transduction was first discovered by Ui and coworkers who showed that the pertussis vaccine suppressed epinephrine-induced hyperglycemia (Sumi and Ui, 1975). Further studies led to the localization of this effect to the pancreatic islet cells, where it was shown that α_2-adrenergic receptor-mediated inhibition of insulin secretion was attenuated by the pertussis vaccine (Katada and Ui, 1977). It was subsequently shown that α_2-adrenergic receptors act by inhibiting adenylyl cyclase and that pertussis toxin attenuates this inhibition (Katada and Ui, 1979). From these observations the concept arose that the effect of pertussis toxin was to uncouple receptors from the effector system. The target of pertussis toxin was found to be a 40–41-kDa membrane protein that served as the acceptor for the pertussis toxin-catalyzed ADP-ribose transfer (Katada and Ui, 1982). Studies of the cyc^- mutants of the murine lymphoma cell line S49 indicated that hormone receptor guanine nucleotide-mediated inhibition of adenylyl cyclase was observable in the absence of the stimulatory pathway (Hildebrandt *et al.*, 1983a; Jakobs *et al.*, 1983), and that pertussis toxin served to disrupt communications in this pathway. This led to the concept that an inhibitory G protein (G_i) was the site of pertussis toxin action. Purification of G proteins also indicated the presence of a heterotrimeric G protein whose α subunit was the pertussis toxin substrate. Hence this protein, similar to the G_s except for differential toxin sensitivity, was also called G_i (Bokoch *et al.*, 1983; Codina *et al.*, 1983).

Recent biochemical and molecular biological studies have shown that there are several substrates for pertussis toxin, including G_i, transducin (Van Dop et al., 1984), and G_o, the major G-protein from brain (Sternweis and Robishaw, 1984; Neer et al., 1984). Indeed most of the G proteins currently identified contain the cysteine in the fourth position from the carboxy terminus which serves as the acceptor for the toxin-catalyzed ADP-ribosylation reaction. Thus pertussis-toxin sensitivity now serves as a general indicator of the involvement of a G protein in a signal transduction pathway, though lack of a pertussis toxin effect does not necessarily mean the lack of G-protein involvement, as is the case in G_s stimulation of adenylyl cyclase. Currently, receptor-regulated stimulation of phosphatidyl inositol-specific phospholipase C, phospholipase A_2, opening of K^+ channels, and both opening and closing of Ca^{2+} channels involve pertussis toxin-sensitive G proteins. In this chapter we describe the various G proteins that are pertussis-toxin substrates and the methods for their identification and purification, as well as functional analysis using both native and bacterially synthesized proteins. The findings lead to an interesting discussion of how signal transduction pathways involving G proteins are set up.

II. PERTUSSIS TOXIN: STRUCTURE AND CONDITIONS FOR TOXIN-CATALYZED ADP-RIBOSYLATION

Pertussis toxin belongs to the A-B class of bacterial toxins (Gill, 1977). The A and B components of the toxin are noncovalently linked. The A component is also called the S1 component (M_r 28,000) and contains the active site of the enzyme ADP-ribosyltransferase. Within the A protomer there are disulfide bonds that have to be reduced to obtain the active form of the enzyme, when the A protomer is dissociated with the B oligomer. Treatment of the toxin with a reducing agent such as 2-mercaptoethanol or dithiothreitol (DTT) is often referred to as activation. Such activation is required if the toxin is to be utilized in a cell-free system. However, if toxin treatment of the intact cell or tissue is required, the toxin is utilized without the activation, since it is the B oligomer that binds to the cell surface to allow the A protomer to enter the cell. The B oligomer is a pentamer consisting of two dimers. Dimer 1 is composed of one S2, (M_r 23,000) and one S4 (M_r 11,700) subunit. Dimer 2 is composed of one S3 (M_r 2200) and one S4 (M_r 11,700) subunit. The S4 subunits are connected to each other through the S5 (M_r 9,300) subunit (Tamura et al., 1982; Ui, 1986). Studies by Ui and coworkers have shown that free amino groups in Dimer 2 are essential for toxin binding to the cell surface, and that modification of Dimer 2 leads to the selective retention of the ADP-ribosylating activity of the toxin while other biologic effects of the toxin such as mitogenicity are abolished. A schematic diagram of pertussis toxin is shown in Fig. 1.

A   A Protomer: S_1
    B Oligomer:
       S_5 + Dimer 1 + Dimer 2
       Dimer 1: S_2 + S_4
       Dimer 2: S_3 + S_4
       S_1= 28K
       S_2= 23K
       S_3= 22K
       S_4= 11.7K
       S_5= 9.3K

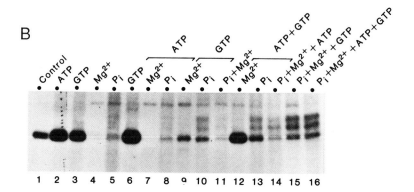

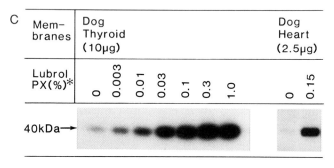

Fig. 1. ADP-ribosylation of membrane proteins by pertussis toxin. (A): Subunit structure of pertussis toxin as deduced by Tamura *et al.* (1982) and Sekura *et al.* (1983). (B): Effects of additives on labeling intensities of pertussis and cholera toxin substrates by cholera toxin plus pertussis toxin in thyroid membranes (adapted from Ribeiro-Neto *et al.*, 1987a). (C): Typical intensification of labeling of pertussis toxin substrates in membrane upon addition of Lubrol PX (adapted from Ribeiro-Neto *et al.*, 1987b). Incubations: 30 min at 32°C. Pertussis toxin: 10 mg/ml; no P_i; no Mg^{2+}; 1 mM EDTA; 1 mM ADP; 0.1 mM GTP. Lubrol PX was added to membranes prior to assay at 5-fold dilution.

The expression of the ADP-ribosylating activity of the toxin requires not only the activation of the A subunit and the appropriate substrates (acceptor protein, NAD^+) but also the presence of ATP (Lim et al., 1985). In cell-free systems the requirement of ATP for the ADP-ribosylation reaction is quite stringent. The ATP requirement appears to be related to the activation (Mattera et al., 1986a) and subunit dissociation (Burns and Manclark, 1986), and is allosteric in nature since ATP hydrolysis is not required.

The substrate for pertussis toxin is the α subunit of the G protein. However the free α subunit is not a substrate. The GDP-liganded $\alpha\beta\gamma$ complex serves as the exclusive substrate. This was first supported by Tsai et al. (1984) and was confirmed by Mattera et al. (1987). It has proven to be true for transducin, G_i and G_o. Thus addition of agents such as Mg^{2+} ions and GTP analogs that promote the activation and hence dissociation of G proteins greatly reduces the capability of the α subunit to serve as an acceptor for pertussis toxin-catalyzed ADP-ribosylation.

In the case of pertussis toxin-catalyzed ADP-ribosylation, the effect is one of uncoupling the G protein from the receptor. The ADP-ribosylated α subunits bind GTP or GTP analogs in a manner indistinguishable from the native protein. The GTPase activity of the α subunit is also unaffected by pertussis toxin treatment (Ui, 1986; Sunyer et al., 1989). Thus contrary to the general impression that pertussis toxin inactivates G proteins, the effect of toxin is specific for receptor interaction.

III. WHAT IS G_i AND WHAT IT MEANS NOW

The original concept of G_i arose from a combination of functional studies in the G_s-deficient cyc^- S49 cell line, where receptor and guanine nucleotide-dependent inhibition of adenylyl cyclase was demonstrated (Hildebrandt et al., 1982, 1984; Hildebrandt and Birnbaumer, 1983), and biochemical studies that led to the characterization of the pertussis toxin substrate as a heterotrimeric G protein whose α subunit was in the 41–40 kDa range (Bokoch et al., 1983; Codina et al., 1983). In contrast to G_s effects on adenylyl cyclase, inhibition of purified adenylyl cyclase has never been obtained with pure G_i. The three cDNAs encoding G_{i1}, G_{i2}, and G_{i3} were isolated from cDNA libraries utilizing synthetic oligonucleotide probes that had been constructed on the basis of partial amino acid sequences (Nukuda et al., 1986, Itoh et al., 1986; Suki et al., 1987) or by the use of oligonucleotide probes encoding the "identity box" region in the GTP binding domain (Jones and Reed, 1987). Thus the name G_i has

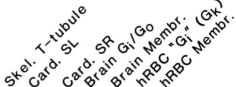

1hr

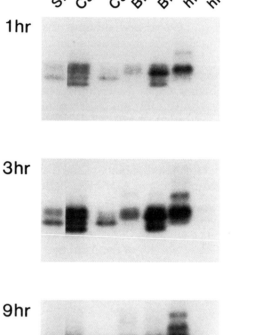

3hr

9hr

36hr

Fig. 2. Membranes and "purified" proteins contain multiple PTX substrates. PTX-labeled membranes (1 μg/lane) or purified bovine brain mixture of G/G_o and human erythrocyte G_i (20 ng/lane) were separated by SDS-PAGE in 8% polyacrylamide gels in the presence of a 4–8 M urea gradient. The gel slabs were fixed, stained with Coomassie blue and destained (not shown) and then autoradiographed for the times indicated. Only the sections of the autoradiograms with molecules between 35 and 45 kDa are shown.

essentially become a generic misnomer since none of the three G_i gene products have been shown to inhibit adenylyl cyclase. However all of them have G_k activity (see below) in that the resolved and purified α subunits, on activation, can open K^+ channels in excised atrial membrane patches. At this stage it is uncertain what other effector systems these proteins will affect. However for historical reasons it is expected that the name G_i will persist, and in this chapter we will continue to refer to them as G_i.

The three α_i cDNAs have strong (ca. 90%) amino acid sequence homology. α_{i1} and α_{i3} are more closely related to each other (95% homology) than either are to α_{i2}. Thus physiological assignment of functions to each of these G proteins appears to be a formidable challenge. Moreover, while molecular cloning has revealed only four pertussis toxin substrates (not counting transducin), labeling experiments with [^{32}P]NAD reveal many more than four bands (Fig. 2), the identity and/or function of which is as yet unclear.

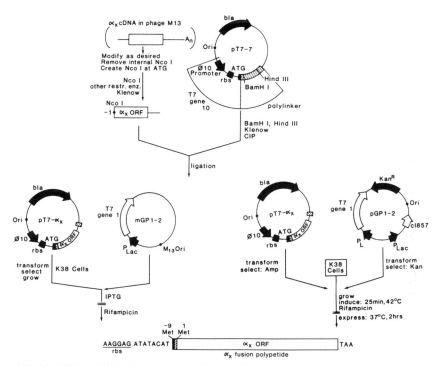

Fig. 3. Tabor–Richardson strategies for bacterial expression of cloned proteins by the plasmid/phage (lower left) and plasmid/plasmid (lower right) methods as applied to α subunits of G proteins. (Tabor and Richardson, 1985; Tabor et al., 1987).

IV. BACTERIAL EXPRESSION OF α SUBUNITS

The multiplicity of G_i proteins and possibility that each of the purified G_i proteins is crosscontaminated with another leads to severe uncertainties in the assignment of effector function to the various G proteins. It is standard practice in the field of organic chemistry that structures of purified active principles deduced by analytical methods need confirmation by *de novo* synthesis. Although with proteins this was impossible in the past, the advent of recombinant DNA techniques and bacterial expression vectors opened the possibility to test and/or confirm the assignment of function of purified polypeptides. A search for an adequate bacterial expression vector to synthesize α subunits encoded in their respective cDNA molecules led

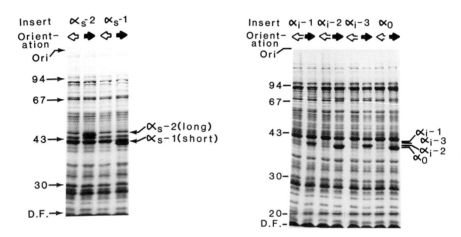

Fig. 4. Bacterial expression of α-subunit fusion proteins by the approach of Tabor and Richardson. cDNAs were inserted into the *Bam*HI site of pT7-7 plasmid and recombinant pT7-α_x plasmids with cDNAs inserted in either the antisense (3' to 5', open arrows) or sense (5' to 3', filled arrows) direction were isolated and transfected into *E. coli* K38 cells. The open reading frame encoded downstream of the T7 promoter was expressed after infection of the cells with phage mGP1-2 and analyzed by SDS-PAGE in 10% polyacrylamide gel slabs. The slabs were stained with Coomassie blue and photographed. Note the appearance of insert-specific and orientation-dependent α subunit bands that constitute between 5 and 8% of the total cell protein. α_{s1}, short human α_s of 380 amino acids with Asp in position 71 and Glu in position 72 (Mattera *et al.*, 1986b) α_{s2}, long human α_s of 394 amino acids with sequence Glu–Gly–Gly–Glu–Glu–Asp–Pro–Gln–Ala–Ala–Arg–Ser–Asn–Ser–Asp–Gly in the place of Asp (Mattera *et al.*, 1986b); α_{i1}, PTX substrate of 355 amino acids cloned from a human brain library (Bray *et al.*, 1987); α_{i2}, PTX substrate of 354 amino acids cloned from a monocyte cDNA library (Sullivan *et al.*, 1986); α_{i3}, 354-amino-acid human liver α_{i3} (Codina *et al.*, 1988); α_o, PTX substrate of 354 amino acids cloned from a rat heart cDNA library identical to rat brain and bovine retina α_o (VanDongen *et al.*, 1988). D.F., dye front.

Graziano *et al.* (1987) as well as our group to the use of pT7 plasmids developed by Tabor and Richardson, (Tabor and Richardson, 1985; Tabor, 1987; Tabor *et al.,* 1987). Several α subunits expressed in *Escherichia coli* were then tested for functions: α_s for reconstitution of the *cyc⁻* adenylyl cyclase system and stimulation of skeletal muscle Ca^{2+} channels in lipid bilayers (Mattera *et al.,* 1988a), α_i molecules for the opening of K^+ channels in excised atrial patches (Mattera *et al.,* 1988b; Yatani *et al.,* 1988a), and α_o molecules for regulation of K^+ channels in excised patches from fetal hippocampal cells (VanDongen *et al.,* 1988).

The pT7-based bacterial expression systems are double-vector systems formed of a plasmid (pT7), into which the cDNA encoding the protein of interest is placed downstream from a phage T7 promoter sequence, and either a phage or a second plasmid that encodes the T7 RNA polymerase, which is required for transcription of the cDNA inserted into the first plasmid. In mGP1-2, a derivative of phage M13, the T7 RNA polymerase gene (gene 1) is under control of the *lac* promoter allowing for induction of the T7 RNA polymerase by isopropylthiogalactoside (IPTG). In pGP1-2, a plasmid carrying the kanamycin resistance gene, the T7 RNA polymerase gene is under control of the leftward λ promoter (P_L) so that induction of the polymerase is contingent on inhibition of the λ repressor (cI gene product). Expression is carried out in *E. coli* K38 cells which are constitutive producers of the *ts* cI857 form of the repressor. Currently a family of the pT7 plasmids exists that are distinguished by suffix 1 through 7. All contain the β-lactamase selection marker, the ColE1 origin of replication for growth in host *E. coli* cells, and the promoter region of gene 10 of the T7 phage. They differ in the nucleotide composition downstream from the gene 10 promoter, containing or not a ribosome-binding site (rbs) and a cloning cassette into which the cDNA of interest is cloned according to a desired reading frame. Feasibility of a α subunit expression was first shown with pT7-5, which lacked a ribosomal binding site as well as a true cloning cassette (Graziano *et al.,* 1987). Experiments obtained with α subunit cDNAs inserted into the pT7-7 plasmid, which has both the T7 gene 10 rbs and its natural ATG initiation codon followed by a multiple cloning cassette, are reported here. After insertion of an α-subunit cDNA, pT7-7 plasmids are referred to as pT7-α_x plasmids. A typical pT7-α_x plasmid is shown in Fig. 3. In initial experiments we used the plasmid–phage system in which T7 RNA polymerase (gene 1 of the T7 phage) is supplied by infecting pT7-α_x-transformed host cells with phage mGP1-2 and polymerase synthesis was induced with IPTG. In later experiments we also used the two-plasmid system in which polymerase, and α subunit expression, is induced by a temperature shift from 30° to 40°C (Tabor *et al.,* 1987).

To construct pT7-α_x plasmids the various α subunit cDNAs were first

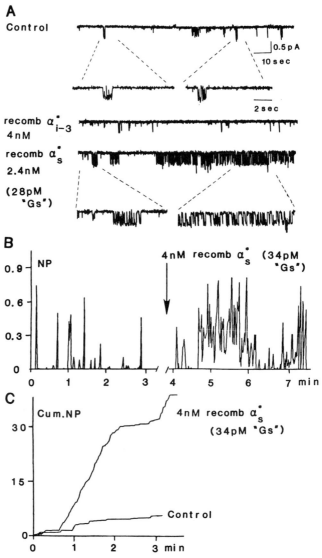

Fig. 5. Stimulation of skeletal muscle T tubule dihydropyridine-sensitive Ca^{2+} channel incorporated into lipid bilayers by the short form of recombinant α_s. Recombinant α_s (Asp^{71}, Ser^{72} version) and recombinant α_{i3} were prepared in their GTPγS-activated form and partially purified by DEAE-Sephacel chromatography, except that the two-plasmid approach was used to express the recombinant proteins (Tabor *et al.*, 1987). (A): single channel Ba^{2+} currents (*trans* to *cis*) before and after addition first of GTPγS-activated recombinant α_{i3} and then of GTPγS-activated recombinant α_s. (B) and (C): NPo diaries of Ca^{2+}-channel activities before and after addition of GTPγS-activated recombinant α_s to the cis chamber (B) and cumulative NPo values as a function of time of the same experiment (C). NPo, product of N, the number of active channels in the phospholipid bilayer, times P_o, the opening probability of each channel as assessed in consecutive 200 ms segments of time.

modified by site-directed mutagenesis in order to generate a unique *Nco*I site at the ATG initiation codon. The engineering was done in replicative forms of M13 phages. The open reading frames of the engineered cDNAs were then excised by digestion with *Nco*I and an appropriate second restriction enzyme. The excised cDNAs were filled with the Klenow fragment of DNA polymerase and subcloned by blunt-end ligation into the *Bam*HI site of the pT7-7 expression vector. This resulted in formation of open reading frames encoding a fusion polypeptide formed of the full sequence encoded by the inserted cDNA preceded by nine extra amino acids (recombinant subunits). Plasmids were isolated containing cDNA inserts in both the correct (sense) and the reverse (anti-sense) orientation with respect to the T7 gene 10 promoter. The cDNAs inserted into the *Bam*HI site of the pT7-7 vector yielded high levels of polypeptide expression (Fig. 4). In each case, the recombinant polypeptides accumulated within 60 min after in-

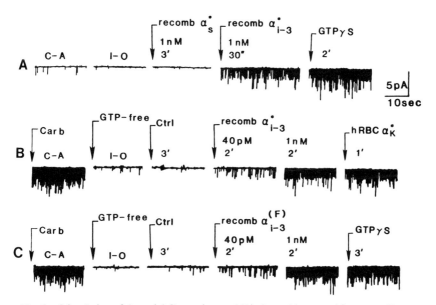

Fig. 6. Stimulation of the atrial G protein-gated K$^+$ channel by recombinant α_{i3}. Recombinant α_{i3} and α_s subunits encoded in their respective cDNAs inserted into pT7-7 vector were expressed in *E. coli* by the one plasmid/one phage method of Tabor and Richardson. The cells were lysed and the recombinant α subunits were activated and partially purified by chromatography over DEAE-Sephacel (0 to 400 mN NaCl gradient). The partially purified recombinant GTPγS-activated α subunits were then added to inside-out membrane patches of guinea pig atria to test for possible K$^+$-channel stimulatory effects. Experiment (A): recombinant α_s does not, but recombinant α_{i3} does, stimulate the atrial K$^+$ channel. Experiments (B) and (C): effects of two different doses of GTPγS-activated or fluoride-activated recombinant α_{i3} (recomb $\alpha_{i3}^{(F)}$).

duction and constituted ca. 5–8% of total bacterial cell protein when the plasmid/phage system was used. Although the absolute amounts of recombinant α subunits made was similar on using the two-plasmid system, they represented a smaller proportion of the total protein because of induction of several heat-shock proteins just prior to the expression of the recombinant α proteins. The various recombinant α subunits migrate on sodium dodecyl sulfate-polyacrylamide gel electrophoresis (SDS-PAGE) in a manner that is consistent with the native α subunits. Thus α_{i1} and α_{i3} migrate at 41,500 kDa while α_{i2} migrates at 40 kDa and α_o at 39 kDa (see Fig. 4).

Even though the levels of α subunit protein synthesized by bacteria using the Tabor–Richardson two-vector system are very high, making purification a relatively easy task, only a fraction (ca. 5 to 10%) can be recovered as a soluble protein after bacterial lysis, the remainder being soluble only in high urea and reducing agents such as DTT or 2-mercaptoethanol. On assaying the soluble portion of α_s molecules for their cyc^--reconstituting activity, they were found to have rather low intrinsic activity. Regardless of whether the recombinant α_s molecules were derived from a cDNA encoding a 394- or a 380-amino acid protein (Robishaw *et al.*, 1986; Mattera *et al.*, 1986b) we obtained specific (intrinsic) activities that are only 3 to 5% of that of native human erythrocyte G_s. The principal advantage of synthesizing α subunits in bacteria is that the product obtained is absolutely free of any other α subunit. In spite of the drawbacks listed above, of which the relatively low intrinsic activity is the more bothersome, the fact that a definable activity is associated with them makes them useful for the purpose of identifying their functional role or roles.

V. RECOMBINANT α SUBUNITS AND EFFECTOR FUNCTIONS

A. α_s Function

As expected, both the long and the short forms of α_s stimulate adenylyl cyclase. In addition to its effects on adenylyl cyclase, it has recently been found that α_s stimulates Ca^{2+} channels directly. This is in addition to the well established effect of cAMP-dependent phosphorylation of the channel, thus implying dual and possibly redundant effects of G_s. In order to rule out the possibility that it was a contaminant in the G_s preparations that opened the channels, we used bacterially synthesized α_s to stimulate the T-tubule Ca^{2+} channel. It was found that the bacterially expressed α_s, but not α_{i3}, stimulated the Ca^{2+} channels (Fig. 5). Both the long (α_{s1}) and short (α_{s4}) forms were able to stimulate the Ca^{2+} channels. Similarly both α_{s1} and

α_{s2}, which differ by a single serine residue, were able to stimulate the Ca^{2+} channels. These experiments demonstrate that in previous experiments (Yatani *et al.*, 1987a, 1988b) it had not been a contaminant in the G_s preparations that stimulated the Ca^{2+} channels but α_s itself.

B. α_i Function

The major pertussis toxin substrate in human erythrocytes had been shown to have G_k activity in that upon application to excised atrial patches it was capable of opening K^+ channels (Yatani *et al.*, 1987b, 1987c; Codina *et al.*, 1987a, 1987b). Molecular cloning studies had indicated the presence of α_{i3} in human erythrocytes and it was predicted that α_{i3} would have G_k activity (Codina *et al.*, 1988). As expected both native and recombinant α_{i3} have G_k activity (Fig. 6, Mattera *et al.*, 1988b). Surprisingly, on further analysis it was found that not only recombinant α_{i3} but also α_{i2} and α_{i1} were capable of opening K^+ channels (Yatani *et al.*, 1988a), albeit all at reduced potency when compared to that of native human erythrocyte α_{i3} (α_k). Hence, it became imperative to test if the native α_{i1} and α_{i2} have the capability of opening K^+ channels as well. As can be seen from the experiments shown in Fig. 7A–C, all native α_is are capable of opening atrial K^+ channels with approximately equal potencies.

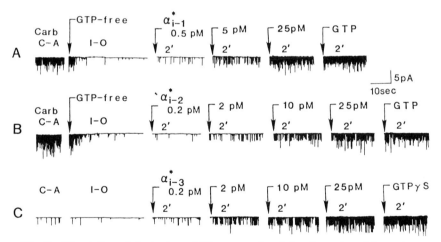

Fig. 7. Stimulation of single-channel K^+ currents in adult guinea pig atrial membrane patches by all three types of native GTPγS-activated (*) α_i subunits. G_{i2} and G_{i3} were purified by DEAE-Toyopearl as shown in Figs. 10 and 11 and G_{i1} was purified by Mono-Q FPLC as shown in Figs. 12 and 13. The purified proteins were activated with GTPγS and their α subunits separated from $\beta\gamma$ dimers by DEAE-Sephacel chromatography (Codina *et al.*, 1987a, b).

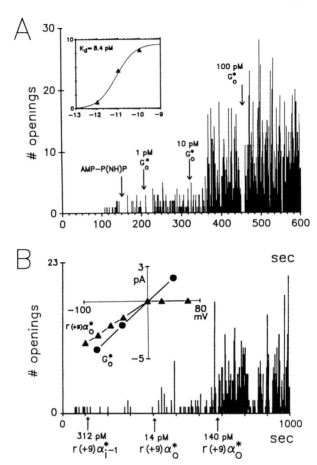

Fig. 8. Bovine brain G_o (A) and *recombinant* α_o, but not *recombinant* α_{i1} (B), gate single K$^+$ channel currents in hippocampal pyramidal cells of neonatal rats. (A) The effect of increasing concentrations of G_o^* on a 55-pS K$^+$ channel is shown by plotting the number of openings per 0.8 seconds as a function of time of continuous recording. Inset: Openings per 0.8 seconds averaged for 1 min as a function of G_o^* concentration. (B): Lack of effects of 312 pM *recombinant* α_{i1}^*, and stimulatory effect of increasing concentrations of *recombinant* α_o^* on a 40-pS K$^+$ channel, plotted as number of opening per 0.2 seconds as a function of the time of continuous recording shown. Single K$^+$-channel currents were recorded from excised inside-out membrane patches as in Fig. 7 in the presence of 0.2 mM AMP-P(NH)P. Holding potential was -80 mV and both pipette and bath solutions contained 140 mM potassium methanesulfonate, 1 mM EGTA, 1 mM MgCl$_2$ and 10 mM HEPES adjusted to pH 7.4 with Tris base (adapted from VanDongen *et al.*, 1988).

C. α_0 Function

The major pertussis toxin substrate in the brain is G_0, and it has been purified and biochemically characterized extensively (see Bockaert *et al.*, Chapter 5 in this volume). Studies have shown that G_0 mediates receptor-regulated closing of Ca^{2+} channels (Heschler *et al.*, 1987; Ewald *et al.*, 1988). It is as yet unclear if this effect of G_0 results from direct interactions

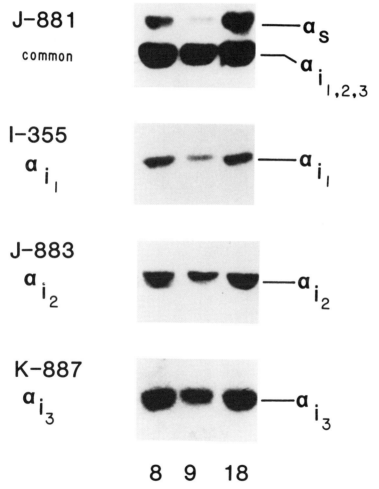

Fig. 9.   Antisera specific for α_{i1} (I-355), α_{i2} (J-883) and α_{i3} (K-887) recognize the presence of each of these α subunits in partially purified human erythrocyte G proteins obtained after the second DEAE-Sephacel chromatography (Codina *et al.*, 1984). Three such preparations (#8, #9 and #18) were resolved on SDS-polyacrylamide gels and immunoblotted with the indicated antibodies.

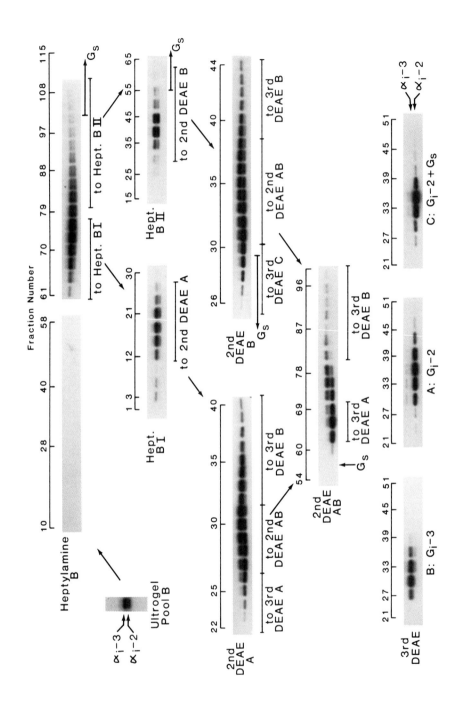

with the Ca^{2+} channel or if the effect is mediated through second messengers. Recently using cell-free patches of hippocampal pyramidal neurons we have found four distinct types of potassium channels that are activated by isolated brain G_o as well as recombinant α_o (VanDongen et al., 1988). These effects are specific for α_o since recombinant α_{i1} was not capable of opening these K^+ channels. An experiment comparing recombinant α_o and α_{i1} is shown in Fig. 8. In contrast to the atrial patches where a single class of K^+ channels was regulated by G proteins, in the pyramidal neuron patches the K^+ channels regulated by G_o could be categorized as four types due to their conductance, mean open time or openings per burst (VanDongen et al., 1988).

VI. IDENTIFICATION OF NATIVE G_i: IMMUNOBLOTTING USING SEQUENCE-SPECIFIC ANTISERA

Since all three α_i subunits are very similar and since all of these are pertussis-toxin substrates, the unequivocal identification of the individual α_i subunits is necessary to develop an understanding of how these proteins

---

Fig. 10. Strategy used to purify human erythrocyte G_{i2} (Fig. 2 "lower" PTX substrate) and G_{i3} (Fig. 2 "upper" PTX substrate) using DEAE-Toyopearl chromatography. The progress of the purification is shown from the point of chromatographing pool B from the Ultrogel AcA 34 column over heptylamine-Sepharose [variant (B)] until "upper" and "lower" PTX substrates were separated and purified to better than 95% purity from other proteins. In the particular purification shown in the figure the aliquots from fractions eluted off the heptylamine-Sepharose column were ADP-ribosylated with PTX and analyzed by urea gradient/SDS-PAGE to give the pattern shown for the heptylamine B column. "Upper" (leading) PTX substrate was partially separated from the "lower" (trailing) PTX substrate, the last part of which in turn overlapped with G_s. Fractions were pooled as shown and rechromatographed under identical conditions (heptylamine B-I and B-II chromatographies). The fractions from these two steps were then subjected to separate second DEAE chromatographies over DEAE-Toyopearl. This resulted in a further partial separation of "lower" (now leading) from "upper" (now trailing) PTX substrate, with the difference that while the leading "lower" PTX substrate eluting from the second DEAE-A was free of G_s, that eluting from the second DEAE-B contained an about equimolar amount of G_s. The eluates from each of the second DEAE columns were combined to give three pools of which the two center pools (ca. 50/50 "upper" and "lower" PTX substrates) were combined and re-chromatographed over a 0.9 × 75 cm DEAE-Toyopearl column (2nd DEAE-AB). The fractions with the leading PTX substrate ("lower") from the second DEAE-AB were combined with those from the second DEAE-A, the fractions with the trailing "upper" PTX substrate were combined with those from both the second DEAE-A and -B. These and the pool of leading "lower" PTX substrate plus contaminating G_s from the second DEAE-B resulting pools were then subjected to a third DEAE chromatography (columns designated A, B, and C as shown on the figure). "Upper" PTX substrate (G_{i3}) elutes from the third DEAE column slightly earlier than the "lower" PTX substrate (G_{i2}), which in turn essentially cochromatographs with G_s. For buffer compositions see Codina et al. (1984).

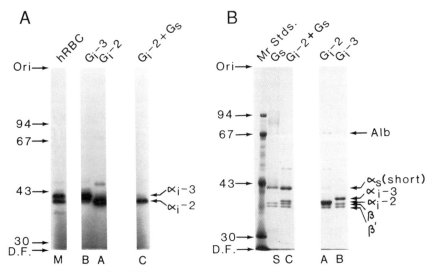

Fig. 11. Urea gradient/SDS-PAGE analysis of fractions from the purification shown in Fig. 3. *Left Panel,* PTX labeling pattern: M, 7 μg human erythrocyte membranes; B, 5 ng of protein obtained from the third DEAE-B, A, 10 ng of protein obtained from the third DEAE-A; C, 4 ng of protein obtained from the third DEAE-B. *Right Panel,* Coomassie blue stain: the same fractions analyzed for PTX labeling were analyzed: B, 500 ng; A, 250 ng; C, 350 ng. Note presence of two β subunits in each of the proteins. D.F., dye front.

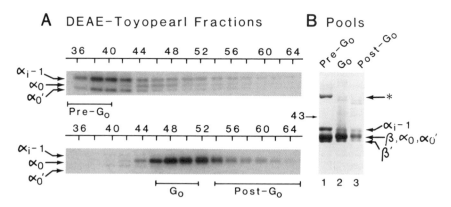

Fig. 12. Separation by DEAE-Toyopearl chromatography (NaCl gradient) of two bovine brain fractions containing different proportions of G_{i1} (upper band); G_{o1} (middle band) and G_{o2} (lower band). The eluates were labeled with PTX and [^{32}P]NAD, electrophoresed as in Fig. 11, and autoradiographed. Note that while it was possible to separate G_{o1}, G_{i1} and G_{i2} cochromatographed.

may have similar or different functions. The approach that has proven most useful is the use of sequence-specific antisera to study G-protein subunits. This approach is discussed in detail in Chapter 6 by Spiegel. In our laboratories the use of sequence-specific antisera has proven useful in identifying α_{i1}, α_{i2} and α_{i3} in human erythrocytes as well as aiding the purification of what appear to be two forms of α_o. For the various α_i subunits we have utilized the antisera developed by Mumby et al., (1986, 1988) utilizing the differences in amino acid sequences in region 113–128 for α_{i1}, α_{i2}, and α_{i3}. The specificity of these antisera was verified by the interaction with the recombinant α_{i1}, α_{i2}, and α_{i3}. On the basis of the specificity of these interactions we have found that partially purified human erythrocyte G proteins preparations contain α_{i1}, α_{i2} and α_{i3}. This is shown in Fig. 9. Such immunoblotting is proving extremely useful in identifying closely related G proteins of known composition during purification.

VII. G-PROTEIN PURIFICATION: RESOLUTION OF CLOSELY RELATED G PROTEINS

Until recently the focus of G-protein purification has been to remove all other contaminating proteins and in the process separate G proteins that were quite distinct from each other. Thus the purification procedure resolved G_s from G_i (Codina et al., 1983, 1984; Bokoch et al., 1983, 1984) or G_i from G_o (Sternweis and Robishaw, 1984). These separations were based on DEAE-Sephacel chromatographies. More closely related G proteins such as G_{i2} and G_{i3} can be separated by repetitive DEAE-Toyopearl chromatography as illustrated in Fig. 10 and 11, but as shown in Fig. 12, this approach is not always successful for it does not separate brain G_{o2} from G_{i1}. The existence of various forms of G_o proteins have been shown in heart upon combining urea gradient gel electrophoresis and immunoblotting with G_o-specific peptide antibody (Scherer et al., 1987). Resolution of these brain proteins can be achieved by Mono Q fast liquid chromatography (Fig. 13). This methodology was brought into the G-protein field by Katada and colleagues (Katada et al., 1987; Oinuma et al., 1987) who applied it to separate rat brain G_{i1} from G_{i2} and HL-60 cell G_{i2} from G_{i3}. As was the case with DEAE-Toyopearl, pure proteins are obtained with Mono Q when partially separated peaks such as shown in Fig. 13 are rechromatographed using a combination of NaCl and cholate gradients. Under these conditions it is thus possible to obtain purified G proteins that are not measurably crosscontaminated with other *known* G proteins. For the case of bovine brain G_{o1}, G_{o2}, and G_{i1}, such crosscontamination is

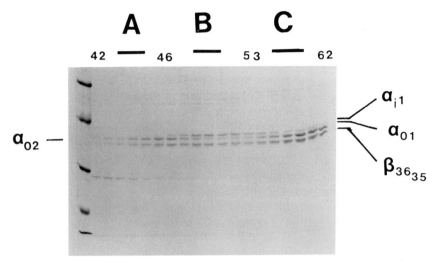

Fig. 13. Resolution of various G proteins by Mono Q chromatography using medium pressure chromatography. Heterotrimeric G proteins from bovine brain after heptylamine-Sepharose chromatography were resolved by use of 150–250 mM NaCl gradient. Three distinct protein pools were made consisting of fractions 42–46 (Pool A), 46–53 (Pool B) and 53–62 (Pool C).

determined by immunoblotting (Fig. 14). The basis for the two distinct forms of G_o is currently not known. They could be distinct gene products, splice variants from the same gene product or a result of posttranslational modification. Further it is not known if the two G_os display similar effector functions such as regulation of K^+ channels and this is a subject of active investigation in our laboratories.

VIII. EFFECTOR FUNCTIONS OF α SUBUNITS: RECOMBINANT VERSUS NATIVE PROTEINS

The singular advantage recombinant α subunits have over native proteins is the certainty with respect to the identity of the protein used. However, the overwhelming drawback in using the recombinant proteins is the low potency with which they activate effector systems. Thus the GTPγS-activated native α_{i3} activates atrial K^+ channels with an EC_{50} of 20 pM while the recombinant activated α_{i3} has an EC_{50} of 500 pM (Fig. 15). Similarly, substantially lower potencies have been observed when the α_s subunits are utilized to activate adenylyl cyclase. Since there is such a large shift in potency, if the differences in the specificity of the three α_is rest on a

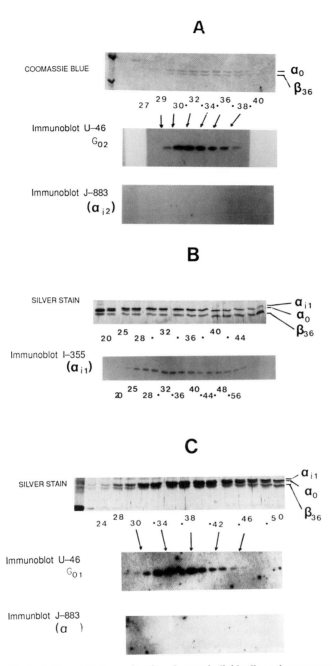

Fig. 14. Pools A, B and C shown in Fig. 13 were individually rechromatographed on Mono Q using a 0–2% cholate gradient. Fractions analyzed by SDS-PAGE followed by staining and immunoblotting are shown. It can be seen that two of the pools contain distinct G_o proteins, while the middle pool contains G_{i1}.

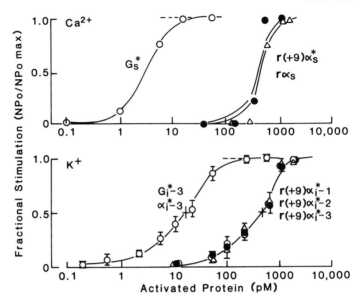

Fig. 15. Dose–response relationships for effects of native and recombinant α subunits on skeletal muscle Ca^{2+}-channel activity (upper panel) and atrial K^+-channel activity (lower panel). Upper panel: G_s^*, GTPγS-activated human erythrocyte G_s (mixture of GTPγS-activated α_s (unknown mix of short forms) plus an equimolar concentration of $\beta\gamma$ dimers); $r(+9)$ α_s^*, GTPγS-activated recombinant α_s biosynthesized in bacteria from α_s placed into the PT7-vector and having a nine-amino acid extension at its amino terminal end (Fig. 2); $r\alpha_s^*$, recombinant α_s molecule prepared by Graziano and Gilman, using a pT5 expression vector that yields α subunits without added amino acids. For details of synthesis see Graziano *et al.* (1987) and for effects of recombinant α_s subunits of Ca^{2+} currents see Mattera *et al.* (1988b). Lower panel: α_{i3}^*, GTPγS-activated α subunit of human erythrocyte G_{i3} after separation from $\beta\gamma$ dimers by DEAE-Toyopearl chromatography; G_i^*, GTPγS-activated human erythrocyte G_{i3} before DEAE-Toyopearl chromatography; the recombinant GTPγS-activated forms of the three types of α_i made in bacteria with the help of the pT7 expression vector are denoted as $r^{(+9)}\alpha_i^*$-x. Adapted from Yatani *et al.* (1988a).

difference in affinity, these specificities could be obscured when recombinant α subunits are used. Hence it is essential to verify the data obtained with recombinant α subunits with those obtained with the native α subunits. We have done this for the α_i effects on the atrial K^+ channels (Yatani *et al.*, 1988a) and the G_o (G_{o1}) effects on hippocampal neuron K^+ channels (VanDongen *et al.*, 1988). However, a careful comparison of potency measurements using receptor to activate the G protein is essential to develop an understanding of the specificity of G protein coupling in the various pathways.

IX. CONCLUSIONS AND FUTURE DIRECTIONS

Studies from our laboratories as well as others have yielded a wealth of information about the existence of multiple G proteins as well as numerous systems where G proteins couple receptors to effectors. From these studies it has become obvious that there are far more G proteins, receptors, and effectors than had been previously envisaged. However, whether each receptor will specifically interact with a unique G protein remains to be determined. Clearly several G proteins are capable of regulating a single effector, such as seen by the effects of G_{i1}, G_{i2} and G_{i3} on the atrial K^+ channel. It is also evident that a single G protein can regulate several distinct effector systems; the best example of this is G_s, which interacts with adenylyl cyclase and Ca^{2+} channels. Thus it is entirely possible that within the various groups of G proteins there is very little specificity in terms of effector regulation. However this is a tentative conclusion. Thus far potency measurements of G protein interactions with effectors has only been done after persistent activation with GTPγS. It is possible that if the activation of the G protein is receptor regulated, one might detect differential specificity. This will have to be determined experimentally. Further if one assumes within the various classes of G protein there is little specificity for effector regulation then it is essential to determine the various effectors that each class of G proteins can regulate. Future studies aimed at discerning the specificity of the various signal transduction pathways as well as cross talk between these pathways should therefore lead to both new and useful information about these complex systems at the cell surface that modulate cellular responsiveness to external signals.

ACKNOWLEDGMENT

Supported in part by United States Public Health Service research grants DK-19318, HD-09581, HL-31164 and HL-37044 to LB; HL-39262 to AMB; EY-06062 to HEH and DK-38761 and CA-44998 to RI, the Baylor College of Medicine Diabetes and Endocrinology Research Center grant DK-27685 (James B. Field, Director), and grant Q1075 from the Welch Foundation.

REFERENCES

Bokoch, G. M., Katada, T., Northup, J. K., Hewlett, E. L., and Gilman, A. G. (1983). Identification of the predominant substrate for ADP-ribosylation by islet-activating protein. *J. Biol. Chem.* **258**, 2072–2075.

Bokoch, G. M., Katada, T., Northup, J. K., Ui, M., and Gilman, A. G. (1984). Purification and Properties of the inhibitory guanine nucleotide binding regulatory component of adenylyl cyclase. *J. Biol. Chem.* **259**, 3560–3567.

Bray, P., Carter, A., Guo, V., Puckett, C., Kamholz, J., Spiegel, A., and Nirenberg, M. (1987). Human cDNA clones for an α subunit of G_i signal-transduction protein. *Proc. Natl. Acad. Sci. U.S.A.* **84**, 5115–5119.

Burns, D. L., and Manclark, C. R. (1986). Adenine nucleotides promote dissociation of pertussis toxin subunits. *J. Biol. Chem.* **261**, 4324–4327.

Codina, J., Hildebrandt, J. D., Iyengar, R., Birnbaumer, L., Sekura, R. D., and Manclark, C. R. (1983). Pertussis toxin substrate, the putative N_i of adenylyl cyclases, is an *alpha/beta* heterodimer regulated by guanine nucleotide and magnesium. *Proc. Natl. Acad. Sci. U.S.A.* **80**, 4276–4280.

Codina, J., Hildebrandt, J. D., Sekura, R. D., Birnbaumer, M., Bryan, J., Manclark, C. R., Iyengar, R., and Birnbaumer, L. (1984). N_s and N_i, the stimulatory and inhibitory regulatory components of adenylyl cyclases. Purification of the human erythrocyte proteins without the use of activating regulatory ligands. *J. Biol. Chem.* **259**, 5871–5886.

Codina, J., Yatani, A., Grenet, D., Brown, A. M., and Birnbaumer, L. (1987a). The *alpha* subunit of the GTP binding protein G_k opens atrial potassium channels. *Science* **236**, 442–445.

Codina, J., Grenet, D., Yatani, A., Birnbaumer, L., and Brown, A. M. (1987b). Hormonal regulation of pituitary GH_3 cell K^+ channels by G_k is mediated by its *alpha* subunit. *FEBS Lett.* **216**, 104–106.

Codina, J., Olate, J., Abramowitz, J., Mattera, R., Cook, R. G., and Birnbaumer, L. (1988). *Alpha*$_i$-3 cDNA encodes the *alpha* subunit of G_k, the stimulation G protein of receptor-regulated K^+ channels. *J. Biol. Chem.* **263**, 6746–6750.

Ewald, D. A., Sternweis, P. C., and Miller, R. J. (1988). Guanine nucleotide-binding protein G_o-induced coupling of neuropeptide Y receptors to Ca^{2+} channels in sensory neurons. *Proc. Natl. Acad. Sci. U.S.A.* **85**, 3633–3637.

Gill, D. M. (1977). Mechanism of action of cholera toxin. *Adv. Cyclic Nucleotide Res.* **6**, 85–118.

Graziano, M. P., Casey, P. J., and Gilman, A. G. (1987). Expression of cDNAs for G proteins in *Escherichia coli.* Two forms of $G_{s\alpha}$ stimulate adenylate cyclase. *J. Biol. Chem.* **262**, 11375–11381.

Hescheler, J., Rosenthal, W., Trautwein, W., and Schultz, G. (1987). The GTP-binding protein, N_o, regulated neuronal calcium channels. *Nature* **325**, 445–447.

Hescheler, J., Rosenthal, W., Hinsch, K.-D., Wulfern, M., Trautwein, W., and Schultz, G. (1988). Angiotensin II-induced stimulation of voltage-dependent Ca^{2+} currents in an adrenal cortical cell line. *EMBO J.* **7**, 619–624.

Hildebrandt, J. D., and Birnbaumer, L. (1983). Inhibitory Regulation of Adenylyl cyclase in the absence of stimulatory regulation. Requirements and kinetics of guanine nucleotide induced inhibition of the *cyc⁻* S49 adenylyl cyclase. *J. Biol. Chem.* **258**, 13141–13147.

Hildebrandt, J. D., Hanoune, J., and Birnbaumer, L. (1982). Guanine nucleotide inhibition of cyc^- S49 mouse lymphoma cell membrane adenylyl cyclase. *J. Biol. Chem.* **257**, 14723–14725.

Hildebrandt, J. D., Sekura, R. D., Codina, J., Iyengar, R., Manclark, C. R., and Birnbaumer, L. (1983). Stimulation and inhibition of adenylyl cyclases is mediated by distinct proteins. *Nature* **302**, 706–709.

Hildebrandt, J. D., Codina, J., and Birnbaumer, L. (1984). Interaction of the stimulatory and inhibitory regulatory proteins of the adenylyl cyclase system with the catalytic component of cyc^- S49 cell membranes. *J. Biol. Chem.* **259**, 13178–13185.

Itoh, H., Kozasa, T., Nagata, S., Nakamura, S., Katada, T., Ui, M., Iwai, S., Ohtsuka, E., Kawasaki, H., Suzuki, K., and Kaziro, Y. (1986). Molecular cloning and sequence determination of cDNAs for the alpha subunits of the guanine nucleotide-binding proteins G_s, G_i, and G_o from rat brain. *Proc. Natl. Acad. Sci. U.S.A.* **83**, 3776–3780.

Jakobs, K. H., Aktories, K., and Schultz, G. (1983). A nucleotide regulatory site for somatostatin inhibition of adenylate cyclase in S49 lymphoma cells. *Nature* **303**, 177–178.

Jones, D. T., and Reed, R. R. (1987). Molecular cloning of five GTP-binding protein cDNA species from rat olfactory neuroepithelium. *J. Biol. Chem.* **262**, 14241–14249.

Katada, T., and Ui, M. (1977). Perfusion of the pancreas isolated from pertussis-sensitized rats: Potentiation of insulin secretory responses due to β-adrenergic stimulation. *Endocrinology* **101**, 1247–1255.

Katada, T., and Ui, M. (1979). Islet-activating protein. Enhanced insulin secretion and cAMP assumulation in pancreatic islets due to activation of native Ca ionophores. *J. Biol. Chem.* **254**, 469–479.

Katada, T., and Ui, M. (1982). Direct modification of the membrane adenylate cyclase system by islet-activating protein due to ADP-ribosylation of a membrane protein. *Proc. Natl. Acad. Sci. U.S.A.* **79**, 3129–3133.

Katada, T., Oinuma, M., Kusakabe, K., and Ui, M. (1987a). A new GTP-binding protein in brain tissues serving as the specific substrate of islet-activating protein, pertussis toxin. *FEBS Lett.* **213**, 353–358.

Lim, L.-K., Sekura, R. D., and Kaslow, H. R. (1985). Adenine nucleotides directly stimulate pertussis toxin. *J. Biol. Chem.* **260**, 2585–2588.

Mattera, R., Codina, J., Sekura, R. D., and Birnbaumer, L. (1986a). The interaction of nucleotides with pertussis toxin. Direct evidence for a nucleotide binding site on the toxin regulating the rate of ADP-ribosylation of N_i, the inhibitory regulatory component of adenylyl cyclase. *J. Biol. Chem.* **261**, 11174–11179.

Mattera, R., Codina, J., Crozat, A., Kidd, V., Woo, S. L. C., and Birnbaumer, L. (1986b). Identification by molecular cloning of two forms of the *alpha* subunit of the human liver stimulatory (G_s) regulatory component of adenylyl cyclase. *FEBS Lett.* **206**, 36–42.

Mattera, R., Codina, J., Sekura, R. D., and Birnbaumer, L. (1987). Guanosine 5'-O-(3-thiotriphosphate) reduces ADP-ribosylation of the inhibitory guanine nucleotide-binding regulatory protein of adenylyl cyclase (N_i) by pertussis toxin without causing dissociation of the subunits of N_i. Evidence of existence of heterotrimeric pt$^+$ and pt$^-$ conformation of N_i. *J. Biol. Chem.* **262,** 11247 – 11251.

Mattera, R., Graziano, M. P., Yatani, A., Zhou, Z., Graf, R., Codina, J., Birnbaumer, L., Gilman, A. G., and Brown, A. M. (1988a). Individual splice variants of the α subunit of the G protein G_s activate both adenylyl cyclase and Ca^{2+} channels. *Science* **243,** 804 – 807.

Mattera, R., Yatani, A., Kirsch, G. E., Graf, R., Olate, J., Codina, J., Brown, A. M., and Birnbaumer, L. (1988b). Recombinant α_i-3 subunit of G protein activates G_k-gated K^+ channels. *J. Biol. Chem.* **264,** 465 – 471.

Mumby, S. M., Kahn, R. A., Manning, D. R., and Gilman, A. G. (1986). Antisera of designed specificity for subunits of guanine nucleotide-binding regulatory proteins. *Proc. Natl. Acad. Sci. U.S.A.* **83,** 265 – 269.

Mumby, S., Pang, I.-H., Gilman, A. G., and Sternweis, P. C. (1988). Chromatographic resolution and immunologic identification of the α_{40} and α_{41} subunits of guanine nucleotide-binding regulatory proteins from bovine brain. *J. Biol. Chem.* **263,** 2020 – 2026.

Neer, E. J., Lok, J. M., and Wolf, L. G. (1984). Purification and properties of the inhibitory guanine nucleotide regulation unit of brain adenylate cyclase. *J. Biol. Chem.* **259,** 14222 – 14229.

Nukada, T., Tanabe, T., Takahashi, H., Noda, M., Haga, K., Haga, T., Ichiyama, A., Kangawa, K., Hiranaga, M., Matsuo, H., and Numa, S. (1986). Primary Structure of the alpha-subunit of bovine adenylate cyclase inhibiting G-protein deduced from the cDNA sequence. *FEBS Lett.* **197,** 305 – 310.

Oinuma, M., Katada, T., and Ui, M. (1987). A new GTP-binding protein in differentiated human leukemic (HL-60) cells serving as the specific substrate of islet-activating protein, pertussis toxin. *J. Biol. Chem.* **262,** 8347 – 8353.

Ribeiro-Neto, F., Mattera, R., Grenet, D., Sekura, R., Birnbaumer, L., and Field, J. B. (1987a). ADP-ribosylation of G proteins by pertussis and cholera toxin in cell membranes. Different requirements for and effects of guanine nucleotides and Mg^{2+}. *Mol. Endocrinol.* **1,** 472 – 481.

Ribeiro-Neto, F., Birnbaumer, L., and Field, J. B. (1987b). TSH-induced desensitization of thyroid slices is associated with a decrease in the ability of pertussis toxin to ADP-ribosylate guanine nucleotide regulatory component(s). *Mol. Endocrinology* **1,** 482 – 490.

Robishaw, J. D., Smigel, M. D., and Gilman, A. G. (1986). Molecular basis for two forms of the G protein that stimulates adenylate cyclase. *J. Biol. Chem.* **261,** 9587 – 9590.

Rosenthal, W., Hescheler, J., Hinsch, K.-D., Spicher, K., Trautwein, W., and Schultz, G. (1988). Cyclic AMP-independent, dual regulation of voltage-dependent Ca^{2+} currents by LHRH and somatostatin in a pituitary cell line. *EMBO J.* **7,** 1627 – 1633.

Scherer, N. M., Toro, M.-J., Entman, M. L., and Birnbaumer, L. (1987). G protein distribution in canine cardiac sarcoplasmic reticulum and sarcolemma. Comparison to rabbit skeletal membranes and brain and erythrocyte G proteins. *Arch. Biochem. Biophys.* **259**, 431–440.

Sekura, R. D., Fish, F., Manclark, C. R., Meade, B., and Zhang, Y. (1983). Pertussis toxin. Affinity purification of a new ADP-ribosyltransferase. *J. Biol. Chem.* **258**, 14647–14651.

Sternweis, P. C., and Robishaw, J. D. (1984). Isolation of two proteins with high affinity for guanine nucleotides from membranes of bovine brain. *J. Biol. Chem.* **259**, 13806–13813.

Suki, W., Abramowitz, J., Mattera, R., Codina, J., and Birnbaumer, L. (1987). The human genome encodes at least three non-allelic G proteins with $alpha_i$ type subunits. *FEBS Lett.* **220**, 187–192.

Sullivan, K. A., Liao, Y.-C., Alborzi, A., Beiderman, B., Chang, F.-H., Masters, S. B., Levinson, A. D., and Bourne, H. R. (1986). Inhibitory and stimulatory G proteins of adenylate cyclase: cDNA and amino acid sequences of the alpha chains. *Proc. Natl. Acad. Sci. U.S.A.* **83**, 6687–6691.

Sumi, T., and Ui, M. (1975). Potentiation of the adrenergic β-receptor-mediated insulin secretion in pertussis-sensitized rats. *Endocrinology* **97**, 352–358.

Sunyer, T., Monastirsky, B., Codina, J., and Birnbaumer, L. (1989). Effect of pertussis toxin on properties of regulatory G proteins. *Mol. Endocrinol.* (in press).

Tabor, S. (1987). Dissection of the bacteriophage T7 DNA replication by the overproduction of its essential genetic elements. Doctoral Thesis, Harvard University School of Medicine.

Tabor, S., and Richardson, C. C. (1985). A bacteriophage T7RNA polymerase/ promote system for controlled exclusive expression of specific genes. *Proc. Natl. Acad. Sci. U.S.A.* **82**, 1074–1078.

Tabor, S., Huber, H. E., and Richardson, C. C. (1987). *Escherichia coli* thioredoxin confers processivity on the DNA polymerase activity of the gene 5 protein of bacteriophage T7. *J. Biol. Chem.* **262**, 16212–16223.

Tamura, M., Nigimori, K., Murai, S., Yajima, M., Ito, K., Katada, T., Ui, M., and Isahi, S. (1982). Subunit structure of islet activating protein, pertussis toxin, in conformity with the A-B model. *Biochemistry* **21**, 5516–5522.

Tsai, S.-C., Adamik, R., Kanaho, Y., Hewlett, E. L., and Moss, J. (1984). Effects of guanyl nucleotides and rhodopsin on ADP-ribosylation of the inhibitory GTP-binding component of adenylate cyclase by pertussis toxin. *J. Biol. Chem.* **259**, 15320–15323.

Ui, M. (1986). Pertussis toxin as a probe of receptor coupling to inositol lipid metabolism. *In* "Phosphoinositides and Receptor Mechanisms" (J. W. Putney, Jr., ed.), pp. 163–195, Alan R. Liss, New York.

Van Dop, C., Yamanaka, G., Steinberg, F., Sekura, R. D., Manclark, C. R., Stryer, L., and Bourne, H. R. (1984). ADP-ribosylation of transducin by pertussis toxin blocks the light-stimulated hydrolysis of GTP and cGMP in retinal photoreceptors. *J. Biol. Chem.* **259**, 23–26.

VanDongen, A., Codina, J., Olate, J., Mattera, R., Joho, R., Birnbaumer, L., and Brown, A. M. (1988). Newly identified brain potassium channels gated by the guanine nucleotide binding (G) protein G_o. *Science* **242**, 1433–1437.

Yatani, A., Codina, J., Imoto, Y., Reeves, J. P., Birnbaumer, L., and Brown, A. M. (1987a). A G protein directly regulates mammalian cardiac calcium channels. *Science* **238,** 1288–1292.

Yatani, A., Codina, J., Brown, A. M., and Birnbaumer, L. (1987b). Direct activation of mammalian atrial muscarinic potassium channels by GTP regulatory protein G_k. *Science* **235,** 207–211.

Yatani, A., Codina, J., Sekura, R. D., Birnbaumer, L., and Brown, A. M. (1987c). Reconstitution of somatostatin and muscarinic receptor mediated stimulation of K^+ channels by isolated G_k protein in clonal rat anterior pituitary cell membranes. *Mol. Endocrinol.* **1,** 283–289.

Yatani, A., Mattera, R., Codina, J., Graf, R., Okabe, K., Padrell, E., Iyengar, R., Brown, A. M., and Birnbaumer, L. (1988a). The G protein-gated atrial K^+ channel is stimulated by three distinct $G_i\alpha$-subunits. *Nature* **336,** 680–682.

Yatani, A., Imoto, Y., Codina, J., Hamilton, S. L., Brown, A. M., and Birnbaumer, L. (1988b). The stimulatory G protein of adenylyl cyclase, G_s, directly stimulates dihydropyridine-sensitive skeletal muscle Ca^{2+} channels. Evidence for direct regulation independent of phosphorylation by cAMP-dependent protein kinase. *J. Biol. Chem.* **263,** 9887–9895.

CHAPTER 13

Structure and Function of Adrenergic Receptors: Models for Understanding G Protein-Coupled Receptors

Marc G. Caron, Mark Hnatowich, Henrik Dohlman, Michel Bouvier, Jeffrey L. Benovic, Brian F. O'Dowd, Brian K. Kobilka, William P. Hausdorff, Robert J. Lefkowitz
Departments of Cell Biology, Biochemistry, and Medicine, Howard Hughes Medical Institute, Duke University Medical Center, Durham, North Carolina 27710

Many hormones and drugs initiate their biological actions by interacting with receptor macromolecules in or on cells. Of the receptors which are located in the plasma membrane of cells, many are coupled to specific effector molecules by the intermediacy of coupling proteins. The latter are termed G proteins because they bind and hydrolyze the nucleotide GTP. Among the biological agents whose actions are mediated by such pathways are the catecholamines epinephrine and norepinephrine, and their many synthetic congeners that are used as therapeutic agents.

As with many other hormones and drugs, several distinct types of receptors for catecholamines have been defined, based both on the distinct physiological actions that they subserve and on their pharmacological specificity. These include the α_1-, α_2-, β_1- and β_2-adrenergic receptor subtypes (Lefkowitz and Caron, 1988). Moreover, these different receptor subtypes are coupled to distinct effector pathways. Thus, both β_1- and β_2-adrenergic receptors stimulate the plasma membrane-bound enzyme adenylyl cyclase leading to the generation of cAMP. G_s is the G protein involved in this pathway. α_2-Adrenergic receptors inhibit adenylyl cyclase via the inhibitory regulatory G protein, G_i. α_2-Adrenergic receptors may also stimulate a Na^+/H^+ antiporter, but the related G protein involved, if any, is not known (Isom et al., 1987). In contrast, α_1-adrenergic receptors activate the enzyme phospholipase C through an, as yet, poorly characterized G protein. Stimulation of this effector system leads to the generation of the second messengers diacylglycerol and inositol trisphosphate which, respectively, activate Ca^{2+}/calmodulin-dependent protein kinase (protein kinase C) and mobilize Ca^{2+} from intracellular stores (Berridge and Irvine, 1984).

Of the many receptors coupled to G proteins, more is known about the family of adrenergic receptors than perhaps any of the others. Thus, all of these receptor subtypes have been purified to homogeneity and reconstituted with G proteins in phospholipid vesicles, their function and regulation by covalent modification has been extensively studied and the genes and/or cDNAs for all of the major subtypes have been isolated and sequenced. From these studies, an ever-sharpening picture of an extremely large gene family is emerging. While this family is exemplified by the adrenergic receptors, it appears to include numerous other receptors for a wide variety of hormones, drugs and neurotransmitters (O'Dowd et al., 1988; Julius et al., 1988; Fargin et al., 1988). This gene family also includes the visual system light-receptor, rhodopsin, which is coupled via the G protein transducin to a cGMP phosphodiesterase (Stryer, 1986). It is also possible that other sensory receptors, e.g., odorant or taste receptors, may be a part of this family of proteins (Lancet and Pace, 1987).

I. COMPONENTS OF HORMONE-SENSITIVE ADENYLYL CYCLASE SYSTEMS

The biochemical properties of hormone receptor-coupled adenylyl cyclase systems have been reviewed (Lefkowitz and Caron, 1988; O'Dowd et al., 1988) and are also the subject of several essays in this volume. The β_2-adrenergic receptor is prototypic of receptors which stimulate this system and the α_2-adrenergic receptor of those which inhibit it. Within the G-protein family of homologous proteins involved in coupling of plasma membrane-bound receptors to different intracellular effector systems, two such proteins, in particular, are involved in coupling adrenergic receptors to the adenylyl cyclase system: G_s, which mediates the effects of stimulatory receptors (e.g., β_2-adrenergic receptors), and G_i, which mediates the effects of inhibitory receptors (e.g., α_2-adrenergic receptors) (Gilman, 1987). These heterotrimeric molecules are composed of α subunits (M_r 39,000–52,000), which bind and hydrolyze GTP and interact with receptors, β subunits (M_r 35,000–36,000) and γ subunits (M_r 5,000–10,000); the latter two are usually tightly associated with each other and thought to regulate the GTPase activity of the α subunits. Although the mechanisms by which G_s and G_i interact with adrenergic receptors to regulate the adenylyl cyclase effector system are reasonably well understood, some details remain controversial, especially in the case of G_i (reviewed in Chapter 1). However, in the case of receptor-mediated adenylyl cyclase stimulation, it appears that agonist occupancy of the receptor induces a conformational change in G_s that promotes GDP dissociation followed by GTP association. Subsequent dissociation of G_s into α_s–GTP and $\beta\gamma$ subunits (Gilman, 1987) appears to cause direct activation of adenylyl cyclase through the action of α_s–GTP on the catalytic unit of the enzyme. The activation cycle is then thought to terminate following the slow hydrolysis of GTP by α_s.

Less is currently known about the molecular properties of the catalytic unit of adenylyl cyclase itself. The enzyme has been purified by affinity chromatography and appears to consist of a single glycoprotein chain of M_r 150,000 (Pfeuffer et al., 1985; Smigel, 1986). The calmodulin-sensitive form of the enzyme appears to have a somewhat smaller M_r of $\sim$ 116,000–120,000 (Mollner and Pfeuffer, 1988). All of these components: receptor, G_s, and catalytic unit, have been successfully reconstituted in phospholipid vesicles and shown to be sufficient to produce catecholamine-responsive adenylyl cyclase activity (May et al., 1985; Cerione et al., 1984; Feder et al., 1986).

II. STRUCTURE OF ADRENERGIC RECEPTORS

All the adrenergic and related G-protein-coupled receptors are integral membrane proteins that require detergents for their solubilization. We have recently deduced the complete amino-acid sequence of all of the mammalian adrenergic receptors, including the β_1 and β_2, α_1 and α_2 subtypes, as well as those for several structurally related receptors (Lefkowitz and Caron, 1988; Cotecchia *et al.*, 1988; Regan *et al.*, 1988; Fargin *et al.*, 1988). A schematic diagram of how these proteins may be arranged in the plasma membrane is shown in Fig. 1. The most striking feature is that each receptor comprises seven stretches of 20–28 hydrophobic amino acids that likely represent membrane-spanning regions. Amino-acid identity concentrated within these membrane-spanning domains ranges from 40–70% among the different members of the adrenergic receptor family. Within the larger family of G-protein-coupled receptors, including the muscarinic, several serotonergic and substance-K receptors, and the various opsins, these identities range from 20–30%. Other regions that are less, but still reasonably, well conserved are the first two cytoplasmic loops. Extracellular domains, the third cytoplasmic loop and the carboxyl-terminal tail are all quite divergent. Other features shared among the adrenergic, muscarinic, serotonergic, and light receptors are one or more potential sites of N-glycosylation near their amino termini, and potential sites of regulatory phosphorylation on cytoplasmic domains which are discussed below.

The biological significance of the proposed seven-membrane-spanning topography of this group of receptors is not presently clear. Moreover, for most of these receptors this model is largely speculative at this point, based primarily on hydropathicity analyses of the sequences, using methods such as those developed by Kyte and Doolittle (1982), and by analogy with the structure of bacteriorhodopsin.

At present, most structural information on such an arrangement for

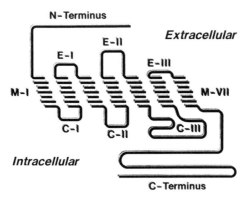

Fig. 1. Schematic representation of the possible organization of G-protein-coupled receptor proteins within the plasma membrane. Segments M-I through M-VII are proposed as spanning the lipid bilayer. With this arrangement, the amino terminus and the connecting loops E-I, E-II and E-III would lie on the extracellular surface of the membrane. Connecting loops C-I, C-II and C-III and the carboxyl terminus would be exposed to the cytoplasmic side of the plasma membrane.

membrane proteins comes from the principal membrane protein of the purple bacterium *Halobacterium halobium* known as bacteriorhodopsin. In this case, the presence of a bundle of seven membrane-spanning α-helices has been established by high-resolution electrodiffraction. Also, with vertebrate rhodopsin, an analogous structure is supported by a variety of physical measurements (reviewed by Applebury and Hargrave, 1986). For the other G-protein-coupled receptors, however, the scarcity of the proteins has, to date, hindered such physical studies. Recently, Dohlman *et al.* (1987a) have assessed various features of the model shown in Fig. 1 for the hamster lung β_2-adrenergic receptor using techniques of limited proteolysis. A schematic diagram of this approach and the findings are shown in Fig. 2. Limited trypsinization of the β_2-adrenergic receptor reconstituted in

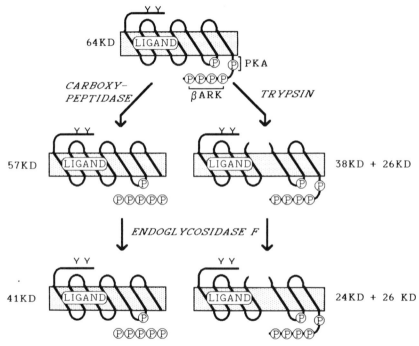

Fig. 2. Schematic diagram outlining a biochemical approach for assessing the topographical organization of the β_2-adrenergic receptor within the plasma membrane. The sites of ligand binding, N-glycosylation, and phosphorylation are represented as existing within the putative membrane-spanning segments, on the amino terminus and on the carboxyl terminus, respectively. Indicated as KD are the sodium dodecyl sulfate-polyacrylamide gel electrophoresis (SDS-PAGE) fragments generated as a result of the various enzymatic treatments shown. The experimental details and supporting evidence for this model are presented in Dohlman *et al.* (1987a). Taken from Dohlman *et al.* (1987a).

lipid vesicles yields two insoluble integral membrane domains of M_r 38,000 and 26,000. Identical results are obtained in intact cells; thus, it can be concluded that the cleavage site of the receptor is accessible at the extracellular surface of the plasma membrane. The amino-terminal 38-kDa domain contains the site of incorporation of the affinity probe ^{125}I-labeled *p*-(bromoacetamidyl)benzylcarazolol, as well as the sites of N-glycosylation as revealed by its sensitivity to endoglycosidase F. The carboxyl-terminal 26-kDa domain, however, contains all the sites of *in vitro* phosphorylation for the cAMP-dependent protein kinase (protein kinase A), as well as for the β-adrenergic receptor kinase (βARK) (see below). With four consensus sites for N-glycosylation, two near the amino- and two near the carboxyl terminus, only those in the amino-terminal domain are utilized and sensitive to endoglycosidase F. Carboxypeptidase Y (serine carboxypeptidase) treatment of reconstituted native $β_2$-adrenergic receptor generates a truncated glycopeptide of M_r ~ 57 K that has lost most of the sites phosphorylated by βARK and one of the sites phosphorylated by protein kinase A. The various features delineated (see Fig. 2), including the length of the carboxypeptidase Y-sensitive region, the extracellular location of the trypsin-sensitive site, and the location of the sites of phosphorylation and N-glycosylation, all constrain the receptor to a rhodopsin-like structure with multiple membrane-spanning segments (Dohlman *et al.*, 1987a).

III. FUNCTIONAL DOMAINS

Now that the genes and/or cDNAs for a number of receptors coupled to G proteins have been sequenced and their amino acid sequences deduced (Dohlman *et al.*, 1987b), a major focus of research is on attempts to determine which structural domains of these proteins are responsible for the known functions of receptors. The structural bases being sought are those for the ligand binding and G-protein coupling functions of the receptors, as well as those involved in the covalent modifications of receptors in processes regulating responsiveness.

IV. LIGAND BINDING

Several lines of evidence suggest that the conserved membrane-spanning domains of the receptors are crucially involved in the binding of adrenergic ligands. These regions appear to contribute to the formation of a ligand binding site. Mutants of the hamster $β_2$-adrenergic receptor in which various hydrophilic regions of the molecule (which presumably are on the internal and external connecting loops between the membrane-spanning segments) were deleted, showed no major alterations in their ligand binding properties (Dixon *et al.,* 1987). However, mutants with truncated amino-acid sequence midway through the seventh or after the fifth putative membrane-spanning regions of the protein do not bind ligands (Kobilka *et al.,* 1987b). Expression of a truncated protein containing only the sixth and seventh membrane-spanning segments is also devoid of ligand binding activity (Kobilka *et al.,* 1988). Interestingly, coexpression of the truncated proteins containing the first five and last two membrane-spanning domains exhibit significant ligand binding activity (Kobilka *et al.,* 1988). Single amino acid changes of highly conserved residues present in membrane-spanning segments II, III, or VII lead to striking changes in the ligand-binding properties of the receptor (Dixon *et al.,* 1987; Strader *et al.,* 1987b). Moreover, limited tryptic digestion of avian β-adrenergic receptor into which an antagonist affinity probe has been covalently incorporated documents that the probe is contained exclusively in the fragment comprising the first four of the seven putative membrane-spanning domains (Rubenstein *et al.,* 1987). Furthermore, covalent incorporation of a similar β-adrenergic receptor probe in the second membrane-spanning segment has been recently documented (Dohlman *et al.,* 1988). These lines of evidence are consistent with the notion that the membrane-spanning regions of receptors determine the characteristics of their ligand-binding specificity. Precisely which residues are involved in determining this specificity, as well as the ligand selectivity of the various receptor subtypes, remains to be examined in detail.

V. RECEPTOR-G PROTEIN COUPLING

The cytoplasmic domains of the various adrenergic and related receptors appear to be involved in coupling to G proteins. Recently, this problem has been approached using both biochemical and mutagenesis techniques. To date, most interest and speculation has focused on the third cytoplasmic loop as being critically involved in coupling of receptors to G proteins. Strader *et al.* (1987a) have constructed several deletion mutations in this loop that markedly impair the ability of the hamster β_2-adrenergic receptor to mediate stimulation of adenylyl cyclase. These include a large 33-residue deletion involving much of the loop (residues 239–272), and two shorter deletions at the amino (222–229) and carboxyl (258–270) ends of the loop. We have found that two mutations in the carboxyl-terminal region of the third cytoplasmic loop of the human β_2-adrenergic receptor cause marked reductions in agonist-dependent adenylyl cyclase activation

(O'Dowd *et al.,* 1988). One of these involves the deletion of seven amino acid residues (267–273) and the other a substitution of four amino acid residues (269–272). Interestingly, however, when the corresponding sequence of the human platelet α_2-adrenergic receptor was substituted into the amino-terminal portion of the third cytoplasmic loop of the human β_2-adrenergic receptor, approximately in the same position where the deletion made by Strader *et al.* (1987a) led to total loss of activity, only a slight impairment (23%) was observed (O'Dowd *et al.,* 1988). These data indicate that the sequences identified by Strader *et al.* (1987a) in the amino terminal portion of this loop cannot be crucially involved in determining the specificity of G-protein binding to the human β_2-adrenergic receptor. Perhaps, instead, this region represents an important structural component of G protein-coupled receptors in determining the overall efficiency of coupling. Further, consistent with these findings are recent observations of Kobilka *et al.* (1988) who constructed a series of α_2/β_2-adrenergic receptor chimeras and expressed these in *Xenopus laevis* oocytes. Several chimeras bound ligands with α_2-adrenergic receptor specificity, but by virtue of incorporation of β_2-adrenergic receptor sequence were now able to mediate activation of adenylyl cyclase, albeit not as well as the wild-type β_2-adrenergic receptor. The minimal regions of the β_2-adrenergic receptor which conferred on the α_2-adrenergic receptor the ability to mediate stimulation of the adenylyl cyclase involved the region extending from the amino terminal region of the fifth membrane-spanning domain through the entire third cytoplasmic loop to the carboxyl-terminal region of the sixth membrane-spanning domain. A second chimera which was similar in structure, but which included only a portion of the fifth membrane-spanning domain of the β_2-adrenergic receptor, also mediated activation of adenylyl cyclase, though less effectively than the former chimera. Taken together, these data are consistent with the notion that sequences in the third cytoplasmic loop, and perhaps adjacent membrane-spanning segments, are involved in coupling of the β_2-adrenergic receptor to G_s.

Several mutations made by O'Dowd *et al.* (1988) in the most amino-terminal region of the carboxyl tail of the human β_2-adrenergic receptor also prevented efficient receptor coupling to G_s. Moreover, results with several mutations in the first and second cytoplasmic loops indicate that these regions may contribute to, but not determine directly, the coupling of the β_2-adrenergic receptor to G_s. Fig. 3 highlights several of the regions and residues that have been delineated by the mutagenesis approach to influence receptor coupling function.

In contrast, however, Rubenstein *et al.* (1987), using limited trypsinization of turkey erythrocyte β-adrenergic receptor, have concluded that most, if not all, of the third cytoplasmic loop, as well as the hydrophilic carboxyl-

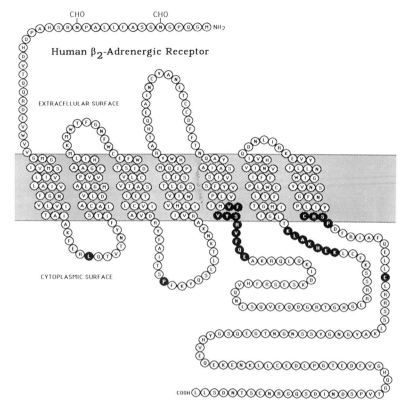

Fig. 3. Putative regions of the human β_2-adrenergic receptor directly or indirectly involved in the interaction of the receptor with G_s. Represented is the sequence of the human β_2-adrenergic receptor (Kobilka *et al.*, 1987a) organized about the plasma membrane according to the seven-membrane-spanning model (Dohlman *et al.*, 1987a). Residues presented in black indicate regions of the third cytoplasmic loop and carboxyl terminus which, when deleted or substituted, were found to affect the ability of the mutated receptors to mediate isoproterenol-stimulation of adenylyl cyclase. Darkened residues in cytoplasmic loops I (Leu[64]) and II (Pro[138]) and the carboxyl tail (Cys[341]) are conserved residues in many G-protein-coupled receptors, which when mutated to Gly[64], Thr[138], and Gly[341] were found to markedly affect the expression and functionality of the expressed receptors (O'Dowd *et al.*, 1988).

terminal region of the receptor, are not necessary for agonist-stimulated regulation of G_s.

It is not presently clear how to reconcile these discrepant findings from biochemical and mutagenesis experiments. One possibility is that the regions of the third cytoplasmic loop removed by the trypsinization protocol of Rubenstein *et al.* (1987) do not encompass the crucial regulatory sequences identified by the deletion or substitution approach. This possibility is, in fact, consistent with the published data. Another possibility is that deletion of amino acids by genetic methods imposes alterations in the tertiary structure of the mutant receptors which are more generalized than those imposed by proteolysis of the wild-type receptors. A third possibility is that the tryptic fragments generated remain functionally associated, albeit noncovalently, with the receptor. While more work will be necessary to resolve these issues, the results of the substitution and chimeric receptor studies strongly buttress the notion that sequences in the third cytoplasmic loop are involved in receptor coupling to G proteins.

VI. REGULATION OF RECEPTOR FUNCTION BY COVALENT MODIFICATIONS

Hormonal responsiveness of target tissues is a dynamic phenomenon. Several different mechanisms have been proposed that may regulate the number and/or function of receptors and other components of signal transduction systems. In the field of the hormone-responsive adenylyl cyclase system, a paradigm which has been extensively studied is that of desensitization. Desensitization appears to be a general biological phenomenon by which the response to a stimulus decreases with time despite the presence of a constant stimulus. In adenylyl cyclase-responsive systems, two major patterns of desensitization have been distinguished and are generally referred to as homologous and heterologous. In homologous desensitization, only responsiveness to the specific desensitizing hormone is attenuated while the efficacy of other hormone activators is unimpaired. Conversely, heterologous desensitization occurs when exposure to one agonist results in the decrease of the response to multiple agonists operating through distinct receptors. In some instances, the pattern of unresponsiveness of adenylyl cyclase in heterologous desensitization may be so broad as to include decreased sensitivity to activators that bypass the receptors, for example, NaF and guanine nucleotides, both of which directly activate G_s. It appears that multiple mechanisms can simultaneously contribute to the development of both heterologous and homologous forms of desensitization. In particular, in the case of heterologous desensitization, available evidence suggests that there may be alterations of each of the components of the adenylyl cyclase system: the receptors, the G proteins and the catalytic moiety of the enzyme (Sibley and Lefkowitz, 1985; Sibley et al., 1987).

Phosphorylation of β-adrenergic receptors accompanies both heterologous and homologous forms of desensitization. Moreover, receptor phosphorylation appears to contribute to the mechanism of desensitization in each case. For heterologous desensitization, most of the evidence has been obtained using avian erythrocyte systems. It has been documented that maximal phosphorylation of β-adrenergic receptors in these cells occurs to a stoichiometry of 2–3 mol of phosphate/mol of receptor during the desensitization process. Under a variety of conditions, the phosphorylation stoichiometry is found to be tightly correlated with the extent of desensitization (Sibley and Lefkowitz, 1985).

It has been demonstrated *in vitro* that protein kinase A can catalyze the addition of up to 2 mol of phosphate/mol of purified hamster lung β_2-adrenergic receptor (Benovic *et al.*, 1985). This correlates with the observation that heterologous desensitization often appears to be cAMP-mediated, and that in several systems desensitization can actually be induced by cAMP analogs (Sibley *et al.*, 1987). Agonist occupancy of the receptor enhances the rate of this phosphorylation reaction, but not the ultimate stoichiometry achieved (Bouvier *et al.*, 1987). Further, it has been demonstrated that β_2-adrenergic receptor phosphorylated by protein kinase A has reduced ability to activate G_s in a reconstituted system (Benovic *et al.*, 1985). Both human and hamster β_2-adrenergic receptors have two consensus sites for protein kinase A. One is on the carboxyl-terminal portion of the third cytoplasmic loop and the other is found on the cytoplasmic carboxyl terminus about 20 amino acid residues removed from the putative seventh membrane-spanning segment. Taken together, these data suggest that one mechanism by which heterologous desensitization can occur at the receptor level is by a feedback regulatory effect mediated by protein kinase A such that agonist occupancy of the receptor leads to the generation of the second messenger cAMP, activation of the kinase and subsequent phosphorylation of the receptors. The consensus sites for protein kinase A phosphorylation of the β_2-adrenergic receptor are also potential sites for protein kinase C phosphorylation. In fact, protein kinase C is also capable of phosphorylating and regulating the β-adrenergic receptor. Tumor-promoting phorbol diesters, which potently activate protein kinase C, stimulate β-adrenergic receptor phosphorylation and induce desensitization of the β-adrenergic-responsive adenylyl cyclase in avian erythrocytes (Sibley *et al.*, 1984). Since a wide variety of hormones, drugs, and neurotransmitters activate protein kinase C, there is at least the potential for regulation of β-adrenergic receptor function by stimulation with these various agents (Sibley *et al.*, 1987).

In vitro, protein kinase C can phosphorylate the purified β_2-adrenergic receptor. However, in contrast to the situation with protein kinase A, the rate of receptor phosphorylation by protein kinase C is not influenced by agonist occupancy. Table I summarizes the agonist dependency for the

Table I   Covalent Modification of β-Adrenergic Receptor by Various Protein Kinases and Dependence on Agonist Occupancy of Receptor[a]

Parameter	PKC	PKA	βARK
$β_2$-Adrenergic receptor phosphorylation	Yes	Yes	Yes
Effect of agonist occupancy	No effect on rate or extent	Increases rate but not extent	Markedly increases both rate and extent

[a] PKC, protein kinase C; PKA, protein kinase A; βARK, β-adrenergic receptor kinase.

phosphorylation of the $β_2$-adrenergic receptor by various kinases (Bouvier *et al.*, 1987; Benovic *et al.*, 1986b). To date, however, no reports assessing the functional consequences of *in vitro* phosphorylation of β-adrenergic receptors by protein kinase C have appeared. In the process of heterologous desensitization, covalent modification of the other components of the adenylyl cyclase transduction system could possibly be involved. It has been shown that the catalytic moiety of adenylyl cyclase is a substrate for protein kinase A and that phosphorylation of the enzyme reduces its responsiveness to guanine nucleotides upon reconstitution with G_s (Yoshimasa *et al.*, 1988). In turkey erythrocytes following exposure to isoproterenol, G_s has been shown to be functionally impaired (Briggs *et al.*, 1983). However, whether this impairment in function is the result of covalent phosphorylation or whether the catalytic moiety of the enzyme can be phosphorylated *in vivo* remains to be determined.

Agonist-specific or homologous desensitization is defined as the loss of adenylyl cyclase responsiveness to a single hormone. At least three distinct phenomena are involved in this process: (1) an initial rapid uncoupling of the receptor from adenylyl cyclase activation, (2) a rapid sequestration or internalization of the receptor away from the cell surface, but without loss of total receptor from the cell, and (3) a slow down-regulation or inactivation of the receptor associated with a loss of total receptor content of the cell.

Substantial evidence indicates that phosphorylation of the β-adrenergic receptor accompanies its homologous desensitization. In every case in which this has been examined, time courses of phosphorylation are at least, if not more, rapid than the desensitization process itself. Both homologous desensitization, as well as the receptor phosphorylation which accompanies it, appear to be *cAMP-independent* processes. Recently, a protein kinase capable of phosphorylating purified β-adrenergic receptor in a *cAMP-independent* fashion has been identified and purified (Strasser *et al.*, 1986b;

Benovic *et al.*, 1986b). Phosphorylation of the receptor by this enzyme, however, is almost totally dependent on agonist occupancy of the receptor (Table I). Initial identification of the kinase activity was made in the *kin⁻* mutant of S49 lymphoma cells which lack protein kinase A (Strasser *et al.*, 1986b). This kinase has been termed the β-adrenergic receptor kinase, or βARK, and is distinct from other known kinases such as protein kinase A, protein kinase C, Ca^{2+}/calmodulin- and Ca^{2+}/phospholipid-dependent protein kinases. βARK does not phosphorylate general kinase substrates such as casein and histones. The enzyme consists of a single subunit of M_r 80,000 and phosphorylates the agonist-occupied form of the receptor to a stoichiometry of ~8 mol of phosphate/mol of receptor *in vitro* (Benovic *et al.*, 1987b).

When β-adrenergic receptor is phosphorylated by pure βARK, the resulting phosphorylated receptor has only a slightly reduced ability to interact with G_s, as assessed by agonist-stimulated GTPase activity of G_s in a reconstituted system. However, when pure β-adrenergic receptor is phosphorylated with a crude βARK preparation, the receptor activation of G_s is very significantly reduced to about 30% of its original activity (Benovic *et al.*, 1987a). This suggests that the crude βARK fraction contains a factor which enhances the uncoupling of phosphorylated β-adrenergic receptor from interaction with G_s. Interestingly, following the addition of pure retinal arrestin (i.e., 48K protein), which is involved in enhancing the inactivating effect of rhodopsin phosphorylation by rhodopsin kinase, to β-adrenergic receptor phosphorylated by pure βARK, the uncoupling of phosphorylated β-adrenergic receptor from G_s is enhanced (41% inactivation) (Fig. 4). These findings suggest that a protein analogous to retinal arrestin may be involved in mediating the functional consequences of receptor phosphorylation by βARK (Benovic *et al.*, 1987a).

Rhodopsin kinase is an enzyme restricted to rod outer segments which phosphorylates only the light-bleached form of rhodopsin. Phosphorylated rhodopsin interacts with arrestin which quenches the ability of rhodopsin to activate transducin and, hence, cGMP phosphodiesterase activity. This series of events essentially turns off the light transduction process (Stryer, 1986). The analogies with βARK are apparent. Both enzymes phosphorylate G-protein-coupled receptors, but only when the receptors have been stimulated either by agonist occupancy in one case or by absorption of a photon of light in the other. In each case, phosphorylation reactions lead to a damping of the ability of the receptor to interact with its respective G protein. Moreover, βARK is capable of phosphorylating light-bleached rhodopsin and, conversely, rhodopsin kinase is capable of phosphorylating agonist-occupied β-adrenergic receptors, albeit weakly (Benovic *et al.*, 1986a). Clearly, each enzyme has a preference for its native substrate.

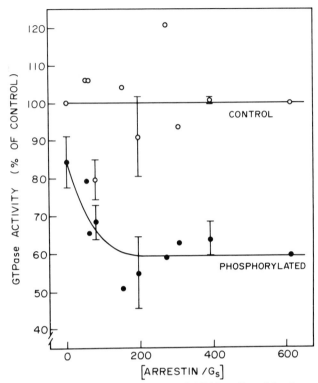

Fig. 4. Ability of retinal arrestin to augment the inhibiting effect of β_2-adrenergic receptor phosphorylation on receptor–G_s coupling. Purified hamster lung β_2-adrenergic receptor in phospholipid vesicles was phosphorylated with purified β-adrenergic receptor kinase to a stoichiometry of ~8 mol of phosphate/mol of receptor. Control receptor was similarly treated except that App(NH)p was used instead of ATP. Receptor preparations (~0.4 pmol) in phospholipid vesicles were then incubated with retinal arrestin (0–1500 pmol) and pure G_s (~2.4 pmol) and assayed for agonist-mediated GTPase activity. Results shown are from Benovic *et al.* (1987a).

βARK is also capable of phosphorylating the agonist-occupied form of the α_2-adrenergic receptor (Benovic *et al.*, 1987c). In fact, the α_2-adrenergic receptor is as good a substrate for βARK as the β_2-adrenergic receptor. Again, phosphorylation is entirely agonist-dependent. In contrast, the α_1-adrenergic receptor, which is coupled to stimulation of phosphatidylinositol phosphate hydrolysis, does not serve as a good substrate for βARK. These results suggest that βARK may be a general adenylyl cyclase-coupled receptor kinase.

At present, little is known about the mechanisms which lead to receptor dephosphorylation and regeneration of biological activity. The process of sequestration, which usually rapidly follows the initial functional uncoupling of receptors from their G proteins, may well play a role in the dephosphorylation mechanism. Sibley *et al.* (1986) have demonstrated that frog erythrocyte β-adrenergic receptors remaining at the cell surface following isoproterenol desensitization are highly phosphorylated, whereas receptors isolated from a sequestered membrane fraction are phosphorylated to a much lower extent. Moreover, the sequestered vesicular compartment contains a phosphatase activity which is capable of rapidly dephosphorylating phosphorylated β-adrenergic receptors (Yang *et al.*, 1988). *In vitro* studies have demonstrated that the Type I, IIa and IIb protein phosphatases do not dephosphorylate βARK-phosphorylated β-adrenergic receptor. However, a latent high-molecular-weight Type II phosphatase does effectively dephosphorylate the receptor (Yang *et al.*, 1988). Further studies will be necessary to determine whether this latent phosphatase acts as a receptor phosphatase *in vivo*.

Similar lines of evidence currently suggest that the major sites for βARK-catalyzed phosphorylation of the β_2-adrenergic receptor are at or near the carboxyl terminus of the protein. First is the analogy with rhodopsin. It has been clearly shown that the cluster of serine and threonine residues found at the carboxyl terminus of rhodopsin represents the major sites of phosphorylation by rhodopsin kinase (Wilden and Kuhn, 1982). Apparently, there is an additional site in the third cytoplasmic loop (Applebury and Hargrave, 1986). Second, as noted above, Dohlman *et al.* (1987a) have demonstrated that carboxypeptidase Y (serine carboxypeptidase) is capable of digesting most or all of the phosphate added to the β-adrenergic receptor by βARK. Finally, compelling data have been recently obtained with two β-adrenergic receptor mutants designed to remove the putative phosphate acceptor sites on the receptor. One is a truncation mutant lacking the last 48 amino acid residues of the protein, including the concentration of serine and threonine residues near the carboxyl terminus. The other is a receptor in which 11 of the serine and threonine residues nearest the carboxyl terminus of the receptor have been substituted with alanine or glycine by site-directed mutagenesis procedures (Bouvier *et al.*, 1988). Fig. 5 illustrates the portion of the carboxyl-terminal sequence of the human β_2-adrenergic receptor which has been mutated in these two receptor constructs. It was documented that both mutants showed markedly reduced agonist-stimulated phosphorylation when compared to the wild-type receptor. This further strengthens the notion that the cluster of serine and threonine residues at the carboxyl terminus of the β_2-adrenergic receptor in fact represents the major site of phosphorylation by βARK *in vivo* (Bouvier *et al.*, 1988). Moreover, as shown in Fig. 6, when both mutants were assessed for their ability to desensitize, they were found

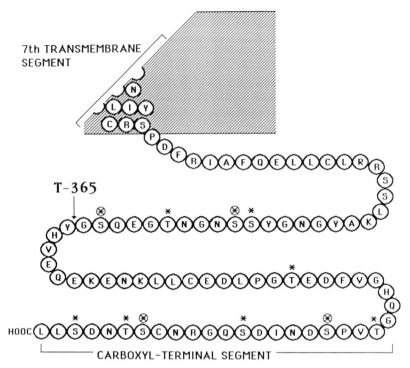

Fig. 5. Carboxyl-terminal sequence of the human β_2-adrenergic receptor showing various sites of mutation. The arrow (T-365) indicates the site of a truncation mutant produced by inserting a stop codon after Gly[365]. The substitution mutant S-351 was produced by substituting the various serine and threonine residues with Ala (*) or Gly ⊕ by site-directed mutagenesis procedures. Both receptor mutants, as well as wild-type β_2-adrenergic receptor, were stably expressed in transformed Chinese hamster fibroblasts (CHW) as described in Bouvier *et al.* (1988).

to exhibit a markedly delayed onset of homologous desensitization. Agonist-promoted phosphorylation of the carboxyl terminus of the receptor, presumably by βARK, therefore appears to be a crucial event in the early stage of homologous desensitization. Interestingly, however, there is no impairment of sequestration of the receptors in either of these mutants, suggesting that phosphorylation of the receptor at the carboxyl terminus is not the triggering signal for sequestration (Bouvier *et al.*, 1988).

Current understanding of the biochemical mechanisms involved in homologous desensitization is depicted in Fig. 7. One of the earliest steps which occurs following agonist occupancy of receptor is G_s activation followed by activation of adenylyl cyclase. Shortly thereafter, agonist occu-

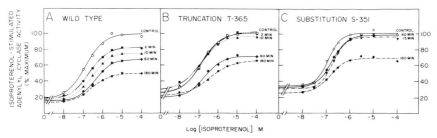

Fig. 6. Effect of the removal of the potential acceptor sites for phosphorylation on the development of agonist-promoted desensitization of the β_2-adrenergic stimulation of adenylyl cyclase. CHW cells expressing the mutants and wild-type β_2-adrenergic receptors were exposed to 2 μM isoproterenol for the time periods indicated in the panels. Cells were then washed, and membranes prepared and assayed for the ability of increasing concentrations of isoproterenol to stimulate adenylyl cyclase. Results shown are from Bouvier *et al.* (1988).

pancy of the receptor induces phosphorylation of the receptor by βARK. This phosphorylation may initially involve translocation of βARK from the cytosol to the plasma membrane (Strasser *et al.*, 1986a). βARK is capable of multiply phosphorylating the agonist-occupied form of the receptor at sites predominantly localized near the carboxyl terminus. βARK-phosphorylation leads to an uncoupling of the receptor from G_s by a mechanism that may involve the interaction of an arrestin-like molecule with the phosphorylated receptor. The receptor is then sequestered from the cell surface by an, as yet, unknown mechanism where it is presumably dephosphorylated and recycled back to the plasma membrane where it recovers full functionality.

Although many of the steps postulated in this process are not fully understood, and their implications not fully proven, it is reasonably well established from the results described above that phosphorylation of the receptor near its carboxyl terminus is implicated in the onset of this form of desensitization. Although βARK is capable of phosphorylating several receptors, the agonist specificity or homologous character of the desensitization process arises from the fact that βARK can only phosphorylate agonist-occupied or activated forms of receptors.

As we have reviewed in this chapter, the β-adrenergic receptor can be phosphorylated by three different kinases: protein kinase A and C, as well as βARK. Phosphorylation by these different kinases shows distinct dependencies on agonist occupancy: independent for protein kinase C; accelerated for protein kinase A and nearly totally dependent for βARK (Table I). It is likely that these processes serve different needs or occur under different circumstances in the cell.

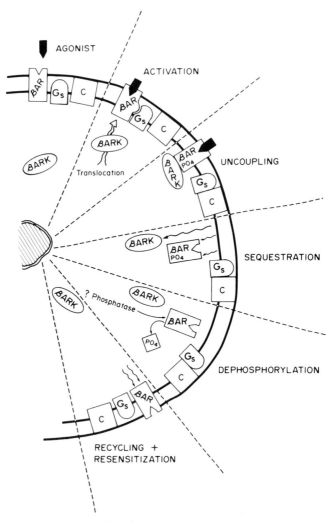

Fig. 7. Schematic representation of the possible molecular events involved in the process of homologous desensitization. βAR, β-adrenergic receptor; G_s, stimulatory guanine nucleotide regulatory protein; C, catalytic unit of adenylyl cyclase; βARK, β-adrenergic receptor kinase; βAR-PO$_4$, phosphorylated β-adrenergic receptor.

REFERENCES

Applebury, M. L., and Hargrave, P. A. (1986). Molecular biology of the visual pigments. *Vision Res.* **26**, 1881–1895.

Benovic, J. L., Pike, L. J., Cerione, R. A., Staniszewski, C., Yoshimasa, T., Codina, J., Birnbaumer, L., Caron, M. G., and Lefkowitz, R. J. (1985). Phosphorylation of the mammalian β-adrenergic receptor by cyclic AMP-dependent protein kinase: Regulation of the rate of receptor phosphorylation and dephosphorylation by agonist occupancy and effects on coupling of the receptor to the stimulatory guanine nucleotide regulatory protein. *J. Biol. Chem.* **260**, 7094–7101.

Benovic, J. L., Mayor, F., Jr., Somers, R. L., Caron, M. G., and Lefkowitz, R. J. (1986a). Light dependent phosphorylation of rhodopsin by the β-adrenergic receptor kinase. *Nature* **322**, 869–872.

Benovic, J. L., Strasser, R. H., Caron, M. G., and Lefkowitz, R. J. (1986b). β-Adrenergic receptor kinase: Identification of a novel protein kinase which phosphorylates the agonist-occupied form of the receptor. *Proc. Natl. Acad. Sci. U.S.A.* **83**, 2797–2801.

Benovic, J. L., Kuhn, H., Weyland, I., Codina, J., Caron, M. G., and Lefkowitz, R. J. (1987a). Functional desensitization of the isolated β-adrenergic receptor by the β-adrenergic receptor kinase: Potential role of an analog of the retinal protein arrestin (48kDa protein). *Proc. Natl. Acad. Sci. U.S.A.* **84**, 8879–8882.

Benovic, J. L., Mayor, F., Jr., Staniszewski, C., Lefkowitz, R. J., and Caron, M. G. (1987b). Purification and characterization of the β-adrenergic receptor kinase. *J. Biol. Chem.* **262**, 9026–9032.

Benovic, J. L., Regan, J. W., Matsui, H., Mayor, F., Jr., Cotecchia, S., Leeb-Lundberg, L. M. F., Caron, M. G., and Lefkowitz, R. J. (1987c). Agonist-dependent phosphorylation of the α_2-adrenergic receptor by the β-adrenergic receptor kinase. *J. Biol. Chem.* **262**, 17251–17253.

Berridge, M. J., and Irvine, R. F. (1984). Inositol trisphosphate, a novel second messenger in cellular signal transduction. *Nature* **312**, 315–321.

Bouvier, M., Leeb-Lundberg, L. M., Benovic, J. L., Caron, M. G., and Lefkowitz, R. J. (1987). Regulation of adrenergic receptor function by phosphorylation. II. Effects of agonist occupancy on phosphorylation of α_1- and β_2-adrenergic receptors by protein kinase C and the cyclic AMP-dependent protein kinase. *J. Biol. Chem.* **262**, 3106–3113.

Bouvier, M., Hausdorff, W. P., De Blasi, A., O'Dowd, B. F., Kobilka, B. K., Caron, M. G., and Lefkowitz, R. J. (1988). Removal of phosphorylation sites from the β_2-adrenergic receptor delays onset of agonist-promoted desensitization. *Nature* **333**, 370–372.

Briggs, M. M., Stadel, J. M., Iyengar, R., and Lefkowitz, R. J. (1983). Functional modification of the guanine nucleotide regulatory protein after desensitization of turkey erythrocytes by catecholamines. *Arch. Biochem. Biophys.* **224**, 142–151.

Cerione, R. A., Sibley, D. R., Benovic, J. L., Caron, M. G., and Lefkowitz, R. J. (1984). Reconstitution of a hormone-sensitive adenylate cyclase system: The pure β-adrenergic receptor and guanine nucleotide regulatory protein confer hormone responsiveness on the resolved catalytic unit. *J. Biol. Chem.* **259**, 9979–9982.

Cotecchia, S., Schwinn, D. A., Randall, R. R., Lefkowitz, R. J., Caron, M. G., and Kobilka, B. K. (1988). Molecular cloning and expression of the cDNA for the hamster α_1-adrenergic receptor. *Proc. Natl. Acad. Sci. U.S.A.* **85**, 7159–7163.

Dixon, R. A. F., Sigal, I. S., Candelore, M. R., Register, R. B., Scattergood, W.,

Rands, E., and Strader, C. D. (1987). Structural features required for ligand binding to the β-adrenergic receptor. *EMBO J.* **6**, 3269.

Dohlman, H. G., Bouvier, M., Benovic, J. L., Caron, M. G., and Lefkowitz, R. J. (1987a). The multiple membrane spanning topography of the β_2-adrenergic receptor. Localization of the sites of binding, glycosylation and regulatory phosphorylation by limited proteolysis. *J. Biol. Chem.* **262**, 14282–14288.

Dohlman, H. G., Caron, M. G., and Lefkowitz, R. J. (1987b). A family of receptors coupled to guanine nucleotide regulatory proteins. *Biochem.* **26**, 2657–2664.

Dohlman, H. G., Caron, M. G., Strader, C. D., Amlaiky, N., and Lefkowitz, R. J. (1988). Identification and sequence of a binding site peptide of the β_2-adrenergic receptor. *Biochemistry* **27**, 1813–1817.

Fargin, A., Raymond, J. R., Lohse, M. J., Kobilka, B. K., Caron, M. G., and Lefkowitz, R. J. (1988). The genomic clone G-21, which resembles a β-adrenergic receptor sequence, encodes the 5-HT$_{1A}$ receptor. *Nature* **335**, 358–360.

Feder, D., Im, M-J, Klein, H. K., Hekman, M., Holzhofer, A., Dees, C., Levitzki, A., Helmreich, E. J. M., and Pfeuffer, T. (1986). Reconstitution of β_1-adreno-ceptor-dependent adenylate cyclase from purified components. *EMBO J.* **5**, 1509–1514.

Gilman, A. G. (1987). G proteins: Transducers of receptor-generated signals. *Annu. Rev. Biochem.* **56**, 615–649.

Isom, L. L., Cragoe, E. J., Jr., and Limbird, L. E. (1987). Multiple receptors linked to inhibition of adenylyl cyclase accelerate Na$^+$/H$^+$ exchange in neuroblastoma X glioma cells via a mechanism other than decreased cAMP accumulation. *J. Biol. Chem.* **262**, 6750–6757.

Julius, D., MacDermott, A. B., Axel, R., and Jessell, T. M. (1988). Molecular characterization of a functional cDNA encoding the serotonin 1C receptor. *Science* **241**, 558–561.

Kobilka, B. K., Dixon, R. A. F., Frielle, T., Dohlman, H. G., Bolanowski, M. A., Sigal, I. S., Yang-Feng, T. L., Francke, U., Caron, M. G., and Lefkowitz, R. J. (1987a). cDNA for the human β_2-adrenergic receptor: A protein with multiple membrane spanning domains and a chromosomal location shared with the PDGF receptor gene. *Proc. Natl. Acad. Sci. U.S.A.* **84**, 46–50.

Kobilka, B. K., MacGregor, C., Daniel, K., Kobilka, T. S., Caron, M. G., and Lefkowitz, R. J. (1987b). Functional activity and regulation of human β_2-adrenergic receptors expressed in xenopus oocytes. *J. Biol. Chem.* **262**, 15796–15802.

Kobilka, B. K., Kobilka, T. S., Regan, J. W., Caron, M. G., and Lefkowitz, R. J. (1988). Chimeric α_2-β_2-adrenergic receptors: Delineation of domains involved in effector coupling and ligand binding specificity. *Science* **240**, 1310–1316.

Kyte, J., and Dotlittle, R. F. (1982). A simple method for displaying the hydropathic character of a protein. *J. Molec. Biol.* **157**, 105–132.

Lancet, D., and Pace, U. (1987). The molecular basis of odor recognition. *Trends Biochem. Sci.* **12**, 63–66.

Lefkowitz, R. J., and Caron, M. G. (1988). The adrenergic receptors: Models for the study of receptors coupled to guanine nucleotide regulatory proteins. *J. Biol. Chem.* **263**, 4993–4996.

May, D. C., Ross, E. M., Gilman, A. G., and Smigel, M. D. (1985). Reconstitution of a catecholamine stimulated adenylate cyclase activity using three purified proteins. *J. Biol. Chem.* **260**, 15829–15833.

Mollner, S., and Pfeuffer, T. (1988). Two different adenylate cyclase in brain distinguished by monoclonal antibodies. *Eur. J. Biochem.* **171**, 265–271.

O'Dowd, B. F., Hnatowich, M., Regan, J. W., Leader, W. M., Caron, M. G., and Lefkowitz, R. J. (1988). Site-directed mutagenesis of the cytoplasmic domains of the human β_2-adrenergic receptor: Localization of regions involved in a G protein-receptor coupling. *J. Biol. Chem.* **263**, 15985–15992.

O'Dowd, B. F., Lefkowitz, R. J., and Caron, M. G. (1989). Structure of the adrenergic and related receptors. *Annu. Rev. Neuroscience* **12**, 67–83.

Pfeuffer, E., Drehev, R-M, Metzger, H., and Pfeuffer, E. (1985). Catalytic unit of adenylate cyclase: Purification and identification by affinity cross-linking. *Proc. Natl. Acad. Sci. U.S.A.* **82**, 3086–3090.

Regan, J. W., Kobilka, T. S., Yang-Feng, T. L., Caron, M. G., and Lefkowitz, R. J. (1988). Cloning and expression of a human kidney cDNA for a novel α_2-adrenergic receptor. *Proc. Natl. Acad. Sci. U.S.A.* **85**, 6301–6305.

Rubenstein, R. C., Wong, S. K.-F., and Ross, E. M. (1987). The hydrophobic tryptic core of the β-adrenergic receptor retains G_s regulatory activity in response to agonists and thiols. *J. Biol. Chem.* **262**, 16655–16662.

Sibley, D. R., and Lefkowitz, R. J. (1985). Molecular mechanisms of receptor desensitization using the β-adrenergic receptor-coupled adenylyl cyclase system as a model. *Nature* **317**, 124–129.

Sibley, D. R., Nambi, P., Peters, J. R., and Lefkowitz, R. J. (1984). Phorbol diesters promote β-adrenergic receptor phosphorylation and adenylyl cyclase desensitization in duck erythrocyte. *Biochem. Biophys. Res. Commun.* **121**, 973–979.

Sibley, D. R., Strasser, R. H., Caron, M. G., and Lefkowitz, R. J. (1985). Homologous desensitization of adenylyl cyclase is associated with phosphorylation of the β-adrenergic receptor. *J. Biol. Chem.* **260**, 3883–3886.

Sibley, D. R., Strasser, R. H., Benovic, J. L., Daniel, K., and Lefkowitz, R. J. (1986). Phosphorylation/dephosphorylation of the β-adrenergic receptor regulates its functional coupling to adenylate cyclase and subcellular distribution. *Proc. Natl. Acad. Sci. U.S.A.* **83**, 9408–9412.

Sibley, D. R., Benovic, J. L., Caron, M. G., and Lefkowitz, R. J. (1987). Regulation of transmembrane signaling by receptor phosphorylation. *Cell* **48**, 913–922.

Smigel, D. (1986). Purification of the catalyst of adenylate cyclase. *J. Biol. Chem.* **261**, 1976–1982.

Strader, C. D., Dixon, R. A. F., Cheung, A. H., Candelore, M. R., Blake, A. D., and Sigal, I. S. (1987a). Mutations that uncouple the β-adrenergic receptor from G_s and increase agonist affinity. *J. Biol. Chem.* **262**, 16439–16443.

Strader, C. D., Sigal, I. S., Register, R. B., Candelore, M. R., Rands, E., and Dixon, R. A. F. (1987b). Identification of residues required for ligand binding to the β-adrenergic receptor. *Proc. Natl. Acad. Sci. U.S.A.* **84**, 4384–4388.

Strasser, R. H., Benovic, J. L., Caron, M. G., and Lefkowitz, R. J. (1986a). β-Agonist and prostaglandin E_1-induced translocation of the β-adrenergic receptor kinase: Evidence that the kinase may act on multiple adenylyl cyclase coupled receptors. *Proc. Natl. Acad. Sci. U.S.A.* **83**, 6362–6366.

Strasser, R. H., Sibley, D. R., and Lefkowitz, R. J. (1986b). A novel catecholamine activated adenosine cyclic 3′,5′-phosphate independent pathway for β-adrenergic receptor phosphorylation in wild type and mutant S49 lymphoma cells: Mechanism of homologous desensitization of adenylyl cyclase. *Biochemistry* **25**, 1371–1377.

Stryer, L. (1986). Cyclic GMP cascade of vision. *Annu. Rev. Neurosci.* **9**, 87–119.

Wilden, U., and Kuhn, H. (1982). Light-dependent phosphorylation of rhodopsin: Number of phosphorylation sites. *Biochemistry* **21**, 3014–3022.

Yang, S.-D., Fong, Y.-L., Benovic, J. L., Sibley, D. R., Caron, M. G., and Lefkowitz, R. J. (1988). Dephosphorylation of the β-adrenergic receptor and rhodopsin by latent phosphatase 2. *J. Biol. Chem.* **263**, 8856–8858.

Yoshimasa, T., Bouvier, M., Benovic, J. L., Amlaiky, N., Lefkowitz, R. J., and Caron, M. G. (1988). Regulation of the adenylyl cyclase signalling pathway: Potential role of the phosphorylation of the catalytic unit by protein kinase A and protein kinase C. *In* "Molecular Biology of Brain and Endocrine Peptidergic Systems" K. W. McKerns and M. Chretien, eds.), pp. 123–139. Plenum, New York.

CHAPTER 14

Muscarinic Receptors and Their Interactions with G Proteins

Michael W. Martin, José Luis Boyer, John M. May
Department of Pharmacology, University of North Carolina, Chapel Hill,
North Carolina 27599

Lutz Birnbaumer
Department of Cell Biology, Baylor College of Medicine, Houston, Texas 77030

T. Kendall Harden
Department of Pharmacology, University of North Carolina, Chapel Hill,
North Carolina 27599

I. Introduction
II. Muscarinic Receptors and Phosphoinositide Metabolism
III. Muscarinic Receptors and Ion Channels
IV. Subtypes of Muscarinic Receptors
V. Molecular Cloning and Structure of mAChR
  A. Molecular Cloning
  B. Structural Features
  C. Signal Transduction Systems Used by Muscarinic Receptors
  References

I. INTRODUCTION

Muscarinic cholinergic receptors (mAChR) are integral membrane proteins expressed by neurons in the peripheral and central nervous systems,
by cardiac and smooth muscle cells, and by various exocrine gland cells.
Acetylcholine interacts with these receptors to produce a myriad of physiological responses including decreased heart rate, contraction of smooth

muscle, increased secretion from glandular cells, and excitation or inhibition of neuronal activity. What mechanisms allow a single endogenous neurotransmitter to convey selective information in such diverse systems? Selectivity of responses is accounted for, at least in part, by the expression of multiple subtypes of mAChR. The mAChR subtypes possess common structural features that are characteristic of a large group of receptor molecules including rhodopsin, β- and α_2-adrenergic receptors, and substance-K receptors (Applebury and Hargrave, 1986; Dohlman et al., 1987; Kerlavage et al., 1987). In addition to similar structural domains, these receptors interact with various guanine nucleotide regulatory proteins (G proteins) to effect trans-plasma-membrane signal transduction (Stryer and Bourne, 1986; Gilman, 1987; Harden, 1989a). A variety of biochemical and electrophysiological responses have been correlated with mAChR activation including: inhibition of adenylyl cyclase activity and subsequent reduction of cyclic AMP levels (Harden, 1989b); stimulation of phospholipase C-catalyzed polyphosphoinositide breakdown resulting in liberation of inositol 1,4,5-trisphosphate and 1,2-diacylglycerol, with subsequent mobilization of intracellular calcium and activation of protein kinase C (Putney, 1986); activation of cardiac potassium channels (Pfaffinger et al., 1985; Breitwieser and Szabo, 1985; Yatani et al., 1987); stimulation of cyclic GMP accumulation (Study et al., 1978; Richelson and El-Fakahany 1981); and opening or closing of chloride, calcium, and potassium channels (Nathanson, 1987). It has been clearly documented that adenylyl cyclase, phospholipase C, and cardiac potassium channels are primary effectors linked to mAChR at the level of the plasma membrane and that G proteins are essential coupling elements of these receptor–effector systems. It remains to be determined if there are additional primary effector enzymes or channels and whether all effector responses to mAChR involve a G protein.

In this review we will discuss evidence documenting the role of G proteins in transducing various mAChR-mediated responses, will consider similarities between mAChR–G protein–effector systems and other receptor systems, and will point to areas of ignorance or controversy where further research is required before definitive answers can be provided concerning components and mechanisms.

Early evidence for the involvement of G proteins in mAChR-mediated responses evolved from studies of two related phenomena, receptor-mediated inhibition of adenylyl cyclase activity and guanine nucleotide effects on receptor affinity for agonists. Murad et al. (1962) first reported that activation of mAChR inhibits production of cyclic AMP in cardiac tissue. This observation was not unambiguously confirmed until the late 1970s when it was demonstrated that mAChR and a variety of other receptors, e.g. α_2-adrenergic, opiate, and adenosine-A_1, inhibit adenylyl cyclase in a variety of target tissues (Jakobs *et al.*, 1976, 1978, 1979, 1985; Limbird, 1981). A G protein, G_s, had been shown to couple β-adrenergic receptors to activation of adenylyl cyclase. Thus, it was not surprising that a G protein was implicated in coupling inhibitory receptors to adenylyl cyclase and that this protein was distinct from G_s: GTP was necessary for observation of receptor-mediated inhibition of adenylyl cyclase (Jakobs *et al.*, 1978; Londos *et al.*, 1978); adenylyl cyclase was still under inhibitory regulation in tissues deficient in G_s (Motulsky *et al.*, 1982; Hildebrandt *et al.*, 1982, 1983; Jakobs *et al.*, 1983); and activation and inhibition of adenylyl cyclase activity were differentially sensitive to inactivation by various reagents (Hoffman *et al.*, 1981; Harden *et al.*, 1982; Jakobs *et al.*, 1982).

Muscarinic receptor–G protein interaction can be observed as a negative heterotropic influence of guanine nucleotides on the apparent binding affinity of agonists to mAChR in washed membrane preparations. Birdsall and colleagues (Birdsall *et al.*, 1978; Birdsall and Hulme, 1983) were the first to demonstrate that muscarinic agonists interact with as many as three populations of binding sites with differing affinities. The possibility that G-protein interactions might contribute to the heterogeneity of agonist affinity states was suggested by the observation that GTP analogs cause a rightward shift of agonist/radiolabeled antagonist competition curves (Berrie *et al.*, 1979; Rosenberger *et al.*, 1979, 1980; Ehlert *et al.*, 1981). The effect of guanine nucleotides generally reflects a decrease in the proportion of binding sites with high agonist affinity with a concomitant increase in the number of lower affinity sites. β-Adrenergic receptors previously had been shown to form a ternary complex of agonist, receptor, and G_s that is not formed in the presence of antagonist, that is stable to solubilization from the membrane, and that dissociates in the presence of guanine nucleotides (Limbird and Lefkowitz, 1978; Limbird, 1981). mAChR from rat heart, prelabeled with an agonist prior to solubilization, also form a guanine nucleotide-sensitive complex larger in apparent molecular size than antagonist prelabeled or unliganded receptors (Berrie *et al.*, 1984).

Direct evidence for a G protein distinct from G_s linking inhibitory receptors to adenylyl cyclase developed rapidly from the demonstration by Ui and colleagues that pertussis toxin (islet-activating protein) catalyzes the ADP-ribosylation of a 41-kDa membrane protein and blocks receptor-mediated inhibition of adenylyl cyclase (Katada and Ui, 1981, 1982; Ui, 1984). The 41-kDa substrate of pertussis toxin was shown subsequently to copurify with a GTP-binding activity (Bokoch *et al.*, 1984; Codina *et al.*, 1983) that reconstituted receptor- and guanine nucleotide-dependent inhibition of adenylyl cyclase (Bokoch *et al.*, 1984; Katada et al., 1984a,b,c). This protein, defined functionally by its activity as an inhibitor of adenylyl cyclase, was termed G_i. The inhibitory effects of mAChR on adenylyl cyclase activity are susceptible to pertussis toxin blockade (Hazeki and Ui,

1981; Kurose et al., 1983; Hughes *et al.*, 1984), as are the effects of guanine nucleotides on agonist affinity in some (Kurose *et al.*, 1983) but not all cell types (Martin *et al.*, 1985a; Evans *et al.*, 1985b).

Activation of G_i by agonist-occupied mAChR obviously does not account for all mAChR-mediated effects. For example, comparison of two cultured cell lines has shown that muscarinic receptor stimulation inhibits the accumulation of cyclic AMP in both cell types, but by two entirely different mechanisms (Harden *et al.*, 1986). NG108–15 neuroblastoma × glioma cells appear to express a typical G_i-regulated response in which synthesis of cyclic AMP is decreased on exposure to muscarinic agonists; this response is readily demonstrable in membrane preparations and is effectively blocked by pertussis toxin (Kurose *et al.*, 1983; Hughes *et al.*, 1984). On the other hand, activation of mAChR on 1321N1 human astrocytoma cells results in a pronounced attenuation of hormone- or forskolin-stimulated cyclic AMP accumulation (Meeker and Harden, 1982, 1983), but direct inhibition of adenylyl cyclase by cholinergic agonists in membrane preparations is not observed (Meeker and Harden, 1982; Hughes *et al.*, 1984). In these cells, activation of mAChR enhances degradation rather than diminishes synthesis of cyclic AMP. That is, stimulation of mAChR in 1321N1 cells results in activation of phospholipase C which liberates inositol 1,4,5-trisphosphate (IP_3), which in turn releases calcium from intracellular stores (Masters *et al.*, 1985; Evans *et al.*, 1985a; Nakahata *et al.*, 1986). One of the cellular components activated by the elevated cytosolic Ca^{2+} is a Ca^{2+}/calmodulin (CaM)-dependent cyclic nucleotide phosphodiesterase that degrades cyclic AMP (Miot *et al.*, 1984; Tanner *et al.*, 1986).

A number of questions arise from these comparisons. For example, do two different mAChR subtypes mediate activation of phospholipase C and inhibition of adenylyl cyclase and are separate G proteins involved in coupling receptors to adenylyl cyclase and phospholipase C? 1321N1 cells express a G_i-like pertussis toxin substrate and the direct inhibition of adenylyl cyclase activity by adenosine analogs (through an A_1-adenosine receptor) and/or guanine nucleotides has been demonstrated, suggesting that the components necessary for receptor-mediated inhibition of adenylyl cyclase are functionally expressed in these cells (Hughes *et al.*, 1984; Hughes and Harden, 1986). Therefore, it appears that the mAChR of 1321N1 cells do not readily couple to G_i and adenylyl cyclase. The converse is true for NG108–15 cells. Bradykinin stimulates phosphoinositide hydrolysis and Ca^{2+} mobilization in these cells (Yano *et al.*, 1985; Hepler *et al.*, 1987); thus, the requisite components for activation of phospholipase C are functionally expressed. However, the muscarinic agonist carbachol is incapable of evoking the phosphoinositide/Ca^{2+} response (Hepler *et al.*, 1987).

This difference in coupling mechanism of the mAChR also can be observed in radioligand binding studies. NG108–15 cells express high affinity agonist binding sites that are converted to lower affinity sites in the presence of GTP and its analogs. Pertussis toxin also converts the receptor population to a guanine nucleotide-insensitive low affinity state, as would be predicted for a G_i-coupled receptor. Guanine nucleotide-sensitive high affinity binding of agonists also occurs to mAChR of 1321N1 astrocytoma cells. However, neither the affinity of agonists nor the effect of guanine nucleotides is modified by pertussis toxin or by treatment with N-ethylmaleimide, a nonspecific reagent that inhibits G_i-dependent interactions (Evans et al., 1985b; Martin et al., 1985a). These data suggest that a G protein, distinct from G_i and insensitive to pertussis toxin, is a component of the mAChR–effector system in 1321N1 cells. Consistent with this interpretation, pertussis toxin does not inhibit formation of inositol phosphates or the mobilization of intracellular Ca^{2+} in these cells (Masters et al., 1985; Nakahata et al., 1986).

Thus, these data obtained from two clonal cell lines suggest that mAChR couple to either inhibition of adenylyl cyclase or to activation of phospholipase C, but not to both. The data also suggest that two different G proteins, one sensitive and one insensitive to pertussis toxin, are involved in transducing receptor activation in the two systems. The simplicity of such a model is attractive. However, in the sections that follow we will present considerations that make it difficult to account for all the existing data with such a simple model.

II. MUSCARINIC RECEPTORS AND PHOSPHOINOSITIDE METABOLISM

The pioneering work of Hokin and Hokin (1954), demonstrated that acetylcholine (ACh), acting at mAChR in pancreatic cells, increased labeling of phosphatidylinositol in tissue that had been preincubated with $[^{32}P]P_i$. Subsequently, it has been shown that ACh, as well as other neurotransmitters and hormones, acting on their specific receptors in a great variety of target tissues, induces the activation of a membrane-bound phospholipase C that hydrolyzes phosphatidylinositol 4,5-bisphosphate (PIP$_2$), yielding the formation of the calcium mobilizing second messenger IP$_3$ and the protein kinase C activator, diacylglycerol (for reviews, see Berridge and Irvine, 1984; Abdel-Latif, 1986; Putney, 1986; Berridge, 1987).

The mechanism of receptor-mediated stimulation of phospholipase C activity is less clearly understood than that for activation or inhibition of adenylyl cyclase, but there is strong evidence that a guanine nucleotide-binding protein is involved. First, agonist binding to receptors known to activate the phosphoinositide pathway, e.g., muscarinic cholinergic (Evans *et al.*, 1985a,b), α_1-adrenergic (Boyer *et al.*, 1984), and H_1-histamine (Nakahata *et al.*, 1986), is modulated by guanine nucleotides in a pertussis toxin-insensitive manner.

Second, experiments with permeabilized cells and membrane preparations demonstrate that guanine nucleotides are required for stimulation of inositol phosphate formation and that agonist stimulation of PIP_2 hydrolysis is potentiated by guanine nucleotides (Haslam and Davidson, 1984; Cockcroft and Gomperts, 1985; Litosch *et al.*, 1985; Smith *et al.*, 1985; Uhing *et al.*, 1985; Harden *et al.*, 1987). The pharmacological and kinetic properties of the guanine nucleotide effects on the phosphoinositide response are quite similar to properties documented for the adenylyl cyclase system.

The nature of the G protein coupled to phospholipase C is not known. In several cell types, pertussis toxin blocks agonist-stimulated phospholipase C activation (Brandt *et al.*, 1985; Smith *et al.*, 1985; Okajima and Ui, 1984; Pfeilschifter and Bauer, 1986, Nakamura and Ui, 1984) and blocks a number of cellular responses that may be activated subsequent to Ca^{2+} and diacylglycerol mobilization. However, it should be emphasized that in most tissues studied, receptor-stimulated activation of phospholipase C is not affected by pertussis toxin (Masters *et al.*, 1985; Nakahata *et al.*, 1986; Burch *et al.*, 1986; Martin *et al.*, 1986; Merritt *et al.*, 1986; Hepler *et al.*, 1987). In all tissues studied to date, the stimulatory effects of mAChR agonists on inositol phosphate formation are not modified by pertussis toxin.

III. MUSCARINIC RECEPTORS AND ION CHANNELS

Considerable evidence has accumulated indicating that mAChR exert control of several ion channels in excitable and nonexcitable cells (North, 1986; Nathanson, 1987; Marty, 1987). In some excitable cells, stimulation of mAChR induces hyperpolarization or depolarization by changes in the permeability of the plasma membrane to K^+, Ca^{2+}, or Cl^- ions. In heart cells, activation of mAChR produces hyperpolarization by activating potassium channels (Giles and Noble, 1976). The same effect has been found in some sympathetic and parasympathetic neurons (Dodd and Horn, 1983; Hartzell *et al.*, 1977), whereas in neurons from the spinal cord (Nowak and MacDonald, 1983), hippocampus (Cole and Nicoll, 1984), and myenteric plexus (Morita *et al.*, 1982), mAChR activation produces depolarization by closing potassium channels. mAChR also are associated with the closing of calcium-dependent potassium channels (Cole and Nicoll, 1984; North and Tokimasa, 1983) and with increases in permeability of Cl^- (Oron et al., 1985) and Ca^{2+} ions (Bigeon and Pappano, 1980).

Receptor-mediated modulation of ion channel activity can be envisaged at different levels in the cascade of activation induced by the binding of the agonist with the receptor: (a) by formation of all or part of the ion channel by the receptor itself; (b) by the direct action of a second messenger; (c) by the action of protein kinases activated by the second messenger (protein kinase C, cAMP-, cGMP- or Ca^{2+}/CaM-dependent protein kinases); or (d) by the interaction of G proteins with ion channels in the membrane. There is no evidence suggesting that the mAChR itself is an ion channel; however, there is some evidence supporting the other possibilities mentioned above (Nathanson, 1987; Marty, 1987).

Recently, the use of techniques for ion channel recordings on "cell-attached," "whole-cell" or "excised patch" cellular membranes have conferred a new dimension in the study of receptor-mediated responses of ion channels. These patch-clamp techniques have proved particularly useful for elucidating the mechanisms underlying mAChR-modulation of K^+ channels of heart cell membranes. When patches of membranes are excised from the cell ("inside-out" excised patches), K^+-channel activity decreases markedly, but still is enhanced when ACh is present in the extracellular face of the patch (Soejima and Noma, 1984). No activity is recorded when ACh is applied in the intracellular face. Addition of GTP to the intracellular face of the patch (but not into the extracellular) reactivates the K^+-channel activity in the presence of ACh inside of the pipette (Kurachi et al., 1986). Moreover, under the same inside-out patch conditions addition of the nonhydrolyzable GTP analogs Gpp(NH)p (Breitwieser and Szabo, 1985) or GTPγS (Kurachi et al., 1986) gradually caused activation of the K^+ channel, even in the absence of ACh. These data strongly support the involvement of GTP-binding proteins in receptor-mediated K^+-channel activity. Interestingly, treatment of rats or chicks with pertussis toxin blocks mAChR-mediated decreases in heart rate (Endoh et al., 1985; Boyer et al., 1986) and hyperpolarization of atrial cells (Sorota et al., 1985). Martin et al. (1985b) demonstrated that treatment of cultured chick heart cells with pertussis toxin blocks muscarinic receptor-mediated increases in K^+ permeability. Pfaffinger et al. (1985) showed, using a whole cell voltage-clamp technique, that mAChR in atrial cells are coupled to the potassium channel through a pertussis toxin-sensitive G protein. This was the first demonstration that a GTP-binding protein couples the action of a receptor to the function of an ion channel without the participation of second messengers. The effect of pertussis toxin was confirmed by Kurachi et al. (1986) using excised inside out patches; treatment of the patches with the active subunit of pertussis toxin (A protomer) gradually eliminated mAChR- and P_1-purinergic receptor-induced K^+-channel activity.

The nature of the G protein involved in mAChR modulation of K^+-channel activity is currently an area of considerable controversy. Yatani *et al.* (1987) and Codina *et al.* (1987), using inside-out patches from guinea pig heart cells, demonstrated that reconstitution of activated α subunit of a G protein purified from human erythrocytes (named G_k) stimulated, in a dose- and time-dependent manner, the opening of K^+ channels. This 40K protein, a pertussis toxin substate, is able to reverse the inhibition of K^+ channel opening produced by treatment of the cells with pertussis toxin. These authors also reported (Codina *et al.*, 1987) that $\beta\gamma$ subunits purified from human erythrocytes were without effect. In contrast, independent studies by Logothetis *et al.* (1987) suggested that the $\beta\gamma$ and not the α subunits are responsible for activating the muscarinic-gated potassium channels. (See Chapter 12 by Codina *et al.* in this volume.)

Other types of muscarinic receptor-regulated ion channels, besides the atrial K^+ channel, may involve a G protein in the activation pathway. For example, there are several different types of Ca^{2+}-dependent conductances that are under the influence of mAChR, including K^+, Cl^-, and monovalent cation-selective channels (Marty, 1987). It is plausible to suggest that the increase in the intracellular calcium ion concentration that modulates these channel activities may be induced by mAChR that activate the phospholipase C-IP_3 pathway via a G protein (see Oron *et al.*, 1985 for example). Likewise, G-protein-linked mAChR may be initial components of pathways that modulate channel activities via second messenger-activated protein kinases.

IV. SUBTYPES OF MUSCARINIC RECEPTORS

Pharmacological data suggesting the existence of multiple subtypes of mAChR are in abundance, but definitive characterization of these subtypes in terms of pharmacological selectivity of biochemical or cellular responses, tissue localization, or mechanism of action has been difficult to achieve. The initial evidence for mAChR heterogeneity was the observation that a series of cholinergic agonists had different orders of potency for causing contraction of smooth muscle versus stimulation of K^+ secretion in the same tissue (Burgen and Spero, 1968). Goyal and Rattan (1978) subsequently proposed that mAChR could be classified as M_1 or M_2 based on the relative activities of McN-A343 and bethanechol in the lower esophageal sphincter of the opossum.

Although results from subsequent radioligand binding studies with agonists were also consistent with the existence of multiple subtypes of mAChR, the contribution of G-protein interactions to the binding heterogeneity has been difficult to assess completely. The use of antagonists has allowed more confident subclassification of mAChR subtypes. Hammer *et al.* (1980) showed that the antagonist pirenzepine inhibited the binding of a nonselective radiolabeled antagonist, [^{3}H] quinuclidinyl benzilate, over a very broad concentration range, suggesting that pirenzepine bound to more than one receptor site. Muscarinic receptors were subsequently classified into those with a high affinity ($\sim$10 nM) and those with a low affinity ($>$100 nM) for pirenzepine, and the original M_1/M_2 classification was adapted with the suggestion that M_1 receptors bound pirenzepine with high affinity and were selectively activated by low concentrations of McN-A343 (Hammer and Giachetti, 1982).

Not all data with pirenzepine were suggestive of drug interaction with two separate mAChR subtypes. However, the introduction of new selective compounds has quieted some of the fear concerning mAChR subclassification based primarily on the use of a single antagonist. As early as 1976, Barlow and co-workers reported that the drug 4-DAMP appeared to be more potent in blocking mAChR on smooth muscle than in cardiac tissue (Barlow *et al.*, 1976). These studies were later confirmed (Barlow *et al.*, 1980), and suggested heterogeneity between peripheral types of mAChR. Peripheral mAChR (except for the ganglionic receptors) were generally considered to be M_2 based on pirenzepine affinities; therefore the work of Barlow and associates suggested that M_2 receptors could be subdivided further (Barlow *et al.*, 1980). Also, the drugs hexahydrosiladifenadol (Mutschler and Lambrecht, 1984), himbascine (Anwar-ul *et al.*, 1986), and AF-DX116 (Hammer *et al.*, 1986) displayed cardioselectivity in antagonizing mAChR-mediated responses, and Nilvebrant and Sparf (1983) reported that several drugs could distinguish between the mAChR of smooth muscle and those in the parotid gland.

Based on these observations new mAChR classification schemes have been suggested (Waelbroeck *et al.*, 1986; Doods *et al.*, 1987). These new attempts at subclassification of mAChR based on functional assays suggest that at least three and possibly four subtypes of mAChR exist. There are currently five known genes encoding different mAChR proteins (see below). A successful classification scheme will incorporate data from molecular cloning with the pharmacology of the receptors using all of the available selective drugs, e.g. see Peralta *et al.*, 1987b.

Attempts have been made to determine whether a particular mAChR subtype selectively mediates a particular biochemical response. Thus, Gil and Wolfe (1985) found that pirenzepine was more potent in antagonizing phosphoinositide hydrolysis than in antagonizing the inhibition of adenylyl cyclase activity using rat cerebral cortex, heart, and parotid as model tissues. However, Lazareno *et al.* (1985) and Fisher (1986) found that receptors possessing both high and low affinities for pirenzepine could initiate the hydrolysis of phosphoinositides. Furthermore, Brown *et al.* (1985) demonstrated that in chick heart cells, pirenzepine was more potent in antagonizing the effects of mAChR stimulation on adenylyl cyclase activity than in blocking the phosphoinositide response. There is currently no consensus concerning the selective role(s) played by particular subtypes of mAChR in the activation of second messenger systems. Reexamination of the antagonism of mAChR-regulated second messenger responses using selective drugs in addition to pirenzepine might provide a clearer picture of the relationship between putative receptor subtypes and the various biochemical signals they control.

V. MOLECULAR CLONING AND STRUCTURE OF mAChR

A. Molecular Cloning

Until recently, little structural data were available for mAChR. mAChR have been covalently labeled with [³H]propylbenzylcholine mustard [³H]PBCM), solubilized from membranes, and visualized by autoradiography after separation by sodium dodecyl sulfate-polyacrylamide gel electrophoresis (SDS-PAGE). They migrate as diffuse bands of radioactivity ranging in size from 70,000–97,000 daltons depending on the tissue of origin (see Kerlavage *et al.*, 1987; Hootman *et al.*, 1985). Using this approach some delineation of receptor heterogeneity has been made. For example, [³H]PBCM-labeled mAChR from 1321N1 human astrocytoma cells is 20,000–25,000 daltons larger than the [³H]PBCM-labeled species from NG108–15 neuroblastoma × glioma cells (Liang *et al.*, 1987). This gives further credence to the notion that two different subtypes of receptors selectively coupling to two different second messenger systems are expressed by the two cell lines.

The pivotal achievement in development of a successful approach for purification of mAChR was made by Haga and Haga (1983) who synthesized a mAChR antagonist, 2-aminobenzyhydroxyltropane (ABT), suitable for affinity chromatography. ABT-affinity chromatography allowed Haga and Haga (1983, 1985) and Petersen *et al.* (1984) to purify to apparent homogeneity the mAChR from porcine cerebral cortex and porcine heart, respectively. The availability of purified receptor made possible the cloning of the mAChR genes.

Several distinct genes encoding mAChR have been isolated by molecular cloning techniques. Following the nomenclature suggested by Bonner *et al.*

(1987), the four initial receptor subtypes encoded by four genes are called the m_1 through m_4 subtypes. Based on partial amino sequences derived from tryptic fragments of the mAChR purified from porcine cerebral cortex (Haga and Haga, 1985), Numa's group was the first to clone, sequence, and express complementary DNA (cDNA) encoding an mAChR (m_1). The cDNA isolated by this group corresponds to a protein composed of 460 amino acids with a calculated M_r of 51,416 (Kubo et al., 1986a). Some peptide fragments from the same receptor preparation (Kubo et al., 1986a) were not encoded by the isolated m_1 cDNA and provided the basis for isolating from a porcine atrial cDNA library a second receptor species, the m_2 subtype, composed of 466 amino acids and having a calculated M_r of 51,670 (Kubo et al., 1986b). At the same time Peralta et al. (1987a) cloned, sequenced, and expressed the m_2 subtype starting from partial amino acid sequences derived from purified porcine atrial mAChR (Peterson et al., 1984).

The rat and human homologs of the porcine m_1 clone, as well as the human homologs of the porcine m_2 clone, were subsequently isolated from rat cerebral cortex cDNA and human genomic libraries utilizing probes corresponding to a highly conserved domain of the putative second transmembrane sequence (Bonner et al., 1987). The cloning strategy used by these investigators allowed them to detect and isolate two additional cDNA sequences encoding what are apparently two unique mAChR, the m_3 and m_4 subtypes (Bonner et al., 1987). The m_3 cDNA codes for a much larger receptor protein of 589 amino acids; ($M_r \sim 66,100$) than the other cDNA clones; the m_4 subtype is only slightly larger than the m_1 and m_2 proteins (478 amino acids; $M_r \sim 53,000$). The human homologs of all four of the receptor subtype cDNAs recently have been isolated and characterized in detail by Peralta et al. (1987b).

These cDNA sequences encode functional muscarinic receptors as demonstrated by expression of the genes in cellular systems devoid of endogenous mAChR (Kubo et al., 1986a,b; Peralta, 1987 a,b; Fukuda et al., 1987; Brann et al., 1987; Braun et al., 1987; Bonner et al., 1987; Ashkenazi et al., 1987). The criterion for successful expression has been based on detection of a specific radiolabeled muscarinic antagonist binding site with the appropriate pharmacological specificity. The high (Kubo et al., 1986a) or low (Peralta et al., 1987a; Fukuda et al., 1987) affinity of pirenzepine has been used to support the assignment of the $M_1 = m_1$ and $M_2 = m_2$ subtype classifications to the cloned receptors. The m_3 and m_4 subtypes displayed apparent affinities for pirenzepine closer to that of M_1 than M_2 (Bonner et al., 1987; Peralta et al., 1987b) and in previous radioligand binding experiments with mammalian tissues probably would have been classified as M_1-type receptors. In addition, the antagonist AF-DX116, which was reported to be relatively selective for cardiac mAChR (Hammer et al., 1986), binds with higher affinity to the m_2 subtype than to the m_1 (Fukuda et al., 1987; Brann et al., 1987; Peralta et al., 1987b) or m_3 and m_4 subtypes (Peralta et al., 1987b).

Cloned mAChR incorporated into the membranes of "foreign" cells retain their capacity to interact with endogenous G proteins. Brann et al. (1987) transfected murine A9L cells with cloned porcine brain (m_1) cDNA and showed that "heterogeneous" agonist binding sites, sensitive to modulation by guanine nucleotides, were expressed. Similarly, Chinese hamster ovary (CHO) cells transfected with the porcine m_2 cDNA expressed high and low affinity agonist binding sites that were modulated by guanine nucleotides (Ashkenazi et al., 1987); pertussis toxin treatment caused conversion of high affinity sites to low affinity sites. Peralta et al. (1987b) expressed each recombinant mAChR subtype (m_1-m_4) in human embryonic kidney cells, and showed that each receptor displayed multiple affinity states for agonists. The proportion of high and low affinity binding sites varied, leading to the suggestion that each receptor expresses variable efficiency of coupling with the endogenous G proteins of these cells.

It has become evident that a number of different receptors have common structural features and predicted topological organization in the lipid bilayer of the membrane (Applebury and Hargrave, 1986; Dohlman et al., 1987). The receptor proteins of this class are similar in size and all contain seven stretches of 20–28 hydrophobic amino acids. Based primarily on comparisons with the well-characterized structure of bacteriorhodopsin and mammalian rhodopsin and on hydropathy profiles, these seven hydrophobic regions are believed to form α-helical membrane-spanning domains (see Fig. 1). The amino- and carboxyl-terminal segments are composed of predominantly hydrophilic amino acid residues, as are the segments between the transmembrane domains. These hydrophilic domains are proposed to extend as loops into the extracellular (o_1, o_2, o_3) and cytoplasmic (i_1, i_2, i_3) spaces. The mAChR subtypes (m_1-m_4) share this type of structural organization with β-adrenergic receptors (Dixon et al., 1986; Yarden et al., 1986; Dohlman et al., 1987; Gocayne et al., 1987), α_2-adrenergic receptors (Kobilka et al., 1987a), bovine rhodopsin (Nathans and Hogness, 1983; 1984; Applebury and Hargrave, 1986), other visual opsins (Nathans et al., 1986), and less well-defined receptors or receptor-like proteins, such as the substance K receptor (Masu et al., 1987), the mas oncogene, (Young et al., 1986), the G-21 protein (Kobilka et al., 1987b), and yeast a and α mating factor receptors (Herskowitz and Marsh, 1987). The overall sequence homology between members of this class of proteins is relatively low, but if only the putative transmembrane regions are considered, there is a high degree of localized sequence homology. Other features also suggest a conserved general pattern of structure. For example,

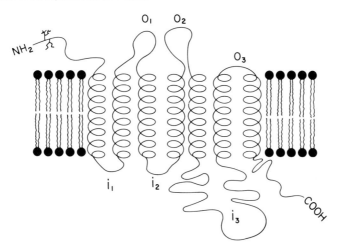

Fig. 1. Seven hydrophobic regions forming α-helical membrane-spanning domains (see text).

the proteins have potential glycosylation sites at the amino terminus, no signal sequence, and serine- and threonine-rich sequences at or near the carboxyl terminus that could serve as potential phosphorylation sites for regulatory kinases. Cholinergic agonist-stimulated phosphorylation of mAChR from chick heart occurs *in vivo* (Kwatra and Hosey, 1986) and after reconstitution of purified porcine heart mAChR and G_i in liposomes (Rosenbaum *et al.*, 1987). Of the kinases tested *in vitro*, only cyclic AMP-dependent protein kinases stimulated phosphorylation of purified heart mAChR (Rosenbaum *et al.*, 1987). However, there are dramatic effects of phorbol esters on mAChR-regulated phosphoinositide turnover (Orellana *et al.*, 1985) and on receptor down-regulation in some systems (Liles *et al.*, 1986). It is not known if these events, presumably mediated by protein kinase C, are due to phosphorylation of receptors or some other component(s), e.g., G proteins, of the system.

The mAChR subtypes are highly conserved between species (Bonner *et al.*, 1987; Peralta *et al.*, 1987b) and have a high degree of homology (65%) within the membrane-spanning segments and short connecting loops (Peralta *et al.*, 1987b). In addition, two negatively charged aspartic acid residues located in the otherwise highly hydrophobic transmembrane domains II and III are conserved among all four subtypes. It has been suggested that the putative transmembrane helices and their corresponding extracellular loops may form a binding pocket for acetylcholine in which the aspartic acid residues participate directly (Peralta *et al.*, 1987a,b). Relative fidelity in the structure of the binding site for acetylcholine seems likely. The regions where the members of this superfamily of receptors display the greatest variability in size and weakest sequence homology are the amino- and carboxyl-terminal segments and the third cytoplasmic loop (i_3) between the putative transmembrane segments V – VI. The four mAChR subtypes diverge strikingly from each other and from the other receptors of similar structure in this i_3 loop. This region varies considerably in length between the subtypes and differs enough in sequence to be subtype-specific (Bonner *et al.*, 1987; Peralta *et al.*, 1987b). In fact, probes derived from the subtype-specific regions of the i_3 loop hybridize with discrete species of mRNA in different tissues and cell types and Southern blots of genomic DNA utilizing these probes suggest that these domains are unique to mAChR (Peralta *et al.*, 1987b).

Now that primary structural data are available for different subtypes of mAChR (Bonner *et al.*, 1987; Peralta *et al.*, 1987; Kerlavage *et al.*, 1987), as well as for many of the known G proteins (Stryer and Bourne, 1986), it is anticipated that progress will be made in identifying the domains of interaction between the two sets of proteins and in determining the relationship of structural determinants to functional specificity. For example, recent work has suggested the carboxyl-terminal segment of G_s as critically important for interaction with β-adrenergic receptor (βAR) (Sullivan *et al.*, 1987). Primary structural data predict that the carboxyl-terminal domains of other G proteins will play similar receptor-coupling roles (Stryer and Bourne, 1986; Masters *et al.*, 1986). The G-protein α subunits examined so far (G_t, G_i, G_s, G_o) have homologous regions that probably represent the GTP-binding domains and possibly domains that interact with the $\beta\gamma$ subunits (Stryer and Bourne, 1986). Highly variable sequences are also evident, and are candidates for effector binding sites.

On the receptor side, some information on the domains responsible for interaction with G proteins is available for rhodopsin and βAR. Proteolytic removal of the carboxyl terminus and part of the i_3 loop of rhodopsin destroys its capacity to couple to transducin (Kuhn and Hargrave, 1981; Kuhn, 1984). The large cytoplasmic loop of the βAR is implicated in effector coupling since mutants with deletions in this portion of the molecule lose capacity to regulate adenylyl cyclase activity (Dixon *et al.*, 1987). The unique character of the large cytoplasmic loop of each mAChR subtype together with a comparison to the analogous region of βAR, rhodopsin, and α_2-receptors suggests that this region may play an important role in G-protein coupling. However, the receptors should show some similarity in structure in the domain that interacts with G proteins. For example, the α_2-adrenergic receptor and M_2 (=m_2) receptors both inhibit adenylyl cyclase, presumably through interactions with the same pertussis toxin-sensitive G_i, and should therefore contain homologous G_i-binding domains.

There are data from reconstitution studies with purified G proteins and receptors that suggest that receptor–G-protein interactions may not be entirely specific. Thus, activated rhodopsin will activate the GTPase activity of transducin as well as G_i and G_o (Cerione et al., 1985; Kanaho et al., 1984; Fung, 1983; Tsai et al., 1984), but not G_s. Similarly, βAR reconstituted into phospholipid vesicles with purified G_i can stimulate the binding of GTPγS and GTPase activity of G_s and G_i to the same extent (Asano et al., 1984). Data from cyc^- S49 lymphoma cells (Abramson and Molinoff, 1985) and adipocytes (Murayama and Ui, 1983) also strongly suggest that βAR can interact with G_i. In addition, α_2-adrenergic receptors only minimally affect the GTPase activity of G_s, but partially activate transducin GTPase and fully activate both G_i and G_o GTPase activities (Cerione et al., 1986).

Sternweis and Robishaw (1984) purified a heterotrimeric ($\alpha\beta\gamma$) GTP-binding pertussis toxin substrate from bovine brain (see also Neer et al., 1984) and later showed that this protein, as well as G_i, reconstituted GTP-sensitive, high-affinity agonist binding to mAChR (Florio and Sternweis, 1985). These results have been extended, using purified mAChR preparations from porcine brain (Haga et al., 1985, 1986; Kurose et al., 1986) and porcine heart (Tota et al., 1987), to show that both of these "pure" mAChR stimulate [^{35}S]GTPγS binding and GTPase activities of G_i and G_o. These studies are important because they demonstrate direct interactions between "purified" mAChR and "purified" G proteins in a defined in vitro system, and confirm mechanistic aspects of the interactions; i.e., agonists increase the affinity for GTP, decrease the affinity for GDP, stimulate GTPase V_{max}. The actual purity of G-protein and mAChR preparations must be qualified, however, based on evidence from molecular cloning experiments. Thus, the so-called M_1 receptor preparation from cerebral cortex could clearly contain m_1, m_3, and m_4 subtypes (Bonner et al., 1987; Peralta et al., 1987b). The $M_2 = m_2$ classification for the heart mAChR is perhaps less tenuous, since heart mRNA hybridizes only with M_2-specific probes (Peralta et al., 1987b). Furthermore, one can clearly ask: what is G_i? (Gilman, 1987). Here too, cloning data strongly support at least three distinct α_i-like gene products (Jones and Reed, 1987; Nakada et al., 1986; Itoh et al., 1986; Graziano and Gilman, 1987). At this point then, we are forced to conclude that little unequivocal data is available concerning interactions between specific mAChR subtypes ($m_1 - m_4$) and specific G-protein α subunits (α_{i1}, α_{i2}, α_{i3}, α_o).

Recently, Ashkenazi et al. (1987) have shown that the recombinant M_2 muscarinic receptor subtype expressed in CHO cells can evoke both adenylyl cyclase inhibition and phospholipase C activation. After mAChR expression in these cells, activation of phospholipase C by carbachol was significantly less efficient than the inhibition of adenylyl cyclase, and required the expression of at least 1.45×10^6 receptors/cell in order to reach a maximal inositol phosphate response. Both responses were inhibited after pertussis toxin treatment of the cells, the inhibition of adenylyl cyclase being about 10-fold more sensitive to pertussis toxin treatment than was activation of phospholipase C. Atropine and pirenzepine exhibited approximately the same affinities for antagonizing the effects of carbachol on inhibition of adenylyl cyclase and activation of phospholipase C. These data suggest that a single receptor subtype (M_2) can be coupled to more than one effector system. The differential coupling efficiency and pertussis toxin sensitivity of the two responses are consistent with the idea that different G proteins are involved in the two mechanisms. Altogether sanguine conclusions cannot be made from this work, however. For example, the appropriateness of the high level of receptor expression (>100 times that of most target cells) required to couple mAChR to phospholipase C is questionable.

As yet, it is not clear which mAChR subtypes are capable of affecting ion channel activity or which of the direct (receptor, G protein) or indirect (second messengers, kinases) mechanisms apply to particular ion conductance changes. Expression of recombinant mAChR proteins provides an approach to answer some of these questions. For example, Fukuda *et al.* (1987) recorded ACh-activated currents in *Xenopus* oocytes injected with mRNA coding for either the porcine m_1 or m_2 subtypes and showed that the responses evoked by the two receptors were clearly different. ACh applied to cells expressing the m_1 subtype stimulated an inward current of large amplitude that was oscillatory in nature, and apparently carried by Cl^- ions. This m_1-mediated response had a relatively long latency (3 sec) and could be abolished completely by intracellular injection of EGTA, suggesting that this Cl^- channel may become activated secondary to mobilization of Ca^{2+} (Kubo *et al.*, 1986a; Fukuda *et al.*, 1987). Cells expressing m_2 receptors responded to ACh with a much (560-fold) lower amplitude inward current that appeared to consist of two components: a smooth current, unaffected by EGTA and attributable to Na^+ and K^+ ion fluxes, and a small oscillatory current, blocked by EGTA.

Brann *et al.* (1987) also have measured electrophysiological responses after transfecting cells (mouse A9L cells) with the rat homolog of the porcine m_1 cDNA. These cells also responded to ACh with a Cl^- conductance after a long (3 sec) delay. Though no further data was obtained as to the mechanism underlying this ACh-induced conductance change, these workers have shown that G_s, G_{i1}, G_{i2}, but not G_o or G_{i3}, are expressed by these cells (Brann *et al.*, 1987). As these authors suggest, perhaps cotransfection of receptors and G proteins in defined cell systems will provide a means of definitively assigning particular receptor subtypes to G-protein subtypes and channel and enzyme effectors.

B. Structural Features

A total of five muscarinic receptors have now been identified by molecular cloning techniques followed by expression in tissue culture cells. As discussed above, they were cloned independently in several laboratories and have unfortunately received different names: mACHR I, II, and III in Japan; m1, m2, m3, m4, and m5 at the National Institutes of Health and Israel; and M1, M2, *M*3, *M*4, and M5 at Genentech, where numbers one, two and five are coherent, but III = m3 (or M3) = *M*4 and m4 (or M4) = *M*3. The equivalence as well as references to the cloning and expression of these receptors can be found in Tables I and II. Southern analysis of rat DNA carried out by Bonner *et al.* (1987) indicates that additional muscarinic receptor forms are possible.

Hydropathy analysis of deduced amino-acid sequences of muscarinic receptors shows that they are fit well by models of molecules that, like opsins and adrenergic receptors, span the plasma membrane seven times. An amino-acid sequence comparison among the five mAChRs (Figs. 2 and 3) shows them to be approximately 40% identical, with areas of much higher sequence identity, especially the proposed membrane-spanning regions where up to 90% sequence identity can be found, and other areas of almost no sequence identity, best exemplified by the amino termini, the carboxy termini and, most of all, the third intracellular loop, all of which vary not only in composition but also in length. The existence of areas of

Table I   Cloning and Naming of Muscarinic Receptors

Research group	Name of molecule					Reference
Kyoto	mAChR I					Kubo *et al.*, 1986a
		mACR II				Kubo *et al.*, 1986b
			mAChR III			Akiba *et al.*, 1988
NIH	ml	m2	m3	m4		Bonner *et al.*, 1987
					m5	Bonner *et al.*, 1988
Genentech		M2				Peralta *et al.*, 1987a
	M1	M2	*M*4	*M*3		Peralta *et al.*, 1987b
Houston					M5	Liao *et al.*, 1989

Table II Expression and Functional Effects of Cloned Muscarinic Receptors

AChR receptor	Recipient cell[a]	Intact cells				Cell-free	Reference
		PI hydrolysis	cAMP levels	Arachidonic acid release	Ca^{2+}-dependent ion channels	adenylyl cyclase	
M1	hEK	Stimulation	Increase	—	—	—	Peralta et al., 1988
	Xenopus oocyte	—	—	—	Stimulation	—	Fukuda et al., 1987
	A9 L	Stimulation	Increase	Stimulation	—	—	Conklin et al., 1988
	A9 L	—	—	—	Stimulation	—	Brann et al., 1987
	A9 L	—	—	—	Stimulation	—	Jones et al., 1988
	NG108-15	Stimulation	—	—	Stimulation	—	Fukuda et al., 1988
	RAT-1	Stimulation	Decrease	—	—	—	Stein et al., 1988
	CHO	Weak stim.	Decrease	—	—	—	Ashkenazi et al., 1987
M2	Xenopus oocyte	—	—	—	No effect	—	Fukuda et al., 1987
	hEK	Weak stim.	Decrease	—	No effect	—	Peralta et al., 1988
	A9 L	—	—	No effect	—	—	Conklin et al., 1988
	A9 L	—	—	—	No effect	—	Jones et al., 1988
	NG108-15	No effect	—	—	Marginal stim.	—	Fukuda et al., 1988

Receptor	Cell line						Reference
M3 (*M4*)	hEK	Stimulation	Increase	—	—	—	Peralta et al., 1988
	A9 L	—	—	Stimulation	—	—	Conklin et al., 1988
	A9 L	Stimulation	No effect	—	Stimulation	—	Jones et al., 1988
	RAT-1	—	—	—	—	—	Pinkas-Kramarski et al., 1988[b]
	NG108-15	Stimulation	—	—	Stimulation	—	Fukuda et al., 1988
M4 (*M3*)	hEK	Weak stim.	Decrease	—	—	—	Peralta et al., 1988
	A9 L	—	—	No effect	—	—	Conklin et al., 1988
	A9 L	—	—	—	No effect	—	Jones et al., 1988
	NG108-15	No effect	—	—	No effect	—	Fukuda et al., 1988
M5	CHO	Stimulation	—	—	—	—	Bonner et al., 1988
	L(*tk*⁻)	Stimulation	—	Stimulation	—	No effect	Liao et al., 1989[c]

[a] hEK cells, human embryonic kidney cells.
[b] Refer to m3 receptor as M3.
[c] Refer to m receptors as M receptors.

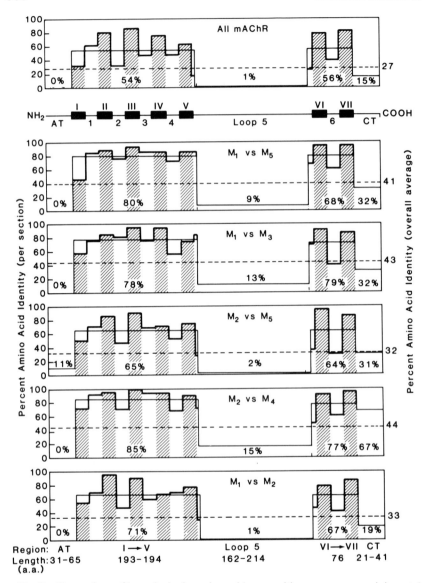

Fig. 2. Comparison of homologies in amino-acid composition among muscarinic acetylcholine receptors in different segments of the molecules. See text for discussion of nomenclature used.

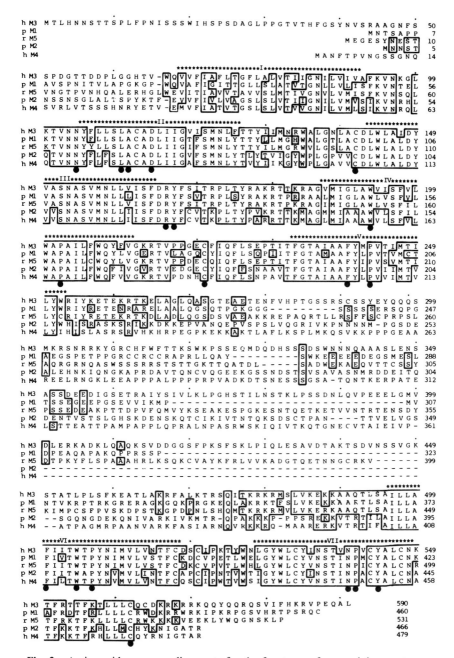

Fig. 3. Amino-acid sequence alignments for the five types of muscarinic receptors as deduced from molecular cloning. Prefixes: h, human; p, porcine; r, rat. Boxes denote amino acid identities. ●, amino acids present in eighteen G-protein-coupled receptors (see text and Fig. 5).

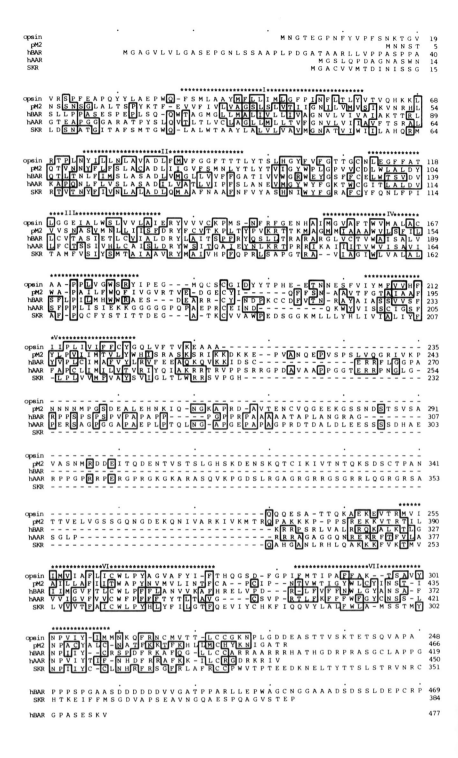

high sequence identity would make it easy to recognize whether or not a newly cloned receptor is of the muscarinic type. For example, after having cloned the cDNA for the M5 receptor and having aligned its predicted amino acid composition with adrenergic, muscarinic, light, and substance-K receptors, we had little doubt that the newly identified molecule would bind muscarinic ligands, as can be seen from comparing Fig. 3 and 4.

At this time 18 separate members of the superfamily of G protein-coupled receptors have been cloned: five muscarinic receptors (Table I), one avian (Yarden *et al.*, 1986) and two mammalian β-adrenergic receptors (Dixon *et al.*, 1986; Frielle *et al.*, 1987), one α_1- and two α_2-adrenergic receptors (Kobilka *et al.*, 1987a; Regan *et al.*, 1988), four mammalian opsins (Nathans and Hogness, 1983; Nathans *et al.*, 1986), one neurokinin (substance K) receptor (Masu *et al.*, 1987) and two serotonin receptors (Julius *et al.*, 1988; Kobilka *et al.*, 1987b, Fargin *et al.*, 1988). As illustrated by Fig. 4, the comparison of their amino acid compositions shows a remarkable variability within the same general architecture predicted by analysis of hydropathy. Thus, while there are ca. 40% identical amino acids among the muscarinic receptors, opsins and adrenergic receptors, there is only a 4% identity when the different families are compared. Interestingly, only 19 amino acids are conserved out of an average of 475 (heavy dots on Fig. 3; summarized in Fig. 5). These include three cysteines, of which one is located in the carboxy terminus and the other two in extracellular loops 1 and 2, respectively, forming a potential extracellular disulfide bridge (Karnik *et al.*, 1988), an asparagine that initiates the second transmembrane region, and a sequence Leu – Ala – X – X – Asp in this same transmembrane region, of which the aspartic acid has been shown to be crucial for ligand binding to the β-adrenergic receptor (Strader *et al.*, 1987a,b; Fraser *et al.*; 1988; Chung *et al.*, 1988). Of the remainder of the identities that are conserved in all of the cloned G-protein-coupled receptors, perhaps the most interesting are four prolines, each of which can be predicted to impart a 30° bend to the putative transmembrane regions IV, V, VI and VII. Except for the initiator methionine, there are no conserved amino-acid identities in the amino terminus or throughout the first transmembrane region and essentially all of the first intracellular loop. In fact, just like the third intracellular loop, which can be as short as 27 amino acids (e.g., bovine rhodopsin) or as long as 282 amino acids (e.g., human M4 receptor), the lengths of both the amino- and the carboxy-terminal regions vary

---

Fig. 4. Amino-acid sequence alignments for bovine opsin, porcine M2 muscarinic acetylcholine receptor, human β_1-adrenergic receptor, human α_{2a}-adrenergic receptor, and substance-K receptor.

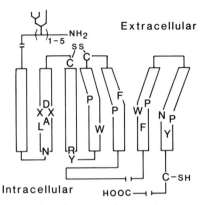

Fig. 5. Schematic representation of transmembrane regions, connecting extracellular and intracellular loops, and amino-terminal and carboxy-terminal segments of G-protein-coupled receptors. Amino acids identical in eighteen cloned G-protein receptors are highlighted as is the general location of 1–5 glycosylation sites on the amino-terminal end of the molecule.

widely (shortest amino terminus: porcine M2 with 22 amino acids, longest amino terminus: human M4 receptor with 70 amino acids; shortest carboxy terminus: human M3 receptor with 22 amino acids; longest carboxy terminus: turkey β-adrenergic receptor with 148 amino acids). Between one (human β-adrenergic receptor) and five (human M4 receptor) glycosylation consensus sequences (Asn–X–Thr/Ser) are found in the amino termini of these receptors.

C. Signal Transduction Systems Used By Muscarinic Receptors

Functional effects of muscarinic receptors have been studied after expression of the individual molecules in a variety of tissue culture cells (Table II). Although this field is relatively new, there are already reports that conflict in their conclusions. It is clear that answers depend on the cell line used to express the receptor, and that one can look forward to new insights into how receptors regulate differing signal transduction pathways.

ACKNOWLEDGMENTS

Work in the authors' laboratory was supported by National Institutes of Health Grants GM 29536, GM 38213 and NS 23019; José Luis Boyer is on leave from the

Instituto Nacional de Cardiologia "Ignacio Chavez," Mexico and is supported by a Fogarty International Fellowship Grant (TWO3737).

REFERENCES

Abdel-Latif, A. A. (1986). Calcium-mobilizing receptors, polyphosphoinositides, and the generation of second messengers. *Pharmacol. Rev.* **38**, 227–272.

Abramson, S., and Molinoff, P. (1985). Properties of β-adrenergic receptors of cultured mammalian cells: Interactions of receptors with a guanine nucleotide-binding protein in membranes prepared from L6 myoblasts and from wild-type and cyc⁻ S49 lymphoma cells. *J. Biol. Chem.* **260**, 14580–14588.

Akiba, I., Kubo, T., Maeda, A., Bujo, H., Nakai, J., Mishina, M., and Numa, S. (1988). Primary structure of porcine muscarinic acetylcholine receptor III and antagonist binding studies. *FEBS Lett.* **235**, 257–261.

Anwar-ul, S., Gilani, H., and Cobbin, L. B. (1986). The cardioselectivity of himbascine: A muscarinic receptor antagonist. *Naunyn Schmiedebergs Arch. Pharmacol.* **332**, 16–20.

Applebury, M. L., and Hargrave, P.A. (1986). Molecular biology of the visual pigments. *Vision Res.* **26**, 1881–1895.

Asano, T., Katada, T., Gilman, A. G., and Ross, E. M. (1984). Activation of the inhibitory GTP-binding protein of adenylate cyclase, G_i, by β-adrenergic receptors in reconstituted phospholipid vesicles. *J. Biol. Chem.* **259**, 9351–9354.

Ashkenazi, A., Winslow, J. W., Peralta, E. G., Peterson, G. L., Schimerlik, M. I., Capon, D. J., and Ramachandran, J. (1987). An M_2 muscarinic receptor subtype coupled to both adenylyl cyclase and phosphoinositide turnover. *Science* **238**, 672–675.

Barlow, R. B., Berry, K. J., Glenton, P. A. M., Nikolaou, N. M., and Soh, K. S. (1976). A comparison of affinity constants for muscarinic-sensitive acetylcholine receptors in guinea-pig atrial pacemaker cells at 29°C and ileum of 29°C and 37°C. *Br. J. Pharmacol.* **58**, 613–620.

Barlow, R. B., Burston, K. N., and Vis, A. (1980). Three types of muscarinic receptors? *Br. J. Pharmac.* **68**, 141–142P.

Berridge, M. J. (1987). Inositol trisphosphate and diacylglycerol: Two interacting second messengers. *Annu. Rev. Biochem.* **56**, 159–193.

Berridge, M. J., and Irvine, R. F. (1984). Inositol trisphosphate, a novel second messenger in cellular signal transduction. *Nature* **312**, 315–321.

Berrie, C. P., Birdsall, N. J. M., Burgen, A. S. V., and Hulme, E. C. (1979). Guanine nucleotides modulate muscarinic receptor binding in the heart. *Biochem. Biophys. Res. Commun.* **87**, 1000–1005.

Berrie, C. P., Birdsall, N. J. M., Hulme, E. C., Keen, M., and Stockton, J. M. (1984). Solubilization and characterization of guanine nucleotide-sensitive muscarinic agonist binding sites from rat myocardium. *Br. J. Pharmacol.* **82**, 853–861.

Bigeon, R. L., and Pappano, A. J. (1980). Dual mechanism for inhibition of calcium-dependent action potentials by acetylcholine in avian ventricular muscle: Relationship to cyclic AMP. *Circ. Res.* **46**, 353–362.

Birdsall, N. J. M., and Hulme, E. C. (1983). Muscarinic receptor subclasses. *Trends Pharmacol Sci.* **4**, 459–463.

Birdsall, N. J. M., Burgen, A. S. V., and Hulme, E. C. (1978). The binding of agonists to brain muscarinic receptors. *Mol. Pharmacol.* **14**, 723–736.

Bokoch, G. M., Katada, T., Northup, J. K., Ui, M., Gilman, A. G. (1984). Purification and properties of the inhibitory guanine nucleotide-binding regulatory component of adenylate cyclase. *J. Biol. Chem.* **259**, 3560–3567.

Bonner, T. I., Buckley, N. J., Young, A. C., and Brann, M. R. (1987). Identification of a family of muscarinic acetylcholine receptor genes. *Science* **237**, 527–532.

Bonner, T. I., Young, A. C., Brann, M. R., and Buckley, N. J. (1988). Cloning and expression of the human and rat m5 muscarinic acetylcholine receptor genes. *Neuron* **1**, 403–410.

Boyer, J. L., Garcia, A., Posadas, C., and Garcia-Sainz, J. A. (1984). Differential effect of pertussis toxin on the affinity state for agonists of renal α_1 and α_2-adrenoceptors. *J. Biol. Chem.* **259**, 8076–8079.

Boyer, J. L., Martinez-Carcamo, M., Monroy-Sanchez, J. A., Juarez-Ayala, J., Pastelin, G., Posadas, C., and Garcia-Sainz, J. A. (1986). Effect of pertussis toxin on the heart muscarinic cholinergic receptors and their function. *Life Sci.* **39**, 603–610.

Brandt, S. J., Dougherty, R. W., Lapetina, E. G., and Niedel, J. E. (1985). Pertussis toxin inhibits chemotactic peptide-stimulated generation of inositol phosphates and lysosomal enzyme secretion in human leukemic (HL-60) cells. *Proc. Natl. Acad. Sci. U.S.A.* **82**, 3277–3280.

Brann, M. R., Buckley, N. J., Jones, S. V. P., and Bonner, T. I. (1987). Expression of a cloned muscarinic receptor in A9 L cells. *Mol. Pharmacol.* **32**, 450–455.

Braun, T., Schofield, P. R., Shivers, B. D., Pritchett, D. B., and Seeburg, P. H. (1987). A novel subtype of muscarinic receptor identified by homology screening. *Biochem. Biophys. Res. Commun.* **149**, 125–132.

Breitwieser, G. E. and Szabo, G. (1985). Uncoupling of cardiac muscarinic and β-adrenergic receptors from ion channels by a guanine nucleotide analogue. *Nature* **317**, 538–540.

Brown, J. H., Goldstein, D., and Masters, S. B. (1985). The putative M_1 muscarinic receptor does not regulate phosphoinositide hydrolysis. *Mol. Pharmacol.* **27**, 525–532.

Burch, R. M., Luini, A., and Axelrod, J. (1986). Phospholipase A_2 and phospholipase C are activated by distinct GTP-binding proteins in response to α_1-adrenergic stimulation in FRTL5 thyroid cells. *Proc. Natl. Acad. Sci. U.S.A.* **83**, 7201–7205.

Burgen, A. S. V., and Spero, L. (1968). The action of acetylcholine and other drugs on the efflux of potassium and rubidium from smooth muscle of the guinea pig intestine. *Br. J. Pharmacol.* **34**, 99–115.

Cerione, R. A., Staniszewski, C., Benovic, J. L., Lefkowitz, R. J., Caron, M. G., Gierschik, P., Somers, R., Spiegel, A. M., Codina, J., and Birnbaumer, L.

(1985). Specificity of the functional interactions of the β-adrenergic receptor and rhodopsin with guanine nucleotide regulatory proteins reconstituted in phospholipid vesicles. *J. Biol. Chem.* **260**, 1493–1500.

Cerione, R. A., Regan, J. W., Nakata, H., Codina, J., Benovic, J. L., Gierschik, P., Somers, R. L., Spiegel, A. M., Birnbaumer, L., Lefkowitz, R. J., and Caron, M. G. (1986). Functional reconstitution of the α_2-adrenergic receptor with guanine nucleotide regulatory proteins in phospholipid vesicles. *J. Biol. Chem.* **261**, 3901–3909.

Chung, F. -Z, Wang, C. -D., Potter, P. C., Venter, J. C. and Fraser, C. M. (1988). Site-directed mutagenesis and continuous expression of human β-adrenergic receptors. Identification of a conserved aspartate residue involved in agonist binding and receptor activation. *J. Biol. Chem.* **263**, 4052–4055.

Cockcroft, S., and Gomperts, B. D. (1985). Role of guanine nucleotide binding protein in the activation of polyphosphoinositide phosphodiesterase. *Nature* **314**, 534–536.

Codina, J., Hildebrandt, J., Iyengar, R., Birnbaumer, L., Sekura, R. D., and Manclark, C. R. (1983). Pertussis toxin substrate, the putative N_i component of adenylyl cyclases, is an $\alpha\beta$ heterodimer regulated by guanine nucleotide and magnesium. *Proc. Natl. Acad. Sci. U.S.A.* **80**, 4276–4280.

Codina, J., Yatani, A., Grenet, D., Brown, A. M., and Birnbaumer, L. (1987). The α subunit of the GTP-binding protein G_K opens atrial potassium channels. *Science* **236**, 442–445.

Cole, A. E., and Nicoll, R. A. (1984). Characterization of a slow cholinergic post synaptic potential recorded in vitro from rat hippocampal pyramidal cells. *J. Physiol.* **352**, 173–188.

Conklin, B. R., Brann, M. R., Buckley, N. J., Ma, A. L., Bonner, T. I., and Axelrod, J. (1988). Stimulation of arachidonic acid release and inhibition of mitogenesis by cloned genes for muscarinic receptor subtypes stably expressed in A9 L cells. *Proc. Natl. Acad. Sci. U.S.A.* **85**, 8698–8702.

Dixon, R. A. F., Kobilka, B. K., Strader, D. J., Benovic, J. L., Dohlman, H. G., Frielle, T., Bolanowski, M. A., Bennett, C. D., Rands, E., Diehl, R. E., Mumford, R. A., Slater, E. E., Sigal, I. S., Caron, M. G., Lefkowitz, R. J., and Strader, C. D. (1986). Cloning of the gene and cDNA for mammalian β-adrenergic receptor and homology with rhodopsin. *Nature* **321**, 75–79.

Dixon, R. A. F., Sigal, I. S., Rands, E., Register, R. B., Candelore, M. R., Blake, A. D., and Strader, C. D. (1987). Ligand binding to the β-adrenergic receptor involves its rhodopsin-like core. *Nature* **326**, 73–76.

Dodd, J., and Horn, J. P. (1983). Muscarinic inhibition of sympathetic C neurones in the bullfrog. *J. Physiol.* **334**, 271–291.

Dohlman, H. G., Caron, M. G., and Lefkowitz, R. J. (1987). A family of receptors coupled to guanine nucleotide regulatory proteins. *Biochemistry* **26**, 2657–2664.

Doods, H. N., Mathy, M. -J., Davidesko, D., Van Charldorp, K. J., DeJonge, A., and Van Zwieten, P. A. (1987). Selectivity of muscarinic antagonists in radioligand and *in vivo* experiments for the putative M_1, M_2, and M_3 receptors. *J. Pharmacol. Exp. Ther.* **242**, 257–262.

Ehlert, F. J., Roeske, W. R., and Yamamura, H. I. (1981). Muscarinic receptor: Regulation by guanine nucleotides, ions, and N-ethylmaleimide. *Fed. Proc. Fed. Am. Soc. Exp. Biol.* **40**, 153–159.

Endoh, M., Maruyama, M., and Iijima, T. (1985). Attenuation of muscarinic cholinergic inhibition by islet-activating protein in the heart. *Am. J. Physiol.* **249**, H309–H320.

Evans, T., Hepler, J. R., Masters, S. B., Brown, J. H., and Harden, T. K. (1985a). Guanine nucleotide regulation of agonist binding to muscarinic cholinergic receptors: Relation to efficacy of agonists for stimulation of phosphoinositide breakdown and Ca^{++} mobilization. *Biochem. J.* **232**, 751–757.

Evans, T., Martin, M. W., Hughes, A. R., and Harden, T. K. (1985b). Guanine nucleotide sensitive, high affinity binding of carbachol to muscarinic cholinergic receptors of 1321N1 astrocytoma cells is insensitive to pertussis toxin. *Mol. Pharmacol.* **27**, 32–37.

Fargin, A., Raymond, J. R., Lohse, M. J., Kobilka, B. K., Caron, M. G., and Lefkowitz, R. J. (1988). The genomic clone G-21 which resembles a β-adrenergic receptor sequence encodes the 5-HT$_{1a}$ receptor. *Nature* **335**, 358–360.

Fisher, S. K. (1986). Inositol lipids and signal transduction at CNS muscarinic receptors. *Trends Pharmacol. Sci. (Suppl.)* **7**, 61–65.

Florio, V. A., and Sternweis, P.C. (1985). Reconstitution of resolved muscarinic cholinergic receptors with purified GTP-binding proteins. *J. Biol. Chem.* **260**, 3477–3483.

Fraser, C. M., Chung, F. -Z., Wang, D. -D., and Venter, J. C. (1988). Site-directed mutagenesis of human β-adrenergic receptors: substitution of aspartic acid-130 by asparagine produces a receptor with high affinity agonist binding that is uncoupled from adenylyl cyclase. *Proc. Natl. Acad. Sci. U.S.A.* **85**, 5478–5482.

Frielle, T., Collins, S., Daniel, K. W., Caron, M. G., Lefkowitz, R. J., and Kobilka, B. K. (1987). Cloning of the cDNA for the human β_1-adrenergic receptor. *Proc. Natl. Acad. Sci. U.S.A.* **84**, 7920–7924.

Fukuda, K., Kubo, T., Akiba, I., Meada, A., Mishina, M., and Numa, S. (1987). Molecular distinction between muscarinic acetylcholine receptor subtypes. *Nature* **327**, 623–625.

Fukuda, K., Higashida, H., Kubo, T., Maeda, A., Akiba, I., Bujo, H., Mishina, M., and Numa, S. (1988). Selective coupling with K^+ currents of muscarinic acetylcholine receptor subtypes in NG108–15 Cells. *Nature* **335**, 355–358.

Fung, B. K-K. (1983). Characterization of transducin from bovine retinal rod outer segments: I. Separation and reconstitution of the subunits. *J. Biol. Chem.* **258**, 10495–10502.

Gierschik, P., Falloon, J., Milligan, G., Pines, M., Gallin, J. I., and Spiegel, A. (1986). Immunological evidence for a novel pertussis toxin substrate in human neutrophils. *J. Biol. Chem.* **261**, 8058–8062.

Gil, D. W., and Wolfe, B. B. (1985). Pirenzepine distinguishes between muscarinic receptor-mediated phosphoinositide breakdown and inhibition of adenylate cyclase. *J. Pharmacol. Exp. Ther.* **232**, 608–616.

Giles, N., and Noble, S. J. (1976). Changes in membrane currents in bullfrog atrium produced by acetylcholine. *J. Physiol.* **261**, 103–123.

Gilman, A. G. (1987). G-proteins: Transducers of receptor-generated signals. *Annu. Rev. Biochem.* **56**, 615–649.

Gocayne, J., Robinson, D. A., Fitzgerald, M. G., Chung, F. Z., Kerlavage, A. R., Lentes, K. U., Lai, J., Wang, C. D., Fraser, C. M., and Venter, J. C. (1987). Primary structure of rat cardiac β-adrenergic and muscarinic cholinergic receptors obtained by automated DNA sequence analysis: Further evidence for a multigene family. *Proc. Natl. Acad. Sci. U.S.A.* **84**, 8296–8300.

Goyal, R. K., and Rattan, S. (1978). Neurohumoral, hormonal, and drug receptors for the lower esophageal sphincter. *Gastroenterology* **74**, 598–619.

Graziano, M. P., and Gilman, A. G. (1987). Guanine nucleotide-binding regulatory proteins: Mediators of transmembrane signalling. *Trends Pharmacol. Sci.* **8**, 478–481.

Haga, K., and Haga, T. (1983). Affinity chromatography of the muscarinic acetylcholine receptor. *J. Biol. Chem.* **258**, 13575–13579.

Haga, K., and Haga, T. (1985). Purification of the muscarinic acetylcholine receptor from porcine brain. *J. Biol. Chem.* **260**, 7927–7935.

Haga, K., Haga, T., Ichiyama, A., Katada, T., Kurose, H., and Ui, M. (1985). Functional reconstitution of purified muscarinic receptors and inhibitory guanine nucleotide regulatory protein. *Nature* **316**, 731–733.

Haga, K., Haga, T., and Ichiyama, A. (1986). Reconstitution of the muscarinic acetylcholine receptor: Guanine nucleotide-sensitive high affinity binding of agonists to purified muscarinic receptors reconstituted with GTP-binding proteins (G_i and G_o). *J. Biol. Chem.* **261**, 10133–10140.

Hammer, R., and Giachetti, A. (1982). Muscarinic receptor subtypes: M_1 and M_2 biochemical and functional characterization. *Life Sci.* **31**, 2991–2998.

Hammer, R., Berrie, C. P., Birdsall, N. J. M., Burgen, A. S. V., and Hulme, E. C. (1980). Pirenzepine distinguishes between different subclasses of muscarinic receptors. *Nature* **283**, 90–92.

Hammer, R., Giraldo, E., Schiavi, G. B., Monterini, E., and Ladinsky, H. (1986). Binding profile of a novel cardioselective muscarinic receptor antagonist, AF-DX 116, to membranes of peripheral tissues and brain in the rat. *Life Sci.* **38**, 1653–1662.

Harden, T. K. (1989a). The role of guanine nucleotide regulatory proteins in receptor-selective direction of second messenger signalling. *In* "Inositol Lipids in Cell Signalling" (R. H. Michell, A. H. Drummond, and C. P. Downes, eds), pp. 113–133. Academic Press, London.

Harden, T. K. (1989b). Muscarinic cholinergic receptor-mediated regulation of cyclic AMP metabolism. *In* "The Muscarinic Receptors" (J. H. Brown, ed.) pp. 223–259. Humana Press, Clifton, New Jersey.

Harden, T. K., Scheer, A. G., and Smith, M. M. (1982). Differential modification of the interaction of cardiac muscarinic cholinergic and β-adrenergic receptors with a guanine nucleotide binding site(s). *Mol. Pharmacol.* **21**, 570–580.

Harden, T. K., Tanner, L. I., Martin, M. W., Nakahata, N., Hughes, A. R., Hepler, J. R., Evans, T., Masters, S. B., and Brown, J. H. (1986). Characteristics of two biochemical responses to stimulation of muscarinic cholinergic receptors. *Trends Pharmacol. Sci.* **7** (Suppl.), 14–18.

Harden, T. K., Stephens, L., Hawkins, P. T., and Downes, C. P. (1987). Turkey erythrocyte membranes as a model for regulation of phospholipase C by guanine nucleotides. *J. Biol. Chem.* **262**, 9057–9061.

Hartzell, H. C., Kueffler, S. W., Strickgold, R., and Yoshikama, D. (1977). Synaptic excitation and inhibition resulting from direct action of acetylcholine on two types of chemoreceptors on individual amphibian parasympathetic neurones. *J. Physiol.* **271**, 317–346.

Haslam, R. J., and Davidson, M. M. L. (1984). Receptor-induced diacylglycerol formation in permeabilized platelets; Possible role for a GTP-binding protein. *J. Recept. Res.* **4**, 605–629.

Hazeki, O., and Ui, M. (1981). Modification by islet-activating protein of receptor-mediated regulation of cyclic AMP accumulation in isolated rat heart cells. *J. Biol. Chem.* **256**, 2856–2862.

Hepler, J. R., Hughes, A. R., and Harden, T. K. (1987). Evidence that muscarinic cholinergic receptors selectively interact with either the cyclic AMP or the inositol phosphate second messenger response systems. *Biochem. J.* **247**, 793–796.

Herskowitz, I., and Marsh, L. (1987). Conservation of a receptor/signal transduction system. *Cell* **50**, 995–996.

Hildebrandt, J. D., Hanoune, J., and Birnbaumer, L. (1982). Guanine nucleotide inhibition of cyc$^-$ S49 mouse lymphoma cell membrane adenylyl cyclase. *J. Biol. Chem.* **257**, 14723–14725.

Hildebrandt, J. D., Sekura, R. D., Codina, J., Iyengar, R., Manclark, C. R., and Birnbaumer, L. (1983). Stimulation and inhibition of adenylyl cyclase mediated by distinct regulatory proteins. *Nature* **302**, 706–709.

Hoffman, B. B., Yim, S., Tsai, B. S., and Lefkowitz, R. J. (1981). Preferential uncoupling by manganese of alpha-adrenergic receptor-mediated inhibition of adenylate cyclase in human platelets. *Biochem. Biophys. Res. Commun.* **100**, 724–731.

Hokin, M. R., and Hokin, L. E. (1954). Effects of acetylcholine on phospholipids in the pancreas. *J. Biol. Chem.* **209**, 549–558.

Hootman, S. R., Picado-Leonard, T. M., and Burnham, D. B. (1985). Muscarinic acetylcholine receptor structure in acinar cells of mammalian exocrine glands. *J. Biol. Chem.* **260**, 4186–4194.

Hughes, A. R., and Harden, T. K. (1986). Adenosine and muscarinic cholinergic receptors attenuate cyclic AMP accumulation by different mechanisms in 1321N1 astrocytoma cells. *J. Pharmacol. Exp. Ther.* **237**, 173–178.

Hughes, A. R., Martin, M. W., and Harden, T. K. (1984). Pertussis toxin differentiates between two mechanisms of regulation of cyclic AMP accumulation by muscarinic cholinergic receptors. *Proc. Natl. Acad. Sci. U.S.A.* **81**, 5680–5684.

Itoh, H., Kozasa, T., Nagata, S., Nakamura, S., Katada, T., Ui, M., Iwai, S., Ohtsuka, E., Kawasaki, A., Suzuki, K., and Kaziro, Y. (1986). Molecular cloning and sequence determination of cDNAs coding for α subunits of G_s, G_i, and G_o proteins from rat brain. *Proc. Natl. Acad. Sci. U.S.A.* **83**, 3776–3780.

Jakobs, K. H., Saur, W., and Schultz, G. (1976). Reduction of adenylate cyclase activity in lysates of human platelets by the alpha-adrenergic component of epinephrine. *J. Cyclic Nucleotide Res.* **2,** 381–392.

Jakobs, K. H., Saur, W., and Schultz, G. (1978). Inhibition of platelet adenylate cyclase by epinephrine requires GTP. *FEBS Lett.* **85,** 167–170.

Jakobs, K. H., Aktories, K., and Schultz, G. (1979). GTP-dependent inhibition of cardiac adenylate cyclase by muscarinic cholinergic agonists. *Naunyn Schmiedeberg's Arch. Pharmacol.* **310,** 113–119.

Jakobs, K. H., Lasch, P., Minuth, M., Aktories, K., and Schultz, G. (1982). Uncoupling of α-adrenoceptor-mediated inhibition of human platelet adenylate cyclase by N-ethylmaleimide. *J. Biol. Chem.* **257,** 2829–2833.

Jakobs, K. H., Aktories, K., and Schultz, G. (1983). A nucleotide regulatory site for somatostatin inhibition of adenylate cyclase in S49 lymphoma cells. *Nature* **303,** 177–178.

Jakobs, K. H., Aktories, K., Minuth, M., and Schultz, G. (1985). Inhibition of adenylate cyclase. *Adv. Cyclic Nucleotide Res.* **19,** 137–150.

Jones, D. T., and Reed, R. R. (1987). Molecular cloning of five GTP-binding protein cDNA species from the olfactory neuroepithelium. *J. Biol. Chem.* **262,** 14241–14249.

Jones, S. V. P., Barker, J. L., Buckley, N. J., Bonner, T. I., Collins, R. M., and Brann, M. R. (1988). Cloned muscarinic receptor subtypes expressed in A9 L cells differ in their coupling to electrical responses. *Mol. Pharmacol.* **34,** 421–426.

Julius, D., MacDermott, A., Jessel, T., and Axel, R. (1988). Molecular characterization of a functional cDNA encoding the serotonin 1c receptor. *Cold Spring Harbor Symp. Quant. Biology* **53** (in press).

Kanaho, Y., Tsai, S. C., Adamik, R., Hewlett, E. L., Moss, J., Vaughan, M. (1984). Rhodopsin-enhanced GTPase activity of the inhibitory GTP-binding protein of adenylate cyclase. *J. Biol. Chem.* **259,** 7378–7381.

Karnik, S. S., Sakmar, T. P., Chen, H. -B., and Khorana, H. G. (1988). Cysteine residues 110 and 187 are essential for the formation of correct structure in bovine rhodopsin. *Biochemistry* **85,** 8459–8463.

Katada, T., and Ui, M. (1981). Islet activating protein: A modifier of receptor-mediated regulation of rat islet adenylate cyclase. *J. Biol. Chem.* **256,** 8310–8317.

Katada, T., and Ui, M. (1982). ADP ribosylation of the specific membrane protein of C6 cells by islet-activating protein associated with modification of adenylate cyclase activity. *J. Biol. Chem.* **257,** 7210–7216.

Katada, T., Bokoch, G. M., Northup, J. K., Ui, M., and Gilman, A. G. (1984a). The inhibitory guanine nucleotide-binding regulatory component of adenylate cyclase. Properties and function of the purified protein. *J. Biol. Chem.* **259,** 3568–3577.

Katada, T., Bokoch, G. M., Smigel, M. D., Ui, M., and Gilman, A. G. (1984b). The inhibitory guanine nucleotide-binding regulatory component of adenylate cyclase: Subunit association and the inhibition of adenylate cyclase in S49 lymphoma cyc⁻ and wild type membranes. *J. Biol. Chem.* **259,** 3586–3595.

Katada, T., Northup, J. K., Bokoch, G. M., Ui, M., and Gilman, A. G. (1984c).

The inhibitory guanine nucleotide-binding regulatory component of adenylate cyclase. Subunit dissociation and guanine nucleotide-dependent hormonal inhibition. *J. Biol. Chem.* **259**, 3578–3585.

Kerlavage, A. R., Fraser, C. M., and Venter, J. C. (1987). Muscarinic cholinergic receptor structure: Molecular biological support for subtypes. *Trends Pharmacol. Sci.* **8**, 426–431.

Kobilka, B. K., Matsui, H., Kobilka, T. S., Yang-Feng, T. L., Francke, U., Caron, M. G., Lefkowitz, R. J., and Regan, J. W. (1987a). Cloning, sequencing, and expression of the gene coding for the human platelet α_2-adrenergic receptor. *Science* **238**, 650–656.

Kobilka, B. K., Frielle, T. Collins, S., Yang-Feng, T., Kobilka, T. S., Francke, U., Lefkowitz, R. J., and Caron, M. G. (1987b). An intronless gene encoding a potential member of the family of receptors coupled to guanine nucleotide regulatory proteins. *Nature* **329**, 75–79.

Kubo, T., Fukuda, K., Mikani, A., Maeda, H., Takahashi, H., Mishina, M., Haga, T., Haga, K., Ichiyama, A., Kanagawa, K., Kojima, M., Matsuo, H., Hirose, T., and Numa, S. (1986a). Cloning, sequencing and expression of complementary DNA encoding the muscarinic acetylcholine receptor. *Nature* **323**, 411–416.

Kubo, T., Maeda, H., Sugimoto, K., Akiba, I., Mikani, A., Takahashi, H., Haga, T., Haga, K., Ichiyama, A., Kanagawa, K., Matsuo, H., Hirose, T., and Numa, S. (1986b). Primary structure of porcine cardiac muscarinic acetylcholine receptor deduced from the cDNA sequence. *FEBS Lett.* **209**, 367–372.

Kuhn, H. (1984). Interactions between photoexcited rhodopsin and light-activated enzymes in rods. *Prog. Retinal Res.* **3**, 123–156.

Kuhn, H., and Hargrave, P. A. (1981). Light-induced binding of guanosine triphosphate to bovine photoreceptor membranes: Effect of limited proteolysis of membranes. *Biochemistry* **20**, 2410–2417.

Kurachi, Y., Nakajima, T., and Sugimoto, T. (1986). On the mechanism of activation of muscarinic K^+ channels by adenosine in isolated atrial cells: Involvement of GTP-binding proteins. *Pfluegers Arch.* **407**, 264–274.

Kurose, H., Katada, T., Amano, T., and Ui, M. (1983). Specific uncoupling by islet-activating protein, pertussis toxin, of negative signal transduction via α-adrenergic, cholinergic, and opiate receptors in neuroblastoma × glioma hybrid cells. *J. Biol. Chem.* **258**, 4870–4875.

Kurose, H., Katada, T., Haga, T., Haga, K., Ichiyama, and Ui, M. (1986). Functional interaction of purified muscarinic receptors with purified inhibitory guanine nucleotide regulatory proteins reconstituted in phospholipid vesicles. *J. Biol. Chem.* **261**, 6423–6428.

Kwatra, M. M., and Hosey, M. M. (1986). Phosphorylation of the cardiac muscarinic receptor of intact chick heart and its regulation by a muscarinic agonist. *J. Biol. Chem.* **261**, 12429–12432.

Lazareno, S., Kendall, D. A., and Nahorski, S. R. (1985). Pirenzepine indicates heterogeneity of muscarinic receptors linked to cerebral inositol phospholipid metabolism. *Neuropharmacology* **24**, 593–595.

Liang, M., Martin, M. W., and Harden, T. K. (1987). [³H]Propylbenzilylcholine mustard-labeling of muscarinic cholinergic receptors that selectively couple to phospholipase C or adenylate cyclase in two cultured cell lines. *Mol. Pharmacol.* **32**, 443–449.

Liao, C. -F., Themmen, A. P. N., Joho, R., Birnbaumer, M. and Birnbaumer, L. (1989). Molecular cloning and expression of a fifth muscarinic acetylcholine receptor (M5-mAChR). *J. Biol. Chem.* (in press).

Liles, W. C., Hunter, D. D., Meier, K. E., and Nathanson, N. M. (1986). Activation of protein kinase C induces rapid internalization and subsequent degradation of muscarinic acetylcholine receptors in neuroblastoma cells. *J. Biol. Chem.* **261**, 5307–5313.

Limbird, L. E. (1981). Activation and attenuation of adenylate cyclase. *Biochem. J.* **195**, 1–13.

Limbird, L. E., and Lefkowitz, R. J. (1978). Agonist-induced increase in apparent β-adrenergic receptor size. *Proc. Natl. Acad. Sci. U.S.A.* **75**, 228–232.

Litosch, I., Wallis, C., and Fain, J. N. (1985). 5-Hydroxytryptamine stimulates inositol phosphate production in a cell-free system from blowfly salivary gland: Evidence for a role of GTP in coupling receptor activation to phosphoinositide breakdown. *J. Biol. Chem.* **260**, 5464–5471.

Logothetis, D. E., Kurachi, Y., Galper, J., Neer, E. J., and Clapham, D. E. (1987). The $\beta\gamma$ subunits of GTP-binding proteins activate the muscarinic K^+ channel in heart. *Nature* **325**, 321–326.

Londos, C., Cooper, D. M. F., Schlegel, W., and Rodbell, M. (1978). Adenosine analogs inhibit adipocyte adenylate cyclase by a GTP-dependent process: Basis for actions of adenosine and methylxanthines on cyclic AMP production and lipolysis. *Proc. Natl. Acad. Sci. U.S.A.* **75**, 5362–5366.

Martin, M. W., Evans, T., and Harden, T. K. (1985a). Further evidence that muscarinic cholinergic receptors of 1321N1 astrocytoma cells couple to a guanine nucleotide regulatory protein that is not N_i. *Biochem. J.* **229**, 539–544.

Martin, J. M., Hunter, D. D., and Nathanson, N. M. (1985b). Islet activating protein inhibits physiological responses evoked by cardiac muscarinic acetylcholine receptors. Role of GTP-binding proteins in regulation of potassium permeability. *Biochemistry* **24**, 7521–7525.

Martin, T. F. J., Bajjaliek, S. M., Lucas, D. O., and Kowalchyk, J. A. (1986). Thyrotropin-releasing hormone stimulation of polyphosphoinositide hydrolysis in GH_3 cell membranes is GTP dependent but insensitive to cholera or pertussis toxin. *J. Biol. Chem.* **261**, 1041–1149.

Marty, A. (1987). Control of ionic currents and fluid secretion by muscarinic agonists in exocrine glands. *Trends Neurosci.* **10**, 373–377.

Masters, S. B., Martin, M. W., Harden, T. K., and Brown, J. H. (1985). Pertussis toxin does not inhibit muscarinic receptor-mediated phosphoinositide hydrolysis or calcium mobilization. *Biochem. J.* **227**, 933–937.

Masters, S. B., Stroud, R. M., and Bourne, H. R. (1986). Family of G protein α chains: Amphipathic analysis and predicted structure of functional domains. *Protein Eng.* **1**, 47–54.

Masu, Y., Nakayama, K., Tamaki, H., Harada, Y., Kuno, M., and Nakanishi, S.

(1987). cDNA cloning of bovine substance-K receptor through oocyte expression system. *Nature* **329**, 836–838.

Meeker, R. B., and Harden, T. K. (1982). Muscarinic receptor-mediated activation of phosphodiesterase. *Mol. Pharmacol.,* **22**, 310–319.

Meeker, R. B., and Harden, T. K. (1983). Muscarinic cholinergic receptor-mediated control of cyclic AMP metabolism: Agonist-induced changes in nucleotide synthesis and degradation. *Mol. Pharmacol.* **23**, 384–392.

Merritt, J. E., Taylor, C. W., Rubin, R. P., and Putney, J. W. (1986). Evidence suggesting that a novel guanine nucleotide regulatory protein couples receptors to phospholipase C in exocrine pancreas. *Biochem. J.* **236**, 337–343.

Miot, F. C., Erneux, C., Wells, J. N., and Dumont, J. E. (1984). The effects of alkylated xanthines on cyclic AMP accumulation in dog thyroid slices exposed to carbamylcholine. *Mol. Pharmacol.* **25**, 261–266.

Morita, K., North, R. A., and Tokimasa, T. (1982). Muscarinic agonists inactivate potassium conductances of guinea pig myenteric neurones. *J. Physiol.* **333**, 125–139.

Motulsky, H. J., Hughes, R. J., Brickman, A. S., Farfel, Z., Bourne, H. R. and Insel, P. A. (1982). Platelets of pseudohypoparathyroid patients: Evidence that distinct receptor-cyclase coupling proteins mediate stimulation and inhibition of adenylate cyclase. *Proc. Natl. Acad. Sci. U.S.A.* **79**, 4193–4197.

Murad, F., Chi, Y. -M., Rall, T. W., and Sutherland, E. W. (1962). Adenyl cyclase. III. The effect of catecholamines and choline esters on the formation of adenosine 3′, 5′-phosphate by preparations of cardiac muscle and liver. *J. Biol. Chem.* **237**, 1223–1238.

Murayama, T., and Ui, M. (1983). Loss of the inhibitory function of the guanine nucleotide regulatory component of adenylate cyclase due to its ADP ribosylation by islet-activating protein, pertussis toxin, in adipocyte membranes. *J. Biol. Chem.* **258**, 3319–3326.

Mutschler, E., and Lambrecht, G. (1984). Selective muscarinic agonists and antagonists in functional tests. *Trends Pharmacol. Sci.* **5** (Suppl.), 39–44.

Nakada, T., Tanabe, T., Takahashi, H., Noda, M., Haga, K., Haga, T., Ichiyama, A., Kanagawa, K., Hiranaga, M., Matsuo, H., and Numa, S. (1986). Primary structure of the α-subunit of bovine adenylate cyclase-inhibiting G-protein deduced from the cDNA sequence. *FEBS Lett.* **197**, 305–310.

Nakahata, N., Martin, M. W., Hughes, A. R., Hepler, J. R., and Harden, T. K. (1986). H_1-histamine receptors on astrocytoma cells. *Mol. Pharmacol.* **29**, 188–195.

Nakamura, T., and Ui, M. (1984). Simultaneous inhibitions of inositol phospholipid breakdown, arachadonic acid release, and histamine secretion in mast cells by islet-activating protein, pertussis toxin. *J. Biol. Chem.* **260**, 3584–3593.

Nathans, J., and Hogness, D. S. (1983). Isolation sequence analysis and intron-exon arrangement of the gene encoding bovine rhodopsin. *Cell* **34**, 807–814.

Nathans, J., and Hogness, D. S. (1984). Isolation and nucleotide sequence of the gene encoding human rhodopsin. *Proc. Natl. Acad. Sci. U.S.A.* **81**, 4851–4855.

Nathans, J., Thomas, D., and Hogness, D. S. (1986). Molecular genetics of human color vision: The genes encoding blue, green, and red pigments. *Science* **232**, 193–202.

Nathanson, N. M. (1987). Molecular properties of the muscarinic acetylcholine receptor. *Annu. Rev. Neurosci.* **10**, 195–236.

Neer, E. J., Lok, J. M., and Wolf, L. G. (1984). Purification and properties of the inhibitory guanine nucleotide regulatory unit of brain adenylate cyclase. *J. Biol. Chem.* **259**, 14222–14229.

Nilvebrandt, L., and Sparf, B. (1983). Differences between binding affinities of some antimuscarinic drugs in the parotid gland and those in the urinary bladder and ileum. *Acta Pharmacol. Tox.* **53**, 304–313.

North, R. A. (1986). Muscarinic receptors and membrane ion conductances. *Trends Pharmacol. Sci.* **7** (suppl.) 19–27.

North, R. A., and Tokimasa, T. (1983). Depression of calcium-dependent potassium conductance of guinea pig myenteric neurones by muscarinic agonists. *J. Physiol.* **342**, 253–266.

Nowak, L. M., and MacDonald, R. L. (1983). Muscarine-sensitive voltage-dependent potassium current in cultured murine spinal cord neurons. *Neurosci. Lett.* **35**, 85–91.

Oinuma, M., Katada, T., and Ui, M. (1987). A new GTP-binding protein in differentiated human leukemic (HL-60) cells serving as the specific substrate of islet-activating protein, pertussis toxin. *J. Biol. Chem.* **262**, 8347–8353.

Okajima, F., and Ui, M. (1984). ADP-ribosylation of a specific membrane protein by islet-activating protein, pertussis toxin, associated with inhibition of a chemotactic peptide-induced arachidonate release in neutrophils: A possible role of the toxin substrate in Ca^{2+}-mobilizing biosignaling. *J. Biol. Chem.* **259**, 13863–13871.

Orellana, S. A., Solski, P. A., and Brown, J. H. (1985). Phorbol ester inhibits phosphoinositide hydrolysis and calcium mobilization in cultured astrocytoma cells. *J. Biol. Chem.* **260**, 5236–5239.

Oron, Y., Dascal, N., Nadler, E., and Lupu, M. (1985). Inositol 1,4,5-trisphosphate mimics muscarinic responses in Xenopus oocytes. *Nature* **313**, 141–143.

Peralta, E. G., Winslow, J. W., Peterson, G. L., Smith, D. H., Ashkenazi, A., Ramachandran, J., Schimerlik, M. I., and Capon, D. J. (1987a). Primary structure and biochemical properties of an M2 muscarinic receptor. *Science* **236**, 600–605.

Peralta, E. G., Ashkenazi, A., Winslow, J. W., Smith, D. H., Ramachandran, J., and Capon, D. J. (1987b). Distinct primary structures, ligand-binding properties, and tissue-specific expression of four human muscarinic acetylcholine receptors. *EMBO J.* **6**, 3923–3929.

Peralta, E. G., Ashkenazi, A., Winslow, J. W., Ramachandran, J., and Capon, D. J. (1988). Differential regulation of PI hydrolysis and adenylyl cyclase by muscarinic receptor subtypes. *Nature* **334**, 434–437.

Petersen, G. L., Herron, G. S., Yamaki, M., Fullerton, D. S., and Schimerlik, M. I. (1984). Purification of the muscarinic acetylcholine receptor from procine atria. *Proc. Natl. Acad. Sci. U.S.A.* **81**, 4993–4997.

Pfaffinger, P. J., Martin, J. M., Hunter, D. D., Nathanson, N. M., and Hille, B. (1985). GTP-binding proteins couple cardiac muscarinic receptors to a K channel. *Nature* **317**, 536–538.

Pfeilschifter, J., and Bauer, C. (1986). Pertussis toxin abolishes angiotensin II-induced phosphoinositide hydrolysis and prostaglandin synthesis in rat renal mesangial cells. *Biochem. J.* **236**, 289–294.

Pinkas-Kramarski, R., Stein, R., Zimmer, Y., and Sokolovsky, M. (1988). Cloned rat M3 muscarinic receptors mediate phosphoinositide hydrolysis but not adenylate cyclase inhibition. *FEBS Lett.* **239**, 174–178.

Putney, J. W. (ed.) (1986). Phosphoinositides and receptor mechanisms. *Receptor Biochem. and Methodol.,* **7.**

Regan, J. W., Kobilka, T. S., Yang-Feng, T. L., Caron, M. G., Lefkowitz, R. J., and Kobilka, B. K. (1988). Cloning and expression of a human kidney cDNA for an α_2-adrenergic receptor subtype. *Proc. Natl. Acad. Sci. U.S.A.* **85**, 6301–6305.

Richelson, E., and El-Fakahany, E. (1981). The molecular basis of neurotransmission at the muscarinic receptor. *Biochem. Pharmacol.* **30**, 2887–2891.

Rosenbaum, L. C., Malencik, D. A., Andersen, S. R., Tota, M. R., and Schimerlik, M. I. (1987). Phosphorylation of the porcine atrial muscarinic acetylcholine receptor by cyclic AMP dependent protein kinase. *Biochemistry* **26**, 8183–8188.

Rosenberger, L. B., Roeske, W. R., and Yamamura, H. I. (1979). The regulation of muscarinic cholinergic receptors by guanine nucleotides in cardiac tissue. *Eur. J. Pharmacol.* **56**, 179–180.

Rosenberger, L. B., Yamamura, H. I., and Roeske, W. R. (1980). Cardiac muscarinic cholinergic binding is regulated by Na^+ and guanyl nucleotides. *J. Biol. Chem.* **255**, 820–823.

Schwinn, D. A., Randall, R. R., Lefkowitz, R. J., Caron, M. G., and Kobilka, B. K. (1988). Molecular cloning and expression of the cDNA for the hamster α_1-adrenergic receptor. *Proc. Natl. Acad. Sci. U.S.A.* **85**, 7159–7163.

Smith, C. D., Lane, B. C., Kusaka, I., Verghese, M. W., and Snyderman, R. (1985). Chemoattractant receptor-induced hydrolysis of phosphatidylinositol 4,5-bisphosphate in human polymorphonuclear leucocyte membranes: Requirement for a guanine nucleotide regulatory protein. *J. Biol. Chem.* **260**, 5875–5878.

Soejima, M., and Noma, A. (1984). Mode of regulation of the ACh-sensitive K-channel by the muscarinic receptor in rabbit atrial cells. *Pfluegers Arch.* **400**, 424–431.

Sorota, S., Tsuji, Y., Tajima, T., and Pappano, A. (1985). Pertussis toxin treatment blocks hyperpolarization by muscarinic agonists in chick atrium. *Circ. Res.* **57**, 748–758.

Stein, R., Pinkas-Kramarski, R., and Sokolovsky, M. (1988). Cloned M1 muscarinic receptors mediate both adenylate cyclase inhibition and phosphoinositide turnover. *EMBO J.* **7**, 3031–3035.

Sternweis, P. C., and Robishaw, J. D. (1984). Isolation of two proteins with high affinity for guanine nucleotides from membranes of bovine brain. *J. Biol. Chem.* **259,** 13806–13813.

Strader, C. D., Dison, R. A. F., Cheung, A. H., Candelore, M. R., Blake, A., and Sigal, I. S. (1987a). Mutations that uncouple the β-adrenergic receptor from G_s and increase agonist affinity. *J. Biol. Chem.* **262,** 16439–16443.

Strader, C. D., Sigal, I. S., Register, R. B., Candelore, M. R., Rands, E., and Dixon, R. A. F. (1987b). Identification of residues required for ligand binding to the β-adrenergic receptor. *Proc. Natl. Acad. Sci. U.S.A.* **84,** 4384–4388.

Stryer, L., and Bourne, H. R. (1986). G Proteins: A family of signal transducers. *Annu. Rev. Cell Biol.* **2,** 391–419.

Study, R. E., Breakfield, X. O., Bartfai, T., and Greengard, P. (1978). Voltage-sensitive calcium channels regulate guanosine 3′, 5′-cyclic monophosphate levels in neuroblastoma cells. *Proc. Natl. Acad. Sci. U.S.A.* **75,** 6295–6299.

Sullivan, K. A., Miller, R. T., Masters, S. B., Beiderman, B., Heideman, W., and Bourne, H. R. (1987). Identification of receptor contact site involved in receptor-G protein coupling. *Nature* **330,** 758–760.

Tanner, L. I., Harden, T. K., Wells, J. N., and Martin, M. W. (1986). Identification of the phosphodiesterase regulated by muscarinic cholinergic receptors of 1321N1 human astrocytoma cells. *Mol. Pharmacol.* **29,** 455–460.

Tota, M. R., Kahler, K. R., and Schimerlik, M. I. (1987). Reconstitution of the purified porcine atrial muscarinic acetylcholine receptor with purified porcine atrial inhibitory guanine nucleotide binding protein. *Biochemistry* **26,** 8175–8182.

Tsai, S. -C., Adamik, R., Kanaho, Y., Hewlett, E. L., and Moss, J. (1984). Effects of guanyl nucleotides and rhodopsin on ADP-ribosylation of the inhibitory GTP-binding component of adenylate cyclase by pertussis toxin. *J. Biol. Chem.* **259,** 15320–15323.

Uhing, R. J., Jiang, H., Prpic, V., and Exton, J. H. (1985). Regulation of a liver plasma membrane phosphoinositide phosphodiesterase by guanine nucleotides and calcium. *FEBS Lett.* **199,** 317–320.

Ui, M. (1984). Islet activating protein, pertussis toxin: A probe for functions of the inhibitory guanine nucleotide regulatory components of adenylate cyclase. *Trends Pharmacol. Sci.* **6,** 277–279.

Waelbroeck, M. M., Gillard, P., Robberecht, P., and Christophe, J. (1986). Kinetic studies of [³H]-N-methylscopolamine binding to muscarinic receptors in the rat central nervous system: Evidence for the existence of three classes of binding sites. *Mol. Pharmacol.* **30,** 305–314.

Waelbroeck, M. M., Gillard, M., Robberecht, P., and Christophe, J. (1987). Muscarinic receptor heterogeneity in rat central nervous system. I. Binding of four selective antagonists to three muscarinic receptor subclasses: A comparison with M_2 cardiac muscarinic receptors of the C type. *Mol. Pharmacol.* **32,** 91–99.

Yano, K., Higashida, H., Hattori, H., and Nozawa, Y. (1985). Bradykinin-induced transient accumulation of inositol trisphosphate in neuron-like cell line NG108–15 cells. *FEBS Lett.* **181,** 403–406.

Yarden, Y., Rodriguez, H., Wong, S. K. -F., Brandt, D. R., May, D. C., Burnier, J., Harkins, R. N., Chen, E. Y., Ramachandran, J., Ullrich, A., and Ross, E. M. (1986). The avian β-adrenergic receptor: Primary structure and membrane topology. *Proc. Natl. Acad. Sci. U.S.A.* **83,** 6795–6799.

Yatani, A., Codina, J., Brown, A. M., and Birnbaumer, L. (1987). Direct activation of mammalian atrial muscarinic K channels by a human erythrocyte pertussis toxin-sensitive G protein, Gk. *Science* **235,** 207–211.

Young, D., Waitches, G., Birchmeier, C., Fasano, O., and Wigler, M. (1986). Isolation and characterization of a new cellular oncogene encoding a protein with multiple potential transmembrane domains. *Cell* **45,** 711–719.

PART III

Systems Regulated by G
Proteins

G Protein- and Protein Kinase C-Mediated Regulation of Voltage-Dependent Calcium Channels

Stanley G. Rane, Kathleen Dunlap
Department of Physiology, Tufts University School of Medicine, Boston, Massachusetts 02111

I. Introduction
   A. Different Types of Calcium Channels
II. Transmitters, G Proteins, and Second Messenger Systems Associated with Inhibition of Calcium Current
   A. Transmitters
   B. GTP-Binding Proteins
   C. Second Messenger Systems Associated with Calcium Current Inhibition
III. Involvement of Protein Kinase C in Calcium Channel Modulation
   A. Transmitter-Induced Inhibition of Calcium Current Mimicked by Exogenously Applied Protein Kinase C Activators
   B. Action of Protein Kinase C on Calcium Channels in *Aplysia* Bag Cell Neurons
   C. Intracellular Injection of Protein Kinase C
   D. Specific Protein Kinase C Inhibitor Blocking Calcium Channel Modulation in Chick Dorsal Root Ganglion Neurons
   E. Problems with Protein Kinase C Hypothesis for Calcium Current Inhibition
IV. Conclusions
   References

I. INTRODUCTION

The instrumental role that calcium channels play in controlling cell function has been acknowledged for over three decades. Until recently, however, the variety and complexity of cellular processes responsible for the regulation of calcium influx across biological membranes have been underappreciated. In the past few years a number of calcium-permeant channels (both voltage-dependent and -independent) have been described. It is clear, not only that these channels provide the means for regulating a number of separate cellular behaviors, but also that their own behavior can be modified by alterations in the biochemical state of the cell.

This fact was first recognized in early studies on cardiac muscle which demonstrated that binding of norepinephrine to β-adrenergic receptors produces an increase in voltage-dependent calcium current, an effect now known to be mediated through activation of a cAMP-dependent protein kinase. Because the literature in this area is extensive, readers wishing an in-depth treatment of the subject are referred to previous reviews (Reuter, 1983; Tsien, 1987). These initial studies on cardiac muscle paved the way for subsequent investigations into the mechanism of neurotransmitter modulation of calcium channel function in neurons. Although the work on cardiac muscle formed an important conceptual foundation, studies of calcium channel modulation in neurons have uncovered certain regulatory mechanisms which are distinct from those found in the heart. In this chapter we will focus on some of these mechanisms.

A wide variety of neurotransmitters have now been described to alter neuronal calcium channel function. While the list of neuroactive substances continues to grow, recent work has focused on the mechanisms by which transmitters and their receptors alter channel function. Two signal transduction events, in particular, are proving integral to calcium channel modulation: (1) activation of GTP-binding proteins (G proteins), the plasma membrane-associated protein heterotrimers described extensively in this volume, and (2) the production of the protein kinase C (PKC) activator, diacylglycerol (DAG), from the cleavage of membrane phospholipids. Increasingly sophisticated biochemical study of these transduction systems, and the development of chemical tools to probe their function, has enabled physiologists to assess their contributions to calcium channel modulation. It now appears clear that both G proteins and activation of protein kinase C are important to transmitter-induced calcium channel modulation. The challenge remains to understand exactly how these signal transduction systems interact with the channel protein to change its function.

A. Different Types of Calcium Channels

It is now generally accepted that there is more than one type of voltage-dependent, neuronal calcium channel (Carbone and Lux, 1984; Nowycky *et al.*, 1985). The distinguishing features of these channel types include differences in their single channel conductances, the voltage ranges over which the channels either activate or inactivate, the kinetics of inactivation, and the pharmacology of agents which can either directly inhibit or potentiate channel activity (Fox *et al.*, 1987a,b). Unambiguous identification of calcium channel subtypes or isolation of a transmitter effect to a particular ionic current requires the use of the voltage clamp technique. In most of the studies reviewed here, calcium currents were recorded from neuronal cell bodies with either one- or two-electrode voltage clamp, or with the whole-cell recording configuration of the patch clamp technique (Hamill *et al.*, 1981). Although the majority of studies to date suggest that the slowly inactivating calcium channel (L type) is a site of transmitter- and second messenger-mediated modulation, single channel recording experiments will be required to unequivocally determine the identity of the calcium channel(s) involved. As will be discussed, a few cases of transmitter-induced modulation of non-L-type calcium channels have also been described, although channel identification has been based only on macroscopic recordings of whole-cell calcium currents. In other experiments, membrane potential recordings of calcium-dependent action potentials were used, rather than voltage clamp techniques, to study transmitter action. Because action potentials can be composed of a variety of currents, the identity of the target currents for transmitter effects can only be tentatively suggested from such experiments.

II. TRANSMITTERS, G PROTEINS, AND SECOND MESSENGER SYSTEMS ASSOCIATED WITH INHIBITION OF CALCIUM CURRENT

A. Transmitters

A number of neurotransmitters have been shown, in a variety of preparations, to cause inhibition of voltage-dependent calcium current or shortening of the calcium-dependent action potential duration (Table I). The inhibitory phenomenon has been extensively studied in dorsal root ganglion (DRG) neurons due both to the relative ease with which they can be grown in primary culture and to the interest in inhibition of calcium

Table I   Examples of Transmitter-Induced Calcium Channel Inhibition

Preparation	Transmitter	Reference[e]
DRG[a] (chick)	norepinephrine	3,7,8,9
	serotonin	3,7,8
	dopamine	3,7,8,19
	GABA (baclofen)	3,7,8
DRG (rat)	GABA (baclofen)	5
	adenosine	6
	neuropeptide Y	24
DRG (mouse)	dynorphin	18
DRG (frog)	serotonin	13
hippocampal neurons	acetylcholine	22
SG[b] (chick)	norepinephrine	19
	dopamine	19
SG (rat)	norepinephrine	10,14
	epinephrine	14
	dopamine	14
	acetylcholine	25
	somatostatin	15
LC[c] (rat)	epinephrine	26
	norepinephrine	26
GH$_3$	somatostatin	21
AtT-20	somatostatin	17
NxG[d]	enkephalin	12
NG108-15	enkephalin	23
	somatostatin	23
Helix		
F5, D2	dopamine	20
F77,D2	cholecystokinin	11
D3, E3	FMRFamide	4
Aplysia		
L10	histamine	16
R2, R15,	FMRFamide	2
L2-4, L6		
Limnaea	dopamine	1

[a]Dorsal root ganglion neurons.
[b]Sympathetic ganglion neurons.
[c]Locus coeruleus neurons.
[d]Neuroblastoma glioma hybrid cells.
[e]Key to references: (1) Akopyan et. al. (1985); (2) Brezina et. al. (1985); (3) Canfield & Dunlap (1984); (4) Colombaioni et. al. (1985); (5) Dolphin & Scott (1987); (6) Dolphin et. al. (1986); (7) Dunlap & Fischbach ((1978); (8) Dunlap & Fischbach (1981); (9) Forscher & Oxford (1985); (10) Galvan & Adams (1982); (11) Hammond et. al. (1987); (12) Hescheler et. al. (1987); (13) Holz et. al. (1986b); (14) Horn & McAfee (1980); (15) Ikeda & Schofield (1989); (16) Kretz et al. (1986); (17) Luini et. al. (1986); (18) Macdonald & Werz (1986); (19) Marchetti et. al. (1986); (20) Paupardin-Tritsch et. al. (1985); (21) Rosenthal et al. (1988); (22) Toselli & Lux (1989); (23) Tsunoo et. al. (1986); (24) Walker et al. (1988); (25) Wanke et. al. (1987); (26) Williams & North (1985).

current as a physiological means of suppressing transmitter release from pain-responsive sensory afferents. As a result, there is more extensive pharmacological information for cultured DRG neurons than for the other preparations.

A variety of amines, including dopamine, serotonin and most notably norepinephrine, inhibit calcium current in cultured embryonic chick DRG neurons (Dunlap and Fischbach, 1978, 1981; Forscher and Oxford, 1985; Marchetti *et al.*, 1986). These transmitters appear to operate through a common α_2-adrenergic-like receptor (Canfield and Dunlap, 1984) in that their actions are blocked by nanomolar concentrations of the selective α_2-adrenergic antagonist yohimbine (and the more general α-antagonist phentolamine), but they are unaffected by the β-blocker propranolol. The α_2 designation is tentative, however, because the amine effects are not mimicked by specific α_2-agonists such as clonidine and xylazine (Canfield and Dunlap, 1984). Calcium-dependent action potentials in other peripheral and central neurons are also affected by stimulation of α receptors. Horn and McAfee (1980) observed adrenergic inhibition of calcium action potentials in rat sympathetic neurons, which was blocked by phentolamine but not by a specific β-blocker, MJ-1999. Further characterization of this receptor was not reported. A more detailed pharmacological description of α receptor-mediated inhibition of calcium action potentials and currents was undertaken in rat locus coeruleus neurons (Williams and North, 1985). Like chick DRG neurons, the actions of norepinephrine and epinephrine were blocked by phentolamine and were unaffected by propranolol. In contrast, yohimbine did not block amine action in the central neurons as it did in DRG cells. Thus, although inhibition of calcium current may be generally associated with α-receptor stimulation, the nature of the receptor subtype seems to vary in different cells.

Other transmitters which inhibit DRG neuron calcium current include γ-aminobutyric acid (GABA) and adenosine. Inhibition of calcium current by the selective $GABA_B$-agonist baclofen has been reported for chick (Dunlap, 1981) and for rat (Dolphin and Scott, 1987) DRG neurons. Canfield and Dunlap (1984) also reported that saturating doses of baclofen and norepinephrine are additive and baclofen-mediated responses are unaffected by yohimbine, suggesting that two distinct receptor types are coupled to calcium-current inhibition. Dolphin and colleagues (1986) showed that the adenosine analog 2-chloroadenosine inhibits rat DRG neuron calcium current at a concentration ($0.5 \mu M$) which suggests activation of the A_1 adenosine receptor subtype. The adenosine antagonists isobutylmethylxanthine and 8-phenyltheophylline blocked the action of 2-chloroadenosine.

Peptides which inhibit neuronal calcium current range from the tripeptide FMRFamide and enkephalin pentapeptide to larger peptides such as dynorphin, somatostatin, neuropeptide Y, and cholecystokinin. The pharmacology of the enkephalin and dynorphin responses has been investigated in an attempt to correlate the specific opioid peptide receptor types responsible for calcium current inhibition with the known effects of the peptides on transmitter release. For mouse DRG neurons, calcium current is inhibited by the selective κ-opioid receptor agonist dynorphin A, while an agonist for both μ and δ receptors, Leu-enkephalin, is without effect (Macdonald and Werz, 1986). Based on macroscopic current data, it has been suggested that dynorphin A inhibits a transient calcium current distinct from the L-type or long-lasting calcium current (Gross and Macdonald, 1987). As mentioned above, single channel data will be necessary to firmly establish the calcium channel type responsive to inhibition by opioid peptides. In contrast to these studies on mouse DRG neurons, calcium current in neuroblastoma-glioma hybrid cells responds to Leu-enkephalin and the specific δ receptor agonist, D-Ala^2,D-Leu^5-enkephalin (Tsunoo *et al.*, 1986; Hescheler *et al.*, 1987). Thus, as with the amine receptors, it seems that the type of opioid receptor which inhibits calcium current may vary with the cell type.

Given the wide variety of neuroactive substances and membrane receptors that inhibit calcium current, it is difficult to make generalizations or predictions as to which substances are most likely to be associated with calcium current inhibition. Rather, the phenomenon will have to be investigated on a case-by-case basis. In addition, it is unlikely that a single mechanism can account for the effects of all of the transmitters known to inhibit neuronal calcium current, although there are striking similarities among some of the examples given in Table I. In the sections below we will discuss some of the mechanisms that may have general importance in mediating transmitter inhibition of neuronal calcium current.

B. GTP-Binding Proteins

It is clear that in all preparations on which the appropriate experiments have been conducted, a role for GTP-binding proteins (G proteins) in transmitter inhibition of calcium current has been confirmed. For the most part, investigations have employed the whole-cell configuration of the patch clamp technique (Hamill *et al.*, 1981) which allows for internal perfusion of neuronal cell bodies with GTP analogs that can alter G protein function. The analogs tested include GDPβS, which blocks activation of G proteins by competing for the GTP binding site and preventing subunit dissociation, and GTPγS, which irreversibly activates G proteins by allowing dissociation of the active and regulatory protein subunits and preventing their reassociation. Thus intracellular application of GDPβS blocks the inhibitory actions of norepinephrine on DRG neurons, D-Ala^2,D-Leu^5-enkephalin on NG108-15 cells, and acetylcholine on sympathetic neurons (Holz et al., 1986a; Hescheler et al., 1987; Wanke et al., 1987). Conversely, GTPγS itself causes inhibition of calcium current when it is intracellularly applied (Lewis et al., 1986; Hescheler et al., 1987; Wanke et al., 1987; Brezina, 1987; Harris-Warrick et al., 1988). Cells which are treated with GTPγS respond poorly, if at all, to transmitter (Lewis et al., 1986; Brezina, 1987; Harris-Warrick et al., 1988). However, Ikeda and Schofield (1989) find that although GTPγS reduces calcium current and blocks subsequent inhibition by somatostatin in sympathetic neurons, application of GDPβS is without effect.

Do all of these inhibitory modulators activate the same G protein? Although this question remains to be answered, it is clear that many of the transmitter-induced effects require the activation of a pertussis toxin-sensitive G protein. Pertussis toxin (PTX), a bacterial endotoxin which prevents receptor interaction with a particular group of G proteins (see Codina et al., this volume), has been demonstrated to block the actions of norepinephrine and baclofen in chick DRG neurons (Holz et al., 1986a), baclofen and neuropeptide Y in rat DRG neurons (Dolphin and Scott, 1987; Ewald et al., 1988), acetylcholine in rat sympathetic and hippocampal neurons (Wanke et al., 1987; Toselli and Lux, 1989), somatostatin in rat sympathetic neurons and in GH$_3$ and AtT-20 pituitary cell lines (Lewis et al., 1986; Rosenthal et al., 1988; Ikeda and Schofield, 1989), enkephalin in a neuroblastoma/glioma hybrid (Hescheler et al., 1987), and dopamine in neurons from the abdominovisceral ganglion of Helix (Harris-Warrick et al., 1988). Hescheler and colleagues have taken these approaches a step further, by showing that intracellular application of the α subunit of G$_o$ reconstitutes enkephalin inhibition of calcium current which had been abolished by prior PTX treatment. Similarly, Ewald and colleagues (1989) have shown reconstitution of neuropeptide Y-induced inhibition in rat DRG neurons, also with the α subunit of G$_o$. While these results demonstrate that G$_o$ is capable of coupling receptors to inhibition of calcium current, it does not rule out the alternative possibility—that another PTX-sensitive G protein is the endogenous transduction molecule. Thus, although G protein reconstitution experiments offer valuable information as to whether a particular molecule transduces a specific signal, additional techniques must be employed before an unequivocal identification can be made. Development of such techniques remains one of the most important future challenges in this area.

Direct Actions of G Proteins on Ion Channels

In most of the preparations discussed above, receptor-mediated inhibition of calcium current is thought to involve G-protein activation of a second messenger pathway. Calcium channel inhibition would, therefore, be dependent on protein phosphorylation due to the second messenger-activated protein kinase (Sections II. C and III). Recently, however, G proteins have been demonstrated to have direct modulatory actions on both calcium and potassium (K^+) channels. The best characterized examples of direct G protein actions are in cardiac muscle, and include muscarinic activation of a voltage-dependent K^+ channel and adrenergic potentiation of voltage-dependent calcium channel activity. Although adrenergic potentiation of cardiac calcium current can proceed through the classic G_s/cAMP/cAMP-dependent protein kinase pathway, Yatani and colleagues (1987) have also demonstrated a direct modulatory influence of the receptor-activated G_s. Thus, application of activated G_s, or its α subunit, to excised membrane patches from ventricular myocytes causes an increase in the opening probability of the calcium channel similar to that observed with the adrenergic agonist isoproterenol, cAMP, or the A-kinase activator forskolin. This effect was observed in the absence of ATP, a condition which precludes activation of kinases which may adhere to the patch membrane. In addition, Yatani *et al.* reconstituted calcium channels into lipid bilayers and showed that these also were subject to direct modulation by G_s.

The only evidence suggesting a direct action of G proteins on neuronal calcium channels has come from neonatal rat sympathetic neurons. Wanke and colleagues (1987) have shown that although acetylcholine inhibition of a transient calcium current in these cells is blocked by both GDPβS and PTX, the effect is not mimicked by application of either cAMP-dependent kinase or protein kinase C activators. Thus the two major second messenger/protein kinase pathways have been ruled out, suggesting the possibility of receptor-to-channel coupling directly through the G protein. Unequivocal proof for this scheme will require excised patch experiments similar to those described in the preceding paragraph. It should be noted that Wanke *et al.* (1987) have described a transient calcium current, which is kinetically quite different from the L-type current modulated via more complex second messenger-driven mechanisms in other preparations.

The physiological significance of direct actions of G proteins on channels remains to be determined. Additionally, the functional significance of two separate mechanisms to increase voltage-dependent calcium current is as yet unclear. Although it is, at present, a finding unique to cardiac muscle, it will be interesting to see whether similar mechanisms underlie calcium channel modulation in other systems. Chapter 3 by Neer and Clapham and Chapter 11 by Yatani, Codina, and Brown will examine these issues in much more detail.

C. Second Messenger Systems Associated with Calcium Current Inhibition

1. cAMP/cAMP-Dependent Kinase

The cAMP/cAMP-dependent protein kinase signal transduction system was demonstrated to mediate modulatory transmitter effects for the vertebrate cardiac calcium channel (Tsien, 1973) and for voltage-dependent channels in *Aplysia* (Kaczmarek *et al.,* 1980; Klein and Kandel, 1978). Fedulova *et al.* (1981) also suggested that proper functioning of the neuronal calcium channel was dependent on maintenance of intracellular cAMP at some minimal level. Galvan and Adams (1982) synthesized this information to hypothesize that receptor-mediated inhibition of adenylyl cyclase, and the subsequent decrease in intracellular cAMP levels, may have been responsible for the norepinephrine-mediated inhibition of calcium current which they observed in rat sympathetic ganglion cells.

Further support for Galvan and Adams' hypothesis comes from the observation that the receptors responsible for modulation of calcium current have been shown, in certain other systems, to cause inhibition of adenylyl cyclase (Sabol and Nirenberg, 1979; Wojcik and Neff, 1984). In addition, PTX, which blocks transmitter-induced inhibition of calcium current in all cases for which G proteins have been implicated (Section II.B), is well-known to block the action of G_i, the G protein responsible for cyclase inhibition. Despite this circumstantial evidence, however, it appears unlikely that decreases in cAMP are responsible for the inhibition of neuronal calcium current, since the transmitter effects are not blocked by application of either membrane-permeable cAMP analogs or the adenylyl cyclase activator forskolin, nor are they affected by internal perfusion of cells with millimolar concentrations of cAMP [Akopyan *et al.,* 1985; Forscher and Oxford, 1985; Paupardin-Tritsch *et al.,* 1985; Holz *et al.,* 1986a; Lewis *et al.,* 1986; Dolphin and Scott, 1987; Wanke *et al.,* 1987; Hammond *et al.,* 1987; Rosenthal *et al.,* 1988; Ikeda and Schofield, 1989; but note that Gray and Johnston (1987) have reported cAMP-dependent potentiation of calcium current in hippocampal granule cells]. The failure of these various treatments, even in the presence of phosphodiesterase inhibitors, argues against the involvement of cAMP and kinase A in transmitter inhibition of calcium current in the majority of cases.

There has been one report of a cGMP/cGMP-dependent kinase-induced modulation of calcium current. This system is proposed to mediate serotonin potentiation of calcium current in neurons D6 and D7 of *Helix* (Paupardin-Tritsch *et al.,* 1986). Although cGMP has also been reported to be involved in the acetylcholine-induced inhibition of calcium current in frog ventricular cells, it is now suggested that the effects of cGMP are indirect, produced via a cGMP-dependent phosphodiesterase. Activation of the enzyme hydrolyzes cAMP and thereby reverses any basal cAMP-mediated enhancement of calcium channel activity (Hartzell and Fischmeister, 1986; Fischmeister and Hartzell, 1987; but see Hescheler *et al.,* 1986 for another interpretation).

2. Diacyglycerol/Protein Kinase C

While it appears unlikely that cAMP mediates the transmitter-induced inhibition of neuronal calcium channels, the diacyglycerol/protein kinase C (DAG/PKC) second messenger system is gaining favor as the signal transduction pathway linking the transmitter/receptor–G-protein interaction with subsequent alterations in calcium-channel function. Free DAG is produced from plasma membrane phosphatidylinositol bisphosphate by the specific action of G protein-regulated phospholipase C (Nishizuka, 1984; see chapter by Majerus). DAG causes translocation of PKC from the cytosol into the membrane, where, in the presence of phosphatidylserine and calcium, it becomes an active kinase.

Exogenously applied synthetic activators of PKC, such as phorbol esters and the DAG analog 1,2-oleoylacetylglycerol (OAG) (Nishizuka, 1984) have been used to determine whether activation of PKC affects calcium channel function. In several preparations, PKC activators have been found to mimic the effects of transmitters which inhibit calcium current (Rane and Dunlap, 1986; Hammond *et al.,* 1987; Lewis and Weight, 1988; Marchetti and Brown, 1988). Studies utilizing intracellular injection of either a specific PKC inhibitor or the kinase itself have provided more convincing evidence that PKC plays a critical physiological role in a number of systems. Details of these studies are given in Section III.

Phorbol esters and OAG have been used to suggest a role for PKC in the modulation of other channels as well. Activation of PKC inhibits: (1) the 'M' type K^+ current, as does bradykinin, in rat sympathetic neurons (Higashida and Brown, 1986), (2) a slow calcium-activated K^+ current, as does acetylcholine, in hippocampal pyramidal cells (Malenka *et al.,* 1986), and (3) voltage-dependent Cl^- currents measured in hippocampal pyramidal cells (Madison *et al.,* 1986; Farley and Auerbach, 1986).

III. INVOLVEMENT OF PROTEIN KINASE C IN CALCIUM CHANNEL MODULATION

A. Transmitter-Induced Inhibition of Calcium Current Mimicked by Exogenously Applied Protein Kinase C Activators

As mentioned above, the action of the endogenous PKC activator, DAG, can be mimicked with the tumor-promoting phorbol esters (Castagna *et al.,* 1982), and synthetic DAG analogs such as OAG (Mori *et al.,* 1982). These compounds intercalate into the cell membrane where they specifically activate PKC (Boni and Rando, 1985; Castagna *et al.,* 1982). Thus, inhibition of current by these compounds can be taken as preliminary evidence that PKC is involved in the modulatory mechanism. Phorbols having the 4-hydroxyl group in the α rather than the β position do not activate PKC but they are otherwise chemically identical to the active compounds. The α-phorbols can, therefore, be used to control for nonspecific effects of phorbol ester treatment (VanDuuren *et al.,* 1979; Nichols *et al.,* 1987). OAG and some of its congeners are amphipathic, allowing for their application in physiological solutions to voltage-clamped neurons either in culture or *in situ.* The structural requirements of these synthetic DAG-like compounds for PKC activation have been well documented (Ganong *et al.,* 1986; Boni and Rando, 1985; Mori *et al.,* 1982). These studies show that there is a greater degree of chemical dissimilarity between the active and inactive DAG analogs than is the case for the phorbols. This dissimilarity makes the inactive DAG analogs poor controls for physiological experiments, in which nonspecific membrane effects of these partially lipophilic compounds must be of some concern. When used with these cautions in mind, however, phorbol esters and OAG offer important information concerning the involvement of PKC in calcium channel modulation.

Application of phorbol esters and OAG from blunt-tipped micropipettes onto cell bodies inhibits calcium current in chick DRG neurons (Rane and Dunlap, 1986; Marchetti and Brown, 1988), identified neurons from *Helix* (Hammond *et al.,* 1987), and cells of the AtT-20 and GH$_3$ lines (Lewis and Weight, 1988; Marchetti and Brown, 1988). The onset time course and reversibility of the OAG effects are similar to those of the transmitters known to inhibit the long-lasting, or L-type, calcium current in these preparations (Table I). OAG inhibits calcium current at a threshold concentration of $0.25-0.6\ \mu M$ and maximal effects are observed at $60\ \mu M$ (Rane and Dunlap, 1986; Hammond *et al.,* 1987). This dose-response relationship is essentially identical to that of OAG concentration and PKC-dependent phosphorylation observed in platelets (Kaibuchi et al., 1983) and in HL-60 cells (Ebeling et al., 1985).

Phorbol esters also decrease action potential duration in mouse DRG neurons (Werz and Macdonald, 1987) and they attenuate voltage-dependent, dihydropyridine-sensitive ^{45}Ca uptake in the PC-12 cell line (DiVirgilio et al., 1986; Harris et al., 1986; Messing et al., 1986). It remains to be seen whether these effects are due to inhibition of the electrophysiologically well characterized L-type calcium current.

The lack of additivity of transmitter effects and those of PKC activators has been cited as evidence for an involvement of PKC in transmitter inhibition of voltage-dependent calcium current. Thus concentrations of OAG or phorbol which maximally inhibit the current also block inhibition by norepinephrine in chick DRG neurons (Fig. 1), and cholecystokinin (CCK) in *Helix* central neurons. These results suggest that both the transmitters, and OAG or phorbols, inhibit calcium current through a common saturable pathway, presumably involving activation of PKC.

B. Action of Protein Kinase C on Calcium Channels in *Aplysia* Bag Cell Neurons

Aplysia bag cell neurons are peptidergic secretory cells which exhibit large, voltage-dependent calcium currents. DeRiemer and colleagues (1985) have shown that the current is greatly enhanced when cells are treated with phorbol ester, and intracellular injection of PKC via a microelectrode causes a change in action potential amplitude consistent with an increase in calcium current. Such increases in calcium current could be observed directly by voltage-clamping cells pretreated with phorbol ester. Currents from these pretreated cells were, on average, larger than those recorded from control cells. Interestingly, DeRiemer et al. observed that phorbol ester was only effective when applied to the bag cell neurons prior to electrical recording with patch pipettes. They concluded that there was a soluble intracellular component essential to PKC action and that this was lost during perfusion with the patch electrode solution (DeRiemer et al., 1985). Thus, the effect of PKC on bag cell neurons is different from the other examples of calcium channel regulation mentioned above in that: (1) stimulation of the kinase results in upmodulation of the current, and (2) the effect can be prevented by cell perfusion, perhaps due to washout of an essential cofactor or intermediate.

The effect of PKC in *Aplysia* bag cell neurons is also unique in that it causes the appearance of a previously quiescent calcium channel. Using

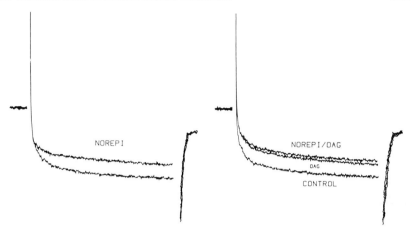

Fig. 1. The protein kinase C activator 1,2-oleoylacetylglycerol (OAG) occludes norepinephrine inhibition of calcium current in chick dorsal root ganglion neurons grown in culture (Dunlap and Fischbach, 1981). Whole-cell patch clamp recordings show voltage-dependent calcium current activated in response to 60 mV depolarizing steps from a holding potential of −60 mV. In the left panel, 10 μM norepinephrine reduced calcium current by about 30%. After recovery of the current, a maximally effective concentration of OAG (60 μM) reduced the current by approximately the same amount. Norepinephrine was then immediately applied in the continued presence of OAG, and the current (NOREPI/OAG) declined by, at most, another 5 to 10%, suggesting that the transmitter and the protein kinase C activator inhibit current through a common saturable pathway. Peak control currents are approximately 1 nA; voltage pulses were 40 msec in duration.

the cell-attached patch recording configuration to study single channel activity, Strong and colleagues (1987) reported that phorbol-treated cells displayed a 23-pS barium-permeant channel which was rare or absent in untreated cells. Control cells had predominantly a 10-pS channel with a voltage dependence similar to the larger conductance channel. This action of PKC has been postulated to underlie long-term changes in the excitability of the bag cell neuron which are caused by certain neuronal inputs, although the endogenous agent responsible for such effects is, at present, unidentified.

C. Intracellular Injection of Protein Kinase C

The large cell bodies of molluscan neurons allow injection of high-molecular-weight proteins via intracellular micropipettes. DeRiemer and colleagues (1985) used intracellular injection of PKC into *Aplysia* bag cell neurons to cause augmentation of calcium current identical to that observed with external application of phorbol ester. The injection of PKC thus provides a very strong corroboration of the results obtained with phorbol.

Likewise, intracellular injection of PKC into cells F77 and D2 of the *Helix* abdominovisceral ganglion has been used to strongly implicate the kinase in mediating CCK-induced inhibition of calcium current (Hammond *et al.*, 1987). Within 10 minutes of injection, PKC caused a large reduction in calcium current similar to that seen in response to either CCK or OAG. Injection of low concentrations of PKC, which were themselves ineffective, enhanced the reduction in calcium current due to CCK. These results suggest that PKC is activated during CCK-induced inhibition of calcium current, and that PKC somehow interacts with the channel to reduce whole-cell calcium current.

A question remains, however, as to how the injected PKC is activated in the absence of agonist-induced DAG production. PKC is thought to exist mainly in an inactive soluble form (Kikkawa *et al.*, 1982), although it may bind to the membrane under certain conditions of elevated internal calcium (Wolf *et al.*, 1985). In either case, it is thought to require free DAG in the presence of calcium and phosphatidylserine for its activation. Under certain purification procedures PKC can be proteolytically converted to a nonregulated form which is constitutively active, independent of DAG, calcium and phosphatidylserine (Takai *et al.*, 1977; Inoue *et al.*, 1977). The studies in both *Aplysia* (DeRiemer *et al.*, 1985) and *Helix* (Hammond *et al.*, 1987), however, used a PKC purification procedure which should have produced only the intact, DAG-sensitive kinase. Indeed, the experiments showing PKC potentiation of the CCK effect in *Helix* suggest that at least some fraction of the injected kinase can respond to agonist-generated second messenger. One possible explanation for the apparently spontaneous activation of PKC is that the molluscan neurons contain a proteolytic enzyme which can cleave and activate the injected enzyme.

Activation of PKC by an endogenous protease may explain why the effects of PKC injection and CCK application are irreversible in *Helix*. During purification from mammalian neural tissue, PKC can be converted to its constitutively active form by endogenous, calcium-dependent, neutral thiol proteases. These proteases can be defeated by very strong calcium buffering (Kikkawa *et al.*, 1982) or by the inhibitor leupeptin (Kishimoto *et al.*, 1983). Chad and Eckert (1986) have shown that when *Helix* neurons are internally dialyzed with solutions containing little or no calcium buffering, there is an irreversible loss of calcium current associated with the entry of calcium from repeated voltage clamp depolarizations. This loss can be prevented by leupeptin, an inhibitor of calcium-dependent protease. It is interesting to speculate that the decrease in calcium channel activity observed by Chad and Eckert might be the result of calcium channel down-modulation by endogenous PKC converted to its nonregulated form by the rise in internal calcium. The irreversibility of CCK, OAG, and PKC effects observed by Hammond *et al.* could, thus, reflect the same basic phenomenon. PKC activation due to increases in calcium at the intracellular opening of the channel may be an important feedback mechanism for regulating calcium influx, although the cell might be expected to have some means of "cleaning up" constitutive PKC.

D. Specific Protein Kinase C Inhibitor Blocking Calcium Channel Modulation in Chick Dorsal Root Ganglion Neurons

As discussed above, initial evidence for the role of PKC in norepinephrine-induced inhibition of chick DRG-neuron calcium current came from experiments showing that OAG and phorbol esters reduce calcium current in a dose-dependent manner (Rane and Dunlap, 1986), and that a maximally effective dose of OAG occludes the action of norepinephrine (Fig. 1). The involvement of cAMP and its kinase were ruled out (Section II.C.1).

To establish that PKC is a necessary intermediate in transmitter-induced inhibition of the calcium current, we have made use of a specific PKC inhibitor (PKCI) isolated from bovine brain. The purification and characterization of this inhibitor by Walsh, McDonald and colleagues (McDonald and Walsh, 1985a,b, 1986; McDonald *et al.*, 1987) has allowed for a direct test of PKC involvement in transmitter-mediated modulation of calcium current. The inhibitor is a 17-kDa protein which, in the micromolar range, inhibits *in vitro* PKC-induced phosphorylation by more than 90%. It has no effect on a variety of other kinases including cAMP-dependent kinase, calcium/calmodulin-dependent myosin light chain kinase, casein kinase, or the epidermal growth factor receptor kinase.

PKCI, intracellularly applied via the patch pipette solution, attenuates norepinephrine-mediated inhibition of DRG neuron calcium current in a dose-dependent fashion (Fig. 2). Calcium-current inhibition due to OAG is also reduced to less than 20% of control by 150 μg/ml PKCI. Intracellular injection of PKCI is ineffective if the inhibitor is digested with trypsin (McDonald and Walsh, 1986) prior to being applied to cells (S. G. Rane, unpublished). Thus, these results strongly suggest that PKC activation is essential to transmitter modulation of calcium current in chick DRG neurons. Experiments using intracellular application of constitutively active PKC will provide convincing corroboration of the inhibitor work. The prediction is that, even in the absence of transmitter, this form of PKC will cause irreversible inhibition of calcium current. Furthermore, the nonreg-

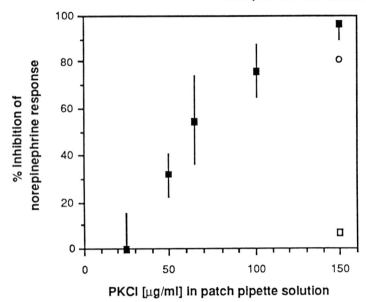

Fig. 2. Norepinephrine- and OAG-induced inhibition of calcium current in cultured chick dorsal root ganglion neurons is blocked by a specific protein kinase C inhibitor (PKCI, see Section III.D). Neurons were voltage clamped using the whole-cell recording configuration of the patch clamp technique and norepinephrine (10 μM, filled squares) or the protein kinase C activator OAG (60 μM, open circle) were externally applied from blunt-tipped pipettes positioned close to the cell. Norepinephrine responses were recorded and averaged from populations of neurons dialyzed with various concentrations of PKCI (4–10 cells/point) and compared to the averaged responses from a population of neurons dialyzed with control solution (4–7 cells/point) without added inhibitor. Attenuation of the OAG response was examined at only one concentration of PKCI (150 $\mu g/ml$). The differences in the values for the PKCI-treated versus the control cells were plotted as a percentage inhibition of the transmitter- or OAG-induced decrease in calcium current vs. PKCI concentration. Trypsin digestion of PKCI lowered its effectiveness in inhibiting the norepinephrine response (open square).

ulated PKC at high concentrations should occlude the action of both norepinephrine and OAG.

In contrast to the direct effects of G-protein subunits reported in other systems (see Chapter 3 by Neer and Clapham and Chapter 11 by Yatani, Codina, and Brown), the results of our experiments, showing that PKCI completely blocks norepinephrine-mediated inhibition of DRG neuron calcium current, argue against a direct action of the PTX-sensitive G protein known to be essential to transmitter effects in these cells (Holz *et al.,* 1986a). The most direct experiment to verify this conclusion would be to test activated G proteins on excised membrane patches containing calcium channels. This technique allows recording from calcium channels in the absence of cytoplasmic components such as soluble second messengers or enzymes. In addition, ATP can be excluded from the bathing solution to prevent phosphorylation-dependent changes in calcium channel activity.

E. Problems with Protein Kinase C Hypothesis for Calcium Current Inhibition

There are some difficulties with the idea that transmitters inhibit calcium current by causing phospholipase-induced breakdown of membrane phosphatidylinositol bisphosphate (PIP_2) to yield free DAG, an endogenous PKC activator. The receptor types known to inhibit calcium current, in particular the $GABA_B$ and α_2-like receptors in chick DRG neurons (Section II.A), have not been associated with PIP_2 breakdown in other systems (Nishizuka, 1984). Indeed, there is some evidence in neuronal tissue that they may inhibit adenylyl cyclase (Section II.C.1). There are now, however, preliminary studies to suggest that these receptors do stimulate PIP_2 breakdown in chick DRG neurons (Harish *et al.*, 1987) and the release of DAG in rat sympathetic neurons (Perney and Miller, 1987) with a time course and pharmacology consistent with their effects on calcium channels. Completion and corroboration of this work is a critical step in defining the biochemical intermediates responsible for the transmitter-induced modulation of calcium channel function.

While it is important that transmitter-stimulated PIP_2 turnover and the release of DAG be examined for each preparation in which PKC is thought to mediate transmitter action, recent work suggests another potential mechanism for triggering the activation of PKC. This involves the transmitter-induced release of compounds other than DAG which are, nonetheless, capable of activating PKC. Specifically, certain fatty acids have been reported to activate PKC *in vitro* (Murakami *et al.*, 1986; McPhail *et al.*, 1984; Hansson *et al.*, 1986; Sekiguchi *et al.*, 1987), and one of these, arachidonic acid, can be liberated by extracellular agonists in a PTX-sensitive fashion (Okajima and Ui, 1984; Bokoch and Gilman, 1984).

Finally, although the importance of a PTX-sensitive G protein in PIP_2 breakdown was at first questioned, it is now clear from a number of cell types that receptor-driven PIP_2 turnover can be inhibited by PTX (see chapters by Deckmyh and Litosch in this volume, and review by Cockcroft, 1987). Worley *et al.* (1986b) have used immunohistochemistry and PTX-ribosylation to localize G_o, a G protein which may regulate PIP_2 turnover, in the brain, and they have correlated its occurrence with that of tritiated phorbol binding, an indicator of PKC activity (Worley *et al.*,

1986a). In addition, antibodies directed against PKC and PKC subspecies have been used in conjunction with electron microscopy to suggest that, in certain brain regions, the enzyme is more prevalent in presynaptic terminals as opposed to postsynaptic specializations (Girard *et al.,* 1985; Saito *et al.,* 1988; Ase *et al.,* 1988). Since voltage-dependent calcium channels are an important component of presynaptic endings where they mediate exocytotic release of transmitter, PKC activation and calcium channel inhibition in nerve terminals would have significant effects on synaptic transmission or the release of neurohormones.

IV. CONCLUSIONS

GTP-binding proteins regulate the activity of a variety of target molecules, including membrane channels. Our understanding of the mechanisms by which these signal transducers carry out their actions on calcium channels is, at present, rudimentary, and many questions remain to be answered. In most systems studied thus far, calcium channel behavior can be modified by activating G proteins, although it remains unclear as to what subspecies of G protein is physiologically important. In addition, the question of whether the calcium channel itself is the target for G-protein regulation or whether alterations in calcium channel function occur indirectly through G protein-mediated activation of secondary target molecules, has only been addressed in a few systems. Recent findings from work on several neuronal preparations demonstrate that an important means of regulating calcium channel function involves not only G proteins but also the activation of protein kinase C. The mechanism by which the kinase is activated is unknown, but may involve a G protein-regulated turnover of membrane phospholipid. This work emphasizes that understanding the complex molecular events underlying the interactions between G proteins and their targets remains one of the most interesting challenges for the future.

REFERENCES

Akopyan, A. R., Chemeris, N. K., and Ilyin, V. I. (1985). Neurotransmitter-induced modulation of neuronal Ca^{2+} is not mediated by intracellular Ca^{2+} or cAMP. *Brain Res.* **326,** 145–148.
Ase, K., Saito, N., Shearman, M. S., Kikkawa, U., Ono, Y., Igarashi, K., Tanaka, C., and Nishizuka, Y. (1988). Distinct cellular expression of βI- and βII-subspecies of protein kinase C in rat cerebellum. *J. Neurosci.* **8,** 3850–3856.
Baraban, J. M., Snyder, S. H., and Alger, B. E. (1985). Protein kinase C regulates

ionic conductance in hippocampal pyramidal neurons: Electrophysiological effects of phorbol esters. *Proc. Natl. Acad. Sci. U.S.A.* **82,** 2538–2542.

Bokoch, G. M., and Gilman, A. G. (1984). Inhibition of receptor-mediated release of arachidonic acid by pertussis toxin. *Cell* **39,** 301–308.

Boni, L. T., and Rando, R. R. (1985). The nature of protein kinase C activation by physically defined phospholipid vesicles and diacylglycerols. *J. Biol. Chem.* **260,** 10819–10825.

Brezina, V. (1987). Non-hydrolyzable analog of GTP mimics modulation of K and calcium currents by FMRF-amide in *Aplysia* neurons. *Biophys. J.* **51,** 251a.

Brezina, V., Erxleben, C., and Eckert, R. (1985). FMRF amide suppresses calcium current in *Aplysia* neurons. *Biophys. J.* **47,** 435a.

Canfield, D. R., and Dunlap, K. (1984). Pharmacological characterization of amine receptors on embryonic chick sensory neurones. *Br. J. Pharmacol.* **82,** 557–563.

Carbone, E., and Lux, H. D. (1984). A low voltage-activated, fully inactivating calcium channel in vertebrate sensory neurones. *Nature* **310,** 501–502.

Castagna, M., Takai, Y., Kaibuchi, K., Sano, K., Kikkawa, U., and Nishizuka, Y. (1982). Direct activation of calcium-activated, phospholipid-dependent protein kinase by tumor-promoting phorbol esters. *J. Biol. Chem.* **257,** 7847–7851.

Chad, J. E., and Eckert, R. (1986). An enzymatic mechanism for calcium current inactivation in dialyzed *Helix* neurones. *J. Physiol.* **378,** 31–51.

Cockcroft, S. (1987). Polyphosphoinositide phosphodiesterase: Regulation by a novel guanine nucleotide binding protein, G_p. *Trends Biochem. Sci.* **12,** 75–78.

Colombaioni, L., Paupardin-Tritsch, D., Vidal, P. P., and Gerschenfeld, H. M. (1985). The neuropeptide FMRF-amide decreases both the Ca^{2+} conductance and a cyclic $3',5'$-adenosine monophosphate-dependent K^+ conductance in identified molluscan neurons. *J. Neurosci.* **5,** 2533–2538.

DeRiemer, S. A., Strong, J. A., Albert, K. A., Greengard, P., and Kaczmarek, L. K. (1985). Enhancement of calcium current in *Aplysia* neurones by phorbol ester and protein kinase C. *Nature* **313,** 313–316.

DiVirgilio, F., Pozzan, T., Wollheim, C. B., Vincentini, L. M., and Meldolesi, J. (1986). Tumor promoter phorbol myristate acetate inhibits Ca^{2+} influx through voltage-gated Ca^{2+} channels in two secretory cell lines, PC 12 and RINm5F. *J. Biol. Chem.* **261,** 32–35.

Dolphin, A. C., and Scott, R. H. (1987). Calcium channel currents and their inhibition by (-)-baclofen in rat sensory neurones: Modulation by guanine nucleotides. *J. Physiol.* **386,** 1–17.

Dolphin, A. C., Forda, S. R., and Scott, R. H. (1986). Calcium-dependent currents in cultured rat dorsal root ganglion neurones are inhibited by an adenosine analogue. *J. Physiol.* **373,** 47–61.

Dunlap, K. (1981). Two types of γ-aminobutyric acid receptor on embryonic sensory neurones. *Br. J. Pharmacol.* **74,** 579–585.

Dunlap, K., and Fischbach, G. D. (1978). Neurotransmitters decrease the calcium component of sensory neurone action potentials. *Nature* **276,** 837–839.

Dunlap, K., and Fischbach, G. D. (1981). Neurotransmitters decrease the calcium conductance activated by depolarization of embryonic chick sensory neurones. *J. Physiol.* **317**, 519–535.

Ebeling, J. G., Vandenbark, G. R., Kuhn, L. J., Ganong, B. R., Bell, R. M., and Neidel, J. E. (1985). Diacylglycerols mimic phorbol diester induction of leukemic cell differentiation. *Proc. Natl. Acad. Sci. U.S.A.* **82**, 815–819.

Ewald, D. A., Sternweis, P. C., and Miller, R. J. (1988). Guanine nucleotide-binding protein G_o-induced coupling of neuropeptide Y receptors to Ca^{2+} channels in sensory neurons. *Proc Natl. Acad. Sci. U.S.A.* **85**, 3633–3637.

Ewald, D. A., Pang, I. -H., Sternweis, P. C., and Miller, R. J. (1989). Differential G protein-mediated coupling of neurotransmitter receptors to Ca^{2+} channels in rat dorsal root ganglion neurons *in vitro. Neuron* **2**, 1185–1193.

Farley, J., and Auerbach, S. (1986). Protein kinase C activation induces conductance changes in *Hermissenda* photoreceptors like those seen in associative learning. *Nature* **319**, 220–223.

Fedulova, S. A., Kostyuk, P. G., and Veselovsky, N. S. (1981). Calcium channels in the somatic membrane of the rat dorsal root ganglion neurons, effect of cAMP. *Brain Res* **214**, 107–128.

Fischmeister, R., and Hartzell, H. C. (1987). Cyclic guanosine 3′,5′-monophosphate regulates the calcium current in single cells from frog ventricle. *J. Physiol.* **387**, 453–472.

Forscher, P., and Oxford, G. S. (1985). Modulation of calcium channels by norepinephrine in internally dialyzed avian sensory neurons. *J. Gen. Physiol.* **85**, 743–763.

Fox, A. P., Nowycky, M. C., and Tsien, R. W. (1987a). Kinetic and pharmacological properties distinguishing three types of calcium currents in chick sensory neurons. *J. Physiol.* **394**, 149–172.

Fox, A. P., Nowycky, M. C., and Tsien, R. W. (1987b). Single channel recordings of three types of calcium channels in chick sensory neurons. *J. Physiol.* **394**, 173–200.

Galvan, M., and Adams, P. R. (1982). Control of calcium current in rat sympathetic neurons by norepinephrine. *Brain Res.* **244**, 135–144.

Ganong, B. R., Loomis, C. R., Hannun, Y. A., and Bell, R. M. (1986). Specificity and mechanism of protein kinase C activation by sn-1,2-diacylglycerols. *Proc. Natl. Acad. Sci. U.S.A.* **83**, 1184–1188.

Girard, P. R., Mazzei, G. J., Wood, J. G., and Kuo, J. F. (1985). Polyclonal antibodies to phospholipid/Ca^{2+}-dependent protein kinase and immunocytochemical localization of the enzyme in rat brain. *Proc. Natl. Acad. Sci. U.S.A.* **82**, 3030–3034.

Gray, R., and Johnston, D. (1987). Noradrenaline and β-adrenoceptor agonists increase activity of voltage-dependent calcium channels in hippocampal neurons. *Nature* **327**, 620–622.

Gross, R. A., and Macdonald, R. L. (1987). Dynorphin A reduces a large transient (N-type) calcium current of mouse dorsal root ganglion neurons in cell culture. *Proc. Natl. Acad. Sci. U.S.A.* **84**, 5469–5473.

Hamill, O. P., Marty, A., Neher, E., Sakmann, B., and Sigworth, F. J. (1981).

Improved patch clamp techniques for high-resolution recording from cells and cell-free membrane patches. *Pfluegers Arch.* **391**, 85–100.

Hammond, C., Paupardin-Tritsch, D., Nairn, A. C., Greengard, P., and Gerschenfeld, H. M. (1987). Cholecystokinin induces a decrease in Ca^{2+} current in snail neurons that appears to be mediated by protein kinase C. *Nature* **325**, 809–811.

Hansson, A., Serhan, C. N., Haeggstrom, J., Ingelman-Sundberg, M., and Samuelsson, B. (1986). Activation of protein kinase C by lipoxin A and other eicosanoids. Intracellular action of oxygenation products of arachidonic acid. *Biochem. Biophys. Res. Commun.* **134**, 1215–1222.

Harish, O. E., Dunlap, K., and Role, L. W. (1987). Phosphoinositide turnover and modulation of calcium current in dorsal root ganglion neurons are mediated by norepinephrine. *Soc. Neurosci. Absr.* **13**, 793.

Harris, K. M., Kongsamut, S., and Miller, R. J. (1986). Protein kinase C mediated regulation of calcium channels in PC-12 pheochromocytoma cells. *Biochem. Biophys. Res. Commun.* **134**, 1298–1305.

Harris-Warrick, R. M., Hammond, C., Paupardin-Tritsch, D., Homburger, V., Rouot, B., Bockaert, J., and Gerschenfeld, H. M. (1988). An α_{40} subunit of a GTP-binding protein immunologically related to G$_o$ mediates a dopamine-induced decrease of Ca^{2+} current in snail neurons. *Neuron* **1**, 27–32.

Hartzell, H. C., and Fischmeister, R. (1986). Opposite effects of cyclic GMP and cyclic AMP on Ca^{2+} current in single heart cells. *Nature* **323**, 273–275.

Hescheler, J., Kameyama, M., and Trautwein, W. (1986). On the mechanism of muscarinic inhibition of the cardiac calcium current. *Pfluegers Arch.* **407**, 182–189.

Hescheler, J., Rosenthal, W., Trautwein, W., and Schultz, G. (1987). The GTP-binding protein, G$_o$, regulates neuronal calcium channels. *Nature* **325**, 445–447.

Higashida, H., and Brown, D. A. (1986). Two polyphosphoinositide metabolites control two K$^+$ currents in a neuronal cell. *Nature* **323**, 333–335.

Holz, G. G., Rane, S. G., and Dunlap, K. (1986a). GTP-binding proteins mediate transmitter inhibition of voltage-dependent calcium channels. *Nature* **319**, 670–672.

Holz, G. G., Shefner, S. A., and Anderson, E. G. (1986b). Serotonin decreases the duration of action potentials recorded from tetraethylammonium-treated bullfrog dorsal root ganglion cells. *J. Neurosci.* **6**, 620–626.

Horn, J. P., and McAfee, D. A. (1980). Alpha-adrenergic inhibition of calcium-dependent potentials in rat sympathetic neurones. *J. Physiol.* **301**, 191–204.

Ikeda, S. R., and Schofield, G. G. (1989). Somatostatin blocks a calcium current in rat sympathetic ganglion neurones. *J. Physiol.* **409**, 221–240.

Inoue, M., Kishimoto, A., Takai, Y., and Nishizuka, Y. (1977). Studies on a cyclic nucleotide-independent protein kinase and its proenzyme in mammalian tissues. II. Proenzyme and its activation by calcium-dependent protease from rat brain. *J. Biol. Chem.* **252**, 7610–7616.

Kaczmarek, L. K., Jennings, K. R., Strumwasser, F., Nairn, A. C., Walter, V., Wilson, F. D., and Greengard, P. (1980). Microinjection of catalytic subunit of cyclic AMP-dependent protein kinase enhances calcium action potentials of bag cell neurones in cell culture. *Proc. Natl. Acad. Sci. U.S.A.* **75**, 3512–3516.

Kaibuchi, K., Takai, Y., Sawamura, M., Hoshijima, M., Fujikura, T., and Nishizuka, Y. (1983). Synergistic functions of protein phosphorylation and calcium mobilization in platelet activation. *J. Biol. Chem.* **258**, 6701–6704.

Kikkawa, U., Takai, Y., Minakuchi, R., Inohara, S., and Nishizuka, Y. (1982). Calcium-activated, phospholipid-dependent protein kinase from rat brain. *J. Biol. Chem.* **257**, 13341–13348.

Kishimoto, A., Kajikawa, N., Shiota, M., and Nishizuka, Y. (1983). Proteolytic activation of calcium-activated, phospholipid-dependent protein kinase by calcium-dependent neutral protease. *J. Biol. Chem.* **258**, 1156–1164.

Klein, M., and Kandel, E. R. (1978). Presynaptic modulation of voltage-dependent Ca^{2+} current: Mechanism for behavioral sensitization in *Aplysia californica. Proc. Natl. Acad. Sci. U.S.A.* **75**, 3512–3516.

Kretz, R., Shapiro, E., Bailey, C. H., Chen, C., and Kandel, E. R. (1986). Presynaptic inhibition produced by an identified presynaptic inhibitory neuron. II. Presynaptic conductance changes caused by histamine. *J. Neurophysiol* **55**, 131–146.

Lewis, D. L., and Weight, F. F. (1988). The protein kinase C activator *1*-oleoyl-2-acetylglycerol inhibits voltage-dependent Ca^{2+} current in the pituitary cell line AtT-20. *Neuroendocrinology* **47**, 169–175.

Lewis, D. L., Weight, F. F., and Luini, A. (1986). A guanine nucleotide-binding protein mediates the inhibition of voltage-dependent calcium current by somatostatin in a pituitary cell line. *Proc. Natl. Acad. Sci. U.S.A.* **83**, 9035–9039.

Luini, A., Lewis, D., Guild, S., Schofield, G., and Weight, F. (1986). Somatostatin, an inhibitor of ACTH secretion, decreases cytosolic free calcium and voltage-dependent calcium current in a pituitary cell line. *J. Neurosci.* **6**, 3128–3132.

Macdonald, R. L., and Werz, M. A. (1986). Dynorphin A decreases voltage-dependent calcium conductance of mouse dorsal root ganglion neurones. *J. Physiol.* **377**, 237–249.

Madison, D. V., Malenka, R. C., and Nicoll, R. A. (1986). Phorbol esters block a voltage-sensitive chloride current in hippocampal pyramidal cells. *Nature* **321**, 695–697.

Malenka, R. C., Madison, D. V., Andrade, R., and Nicoll, R. A. (1986). Phorbol esters mimic some cholinergic actions in hippocampal pyramidal neurons. *J. Neurosci.* **6**, 475–480.

Marchetti, C., and Brown, A. M. (1988). Protein kinase activator 1-oleoyl-2-acetyl-*sn*-glycerol inhibits two types of calcium currents in GH_3 cells. *Am. J. Physiol.* **254**, C206–C210.

Marchetti, C., Carbone, E., and Lux, H. D. (1986). Effects of dopamine and noradrenaline on calcium channels of cultured sensory and sympathetic neurons of chick. *Pfluegers Arch.* **406**, 104–111.

McDonald, J. R., and Walsh, M. P. (1985a). Inhibition of the Ca^{2+}- and phospholipid-dependent protein kinase by a novel M_r 17,000 Ca^{2+}-binding protein. *Biochem. Biophys. Res. Commun.* **129**, 603–610.

McDonald, J. R., and Walsh, M. P. (1985b). Ca^{2+}-binding proteins from bovine brain including a potent inhibitor of protein kinase C. *Biochem. J.* **232**, 559–567.

McDonald, J. R., and Walsh, M. P. (1986). Regulation of protein kinase C activity by natural inhibitors. *Biochem. Soc. Trans.* **14**, 585–586.

McDonald, J. R., Groschel-Stewart, U., and Walsh, M. P. (1987). Properties and distribution of the protein inhibitor (M_r 17,000) of protein kinase C. *Biochem. J.* **242**, 695–705.

McPhail, L. C., Clayton, C. C., and Snyderman, R. (1984). A potential second messenger role for unsaturated fatty acids: Activation of Ca^{2+}-dependent protein kinase. *Science* **224**, 622–625.

Messing, R. O., Carpenter, C. L., and Greenberg, D. A. (1986). Inhibition of calcium flux and calcium channel antagonist binding in the PC12 neuronal cell line by phorbol esters and protein kinase C. *Biochem. Biophys. Res. Commun.* **136**, 1049–1056.

Mori, T., Takai, Y., Yu, B., Takahashi, J., Nishizuka, Y., and Fujikura, T. (1982). Specificity of the fatty acid moieties of diacylglycerol for the activation of calcium-activated, phospholipid-dependent protein kinase. *J. Biochem.* **91**, 427–431.

Murakami, K., Chan, S. Y., and Routtenberg, A. (1986). Protein kinase C activation by *cis*-fatty acid in the absence of Ca^{2+} and phospholipids. *J. Biol. Chem.* **261**, 15424–15429.

Nichols, R. A., Haycock, J. W., Wang, J. K. T., and Greengard, P. (1987). Phorbol ester enhancement of neurotransmitter release from rat brain synaptosomes. *J. Neurochem.* **48**, 615–621.

Nishizuka, Y. (1984). The role of protein kinase C in cell surface signal transduction and tumor production. *Nature* **308**, 693–698.

Nowycky, M. C., Fox, A. P., and Tsien, R. W. (1985). Three types of neuronal calcium channels with different calcium agonist sensitivity. *Nature* **316**, 440–443.

Okajima, F., and Ui, M. (1984). ADP-ribosylation of the specific membrane protein by islet-activating protein, pertussis toxin, associated with inhibition of a chemotactic peptide-induced arachidonate release in neutrophils. *J. Biol. Chem.* **259**, 13863–13871.

Paupardin-Tritsch, D., Colombaioni, L., Deterre, P., and Gerschenfeld, H. M. (1985). Two different mechanisms of calcium spike modulation by dopamine. *J. Neurosci.* **5**, 2522–2532.

Paupardin-Tritsch, D., Hammond, C., Gerschenfeld, H. M., Nairn, A. C., and Greengard, P. (1986). cGMP-dependent protein kinase enhances Ca^{2+} current and potentiates the serotonin-induced Ca^{2+} current increase in snail neurons. *Nature* **323**, 812–814.

Perney, T. M., and Miller, R. J. (1987). Neurotransmitter regulation of lipid metabolism in cultured rat sensory neurons. *Soc. Neurosci. Absr.* **13**, 1135.

Rane, S. G., and Dunlap, K. (1986). Kinase C activator 1,2-oleoylacetylglycerol attenuates voltage-dependent calcium current in sensory neurons. *Proc. Natl. Acad. Sci. U.S.A.* **83**, 184–188.

Reuter, H. (1983). Calcium channel modulation by neurotransmitters, enzymes and drugs. *Nature* **301,** 569–574.

Rosenthal, W., Hescheler, J., Hinsch, K. -D., Spicher, K., Trautwein, W., and Schultz, G. (1988). Cyclic AMP-independent, dual regulation of voltage-dependent Ca²⁺ currents by LHRH and somatostatin in a pituitary cell line. *EMBO* **7,** 1627–1633.

Sabol, S. L., and Nirenberg, M. (1979). Regulation of adenylate cyclase of neuroblastoma x glioma hybrid cells by α-adrenergic receptors. *J. Biol. Chem.* **254,** 1913–1920.

Saito, N., Kikkawa, U., Nishizuka, Y., and Tanaka, C. (1988). Distribution of protein kinase C-like immunoreactive neurons in rat brain. *J. Neurosci.* **8,** 369–382.

Sekiguchi, K., Tsukuda, M., Ogita, K., Kikkawa, U., and Nishizuka, Y. (1987). Three distinct forms of rat protein kinase C: Differential response to unsaturated fatty acids. *Biochem. Biophys. Res. Commun.* **145,** 797–802.

Strong, J. A., Fox, A. P., Tsien, R. W., and Kaczmarek, L. K. (1987). Stimulation of protein kinase C recruits covert calcium channels in *Aplysia* bag cell neurons. *Nature* **325,** 714–717.

Takai, Y., Kishimoto, A., Inoue, M., and Nishizuka, Y. (1977). Studies on a cyclic nucleotide-independent protein kinase and its proenzyme in mammalian tissues. I. Purification and characterization of an active enzyme from bovine cerebellum. *J. Biol. Chem.* **252,** 7603–7609.

Toselli, M., and Lux, H. D. (1989). GTP-binding proteins mediate acetylcholine inhibition of voltage dependent calcium channels in hippocampal neurons. *Pflugers Arch.* **413,** 319–321.

Tsien, R. W. (1973). Adrenaline-like effects of intracellular iontophorosis of cyclic AMP in cardiac Purkinje fibers. *Nature (New Biol.)* **245,** 120–122.

Tsien, R. W. (1987). Calcium currents in heart cells and neurons. *In* "Neuromodulation" (L. K. Kaczmarek and I. B. Levitan, eds.), pp. 206–242. Oxford Univ. Press, New York.

Tsunoo, A., Yoshi, M., and Narahashi, T. (1986). Block of calcium channels by enkephalin and somatostatin in neuroblastoma-glioma hybrid NG108-15 cells. *Proc. Natl. Acad. Sci. U.S.A.* **83,** 9832–9836.

VanDuuren, B. L., Tseng, S. -S., Segal, A., Smith, A. C., Melchiorre, S., and Seidman, I. (1979). Effects of structural changes on the tumor-promoting activity of phorbol myristate acetate on mouse skin. *J. Neurochem.* **39,** 2644–2646.

Walker, M. W., Ewald, D. A., Perney, T. M., and Miller, R. J. (1988). Neuropeptide Y modulates neurotransmitter release and Ca²⁺ currents in rat sensory neurons. *J. Neurosci.* **8,** 2438–2446.

Wanke, E., Ferroni, A., Malgaroli, A., Ambrosini, A., Pozzan, T., and Meldolesi, J. (1987). Activation of a muscarinic receptor selectively inhibits a rapidly inactivated Ca²⁺ current in rat sympathetic neurons. *Proc. Natl. Acad. Sci. U.S.A.* **84,** 4313–4317.

Werz, M. A., and Macdonald, R. L. (1987). Phorbol esters: Voltage-dependent effects on calcium-dependent action potentials of mouse central and peripheral neurons in cell culture. *J. Neurosci.* **7,** 1639–1647.

Williams, J. T., and North, R. A. (1985). Catecholamine inhibition of calcium action potentials in rat locus coeruleus neurones. *Neuroscience* **14,** 103–109.

Wojcik, W. J., and Neff, N. H. (1984). γ-Aminobutyric acid and β receptors are negatively coupled to adenylate cyclase in brain, and in the cerebellum these receptors may be associated with granule cells. *Mol. Pharmacol.* **251,** 24–28.

Wolf, M., LeVine, H., III, May, W. S., Jr., Cuatrecasas, P., and Sahyoun, N. (1985). A model for intracellular translocation of protein kinase C involving synergism between Ca^{2+} and phorbol esters. *Nature* **317,** 546–549.

Worley, P. F., Baraban, J. M., DeSouza, E. B., and Snyder, S. H. (1986a). Mapping second messenger systems in the brain: Differential localizations of adenylate cyclase and protein kinase C. *Proc. Natl. Acad. Sci. U.S.A.* **83,** 4053–4057.

Worley, P. F., Baraban, J. M., Van Dop, C., Neer, E. J., and Snyder, S. H. (1986b). G_o, a guanine nucleotide-binding protein: Immunohistochemical localization in rat brain resembles distribution of second messenger systems. *Proc. Natl. Acad. Sci. U.S.A.* **83,** 4561–4565.

Yatani, A., Codina, J., Imoto, Y., Reeves, J. P., Birnbaumer, L., and Brown, A. M. (1987). A G protein directly regulates mammalian cardiac calcium channels. *Science* **238,** 1288–1292.

CHAPTER 16

Receptor – Ion Channel Coupling through G Proteins

Jürgen Hescheler, Walter Rosenthal, Wolfgang Trautwein, Günter Schultz
Physiologisches Institut der Universität des Saarlandes, D-6650 Homburg/Saar,
Federal Republic of Germany

I. INTRODUCTION

The combination of electrophysiological and biochemical approaches in the investigation of plasmalemmal ionic channels has revealed that they are not only regulated by voltage gates, but also by "biochemical gates." While the characteristics of channel voltage dependency make possible self-generating electrical responses, such as action potentials, biochemical modification is thought to modulate membrane potential changes and thus to coordinate the electrical with the chemical signaling system.

The long-term (range of hours) modulation of channel activity is, of course, the synthesis and degradation of the channel proteins. However, little is known about these regulatory processes (Scott *et al.*, 1986; Yaari *et al.*, 1987, Belles *et al.*, 1988). Cells also possess several other mechanisms for fast (range of seconds to minutes) modifications of voltage-dependent channels. Classification of these fast modifications may include the processes (see also Rosenthal and Schultz, 1987) outlined below.

A. Interaction of Intracellular Signal Molecules with Ion Channels

This direct regulatory principle seems to be important in primary sensory cells. For instance, plasma membranes and disc membranes of rod photoreceptors contain cation channels weakly selective for monovalent cations (Cook *et al.*, 1987), which could be demonstrated to be directly stimulated by cGMP. G proteins are indirectly involved in the regulation of these channels, since the cGMP level in photoreceptor cells is controlled by a phosphodiesterase which is coupled to the light-sensitive rhodopsin via a G protein (transducin; for review see Stryer, 1986, Cook *et al.*, 1987). In olfactory cilia, some odorants increase adenylyl cyclase activity by G protein-coupled receptors (similar to G_s). Since cAMP increased a cation conductance of excised patches (Nakamura and Gold, 1987), it was concluded that this channel is directly regulated by the nucleotide, similar to the cGMP-dependent channel in photoreceptors. This first group of direct regulatory mechanisms may also include the ATP-dependent K^+ channel seen in β-pancreatic cells as well as those observed in cardiac and skeletal muscle cells (for review see Ashcroft, 1988). Furthermore, we can include K^+ channels which are directly activated by intracellular Ca^{2+}; these have been found in a wide range of cells and are proposed as a mechanism for electrical inactivation of voltage-dependent Ca^{2+} channels (see Cook, 1988).

B. Phosphorylation of Channel Proteins Stimulated by Intracellular Messengers

Protein kinases, including cAMP-dependent protein kinase, cGMP-dependent protein kinase, and the Ca^{2+}/phospholipid-dependent protein kinase (protein kinase C), have been shown to stimulate or to inhibit the activity of a variety of voltage-dependent ion channels. A partial list of phosphorylation-modulated voltage-dependent K^+ as well as Ca^{2+} channels is given in Tables I and II. Their common factor is that G proteins exert an indirect role in the regulatory cascade, since the concentration of intracellular messenger molecules is controlled by G protein-dependent membranous effectors such as adenylyl cyclase and phospholipase C. The best understood regulatory cascade is the cAMP-dependent phosphorylation of Ca^{2+} channels (for cardiac cells see Reuter, 1983; Kameyama et al., 1985; Kameyama et al., 1986, Hofmann et al., 1987).

The dihydropyridine receptor, purified from skeletal muscle and reconstituted into phospholipid bilayers, exhibits characteristics similar to functional Ca^{2+} channels. Since channel activity increases after addition of cAMP-dependent protein kinase and MgATP, it was concluded that the channel protein is directly phosphorylated (see Flockerzi et al., 1986). Additional data from Tanabe et al. (1987), who analyzed the amino acid sequence of the dihydropyridine receptor, suggest a total of seven different phosphorylation sites, most of them situated at the carboxy end, although it remains unclear which phosphorylation site(s) may be functionally relevant. In line with the view that the C terminus may be important for regulating the open probability of the Ca^{2+} channel, it was found that intracellular application of carboxypeptidase A mimics the effect of cAMP-dependent phosphorylation in cardiac cells (Hescheler and Trautwein, 1988).

C. Membrane-Confined Interactions of G Proteins with Ion Channels

G proteins functionally couple activated cell surface receptors to effectors, generating intracellular messengers. In analogy to the direct G-protein actions on established enzymatic effectors (for reviews, see Chapters 3, 4, 20), there are also direct G-protein effects on voltage-dependent ion channels (summarized in Tables I and II). In the case of voltage-dependent Ca^{2+} channels, the regulated intracellular messenger is the Ca^{2+} concentration itself. In the case of K^+ channels, Ca^{2+} channels are indirectly influenced via changes of the membrane potential (Cook, 1988; see Fig. 5).

Table I Hormonal Modification of K[+] Channels in Various Cells[a]

Cell type	Hormone	Signal/kinase	Effect	Channel	Reference[b]
Phosphorylation					
Muscle cells					
Cardiac	β-adrenergic	cAMP kinase (?)	+	Del. outw.	*AJP*:**253**,1321
Smooth	β-adrenergic	cAMP	+	M-current	*Sci*:**239**,190
	Acetylcholine	ND	−	M-current	*Sci*:**239**,190
Neuronal cells					
Aplysia sensory	5-HT, serotonin	cAMP kinase	−	S-current	*Nat*:**299**,413
Aplysia sensory	FMRFamide	Arachidonic acid	+	S-current	*Nat*:**328**,38
Aplysia bag cell	—	cAMP	−	Del. outw.	*JNS*:**8**,814
Helix ganglia	—	cAMP-kinase	+	Ca^{2+}-dependent	*Nat*:**315**,503
Snail neurons	—	cAMP-kinase	+	Ca^{2+}-dependent	*PNAS*:**79**,4207
Chick sensory	—	cAMP	−	ND	*PA*:**403**,170
Mouse sensory	Prostaglandin E	cAMP	−	ND	*JNS*:**7**,700
Cultured neurons	—	cAMP, PKC	−	ND	*Sci*:**235**,345
Sympathetic neurons	Substance P, LHRH bradykinin	PKC	−	M-current	*AJP*:**251**,580 *Nat*:**323**,333

386

Cell type	Agonist/receptor	Mechanism		Current	Reference
Hippocampal	SST	ND	+	M-current	Sci:**239**,278
Other cells					
Taste receptor	—	cAMP-kinase	−	44pS	Nat:**331**,351
Murine B cells	—	cAMP	−	Transient	Sci:**335**,1211
Photoreceptor	—	cAMP-kinase	−	A-current	Sci:**219**,303
Rat lacrimal cells	—	PKC	−	Ca^{2+}-dependent	JP:**394**,239
Control by G proteins					
Cardiac cells	Acetylcholine	G_k	+	Inw. rect.	Nat:**317**,536/8 Sci:**235**,207
Neuronal cells					
Hippocampal pyramidal	Serotonin, $GABA_B$	PT-sensitive	+	ND	Sci:**234**,1261
Aplysia neurons	Dopamine, histamine, acetylocholine	ND	+	ND	Nat:**325**,259
Brain neurons	Substance P	PT-insensitive	−	Inw. rect.	PNAS:**85**,3643
Endocrine cells					
Pituitary GH_3	SST, acetylcholine	G_k	+	ND	Endo:**1**,283
Pituitary AtT-20	SST	G_k	+	ND	BR:**444**,346

[a]PKC, protein kinase C; SST, somatostatin; PT, pertussis toxin; del.outw., delayed outward rectifying current; inw.rect., inward rectifying current; ND, not determined.

[b]For every type of regulation, one representative reference is given, cited as journal:volume,first page. The following abbreviations for journals are used: *JNS*, Journal of Neuroscience; *AJP*, American Journal of Physiology; *PA*, Pflügers Archiv; *JP*, Journal of Physiology; *PNAS*, Proceedings of the National Academy of Science, *Nat*, Nature, *Sci*, Science, *Endo*, Endocrinology, *BR*, Brain Research.

Table II Hormonal Modification of Ca^{2+} Channels in Various Cells[a]

Cell type	Hormone	Signal/kinase	Effect	Channel	Reference[b]
Phosphorylation					
Muscle cells					
cardiac	β-adrenergic	cAMP-kinase	+	L-type	PA:**407**,123
skeletal	β-adrenergic	cAMP-kinase	+	L-type	JBC:**260**,13041
smooth (stomach)	Acetylcholine	DAG, PKC (?)	+	DHP-sensitive	FASEB:**2**,2497
endocrine cells					
pituitary GH$_3$	—	cAMP-kinase	Stable	ND	PNAS:**84**,2518
neuronal cells					
Helix aspersa	—	cAMP-kinase	Stable	ND	JP:**378**,31 NS:**11**,263
	5-HT	cGMP-kinase	+	N-type (?)	Nat:**323**,812
	Cholecystokinin	PKC	–	N-type (?)	Nat:**325**,809
chick sensory	Noradrenaline	PKC	–	L-type	Nat:**319**,670
Aplysia neurons	—	PKC	+	ND	Nat:**313**,313
rat hippocampal	β-adrenergic	cAMP-kinase	+	N-type	Nat:**327**,620
other cells					
fibroblasts	—	PKC	+	ND	Biop J:**51**,226

Control by G protein

neuronal cells					
DRG cells (chick)	α-adrenergic,GABA$_B$,5-HT	G$_o$ (?)	−	N-type (?)	*JP*:**317**,519
	Dopamine				*PA*:**406**,104
DRG cells (rat)	A$_1$-adenosine	G$_o$ (?)	−	N-type (?)	*NSL*:**69**,59
	Neuropeptide Y	PT-sensitive, PKC	−	ND	*JNS*:**8**,2438
DRG cells (mouse)	κ-opioid	G$_o$ (?)	−	N-type	*PNAS*:**84**,5469
sympathetic					
ganglion	Muscarinic	G$_o$ (?)	−	N-type (?)	*PNAS*:**84**,4313
NxG hybrid cells	δ-opioid(DADLE), SST	G$_o$	−	N-type (?)	*PNAS*:**83**,9832
	δ-opioid(DADLE)				*Nat*:**325**,445
endocrine cells					
pituitary AT20	SST	ND	−	ND	*PNAS*:**83**,9035
pituitary GH$_3$	SST	G$_o$ (?)	−	ND	*EMBO*:**7**,1627
	LHRH, angioII	G$_i$ (?)	+	ND	*EMBO*:**7**,1627
adrenal cortex Y1	AngioII	G$_i$ (?)	+	ND	*EMBO*:**7**,619
glomerulosa (bovine)	—				*PNAS*:**85**,2412
cardiac myocytes	—	G$_s$	+	L-type	*Sci*:**238**,1288

[a]DAG, Diacylglycerol; PKC, protein kinase C; SST, somatostatin; LHRH, luteotropic hormone-releasing hormone; angioII, angiotensin II; ND, not determined.

[b]For every type of regulation, one representative reference is given, cited as journal:issue,first page. The following abbreviations for journals are used: *JNS*, Journal of Neuroscience; *PA*, Pflügers Archiv; *JBC*, Journal of Biological Chemistry; *JP*, Journal of Physiology; *NS*, Neuroscience; *Biop J*, Biophysical Journal; *NSL*, Neuroscience Letters; *PNAS*, Proceedings of the National Academy of Science; *Nat*, Nature; *Sci*, Science; *FASEB*, Fed Proc., Fed Am. Soc. Exp. Biol.; *EMBO*, European Molecular Biology Organization Journal.

This chapter will review some studies on G-protein effects on voltage-dependent K^+ and Ca^{2+} channels which are apparently independent of intracellular messenger molecules or kinases (see also Chapters 10, 15, and 16). The experimental approach is the patch-clamp technique (Hamill *et al.*, 1981) which provides a convenient method for measurement of whole-cell transmembranous current. This technique also allows the study of currents through single-ion channels in cell-attached or isolated membrane patch configuration or of ion channels incorporated into phospholipid bilayers. The whole-cell clamp configuration allows — besides superfusion of cells with various bathing solutions — the intracellular application of compounds via the patch pipette. Single-channel recording in the inside-out configuration also provides free access to the cytoplasmic face of an isolated plasma membrane patch.

II. G-PROTEIN-CONTROLLED K^+ CURRENT IN CARDIAC CELLS

The classic example for direct G-protein control of voltage-dependent channels is the acetylcholine-sensitive K^+ current primarily found in pace-making or atrial cells of the heart. The voltage dependency of the whole cell as well as of the single-channel current showed inward rectification with chord conductance being 25 pS for inward current flow and 5 pS for outward current flow (with 20 mM K^+ in the pipette). The reversal potential corresponded to the K^+ Nernst potential (Noma and Trautwein, 1978; Sakmann *et al.*, 1983). In a recent study, Codina *et al.* (1987a) described a similar type of K^+ channel in the pituitary GH$_3$ cell line.

From a functional point of view, K^+ current is regarded as the main mechanism for the regulation of the pacemaker frequency in cardiac cells. Superfusion of spontaneously beating cells with acetylcholine (about 1 μM) or with adenosine (about 0.1 mM) resulted in an increased channel open probability, a larger whole cell K^+ conductance, and, consequently, to a hyperpolarization and a slowing of the diastolic depolarization (Trautwein and Dudel, 1958; Hartzell, 1979). A typical voltage clamp experiment (voltage pulses from -40 to 0 mV) on a frog atrial myocyte is demonstrated in Fig. 1. The cell was infused with 10 μM cAMP which resulted in an increment of the Ca^{2+} current (seen as negative directed peak inward current). In contrast, cAMP did not change the K^+ current (seen as positive directed outward current), demonstrating that these channels are not sensitive to this second messenger. When 1 μM acetylcholine was applied to the bath solution, there was a shift of the current to the outward direction

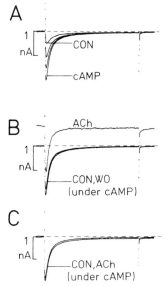

Fig. 1. Effect of acetylcholine on a cardiocyte isolated from frog heart atrium. (A) In contrast to mammalian cardiocytes, the basal Ca^{2+} current is small and can be only seen after intracellular application of 10 μM cAMP (compare with cultured cardioballs, Bechem and Pott, 1985). (B) Superfusion of the cAMP-infused cell with 1 μM acetylcholine causes a shift of the membrane current to the positive direction, whereby the Ca^{2+} current remains unchanged. (C) The effect of acetylcholine of the K^+ outward current can be suppressed by previous incubation of the cardiocytes for six hours in 0.5 μg/ml pertussis toxin-containing medium. Whole-cell current traces in A,B,C were recorded during voltage clamp pulses from -40 to 0 mV. CON, control, WO, current after washing out of acetylcholine. The dotted lines represent the zero current. Duration of test-pulses = 300 msec.

(See Fig. 1, B), which corresponded to the acetylcholine-sensitive K^+ current. Although the current traces in Fig. 1 showed no apparent variation with time, a more appropriate investigation of the time dependency, by subtracting the control current traces from the acetylcholine-modified traces, could demonstrate an exponential relaxation process with a time constant of around 120 msec (Noma and Trautwein, 1978).

The first assumption about the mechanism underlying the acetylcholine-induced increment of K^+ current was a direct junction of the muscarinic receptor with the channel (ligand-operated ion channel; Osterrieder *et al.*, 1981), similar to that proposed for the nicotinic receptor of the neuromuscular junction (Katz and Miledi, 1973; Changeux *et al.*, 1987) or the GABA- or glycine-stimulated Cl⁻ channels in neuronal cells (Schofield *et al.*, 1987; Grenningloh *et al.*, 1987; Betz, 1987). However, this concept did not fit to the experimental result; the intrinsic delay for activation of the muscarinic K^+ channel after a step application of acetylcholine (ionophoresis) was about 30 to 100 msec (Osterrieder et al., 1981; Nargeot et al., 1983) and thereby about 100 times longer than the nicotinic endplate current (Dreyer and Peper, 1973). Therefore it was suggested that some intermediate steps between receptor and channel may be involved. Since neither cAMP nor cGMP mimicked the acetylcholine effects (see Fig. 1, Trautwein et al., 1982), it was more likely that the signal-transduction process may be membrane-confined. This concept was also supported by the observation that single acetylcholine-sensitive K^+ channels in the cell-attached patch did not respond to bath application of the hormone but responded to application through the patch pipette (Soejima and Noma, 1984). Therefore, the receptor may be close to the channel protein and no intracellular signal molecule may be involved. At least three main experimental approaches hint at the participation of G proteins in this transduction process (Pfaffinger et al., 1985; Breitwieser and Szabo, 1985).

A. Effect of Guanyl Nucleotides on K^+ Channel Activity

The whole-cell configuration of the patch clamp method provides free access to the cytoplasm. Diffusion results in an exchange between the intracellular and intrapipette compartments, thus allowing the infusion of substances into the cell, or, vice versa, to dilute cytoplasmic compounds out of the cell. It was found by Pfaffinger and co-workers (1985) that the acetylcholine effect on the K^+ current in pacemaker cells depended on the intracellular GTP concentration. If this nucleotide was absent in the pipette solution and thus diffused out of the cytoplasm, the effect of acetylcholine was lost, but could be restored by intracellular application of GTP. The GTP-dependence of the signal-transduction process was also confirmed by the use of GTP- or GDP analogs (Breitwieser and Szabo, 1985). Guanosine 5' -O-(3-thiotriphosphate) (GTPγS), guanylyl imidodiphosphate (Gpp(NH)p) as well as guanylyl (β, γ-methylene) diphosphate (Gpp(CH_2)p) are analogs of GTP which irreversibly bind to G proteins. In consequence, intracellular application of these derivatives resulted in an amplification of the hormonal effect and a permanent activation of the K^+ current. The action of the analogs depended solely on the intracellular analog/GTP ratio and not on the absolute concentration of the nucleotides, suggesting that they competed with GTP for the same binding site on the G protein. The relative affinities were as follows: GTPγS>GTP>Gpp(NH)p>Gpp(CH_2)p (Breitwieser and Szabo, 1988). On the other hand, G proteins are inactive if they have bound GDP. Guanosine 5'-O-(2-thiodiphosphate) (GDPβS) is an analog which retains the G proteins in that inactive form. As a consequence, it was found in the experiment that intracellular infusion of GDPβS blocked the acetylcholine effect on the K^+ current (Breitwieser and Szabo, 1988).

B. Acetylcholine Effect Inhibited by Preincubation of Cells in Pertussis Toxin-Containing Medium

The main exotoxin of *Bordetella pertussis*, pertussis toxin, is another important tool to study G-protein control of intracellular signal-generating effectors since it prevents functional coupling of activated receptors and a number of G proteins (e.g. G_i-like and G_o) by ADP-ribosylation of G-protein α subunits (Ui *et al.*, 1984; Graziano and Gilman, 1987). In line with that property, it was found that preincubation of cardiac pacemaker cells in pertussis toxin-containing medium prevented the activation of the K^+ current by acetylcholine (Pfaffinger *et al.*, 1985). An experimental example is demonstrated in Fig. 1, C. An atrial cell of the frog heart was incubated for five hours in 0.5 μg/ml pertussis toxin-containing solution. In contrast to nonincubated cells Fig. 1, A and B, there was no change of the outward current after superfusion of the same acetylcholine concentration (1 μM). Similar results were also obtained in the case of adenosine activation of the sinoatrial K^+ current (Isenberg *et al.*, 1987). Taken together, these results were in line with the hypothesis that muscarinic or purinergic receptors activate a pertussis toxin-dependent G protein and that this activated G protein may have a direct effect on the K^+ channels.

C. Reconstitution Experiments

The most direct approach to demonstrate the participation of G proteins in the regulation of K^+ channels was to incorporate preactivated G proteins (as well as their subunits) into membrane patches containing only a few K^+ channels (inside-out configuration of the patch-clamp technique). These experiments, which are reviewed in more detail in Chapters 3 (Neer and Clapham) and 11 (Yatani, Codina, and Brown), also gave the possibility of determining which G protein, or even which subunit, might be involved in the receptor–channel coupling. Unexpectedly, Logothetis and co-workers failed to demonstrate that the acetylcholine-sensitive K^+ channel is activated by the α subunit of any G protein, but they were able to show K^+-channel opening with nanomolar concentrations by the $\beta\gamma$ dimer apparently common to G_s, G_i, and G_o (Logothetis *et al.*, 1987a; Neer and Clapham, 1988). Conversely, Brown and co-workers reported activation of cardiac muscarinic K^+ channels by picomolar concentrations of the α subunit (α_k) isolated from human erythrocytes (Yatani *et al.*, 1987a,b;

Codina et al., 1987b). For the discussion of which subunit may be functionally relevant see Birnbaumer and Brown (1987), Logothetis et al. (1987b), and Cerbai et al. (1988).

III. G-PROTEIN CONTROL OF Ca^{2+} CURRENT IN NEURONAL AND ENDOCRINE CELLS

Voltage-dependent Ca^{2+} currents have been postulated to play an important role in signal transduction for nearly all tissues. They emerged early in evolution and are found throughout eukaryotes, including protozoa, algae, higher plants, fungi, and animals. There are at least three different types of voltage-dependent Ca^{2+} currents, classified by Tsien and co-workers as L- (Long lasting), T- (Transient) and N- (Neuronal) type currents (Carbone and Lux, 1984; Nowycky, et al., 1985). They all have a common feature, that the whole-cell current–voltage relation is U-shaped, i.e., Ca^{2+} current increases above a certain threshold, reaches a maximum and then decreases. The main differences between the three types of current include their threshold potential (T-type: low-voltage activated; L-, N-type: high-voltage activated), their maximum potential, their time course of inactivation (T-type: fast, N-type: moderate, L-type: very slow inactivation) and their sensitivity toward different channel blockers. Single channel data also reveal differences in opening kinetics as well as ionic conductances, which are 25, 13 and 8 pS for L-, N-, and T-type channels, respectively (under 110 mM Ba^{2+} outside, Nowycky et al., 1985).

L- and T-type channels have been identified in variable densities in cardiac, skeletal, and smooth muscle cells as well as neuronal and neurosecretory cells. However, N-type channels have been described only for neurons (literature on the Ca^{2+} currents in the cell types used: Carbone and Lux, 1984; Hering et al., 1985; Tsunoo et al., 1986; Hescheler et al., 1988b; Cohen et al., 1987; Armstrong and Mattison, 1985; DeRiemer and Sakmann, 1985; Nilius et al., 1985). Functionally, Ca^{2+} channels may play a prominent role, since within a few milliseconds Ca^{2+} inflow leads to an approximately 100-fold increase in the intracellular Ca^{2+} concentration and activation of many intracellular Ca^{2+}-dependent enzymes. Thus, Ca^{2+} channels may be specially determined at strategically important points to act as link between electrical events and intracellular regulation. Examples include electromechanical (smooth, cardiac, skeletal muscle) and electrosecretory (neurons, endocrine cells, B lymphocytes) coupling.

According to these functions of Ca^{2+} channels, the intriguing task is to find out the regulatory features of the channel protein. Obviously, there are large differences in the regulation of Ca^{2+} current in different tissues. While

L-type channels in cardiac cells (as well as in skeletal muscle) are the classical example for regulation by cAMP dependent phosphorylation (Reuter, 1983; Flockerzi *et al.*, 1986; Hofmann *et al.*, 1987), it is interesting to note that the Ca^{2+} current in neuronal as well as in endocrine cells could neither be stimulated nor inhibited by intracellular infusion of cAMP or superfusion with the adenylyl cyclase stimulator forskolin (Hescheler *et al.*, 1988b; Lewis *et al.*, 1986; Wanke *et al.*, 1987; Holz *et al.*, 1986). In these cells, other possible regulatory mechanisms may prevail, among them G protein or protein kinase C-mediated interactions (see Rane and Dunlap, Chapter 15).

A. Cyclic AMP-Independent Hormonal Modulations of Ca^{2+} Currents

Hormonal inhibition of Ca^{2+} currents has been observed in various cell types (see Table II), e.g. in dorsal root ganglia of chick (Dunlap and Fischbach, 1981; Holz *et al.*, 1986; Forscher *et al.*, 1986), of rat (Scott and Dolphin, 1986; Dolphin *et al*, 1986), and of mouse (Gross and Macdonald, 1987); in sensory and sympathetic ganglia of chick (Deisz and Lux, 1985; Marchetti *et al.*, 1986); in rat sympathetic neurons (Wanke *et al.*, 1987); in neuroblastoma × glioma hybrid cells (N×G, Tsunoo *et al.*, 1986; Hescheler *et al.*, 1987; Hescheler *et al.*, 1988b); and in a murine pituitary cell line (AtT-20, Lewis *et al.*, 1986). In contrast, there are only a few examples of hormones that appear to act by stimulation of Ca^{2+} channels with evidence that activation of adenylyl cyclase is not involved. Biochemical evidence has been provided that angiotensin II-induced secretion of aldosterone involves activation of voltage-dependent Ca^{2+} channels. Both angiotensin II- and K^+-induced aldosterone secretions strictly depend on extracellular Ca^{2+} (Fakunding *et al.*, 1979), are inhibited by various Ca^{2+}-channel blockers (Fakunding and Catt, 1980; Foster *et al.*, 1981; Kojima *et al.*, 1984; Aguilera and Catt, 1986), and are stimulated by the Ca^{2+} channel agonist, Bay K 8644 (Kojima *et al.*, 1984; Hausdorff *et al.*, 1986). In addition, angiotensin II and K^+ have been shown to stimulate Ca^{2+} influx in a dihydropyridine-sensitive manner (Kojima *et al.*, 1985; Kojima *et al.*, 1986).

Figure 2 summarizes some of the hormonal effects on Ca^{2+} channels, observed in three different cell lines. Superfusion of N×G (108CC15) cells with 1 μM D-Ala-D-Leu-encephalin (DADLE) reduced the whole-cell Ca^{2+} inward current by about 70% (Hescheler *et al.*, 1987; Hescheler *et al.*, 1988b). In contrast, the adrenocortical cell line Y1 responded to superfusion with 0.1 μM angiotensin II with an approximately 50% increase in the slowly inactivating Ca^{2+} current (see also Hescheler *et al.*, 1988a). Finally,

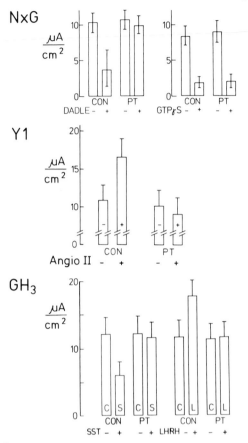

Fig. 2. Mean Ca^{2+}-current densities ($\pm$ standard deviation) and their hormonal modifications of control in pertussis toxin-incubated N$\times$G, Y1, and GH$_3$ cells. Hormonal concentrations used were 1 μM D-Ala^2, D-Leu^5-enkephalin (DADLE), 0.1 μM angiotensin II (AngioII), 0.1 μM somatostatin (SST), and 0.1 μM luteinizing hormone-releasing hormone (LHRH). While the hormones inhibited or stimulated the Ca^{2+} current in control cells, the effect was absent in cells previously incubated in pertussis toxin media.

the pituitary cell line GH$_3$ appears to be an example of bidirectionally regulated Ca^{2+} current. The secretion-inhibiting hormone, somatostatin (0.1 μM) was found to inhibit the Ca^{2+} current whereas the secretion-stimulating hormone, LHRH, increased it (see also Marian and Conn, 1979; Conn *et al.*, 1983; Rosenthal *et al.*, 1988). When GH$_3$ cells were superfused with a mixture of LHRH and somatostatin at maximally effective concentrations of each hormone (1 μM), the inhibitory effect prevailed (not shown). In all cases, the effects of the oligopeptide hormones occurred rapidly (within about 30 sec) and could be completely reversed by washing off the agonist.

There is good evidence that, at least in neuronal cells, the N-type Ca^{2+} current is inhibited by receptor agonist (Tsunoo et al., 1986; Gross and Macdonald, 1987; Wanke et al., 1987). In both endocrine cell lines tested (GH_3, Y1), the hormone-stimulated Ca^{2+} current was only partially sensitive to organic channel agonists and antagonists of the dihydropyridine type, suggesting fluxes through L-type Ca^{2+} channels. Slowly inactivating Ca^{2+} currents of GH_3 cells were also inhibited by the snake venom, ω-conotoxin (Suzuki and Yoshioka, 1987), which affects both L- and N-type Ca^{2+} channels in neuronal tissues (McCleskey et al., 1987). Thus, at present it is not clear whether hormones modulate N- or L-type Ca^{2+} channels in endocrine cells.

B. Effect of Guanyl Nucleotides on Ca^{2+} Channel Activity

In analogy to experiments on the muscarinic K^+ current, GTP or GDP analogs have been used to demonstrate the possible direct involvement of G proteins in the hormonal modulations of Ca^{2+} current. Intracellular infusion of GTPγS at high concentrations (100 to 500 μM) drastically reduced Ca^{2+} currents in neuronal (Holz et al., 1986; Scott and Dolphin, 1986; Wanke et al., 1987; Hescheler et al., 1987; Hescheler et al., 1988b) and pituitary cells (Lewis et al., 1986). Dolphin and co-workers (1988) have used caged GTPγS or Gpp(NH)p [1-(2-nitrophenyl)ethyl p^3-ester derivatives] to effect millisecond step changes in the concentration of GTP analogs by light pulses. They found a relatively slow time course of Ca^{2+} current inhibition with a mean time course of 1.5 min. In N×G cells, GTPγS at low concentrations (1 μM) had no effect on the Ca^{2+} current in the absence of DADLE but caused a marked and irreversible inhibition after extracellular application of the opioid (Hescheler et al., 1987; see also Breitwieser and Szabo, 1985). According to the capacity of GDPβS to inhibit the activation of G proteins, intracellular application of this GDP analog prevented the inhibition of Ca^{2+} current by receptor agonist (Holz et al., 1986; Scott and Dolphin, 1986; Wanke et al., 1987). Thus, the effects of guanyl-nucleotide analogs on Ca^{2+} current resemble their effects on hormone-sensitive adenylyl cyclase (see Birnbaumer et al., 1985). G proteins appear to be not only involved in the hormonal modulation of Ca^{2+} channels but also in their interaction with ligands of the phenylalkylamine and dihydropyridine type (Scott and Dolphin, 1988). This assumption is based on the finding that in dorsal root ganglia the antagonistic effect of Ca^{2+}-channel ligands was turned into an agonistic effect by intracellular infusion of high concentrations (500 μM) of GTPγS.

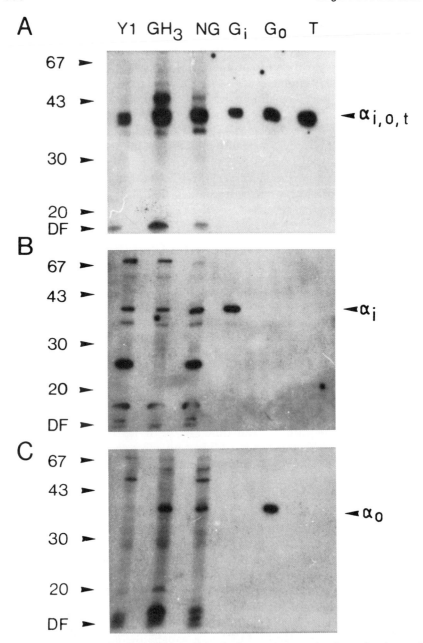

Fig. 3. Identification of G-protein α subunits in membranous fractions of various cell types. Shown are autoradiographs of immunoblots obtained with antisera raised against synthetic peptides corresponding to confined regions of G-protein α subunits. Membranous

C. Effects of Pertussis Toxin on Hormonal Modifications of Ca²⁺ Current

As outlined in Fig. 2 (top), hormonal inhibition of Ca^{2+} current in NXG cells was abolished when cells were pretreated with pertussis toxin. In analogy with the data obtained for inhibition of adenylyl cyclase (Jakobs *et al.*, 1984), GTPγS-induced inhibition of Ca^{2+} current in NXG cells was not affected by preincubation with the toxin (Hescheler *et al.*, 1988b). The suppression of the inhibitory effect of receptor agonists by pertussis toxin has also been described for neuronal (Holz *et al.*, 1986) and pituitary cells (Lewis *et al.*, 1986). The toxin action depended on the incubation time and temperature (Holz *et al.*, 1986), and was irreversible at least 48 hours after washout. In addition, pertussis toxin completely blocked the hormonal stimulation of Ca^{2+} current in Y1 cells (Fig. 2, middle). This observation is consistent with the biochemical finding that the angiotensin II-induced stimulation of Ca^{2+} influx was abolished by the toxin (Kojima *et al.*, 1986). Moreover, the bidirectional hormonal modulation of Ca^{2+} current in GH₃ cells was completely suppressed in cells pretreated with the toxin (Fig. 2, bottom). The occurrence of both pertussis toxin-sensitive hormonal stimulation and inhibition of voltage-dependent Ca^{2+} current in one cell type suggested that these opposite regulations are mediated by distinct G proteins (Rosenthal *et al.*, 1988).

D. Possible Identity of G Proteins Involved in Ca²⁺ Current Modulations

In order to identify the pertussis toxin-sensitive G proteins possibly involved in the signal-transduction process of NXG, Y1, and GH₃ cells, we probed membranous fractions with polyclonal antisera raised against synthetic peptides corresponding to confined regions of G-protein α subunits (Fig. 3, Rosenthal *et al.*, 1988). An antiserum raised against a sequence

---

fractions of each indicated cell type (75 μg) or purified G proteins (1 μg with respect to G-protein α subunit) were subjected to sodium dodecylsulfate-polyocrylamide gel electrophoresis (SDS-PAGE) and transferred to nitrocellulose filters which were incubated with antiserum (diluted 1:300) raised against a peptide common to all sequenced α subunits (A). Antiserum (diluted 1:300) raised against a peptide specific for the α subunit of G₁-like G proteins, or (B) antiserum (diluted 1:300) raised against a peptide specific for the α subunit of G₀. (C) Immunoreactive peptides were identified by autoradiography following incubation of filters with ¹²⁵I-labeled protein A. Figures on the left panel margins indicate relative molecular masses (in kDa). T, transducin; NG, NXG cells. Adapted from Rosenthal *et al.* (1988).

common to all G-protein α subunits (α_{common} peptide) reacted with membranous peptides of about 39 to 41 kDa in all tested cell types. Similarly, an antiserum against the α subunits of G_i-like G proteins (α_i peptide) recognized peptides around 40 kDa in all tested membranes. In contrast, the α_o antiserum specifically reacted with peptides of 39 to 40 kDa in membranes

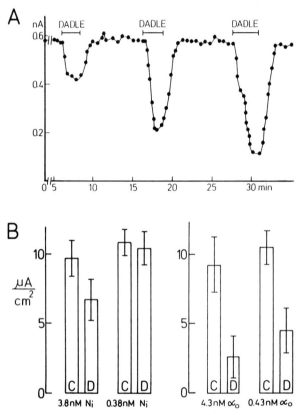

Fig. 4. (A) Reconstitution of the DADLE effect on Ca^{2+} currents by intracellular infusion of a mixture of G_i-like G proteins and G_o in a pertussis toxin-pretreated N×G cell. Ca^{2+} currents were repetitively measured during voltage-clamp pulses from −40 to 0 mV (frequency of 0.2 Hz). DADLE (1 μM) was present at time intervals indicated by the bars. G proteins were purified from porcine brain (Rosenthal et al., 1986). The concentration of each G_i-like G proteins and G_o was 15 μM. Adapted from Hescheler *et al.* (1987). (B) Effect of intracellular infusion of various concentrations of G_i-like G protein (N_i) as well as of the α subunit of G_o (α_o). Experimental procedures were as in panel A. Currents were measured about 15 min after disruption of membrane patches in the absence (C) and presence (D) of 1 μM DADLE. Shown are mean current densities ± standard deviation (numbers of experiments varied between 4 and 5). Adapted from Hescheler *et al.* (1988).

of GH_3 cells and NXG cells but *not* of Y1 cells. The identification of the α subunit of a G_i-like G protein and the absence of the α subunits of G_o in membranes of Y1 cells indicates that the stimulatory effect of angiotensin II on the Ca^{2+} current may be mediated by a G_i-like G protein and not by G_o. On the other hand, the abundance of G_o in NXG as well as in GH_3 cells (see also Huff *et al.*, 1985; Milligan *et al.*, 1986) suggests that the hormonal inhibition of Ca^{2+} current may be mediated by G_o.

This assumption was further tested by reconstitution experiments. We infused G proteins purified from porcine cerebral cortex into NXG cells which had become insensitive to DADLE following pretreatment with pertussis toxin. As shown in Fig. 4, A, intracellular infusion of a mixture of both G_i-like G proteins and G_o completely restored the DADLE effect within approximately 30 min. In order to elucidate which pertussis toxin-sensitive G protein is involved in functional coupling of opioid receptors and neuronal Ca^{2+} channels, individual G proteins or isolated subunits were applied (Fig. 4, B). It was found that the α subunit of G_o was still effective at a concentration of 0.4 nM and thus was about ten times more potent than a G_i preparation from brain, probably representing mainly G_{i1}. High concentrations of the bovine retinal G protein, transducin, and the $\beta\gamma$ complex of G proteins purified from porcine cerebral cortex were not effective (Hescheler *et al.*, 1987). These results were confirmed in rat dorsal root ganglion cells, where infusion of the α subunit of G_o restored the inhibition of Ca^{2+}-current modulation by neuropeptide Y (Ewald *et al.*, 1988).

Although a physiological role of G_i-type G proteins in the inhibitory modulations cannot be excluded, the findings described above as well as the high abundance of G_o in cell types exhibiting inhibitory Ca^{2+}-current modulation favor a physiological role for G_o in functional coupling of inhibitory receptors to voltage-dependent Ca^{2+} channels. This hypothesis is supported by the finding that in rat pituitary tumor cells, which possess G_i-type G proteins but little or no α subunit of G_o, dopamine does not inhibit prolactin release, although it does so in G_o-containing pituitary tumor cells (Collu *et al.*, 1988).

IV. CONCLUSIONS

The three-component concept of a signal transduction system, originally developed to explain hormonal modulations of the adenylyl cyclase, has proven to cover a much wider range of receptor-controlled processes in cells. Thus, it is found that G proteins are not only linked to many enzymatic systems but presumably also to membranous channel proteins. In regulatory pathways which involve second messenger-controlled protein kinases (see Tables I and II), G proteins participate indirectly in channel modulation as they mediate hormonal regulation of intracellular signal-generating enzymes. Examples include the cAMP-dependent phosphorylation of the cardiac Ca^{2+} channel and the diacylglycerol-dependent phosphorylation of the neuronal Ca^{2+} channel (see Fig. 5).

On the other hand, evidence is growing for a close stimulatory and inhibitory control of channels by G proteins that appears to be membrane-confined since neither cytosolic signal molecules nor protein kinases are found to be involved. Examples discussed here are the muscarinic K^+ channel in cardiac cells and voltage-dependent Ca^{2+} channels in neuronal and endocrine cells. The abundance of G_o in neuronal cells and pituitary cells as well as experimental evidence derived from reconstitution experiments suggests that this G protein may be involved in hormonal inhibition of Ca^{2+} channels in these two cell types. The pertussis toxin-sensitive hormonal stimulation of Ca^{2+} current in an adrenocortical cell line (Y1) which possesses a G_i-like G protein but no G_o argues for a role of a G_i-like G protein in hormonal stimulation of Ca^{2+} channels in this and possibly other endocrine cells. A second stimulatory, membrane-confined G protein control of cardiac Ca^{2+} channels has been proposed by Yatani and co-workers (1987c) who described a direct stimulatory effect of the isolated α subunit of G_s in isolated membrane patches from cardiac ventricular

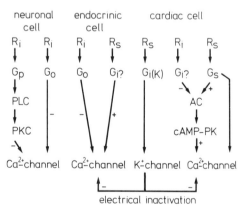

Fig. 5. Proposed mechanisms for the hormonal modulations of voltage-dependent Ca^{2+} and K^+ channels (examples). R_i and R_s, inhibitory and stimulatory receptors, respectively; $G_{i?}$, not yet identified G_i-subtype; PLC, phospholipase C; AC, adenylyl cyclase; PKC, protein kinase C; cAMP-PK, cAMP-dependent protein kinase. (+) stimulation; (-) inhibition. For further explanations see text.

cells (see Fig. 5). Thus, G_s may control cardiac Ca^{2+}-channel function via two pathways (see above).

Several reasons can be envisaged for the variety of mechanisms in Ca^{2+}-channel regulation. (1) Different forms of Ca^{2+}-channel proteins with altered primary structure, subunit composition or posttranslational modifications (e.g. phosphorylation, glycosylation, acylation) may coexist. (2) A tissue-specific "environment" may modify ubiquitous Ca^{2+}-channel proteins. The latter assumption is supported by the finding that ω-conotoxin, a snake venom, blocks L- and N-type Ca^{2+} currents in neuronal cells but not in cardiac myocytes (McCleskey et al., 1987). In contrast to enzymatic effectors (e.g. adenylyl cyclase, retinal cGMP phosphodiesterase; see Graziano and Gilman, 1987), a direct, reversible interaction of G proteins with Ca^{2+} channels has not yet been demonstrated. Although the involvement of cytosolic signal molecules is unlikely, as-yet unknown components, such as directly regulated membrane-associated protein kinases or phosphatases, may be required for functional coupling of G proteins to Ca^{2+} channels.

While the regulation of Ca^{2+} channels seems to be rather complex and involves multiple G protein-controlled mechanisms, only one G protein, termed G_k, has been found to regulate K^+ channels in cardiac pacemaker cells and GH_3 cells. Since both the α subunit as well as the $\beta\gamma$ complex have been demonstrated to affect the K^+ channel in isolated membrane patches, it is at present not clear which subunit of G_k is physiologically relevant. However, because of the specific structure of the α subunit, one may assume that this subunit is involved in receptor–channel coupling, whereas the $\beta\gamma$ complex is responsible for maintenance of the basal activity. Since G_k belongs to the G_i-type G proteins, it is an intriguing task to find out whether G_k is identical to the G protein responsible for the G protein-controlled stimulation of voltage-dependent Ca^{2+} channels in endocrine cells or whether another member of the G_i family serves this function.

REFERENCES

Aguilera, G., and Catt, K. J. (1986). Participation of voltage-dependent calcium channels in the regulation of adrenal glomerulosa function by angiotensin II and potassium. *Endocrinology* **118**, 112–118.

Armstrong, C. M., and Mattison, D. R. (1985). Two distinct populations of calcium channels in a clonal line of pituitary cells. *Science* **227**, 65–67.

Ashcroft, F. (1988). Adenosine 5'-triphosphate-sensitive potassium channels. *Annu. Rev. Neurosci.* **11**, 97–118.

Bechem, M., and Pott, L. (1985). Removal of Ca current inactivation in dialysed guinea-pig atrial cardioballs by Ca chelators. *Pflügers Arch.* **404,** 10–20.

Belles, B., Hescheler, J., Trautwein, W., Blomgren, K., and Karlsson, J. (1988). A possible physiological role of the Ca-dependent protease calpain and its inhibitor calpastatin on the Ca current in guinea pig myocytes. *Pflügers Arch.* **412,** 554–556.

Betz H. (1987). Biology and structure of the mammalian glycine receptor. *Trends Neuro Sci.* **10,** 113–117.

Birnbaumer, L., and Brown, A. (1987). G protein opening of K^+ channels. *Nature* **327,** 21–22.

Birnbaumer, L., Codina, J., Mattera, R., Cerione, R. A., Hildebrandt, J. D., Sunyer, T., Rojas, F. J., Caron, M. G., Lefkowitz, R. J., and Iyengar, R. (1985). Regulation of hormone receptors and adenylyl cyclases by guanine nucleotide binding N proteins. *Rec. Progr. Hormone Res.* **41,** 41–99.

Breitwieser, G., and Szabo, G. (1985). Uncoupling of cardiac muscarinic and β-adrenergic receptors from ion channels by a guanine nucleotide analogue. *Nature* **317,** 538–540.

Breitwieser, G., and Szabo, G. (1988). Mechanism of muscarinic receptor-induced K^+ channel activation as revealed by hydrolysis-resistant GTP analogues. *J. Gen. Physiol.* **91,** 469–493.

Carbone, E., and Lux, H. D. (1984). A low voltage-activated, fully inactivating Ca channel in vertebrate sensory neurones. *Nature* **310,** 501–502.

Cerbai, E., Klöckner, U., and Isenberg, G. (1988). The o subunit of the GTP binding protein activates muscarinic potassium channels of the atrium. *Science* **240,** 1782–1783.

Changeux, J., Giraudat, J., and Dennis, M. (1987). The nicotinic acetylcholine receptor: Molecular architecture of a ligand-regulated ion channel. *Trends Pharmacol. Sci.* **8,** 459–465.

Codina, J., Grenet, D., Yatani, A., Birnbaumer, L., and Brown, A. M. (1987a). Hormonal regulation of pituitary GH_3 cell K^+ channels by G_K is mediated by its o-subunit. *FEBS Lett.* **216,** 104–106.

Codina, J., Yatani, A., Grenet, D., Brown, A. M., and Birnbaumer, L. (1987b). The α-subunit of the GTP binding protein G_K opens atrial potassium channels. *Science* **236,** 442–445.

Cohen, C. J., McCarthy, R. T., Barrett, P. Q., and Rasmussen, H. (1987). Two populations of Ca channels in bovine glomerulosa cells. *Biophys. J.* **51,** T-PM-B5 (abstract).

Collu, R., Bouvier, C., Lagacé, G., Unson, C., Milligan, G., Goldsmith, P., and Spiegel, A. (1988). Selective deficiency of guanine nucleotide-binding protein G_o in two dopamine-resistant pituitary tumors. *Endocrinology* **122,** 1176–1178.

Conn, P. M., Rogers, D. C., and Seay, S. G. (1983). Structure-function relationship of calcium ion channel antagonists at the pituitary gonadotrope. *Endocrinology* **113,** 1592–1595.

Cook, N. (1988). The pharmacology of potassium channels and their therapeutic potential. *Trends Pharmacol. Sci.* **9,**: 21–28.

Cook, N., Hanke, W., and Kaupp, B. (1987). Identification, purification, and functional reconstitution of the cyclic GMP-dependent channel from rod photoreceptors. *Proc. Natl. Acad. Sci. U.S.A.* **84**, 585–589.

Deisz, R. A., and Lux, H. D. (1985). γ-Aminobutyric acid-induced depression of calcium currents of chick sensory neurons. *Neurosci. Lett.* **56**, 205–210.

DeRiemer, S. A., and Sakmann, B. (1985). Two calcium currents in normal rat anterior pituitary cells identified by a plaque technique. *Exp. Brain Res.* **14**, 139–154.

Dolphin, A. A., Forda, S. R., and Scott, R. H. (1986). Calcium-dependent currents in cultured rat dorsal root ganglion neurones are inhibited by an adenosine analogue. *J. Physiol.* **373**, 47–61.

Dolphin, A., Wootton, J., Scott, R., and Trentham, D. (1988). Photoactivation of intracellular guanosine triphosphate analogues reduces the amplitude and slows the kinetics of voltage-activated calcium channel currents in sensory neurones. *Pflügers Arch.* **411**, 628–636.

Dreyer, F., and Peper, K. (1973). Iontophoretic application of acetylcholine: Advantages of high resistance micropipettes in connection with an electronic current pump. *Pflügers Arch.* **348**, 263–272.

Dunlap, K., and Fischbach, G. D. (1981). Neurotransmitters decrease the calcium conductance activated by depolarization of embryonic chick sensory neurones. *J. Physiol.* **317**, 519–535.

Ewald, D., Miller, R., and Sternweis, P. (1988). C-kinase and G proteins mediate inhibition of Ca^{2+} currents by neuropeptide γ in rat dorsal root ganglion neurons. *Biophys. J.* **53**, 234a.

Fakunding, J. L., and Catt, K. J. (1980). Dependence of aldosterone stimulation in adrenal glomerulosa cells on calcium uptake: Effects of lanthanum and verapamil. *Endocrinology* **107**, 1345–1353.

Fakunding, J. L., Chow, R., and Catt, K. J. (1979). The role of calcium in the stimulation of aldosterone production by adrenocorticotropin, angiotensin II, and potassium in isolated glomerulosa cells. *Endocrinology* **105**, 327–333.

Flockerzi, V., Oeken, H., Hofmann, F., Pelzer, D., Cavalie, A., and Trautwein, W. (1986). Purified dihydropyridine-binding site from skeletal muscle t-tubules is a functional calcium channel. *Nature* **323**, 66–68.

Forscher, P., Oxford, G. S., and Schulz, D. (1986). Noradrenaline modulates calcium channels in avian dorsal root ganglion cells through tight receptor-channel coupling. *J. Physiol.* **379**, 131–144.

Foster, R., Lobo, M. V., Rasmussen, H., and Marusic, E. T. (1981). Calcium: Its role in the mechanism of action of angiotensin II and potassium in aldosterone production. *Endocrinology* **109**, 2196–2201.

Graziano, M. P., and Gilman, A. G. (1987). Guanine nucleotide-binding regulatory proteins: Mediators of transmembrane signaling. *Trends Pharmacol. Sci.* **8**, 478–481.

Grenningloh, G., Rienitz, A., Schmitt, B., Methfessel, C., Zensen, M., Beyreuther, K., Gundelfinger, E., and Betz, H. (1987). The strychnine-binding subunit of the glycine receptor shows homology with nicotinic acetylcholine receptors. *Nature* **328**, 215–220.

Gross, R. A., and Macdonald, R. L. (1987). Dynorphin A selectively reduces a large transient (N-type) calcium current of mouse dorsal root ganglion neurons in cell culture. *Proc. Natl. Acad. Sci. U.S.A.* **84,** 5469–5473.

Hamill, O. P., Marty, A., Neher, E., Sakman, B., and Sigworth, F. J. (1981). Improved patch-clamp techniques for high-resolution current recordings from cells and cell-free membrane patches. *Pflügers Arch.* **391,** 85–100.

Hartzell, C. (1979). Adenosine receptors in frog sinus venosus: Slow inhibitory potentials produced by adenine compounds and acetylcholine. *J. Physiol.* **293,** 23–49.

Hausdorff, W. P., Aguilera, G., and Catt, K. J. (1986). Selective enhancement of angiotensin II- and potassium-stimulated aldosterone secretion by the calcium channel agonist BAY K 8644. *Endocrinology* **118,** 869–874.

Hering, S., Bodewei, R., Schubert, B., Rhode, K., and Wollenberger, A. (1985). A kinetic analysis of the inward calcium current in 108CC15 neuroblastoma × glioma hybrid cells. *Gen. Physiol. Biophys.* **4,** 129–142.

Hescheler, J., and Trautwein, W. (1988). Modification of L-type calcium current by intracellularly applied trypsin in guinea-pig ventricular myocytes. *J. Physiol.* **404,** 259–274.

Hescheler, J., Rosenthal, W., Trautwein, W., and Schultz, G. (1987). The GTP-binding protein, G_o, regulates neuronal calcium channels. *Nature* **325,** 445–447.

Hescheler, J., Rosenthal, W., Hinsch, K. -D., Wulfern, M., Trautwein, W., and Schultz, G. (1988a). Angiotensin II-induced stimulation of voltage-dependent Ca^{2+} currents in adrenal cortical cell line. *EMBO J.* **7,** 619–624.

Hescheler, J., Rosenthal, W., Wulfern, M., Tang, M., Yajima, M., Trautwein, W., and Schultz, G. (1988b). Involvement of the guanine nucleotide-binding protein, N_o, in the inhibitory regulation of neuronal calcium channels. *Adv. Second Messenger Phosphoprot. Res.* **21,** 165–174.

Hofmann, F., Nastainczyk, W., Röhrkasten, A., Schneider, T., and Sieber, M. (1987). Regulation of the L-type calcium channel. *Trends Pharmacol. Sci.* **8,** 393–398.

Holz, G. G., IV, Rane, S. G., and Dunlap, K. (1986). GTP-binding proteins mediated transmitter inhibition of voltage-dependent calcium channels. *Nature* **319,** 670–672.

Huff, R. M., Axton, J. M., and Neer, E. J. (1985). Physical and immunological characterization of a guanine nucleotide-binding protein purified from bovine cerebral cortex. *J. Biol. Chem.* **260,** 10864–10871.

Isenberg, G., Cerbai, E., and Klöckner, U. (1987). Ionic channels and adenosine in isolated heart cells. In "Topics and Perspectives in Adenosine Research" (E. Gerlach and B. Becker, eds.) Springer Verlag, Berlin.

Jakobs, K. H., Aktories, K., and Schultz, G. (1984). Mechanism and components involved in adenylate cyclase inhibition by hormones. *Adv. Cyclic Nucleotide Protein Phosphorylation Res.* **17,** 135–143.

Kameyama, M., Hofmann, F., and Trautwein, W. (1985). On the mechanism of the Ca channels in the guinea-pig heart. *Pflügers Arch.* **405,** 285–293.

Kameyama, M., Hescheler, J., Hofmann, F., and Trautwein, W. (1986). Modulation of Ca current during the phophorylation cycle in the guinea pig heart. *Pflügers Arch.* **407,** 123–128.

Katz, B., and Miledi, R. (1973). The characteristics of 'end-plate noise' produced by different depolarizing drugs. *J. Physiol.* **230,** 707–717.

Kojima, K., Kojima, I., and Rasmussen, H. (1984). Dihydropyridine calcium agonist and antagonist effects on aldosterone secretion. *Am. J. Physiol.* **247,** 645–650.

Kojima, I., Kojima, K., and Rasmussen, H. (1985). Characteristics of angiotensin II-, K+- and ACTH-induced calcium influx in adrenal glomerulosa cells. Evidence that angiotensin II, K+- and ACTH may open a common calcium channel. *J. Biol. Chem.* **260,** 9171–9176.

Kojima, I., Shibata, H., and Ogata, E. (1986). Pertussis toxin blocks angiotensin II-induced calcium influx but not inositol triphosphate production in adrenal glomerulosa cell. *FEBS Lett.* **204,** 347–351.

Lewis, D. L., Weight, F. F., and Luini, A. (1986). A guanine nucleotide-binding protein mediates the inhibition of voltage-dependent calcium current by somatostatin in a pituitary cell line. *Proc. Natl. Acad. Sci. U.S.A.* **83,** 9035–9039.

Logothetis, D., Kurachi, Y., Galper, J., Neer, E., and Clapham, D. (1987a). The βv subunits of GTP-binding proteins activate the muscarinic K+ channel in heart. *Nature* **325,** 321–326.

Logothetis, D., Kurachi, Y., Galper, J., Neer, E., and Clapham, D. (1987b). G protein opening of K+ channels. *Nature* **327,** 21–22.

Marchetti, C., Carbone, E., and Lux, H. D. (1986). Effects of dopamine and noradrenaline on Ca channels of cultured sensory and sypoathetic neurons of chick. *Pflügers Arch.* **406,** 104–111.

Marian, J., and Conn, P. M. (1979). Gonadotropin releasing hormone stimulation of cultured pituitary cell requires calcium. *Mol. Pharmacol.* **16,** 196–201.

McCleskey, E. W., Fox, A. P., Feldman, D. H., Cruz, L. J., Olivera, B. M., Tsien, R. W., and Yoshikami, D. (1987). W-conotoxin: Direct and persistent blockade of specific types of calcium channels in neurons but not muscle. *Proc. Natl. Acad. Sci. U.S.A.* **84,** 4327–4331.

Milligan, G., Gierschik, P., Spiegel, A. M., and Klee, W. A. (1986). The GTP-binding regulatory proteins of neuroblastoma × glioma, NG108–15, and glioma, C6, cells. *FEBS Lett.* **195,** 225–230.

Nakamura, T., and Gold, G. (1987). A cyclic, nucleotide-gated conductance in olfactory receptor cilia. *Nature* **325,** 442–444.

Nargeot, J., Nerbonne, J., Engels, J., and Lester, H. (1983). Time course of increase in the myocardial slow inward current after a photochemically generated concentration jump of intracellular cAMP. *Proc. Natl. Acad. Sci. U.S.A.* **80,** 2395–2399.

Neer, E., and Clapham, D. (1988). Roles of G protein subunits in transmembrane signalling. *Nature* **333,** 129–134.

Nilius, B., Hess, P., Lansman, J. B., and Tsien, R. W. (1985). A novel type of cardiac calcium channel in ventricular cells. *Nature* **316,** 443–446.

Noma, A., and Trautwein, W. (1978). Relaxation of the ACh-induced potassium current in the rabbit sinoatrial node cell. *Pflügers Arch.* **377**, 193–200.

Nowycky, M. C., Fox, A. P., and Tsien, R. W. (1985). Three types of neuronal calcium channel with different calcium agonist sensitivity. *Nature* **316**, 440–443.

Osterrieder, W., Yang, Q., and Trautwein, W. (1981). The time course of the muscarinic response to ionophoretic acetylcholine application to the S-A node of the rabbit heart. *Pflügers Arch.* **389**, 283–291.

Pfaffinger, P. J., Martin, J. M., Hunter, D. D., Nathanson, N. M., and Hille, B. (1985). GTP-binding proteins couple cardiac muscarinic receptors to a K$^+$ channel. *Nature* **317**, 536–538.

Reuter, H. (1983). Calcium channel modulation by neurotransmitters, enzymes and drugs. *Nature* **301**, 569–574.

Rosenthal, W., and Schultz, G. (1987). Modulation of voltage-dependent ion channels by extracellular signals. *TIPS* **8**, 351–354.

Rosenthal, W., Koesling, D., Rudolph, U., Kleuss, C., Pallast, M., Yajima, M., Schultz, G. (1986). Identification and characterization of the 35-kDa β subunit of guanine nucleotide-binding proteins by an antiserum raised against transducin. *Eur. J. Biochem.* **158**, 255–263.

Rosenthal, W., Hescheler, J., Hinsch, K. -D., Spicher, K., Trautwein, W., and Schultz, G. (1988). Cyclic AMP-independent, dual regulation of voltage-dependent Ca^{2+} currents by LHRH and somatostatin in a pituitary cell line. *EMBO J.* **7**, 1627–1633.

Sakmann, B., Noma, A., and Trautwein, W. (1983). Acetylcholine activation of single muscarinic K$^+$ channels in isolated pacemaker cells of the mammalian heart. *Nature* **303**, 250–253.

Schofield, P., Darlison, M., Fujita, N., Burt, D., Stephenson, A., Rodriguez, H., Rhee, L., Ramachandran, J., Reale, V., Glencorse, T., Seeburg, P., and Barnard, E. (1987). Sequence and functional expression of the GABA$_A$ receptor shows a ligand-gated receptor super-family. *Nature* **328**, 221–227.

Scott, R. H., and Dolphin, A. C. (1986). Regulation of calcium currents by a GTP anlogue: Potentiation of (-)baclofen-mediated inhibition. *Neurosci. Lett.* **69**, 59–64.

Scott, R. H., and Dolphin, A. C. (1988). Activation of a G protein promotes agonist responses to calcium channel ligands. *Nature* **330**, 760–762.

Scott, I., Åkerman, K., Heikkilä, J., Kaila, K., and Andersson, L. (1986). Development of a neural phenotype in differentiating ganglion cell-derived human neuroblastoma cells. *J. Cell. Physiol.* **128**, 258–292.

Soejima, M., and Noma, A. (1984). Mode of regulation of the ACh-sensitive K-channel by the muscarinic receptor in rabbit atrial cells. *Pflügers Arch.* **400**, 424–431.

Stryer, L. (1986). Cyclic GMP cascade of vision. *Annu. Rev. Neurosci.* **9**, 87–119.

Suzuki, N., and Yoshioka, T. (1987). Differential blocking action of synthetic-conotoxin on components of Ca^{2+} channel current in clonal GH$_3$ cells. *Neurosci. Lett.* **75**, 235–239.

Tanabe, T., Takeshima, H., Mikami, A., Flockerzi, V., Takahashi, H., Kangawa, K., Kojima, M., Matsuo, H., Hirose, T., and Numa, S. (1987). Primary structure of receptor for calcium channel blockers from skeletal muscle. *Nature* **328**, 313–318.

Trautwein, W., and Dudel, J. (1958). Zum Mechanismus der Membranwirkung des Acetylcholin an der Herzmuskelfaser. *Pflügers Arch.* **266**, 324–334.

Trautwein, W., Taniguchi, J., and Noma, A. (1982). The effect of intracellular cyclic nucleotides and calcium on action potential and acetylcholine response of isolated cardiac cells. *Pflügers Arch.* **392**, 307–314.

Tsunoo, A., Yoshii, M., and Narahashi, T. (1986). Block of calcium channels by enkephalin and somatostatin in neuroblastoma-glioma hybrid NG108–15 cells. *Proc. Natl. Acad. Sci. U.S.A.* **83**, 9832–9836.

Ui, M., Katada, T., Murayama, T., Kurose, H., Yajima, M., Tamura, M., Nakamura, T., and Nogimori, K. (1984). Islet-activating protein, pertussis toxin: A specific uncoupler of receptor-mediated inhibition of adenylate cyclase. *Adv. Cyclic Nucleotide Protein Phosphorylation Res.* **17**, 145–151.

Wanke, E., Ferroni, A., Malgaroli, A., Ambrosini, A., Pozzan, T., and Meldolesi, J. (1987). Activation of a muscarinic receptor selectively inhibits a rapidly inactivated Ca^{2+} current in rat sympathetic neurons. *Proc. Natl. Acad. Sci. U.S.A.* **84**, 4313–4317.

Yaari, Y., Hamon, B., and Lux, H. (1987). Development of two types of calcium channels in cultured mammalian hippocampal neurons. *Science* **235**, 680–682.

Yatani, A., Codina, J., Brown, A. M., and Birnbaumer, L. (1987a). Direct activation of mammalian atrial muscarinic potassium channels by GTP regulatory protein G_K. *Science* **235**, 207–211.

Yatani, A., Codina, J., Sekura, R. D., Birnbaumer, L., and Brown, A. M. (1987b). Reconstitution of somatostatin and muscarinic receptor mediated stimulation of K^+ channels by isolated G_K protein in clonal rat anterior pituitary cell membranes. *Mol. Endocrin.* **1**, 283–286.

Yatani, A., Codina, J., Imoto, Y., Reeves, J. P., Birnbaumer, L., and Brown, A. M. (1987c). A G protein directly regulates mammalian cardiac calcium channels. *Science* **238**, 1288–1292.

CHAPTER 17

Signal Transduction in Olfaction and Taste

Richard C. Bruch

Monell Chemical Senses Center, Philadelphia, Pennsylvania 19104

I. INTRODUCTION

Historically, the chemical senses of taste (gustation) and smell (olfaction) have been less intensively studied than other sensory systems such as vision and audition. However, within the past decade, considerable progress has been made toward a more complete understanding of the molecular and cellular mechanisms underlying chemosensation. Chemosensory perception ultimately depends on detection and discrimination of chemical stimuli by peripheral chemoreceptive membranes and the transduction of stimulus-encoded information into neuronal activity. In vertebrates, primary chemosensory neurons detect odorants and transmit this information directly to the central nervous system. In contrast, stimulus interaction with the apical surface of taste cells leads to neurotransmitter release at the basolateral membrane thereby activating primary sensory neurons.

However, definition of the sequence of molecular events between the initial interaction of stimulus with the receptor cell membrane and synaptic transmission remains incompletely characterized and controversial. For example, the neurotransmitter(s) involved in taste stimulus–response coupling has not been identified, nor have the ionic mechanisms that mediate neurotransmitter release been unambiguously characterized in taste cells. A recent review has also emphasized the lack of consensus among laboratories regarding the ionic events mediating stimulus activation of taste cells (Teeter *et al.,* 1987).

It is generally, but not universally, accepted that the initial events in chemoreception involve the reversible interaction of stimuli with specific receptor proteins in the apical membranes of taste cells and olfactory neurons. However, the receptor hypothesis has not been rigorously established since the appropriate molecular species have not been identified or isolated, due in large part to the generally low affinity of stimulus interaction with the binding sites. Recently, with the identification of G proteins and the demonstration of stimulus enhancement of GTP-dependent second messenger formation in chemosensory membrane preparations, additional evidence has accumulated to support the receptor hypothesis. This chapter summarizes recent evidence regarding the interaction of stimuli with the putative receptor sites, the interaction of the receptor sites with G proteins, and stimulus activation of G protein-linked second messenger systems. Since space does not permit a detailed treatment of the chemical senses literature, selected examples will be discussed to illustrate current concepts of the molecular events underlying chemosensory signal transduction. The reader is referred to several recently published reviews that discuss many of the historical concepts and additional aspects of the chemical senses that are not covered here (Bruch *et al.,* 1988; Getchell *et al.,* 1985; Getchell, 1986; Lancet, 1986; Teeter and Brand, 1987a; Teeter and Cagan, 1988; Teeter *et al.,* 1987).

II. RECEPTOR HYPOTHESIS

The peripheral tissues subserving taste and olfaction in vertebrates contain a variety of cell types in addition to the chemosensory receptor cells. Isolated or cultured receptor cell preparations are not yet routinely available for biochemical studies of chemoreception due primarily to the intrinsic cellular heterogeneity of these tissues and their continuous turnover even in mature animals. The olfactory epithelium contains primary chemosensory neurons, specialized for the detection of odorants, as well as sustentacular cells, basal cells, and secretory glands (Getchell *et al.,* 1985; Getchell, 1986). Olfactory receptor cells are bipolar neurons with a single, unbranched, and nonmyelinated axon projecting to the olfactory bulb. The receptor cells also have a single dendrite projecting toward the external environment, which generally terminates at the apical end with cilia, although a second receptor cell type terminates in microvilli (Moran et al., 1982; Yamamoto, 1982). The cilia and microvillar extensions of the dendritic membrane are generally considered to be the site of the initial interaction of stimuli with the chemoreceptive membrane (Rhein and Cagan, 1981). The cilia are readily isolated and have been shown to be an appropriate membrane preparation for biochemical studies of the initial events of olfactory reception and signal transduction (Anholt et al., 1986; Boyle et al., 1987; Chen et al., 1986; Rhein and Cagan, 1981). The isolated cilia preparations may thus be regarded as the olfactory equivalent of the rod outer segment preparation of vision.

Taste receptor cells are localized in multicellular structures (taste buds) that also contain additional cell types. In mammalian taste buds, several cell types have been described, including dark, light, gustatory, and basal cells (Kinnamon, 1987). However, the number of cell types in taste buds and their roles in chemoreception continue to be investigated. Identification of the receptor cells and characterization of their properties have been complicated by the observations that many of the cell types in the taste bud appear to be suitably innervated and nearly all cells within the taste bud respond to common taste stimuli. Occluding junctions near the apical ends of the cells within the taste bud divide the receptor cell membrane into an apical portion that terminates in microvilli and is exposed to the external environment, and a basolateral portion facing the interstitial fluid. The initial interaction of stimulus with the apical receptor cell membrane presumably initiates the transduction events that lead to neurotransmitter release from the basolateral membrane.

The initial interaction of stimuli with the exposed apical membrane of chemosensory cells is generally considered to be mediated by specific interactions of chemical stimuli with macromolecular receptors. Biochemical and electrophysiological evidence is consistent with the hypothesis of receptor involvement in detection and discrimination of olfactory stimuli (Bruch et al., 1988; Getchell, 1986; Lancet, 1986; Price, 1981; Rhein and Cagan, 1981). The ligand-binding properties of olfactory receptors have been studied most extensively in fish, for which amino acids are effective olfactory stimuli for many species (reviewed in Brown and Hara, 1982; Bruch, 1989; Bruch et al., 1988). The interaction of stimulus amino acids with the binding sites exhibited the appropriate specificity, saturability, and reversibility properties expected for the interaction of a ligand with a membrane-associated receptor. In isolated cilia preparations, the ligand-binding properties of the amino acid-binding sites have correlated well with specificity, potency, and stereoselectivity characteristics expected for these sites from electrophysiological data. The agreement between biochemical binding studies and neurophysiological data further supports the hypothesis that stimulus amino acid detection and recognition involves specific stimulus interaction with physiologically relevant receptors in the ciliary membrane. In contrast, a model for olfactory chemoreception has also been proposed that does not invoke stimulus interaction with receptor proteins. In this model, membrane lipid composition was hypothesized to vary from cell to cell, thereby providing selective adsorption sites for odorants (Kurihara et al., 1986; Nomura and Kurihara, 1987a,b). Thus, in addition to stimulus–receptor interaction, the interaction of some odorants, particularly those that are hydrophobic, with membrane lipids may also contribute to the overall mechanism of olfactory neuron activation.

As in olfaction, taste receptor-cell activation by at least some stimuli involves receptor-mediated recognition at the apical chemoreceptive membrane (Bruch et al., 1988; Cagan, 1981; Teeter and Brand, 1987a). Although ligand-binding studies support the receptor hypothesis for amino acid and sweet-tasting stimuli, alternative mechanisms, independent of specific ligand–receptor interaction, have been proposed for receptor cell activation by inorganic ions and bitter-tasting compounds (Bruch et al., 1988; Cagan, 1981; Teeter and Brand, 1987a). For example, taste responses to sodium and lithium, but not potassium or rubidium, salts have been shown to be mediated in part by amiloride-sensitive sodium channels (Brand et al., 1985; Heck et al., 1984). A variety of mechanisms have also been proposed to account for receptor-cell activation by bitter-tasting stimuli, including electrostatic interactions at the membrane surface (Kurihara et al., 1986), nonreceptor-mediated penetration of lipophilic stimuli across the apical membrane and subsequent inhibition of cyclic AMP phosphodiesterase (Kurihara, 1972; Law and Henkin, 1982), and receptor-mediated increase of intracellular calcium (Akabas et al., 1987). Since bitter-tasting stimuli comprise a large group of structurally diverse compounds, are often positively charged at neutral pH, and are generally lipophilic, it is not unlikely that each of these mechanisms may be involved in bitter chemoreception.

The most extensive evidence supporting the receptor hypothesis in vertebrate taste has come from ligand-binding studies in aquatic animals. Amino acids are also effective olfactory stimuli for many fish. Although these stimuli activate both olfactory and taste receptor cells, the two chemosensory systems differ electrophysiologically in their specificity, rank order of stimulus potency, sensitivity, and response characteristics (Brown and Hara, 1982; Caprio, 1977, 1978; Caprio and Byrd, 1984). Ligand-binding studies in isolated membrane preparations derived from taste epithelium and olfactory cilia have paralleled these electrophysiological distinctions between gustatory and olfactory responses to amino acids. Biochemically, taste and olfactory amino-acid receptors have been distinguished by binding affinity and selectivity, differential lectin inhibition, and mode of coupling to G proteins (Brown and Hara, 1982; Bruch *et al.*, 1988; Kalinoski *et al.*, 1987a, 1989; Bruch and Kalinoski, 1987). Taken together, biochemical and neurophysiological evidence is consistent with the hypothesis that, although gustatory and olfactory amino-acid receptors recognize similar ligands, the receptors are probably not identical molecular species in the two chemosensory systems.

III. G PROTEINS IDENTIFIED IN CHEMOSENSORY MEMBRANES

A significant conceptual advance in chemosensory signal transduction emerged with appreciation of the central role of transducin in mediating sensory transduction in visual cells (Stryer, 1986). Several laboratories therefore investigated the possibility that transmembrane signaling in chemosensory cells was also mediated by G proteins. Members of the G-protein family were first identified in isolated olfactory cilia preparations from frog and were subsequently identified in the olfactory epithelium of several other vertebrate species (reviewed in Bruch *et al.*, 1988; Bruch, 1989). Based on the familiar criteria of electrophoretic mobility, bacterial toxin-catalyzed ADP-ribosylation, and immunoreactivity with subunit-specific antisera, the common β subunit and the α subunits of G_s, G_i, and G_o were identified in the olfactory epithelia of fish, amphibia, and rodents. G_s was found to be particularly abundant in olfactory cilia and absent from respiratory cilia. In contrast, G_o and G_i were shown to be localized in membranes derived from deciliated neuroepithelium as well as in both olfactory and respiratory cilia. These observations encouraged speculation that adenylyl cyclase was involved in olfactory signal transduction (Section IV. A).

Although G_s was invariably found in olfactory cilia in all species examined, the distribution of pertussis-toxin substrates across vertebrate species appears to be more variable. In isolated cilia from catfish, a single 40-kDa pertussis-toxin substrate was identified that crossreacted with antiserum to a common amino-acid sequence of G-protein α subunits. This pertussis-toxin substrate, tentatively designated G_{40} (Bruch, 1989), did not crossreact with antisera to G_o and did not comigrate in polyacrylamide gels with the purified 41-kDa subunit of G_i (Bruch and Kalinoski, 1987). Although the identity and function of G_{40} have not been established, it probably corresponds to the 40-kDa form of G_i, designated $G_{i2}\alpha$ (Gilman, 1987) and may be homologous to the 40-kDa pertussis-toxin substrate recently identified in neutrophils (Gierschik *et al.*, 1986), brain (Katada *et al.*, 1987), and differentiated HL-60 cells (Oinuma *et al.*, 1987). It is also of interest to note that cDNA clones encoding G_s, G_o, and three forms of G_i have been isolated from rat olfactory epithelium (Jones and Reed, 1987). Two of the G_i-encoding clones correspond to $G_{i1}\alpha$ and $G_{i2}\alpha$ (Gilman, 1987), while the remaining clones apparently encode a previously unrecognized form of G_i.

In contrast to olfaction, G proteins have been identified thus far in a single vertebrate taste system. ADP-ribosylation and immunoblotting analysis with subunit-specific antisera in purified plasma membranes derived from cutaneous taste epithelium from catfish revealed the presence of the common β subunit and the α subunits of G_s and the 41-kDa form of G_i (Bruch and Kalinoski, 1987). Immunoreactive G_o and G_{40} were not detected in membranes derived from taste epithelium in this animal model. Coupling of gustatory and olfactory amino-acid receptors and G proteins in this model system was also investigated. In isolated olfactory cilia, the affinities of two amino-acid receptors for their ligands were decreased in the presence of GTP or a hydrolysis-resistant analog. In contrast, the corresponding gustatory receptor affinities for the same ligands were unaffected by guanine nucleotides when tested under a variety of conditions shown previously to uncouple a number of hormone and neurotransmitter receptors (Bruch and Kalinoski, 1987). Thus, these results provided direct evidence for functional coupling of olfactory receptors with G proteins at the level of the initial binding event. Based on the observation of stimulus activation of second-messenger formation in isolated membrane preparations, indirect evidence has been presented recently that indicates that the gustatory amino-acid receptors are also coupled to G proteins (Section IV).

IV. SECOND MESSENGERS IN CHEMOSENSORY TRANSDUCTION

A. Cyclic Nucleotides

A role for adenylyl cyclase in signal transduction in vertebrate taste and olfaction was suspected for many years based on the observation of unusually high basal levels of the enzyme in chemosensory tissues (Kurihara and Koyama, 1972). These initial observations were not pursued further until Menevse *et al.* (1977) showed that membrane-permeable cyclic AMP analogs and cyclic nucleotide phosphodiesterase inhibitors reversibly reduced the amplitude of electrophysiological responses to odorants. An additional eight years elapsed before Lancet and co-workers showed that odorants stimulated the formation of cyclic AMP in isolated cilia preparations from frog (Pace *et al.,* 1985). During the next two years, three additional laboratories confirmed and further extended the observations of Pace *et al.* (1985). Stimulus activation of adenylyl cyclase in isolated cilia preparations and olfactory epithelium homogenates has been demonstrated in amphibia, fish, and rodents (reviewed in Bruch, 1989; Bruch *et al.,* 1988; Lancet, 1986). The biochemical evidence implicating the involvement of adenylyl cyclase in olfactory signal transduction may be summarized as follows:

1. Nearly half of the total adenylyl cyclase activity of the olfactory epithelium is associated with the isolated cilia. The olfactory enzyme exhibits many of the familiar characteristics of the classical hormone-regulated enzyme, including sensitivity to cholera toxin, guanine nucleotides, and fluoride ion. Basal, guanine nucleotide-stimulated, and odorant-stimulated adenylyl cyclase activities are markedly inhibited by micromolar concentrations of calcium.
2. G_s is apparently localized in olfactory, but not respiratory, cilia (Section III). Odorants stimulate cyclic AMP formation in a tissue-specific, GTP-dependent manner.
3. Cyclic AMP stimulates cyclic nucleotide-dependent protein kinase activity in isolated olfactory cilia preparations (Heldman and Lancet, 1986; Kropf *et al.,* 1987). Tissue-specific protein substrates for the kinase that exhibit specifically enhanced phosphorylation in response to cyclic AMP were also identified in these preparations, although the identities of these substrates and their possible roles in olfactory signal transduction were not determined.
4. Humans with type 1a pseudohypoparathyroidism exhibit deficits in erythrocyte G_s activity, resistance to the cyclic AMP-mediated actions of several hormones, and impaired ability to correctly identify several common odorants. In contrast, type 1b patients exhibit normal G_s activity in cyc⁻ membranes and normal olfactory ability (Weinstock *et al.,* 1986).

Based on the combined biochemical evidence, it has been proposed that cyclic AMP mediates olfactory signal transduction, presumably by the classical G_s-mediated mechanism (Pace *et al.,* 1985; Lancet, 1986). Neurophysiological evidence also supports the hypothesis of adenylyl cyclase involvement in olfactory signal transduction (reviewed in Bruch and Teeter, 1989; Bruch, 1989; Bruch *et al.,* 1988). Incorporation of rat olfactory epithelium homogenates into planar phospholipid bilayers conferred odorant sensitivity to the bilayers that was ATP- and GTP-dependent and mimicked by cyclic AMP (Vodyanoy and Vodyanoy, 1987a). Since protein kinase inhibitor antagonized the cyclic AMP-induced modulation of membrane conductance in these preparations, it was concluded that cyclic AMP-activated ion channels were deactivated by protein phosphorylation (Vodyanoy and Vodyanoy, 1987b). Modulation of membrane conductance by cyclic nucleotides was also observed in excised membrane patches from individual cilia on dissociated olfactory neurons (Nakamura and Gold, 1987) and in isolated ciliary membranes reconstituted into phospholipid bilayers on the tips of patch-clamp electrodes (Bruch and Teeter, 1989). Cyclic-nucleotide modulation of membrane conductance was mediated by nonselective cation channels of about 40 pS conductance that did not discriminate between sodium and potassium ions and exhibited voltage-dependent blockade by divalent cations. Half-maximal increases in membrane conductance were obtained with $1-5$ μM of either cyclic AMP or cyclic GMP with an average Hill coefficient of 1.5 in excised ciliary membrane patches (Nakamura and Gold, 1987). Since exogenous nucleotide triphosphates were not required to observe the reversible modulation of membrane conductance induced by cyclic nucleotides in these experiments, it was concluded that the cation channels mediating the conductance were gated directly by cyclic nucleotides without intermediary protein phosphorylation. Thus, taken together with the biochemical evidence, neurophysiological evidence is consistent with the hypothesis of cyclic-AMP mediation of olfactory responses by regulation of cation-channel activity by direct gating and/or by protein phosphorylation (Fig. 1).

Although a substantial amount of evidence implicates a second-messenger role for cyclic AMP in olfaction, additional considerations indicate that it is probably not the sole mediator of stimulus–response coupling in olfactory neurons. While a large number of odorants stimulate adenylyl cyclase in isolated cilia, a number of other odorants do not elicit increased cyclic AMP formation (Sklar *et al.,* 1986), suggesting that at least one additional transduction pathway must be available to account for responses to these stimuli. Further, in isolated cilia from catfish, amino-acid regulation of adenylyl cyclase was investigated under conditions that maintained receptor-binding activity and receptor–G protein coupling (Bruch and Teeter, 1989). In this study, significant enhancement of cyclic AMP formation was obtained only at high receptor occupancy after prolonged (greater than 2 min) exposure to stimulus. The ability of ten amino acids, representative of each olfactory receptor class in this animal model and covering a wide range of electrophysiological potency (Caprio, 1978; Caprio and Byrd, 1984), was poorly correlated with neurophysiological potency when tested at equimolar concentrations. In addition, stimulus amino acids did not accelerate the rate of cyclic AMP formation obtained by guanine nucleotide alone. This observation contrasts sharply with other

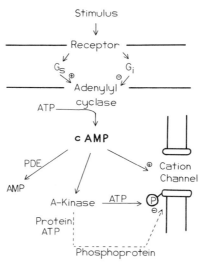

Fig. 1. Second messenger role of cyclic AMP in chemosensory signal transduction. Stimulation of adenylyl cyclase following stimulus–receptor interaction, presumably mediated by G_s, leads to elevation of intracellular cyclic AMP levels. Three metabolic fates of cyclic AMP are indicated: (1) direct activation of cation channels in olfactory cilia; (2) stimulation of cyclic AMP-dependent protein kinase (A-Kinase); and (3) degradation catalyzed by cyclic nucleotide phosphodiesterase (PDE). In olfactory cilia, stimulation of A-Kinase may deactivate cyclic AMP-gated cation channels by phosphorylation of an intermediary modulatory protein that subsequently regulates channel activity (dashed line). Direct gating of ion channels by cyclic nucleotides has not been reported in gustatory cells. Stimulus-dependent inhibition of adenylyl cyclase, mediated by G_i, has also not been reported in chemosensory tissues.

G_s-linked receptors, such as the β-adrenoreceptor (Cerione *et al.,* 1985), suggesting that the olfactory adenylyl cyclase may be regulated by a mechanism distinct from the presumed G_s-mediated pathway. The slow rate of odorant stimulation of cyclic AMP formation also suggests that this second messenger may mediate tonic or adaptive responses rather than the initial rapid (less than 1 sec, Getchell, 1986) phasic response.

The possible second-messenger role of cyclic GMP in olfactory signal transduction has not been extensively investigated. Although membrane-permeable cyclic AMP analogs reduced olfactory electrophysiological responses, the corresponding cyclic GMP analogs were ineffective (Menevse *et al.,* 1977). Cyclic GMP activated cyclic nucleotide-dependent protein kinase in isolated cilia from frog, although about tenfold more cyclic GMP than cyclic AMP was required for half-maximal stimulation (Heldman and Lancet, 1986). In contrast, cyclic GMP and cyclic AMP were equally effective in modulating cation-channel activity in excised ciliary membrane patches (Nakamura and Gold, 1987) and in reconstituted ciliary membranes in artificial phospholipid bilayers (Bruch and Teeter, 1989). Guanylyl cyclase activity was identified in soluble and particulate fractions from the olfactory epithelium, although only a small percentage of the total activity was associated with the isolated cilia (Bruch and Teeter, 1989). The enzyme in all isolated fractions from this tissue exhibited a marked cofactor preference for Mn^{2+} over Mg^{2+}. In isolated cilia, Mn^{2+}-dependent cyclic GMP formation was stimulated by exogenous Ca^{2+}, but only at high (millimolar) concentrations. Ca^{2+}-stimulated cyclic GMP formation was markedly potentiated by the calcium ionophore A23187, indicating that the Ca^{2+}-dependent enhancement of guanylyl cyclase activity was a specific effect. Not unexpectedly, Mn^{2+}-dependent cyclic GMP formation in isolated cilia was unaffected by stimuli, irrespective of the presence or absence of stimulatory levels of exogenous Ca^{2+}. Thus, since receptor activation of guanylyl cyclase has thus far been demonstrated only in intact cells, further evaluation of the role of cyclic GMP in olfactory stimulus–response coupling will depend on the availability of isolated receptor cells.

In contrast to olfaction, less information is currently available regarding the potential involvement of cyclic nucleotides in taste-signal transduction (reviewed in Bruch *et al.,* 1988; Teeter and Brand, 1987a). Adenylyl cyclase and guanylyl cyclase, as well as cyclic nucleotide phosphodiesterase, were identified histochemically in taste bud-containing folliate papillae in mammals (Nomura, 1978, 1980). In these studies, salt stimuli inhibited adenylyl cyclase, suggesting that the enzyme was not involved in salt taste responses. Enhancement of cyclic AMP levels by sweet and bitter stimuli was also not observed in bovine taste preparations (Cagan, 1976). However, perfusion of the lingual artery with cyclic nucleotides affected electrophysiological responses to some stimuli in frog (Kurihara *et al.,* 1986). However, the significance of these observations may be limited since the cellular site of action of these potential second messengers was not identified. In contrast, stimulus enhancement of GTP-dependent cyclic AMP formation by sweet-tasting stimuli was reported in membrane preparations from rat tongue (Striem *et al.,* 1986; Lancet *et al.,* 1987). Thus, due to the contradictory nature of these combined observations, the role of cyclic AMP in mediating gustatory responses to sweet and bitter stimuli remains ambiguous.

Recent evidence has been presented, however, indicating that adenylyl cyclase may be involved in gustatory responses to some amino acids in the channel catfish. In isolated particulate fractions derived from cutaneous taste epithelium from this animal model, cyclic AMP accumulation was stimulated by guanine nucleotides and fluoride ion, indicating functional interaction between adenylyl cyclase and G protein (Kalinoski *et al.,* 1987b). The potent taste stimulus L-alanine enhanced guanine nucleotide-stimulated cyclic AMP formation, suggesting that this receptor was coupled to adenylyl cyclase through G_s (Kalinoski *et al.,* 1988). Time-course studies indicated that the rate of cyclic AMP formation elicited by L-alanine was rapid. Stimulus enhancement of cyclic AMP levels could be observed within 20 seconds of exposure to stimulus, suggesting that the rate of stimulus activation of adenylyl cyclase was sufficiently rapid to mediate gustatory responses. However, in contrast to amino-acid activation of adenylyl cyclase in the olfactory system of this animal model, activation of the gustatory L-arginine receptor did not affect guanine nucleotide-stimulated cyclic AMP accumulation. L-Arginine, which interacts with a distinct gustatory receptor from L-alanine (Kalinoski *et al.,* 1989), modulated cation-channel activity in purified plasma membrane preparations derived from taste epithelium that were incorporated into phospholipid bilayers on the tips of patch-clamp electrodes (Teeter and Brand, 1987b). Since L-arginine modulation of ion-channel activity was observed in the absence of exogenous second messengers or precursors, stimulus activation of ion channels in these preparations was proposed to be directly regulated by receptor occupation. The hypothesis of direct receptor-mediated regulation of ion channels was further supported by the observation that L-alanine did not affect ion channel activity. In addition, D-arginine, which inhibits L-arginine binding but is a poor electrophysiological stimulus (Kalinoski *et al.,* 1989), did not affect cation-channel activity (Teeter and Brand, 1987b).

A role for cyclic nucleotides in taste signal transduction has also been suggested by electrophysiological evidence that indicates that these second messengers regulate at least some potassium channels in taste receptor cells. Partial blocking of potassium currents by cyclic AMP was demonstrated by patch-clamp studies with dissociated taste receptor cells from frog (Avenet and Lindemann, 1987). Similar responses were obtained by stimulation of adenylyl cyclase with forskolin, and inhibition of cyclic nucleotide phosphodiesterase enhanced the effects of cyclic AMP. Since the cyclic AMP-mediated response was observed only in the presence of ATP, it was proposed that these channels were closed by phosphorylation resulting from activation of cyclic AMP-dependent protein kinase. Additional evidence supporting this proposal was subsequently obtained from whole-cell recordings and excised membrane patches that demonstrated that inhibition of the protein kinase and addition of the cyclic AMP-activated subunit of the kinase affected the cyclic AMP-sensitive channels in the expected manner (Avenet *et al.,* 1988). In contrast, potassium conductance was decreased in mouse taste cells following intracellular injection of cyclic GMP (Tonosaki and Funakoshi, 1988). However, similar results were also observed following injection of cyclic AMP, although the response magnitudes were smaller than those elicited by cyclic GMP.

B. Phosphoinositide-Derived Messengers

The potential second messengers derived from the phospholipase C-catalyzed hydrolysis of phosphoinositide lipids may also be involved in signal transduction in both olfaction and taste (Fig. 2). Biochemical evidence supporting this hypothesis has been reported thus far only in a single vertebrate species. Phospholipase C activity toward exogenous phophatidylinositol-4,5-bisphosphate (PIP_2) was identified and partially characterized in isolated cilia and membrane preparations derived from taste epithelium from the channel catfish (Boyle *et al.*, 1987; Bruch *et al.*, 1987a; 1988; Huque *et al.*, 1987; Huque and Brand, 1988). Coupling of the enzyme to G proteins was implicated in these preparations by stimulation of inositol phosphate formation in the presence of guanine nucleotides and fluoride ion (Bruch *et al.*, 1987a; Huque *et al.*, 1987; Huque and Brand, 1988;

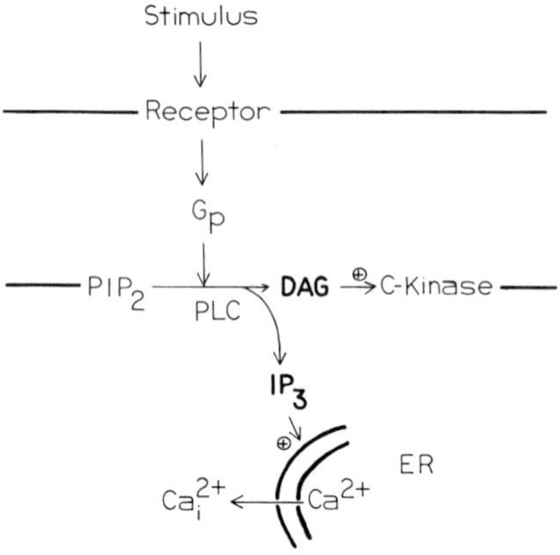

Fig. 2. Phosphoinositide-derived second messengers in chemosensory signal transduction. Stimulus–receptor interaction stimulates phospholipase C (PLC) by a G protein (G_p)-dependent mechanism. Hydrolysis of phosphatidylinositol 4,5-bisphosphate (PIP_2) generates the second messengers diacylglycerol (DAG) and inositol 1,4,5-trisphosphate (IP_3). Although protein kinase C (C-Kinase) has been identified in olfactory cilia (Anholt *et al.*, 1987), stimulus-dependent activation of the enzyme has not been reported. IP_3 releases Ca^{2+} from isolated olfactory microsomes (ER) thereby increasing intracellular calcium (Ca^{2+}) and would be expected to act similarly in gustatory cells, although this has not yet been confirmed.

Huque and Bruch, 1986). Stimulus activation of inositol phosphate production was also demonstrated in both chemosensory membranes. The potent olfactory and gustatory stimulus L-alanine enhanced guanine nucleotide- and fluoride-dependent inositol phosphate release (Bruch *et al.*, 1987a; Huque and Brand, 1988). In both chemosensory systems, stimulus activation of phospholipase C was very rapid with significant formation of inositol phosphate products observed within several seconds. Inositol 1,4,5-trisphosphate (IP_3) was the major inositol phosphate formed under basal and stimulated conditions in both systems. IP_3 was also shown to rapidly and transiently release calcium in isolated microsomes from the olfactory epithelium (Bruch, 1989), suggesting a role for intracellular calcium flux in olfactory stimulus–response coupling. Taken together, these initial studies suggest that at least some chemosensory receptors are coupled to a G protein-linked phospholipase C. The rapid time course of stimulus activation of this pathway also suggests that phosphoinositide hydrolysis may mediate rapid phasic chemosensory responses. Confirmation of this hypothesis and demonstration of stimulus activation of phospholipase C in other vertebrate model systems may be expected to lead to a clearer understanding of the molecular events underlying chemosensory signal transduction.

V. EDITOR'S COMMENTS (BY RAVI IYENGAR)

The central role of cAMP production in the transduction of the olfactory signal has gained further credence from the recent studies of Jones and Reed (1989), who have isolated an olfactory neuroepithelium-specific cDNA encoding a $G_s\alpha$-like protein which they have named $G_{olf}\alpha$. The predicted α_{olf} protein has 381 amino acid residues, and shares 88% amino acid identity with α_{s3}. α_{olf} does not appear to be a product of the α_s gene, and expression of the α_{olf} mRNA appears limited to olfactory neurons. Expression of G_{olf} in cyc^- kin^- S49 lymphoma cells restored β-adrenergic/ GTPγS and NaF stimulation of adenylyl cyclase to isolated cell membranes, indicating that G_{olf} can function as a G_s. Immunohistochemical localization of the G_{olf} protein indicated that expression is highest in the ciliated surface of sensory neurons in the olfactory neuroepithelium, and this immunoreactivity is reduced following bulbectomy-induced degeneration of the neuroepithelium. The exclusive distribution G_{olf} implies that it may play a specialized role in olfactory transduction analogous to that of the G_ts (rod and cone transducins) in visual transduction.

Reference

Jones, D. T., and Reed, R. R. (1989). G_{olf}: An olfactory neuron specific-G protein involved in odorant signal transduction. *Science* **244**, 790–795.

REFERENCES

Akabas, M. H., Dodd, J., and Al-Awgati, Q. (1987). Mechanism of transduction of bitter taste in rat taste bud cells. *Soc. Neurosci. Abstr.* **13**, Part 1, 361.

Anholt, R. R. H., Aebi, U., and Snyder, S. H. (1986). A partially purified preparation of isolated chemosensory cilia from the olfactory epithelium of the bullfrog, *Rana catesbeiana. J. Neurosci.* **6**, 1962–1969.

Anholt, R. R. H., Mumby, S. M., Stoffers, D. A., Girard, P. R., Kuo, J. F., and Snyder, S. H. (1987). Transduction proteins of olfactory receptor cells: Identification of guanine nucleotide binding proteins and protein kinase C. *Biochemistry* **26**, 788–795.

Avenet, P., and Lindemann, B. (1987). Patch-clamp study of isolated taste receptor cells of the frog. *J. Membr. Biol.* **97**, 223–240.

Avenet, P., Hofmann, F., and Lindemann, B. (1988). Transduction in taste receptor cells requires cAMP-dependent protein kinase. *Nature* **331**, 351–354.

Boyle, A. G., Park, Y. S., Huque, T., and Bruch, R. C. (1987). Properties of phospholipase C in isolated olfactory cilia from the channel catfish *(Ictalurus punctatus). Comp. Biochem. Physiol.* **88B**, 767–775.

Brand, J. G., Teeter, J. H., and Silver, W. L. (1985). Inhibition by amiloride of chorda tympani responses evoked by monovalent salts. *Brain Res.* **334**, 207–214.

Brown, S. B., and Hara, T. J. (1982). Biochemical aspects of amino acid receptors in olfaction and taste. *In* "Chemoreception in Fishes" (T. J. Hara, ed.), pp. 159–180. Elsevier, New York.

Bruch, R. C. (1989). G-proteins in olfactory neurons. *In* "G-proteins and Calcium Mobilizing Hormones" (P. H. Naccache, ed.), in press. CRC Press, Boca Raton.

Bruch, R. C., and Kalinoski, D. L. (1987). Interaction of GTP-binding regulatory proteins with chemosensory receptors. *J. Biol. Chem.* **262**, 2401–2404.

Bruch, R. C., and Teeter, J. H. (1989). Role of cyclic nucleotides in olfactory signal transduction. Submitted for publication.

Bruch, R. C., Kalinoski, D. L., and Huque, T. (1987a). Role of GTP-binding regulatory proteins in receptor-mediated phosphoinositide turnover in olfactory cilia. *Chem. Senses* **12**, 173.

Bruch, R. C., Rulli, R. D., and Boyle, A. G. (1987b). Olfactory L-amino acid receptor specificity and stimulation of potential second messengers. *Chem. Senses* **12**, 642–643.

Bruch, R. C., Kalinoski, D. L., and Kare, M. R. (1988). Biochemistry of vertebrate olfaction and taste. *Annu. Rev. Nutr.* **8**, 21–42.

Cagan, R. H. (1976). Biochemical studies of taste sensation. II. Labeling of cyclic AMP of bovine taste papillae in response to sweet and bitter stimuli. *J. Neurosci. Res.* **7**, 37–43.

Cagan, R. H. (1981). Recognition of taste stimuli at the initial binding interaction. *In* "Biochemistry of Taste and Olfaction" (R. H. Cagan and M. R. Kare, eds.), pp. 175–203. Academic Press, New York.

Caprio, J. (1977). Electrophysiological distinctions between the taste and smell of amino acids in catfish. *Nature* **266**, 850–851.

Caprio, J. (1978). Olfaction and taste in the channel catfish: An electrophysiological study of the responses to amino acids and derivatives. *J. Comp. Physiol.* **123**, 357–371.

Caprio, J., and Byrd, R. P., Jr. (1984). Electrophysiological evidence for acidic, basic, and neutral amino acid olfactory receptor sites in the catfish. *J. Gen. Physiol.* **84**, 403–422.

Cerione, R. A., Staniszewski, C., Benovic, J. L., Lefkowitz, R. J., Caron, M. G., Gierschik, P., Somers, R., Spiegel, A. M., Codina, J., and Birnbaumer, L. (1985). Specificity of the functional interactions of the β-adrenergic receptor and rhodopsin with guanine nucleotide regulatory proteins reconstituted in phospholipid vesicles. *J. Biol. Chem.* **260**, 1493–1500.

Chen, Z., Pace, U., Heldman, J., Shapira, A., and Lancet, D. (1986). Isolated frog olfactory cilia: A preparation of dendritic membranes from chemosensory neurons. *J. Neurosci.* **6**, 2146–2154.

Getchell, T. V. (1986). Functional properties of vertebrate olfactory receptor neurons. *Physiol. Rev.* **66**, 772–818.

Getchell, T. V., Margolis, F. L., and Getchell, M. L. (1985). Perireceptor and receptor events in vertebrate olfaction. *Prog. Neurobiol.* **23**, 317–345.

Gierschik, P., Falloon, J., Milligan, G., Pines, M., Gallin, J. I., and Spiegel, A. (1986). Immunochemical evidence for a novel pertussis toxin substrate in human neutrophils. *J. Biol. Chem.* **261**, 8058–8062.

Gilman, A. G. (1987). G-proteins: Transducers of receptor-generated signals. *Annu. Rev. Biochem.* **56**, 615–649.

Heck, G. L., Mierson, S., and DeSimone, J. A. (1984). Salt taste transduction occurs through an amiloride-sensitive sodium transport pathway. *Science* **223**, 403–405.

Heldman, J., and Lancet, D. (1986). Cyclic AMP-dependent protein phosphorylation in chemosensory neurons: Identification of cyclic nucleotide-regulated phosphoproteins in olfactory cilia. *J. Neurochem.* **47**, 1527–1533.

Huque, T., and Brand, J. G. (1988). Phosphatidylinositol-4,5-bisphosphate phosphodiesterase (PIP2-PDE) activity of catfish taste tissue. *Chem. Senses* **13**, 698.

Huque, T., and Bruch,, R. C. (1986). Odorant- and guanine nucleotide-stimulated phosphoinositide turnover in olfactory cilia. *Biochem. Biophys. Res. Commun.* **137**, 36–42.

Huque, T., Brand, J. G., Rabinowitz, J. L., and Bayley, D. L. (1987). Phospholipid turnover in catfish barbel (taste) epithelium with special reference to phosphatidylinositol-4,5-bisphosphate. *Chem. Senses* **12**, 666–667.

Jones, D. T., and Reed, R. R. (1987). Molecular cloning of five GTP-binding protein cDNA species from rat olfactory neuroepithelium. *J. Biol. Chem.* **262,** 14241–14249.

Kalinoski, D. L., Bruch, R. C., and Brand, J. G. (1987a). Differential interaction of lectins with chemosensory receptors. *Brain Res.* **418,** 34–40.

Kalinoski, D. L., LaMorte, V., and Brand, J. G. (1987b). Characterization of a taste stimulus-sensitive adenylate cyclase from the gustatory epithelium of the channel catfish, *Ictalurus punctatus. Soc. Neurosci. Abstr.* **13,** Part 2, 1405.

Kalinoski, D. L., LaMorte, V. J., Johnson, L. C., and Brand, J. G. (1988). L-Amino acid stimulated adenylate cyclase in catfish barbel epithelium: Specificity and characterization. *Chem. Senses* **13,** 700.

Kalinoski, D. L., Bryant, B. P., Shaulsky, G., Brand, J. G., and Harpaz, S. (1989). Specific L-arginine taste receptors in the catfish, *Ictalurus punctatus:* Biochemical and neurophysiological characterization. *Brain Res.* **488,** 163–173.

Katada, T., Oinuma, M., Kusakabe, K., and Ui, M. (1987). A new GTP-binding protein in brain tissues serving as the specific substrate of islet-activating protein, pertussis toxin. *FEBS Lett.* **213,** 353–358.

Kinnamon, J. C. (1987). Organization and innervation of taste buds. *In* "Neurobiology of Taste and Smell" (T. E. Finger and W. L. Silver, eds.), pp. 277–297. Wiley, New York.

Kropf, R., Lancet, D., and Lazard, D. (1987). A bovine olfactory cilia preparation: Specific transmembrane glycoproteins and phosphoproteins. *Soc. Neurosci. Abstr.* **13,** Part 2, 1410.

Kurihara, K. (1972). Inhibition of cyclic 3′:5′-nucleotide phosphodiesterase in bovine taste papillae by bitter stimuli. *FEBS Lett.* **27,** 279–281.

Kurihara, K., and Koyama, N. (1972). High activity of adenyl cyclase in olfactory and gustatory organs. *Biochem. Biophys. Res. Commun.* **48,** 30–34.

Kurihara, K., Yoshii, K., and Kashiwayanagi, M. (1986). Transduction mechanisms in chemoreception. *Comp. Biochem. Physiol.* **85A,** 1–22.

Lancet, D. (1986). Vertebrate olfactory reception. *Annu. Rev. Neurosci.* **9,** 329–355.

Lancet, D., Striem, B. J., Pace, U., Zehavi, U., and Naim, M. (1987). Adenylate cyclase and GTP-binding protein in rat sweet taste transduction. *Soc. Neurosci. Abstr.* **13,** Part 1, 361.

Law, J. S., and Henkin, R. I. (1982). Taste bud adenosine 3′:5′-monophosphate phosphodiesterase: Activity, subcellular distribution, and kinetic parameters. *Res. Commun. Chem. Pathol. Pharmacol.* **38,** 439–452.

Menevse, A., Dodd, G., and Poynder, T. M. (1977). Evidence for the specific involvement of cyclic AMP in the olfactory transduction mechanism. *Biochem. Biophys. Res. Commun.* **77,** 671–677.

Moran, D. T., Rowley, J. C., and Jafek, B. W. (1982). Electron microscopy of human olfactory epithelium reveals a new cell type: The microvillar cell. *Brain Res.* **253,** 39–46.

Nakamura, T., and Gold, G. H. (1987). A cyclic nucleotide-gated conductance in olfactory receptor cilia. *Nature* **325,** 442–444.

Nomura, H. (1978). Histochemical localization of adenylate cyclase and phospho-diesterase activities in the folliate papillae of the rabbit. I. Light microscopic observations. *Chem. Senses Flavour* **3**, 319–324.

Nomura, H. (1980). Is cyclic nucleotide involved in transduction process of mammalian taste receptor cells? *In* "Olfaction and Taste VII" (H. van der Starre, ed.), p. 219. IRL Press, London.

Nomura, T., and Kurihara, K. (1987a). Liposomes as a model for olfactory cells: Changes in membrane potential in response to various odorants. *Biochemistry* **26**, 6135–6140.

Nomura, T., and Kurihara, K. (1987b). Effects of changed lipid composition on response of liposomes to various odorants: Possible mechanism of odor discrimination. *Biochemistry* **26**, 6141–6145.

Oinuma, M., Katada, T., and Ui, M. (1987). A new GTP-binding protein in differentiated human leukemic (HL-60) cells serving as the specific substrate of islet-activating protein, pertussis toxin. *J. Biol. Chem.* **262**, 8347–8353.

Pace, U., Hanski, E., Salomon, Y., and Lancet, D. (1985). Odorant-sensitive adenylate cyclase may mediate olfactory reception. *Nature* **316**, 255–258.

Price, S. (1981). Receptor proteins in vertebrate olfaction. *In* "Biochemistry of Taste and Olfaction" (R. H. Cagan and M. R. Kare, eds.), pp. 69–84. Academic Press, Orlando.

Rhein, L. D., and Cagan, R. H. (1981). Role of cilia in olfactory recognition. *In* "Biochemistry of Taste and Olfaction" (R. H. Cagan and M. R. Kare, eds.), pp. 47–68. Academic Press, Orlando.

Sklar, P. B., Anholt, R. R. H., and Snyder, S. H. (1986). The odorant-sensitive adenylate cyclase of olfactory receptor cells: Differential stimulation by distinct classes of odorants. *J. Biol. Chem.* **261**, 15538–15543.

Striem, B. J., Pace, U., Zehavi, U., Naim, M., and Lancet, D. (1986). Is adenylate cyclase involved in sweet taste transduction? *Chem. Senses* **11**, 669.

Stryer, L. (1986). Cyclic GMP cascade of vision. *Annu. Rev. Neurosci.* **9**, 87–119.

Teeter, J. H., and Brand, J. G. (1987a). Peripheral mechanisms of gustation: Physiology and biochemistry. *In* "Neurobiology of Taste and Smell" (T. E. Finger and W. L. Silver, eds.), pp. 299–329. Wiley, New York.

Teeter, J. H., and Brand, J. G. (1987b). L-Arginine-activated cation channels from the catfish taste epithelium. *Soc. Neurosci. Abstr.* **13**, Part 1, 361.

Teeter, J. H., and Cagan, R. H. (1989). Mechanisms of taste transduction. *In* "Neural Mechanisms in Taste" (R. H. Cagan, ed.), in press. CRC Press, Boca Raton.

Teeter, J., Funakoshi, M., Kurihara, K., Roper, S., Sato, T., and Tonosaki, K. (1987). Generation of the taste cell potential. *Chem. Senses* **12**, 217–234.

Teeter, J. H., Brand, J. G., Kalinoski, D. L., and Bryant, B. P. (1988). Cation channels in reconstituted catfish taste epithelial membrane preparations activated by arginine in an enantiomerically specific manner. *Chem. Senses* **13**, 740.

Tonosaki, K., and Funakoshi, M. (1988). Cyclic nucleotides may mediate taste transduction. *Nature* **331**, 354–356.

Vodyanoy, V., and Vodyanoy, I. (1987a). ATP and GTP are essential for olfactory response. *Neurosci. Lett.* **73**, 253–258.

Vodyanoy, V., and Vodyanoy, I. (1987b). Ion channel modulation by cAMP and protein kinase inhibitor. *Soc. Neurosci. Abstr.* **13**, Part 2, 1410.

Weinstock, R. S., Wright, H. N., Spiegel, A. M., Levine, M. A., and Moses, A. M. (1986). Olfactory dysfunction in humans with deficient guanine nucleotide-binding protein. *Nature* **322**, 635–636.

Yamamoto, M. (1982). Comparative morphology of the peripheral olfactory organs in teleosts. *In* "Chemoreception in Fishes" (T. J. Hara, ed.), pp. 39–59. Elsevier, New York.

CHAPTER 18

Phosphatidylinositol Phospholipase C

Hans Deckmyn*, Brian J. Whiteley, Philip W. Majerus
Division of Hematology–Oncology, Departments of Internal Medicine and
Biological Chemistry, Washington University School of Medicine, St. Louis,
Missouri 63110

Phosphatidylinositol (PI)-specific phospholipase C (PLC) was first de-scribed by Dawson (1959) and Kemp *et al.* (1961) and is a ubiquitous enzyme found in both prokaryotes and eukaryotes (Shukla, 1982). This review concerns mammalian PLC which plays a key role in the response of cells to extracellular agonists.

When certain cell surface receptors are occupied by agonists, activation of PLC follows. PLC cleaves the different phosphatidylinositols giving rise to a number of inositol phosphates and 1,2-diacylglycerol. The latter acti-vates protein kinase C, leading to phosphorylation reactions, or is degraded

---

* Current address: Leuvin University, Leuvin, Belgium.

by lipases to liberate arachidonic acid for eicosanoid production. The inositol phosphates elevate intracellular $[Ca^{2+}]$, which itself facilitates a number of reactions.

I. ACTION OF PHOSPHOLIPASE C

A. Substrates

There are multiple forms of PLC (see Section II), all of which use phosphatidylinositol (PI), phosphatidylinositol 4-phosphate (PIP), and phosphatidylinositol 4,5-bisphosphate (PIP_2) as substrates (Wilson et al., 1984; Ryu et al., 1987b; Rebecchi and Rosen, 1987b). They do not hydrolyze other phospholipids, with the exception of phosphatidylglycerol, which is utilized 0.001 times as well as PI (Hofmann and Majerus, 1982a). When PI, PIP, and PIP_2 are present together in unilamellar vesicles, they compete with each other for PLC. At equimolar proportions, hydrolysis of polyphosphoinositides is favored, while increasing the Ca^{2+} concentration favors PI breakdown (Wilson et al., 1984).

B. Products

The products of the reaction of PLC in vitro are 1,2-diacylglycerol, inositol 1-phosphate (I1P), inositol 1,4-bisphosphate (IP_2), inositol 1,4,5-trisphosphate (IP_3), and the inositol 1,2-cyclic phosphate [cI(1:2)P] counterparts of each of these inositol phosphates when PI, PIP, and PIP_2 are used as substrate, respectively (Fig. 1).

The ratio between cyclic and noncyclic products of PLC depends on pH (Dawson et al., 1971), with relatively less cyclic product formed at higher pH. This is consistent with a competition between free OH^- and the 2-position hydroxyl on the inositol ring as alternative nucleophiles to attack the phosphorus. At pH 5.3, 60–70% cyclic product was formed from PI, 40–50% from PIP, and 30–40% from PIP_2 (Wilson et al., 1985). Carter et al. (1986) indicated that different forms of PLC may produce specific ratios of inositol cyclic versus noncyclic phosphates. Although the formation of inositol cyclic phosphates has been documented in a number of cells (reviewed in Majerus et al., 1988), the ratio between the cyclic and noncyclic products is uncertain due to technical difficulties in measuring the cyclic inositol phosphates which are acid-labile and subject to methanolysis (Lips et al., 1988).

The further metabolism of the six inositol phosphates produced by PLC occurs through a complex pathway catalyzed by a series of kinases and

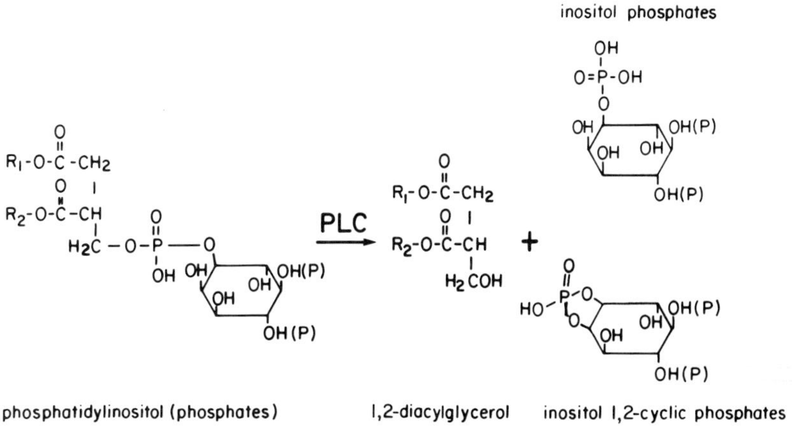

inositol phosphates

phosphatidylinositol (phosphates)   1,2-diacylglycerol   inositol 1,2-cyclic phosphates

Fig. 1. The general reaction catalyzed by phosphatidylinositol-specific phospholipase C (PLC). The (P) is used to denote that PLC utilizes substrates with or without phosphate groups at those positions and that the products are the corresponding inositol phosphate or inositol 1,2-cyclic phosphate compounds.

phosphatases, as reviewed by Majerus *et al.,* 1986, 1988. Presently, 15 inositol phosphates are known to be formed in one or more cell types following PLC activation, four of which, IP_3, $cI(1:2,4,5)P_3$, $I(1,3,4)P_3$, and $I(1,3,4,5)P_4$ may increase the Ca^{2+} levels in cells (reviewed by Berridge, 1987).

1,2-Diacylglycerol stimulates protein kinase C (Nishizuka, 1986), and, further, can be cleaved by diacyl- and monoacylglycerol lipases resulting in the liberation of arachidonic acid (Majerus, 1983). Arachidonic acid itself is used by cyclooxygenase and lipoxygenase enzymes to produce a series of prostaglandins, thromboxanes, leukotrienes, and lipoxins, each of which again can trigger specific cellular responses.

II. PURIFICATION OF PHOSPHOLIPASE C

A. Molecular Heterogeneity

A single tissue may contain multiple, immunologically distinct PLC forms. Hofmann and Majerus (1982a) found two PLC forms, which they termed I and II, in sheep seminal vesicle glands. These two cytosolic enzymes had apparent molecular weights of 65,000 and 85,000, respectively, and although immunologically and chromatographically distinct, were similar with respect to substrate specificity and enzyme kinetics. Rabbit antibodies raised against PLC-I were used to determine that sheep organs had varying proportions of PLC-I. The PLC from cultured human skin fibroblasts was 50%-inhibited by anti-PLC-I antibodies implying that a single cell contains more than one PLC form.

Most tissues contain multiple forms of chromatographically distinguishable PLC as listed in Table I. There are many different molecular weights reported including variations within a single tissue from different species. Molecular weight has been estimated from gel permeation of unpurified cytosolic PLC activity in some studies, while in others molecular weight was estimated using homogeneous protein. It is not known whether the PLC forms of different sizes exhibit any immunological crossreactivity. It is therefore currently impossible to estimate the number of distinct PLC enzymes; furthermore, the nomenclature is ambiguous with respect to PLC forms.

Partially purified PLC from bovine brain exists as a multimer when analyzed by PAGE under nondenaturing conditions (Ryu et al., 1987a) and these authors showed that oligomerization was more extensive following storage at $-20°$. It is unclear whether PLC exists as a monomer or oligomer in intact cells. Low et al. (1984) suggests that supernatant fractions from various fresh tissues contain a small fraction of dimerized PLC. Bennett and Crooke (1987) reported that PLC was phosphorylated in rat basophilic leukemia (RBL-1) cells treated with phorbol myristate acetate. No effect of phosphorylation on PLC activity or subcellular distribution has been reported.

B. Subcellular Distribution

PI-specific PLC activity is found predominantly in the cytosol of mammalian cells. PLC associated with the particulate fraction of brain was initially thought to be a distinct form of the enzyme (Lapetina and Michell, 1973) but others concluded that there was no distinct membrane form of PLC (Irvine and Dawson, 1978). Lysosomal PLC activity is not believed to play a role in receptor-mediated events (Irvine and Dawson, 1978).

PLC has been purified from the supernatant fraction of homogenates of various tissues and also from a bovine brain particulate fraction after extraction of protein with 2 M KCl (Lee et al., 1987). Two membrane-associated PLC activities were resolved, both of which reacted with antibodies directed against a cytosolic PLC that they termed PLC-I. The two membrane-bound forms were estimated to have molecular weights of 140,000 and 150,000 by sodium dodecyl sulfate-polyacrylamide gel electrophoresis (SDS-PAGE) and the authors theorized, based on tryptic peptide analysis, that the 140K form was a degradation product of the 150K enzyme. They found no physical differences between soluble and membrane PLC-I. Katan and Parker (1987) found membrane-associated PLC in bovine brain. Those authors separated brain homogenates into soluble and particulate fractions and then extracted the pellet with 2% cholate/0.4 M NaCl. They isolated PLC and estimated the molecular weight to be 154K by gel permeation and SDS-PAGE. Bennett and Crooke (1987) isolated a PLC, which they termed PLC-I, from the cytosol of guinea pig uterus. Rabbit polyclonal antibody raised against the purified cytosolic enzyme reacted with the membrane-associated and soluble forms of identical molecular weights, 62K based on SDS-PAGE. The authors found very little salt-extractable, membrane-associated PLC activity using 1M KCl. It is unclear whether 2 M KCl would have extracted all of the membrane-bound PLC activity, as was found in a study of bovine brain enzyme (Lee *et al.,* 1987).

III. REGULATION OF PHOSPHOLIPASE C

It is not known whether the different forms of PLC serve different functions or are regulated by separate factors. The activity of the enzymes studied so far is affected by [Ca^{2+}], pH, lipid environment, and the presence of GTP-binding proteins.

A. Ca^{2+} and pH

Ca^{2+} ions are absolutely required in order to see PLC activity using PI as a substrate. PIP- and PIP_2 hydrolysis is detectable in the absence of Ca^{2+} with some PLC forms (Wilson *et al.,* 1984), but not with others (Ryu *et al.,* 1987b; Rebecchi and Rosen, 1987b). Increasing the free Ca^{2+} concentration stimulates the PLC activity. The breakdown of PIP_2 is obtained with Ca^{2+} concentrations in the range of 10^{-7} to 10^{-4} M; above 10^{-3} M, inhibition by Ca^{2+} is observed, with virtually no remaining activity at 10^{-2} M Ca^{2+} (e.g., Ryu *et al.,* 1987b). PI (Nakamura *et al.,* 1985; Ryu *et al.,* 1987b) or PIP (Low *et al.,* 1986) hydrolysis is stimulated by Ca^{2+} at 10^{-5} to 10^{-4} M, however, millimolar concentrations are required for maximum activity.

The pH optimum is pH 5.2 to 5.5 and is the same whether PI or PIP_2 is used as substrate (Carter and Smith, 1987; Manne and Kung, 1987; Melin *et al.,* 1986). The presence of the ionic detergent deoxycholate increases the pH optimum to about pH 7.

Table I Partially and Homogeneously Purified Phospholipases C, Estimated Molecular Weights, Purification over Starting Material, and Final Specific Activity[a]

Source	Form	MW (10^{-3})	Purification (-fold)	Specific activity (μmol/min/mg)	Reference
Rat liver cytosol		68	4310[c]	0.15[d]	Takenawa and Nagai (1981)
Sheep seminal vesicles	I	65	5310	29	Hofmann and Majerus (1982a)
	II	85	730	7	
Bovine platelets	I	143	290	0.6	Hakata et al. (1982)
Human platelets	I	—	10	0.06[d]	Chau and Tai (1982)
	II	—	5	0.05[d]	
Bovine heart	I	~110	—	0.006	Low and Weglicki (1983)
	II	~42	—	0.152	
	III	~64	—	0.336	
	IV	~256	—	—	
Sheep seminal vesicles	II	85	2625	10.5	Wilson et al. (1984)
Rat brain	I	300	100	—	Nakanishi et al. (1985)
	II	250	100	—	
Rat liver	I	200	—	—	Nakamura et al. (1985)
Human platelets	I	—	110	0.006	
Human platelets	I	67	159	0.35	Banno et al. (1986a)
	IIa	120	181	0.40	
	IIb	70	223	0.49	
Human platelets	I/II/III	140[b]	200–500	—	Low et al. (1986)
	IV	95	200–500	—	

Source	Form					Reference
Bovine brain	I	150[b]	—	11–15		Ryu et al. (1986)
	II	145[b]	—	21–26		
Human platelets	1	38–45–58(?)	189	0.68		Manne and Kung (1987)
Porcine lymphocytes	2A	300	—	—		Carter and Smith (1987)
	2B	175	—	—		
	4	78	—	—		
		11				
Bovine brain	I	150	1840[c]	12.9		Ryu et al. (1987a)
	II	145	1240[c]	23.5		
Bovine brain	II	88	58000[c]	0.7	29[e]	Rebecchi and Rosen (1987b)
Bovine brain membranes	II M_1	140	200	20–21		Lee et al. (1987)
	II M_2	150	200	20–26		
Bovine brain	III	85	6800[c]	34		Ryu et al. (1987b)
Bovine brain membranes		154	1154	150[f]		Katan and Parker (1987)
Guinea pig uterus	I_a	62	650	0.32		Bennett and Crooke (1987)
	I_b	62	1436	0.72		
	II	87–73–56(?)	300	—		

[a] Specific activities are difficult to compare in some cases because of differences in substrate, detergent, or Ca^{2+} concentration.
[b] Multimers were detected under nondenaturing conditions.
[c] Apparently homogeneous.
[d] No deoxycholate in assay.
[e] Substrate 100 μM PIP$_2$ in 1% octylglucoside.
[f] Substrate 220 μM PIP$_2$ in 0.6% sodium cholate.

B. Lipids and Detergents

Hydrolysis of PI is inhibited by phosphatidylcholine (PC), which apparently shields the substrate and does not allow PLC to bind to PC-containing PI vesicles (Irvine et al., 1979a,b; Dawson et al., 1980; Hofmann and Majerus, 1982b; Wilson et al., 1984). Phosphatidylserine, diacylglycerol, and free fatty acids can partially reverse this inhibition. Cell membranes, vesicles made from lipids extracted from cell membranes, or vesicles composed of lipids in proportions occurring in cell membranes are all poor substrates because of the amount of PC present.

The ionic detergent deoxycholate stimulates PI hydrolysis by PLC when present in concentrations ranging from 0.5 to 1.5 mg/ml (Hofmann and Majerus, 1982a; Ryu et al., 1987a). Deoxycholate stimulates PLC only at pH values above pH 6 and therefore results in an apparent shift in the pH optimum of the enzyme from pH 5.5 to pH 7.0. Since the pK_a of deoxycholate is 6.5, it is only an effective detergent as pH rises to 6.0 or above (Irvine and Dawson, 1978). Octylglucoside, a nonionic detergent, also stimulates PLC activity, especially using phosphatidylinositol polyphosphates as substrates (Rebecchi and Rosen, 1987b). Triton X-100, on the other hand, inhibits PLC (Takenawa and Nagai, 1981).

C. G Proteins

The activation of adenylyl cyclase, as discussed in other chapters of this volume, involves binding of an agonist to its receptor followed by the association of the receptor with a G protein. This leads to the exchange of GDP for GTP on the G protein, which reduces the affinity of the receptor for the agonist and causes the dissociation of the G protein into its α–GTP and $\beta\gamma$ subunit. α–GTP activates the catalytic unit leading to the formation of cAMP. The activation is terminated by the GTPase activity of the α subunit, after which α–GDP and $\beta\gamma$ reassociate.

PLC activation may occur through a parallel scheme but proof requires the identification, isolation, and reconstitution of the constitutive elements. A number of observations have given indirect evidence for a G protein that activates PLC, which we designate as "G_p".

1. Coupling of PLC-Activating Receptors to G Proteins

GTP and nonhydrolyzable analogs of GTP reduce the affinity of PLC-coupled receptors for their agonists. This has been shown for angiotensin II receptors in adrenal cortex (Glossman et al., 1974), vasopressin (Cantau et al., 1980), and angiotensin binding to liver plasma membranes (Crane et al., 1982); norepinephrine interaction with β_1-receptors (Goodhardt et al.,

1982); carbachol binding to muscarinic receptors (Florio and Sternweiss, 1985); and platelet-activating factor binding to platelet membranes (Hwang *et al.*, 1986). The apparent size of solubilized angiotensin receptor from liver (Guillemette *et al.*, 1984) or adrenal gland (DeLean *et al.*, 1984), or the vasopressin receptor from rat liver (Bojanic and Fain, 1986) is larger when liganded with an agonist than with an antagonist and the dissociation of the agonist from this larger form is enhanced by Gpp(NH)p (DeLean *et al.*, 1984).

In addition, PLC-activating receptors stimulate a GTPase activity, as seen when thrombin, platelet activating factor, the thromboxane mimetic U44069, and vasopressin bind to their respective receptors on platelets (Houslay *et al.*, 1986a,b; Grandt *et al.*, 1986; Hwang *et al.*, 1986). A similar increase in GTPase activity is found when neutrophil receptors are occupied by agonists including formyl-Met–Leu–Phe, platelet activating factor, and leukotriene B_4 (Okajima *et al.*, 1985; Matsumoto *et al.*, 1987); when vasopressin binds to liver plasma membranes (Fain *et al.*, 1985; Fitzgerald *et al.*, 1986); or when thyrotropin-releasing hormone binds to GH_3 cells (Wojcikiewicz *et al.*, 1986) or to GH_4C_1 pituitary tumor cells (Hinkle and Phillips, 1984).

Most of the PLC-activating agonists decrease adenylyl cyclase activity through coupling to G_i. Whether two G proteins are activated by a single receptor or whether stimulated G_i is able to exert two functions is uncertain. An indication for a putative G_p, distinct from G_i, has been inferred by showing additive GTPase activity when agonists linked to G_i and to PLC-activating receptors are combined (Houslay *et al.*, 1986a). Pertussis toxin blocks G_i while leaving G_p activation intact in some cases (Houslay *et al.*, 1986a,b; Grandt *et al.*, 1986; Wojcikiewicz *et al.*, 1986).

2. Coupling of a G Protein to Phospholipase C

It has been difficult to show coupling of a G protein to phospholipase C because of the inability to manipulate intracellular GTP- or stable GTP analog levels in intact cells. Treatment of intact platelets, hepatocytes, neutrophils, and fibroblasts with AlF_4^- yields increased inositol phosphate formation and increases in free $[Ca^{2+}]$ levels, protein phosphorylation, and secretion. AlF_4^-, by mimicking the γ-phosphate of GTP (Bigay *et al.*, 1985), directly activates G proteins, and therefore may cause the above effects by activation of phospholipase C (Kienast *et al.*, 1987).

Pertussis toxin ADP-ribosylates G_i, G_o, and transducin, whereas cholera toxin modifies G_s and transducin. Some authors report inhibition of PLC-activating receptor/agonist reactions by cholera toxin (Imboden *et al.*, 1986; Xuan *et al.*, 1987; Lo and Hughes, 1987). Others find either toxin-

Table II G Protein-Mediated Activation of PLC in Intact Cells, Permeabilized Cells, and Membranes Using Either Endogenous or Exogenous Labeled Phosphatidylinositols

Cell type	Source	Stimulus[b]	Inhibition by pertussis toxin	Reference[a]
Intact cells				
Platelets	Human	AlF_4^-	—	(Mürer, 1976); (Poll et al., 1986); Brass et al., 1986; Fuse and Tai, 1987; Kienast et al., 1987
Hepatocytes	Rat	AlF_4^-	—	Blackmore et al., 1985; Blackmore and Exton, 1986
Neutrophils	Human	AlF_4^-	—	Strnad et al. 1986
Fibroblasts	Hamster	AlF_4^-	Yes	Paris and Pouysségur, 1987
		Thrombin	Yes	Paris and Pouysségur, 1986
Mast cells	Rat	Compound 48/80	Yes	Nakamura and Ui, 1985
Neutrophils	Human	fMLP	Yes	(Lad et al., 1985); Verghese et al., 1985; Krause et al., 1985; Strnad et al., 1986
	Rabbit	fMLP	Yes	Shefcyk et al., 1985; (Molski et al., 1984); Bradford and Rubin, 1985; Volpi et al., 1985
	Guinea pig	LTB_4	Yes	Molski et al., 1984; Bradford and Rubin, 1985
HL-60	Human	fMLP	Yes	Ohta et al. 1985
	Human	fMLP	Yes	Krause et al., 1985; Brandt et al., 1985
WBC 264-9c	Human–mouse	fMLP	Yes	(Backlund et al., 1985)
Hepatocytes		EGF	Yes	Johnson et al., 1986
Endothelial cells	Pig	LTB_4,LTD_4	Yes	(Clark et al., 1986)
Renal mesangial cells	Rat	Angiotensin II	Yes	Pfeilschifter and Bauer, 1986
Permeabilized cells				
Platelets	Human	Thrombin–GTP	—	(Haslam and Davidson, 1984)
		GTP–Ca^{2+}	—	(Knight and Scrutton, 1985)
		thrombin	Yes	Brass et al., 1986, (1987)
		U46619	No	(Brass et al. 1987)
Neutrophils	Rabbit	fMLP–GTP	Yes	Bradford and Rubin, 1986
Mast cells	Human	$GTP\gamma S$	—	(Cockcroft and Gomperts, 1985)
	Rat	Gpp(NH)p	Yes	(Nakamura and Ui, 1984)

Tissue	Species	Ligand		Reference
FRTL5 thyroid	Rat	Norepinephrine–GTPγS	No	Burch et al., 1986
Pancreatic acinar	Rat	Caerulein–GTP	No	Merritt et al., 1986
		Carbachol–GTP	No	Martin et al., 1986a
GH_3 pituitary		TRH–GTPγS	No	
Membranes				
Endogenous-labeled substrate				
Salivary gland	Blowfly	5HT-GTP	—	Litosch et al., 1985
Liver	Rat	Vasopressin-GTP	Yes	Uhing et al., 1985, 1986
Neutrophil	Human	GTP	—	Cockcroft and Gomperts, 1985
		fMLP-GTP	Yes	Smith et al., 1985, 1986
Cortical	Rat	GTP	—	Gonzales and Crews, 1985
Platelet	Human	Thrombin-GTP	—	Baldassare and Fischer, 1986
Astrocytoma	Human	Carbachol-GTP	No	Hepler and Harden, 1986
Pituitary	Rat	TRH-GTP	Yes	Straub and Gershengorn, 1986
WRK1		AlF_4^-	—	Guillon et al., 1986
GH_3 pituitary	Mink	TRH-GTP	No	Martin et al., 1986b
Lung epithelial	Human	GTP	—	Jackowski et al., 1986
HL-60	Human	fMLP-GTP	Yes	Anthes et al., 1987
		GTP, GTPγS	Yes	Rebecchi and Rosen, 1987a
WI38 fibroblast	Human	Thrombin GTP	—	
T-cell leukemic	Human	GTPγS, AlF_4^-	—	Sasaki and Hasegawa-Sasaki, 1987
Exogenous-labeled substrate				
Salivary gland	Blowfly	5HT-GTPγS	—	Litosch and Fain, 1985
Smooth muscle	Pig	Ach-GTP	—	Sasaguri et al., 1985
Liver	Rat	GMP-PCP	—	Melin et al., 1986
Olfactory cilia	Catfish	L-Ala-GTP	—	Hugue and Burch, 1986
Cerebral cortex	Rat	Gpp(NH)p, AlF_4^-	—	Litosch, 1987
DDT_1-MF_2		adrenaline GTP	—	Fulle et al., 1987

[a] References given in parentheses indicate papers where no direct measurement of phosphatidylinositol turnover was undertaken.

[b] Abbreviations used are: fMLP, fMet-Leu-Phe; LTB_4, leukotriene B_4; EGF, epidermal growth factor; LTD_4, leukotriene D_4; TRH, thyrotropin-releasing hormone; 5HT, 5-hydroxytryptamine; Ach, acetylcholine; GMP-PCP, guanylyl (β,γ-methylene)-diphosphonate.

insensitive or pertussis toxin-inhibited systems. Inhibition of agonist-induced formation of inositol phosphates after treatment of intact cells with pertussis toxin has been found in several cell types (Table II). Lack of inhibition may result from the presence of a pertussis toxin-insensitive pathway or because a number of cells such as platelets are impermeable to the toxin. The pertussis toxin holoenzyme itself may cause stimulation of cells (Banga *et al.*, 1987). However, using permeabilized cells or cell membranes, activation of phospholipase C is inhibited by pertussis toxin in only some instances. For example, in permeabilized platelets, phosphatidic acid formation in response to thrombin is inhibited whereas the activation by the thromboxane mimetic U46619 is not (Brass *et al.*, 1986, 1987). Pertussis toxin pretreatment in hepatocytes inhibits the Ca^{2+} increase induced by subsequently added epidermal growth factor but not that induced by angiotensin II (Johnson *et al.*, 1986). In endothelial cells the toxin blocks leukotriene D_4- and C_4-induced but not bradykinin-induced prostaglandin release (Clark *et al.*, 1986). Phospholipase C may interact with G_i, a pertussis toxin substrate, in some cases, and with an undefined G_p, resistant to the toxin, in other cases.

Synergistic action has been observed between PLC-activating agonists and GTP or GTP analogs. Stable GTP analogs and AlF_4^- activate phosphatidylinositol turnover (Table II) without need for a stimulatory agonist, and in most cases GDP or GDPβS inhibits the GTP-mediated activation. Stable GTP analogs are more potent than GTP itself, implying the presence of a GTPase, whereas other nucleotide triphosphates are much less active (see Cockcroft and Gomperts, 1985; Gonzales and Crews, 1985; Uhing *et al.*, 1985; Hepler and Harden, 1986; Martin *et al.*, 1986b, Smith *et al.*, 1986, but see also Rock and Jackowski, 1987).

The addition of purified G_i or G_o to pertussis toxin-pretreated membranes from neutrophils and HL-60 cells restored the formyl-Met–Leu–Phe binding affinity and the GTPase activity and PLC stimulation induced by formyl-Met–Leu–Phe (Okajima *et al.*, 1985; Kikuchi *et al.*, 1986). However, the total PLC stimulation achievable in isolated membranes was only modest when compared to intact cells. It is not clear that this membrane system is representative of all the events occurring in the cell. Subsequently, Banno *et al.* (1987) extracted, separated, and successfully reconstituted PLC and the GTP-binding fraction from platelet membranes, thereby restoring the GTPγS-induced PLC stimulation. In this same study, stimulation of PLC was also achieved by addition of GTPγS-preactivated G_i or G_o. However, the reconstitution of PLC and GTP-binding fractions was not performed in the absence of GTPγS to show an actual stimulation by that compound. There were also no controls to demonstrate that pertussis toxin treatment could eliminate the effects of the purified rat-brain G protein. It is possible that the purified G proteins may not have been completely homogeneous since multiple forms of $G_i\alpha$ have recently been detected in a single cell type (Murphy et al., 1987; Beals et al., 1987), one of which possibly could be linked to PLC rather than to adenylyl cyclase. Alternatively, PLC may be activated by interaction with the $\beta\gamma$ subunits common to several G proteins.

From these results one can postulate a scheme for PLC activation. Upon stimulation of cells with an appropriate agonist, PLC is activated through the interaction with a G protein and acts upon PIP_2 at basal cellular Ca^{2+} levels. Degradation of PIP_2 yields IP_3 and cIP_3, which release Ca^{2+} from intracellular stores. Once Ca^{2+} is elevated, a number of other reactions occur, including breakdown of PI with release of major amounts of diacylglycerol leading to protein kinase C activation and consequent protein phosphorylation, and activation of phospholipase A_2 with release of arachidonic acid and formation of its products.

The receptors for PLC activation are on the plasma membrane, whereas the majority of the PLC is cytosolic (see Table II). Cytosolic PLC is active when attached to a membrane since the substrate is within the membrane bilayer. Translocation of PLC is a potential physiological mechanism for activation by G proteins, but such has not yet been demonstrated. Several studies have demonstrated GTP-dependent and agonist-dependent stimulation of membrane-bound PLC using exogenous substrates (Table II), implying that activated PLC can leave the membrane to find its substrate. We and others found that PLC from platelet cytosol (Deckmyn et al., 1986; Banno et al., 1986b; Baldassare and Fischer, 1986; Rock and Jackowski, 1987), or PLC partially purified from bovine brain (Deckmyn et al., 1986), or a PLC purified to apparent homogeneity from bovine brain cytosol (Ryu et al., 1987a), is stimulated by GTP and GTPγS, whereas purified PLC from ram seminal vesicles (Deckmyn et al., 1986) and another PLC from bovine brain (Ryu et al., 1987a) were not stimulated. AlF_4^- stimulated PLC in platelet cytosol. GDP (Banno et al., 1986b), ATP (Ryu et al., 1987a) and ATP, App(NH)p, CTP, and UTP (Rock and Jackowski, 1987) also stimulate cytosolic PLC, but half-maximal stimulation was obtained with much lower concentrations of GTP and GTP analogs than concentrations of GDP, GDPβS, UTP, CTP, and ATP (Deckmyn et al., 1986). These observations suggest either the presence of a cytosolic G protein associated with the PLC, or, alternatively, direct stimulation of a certain form of PLC by guanine nucleotides. The homogenous cytosolic PLC I that is apparently directly stimulated by GTP (Ryu et al., 1987a) is the one that is also found to be associated with membranes (Lee et al., 1987).

3. Inhibition of Phospholipase C

There may also be systems for G protein-mediated inhibition of PLC. Glutamate inhibits PI turnover stimulated by norepinephrine (Nicoletti *et al.*, 1986), carbachol, and histamine (Baudry *et al.*, 1986) in rat hippocampal slices, and dopamine prevents the formation of inositol phosphates in pituitary cells stimulated by angiotensin II (Enjalbert *et al.*, 1986) and thyrotropin-releasing hormone (Simmonds and Strange, 1985; Journot *et al.*, 1987). The inhibition by dopamine is blocked by pretreatment of the cells with pertussis toxin (Journot *et al.*, 1987), implying the involvement of a G protein and therefore possibly further completing the parallel between the adenylyl cyclase and the PLC systems.

ACKNOWLEDGMENT

This research was supported by Grants HLBI 14147 (Specialized Center for Research in Thrombosis), HLBI 16634, and Training Grant T32 HLBI 07088 from the National Institutes of Health, a NATO Research Fellowship and a Fulbright Award (to H.D.).

REFERENCES

Anthes, J. C., Billah, M. M., Cali, A., Egan, R. W., and Siegel, M. I. (1987). Chemotactic peptide, calcium and guanine nucleotide regulation of phospholipase C activity in membranes from DMSO differentiated HL-60 cells. *Biochem. Biophys. Res. Commun.* **145,** 825–833.

Backlund, P. S., Meade, B. D., Manclark, C. R., Cantoni, G. L., and Aksamir, R. R. (1985). Pertussis toxin inhibition of chemotaxis and the ADP ribosylation of a membrane protein in a human-mouse hybrid cell line. *Proc. Natl. Acad. Sci. U.S.A.* **82,** 2637–2641.

Baldassare, J. J., and Fischer, G. J. (1986). Regulation of membrane-associated and cytosolic phospholipase C activities in human platelets by guanosine triphosphate. *J. Biol. Chem.* **261,** 11942–11944.

Banga, H. S., Walker, R. K., Winberry, L. K., and Rittenhouse, S. E. (1987). Pertussis toxin can activate human platelets. Comparative effects of holotoxin and its ADP-ribosylating S_1 subunit. *J. Biol. Chem.* **262,** 14871–14874.

Bannon, Y., Nakashima, S., and Nozawa, Y. (1986a). Partial purification of phosphoinositide phospholipase C from human platelet cytosol; characterization of its three forms. *Biochem. Biophys. Res. Commun.* **136,** 713–721.

Banno, Y., Nakashima, S., Tohmatsu, T., Nozawa, Y., and Lapetina, E. G. (1986b). GTP and GDP will stimulate platelet cytosolic phospholipase C independent of Ca^{2+}. *Biochem. Biophys. Res. Commun.* **140,** 728–734.

Banno, Y., Nagao, S., Katada, T., Nagata, K., Ui, M., and Nozawa, Y. (1987). Stimulation by GTP-binding proteins (G_i, G_o) of partially purified phospholipase C activity from human platelet membranes. *Biochem. Biophys. Res. Commun.* **146,** 861–869.

Baudry, M., Evans, J., and Lynch, G. (1986). Excitatory amino acids inhibit stimulation of phosphatidylinositol metabolism by aminergic agonists in hippocampus. *Nature* **319,** 329–331.

Beals, C. R., Wilson, C. B., Perlmutter, R. M. (1987). A small multigene family encodes Gi signal-transduction proteins. *Proc. Natl. Acad. Sci. U.S.A.* **84,** 7886–7890.

Bennett, C. F., and Crooke, S. T. (1987). Purification and characterization of a phosphoinositide-specific phospholipase C from guinea pig uterus. Phosphorylation by protein kinase C in vivo. *J. Biol. Chem.* **262,** 13789–13797.

Berridge, M. J. (1987). Inositol trisphosphate and diacylglycerol: two interacting second messengers. *Annu. Rev. Biochem.* **56,** 159–193.

Bigay, J., Deterre, P., Pfister, C., and Chabre, M. (1985). Fluoroaluminates activate transducin-GDP by mimicking the γ-phosphate of GTP in its binding site. *FEBS Lett.* **191,** 181–185.

Blackmore, P. F., and Exton, J. H. (1986). Studies on the hepatic calcium-mobilizing activity of aluminum fluoride and glucagon. Modulation by cAMP and phorbol myristate acetate. *J. Biol. Chem.* **261,** 11056–11063.

Blackmore, P. F., Bocckino, S. R., Waynick, L. E., and Exton, J. H. (1985). Role of guanine nucleotide binding regulatory protein in the hydrolysis of hepatocyte phosphatidylinositol 4,5-bisophosphate by calcium mobilizing hormones and the control of cell calcium. Studies utilizing aluminum fluoride. *J. Biol. Chem.* **260,** 14477–14483.

Bojanic, D., and Fain, J. N. (1986). Guanine nucleotide regulation of [³H]vasopressin binding to liver membranes and solubilized receptors: Evidence for the involvement of a guanine nucleotide regulatory protein. *Biochem. J.* **240,** 361–365.

Bradford, P. G., and Rubin, R. P. (1985). Pertussis toxin inhibits chemotactic factor-induced phospholipase C stimulation and lysosomal enzyme secretion in rabbit neutrophils. *FEBS Lett.* **183,** 317–320.

Bradford, P. G., and Rubin, R. P. (1986). Guanine nucleotide regulation of phospholipase C activity in permeabilized rabbit neutrophils. Inhibition by pertussis toxin and sensitization to submicromolar calcium concentrations. *Biochem J.* **239,** 97–102.

Brandt, S. J., Dougherty, R. W., Lapetina, E. G., and Niedel, J. E. (1985). Pertussis toxin inhibits chemotactic peptide-stimulated generation of inositol phosphates and lysosomal enzyme secretion in human leukemic (HL-60) cells. *Proc. Natl. Acad. Sci. U.S.A.* **82,** 3277–3280.

Brass, L. F., Laposata, M., Banja, H. S., and Rittenhouse, S. E. (1986). Regulation of the phosphoinositide hydrolysis pathway in thrombin-stimulated platelets by a pertussis toxin-sensitive guanine nucleotide binding protein. *J. Biol. Chem.* **261,** 16838–16847.

Brass, L. F., Shaller, C. C., and Belmonte, E. J. (1987). Inositol 1,4,5-trisphosphate-induced granule secretion in platelets. Evidence that the activation of phospholipase C mediated by platelet thromboxane receptors involves a guanine nucleotide binding protein-dependent mechanism distinct from that of thrombin. *J. Clin. Invest.* **79,** 1269–1275.

Burch, R. M., Luini, A., and Axelrod, J. (1986). Phospholipase A_2 and phospholipase C are activated by distinct GTP-binding proteins in response to α_1-adrenergic stimulation in FRTL5 thyroid cells. *Proc. Natl. Acad. Sci. U.S.A.* **83,** 7201–7205.

Cantau, B., Keppens, S., DeWulf, H., and Jard, S. (1980). [^{3}H]-Vasopressin binding to isolated rat hepatocytes and liver membranes: Regulation by GTP and relation to glycogen phosphorylase activation. *J. Recept. Res.* **1,** 137–168.

Carter, H. R., and Smith, A. D. (1987). Resolution of the phosphoinositide-specific phospholipase C isolated from porcine lymphocytes into multiple species. Partial purification of two isoenzymes. *Biochem. J.* **244,** 639–645.

Carter, H. R., Bird, I. M., and Smith, A. D. (1986). Two species of phospholipase C isolated from lymphocytes produce specific ratios of inositol phosphate products. *FEBS Lett.* **204,** 23–27.

Chau, L. Y., and Tai, H. H. (1982). Resolution into two different forms and study of the properties of phosphatidylinositol specific phospholipase C from human platelet cytosol. *Biochim. Biophys. Acta* **713,** 344–351.

Clark, M. A., Conway, T. M., Bennett, F., Cooke, S. T., and Stadel, J. M. (1986). Islet activating protein inhibits leukotriene D4- and leukotriene C4- but not bradykinin or calcium ionophore-induced prostacyclin synthesis in bovine endothelial cells. *Proc. Natl. Acad. Sci. U.S.A.* **83,** 7320–7324.

Cockcroft, S., and Gomperts, B.D. (1985). Role of guanine nucleotide binding protein in the activation of polyphosphoinositide phosphodiesterase. *Nature* **314,** 534–535.

Crane, J. K., Campanile, C. P., and Garrison, J. C. (1982). The hepatic angiotensin II receptor II. Effect of guanine nucleotides and interaction with cyclic AMP production. *J. Biol. Chem.* **257,** 4959–4965.

Dawson, R. M. C. (1959). Studies on the enzymic hydrolysis of monophosphoinositides by phospholipase preparations from P. notatum and ox pancreas. *Biochim. Biophys. Acta* **33,** 68–77.

Dawson, R. M., Freinkel, N., Jungalwala, F. B., and Clarke, N. (1971). The enzymic formation of myoInositol 1:2 cyclic phosphate from phosphatidylinositol. *Biochem. J.* **122,** 605–607.

Dawson, R. M., Hemington, N., and Irvine, R. F. (1980). The inhibition and activation of Ca^{2+}-dependent phosphatidylinositol phosphodiesterase by phospholipids and blood plasma. *Eur. J. Biochem.* **112,** 33–38.

Deckmyn, H., Tu, S.-M., and Majerus, P. W. (1986). Guanine nucleotides stimulate soluble phosphoinositide-specific phospholipase C in the absence of membranes. *J. Biol. Chem.* **261,** 16553–16558.

DeLean, A., Ong, H., Gutkowska, J., Schiller, P. W., and McNicoll, N. (1984). Evidence for agonist-induced interaction of angiotensin receptor with guanine nucleotide-binding protein in bovine adrenal zona glomerula. *Mol. Pharmacol.* **26**, 498–508.

Enjalbert, A., Sladeczek, F., Guillon, G., Bertrand, P., Shu, C., Epelbaum, J., Garcia-Sainz, A., Jard, S., Lombard, C., Kordon, C., and Bockaert, J. (1986). Angiotensin II and dopamine modulate both cAMP and inositol phosphate production in anterior pituitary cells. Involvement in prolactin secretion. *J. Biol. Chem.* **261**, 4071–4075.

Fain, J. N., Brindley, D. N., Pittner, R. A., and Hawthorne, J. N. (1985). Stimulation of specific GTPase activity by vasopressin in isolated membranes from cultured rat hepatocytes. *FEBS Lett.* **192**, 251–259.

Fitzgerald, T. J., Uhing, R. J., and Exton, J. N. (1986). Solubilization of the vasopressin receptor from rat liver plasma membranes. Evidence for a receptor GTP-binding protein complex. *J. Biol. Chem.* **261**, 16871–16877.

Florio, V. A., and Sternweis, P. C. (1985). Reconstitution of resolved muscarinic cholinergic receptors with purified GTP-binding proteins. *J. Biol. Chem.* **260**, 3477–3483.

Fulle, H.-J., Hoer, D., Lache, W., Rosenthal, W., Schultz, G., and Oberdisse, E. (1987). In vitro synthesis of ^{32}P-labeled phosphatidylinositol 4,5-bisphosphate and its hydrolysis by smooth muscle membrane-bound phospholipase C. *Biochem. Biophys. Res. Commun.* **145**, 673–679.

Fuse, I., and Tai, H. H. (1987). Stimulation of arachidonate release and inositol 1,4,5-trisphosphate formation are mediated by distinct G proteins in human platelets. *Biochem. Biophys. Res. Commun.* **146**, 659–665.

Glossman, H., Baukal, A., and Catt, K. J. (1974). Angiotensin II receptors in bovine adrenal cortex. Modification of angiotensin II binding by guanyl nucleotides. *J. Biol. Chem.* **249**, 664–666.

Gonzales, R. A., and Crews, F. T. (1985). Guanine nucleotides stimulate production of inositol trisphosphate in rat cortical membranes. *Biochem. J.* **232**, 799–804.

Goodhardt, M., Ferry, N., Geynet, P., and Hanoune, J. (1982). Hepatic α1-adrenergic receptors show agonist-specific regulation by guanine nucleotide. Loss of nucleotide effect after adrenalectomy. *J. Biol. Chem.* **257**, 11577–11583.

Grandt, R., Aktories, K., and Jakobs, K. H. (1986). Evidence for two GTPases activated by thrombin in membranes of human platelets. *Biochem. J.* **237**, 669–674.

Guillemette, G., Guillon, G., Marie, J., Pantaloni, C., Balestre, M. N., Escher, E., and Jard, S. (1984). Angiotensin-induced changes in the apparent size of rat liver angiotensin receptors. *J. Recept. Res.* **4**, 267–281.

Guillon, G., Mouillac, B., and Balestre, M. N. (1986). Activation of polyphosphoinositide phospholipase C by fluoride in WRK1 cell membranes. *FEBS Lett.* **204**, 183–188.

Hakata, H., Kambayashi, J., and Kosaki, G. (1982). Purification and characterization of phosphatidylinositol-specific phospholipase C from bovine platelet. *J. Biochem.* **92**, 929–935.

Haslam, R. J., and Davidson, M. M. L. (1984). Guanine nucleotides decrease the free [Ca^{2+}] required for secretion of serotonin from permeabilized blood platelets. Evidence of a role for GTP-binding protein in platelet activation. *FEBS Lett.* **174,** 90–95.

Hepler, J. R., and Harden, T. K. (1986). Guanine nucleotide-dependent pertussis toxin insensitive stimulation of inositol phosphate formation by carbachol in a membrane preparation from human astrocytoma cells. *Biochem. J.* **239,** 141–146.

Hinkle, P. M., and Phillips, W. J. (1984). Thyrotropin-releasing hormone stimulates GTP hydrolysis by membranes from $GH4C_1$ rat pituitary tumor cells. *Proc. Natl. Acad. Sci. U.S.A.* **81,** 6183–6187.

Hofmann, S. L., and Majerus, P. W. (1982a). Identification and properties of two distinct phosphatidylinositol-specific phospholipase C enzymes from sheep seminal vesicular glands. *J. Biol. Chem.* **257,** 6461–6469.

Hofmann, S. L., and Majerus, P. W. (1982b) Modulation of phosphatidylinositol-specific phospholipase C activity by phospholipid interactions, diglycerides, and calcium ions. *J. Biol. Chem.* **257,** 14359–14364.

Houslay, M. D., Bojanic, D., and Wilson, A. (1986a). Platelet activating factor and U44069 stimulate a GTPase activity in human platelets which is distinct from the guanine nucleotide regulatory proteins, N_s and N_i. *Biochem. J.* **234,** 737–740.

Houslay, M. D., Bojanic, D., Gawler, D., O'Hagan, S., and Wilson, A. (1986b). Thrombin, unlike vasopressin, appears to stimulate two distinct guanine nucleotide regulatory proteins in human platelets. *Biochem. J.* **238,** 109–113.

Hugue, T., and Burch, R. C. (1986). Odorant- and guanine nucleotide-stimulated phosphoinositide turnover in olfactory cilia. *Biochem. Biophys. Res. Commun.* **137,** 36–42.

Hwang, S. B., Lam, M. H., and Pong, S. S. (1986). Ionic and GTP regulation of binding of platelet activating factor to receptors and platelet activating factor-induced activation of GTPase in rabbit platelet membranes. *J. Biol. Chem.* **261,** 532–537.

Imboden, J. B., Shoback, D. M., Pattison, G., and Stobo, J. D. (1986). Cholera toxin inhibits the T-cell antigen receptor-mediated increases in inositol trisphosphate and cytoplasmic-free calcium. *Proc. Natl. Acad. Sci. U.S.A.* **83,** 5673–5677.

Irvine, R. F., and Dawson, R. M. C. (1978). The distribution of calcium-dependent phosphatidylinositol-specific phosphodiesterase in rat brain. *J. Neurochem.* **31,** 1427–1434.

Irvine, R. F., Letcher, A. J., and Dawson, R. M. (1979a). Fatty acid stimulation of membrane phosphatidylinositol hydrolysis by brain phosphatidylinositol phosphodiesterase. *Biochem. J.* **178,** 497–500.

Irvine, R. F., Hemington, N., and Dawson, R. M. (1979b). The calcium-dependent phosphatidylinositol phosphodiesterase of rat brain, mechanisms of suppression and stimulation. *Eur. J. Biochem.* **99,** 525–530.

Jackowski, S., Rettenmier, C. W., Sherr, C. J., and Rock, C. O. (1986). A guanine nucleotide-dependent phosphatidylinositol 4,5-diphosphate phospholipase C in cells transformed by the v-fms and v-fes oncogenes. *J. Biol. Chem.* **261,** 4978–4985.

Johnson, R., Connelly, P., Sisk, R., Probiner, B., Hewlett, E., and Garrison, J. (1986). Pertussis toxin or phorbol 12-myristate 13-acetate can distinguish between epidermal growth factor- and angiotensin-stimulated signals in hepatocytes. *Proc. Natl. Acad. Sci. U.S.A.* **83,** 2032–2036.

Journot, L., Homburger, V., Pantaloni, C., Priam, M., Bockaert, J., and Enjalbert, A. (1987). An islet activating protein-sensitive G protein is involved in dopamine inhibition of angiotensin and thyrotropin-releasing hormone-stimulated inositol phosphate production in anterior pituitary cells. *J. Biol. Chem.* **262,** 15106–15110.

Katan, M., and Parker, P. J. (1987). Purification of phosphoinositide-specific phospholipase C from a particulate fracture of bovine brain. *Eur. J. Biochem.* **168,** 413–418.

Kemp, P., Hubscher, G., and Hawthorne, J. N. (1961). Phosphoinositides 3. Enzymic hydrolysis of inositol-containing phospholipids. *Biochem. J.* **79,** 193–200.

Kienast, J., Arnout, J., Pfliegler, G., Deckmyn, H., Hoet, B., and Vermylen, J. (1987). Sodium fluoride mimics effects of both agonists and antagonists on intact human platelets by simultaneous modulation of phospholipase C and adenylate cyclase activity. *Blood* **69,** 859–866.

Kikuchi, A., Kozawa, O., Kaibuchi, K., Katada, T., Ui, M., and Takai, Y. (1986). Direct evidence for involvement of a guanine nucleotide-binding protein in chemotactic peptide-stimulated formation of inositol bisphosphate and trisphosphate in differentiated human leukemic (HL-60) cells. Reconstitution with G_i or G_o of the plasma membranes ADP ribosylated by pertussis toxin. *J. Biol. Chem.* **261,** 11558–11562.

Knight, D. E., and Scrutton, M. C. (1985). Effect of various excitatory agonists on the secretion of 5-hydroxytryptamine from permeabilized human platelets induced by Ca^{2+} in the presence or absence of GTP. *FEBS Lett.* **183,** 417–422.

Krause, K. H., Schlegel, W., Wollheim, C. B., Andersson, T., Waldvogel, F. A., and Lew, P. D. (1985). Chemotactic peptide activation of human neutrophils and HL-60 cells. Pertussis toxin reveals correlation between inositol trisphosphate generation, calcium ion transients and cellular activation. *J. Clin. Invest.* **76,** 1348–1354.

Lad, P. M., Olson, C. V., and Smiley, P. A. (1985). Association of the N-formyl-Met-Leu-Phe receptor in human neutrophils with a GTP-binding protein sensitive to pertussis toxin. *Proc. Natl. Acad. Sci. U.S.A.* **82,** 869–873.

Lapetina, E. G., and Michell, R. H. (1973). A membrane-bound activity catalyzing phosphatidylinositol breakdown to 1,2-diacylglycerol, D-myoinositol 1 : 2-cyclic phosphate and D-myoinositol 1-phosphate. *Biochem. J.* **131,** 433–442.

Lee, K. Y., Ryu, S. H., Suh, P. G., Choi, W. C., and Rhee, S. G. (1987). Phospholipase C associated with particulate fractions of bovine brain. *Proc. Natl. Acad. Sci. U.S.A.* **84,** 5540–5544.

Lips, D. L., Bross, T. E., and Majerus, P. W. (1988). Isolation of 1-monomethylphosphorylinositol 4,5-bisphosphate (a product of methanolysis of inositol 1:2(cyclic)4,5-trisphosphate) from Swiss mouse 3T3 cells. *Proc. Natl. Acad. Sci. U.S.A.* **85**, 88–92.

Litosch, I. (1987). Guanine nucleotide and NaF stimulation of phospholipase C activity in rat cerebral cortical membranes. Studies on substrate specificity. *Biochem. J.* **244**, 35–40.

Litosch, I., and Fain, J. N. (1985). 5-Methyltryptamine stimulates phospholipase C-mediated breakdown of exogenous phosphoinositides by blowfly salivary gland membranes. *J. Biol. Chem.* **260**, 16052–16055.

Litosch, I., Wallis, C., and Fain, J. N. (1985). 5-Hydroxytryptamine stimulates inositol phosphate production in a cell-free system from blowfly salivary glands. Evidence for a role of GTP in coupling receptor activation to phosphoinositide breakdown. *J. Biol. Chem.* **260**, 5464–5471.

Lo, W. W. Y., and Hughes, J. (1987). A novel cholera toxin-sensitive G-protein (G_c) regulating receptor-mediated phosphoinositide signalling in human pituitary clonal cells. *FEBS Lett.* **220**, 327–331.

Low, M. G., and Weglicki, W. B. (1983). Resolution of myocardial phospholipase C into several forms with distinct properties. *Biochem. J.* **215**, 325–334.

Low, M. B., Carroll, R. C., and Weglicki, W. B. (1984). Multiple forms of phosphoinositide-specific phospholipase C of different relative molecular masses in animal tissues. Evidence for modification of the platelet enzyme by Ca^{2+}-dependent proteinase. *Biochem. J.* **221**, 813–820.

Low, M. G., Carroll, R. C., and Cox, A. C. (1986). Characterization of multiple forms of phosphoinositide-specific phospholipase C purified from human platelets. *Biochem. J.* **237**, 139–145.

Majerus, P. W., (1983). Arachidonic metabolism in vascular disorders. *J. Clin. Invest.* **72**, 1521–1525.

Majerus, P. W., Connolly, T. M., Deckmyn, H., Ross, T. S., Bross, T. E., Ishii, H., Bansal, V. S., and Wilson, D. B. (1986). The metabolism of phosphoinositide-derived messenger molecules. *Science* **234**, 1519–1526.

Majerus, P. W., Connolly, T. M., Bansal, V. S., Inhorn, R. C., Ross, T. S., and Lips, D. L. (1988). Inositol phosphates: Synthesis and degradation. *J. Biol. Chem.* **263**, 3051–3054.

Manne, V., and Kung, H. F. (1987). Characterization of phosphoinositide-specific phospholipase C from human platelets. *Biochem. J.* **243**, 763–771.

Martin, T. F. J., Lucas, D. O., Bajjalieh, S. M., and Kowalchyk, J. A. (1986a). Thyrotropin-releasing hormone activates a Ca^{2+}-dependent polyphosphoinositide phosphodiesterase in permeable GH_3 cells. GTPγS potentiation by a cholera and pertussis toxin-insensitive mechanism. *J. Biol. Chem.* **261**, 2918–2927.

Martin, T. F. J., Bajjalieh, S. M., Lucas, D. O., and Kowalchyk, J. A. (1986b). Thyrotropin-releasing hormone stimulation of polyphosphoinositide hydrolysis in GH_3 cell membranes is GTP-dependent but insensitive to cholera or pertussis toxin. *J. Biol. Chem.* **261**, 10041–10049.

Matsumoto, T., Molski, T. F. P., Kanaho, Y., Becker, E. L., and Sha'afi, R. I.

(1987). G-protein dissociation, GTP-GDP exchange and GTPase activity in control and PMA-treated neutrophils stimulated by fMet-Leu-Phe. *Biochem. Biophys. Res. Commun.* **143**, 489–498.

Melin, P. M., Sundler, R., and Jergil, B. (1986). Phospholipase C in rat liver plasma membranes. Phosphoinositide specificity and regulation by guanine nucleotides and calcium. *FEBS Lett.* **198**, 85–88.

Merritt, J. E., Taylor, C. W., Rubin, R. P., and Putney, J. W. (1986). Evidence suggesting that a novel guanine nucleotide regulatory protein couples receptor to phospholipase C in exocrine pancreas. *Biochem. J.* **236**, 337–343.

Molski, T. F. P., Naccache, P. N., Marsh, M. L., Kermode, J., Becker, E. L., Sha'afi, R. I. (1984). Pertussis toxin inhibits the rise in the intracellular concentration of free calcium that is induced by chemotactic factors in rabbit neutrophils: Possible role of the G-proteins in calcium mobilization. *Biochem. Biophys. Res. Commun.* **124**, 644–650.

Mürer, E. H. (1976). Effects of fluoride on blood platelets. *Fluoride* **9**, 173–184.

Murphy, P. M., Eide, B., Goldsmith, P., Brann, M., Gierschik, P., Spiegel, A., and Malech, H. L. (1987). Detection of multiple forms of $G_{i\alpha}$ in HL-60 cells. *FEBS Lett.* **221**, 81–86.

Nakamura, K., Kambayashi, J., Suga, K., Hakata, H., and Mori, T. (1985). Hydrolysis of polyphosphoinositides in human platelets. *Thromb. Res.* **38**, 513–525.

Nakamura, T., and Ui, M. (1984). Islet-activating protein, pertussis toxin, inhibits Ca^{2+}-induced and guanine nucleotide-dependent releases of histamine and arachidonic acid from mast cells. *FEBS Lett.* **173**, 414–418.

Nakamura, T., and Ui, M. (1985). Simultaneous inhibitions of inositol phospholipid breakdown, arachidonic acid release, and histamine secretion in mast cells by islet activating protein, pertussis toxin. A possible involvement of the toxin-specific substrate in the Ca^{2+} mobilizing receptor-mediated biosignaling system. *J. Biol. Chem.* **260**, 3584–3593.

Nakanishi, H., Nomura, H., Kikkawa, U., Kishimoto, A., and Nishizuka, Y. (1985). Rat brain and liver soluble phospholipase C: Resolution of two forms with different requirements for calcium. *Biochem. Biophys. Res. Commun.* **132**, 582–590.

Nicoletti, F., Iadarola, M. J., Wroblewski, and Costa, E., (1986). Excitatory amino acid recognition sites coupled with inositol phospholipid metabolism: Developmental changes and interactions with α_1-adrenoceptors. *Proc. Natl. Acad. Sci. U.S.A.* **83**, 1931–1935.

Nishizuka, Y. (1986). Studies and perspectives of protein kinase C. *Science* **233**, 305–312.

Ohta, H., Okajima, F., and Ui, M. (1985). Inhibition by islet-activating protein of a chemotactic peptide-induced early breakdown of inositol phospholipids and Ca^{2+} mobilization in guinea pig neutrophils. *J. Biol. Chem.* **260**, 15771–15780.

Okajima, F., Katada, T., and Ui, M. (1985). Coupling of the guanine nucleotide regulatory protein to chemotactic peptide receptors in neutrophil membranes and its uncoupling by islet-activating protein, pertussis toxin. A possible role of the toxin substrate in Ca^{2+} mobilizing receptor-mediated signal transduction. *J. Biol. Chem.* **260**, 6761–6768.

Paris, S., and Pouysségur, J. (1986). Pertussis toxin inhibits thrombin-induced activation of phosphoinositide hydrolysis and Na^+/H^+ exchange in hamster fibroblasts. *EMBO J.* **5**, 55–60.

Paris, S., and Pouysségur, J. (1987). Further evidence for a phospholipase C-coupled G-protein in hamster fibroblasts. Induction of inositol phosphate formation by fluoroluminate and vanadate and inhibition by pertussis toxin. *J. Biol. Chem.* **262**, 1970–1976.

Pfeilschifter, J., and Bauer, C. (1986). Pertussis toxin abolishes angiotensin II-induced phosphoinositide hydrolysis and prostaglandin synthesis in rat renal mesangial cells. *Biochem. J.* **236**, 289–294.

Poll, C., Kyrle, P., and Westwick, J. (1986). Activation of protein kinase C inhibits sodium fluoride-induced elevation of human platelet cytosolic-free calcium and thromboxane B_2 generation. *Biochem. Biophys. Res. Commun.* **136**, 381–389.

Rebecchi, M. J., and Rosen, O. M. (1987a). Stimulation of polyphosphoinositide hydrolysis by thrombin in membranes from human fibroblasts. *Biochem. J.* **245**, 49–57.

Rebecchi, M. J., and Rosen, O. M. (1987b). Purification of a phosphoinositide-specific phospholipase C from bovine brain. *J. Biol. Chem.* **262**, 12526–12532.

Rock, C. O., and Jackowski, S. (1987). Thrombin- and nucleotide-activated phosphatidylinositol 4,5-bisphosphate phospholipase C in human platelet membranes. *J. Biol. Chem.* **262**, 5492–5498.

Ryu, S. H., Cho, K. S., Lee, K. Y., Suh, P. G., and Rhee, S. G. (1986). Two forms of phosphatidylinositol-specific phospholipase C from bovine brain. *Biochem. Biophys. Res. Commun.* **141**, 137–144.

Ryu, S. H., Cho, K. S., Lee, K. Y., Suh, P. G., and Rhee, S. G. (1987a). Purification and characterization of two immunologically distinct phosphoinositide-specific phospholipase C from bovine brain. *J. Biol. Chem.* **262**, 12511–12518.

Ryu, S. H., Suh, P. G., Cho, K. S., Lee, K. Y., and Rhee, S. G. (1987b). Bovine brain cytosol contains three immunologically distinct forms of inositol phospholipid-specific phospholipase C. *Proc. Natl. Acad. Sci. U.S.A.* **84**, 6649–6653.

Sasaguri, T., Hirata, M., and Kuriyama, H. (1985). Dependence on Ca^{2+} of the activities of phosphatidylinositol 4,5-bisphosphate phosphodiesterase in inositol 1,4,5-trisphosphate phosphatase in smooth muscles of the porcine coronary artery. *Biochem J.* **231**, 497–503.

Sasaki, T., and Hasegawa-Sasaki, H. (1987). Activation of polyphosphoinositide phospholipase C by guanosine 5′-0-(3-thio)triphosphate and fluoroaluminate in membranes prepared from a human T cell leukemia line, JURKAT. *FEBS Lett.* **218**, 87–92.

Shefcyk, J., Yassin, R., Volpi, M., Molski, T. F. P., Naccache, P. H., Munoz, J. J., Becker, E. L., Feinstein, M. B., and Sha'afi, R. I. (1985). Pertussis but not cholera toxin inhibits the stimulated increase in actin association with the cytoskeleton in rabbit neutrophils: Role of the "G proteins" in stimulus-response coupling. *Biochem. Biophys. Res. Commun.* **126,** 1174–1181.

Shukla, S. D. (1982). Phosphatidylinositol-specific phospholipase C. *Life Sci.* **30,** 1323–1335.

Simmonds, S. H., and Strange, P. G. (1985). Inhibition of inositol phospholipid breakdown by D_2-dopamine receptors in dissociated bovine anterior pituitary cells. *Neurosci. Lett.* **60,** 267–277.

Smith, C. D., Lane, B. C., Kusaka, I., Verghese, M. W., and Snyderman, R. (1985). Chemoattractant receptor-induced hydrolysis of phosphatidylinositol 4,5-bisphosphate in human polymorphonuclear leukocyte membranes. Requirement for a guanine nucleotide regulatory protein. *J. Biol. Chem.* **260,** 5875–5878.

Smith, C. D., Cox, C. C., and Snyderman, R. (1986). Receptor-coupled activation of phosphoinositide-specific phospholipase C by an N-protein. *Science* **232,** 97–100.

Straub, R. E., and Gershengorn, M. C. (1986). Thyrotropin-releasing hormone and GTP activate inositol trisphosphate formation in membranes isolated from rat pituitary cells. *J. Biol. Chem.* **261,** 2712–2717.

Strnad, C. F., Parente, J. E., and Wong, K. (1986). Use of fluoride ion as a probe for the guanine nucleotide-binding protein involved in the phosphoinositide-dependent neutrophil transduction pathway. *FEBS Lett.* **206,** 20–24.

Takenawa, T., and Nagai, Y. (1981). Purification of phosphatidylinositol-specific phospholipase C from rat liver. *J. Biol. Chem.* **256,** 6769–6775.

Uhing, R. J., Jiang, H., Prpic, V., and Exton, J. H. (1985). Regulation of a liver plasma membrane phosphoinositide phosphodiesterase by guanine nucleotides and calcium. *FEBS Lett.* **188,** 317–320.

Uhing, R. J., Prpic, V., Jiang, H., and Exton, J. H. (1986). Hormone-stimulated polyphosphoinositide breakdown in rat liver plasma membranes. Roles of guanine nucleotides and calcium. *J. Biol. Chem.* **261,** 2140–2146.

Verghese, M. W., Smith, C. D., and Snyderman, R. (1985). Potential role for a guanine nucleotide regulatory protein in chemoattractant receptor-mediated polyphosphoinositide metabolism, Ca^{++} mobilization and cellular responses by leukocytes. *Biochem. Biophys. Res. Commun.* **127,** 450–457.

Volpi, M., Naccache, P. H., Molski, T. F. P., Shefcyk, J., Huang, C. K., Marsh, M. C, Munoz, J., Becker, E. L., and Sha'afi, R. I. (1985). Pertussis toxin inhibits fMet-Leu-Phe, but not phorbol ester-stimulated changes in rabbit neutrophils: Role of G proteins in excitation response coupling. *Proc. Natl. Acad. Sci. U.S.A.* **82,** 2708–2712.

Wilson, D. B., Bross, T. E., Hofmann, S. L., and Majerus, P. W. (1984). Hydrolysis of polyphosphoinositides by purified sheep seminal vesicle phospholipase C enzymes. *J. Biol. Chem.* **259,** 11718–11724.

Wilson, D. B., Connolly, T. M., Bross, T. E., Majerus, P. W., Sherman, W. R., Tyler, A. N., Rubin, L. J., and Brown, J. E. (1985). Isolation and characterization of the inositol cyclic phosphate products of polyphosphoinositide cleavage by phospholipase C. I. Physiological effects of permeabilized platelets and Limulus photoreceptor cells. *J. Biol. Chem.* **260,** 13496–13501.

Wojcikiewicz, R. J. H., Kent, P. A., and Fain, J. N. (1986). Evidence that thyrotropin-releasing hormone-induced increases in GTPase activity and phosphoinositide metabolism in GH$_3$ cells are mediated by a guanine nucleotide-binding protein other than G$_s$ or G$_i$. *Biochem. Biophys. Res. Commun.* **138,** 1383–1389.

Xuan, Y. T., Su, Y. F., Chang, K. J., and Watkins, W. D. (1987). A pertussis/cholera toxin-sensitive G-protein may mediate vasopressin-induced inositol phosphate formation in smooth muscle cell. *Biochem Biophys. Res. Commun.* **146,** 898–906.

CHAPTER 19

Receptor Modulation of Phospholipase C Activity

Irene Litosch

Department of Pharmacology, University of Miami School of Medicine
Miami, Florida 33101

I. INTRODUCTION

The phosphoinositide receptor system has long been recognized to constitute an important transmembrane signaling pathway utilized by hormones, neurotransmitters, and growth factors which promote cellular activation through elevation of cytosolic Ca^{2+} (Berridge and Irvine, 1984; Hokin, 1985; Litosch and Fain, 1986a). Agonists stimulate a rapid phospholipase C-mediated hydrolysis of phosphatidylinositol 4,5-bisphosphate (PIP_2) with generation of diacylglycerol and inositol 1,4,5-trisphosphate. Inositol 1,2-cyclic,4,5-trisphosphate is also produced (Wilson *et al.*, 1985a). Diacylglycerol activates protein kinase C while both inositol 1,4,5-trisphosphate and inositol 1,2-cyclic,4,5-trisphosphate trigger the release of Ca^{2+} from endoplasmic reticulum (Nishizuka, 1984; Wilson *et al.*, 1985a;

Irvine *et al.,* 1986); an event which initiates the cellular activation process.

Despite the importance of the phosphoinositide system in cellular activation, the mechanism underlying agonist-stimulated hydrolysis of phosphoinositides remained undetermined for a considerable length of time. Data from radioligand binding studies had suggested that a GTP-dependent mechanism was involved in activation of the phospholipase C system. GTP and hydrolysis-resistant GTP analogs modulated agonist binding to receptors linked to phospholipase C activation (Cantau *et al.,* 1980; Snavely and Insel, 1982; Kuo *et al.,* 1983). The demonstration of agonist activation of phospholipase C activity in cell-free systems provided additional direct biochemical evidence that a regulatory GTP-binding protein (G protein) functioned in the phosphoinositide–phospholipase C system. Studies in permeabilized platelets showed that the hydrolysis-resistant guanine nucleotide, GTPγS, stimulated phospholipase C activity. An enhancement of thrombin action by GTPγS was observed, although thrombin appreciably activated phospholipase C activity in the absence of added nucleotides (Haslam and Davidson, 1984). In neutrophil membranes, hydrolysis-resistant guanine nucleotides stimulated the loss of labeled PIP_2 (Cockcroft and Gomperts, 1985). Studies in blowfly salivary gland membranes demonstrated that guanine nucleotides were essential to promote agonist activation of phospholipase C activity. The rank-order potency of guanine-nucleotide analogs in promoting the effect of agonist was GTPγS > Gpp(NH)p >>GTP (Litosch *et al.,* 1985). The maximal degree of phospholipase C stimulation due to agonist was greater in the presence of the hydrolysis resistant analogs GTPγS or Gpp(NH)p than with GTP, which is rapidly hydrolyzed.

Studies by several laboratories have now shown that guanine nucleotides are required for agonist-induced activation of phospholipase C activity in a variety of cell-free systems (Smith *et al.,* 1985; Uhing *et al.,* 1986; Straub and Gershengorn, 1986; Martin *et al.,* 1986; Guillon *et al.,* 1986; Merritt *et al.,* 1986; Baldassare and Fisher, 1986). Hydrolysis-resistant guanine nucleotides, Gpp(NH)p and GTPγS, activate phospholipase C activity in the absence of agonists, indicating that activation of the putative G protein (G_{PLC}) is sufficient to promote an increase in phospholipase C activity (Haslam and Davidson, 1984; Cockcroft and Gomperts, 1985; Litosch *et al.,* 1985; Smith *et al.,* 1985; Uhing *et al.,* 1986; Straub and Gershengorn, 1986; Martin *et al.,* 1986; Guillon *et al.,* 1986; Merritt *et al.,* 1986; Baldassare and Fisher, 1986; Wallace and Fain, 1985; Gonzales and Crews, 1985; Litosch, 1987a; Cockcroft and Taylor, 1987; Harden *et al.,* 1987). Fluoride, which activates G proteins linked to regulation of adenylyl cyclase or cyclic GMP phosphodiesterase, promotes an increase in phospholipase C activity in membranes (Martin *et al.,* 1986; Guillon *et al.,* 1986; Litosch,

1987a; Cockcroft and Taylor, 1987) and cells (Blackmore *et al.,* 1985) lending further credence to the view that a stimulatory G protein functions in the phospholipase C system (Strnad *et al.,* 1986; Litosch, 1987a).

Despite the general agreement that a G protein is involved in the signal transduction process leading to phospholipase C activation, there is still a considerable lack of information, and even confusion, concerning the identity of the G protein or the phospholipase C involved in the process. This may be due, in part, to the complexity of the system under study. The enzyme is a phospholipase and interacts with a lipid substrate. The properties and regulation of phospholipases are not as well understood as those of other enzymes. The substrate is a lipid which is an integral part of the membrane and whose physicochemical state is affected by a number of factors including ionic strength, pH, proteins, phospholipids, and detergents.

Levels of PIP_2 are regulated, as shown in Fig. 1, by metabolic processes which result in the synthesis or catabolism of PIP_2. Synthesis of PIP_2 occurs through Mg^{2+}-dependent kinases (Fig. 1a,b) which catalyze the phosphorylation of phosphatidylinositol (PI) to phosphatidylinositol 4-phosphate (PIP) and subsequently PIP_2 (Hawthorne and Kemp, 1964; Colodzin and Kennedy, 1965; Jergil and Sundler, 1983). Loss of PIP_2 occurs through Mg^{2+}-activated lipid phosphatases (Fig. 1c,d,), (Dawson and Thompson, 1964; Nijjar and Hawthorne, 1977; Irvine, 1982; Irvine *et al.,* 1984), the receptor–G protein-regulated phospholipase C and possibly cytosolic phospholipase C activities which may hydrolyze PIP_2 as well as other phosphoinositides under appropriate conditions, i.e., elevated Ca^{2+} levels. Unfortunately, little is known concerning the identity or regulation of kinases and phosphatases involved in PIP_2 metabolism. Both Mg^{2+} and Ca^{2+}, which stimulate phospholipase C activity, may also stimulate PIP_2 phosphatase activity resulting in degradation of PIP_2 (Hawthorne and Kemp, 1964; Colodzin and Kennedy, 1965; Jergil and Sundler, 1983). Since the specific activity of phospholipase C is dependent on substrate levels, possible secondary effects of added agents on the rate of synthesis or catabolism of PIP_2 may affect the determination of phospholipase C activ-

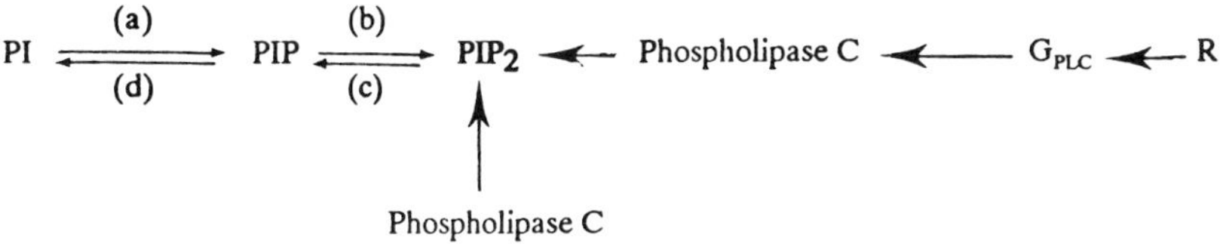

Fig. 1. Metabolic regulation of phosphatidylinositol 4,5-bisphosphate levels in membranes.

ity and thereby complicate interpretation of the kinetics of phospholipase C activation.

II. STUDIES IN CELL-FREE SYSTEMS

A. Guanine-Nucleotide Sensitivity

Several membrane systems have been used to study the regulation of phospholipase C by guanine nucleotides and agonists. The responsiveness of some membrane systems to guanine nucleotides is shown in Table I. Values have been derived from studies in which complete saturable dose–response curves to GTPγS or Gpp(NH)p have been published.

For many mammalian systems, GTPγS appears to be slightly more

Table I   Guanine-Nucleotide Sensitivity of Membrane Phospholipase C

Membrane	Method of assay	Concentration (μM) for half-maximal activation[a]		Reference[b]
		Basal	+ Agonist	
Blowfly salivary gland	Endogenous [³H]inositol substrate	GTPγS = 0.33 Gpp(NH)p = 10 GTP = ineffective	0.01 1	(1)
	Exogenous [³H]PIP₂ substrate	GTPγS = 0.33	0.05	(2)
Neutrophil	Endogenous [³²P] phosphoinositide substrate	GTPγS = 10	N.D.	(3)
Hepatocyte	Endogenous [³²P]phosphoinositide substrate	GTPγS = 0.1 GTP = partially effective	N.D.	(4)
Cerebral cortex	Endogenous [³H]inositol substrate	GTPγS = 27 Gpp(NH)p = 36 GTP = partially effective	N.D.	(5)
	Endogenous [³H]PIP₂substrate	Gpp(NH)p = 33 GTP = ineffective	N.D.	(6)
Turkey	Endogenous [³H]inositol substrate	GTPγS = 0.6 Gpp(NH)p = 36 GTP = partially effective	N.D.	(7)

[a] ND, Not determined.

[b] Key to references: (1) Litosch et al., 1985; (2) Litosch and Fain, 1986b; (3) Cockcroft and Gomperts, 1985; (4) Wallace and Fain, 1985; (5) Gonzales and Crews, 1985; (6) Litosch, 1987a; (7) Harden et al., 1987.

effective or equivalent to Gpp(NH)p in stimulating phospholipase C activity. GTP is relatively ineffective in stimulating basal phospholipase C activity. The apparent reported similarity in potency of GTPγS and Gpp(NH)p in activating phospholipase C activity in some systems contrasts with the adenylyl cyclase system where GTPγS is more potent than Gpp(NH)p (Gilman, 1987; Birnbaumer *et al.*, 1989). The reason for this is not apparent.

Some membrane systems retain the capacity to be stimulated by agonists and a number of studies have documented the essential role of guanine nucleotides in agonist-mediated stimulation of phospholipase C activity (Smith *et al.*, 1985; Uhing *et al.*, 1986; Straub and Gershengorn, 1986; Martin *et al.*, 1986; Guillon *et al.*, 1986; Merritt *et al.*, 1986; Baldassare and Fisher, 1986). Agonist appears to reduce the concentration of guanine nucleotides required for half-maximal activation of phospholipase C activity (Litosch *et al.*, 1985; Litosch and Fain, 1986a). However, this cannot be taken as a universal effect, given the limited number of systems which have been studied in this manner.

Thus, although similarities exist between phospholipase C and adenylyl cyclase with regard to the requirement of guanine nucleotides for activation of enzymatic activity, there are some apparent differences, particularly with regard to guanine-nucleotide potency. Whether this reflects a true difference in the regulation of the respective G proteins, a deficiency in our current understanding of phospholipase C regulation, or a difficulty in the assay remains to be determined.

B. Method of Assay

Two major approaches have been used to study phospholipase C activation in membranes, as illustrated by the data in Table I. These include the use of endogenous substrate, in which labeled PIP$_2$ substrate is provided by the membrane, or exogenously supplied substrate, in which labeled substrate is added to an unlabeled membrane. Phospholipase C activity is monitored by measuring the production of labeled water-soluble products. In some cases, loss of labeled lipid is measured (Cockcroft and Gomperts, 1985; Litosch *et al.*, 1985; Smith *et al.*, 1985; Wallace and Fain, 1985). The advantages and disadvantages of both approaches are discussed in the following sections.

1. Endogenous Substrate

In the endogenous substrate system, the membrane synthesizes the labeled substrate, which is subsequently hydrolyzed by phospholipase C. All the components of the phosphoinositide system including substrate, phospholipase C, receptor, and putative regulatory G protein are therefore present in the membrane under study. In the typical protocol, cells or tissue are incubated with [^{3}H]inositol or [^{32}P]P$_i$ for several hours. During this time, label is incorporated into the phosphoinositides. Homogenates and membranes are subsequently prepared for study. The apparent phospholipase C activity, measured in different tissues, varies, but a strict comparison of phospholipase C activity between tissues is difficult since the specific activity of the pool of phosphoinositides hydrolyzed is not known. In general, a mixture of inositol monophosphate, inositol bisphosphate and inositol trisphosphate is obtained. The concurrent appearance of inositol mono- and bisphosphate with trisphosphate is thought to be a consequence of phosphatase-mediated dephosphorylation of inositol trisphosphate. However, hydrolysis of PI and PIP may also contribute to the inositol mono- and bisphosphate produced.

The successful use of prelabeled membranes depends on three factors: (1) the cell must incorporate label into the hormone-sensitive pool of phosphoinositides, (2) the labeled lipids should survive the protocols designed to isolate the membranes, and (3) activation of phospholipase C must be measured without changes in substrate levels which may occur due to activation of membrane kinases or phosphatases. Concerning the first point, cells or tissues which do not incorporate adequate levels of isotopes into phosphoinositides cannot be studied. This approach is therefore restrictive and may completely preclude some systems from study. Furthermore, comparative studies are difficult since absolute specific activities of the enzymatic process cannot be derived from this approach. Studies on the kinetics of incorporation of label into phosphoinositides indicate that cells contain multiple pools of PIP$_2$ (Fain and Berridge, 1979; Quist, 1982; Vickers and Mustard, 1986; Monaco, 1987). In some cells, only a small fraction of the total label appears to be incorporated into the pool of labeled lipid which is hydrolyzed upon receptor activation. The basis for the existence of these apparent discrete pools of cellular lipids is not known. Since the size of the specific pool involved in hormone action may vary with cell type, simple expression of specific activity based on total lipid content may be inaccurate. Second, the availability of labeled lipid is determined by the recovery of substrate which survived the preparative procedure. This may amount to a full or partial recovery of substrate. Since substrate levels affect the specific activity of phospholipase C (Litosch and Fain, 1986b), the measured phospholipase C activity will vary with recovery of substrate. Clearly, in the ideal situation, assays for phospholipase C activity should be conducted at saturating levels of substrate. Unfortunately, this is not likely to be the case in most studies. PIP_2 is rapidly hydrolyzed during receptor activation. The assay is therefore carried out at constantly decreasing levels of substrate unless a mechanism exists for replenishment of PIP_2 levels. This effect may be translated into a rapidly decreasing specific activity of the enzyme. This problem can be partially overcome if the membrane replenishes its store of PIP_2 during the assay; this occurs to some extent. In blowfly salivary gland membranes, synthesis of PIP_2 occurs during stimulation of PIP_2 hydrolysis by agonist (Litosch *et al.*, 1985; Litosch *et al.*, 1986). The newly synthesized PIP_2 is available for phospholipase C action and contributes to the pool of inositol trisphosphate generated during agonist stimulation. However, a linear rate of inositol phosphate production is not maintained, suggesting that replenishment of PIP_2 levels does not occur at a rate equivalent to the hydrolysis rate. Similar observations have been reported in GH_3 cell membranes where inclusion of ATP in the assay resulted in an enhanced net production of inositol trisphosphate. This effect could be attributed to a net increased synthesis of PIP_2 (Straub and Gershengorn, 1986; Martin *et al.*, 1986). Since the rate of PIP_2 hydrolysis generally exceeds the rate of synthesis, levels of PIP_2 decrease during agonist stimulation (Litosch *et al.*, 1986). Thus, under conditions which promote metabolism of PIP_2 (synthesis or hydrolysis), secondary effects on metabolism may alter substrate levels and affect the apparent measured phospholipase C activity.

Data from several laboratories are in good agreement concerning the general properties of the phosphoinositide–phospholipase C system. Chelators (EGTA) markedly inhibit phospholipase C activity. The agonist-regulated phospholipase C requires a critical amount of Ca^{2+} for activity (Litosch *et al.*, 1985; Uhing *et al.*, 1986; Straub and Gershengorn, 1986; Martin *et al.*, 1986; Litosch, 1987a). In neutrophil membranes, guanine nucleotides increased the Ca^{2+} affinity of phospholipase C suggesting an allosteric effect of guanine nucleotides (Smith *et al.*, 1986). In hepatocyte membranes, a shift in the Ca^{2+} affinity was observed only in the presence of 10 mM $MgCl_2$ (Uhing *et al.*, 1986). Guanine nucleotides clearly reduce the Ca^{2+} required to produce a given degree of phospholipase C activity. However, the mechanism underlying this effect remains to be ascertained.

Stimulation of phospholipase C activity by a maximal effective dose of Gpp(NH)p and Ca^{2+} is not additive, suggesting that Gpp(NH)p and Ca^{2+} activate the same enzyme (Straub and Gershengorn, 1986; Blackmore *et al.*, 1985). Alternatively, Ca^{2+} may activate more than one phospholipase C activity. The same pool of substrate may be degraded by both the Ca^{2+} and G protein–Ca^{2+} regulated phospholipase C. It is not possible to distinguish between these two possibilities at the present time.

Gpp(NH)p ─────────➤ G$_{PLC}$ ─────────➤ Phospholipase C ◄───────── Ca^{2+}

Fig. 2. Model for activation of phospholipase C by Ca^{2+} and Gpp(NH)p.

In summary, most studies suggest the following model of phospholipase C activation as outlined in Fig. 2. Gpp(NH)p activates phospholipase C through the interaction with the G protein (G$_{PLC}$), while Ca^{2+} activates the enzyme directly. One concern with this model, however, is that in intact cells, elevation of Ca^{2+} by treatment with ionophores such as A23187 does not stimulate PIP$_2$ hydrolysis (Litosch and Fain, 1986a). The enzyme *in vivo* may have sufficient Ca^{2+} to satisfy its requirement for catalytic activity and thus activation by G$_{PLC}$ appears to be the rate limiting event during receptor activation.

2. Exogenous Substrate

Exogenously supplied substrate has also been used to study phospholipase C activation. In this approach, unlabeled membranes are incubated with substrate which is labeled with [³H]inositol or [³²P]Pi. Activity is measured by monitoring the formation of labeled inositol phosphates. One advantage of this approach is that cells which do not incorporate [³H]inositol into phosphoinositides or whose labeled pool of phosphoinositides is unstable during membrane isolation can be used to measure phospholipase C activity. Furthermore, with exogenous substrate it is not essential to incubate cells for a prolonged period with isotope. Exogenously supplied substrate affords the opportunity to regulate substrate levels, and thereby measure phospholipase C activity at a constant substrate level, provided the assay conditions are established to minimize degradation of PIP$_2$ through other metabolic processes.

The results shown in Table II demonstrate the basal and Gpp(NH)p-activated phospholipase C activity measured with exogenously added substrate in various tissues obtained from rat. All studies were carried out under similar conditions with no attempt to optimize incubation parameters for the individual membranes. Furthermore, the results do not take into account the amount of phospholipid contributed by the membrane that may be concurrently hydrolyzed during the assay.

The results in Table II demonstrate that membranes prepared from different tissues of rat exhibit a range of phospholipase C activity when assayed with added substrate. Of the four tissues tested, both lung and kidney have a high basal phospholipase C activity, which is stimulated by Gpp(NH)p. Brain membranes have an intermediate activity while liver membranes exhibit the lowest basal and Gpp(NH)p-stimulated activity. Clearly, the ability to monitor phospholipase C activity using added sub-

Table II Basal and Gpp(NH)p-Stimulated
Phospholipase C Activity in Membranes Measured
With Exogenously Supplied Substrate[a]

| Source of membranes | Phospholipase C activity (cpm) | |
	Basal	+Gpp(NH)p
Brain	40	220
Kidney	160	232
Liver	10	90
Lung	182	521

[a] Membranes were prepared from the indicated tissue and assayed for phospholipase C activity in incubation buffer containing 25 mM HEPES (pH 6.75), 50 μM CaCl$_2$, 12 mM LiCl, 0.5 mM MgCl$_2$ and 0.4 μM PIP$_2$. Assay was at 35°C for 2 min. Membrane protein was 15 μg/tube. Gpp(NH)p was present at 10 μM.

strate is not equivalent for all membranes. The basis for this is not apparent but may reflect the presence of different levels of phospholipase C or inhibitors and activators of the enzyme.

Exogenously added substrate has been used in a few studies. In blowfly salivary gland membranes, phospholipase C activity was measured with exogenously added substrate in the absence of added detergent (Litosch and Fain, 1986b). Often, however, it is necessary to add a trace amount of deoxycholate to facilitate suspension of commercially available PIP$_2$ substrate. However, a minimal amount of detergent is used for this and the resulting final detergent concentration is 0.00025% in the assay buffer (Litosch, 1987a). Analysis of product formation is done by Dowex formate chromatography to ensure that phospholipase C activity is measured under all experimental conditions. The data derived from the studies with exogenous substrate compares favorably with data derived from membrane studies employing endogenous substrate in blowfly salivary gland membranes (Litosch *et al.*, 1985) or cerebral cortex membranes (Gonzales and Crews, 1985).

In blowfly salivary gland membranes, hydrolysis of added substrate resulted in a time-dependent production of inositol phosphates (Litosch and Fain, 1986b). Agonist promoted a marked increase in phospholipase C activity in the presence of guanine nucleotides, indicating a functional interaction of the receptor system with the added substrate. Similarly in DDT-MF-2 cell membranes, GTPγS-stimulated phospholipase C activity was enhanced by norepinephrine (Fülle *et al.*, 1987). Thus, exogenous substrate can be used to monitor receptor-promoted increases in phospholipase C activity provided suitable precautions for nonspecific effects are employed and appropriate guanine nucleotide specificity is demonstrated.

Exogenous substrate has also been used in studies with porcine pig coronary artery homogenates (Sasaguri et al., 1985). Divalent chelators, EGTA and EDTA, inhibited phospholipase C activity while Ca^{2+} activated hydrolysis of PIP_2. In the presence of 90 mM KCl, Mg^{2+} inhibited the activation of phospholipase C by Ca^{2+}. This inhibitory effect of Mg^{2+} on Ca^{2+}-activation may reflect partial effects on substrate due to the inclusion of 90 mM KCl in the assay buffer. KCl, at 90 mM, has been shown to facilitate an apparent Mg^{2+}-inhibition of the Ca^{2+}-activated brain cytosolic phospholipase C (Irvine et al., 1984) (see discussion in Section III). GTPγS stimulated PIP_2 hydrolysis and potentiated the effects of carbachol as measured in the presence of 1 mM $MgCl_2$. At least 1 μM GTPγS was needed to facilitate the effect of hormone.

In rat aorta homogenates, Gpp(NH)p stimulated the hydrolysis of exogenously added PIP_2 with half-maximal activation at approximately 33 μM (Roth, 1987). Inclusion of EGTA inhibited, while Ca^{2+}-stimulated, phospholipase C activity.

Two studies used 9 mM cholate in the assay to promote phospholipase C activity in membranes (Jackowski et al., 1986; Rock and Jackowski, 1987). The results of these two studies are difficult to evaluate. Although the studies were done by the same laboratory and employed similar assay conditions, the results with mink lung epithelial cell membranes (Jackowski et al., 1986) and platelet membranes (Rock and Jackowski, 1987), are dramatically different from each other and from other published studies. In lung epithelial cell membranes (Rock and Jackowski, 1987), the divalent chelator, EDTA, inhibited both the basal and Gpp(NH)p-stimulated phospholipase C activity. Ca^{2+} did not activate phospholipase C activity. Although the sensitivity to guanine nucleotide activation was low, the membranes responded to Gpp(NH)p and were not appreciably activated by GTP or ATP. In platelet membranes, however, the chelators, EDTA and EGTA, markedly stimulated phospholipase C activity while Ca^{2+} was ineffective (Rock and Jackowski, 1987). Guanine nucleotides, GTP and Gpp(NH)p were equally effective in stimulating hydrolysis as were all triphosphates. Furthermore, App(NH)p was reported to activate phospholipase C activity. The logic of the reasoning in the platelet studies, that since chelators activate phospholipase C, Gpp(NH)p activates phospholipase C and therefore Gpp(NH)p activates through chelation, is difficult to follow since the two events have no causal relationship to each other. Furthermore, the reported effects of App(NH)p should be taken with some reservation since commercial preparations of App(NH)p may be contaminated with a GTP-like substance and are also capable of promoting transphosphorylation reactions (Clark, 1978; Kimura *et al.*, 1985). The lack of a stimulatory effect of Ca^{2+} on phospholipase C activity and stimulation of phospholipase C activity by divalent chelators is contrary to other published work using either endogenous or exogenous substrate.

The reason for these results remains to be determined; this emphasizes the need to consider that phospholipase C:substrate interactions may be complex. It is possible that the reported activity did not represent phospholipase C. The only evidence for phospholipase C activity was an analysis of the total water-soluble fraction of a pooled aqueous methanol:chloroform extract derived from one assay. In this fraction only inositol trisphosphate was detected and other dephosphorylated inositol phosphates were not evident. This is different from a number of studies in which inositol monophosphate and inositol bisphosphate are also generated (Litosch and Fain, 1986b; Fülle *et al.*, 1987; Litosch *et al.*, 1985; Straub and Gershengorn, 1986; Martin *et al.*, 1986, to cite a few). Subsequent data were based on measurement of radioactivity in the total aqueous extract. The time-course of production of inositol trisphosphate or its formation in the presence of EGTA, guanine nucleotides, or other additions was not documented.

The use of cholate or any detergent in the phospholipase C assay with membranes requires special consideration with regard to (1) possible effects of detergent to solubilize membrane components, (2) possible effects of detergent on extraction of inositol phosphates and lipids, (3) effects of incubation conditions on the solubility and ionization state of the detergent, as well as (4) activation of other phospholipase activities during the experiment. With regard to the first point, the guanine nucleotide-regulated phospholipase C activity is solubilized with detergent. Incubation of membranes with detergent will result in a progressive solubilization of membrane phospholipase C activity. Since the stability and properties of the solubilized enzyme may differ from that of the membrane-bound enzyme, any alteration in the assay conditions which affect the solubilization efficiency of cholate or stability of the solubilized enzyme will complicate interpretation of data. Second, detergents alter the efficiency of lipid partitioning during a methanol:chloroform extraction. The efficiency of this partitioning should be determined and inositol phosphate production should be quantitated to ensure that the measured radioactivity corresponds to inositol phosphates. Third, if detergent is needed to promote activation of the enzyme under study, then clearly any change in the concentration of the effective form of the detergent, due to altered ionic composition of the buffer, will indirectly affect the measured enzymatic activity. Finally, detergent activation of other phospholipase activities which hydrolyze PIP_2 may contribute to the total measured activity and thus it is essential to ascertain that other phospholipase activities do not contribute to the total activity.

Possible problems with use of exogenous substrate are similar to those which may be encountered with endogenous substrate. It is essential that substrate levels are maintained during the experiment. Mg^{2+}-dependent activation of PIP_2 phosphatase activity has been shown in rat aorta homogenates (Roth, 1987) and mink lung epithelial membranes (Jackowski et al., 1986). In rat aorta homogenates, 33% of the PIP_2 was converted to PIP after a 10-min incubation in the presence of 100 μM Ca^{2+}. At 10 mM Mg^{2+}, approximately 80% of the substrate was converted to PIP within 10 min. Loss of substrate will result in an apparent inhibition of measured enzyme activity. In the absence of a specific inhibitor of the phosphatases, some of these problems may be partially circumvented through the use of purified membranes and minimizing the time of incubation.

Other considerations in the use of exogenous substrate include effects of incubation conditions on the physicochemical state of the substrate. However, similar considerations may apply to studies using endogenous substrate (see Section III.A). The contribution of unlabeled membrane phospholipid to the measured phospholipase C may result in an underestimation of the true phospholipase C activity. This assay, however, may be the only way to assess phospholipase C activity in some membranes and thus provides a valuable complimentary approach to studies with endogenous substrate.

III. CHARACTERIZATION OF G PROTEIN– PHOSPHOLIPASE C COMPLEX

A. Phospholipase C

The most thoroughly studied phospholipase C activity to date has been the cytosolic phospholipase C activity. Its relationship to the agonist-regulated activity has not been defined. A considerable amount of information has been derived from studies with soluble phospholipase C and it is likely that general principles will also apply, to some extent, to the guanine nucleotide-regulated phospholipase C. The activity of the enzyme is usually measured with PI substrate, and deoxycholate is often included in the assay buffer. Cytosolic phospholipase C activity has a pH optimum of approximately 5.5 (Irvine et al., 1984; Hofmann and Majerus, 1982). When the enzyme is assayed in the presence of a detergent such as deoxycholate, a pH optimum at approximately 7.0 is also observed. The pK of deoxycholic acid is 6.5. Since only the deoxycholate anion form is the active form of the detergent (Irvine and Dawson, 1978), the activity measured at pH 7.0 may reflect predominately a pH effect on the concentration of the effective form of detergent. PIP_2 and PIP are the preferred substrates for cytosolic phospholipase C, although PI is also hydrolyzed (Irvine and Dawson, 1978; Hofmann and Majerus, 1982; Takenawa and Nagai, 1981; Low and Weglicki, 1983; Carter and Smith, 1987; Wilson et al., 1984; Rittenhouse-Simmons, 1979; Graff et al., 1984; Banno et al., 1986; Downes and Michell, 1982). Ca^{2+} activates the cytosolic phospholipase C and affects the ability of the enzyme to hydrolyze PI. At $1-10$ μM Ca^{2+}, PIP_2 and PIP are hydrolyzed. Higher Ca^{2+} concentrations result in PI hydrolysis (Irvine et al., 1984; Irvine and Dawson, 1978; Hofmann and Majerus, 1982; Takenawa and Nagai, 1981; Low and Weglicki, 1983; Carter and Smith, 1987; Wilson et al., 1984; Rittenhouse-Simmons, 1979; Graff et al., 1984; Banno et al., 1986).

It is possible that the Ca^{2+} requirement of phospholipase C reflects an effect on both the substrate and on the enzyme. PIP_2 is ionized at physiological pH and Ca^{2+} may neutralize the substrate charge, thereby allowing an effective interaction of enzyme with substrate (Irvine et al., 1984). In vivo, the substrate may have sufficient bound Ca^{2+} to satisfy its divalent requirements (Irvine et al., 1984). When assayed with exogenous substrate, Mg^{2+} activates the soluble phospholipase C as does Ca^{2+}. Stimulatory effects of Ca^{2+} are observed at 1 μM while Mg^{2+} stimulation is observed at concentrations of 1 mM or greater (Irvine et al., 1984). The ionic strength of the incubation buffer markedly affects the Ca^{2+} dependency of the enzyme. The soluble enzyme expresses considerable activity at 10 nM Ca^{2+}. However, in the presence of 80 mM KCl and Mg^{2+}, little activity is expressed below 100 μM Ca^{2+} (Irvine et al., 1984). A similar situation occurs in human red blood cell membranes where hydrolysis of endogenous membrane substrate is inhibited by KCl (Downes and Michell, 1982). This inhibitory effect of KCl may reflect, in part, an effect on the physical state of the substrate (Irvine et al., 1984; Downes and Michell, 1982).

The physicochemical state of PIP_2 is markedly affected by the ionic strength of the buffer. In aqueous solution, PIP_2 exists primarily as small micelles. Addition of 10 mM $MgCl_2$ results in vesicle fusion and formation of large vesicles (Dawson, 1965). This event is accelerated by high KCl concentrations (Janmey et al., 1987). Thus, perturbation of the environment may profoundly affect the state of the lipid and this may affect the apparent measured enzymatic activity. Although these types of changes are readily apparent with pure substrate or substrate presented in mixed lipid vesicles, it is possible that similar considerations may also apply to the lipid present in the native membrane. Unfortunately, the latter is more difficult to document and possible regulation of membrane substrate by similar factors can only be inferred from studies with pure lipid substrate.

Apart from ionic strength, the lipid and protein environment may affect the measured phospholipase C activity. Inclusion of phosphatidylcholine to lipid vesicles results in a marked inhibition of phospholipase C activity while phosphatidylethanolamine and phosphatidylserine are stimulatory. Basic proteins such as histones are inhibitory (Irvine *et al.*, 1984). A similar regulation may also occur to some extent in native membranes (Irvine *et al.*, 1984).

Purification studies indicate that multiforms of soluble phospholipase C exist. Four forms of soluble phospholipase C exist in heart and these exhibit a range of apparent molecular weights from 40kDa to 120kDa (Low and Weglicki, 1983). Lymphocytes contain multiforms of soluble phospholipase C activities which range from 300kDa to 11kDa (Carter and Smith, 1987). Sheep seminal vesicles contain two immunologically distinct forms of phospholipase C with apparent molecular weights of 68kDa (Wilson *et al.*, 1984). A soluble phospholipase C activity with an apparent molecular weight of 63kDa has been purified from guinea pig uterus (Bennett and Crooke, 1987). Several groups have purified soluble phospholipase C activity from the supernatant of bovine brain (Rebecchi and Rosen, 1987; Lee *et al.*, 1987). The relationship between the various multiforms of soluble phospholipase C remains to be defined.

In contrast to the cytosolic phospholipase C, the agonist-regulated phospholipase C activity is membrane associated (Cockcroft and Gomperts, 1985; Litosch *et al.*, 1985; Smith *et al.*, 1985; Uhing *et al.*, 1986; Straub and Gershengorn, 1986; Martin *et al.*, 1986; Guillon *et al.*, 1986; Merritt *et al.*, 1986; Baldassare and Fisher, 1986), is activated by guanine nucleotides (Cockcroft and Gomperts, 1985; Litosch *et al.*, 1985; Smith *et al.*, 1985; Uhing *et al.*, 1986; Straub and Gershengorn, 1986; Martin *et al.*, 1986; Guillon *et al.*, 1986; Merritt *et al.*, 1986; Baldassare and Fisher, 1986; Wallace and Fain, 1985; Gonzales and Crews, 1986; Litosch, 1987a; Cockcroft and Taylor, 1987; Harden *et al.*, 1987), and has a pH optimum of 6.7 in the presence of guanine nucleotides and absence of detergents (Martin *et al.*, 1986; Litosch, 1987a). A second pH optimum of 5.5 is detected in the presence of NaF (Litosch, 1987a). Apart from this scant information, little else is known. A guanine nucleotide-stimulated cytosolic activity has been described in calf brain (Deckmyn *et al.*, 1986). Hydrolysis-resistant guanine nucleotides and fluoride promoted activation of this enzyme. Maximal activity was measured at pH 5.5 with little detectable activity at neutral pH. Whether this activity represents a membrane-dissociated activity complexed with its regulatory GTP-binding component has not yet been established. However, the acidic pH optimum of this measured enzyme activity is difficult to reconcile with the neutral pH optimum of the agonist-regulated phospholipase C activity measured in membranes.

Little is also known concerning the regulation of phospholipase C activity. Phorbol esters which activate protein kinase C (Castagna, 1982) interfere with guanine nucleotide-, but not Ca^{2+}-activation of phospholipase C activity (Smith et al., 1987). These studies suggest that protein kinase C activation inhibits the ability of the G protein to interact with phospholipase C. Recently, phorbol esters have been shown to induce the phosphorylation of a soluble phospholipase C activity purified from guinea pig uterus (Bennett and Crooke, 1987) but the functional consequences of this phosphorylation have not been addressed.

B. G Protein

The identity of G_{PLC} has not been ascertained and thus its relationship to the presently characterized G proteins is not known. G_{PLC} may resemble other functionally characterized G proteins which are oligomeric and consist of α, β, and γ subunits. Four G proteins have been characterized. G_s is involved in the activation of adenylyl cyclase, G_i is involved in the inhibition of adenylyl cyclase, G_k activates potassium channels, and transducin activates cyclic GMP phosphodiesterase. However, these G proteins may have additional functions (Gilman, 1987; Birnbaumer et al., 1989).

The current scheme for G-protein activation (subunit dissociation model) depicts the event as occurring through a receptor-promoted exchange of GTP for the GDP bound on the α subunit, resulting in activation of the α subunit. The oligomer dissociates into an activated α subunit and $\beta\gamma$ subunit. The α subunits range in molecular weights from 39–52kDa. The α subunit contains the GTP binding site, GTPase activity, receptor and target specificity. These subunits are ADP-ribosylated by specific toxins (Gilman, 1987; Birnbaumer et al., 1989). Two types of β subunits are present in cells. On sodium dodecyl sulfate-polyacrylamide gel electrophoresis (SDS-PAGE), a doublet at 36/35kDa is observed. The ratio of these peptides varies in different tissues (Sternweis and Robishaw, 1984; Carter and Smith, 1987). In rod outer segment, only a 36kDa form is observed (Fong et al., 1986). Isolated γ subunits have apparent molecular weights of approximately 10K (Gilman, 1987; Birnbaumer et al., 1989). The role of $\beta\gamma$ subunits in G-protein function has not yet been completely established. $\beta\gamma$ subunits copurify with α subunits and are functionally interchangeable (Cerione et al., Cerione et al., 1987). They function in reversal of α-activated events (Cerione et al., 1985; Cerione et al., 1987; Katada et al., 1984a,b), promote effective guanine-nucleotide exchange on the α subunit (Fung, 1983; Correze et al., 1987), allow efficient ADP-ribosylation of α-subunits by pertussis toxin (Neer *et al.*, 1984; Katada *et al.*, 1986), stabilize α subunits against thermal denaturation (Neer *et al.*, 1984; Katada *et al.*, 1986), and provide a membrane-anchoring function (Sternweis, 1986).

The bacterial toxins, cholera toxin and pertussis toxin, were a major tool which helped identify the G proteins linked to regulation of adenylyl cyclase. These toxins promoted a functional modification of the G protein-regulated system and this alteration could be correlated with the degree of ADP-ribosylation of the α subunit. This allowed the establishment of protocols aimed at purification and identification of the respective G proteins. Cholera toxin helped identify G_s (Northup *et al.*, 1980; Hildebrandt *et al.*, 1984). Pertussis toxin has been used to identify G_i (Codina *et al.*, 1984; Bokoch *et al.*, 1983) and G_{K+} (Yatani *et al.*, 1987).

The use of these toxins in the phospholipase C system has generated some confusion. Pertussis toxin inhibits agonist-induced phospholipase C activation in neutrophils (Bradford and Rubin, 1985; Shefcyk *et al.*, 1985; Verghese *et al.*, 1985), leukocytes (Ohta *et al.*, 1985), HL-60 leukemic cells (Brandt *et al.*, 1985), and cultured endothelial (Lambert *et al.*, 1986) and rat aortic smooth muscle cells (Xuan *et al.*, 1987). However, in hepatocytes (Uhing *et al.*, 1986; Pobiner *et al.*, 1985), brown fat cells (Schimmel and Elliot, 1986), and FRT1-5 thyroid cells (Albert and Tashjian, 1984), pertussis toxin does not affect phosphoinositide hydrolysis. The basis for this differential susceptibility to pertussis toxin is not clear. In some cases, the toxin was shown to catalyze the complete ADP-ribosylation of G_i, indicating that G_i was not the G protein involved in phospholipase C activation (Uhing *et al.*, 1986; Pobiner *et al.*, 1985).

Recent studies indicate that cells contain a number of pertussis toxin-sensitive substrates with apparent molecular weights of 41kDa, 40kDa, and 39kDa (Gilman, 1987; Birnbaumer *et al.*, 1989). Two forms of G_i exist (Bray *et al.*, 1987). In cardiac tissue, two G_o-type pertussis toxin-sensitive substrates are present (Scherer *et al.*, 1987). G_i and G_o differ in their ability to be ADP-ribosylated by pertussis toxin and this may be due to a difference in their affinity for $\beta\gamma$ subunits which are necessary for optimal ADP-ribosylation of G_i and G_o (Neer *et al.*, 1984; Katada *et al.*, 1986). The efficiency of toxin-mediated ADP-ribosylation of other G proteins has not been thoroughly characterized.

There is little evidence for the involvement of a cholera toxin substrate in the phospholipase C system. However, it has been reported that cholera toxin activates phosphoinositide hydrolysis in flow 9000 cloned pituitary cells (Lo and Hughes, 1987).

Few reconstitution studies have been published on the effects of added resolved α on phospholipase C. In neutrophil membranes, a pertussis toxin-sensitive phospholipase C system, addition of a mixture of the pertussis toxin substrates G_i/G_o restored agonist-stimulated phospholipase C activity in pertussis toxin-treated membranes (Kikuchi *et al.*, 1986). Other G proteins and resolved α subunits however, were not tested.

IV. FUTURE DIRECTIONS

A current view of the regulation of phospholipase C by agonists and guanine nucleotides, patterned after the scheme for currently characterized functional G proteins (Gilman, 1987; Birnbaumer *et al.*, 1989), is summarized in Fig. 3.

The receptor–phopholipase C system is depicted as a tertiary complex consisting of receptor, G protein (G_{PLC+}) and phospholipase C. Stimulatory receptors induce a conformational change in G_{PLC+}, enhancing its ability to interact with phospholipase C. The G protein is depicted as an oligomer consisting of α, β, and γ subunits. Activation of G_{PLC+} results in the dissociation of the oligomer into activated α subunits and $\beta\gamma$ subunits. In this sequence, the α–GDP is converted to α–GTP by an acceleration of the nucleotide exchange reaction. The activation state of the α subunit is regulated by the stability of the GTP-liganded state. The activated α subunit interacts with phospholipase C and stimulates enzymatic activity. Hydrolysis of GTP by a GTPase activity on the α subunit results in an attenuation of the stimulatory signal and restoration of basal phospholipase C activity. $\beta\gamma$ subunits regulate phospholipase C indirectly through reassociation with α subunits, resulting in a reversal of α-mediated events.

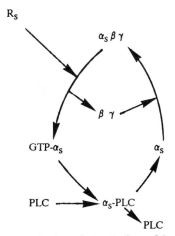

Fig. 3. Proposed scheme for activation of phospholipase C by regulatory G Proteins. R_s, stimulatory receptor; PLC, phospholipase C.

The realization that cells contain multiple forms of α, β, and γ subunits, as well as the realization that a G protein is able to regulate various cellular processes, emphasizes the complexity of G protein-regulated systems. The challenge of the present is to identify the G protein involved in regulation of the phospholipase C system and to integrate this regulation into the current scheme of G-protein function.

ACKNOWLEDGMENTS

I.L. is an Established Investigator of the American Heart Association. This work was supported by NIH grant DK 37007 and by funds contributed by the American Heart Association, Florida Affiliate. Excellent secretarial assistance was provided by Gerry Trebilcock.

REFERENCES

Albert, P. R., and Tashjian, A. H. (1984). Thyrotropin-releasing hormone-induced spike and plateau in cytosolic free Ca^{2+} concentrations in pituitary cells. *J. Biol. Chem.* **259**, 5827–5832.

Baldassare, J. J., and Fisher, G. J. (1986). GTP and cytosol stimulate phosphoinositide hydrolysis in isolated platelet membranes. *Biochem. Biophys. Res. Commun.* **137**, 801–805.

Banno, Y., Nakashima, S., and Nozawa, Y. (1986). Partial purification of phosphoinositide phospholipase C from human platelet cytosol; Characterization of its three forms. *Biochem. Biophys. Res. Commun.* **136**, 713–721.

Baudry, M., Evans, J., and Lynch, G. (1986). Excitatory amino acids inhibit stimulation of phosphatidylinositol metabolism by aminergic agonists in hippocampus. *Nature (London)* **319**, 329–331.

Bennett, C. F., and Crooke, S. T. (1987). Purification and characterization of a phosphoinositide-specific phospholipase C from guinea pig uterus. *J. Biol. Chem.* **262**, 13789–13797.

Berridge, M. J., and Irvine, R. F. (1984). Inositol trisphosphate: A novel second messenger in cellular signal transduction. *Nature* **312**, 315–321.

Birnbaumer, L., Codina, J., Mattera, R., Yatani, A., Scherer, N., Toro, M-J, and Brown, A. M. (1989). Signal transduction by G proteins. *Kidney Int.* (in press).

Blackmore, P. F., Bocckino, S. B., Waynick, L. E., and Exton, J. H. (1985). Role of a guanine nucleotide-binding regulatory protein in the hydrolysis of hepatocyte phosphatidylinositol 4,5-bisphosphate by calcium-mobilizing hormones and the control of cell calcium. *J. Biol. Chem.* **260**, 14477–14483.

Bokoch, G. M., Katada, T., Northup, J. K., Hewlett, E. L., and Gilman, A. G. (1983). Identification of predominant substrate for ADP-ribosylation by islet activating protein. *J. Biol. Chem.* **258**, 2072–2075.

Bradford, P. G., and Rubin, R. P. (1985). Pertussis toxin inhibits chemotactic factor induced phospholipase C stimulation and lysosomal enzyme secretion in rabbit neutrophils. *FEBS Lett.* **183**, 317–320.

Brandt, S. J., Dougherty, R. W., Lapetina, E. G., and Niedel, J. E. (1985). Pertussis toxin inhibits chemotactic peptide-stimulated generation of inositol phosphates and lysosomal enzyme secretion in human leukemic (HL-60) cells. *Proc. Natl. Acad. Sci. U.S.A.* **82**, 3277–3280.

Bray, P., Carter, A., Guo, V., Puckett, C., Kamholz, J., Spiegel, A., and Nirenberg, M. (1987). Human cDNA clones for an α subunit of G_i signal-transduction protein. *Proc. Natl. Acad. Sci. U.S.A.* **84**, 5115–5119.

Cantau, B., Keppens, S., de Wulf, H., and Jard, S. (1980). (^{3}H)-vasopressin binding to isolated rat hepatocytes and liver membranes: Regulation by GTP and relation to glycogen phosphorylase activation. *J. Recept. Res.* **1**, 137–168.

Carter, H. R., and Smith, A. D. (1987). Resolution of the phosphoinositide-specific phospholipase C isolated from porcine lymphocytes into multiple species. *Biochem. J.* **244**, 639–645.

Castagna, M., Takai, Y., Kaibuchi, K., Sano, K., Kikkawa, U., and Nishizuka, Y. (1982). Direct activation of calcium-activated, phospholipid-dependent protein kinase by tumor-promoting phorbol esters. *J. Biol. Chem.* **257**, 7847–7851.

Cerione, R. A., Codina, J., Kilpatrick, B. F., Staniszewski, C., Gierschik, P., Somers, R. L., Spiegel, A. M., Birnbaumer, L., Caron, M. G., and Lefkowitz, R. J. (1985). Transducin and the inhibitory nucleotide regulatory protein inhibit the stimulatory nucleotide regulatory protein mediated stimulation of adenylate cyclase in phospholipid vesicle systems. *Biochemistry* **24**, 4499–4503.

Cerione, R. A., Gierschik, P., Staniszewski, C., Benovic, J. L., Codina, J., Somers, R., Birnbaumer, L., Spiegel, A. M., Lefkowitz, R. J., and Caron, M. G. (1987). Functional differences in the $\beta\gamma$ complexes of transducin and the inhibitory guanine nucleotide regulatory protein: Possible role of the γ subunit in modulating inhibition of adenylate cyclase. *Biochemistry* **26**, 1485–1491.

Clark, R. B. (1978). Endogenous GTP and the regulation of epinephrine stimulation of adenylate cyclase. *J. Cyclic Nucleotide Res.* **4**, 259–270.

Cockcroft, S., and Gomperts, B. D. (1985). Role of guanine nucleotide binding protein in the activation of polyphosphoinositide phosphodiesterase. *Nature* **314**, 534–536.

Cockcroft, S., and Taylor, J. A. (1987). Fluoroaluminate mimics guanosine 5'-[γ-thio]triphosphate in activating the polyphosphoinositide phosphodiesterase of hepatocyte membranes. Role for the guanine nucleotide regulatory protein G_p in signal transduction. *Biochem. J.* **241**, 409–414.

Codina, J., Hildebrandt, J. D., Sekura, R. D., Birnbaumer, M., Bryan, J., Manclark, C. R., Iyengar, R., and Birnbaumer, L. (1984). N_s and N_i, the stimulatory and inhibitory regulation components of adenylyl cyclase. *J. Biol. Chem.* **259**, 5871–5886.

Colodzin, M., and Kennedy, E. P. (1965). Biosynthesis of diphosphoinositide in brain. *J. Biol. Chem.* **240**, 3771–3781.

Correze, C., d'Alayer, J., Coussen, F., Berthillier, G., Deterre, P., and Monneron, A. (1987). Antibodies directed against transducin β-subunits interfere with the regulation of adenylate cyclase activity in brain membranes. *J. Biol. Chem.* **262**, 15182–15187.

Dawson, R. M. C. (1965). Phosphatido-peptide-like complexes formed by the interaction of calcium triphosphoinositide with protein. *Biochem. J.* **97**, 134–138.

Dawson, R. M. C., and Thompson, W. (1964). The triphosphoinositide phosphomonesterase. *Biochem. J.* **91**, 244–250.

Deckmyn, H., Tu, S-M, and Majerus, P. W. (1986). Guanine nucleotides stimulate soluble phosphoinositide-specific phospholipase C in the absence of membranes. *J. Biol. Chem.* **261**, 16553–16558.

Downes, C. P., and Michell, R. H. (1982). The control by Ca^{2+} of the polyphosphoinositide phosphodiesterase and the Ca^{2+}-pump ATPase in human erythrocytes. *Biochem. J.* **202**, 53–58.

Enjalbert, A., Sladeczek, F., Guillon, G., Bertrand, P., Shu, C., Epelbaum, J., Garcia-Sainz, A., Jard, S., Lombard, C., Kordon, C., and Bockaert, J. (1986). Angiotensin II and dopamine modulate both cAMP and inositol phosphate productions in anterior pituitary cells. Involvement in prolactin secretion. *J. Biol. Chem.* **261**, 4071–4075.

Fain, J. N., and Berridge, M. J. (1979). Relationship between hormonal activation of phosphatidylinositol hydrolysis, fluid secretion and calcium flux in the blowfly salivary gland. *Biochem. J.* **178**, 45–58.

Fong, H. K. W., Hurley, J. B., Hopkins, R. S., Miake-Lye, R., Johnson, M. S., Doolittle, R. F., and Simon, M. I. (1986). Repetitive segmental structure of the transducin β subunit: Homology with the CDC4 gene and identification of related mRNAs. *Proc. Natl. Acad. Sci. U.S.A.* **83**, 2162–2166.

Fülle, H-J., Hoer, D., Lache, W., Rosenthal, G., Schultz, G., and Oberdisse, E. (1987). In vitro synthesis of ^{32}P-labelled phosphatidylinositol 4,5-bisphosphate and its hydrolysis by smooth muscle membrane-bound phospholipase C. *Biochem. Biophys. Res. Commun.* **145**, 673–679.

Fung, B., K-K. (1983). Characterization of transducin from bovine retinal rod outer segments. *J. Biol. Chem.* **258**, 10495–10502.

Gilman, A. G. (1987). G proteins: Transducers of receptor-generated signals. *Annu. Rev. Biochem.* **56**, 615–649.

Gonzales, R. A., and Crews, F. T. (1985). Guanine nucleotides stimulate production of inositol trisphosphate in rat cortical membranes. *Biochem. J.* **232**, 799–804.

Graff, G., Nahas, N., Nikolopoulon, M., Natarajan, V., and Schmid, H. H. O. (1984). Possible regulation of phospholipase C activity in human platelets by phosphatidylinositol 4,5-bisphosphate. *Arch. Biochem. Biophys.* **228**, 299–308.

Guillon, G., Mouillac, B., and Balestre, M-N. (1986). Activation of membrane phospholipase C by vasopressin. A requirement for guanyl nucleotides. *FEBS Lett.* **196**, 155–159.

Harden, T. K., Stephens, L., Hawkins, P. T., and Downes, C. P. (1987). Turkey erythrocyte membranes as a model for regulation of phospholipase C by guanine nucleotides. *J. Biol. Chem.* **262,** 9057–9061.

Haslam, R. J., and Davidson, M. M. L. (1984). Receptor-induced diacylglycerol formation in permeabilized platelets: Possible role for a GTP-binding protein. *J. Recept. Res.* **4,** 605–629.

Hawthorne, J. N., and Kemp, P. (1964). The brain phosphoinositides. *Adv. Lipid Res.* **2,** 127–166.

Hildebrandt, J. D., Codina, J., Risinger, R., and Birnbaumer, L. (1984). Identification of a γ subunit associated with the adenylyl cyclase regulatory proteins N_s and N_i. *J. Biol. Chem.* **259,** 2039–2042.

Hofmann, S. L., and Majerus, P. W. (1982). Identification and properties of two distinct phosphatidylinositol-specific phospholipase C enzymes from sheep seminal vesicular glands. *J. Biol. Chem.* **257,** 6461–6469.

Hokin, L. E. (1985). Receptors and phosphoinositide-generated second messengers. *Annu. Rev. Biochem.* **54,** 205–235.

Irvine, R. F. (1982). The enzymology of stimulated inositol lipid turnover. *Cell Calcium* **3,** 295–309.

Irvine, R. F., and Dawson, R. M. C. (1978). The distribution of calcium dependent phosphatidylinositol phosphodiesterase in rat brain. *J. Neurochem.* **31,** 1427–1434.

Irvine, R. F., Letcher, A. J., and Dawson, R. M. C. (1984). Phosphatidylinositol-4,5-bisphosphate phosphodiesterase and phosphomonoesterase activities of rat brain. *Biochem. J.* **218,** 177–185.

Irvine, R. F., Letcher, A. J., Lander, D. J., and Berridge, M. J. (1986). Specificity of inositol phosphate-stimulated Ca^{2+} mobilization from Swiss-mouse 3T3 cells. *Biochem. J.* **240,** 301–304.

Jackowski, S., Rettenmier, C. V., Sherr, C. J., and Rock, C. O. (1986). A guanine nucleotide-dependent phosphatidylinositol 4,5-diphosphate phospholipase C in cells transformed by the v-fms and v-fes oncogenes. *J. Biol. Chem.* **261,** 4978–4985.

Janmey, P. A., Iida, K., Yin, H. L., and Stossel, T. P. (1987). Polyphosphoinositide-containing vesicles dissociate endogenous gelsolin-actin complexes and promote actin assembly from the fast-growing end of actin filaments blocked by gelsolin. *J. Biol. Chem.* **262,** 12228–12236.

Jergil, B., and Sundler, R. (1983). Phosphorylation of phosphatidylinositol in rat liver Golgi. *J. Biol. Chem* **258,** 7968–7973.

Journot, L., Homburger, V., Pantaloni, C., Priam, M., Bockaert, J., and Enjalbert, A. (1987). An islet activating protein-sensitive G protein is involved in dopamine inhibition of angiotensin and thyrotropin-releasing hormone-stimulated inositol phosphate production in anterior pituitary cells. *J. Biol. Chem.* **262,** 15106–15110.

Katada, T., Bokoch, G. M., Smigel, M. D., Ui, M., and Gilman, A. G. (1984a). The inhibitory guanine nucleotide-binding regulatory component of adenylate cyclase. Subunit dissociation and the inhibition of adenylate cyclase in S49 lymphoma Cyc- and wild type membranes. *J. Biol. Chem.* **259,** 3586–3595a.

Katada, T., Northup, J. K., Bokoch, G. M., Ui, M., and Gilman, A. G. (1984b).

The inhibitory guanine nucleotide-binding regulatory component of adenylate cyclase. Subunit dissociation and guanine nucleotide-dependent hormonal inhibition. *J. Biol. Chem.* **259,** 3578–3585b.

Katada, T., Oinuma, M., and Ui, M. (1986). Two guanine nucleotide-binding proteins in rat brain serving as the specific substrate of islet-activating protein, pertussis toxin. Interaction of the α-subunits with $\beta\gamma$-subunits in development of their biological activities. *J. Biol. Chem.* **261,** 8182–8191.

Kikuchi, A., Kozawa, O., Kaibuchi, K., Katada, T., Ui, M., and Takai, Y. (1986). Direct evidence for involvement of a guanine nucleotide-binding protein in chemotactic peptide-stimulated formation of inositol bisphosphate and tris-phosphate in differentiated human leudemic (HL-60) cells. Reconstitution with G_i or G_o of the plasma membranes ADP-ribosylated by pertussis toxin. *J. Biol. Chem.* **261,** 11558–11562.

Kimura, N., Shimada, N., and Tsubokura, M. (1985). Adenosine 5'-(β, γ-imino) triphosphate and guanosine 5'-0-(2-thiodiphosphate) do not necessarily pro-vide non-phosphorylating conditions in adenylate cyclase studies. *Biochem. Biophys. Res. Commun.* **126,** 983–991.

Kuo, C., Lefkowitz, R. J., and Snyderman, R. (1983). Guanine nucleotides modu-late the binding affinity of the oligopeptide chemoattractant receptor on human polymorphonuclear leukocytes. *J. Clin. Invest.* **72,** 748–758.

Lambert, T. L., Kent, R. S., and Whorton, A. R. (1986). Bradykinin stimulation of inositol polyphosphate production in porcine aortic endothelial cells. *J. Biol. Chem.* **261,** 15288–15293.

Lee, K-Y., Ryu, S. H., Suh, P-G., Choi, W. C., and Rhee, S. G. (1987). Phospholi-pase C associated with particulate fractions of bovine brain. *Proc. Natl. Acad. Sci. U.S.A.* **84,** 5540–5544.

Litosch, I. (1987a). Guanine nucleotide and NaF stimulation of phospholipase C activity in rat cerebral-cortical membranes. *Biochem. J.* **244,** 35–40.

Litosch, I. (1987b). Regulatory GTP-binding proteins: Emerging concepts on their role in cell function. *Life Sci.* **41,** 251–258.

Litosch, I., and Fain, J. N. (1986a). Regulation of phosphoinositide breakdown by guanine nucleotides. *Life Sci.* **39,** 187–194.

Litosch, I., and Fain, J. N. (1986b). 5-Methyltryptamine stimulates phospholipase C-mediated breakdown of exogenous phosphoinositides by blowfly salivary gland membranes. *J. Biol. Chem.* **260,** 16052–16055.

Litosch, I., Wallis, C., and Fain, J. N. (1985). 5-Hydroxytryptamine stimulates inositol phosphate production in a cell-free system from blowfly salivary glands. *J. Biol. Chem.* **260,** 5464–5471.

Litosch, I., Calista, C., Wallis, C., and Fain, J. N. (1986). 5-Methyltryptamine decreases net accumulation of ^{32}P into the polyphosphoinositides from [γ-^{32}P]ATP in a cell-free system from blowfly salivary glands: Activation of breakdown of the newly synthesized [^{32}P]polyphosphoinositides. *J. Biol. Chem.* **261,** 638–643.

Lo, W. W. Y., and Hughes, J. (1987). A novel cholera toxin-sensitive G-protein (G_c) regulating receptor-mediated phosphoinositide signaling in human pitui-tary clonal cells. *FEBS Lett.* **220,** 327–331.

Low, M. G., and Weglicki, W. B. (1983). Resolution of myocardial phospholipase C into several forms with distinct properties. *Biochem. J.* **215**, 325–334.

Majerus, P. W., Connolly, T. M., Deckymyn, H., Ross, T. S., Boss, T. E., Ishii, H., Bansal, U. S., and Wilson, D. B. (1986). The metabolism of phosphoinositide-derived messenger molecules. *Science* **234**, 1519–1526.

Martin, T. F. J., Bajjalieh, S. M., Lucas, D. O., and Kowalchyk, J. A. (1986). Thyrotropin-releasing hormone stimulation of polyphosphoinositide hydrolysis in GH$_3$ cell membranes is GTP dependent but insensitive to cholera or pertussis toxin. *J. Biol. Chem.* **261**, 10141–10149.

Merritt, J. E., Taylor, C. W., Rubin, R. P., and Putney, Jr. J. W. (1986). Evidence suggesting that a novel guanine nucleotide regulatory protein couples receptors to phospholipase C in exocrine pancreas. *Biochem. J.* **236**, 337–343.

Monaco, M. (1987). Inositol metabolism in WRK-1 cells. Relationship of hormone-sensitive to -insensitive pools of phosphoinositides. *J. Biol. Chem.* **262**, 13001–13006.

Neer, E. J., Lok, J. M., and Wolf, L. G. (1984). Purification and properties of the inhibitory guanine nucleotide regulatory unit of brain adenylate cyclase. *J. Biol. Chem.* **259**, 14222–14229.

Nijjar, M. S., and Hawthorne, J. N. (1977). Purification and properties of polyphosphoinositide phosphomonoesterase from rat brain. *Biochem. Biophys. Acta.* **480**, 390–402.

Nishizuka, Y. (1984). The role of protein kinase C in cell surface signal transduction and tumor promotion. *Nature* **308**, 693–698.

Northup, J. K., Sternweis, P. C., Smigel, M. D., Schleifer, L. S., Ross, E. M., and Gilman, A. G. (1980). Purification of the regulatory component of adenylate cyclase. *Proc. Natl. Acad. Sci. U.S.A.* **77**, 6516–6520.

Ohta, H., Okajima, F., and Ui, M. (1985). Inhibition by islet-activating protein of a chemotactic peptide-induced early breakdown of inositol phospholipids and Ca^{2+} mobilization in guinea pig neutrophils. *J. Biol. Chem.* **260**, 15771–15780.

Pobiner, B. F., Hewlett, E. L., and Garrison, J. C. (1985). Role of N$_i$ in coupling angiotensin receptors in inhibition of adenylate cyclase in hepatocytes. *J. Biol. Chem.* **260**, 16200–16209.

Quist, E. E. (1982). Polyphosphoinositide synthesis in rabbit erythrocyte membranes. *Arch. Biochem. Biophys.* **219**, 58–60.

Rebecchi, M. J., and Rosen, O. M. (1987). Purification of a phosphoinositide-specific phospholipase C from bovine brain. *J. Biol. Chem.* **262**, 12526–12532.

Rittenhouse-Simmons, S. (1979). Production of diglyceride from phosphatidylinositide in activated human platelets. *J. Clin. Invest.* **63**, 580–587.

Rock, C. O., and Jackowski, S. (1987). Thrombin- and nucleotide-activated phosphatidylinositol 4,5-bisphosphate phospholipase C in human platelet membranes. *J. Biol. Chem.* **262**, 5492–5498.

Roth, B. L. (1987). Modulation of phosphatidylinositol-4,5-bisphosphate hydrolysis in rat aorta by guanine nucleotides, calcium and magnesium. *Life Sci.* **41**, 629–634.

Sasaguri, T., Hirata, M., and Kuriyama, H. (1985). Dependence on Ca^{2+} of the activities of phosphatidylinositol 4,5-bisphosphate phosphodiesterase and inositol 1,4,5-trisphosphate phosphatase in smooth muscle of the porcine coronary artery. *Biochem. J.* **231,** 497–503.

Scherer, N. M., Toro, M-J., Entman, M. L., and Birnbaumer, L. (1987). G-protein distribution in canine cardiac sarcoplasmic reticulum and sarcolemma: Comparison to rabbit skeletal muscle membranes and to brain and erythrocyte G proteins. *Arch. Biochem. Biophys.* **259,** 431–440.

Schimmel, R. J., and Elliott, M. E. (1986). Pertussis toxin does not prevent α-adrenergic stimulated breakdown of phosphoinositides or respiration in brown adipocytes. *Biochem. Biophys. Res. Commun.* **135,** 823–829.

Shefcyk, J., Yassin, R., Volpi, M., Molski, T. F. P., Naccache, P. H., Munoz, J. J., Becker, E. L., Feinstein, M. B., and Shaafi, R. I. (1985). Pertussis but not cholera toxin inhibits the stimulated increase in actin association with the cytoskeleton in rabbit neutrophils: Role of the 'G proteins' in stimulus-response coupling. *Biochem. Biophys. Res. Commun.* **126,** 1174–1181.

Smith, C. D., Lane, B. C., Kusaka, I., Verghese, M. W., and Snyderman, R. (1985). Chemoattractant receptor-induced hydrolysis of phosphatidylinositol 4,5-bisphosphate in human polymorphonuclear leukocyte membranes. *J. Biol. Chem.* **260,** 5875–5878.

Smith, C. D., Cox, C. C., and Snyderman, R. (1986). Receptor-coupled activation of phosphoinositide-specific phospholipase C by an N protein. *Science* **232,** 97–100.

Smith, C. D., Uhing, R. J., and Snyderman, R. (1987). Nucleotide regulatory protein-mediated activation of phospholipase C in human polymorphonuclear leukocytes is disrupted by phorbol esters. *J. Biol. Chem.* **262,** 6121–6127.

Snavely, M. D., and Insel, P. A. (1982). Characterization of α-adrenergic receptor subtypes in the rat renal cortex. Differential regulation of α_1- and α_2-adrenergic receptors by guanyl nucleotides and Na^+. *Mol. Pharmacol.* **22,** 532–546.

Sternweis, P. (1986). The purified α-subunits of G_o and G_i from bovine brain require $\beta\gamma$ for association with phospholipid vesicles. *J. Biol. Chem.* **261,** 631–637.

Sternweis, P. C., and Robishaw, J. D. (1984). Isolation of two proteins with high affinity for guanine nucleotides from membranes of bovine brain. *J. Biol. Chem.* **259,** 13806–13813.

Sternweis, P. C., Northup, J. K., Murray, D., Smigel, M. D., and Gilman, A. G. (1981). The regulatory component of adenylate cyclase. Purification and Properties. *J. Biol. Chem.* **256,** 11517–11526.

Straub, R. E., and Gershengorn, M. C. (1986). Thyrotropin-releasing hormone and GTP activate inositol trisphosphate formation in membranes isolated from rat pituitary cells. *J. Biol. Chem.* **261,** 2712–2717.

Strnad, C. F., Parente, J. E., and Wong, K. (1986). Use of fluoride ion as a probe for the guanine nucleotide-binding protein involved in the phosphoinositide-dependent neutrophil transducin pathway. *FEBS Lett.* **206,** 20–24.

Takenawa, T., and Nagai, Y. (1981). Purification of phosphatidylinositol-specific phospholipase C from rat liver. *J. Biol. Chem.* **256,** 6769–6775.

Uhing, R. J., Prpic, V., Jiang, H., and Exton, J. H. (1986). Hormone-stimulated polyphosphoinositide breakdown in rat liver plasma membranes. *J. Biol. Chem.* **261,** 2140–2146.

Verghese, M. W., Smith, C. D., and Snyderman, R. (1985). Potential role for a guanine nucleotide regulatory protein in chemoattractant receptor mediated polyphosphoinositide metabolism, Ca^{2+} mobilization and cellular responses by leukocytes. *Biochem. Biophys. Res. Commun.* **127,** 450–457.

Vickers, J. D., and Mustard, J. F. (1986). The phosphoinositides exist in multiple metabolic pools in rabbit platelets. *Biochem. J.* **238,** 411–417.

Wallace, M. A., and Fain, J. N. (1985). Guanosine 5'-0-thiotriphosphate stimulates phospholipase C activity in plasma membranes of rat hepatocytes. *J. Biol. Chem.* **260,** 9527–9530.

Wilson, D. B., Bross, T. E., Hofmann, S. L., and Majerus, P. W. (1984). Hydrolysis of polyphosphoinositides by purified sheep seminal vesicle phospholipase C enzymes. *J. Biol. Chem.* **259,** 11718–11724.

Wilson, D. B., Connolly, T. M., Bross, T. E., Majerus, P. W., Sherman, W. R., Tyler, A. N., Rubin, L. J., and Brown, J. E. (1985a). Isolation and characterization of the inositol cyclic phosphate products of polyphosphoinositide cleavage by phospholipase C. Physiological effects in permeabilized platelets and Limulus photoreceptor cells. *J. Biol. Chem.* **260,** 13496–13501.

Wilson, D. B., Bross, T. E., Sherman, W. R., Berger, R. A., and Majerus, P. W. (1985b). Inositol cyclic phosphates are produced by cleavage of phosphatidyl-phosphoinositols (polyphosphoinositides) with purified sheep seminal vesicle phospholipase C enzymes. *Proc. Natl. Acad. Sci. U.S.A.* **82,** 4013–4017.

Xuan, X-T., Su, Y-F., Chang, K. J., and Watkins, W. D. (1987). A pertussis/cholera toxin sensitive G-protein may mediate vasopressin-induced inositol phosphate formation in smooth muscle cell. *Biochem. Biophys. Res. Commun.* **146,**898–906.

Yatani, A., Codina, J., Brown, A. M., and Birnbaumer, L. (1987). Direct activation of mammalian atrial muscarinic potassium channels by GTP regulatory protein G_k. *Science* **235,** 207–211.

CHAPTER 20

Xenopus Oocyte as Model System to Study Receptor Coupling to Phospholipase C

Thomas M. Moriarty

Department of Psychiatry, Mount Sinai School of Medicine of the City University of New York, New York, New York 10029

Emmanuel M. Landau

Departments of Psychiatry and Pharmacology, Mount Sinai School of Medicine of the City University of New York, New York, New York 10029 and Department of Psychiatry, Veterans Administration Medical Center, Bronx, New York 10468

I. INTRODUCTION

The immature oocyte of the African clawed frog *Xenopus laevis* has emerged as a major tool for the elucidation of the molecular basis of cellular communication (Dascal, 1987; Lester, 1988; Levitan, 1988; Snutch, 1988). Gurdon's observations that the oocyte can be used as a

G Proteins   Copyright © 1990 by Academic Press, Inc. All rights of reproduction in any form reserved.

high-fidelity *in vivo* translation system (Gurdon *et al.,* 1971; Lane *et al.,* 1971) has led to the development of a variety of uses for the oocyte. Perhaps the most powerful is the use of the oocyte as a membrane source that is functionally coupled to cellular effectors. Structurally defined foreign proteins can be introduced into the oocyte, by translation of mRNA or by direct injection of pure protein, then analyzed for function in a simple and well-controlled environment.

The oocyte is particularly suited to the study of signal transduction. Endogenous receptors which couple to the two primary second messenger systems, adenylyl cyclase and phospholipase C, have been identified and characterized (Kusano *et al.,* 1977; Dascal and Landau, 1980; Lotan *et al.,* 1982; VanRhenterghem *et al.,* 1984; for review see Dascal, 1987). Foreign receptors that act on both of these systems can be transplanted into the oocyte by the expression of exogenous RNA (Gundersen *et al.,* 1984). Most receptor types have been functionally demonstrated in the oocyte including voltage-activated channels, ligand-gated ion channels, and the receptors which couple to guanine nucleotide-binding regulatory proteins (G proteins) (Dascal, 1987; Lester, 1988). Second messenger-mediated pathways can be studied using conventional biochemical techniques; however, the function of ligand-gated channels, such as $GABA_A$, or the receptor–G protein-regulated channels, such as the muscarinic-regulated K^+ channels, can only be studied using sensitive electrophysiological methods. In that the oocyte can be readily analyzed biochemically *in vitro* and can be studied as an isolated *in vivo* electrophysiological preparation, it is uniquely suited to the study of direct and second messenger-coupled receptor–effector pathways as well as the complex interactions of multiple receptor systems.

The pathways from receptor to activation of distal effectors, via cAMP or inositol 1,4,5-trisphosphate (IP_3), have been well characterized in the oocyte. This is especially true for the phosphatidylinositol system (Fig. 1). Activation of endogenous muscarinic receptors in the oocyte evokes a complex depolarization of the plasma membrane. A number of expressed exogenous receptors of the G protein-linked family have been demonstrated to activate this same depolarization event. Using both electrophysiology and biochemistry, it has been shown that the depolarization is due to an agonist-induced turnover of phosphatidylinositol 4,5-bisphosphate to IP_3. This evoked release of IP_3 is mediated by a G protein(s). The IP_3 formed causes a release of Ca^{2+} from intracellular stores. Ca^{2+}-sensitive Cl^- channels respond to this surge in intracellular Ca^{2+} by opening. It has been shown that the magnitude of the macroscopic Cl^- current (the sum of the individual channel conductances) is proportional to the amount of cytosolic IP_3 released. Therefore, in a single cell under voltage clamp, the Cl^-

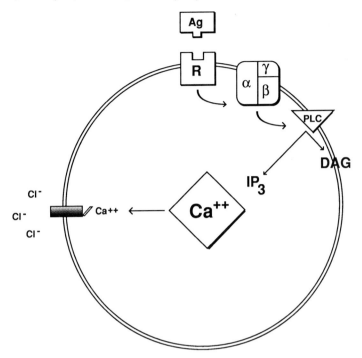

Fig. 1. The receptor-activated IP$_3$-mediated Cl⁻ conductance in *Xenopus* oocyte: general pathway. This schematic represents a single oocyte. The agonist (Ag)-bound receptor (R) (either the native muscarinic receptor or a transplanted receptor) will couple to phospholipase C (PLC) via a heterotrimeric G protein ($\alpha\beta\gamma$). Phospholipase C will break down phosphatidyl-inositol 4,5-bisphosphate (PIP$_2$) to inositol 1,4,5-trisphosphate (IP$_3$) and diacylglycerol (DAG). The IP$_3$ will bind to receptors on endoplasmic reticulum and evoke a release of Ca^{2+} into the cytosol. Ca^{2+}-sensitive Cl⁻ channels in the plasma membrane will open in response to the increased cytosolic Ca^{2+}. The coordinated opening and closing of Cl⁻ channels is responsible for the complex electrical response of the membrane.

current can be used as a sensitive and easily monitored measure of IP$_3$ production and phospholipase C activity. Given that the oocyte (1) has an intact IP$_3$-mediated receptor pathway, (2) has the ability to express various other receptors which utilize this pathway, and (3) supports a variety of powerful experimental interventions (such as direct injection of substances into the cell), the oocyte affords a prime opportunity to discover the role of G proteins in activation of phospholipase C.

In this chapter we will consider how the oocyte can be used to study the role of G proteins in coupling receptors to phospholipase C. We will review some of the basic properties of the oocyte essential to understanding electrophysiological experimentation in this preparation, and review the literature on the receptor-activated IP_3-mediated response in the oocyte and the role of G proteins in this response.

II. OOCYTE MORPHOLOGY, MEMBRANE PROPERTIES, AND ELECTROPHYSIOLOGY

The oocytes used in these studies are harvested surgically from mature females under tricaine anesthesia. Each of the ovarian lobes bears numerous oocytes arrested in first meiotic prophase (Maller and Krebs, 1980). These oocytes are in various stages of growth, termed stage one through six by the convention of Dumont (1972). Only the largest, stage V and VI, are used for experiments. These cells have an average diameter of 1.2–1.3 mm and a volume of approximately 1 μl.

Fully grown oocytes are surrounded by a number of cellular and connective tissue layers (Dumont and Brummett, 1978). Closest to the oocyte surface is the fibrous vitelline membrane. This is surrounded by a layer of follicle cells. The follicle cells and the oocyte are connected by numerous gap junctions and form a functional syncytium (Brown et al., 1979). The gap junctions allow for the intercellular transmission of small (<1000 Da) constituents and the direct electrical coupling of the oocyte and follicle cells. The follicular layer is surrounded by the theca, which contains blood vessels, nerve fibers, and smooth muscle cells. An epithelial cell layer, which is a continuation of the surface of the ovary, forms the outermost layer. The oocyte with the external cells intact is known as a follicular cell or a follicle. These layers, except the vitelline, can be removed by treatment with collagenase. Such collagenase-treated oocytes are known as denuded oocytes or simply collagenase-treated oocytes. Most electrophysiological studies are performed in either follicles or denuded oocytes. If multiple penetrations by microelectrodes and injector pipettes are anticipated, denuded oocytes are typically used. The vitelline layer can be removed with microforceps after incubating denuded oocytes in a hyperosmolar medium. This treatment is necessary for patch-clamp studies, but is not necessary for the more usual whole-cell voltage clamping.

Fully grown oocytes are remarkable for their bipolar coloration with one hemisphere nearly black and the other light yellow or beige. This pigment separation demarcates a morphological and functional polarization (for review see Dascal, 1987). The dark pole, the animal pole, is highest in melanin-containing granules, Ca^{2+}-dependent Cl^- channels and specialized plasma membrane–endoplasmic reticulum membrane junctions. It also demonstrates a higher sensitivity to some iontophoretically applied transmitter substances (Kusano et al., 1982; Oron et al., 1988a). The light colored hemisphere, the vegetal pole, has the highest concentrations of yolk proteins and RNA.

To do single-cell electrophysiological experiments, it is necessary to know the concentrations and activities of relevant ions inside and outside of the cell. The constitution of the extracellular fluid is always known because the various bathing media are made in the laboratory and applied to the cell in a manifold-controlled perfusion apparatus. Several examples of physiological bathing solutions are summarized in Table I. The activities of ions inside of the cell are known from the literature which is reviewed extensively by Dascal (1987). He summarizes the approximate values as: Ca^{2+}, $0.1-0.4\ \mu M$; Na^+, $1-6$ mM; K^+, $80-120$ mM; and Cl^-, $44-62$ mM.

The electrophysiological experiments described in this review utilize the two-electrode voltage clamp or the patch-clamp technique. Both methods are based on the same principle: the function of the membrane can be analyzed by uncoupling the mutually dependent current and voltage properties of the membrane. The voltage clamp monitors macroscopic currents which are the sum of the unit conductances through all open ion channels. The patch clamp isolates a patch of membrane and allows for the monitoring of the unit conductances of one, or just a few, ion channels.

The oocyte is an ideal cell for voltage-clamp experiments (Fig. 2). Its spherical shape simplifies the mathematical assumptions of the theoretical

Table I Representative Extracellular Bathing Media (mM)

Components		Source			
	ND96[a]	Modified Barth's solution[b]	Berridge[c]	Miledi[d]	Barish[e]
NaCl	96	88	115	115.6	81
KCl	2	1	2	2	2.5
$CaCl_2$	1.8	0.41	1.8	1.8	1
$MgCl_2$	1	—	1	—	1
$Ca(NO_3)_2$	—	0.33	—	—	—
$MgSO_4$	—	0.82	—	—	—
$NaHCO_3$	—	2.4	—	2.4	2.5
HEPES	5	—	—	—	5
Tris	—	7.5	7.5	—	—
pH	7.5	7.6	7.6	7.2	7.4

[a] Dascal *et al.* (1986).
[b] Kaneko *et al.* (1987).
[c] Berridge (1988).
[d] Kusano *et al.* (1982).
[e] Barish (1983).

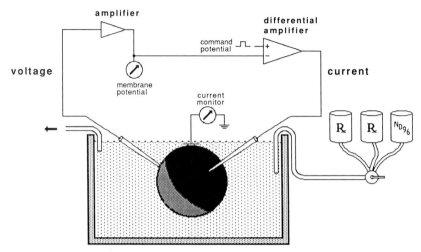

Fig. 2. Electrophysiology experimental apparatus. Most electrophysiology experiments described in this review use the two-electrode voltage clamp technique. Single oocytes are placed in a small chamber which is perfused with frog Ringer's solution (ND96). The content of the bathing medium is controlled using a stopcock and various reservoirs. Two glass microelectrodes are pulled to very fine points and filled with 3 M KCl (1–5 MΩ resistance). These electrodes are mounted on micromanipulators and inserted into the oocyte. One electrode (voltage) is connected to an amplifier and records the membrane potential. This potential is fed into one terminal of a differential feedback amplifier. The experimenter sets the membrane potential at which the experiment will be performed. This "command potential" is fed into the other pole of the differential amplifier. Any difference between the recorded membrane potential and the command potential will be pumped as current into the cell via the other (current) electrode to "clamp" the membrane at the desired potential. The magnitude of this current is measured at an ammeter (current monitor).

circuit. Space clamping is easily achieved. It is large enough for the two-electrode voltage clamp configuration. Its size also allows for experimental control of the intracellular environment by direct injection of substances into the cell. The extracellular environment, the bathing medium, is very easily manipulated. The powerful tool of patch clamping can also be used with oocytes. Both the intracellular and extracellular faces of the patch are under direct experimental control. The advantage of the whole cell voltage clamp is that one may study multiple-component pathways in an intact living cell. Direct injection of carefully chosen biologicals into oocytes allows one to dissect and understand such complex pathways. The power of patch clamping is in analyzing the function of single ion channels.

The literature on the properties of the oocyte membrane at rest are also reviewed by Dascal (1987). Resting potentials from -20 to -90 mV have been reported, however several sources of error contribute to this wide range. Each electrode penetration introduces a depolarization which partially recovers over several minutes. Therefore, any potential recorded by electrode will underestimate the actual resting potential. Manual or enzymatic defolliculation induces a hyperpolarization which will recover somewhat over several hours. Therefore, healthy defolliculated oocytes have resting membrane potential of approximately -45 to -65 mV. Resting potentials of healthy follicles are somewhat more positive. The input resistance, also underestimated by electrode penetration, is about $0.1-3$ MΩ. The total capacitance of the oocyte is 230 nF with a specific capacitance of $4-7$ μF/cm^2. The membrane selectivity of the oocyte at rest has also been investigated. At rest, the major monovalent ions are not at equilibrium. The approximate equilibrium potentials for these are: K$^+$, -100 mV; Na$^+$, $+80$ mV; and Cl$^-$, -25 mV.

III. RECEPTOR-ACTIVATED INOSITOL 1,4,5-TRISPHOSPHATE MEDIATED Cl$^-$ CONDUCTANCE IN *XENOPUS* OOCYTE: NATIVE MUSCARINIC RECEPTOR, TRANSPLANTED RECEPTORS, AND GENERAL PATHWAY

In 1977, Miledi and co-workers, in a study on the neurotransmitter sensitivity of undifferentiated progenitor cells, described an electrophysiological responsiveness of oocytes to acetylcholine (ACh) (Kusano *et al.*, 1977). They found that bath application of ACh to individual voltage-clamped oocyte follicles resulted in a complex depolarization of the membrane (Fig. 3A). In this brief seminal report the authors addressed several key issues concerning this response. They showed that the response was sensitive to ACh in a dose-dependent manner, was evoked by muscarinic and mixed cholinergic agents and was antagonized by atropine, but not curare, α-bungarotoxin, or tetrodotoxin. They concluded that the receptors involved are muscarinic in nature. Treatment of the oocyte follicle with collagenase, which removes the enveloping follicular layers, did not abolish the response to ACh in most cells, suggesting that the cholinergic receptors are located in the oocyte membrane. Morphological examination of ACh-sensitive oocytes after collagenase indeed showed that the follicle cells had all been removed. Voltage clamp analysis of the ionic basis of the depolarization revealed that ACh caused a fall in membrane resistance and an inward flux of current. Ion substitution experiments and reversal potential determinations suggested that the current flux was carried primarily by Cl$^-$ ions. Iontophoretic application of ACh showed a marked delay in activation of

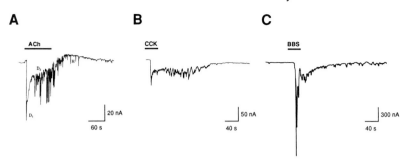

Fig. 3. The receptor-activated IP$_3$-mediated Cl$^-$ conductance: representative current traces from voltage-clamped oocytes. (A) The native muscarinic response in a follicle-enclosed oocyte. The cell was clamped at -60 mV. ACh was at 0.1 μM. Note the four components of the response: D$_1$, D$_2$, and H, and the rapid, superimposed fluctuations (F). (B) The response to cholecystokinin (0.1 μM) and (C) to bombesin (0.1 μM) in two different RNA-injected oocytes clamped at -70 mV. These oocytes had been defolliculated with collagenase, then injected with 50 nl of rat brain RNA (5 mg/ml) three days prior to assay. Note that the D$_1$, D$_2$, and F components are present, but the H component is not seen in these collagenase-treated oocytes. The bar above each response represents the duration of application of transmitter. CCK, cholecystokinin-8 (sulfated); BBS, bombesin. (A) Modified from Dascal and Landau (1982). (B) and (C) Modified from Moriarty *et al.* (1988).

the response, suggesting that either a number of receptors must bind ligand and interact in some fashion to initiate the response, or that the depolarization is mediated by an intracellular second messenger.

Oocytes injected with exogenous RNA will express sensitivity to various transmitter substances. The oocyte readily expresses receptors which activate a Cl$^-$ conductance similar to that seen for the native muscarinic receptor (Fig. 3). These receptors include 5-HT$_{1c}$, (Lübbert *et al.*, 1987; Julius *et al.*, 1988) muscarinic cholinergic (Kubo *et al.*, 1986), glutamate (Sugiyama *et al.*, 1987), and several peptidergic transmitters (Hirono *et al.*, 1987; Masu *et al.*, 1987; McIntosh and Catt, 1987; Oron *et al.*, 1987; Moriarty *et al.*, 1988a; Myerhoff *et al.*, 1988; Williams *et al.*, 1988). The accumulated evidence strongly favors a theory that such transplanted receptors couple to native Cl$^-$ channels via the same biochemical pathway as the native muscarinic receptor. The remainder of this review will treat the native muscarinic response and the response to transmitters seen in RNA-injected cells as two manifestations of the same phenomenon. Examples will be drawn from the literature on both native and expressed receptors such that a general mechanism can be outlined.

Further studies on the muscarinic receptor-evoked depolarization were aimed at understanding the underlying ionic and biochemical mediators of the response. The complex waveform was reduced to constituent components by Dascal and Landau (1980). They reported that not all components were seen in all oocytes; however, the response could be generally broken down into four components (Fig. 3A). Application of agonist will, after a few-second delay, evoke a fast transient inward current (D_1), followed by, or perhaps concurrent with, a slower inward conductance (D_2), which is succeeded by a slow outward hyperpolarizing current (H). All of these components may be superimposed with rapid fluctuational currents (F) of variable magnitude and duration. The slow outward H current is often overshadowed by the more substantial inward D_2 response, and therefore not readily apparent. There is a seasonal variation in the appearance of the various components. The D_1 may, in the southern hemisphere at any rate, disappear for the entire winter. However, with proper care of the frogs, including careful monitoring of light/dark cycle, water temperature and composition, and feed, these seasonal variations can be reduced (Dascal *et al.*, 1984). If the oocytes have been treated with collagenase, the response to ACh will most likely disappear, but can reappear after two or three days rest. The reasons for this are unclear, but it may be due to nonspecific proteolysis of existing muscarinic receptors during incubation with impure collagenase. The H response is always abolished by defolliculation suggesting that either the intracellular or membrane mediators of the response lie in the follicular cells. Iontophoretic application of ACh to various parts of the oocyte has revealed that these cells exhibit a marked polarity with either the animal pole (Kusano *et al.*, 1982) or vegetal pole (Oron *et al.*, 1988a) being considerably more sensitive to ACh than the opposite hemisphere.

In their original study, Kusano *et al.*, (1977) suggested that Cl^- was the ion primarily responsible for the depolarization. Dascal and Landau demonstrated that the reversal potential of the D_1 component was about -22 mV (Dascal and Landau, 1982), which was very close to the equilibrium potential of Cl^- in the oocyte (Barish, 1983). They also showed that this reversal potential depends on $[Cl^-]_{out}$ in a Nernstian fashion (Dascal *et al.*, 1984). Varying $[K^+]_{out}$ or $[Na^+]_{out}$ did not affect the reversal potential. Taken together, these data strongly suggest that the D_1 component is carried almost exclusively by Cl^- ions. The D_2 component was shown to have a somewhat more negative reversal potential and a dependence on $[K^+]_{out}$. (Dascal *et al.*, 1984). This is consistent with a simultaneously increased conductance to Cl^- and K^+ ions due to an overlapping of the D_2 and H components. Intracellular injection of (TEA) or removal of the surrounding follicular cells will eliminate the K^+-conductance contamination of the D_2 response. Therefore, in collagenase-treated cells, both the D_1 and D_2 components are strictly carried by Cl^-. The H component has been shown to be primarily an outward K^+ current (Dascal *et al.*, 1984). The fluctuational F component disappears if the cell is voltage-clamped at the Cl⁻ equilibrium potential, which indicates that these fluctuations are Cl⁻ currents. In summary, the response to ACh is a complex, time-restricted increase in conductance to Cl⁻ and K⁺ ions. If collagenase-treated cells are used, the entire waveform is due to the opening and closing of Cl⁻ channels (Fig. 4).

The receptors responsible for the ACh response are muscarinic choliner-

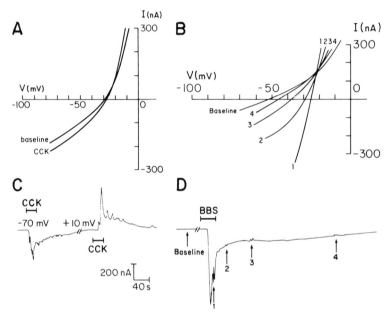

Fig. 4. Current–voltage characteristics (I–V) of the receptor-activated, IP$_3$-mediated Cl⁻ conductance. Examples from rat brain RNA-injected oocytes (50 nl of 5 mg/ml total RNA per oocyte). (A) I–V characteristics of the response to CCK shown in a single cell by the ramp method. I–V curves were recorded automatically on an X–Y plotter. The baseline curve was made one minute before application of the peptide. The CCK curve was made at the peak of the fast response to the peptide. The reversal potential is the point of intersection of the two curves. (B) A voltage ramp study of the response to BBS. Voltage ramps similar to the ones shown in part A of this figure were obtained at baseline and at various times after application of BBS [see part (D)]. Note that all the voltage ramps cross near the same voltage (−23 mV). (C) Application of CCK in a different cell voltage clamped at −70 mV and then at +10 mV after a 15-minute washout period. Note that both the early peak and the later part of the response are inverted at +10 mV indicating that both have a similar reversal potential. (D) The time course of the current induced by application of BBS in the same cell as in part (B). The times when the ramps were applied are shown by arrows. The actual ramps were removed from the record for clarity. CCK, cholecystokinin; BBS, bombesin. Modified from Moriarty *et al.* (1988).

gic (Kusano *et al.,* 1977; Dascal and Landau, 1982). Muscarinic and mixed muscarinic agents will evoke the response; atropine will antagonize the response. VanWezenbeek *et al.* (1988) show that the endogenous muscarinic response is antagonized by muscarinic antagonists in the following rank order of potency: 4-DAMP > pirenzipine > AF-DX 116, which identifies the receptor as the M_3 subtype by the convention of Doods *et al.* (1987).

Dascal and Landau (1982) analyzed the dose-response characteristics of the D_1 component and found that either a cooperativity of binding or a cooperation of binding with a downstream biochemical event was necessary. They demonstrated two categories of binding site, which they called the low sensitivity site, with an approximate K_d of about 0.2 μM, and the high sensitivity with approximate K_d of 0.03 μM. Individual cells expressed one of the two receptor types with the low sensitivity type being most common. The existence of two apparent binding affinities is similar to that seen in the "superhigh" and "high" affinity muscarinic cholinergic binding sites in brain (Birdsall *et al.,* 1980). Later studies by Dascal and Cohen (1987) on the muscarinic response found that the apparent K_d is also approximately 0.4 μM for the D_2 and H components. They concluded that all major components of the complex response are a result of ACh binding to the same receptor, but are mediated by different postreceptor effectors.

In 1977, Miledi and co-workers suggested, on the basis of time course after iontophoresis, that the muscarinic response might be mediated by a second messenger (Kusano *et al.,* 1977). Since then, the biochemical steps between binding of agonist and the activation of current have been examined in detail. The following pathway is presently understood for both the native muscarinic receptor and certain expressed receptors (Fig. 1). After binding agonist, the receptor couples to a G protein to activate phospholipase C. The lipase breaks down PIP_2 to inositol 1,4,5-trisphosphate (IP_3) and diacylglycerol. The liberated IP_3 binds to receptors on endoplasmic reticulum to cause a release of Ca^{2+} into the cytosol. The coordinated release of Ca^{2+} opens populations of Ca^{2+}-sensitive Cl^- channels in the plasma membrane and gives rise to the complex macroscopic current seen in voltage clamp. A possible role for calmodulin has also been suggested. Each step in this pathway will now be considered in detail. The role of G proteins in this response will be discussed in a separate section.

The function of phospholipase C in the muscarinic response was first addressed in a paper by Oron *et al.* (1985) in which a direct link was demonstrated between phosphoinositide metabolism and a transmitter-induced physiological response. Oocytes prelabeled with [³H]inositol were used to demonstrate cholinergic stimulation of PIP_2 breakdown to IP_3, inositol 1,4-bisphosphate, and inositol 1-phosphate within 30–120 sec after challenge with agonist. Similar hydrolysis of PIP_2 was demonstrated in RNA-injected cells in response to ACh (Nomura et al., 1987), 5-HT (Nomura et al., 1987), thyrotropin-releasing hormone (McIntosh and Catt, 1987; Oron et al., 1988b), and angiotensin II (McIntosh and Catt, 1987).

Direct injection of IP_3 into the oocyte evokes a depolarization with the characteristic waveform and current–voltage properties of the native muscarinic response (Oron et al., 1985; Nadler et al., 1986; Parker and Miledi, 1986; Gillo et al., 1987) (Fig. 5). Studies on injection of IP_3 show that the site of injection can alter the response (Gillo et al., 1987). Shallow injections of IP_3 give a response with prominent D_1 and D_2 components. Deeper injections give a response with a diminished D_1 and a more marked D_2. A study by Gillo et al. (1987), with careful control of the location of the injector pipette, demonstrated that the response to injected IP_3 is dose-dependent. It has also been shown that the animal hemisphere is more sensitive to IP_3 injections (Berridge, 1988).

IP_3 activates the depolarization by mobilizing intracellular stores of Ca^{2+}. Application of ACh to native oocytes, or injection of IP_3, will evoke a release of $^{45}Ca^{2+}$ from preloaded cells with a time course similar to the membrane electrical response (Nadler et al., 1986; Dascal et al., 1985a). This is also true of transplanted receptors (Oron et al., 1988b, Williams et al., 1988). A rise in intracellular Ca^{2+} has been demonstrated, after injection of IP_3 with Ca^{2+}-sensitive electrodes (Busa et al., 1985) and after application of agonist, by fura-2 fluorescence (Takahashi et al., 1987). Depletion of stored Ca^{2+} with ionophore A23187 prevents transmitter and IP_3-evoked membrane responses (Gillo et al., 1987). The responses to ACh or IP_3 are not abolished by depleting the extracellular fluid of Ca^{2+}, which indicates that the response is not exclusively due to an influx of Ca^{2+} (Dascal et al., 1985a; Parker and Miledi, 1986; Gillo et al., 1987). Some reduction in amplitude of both the D_1 and D_2 components has been observed in Ca^{2+}-free media, with the D_2 component being more sensitive or even abolished (Dascal et al., 1984; Dascal et al., 1985a; Snyder et al., 1988). Intracellular injection of the Ca^{2+} chelator EGTA will block the response to agonist (Dascal et al., 1985a) or IP_3. Injection of Ca^{2+} directly into the cell evokes either a single fast D_1-like response or a biphasic (D_1 and D_2) response carried by Cl^- ions (Miledi and Parker, 1984; Dascal et al., 1985a; Gillo et al., 1987). The magnitude of the Ca^{2+}-evoked current is dose-dependent (Dascal et al., 1985a). A role for calmodulin in mediating the Ca^{2+}-evoked Cl^- current has been suggested by the observed inhibition of both the muscarinic response and the response to injected Ca^{2+} by the calmodulin inhibitor trifluoperazine. (Dascal et al., 1985a; Ito et al., 1988).

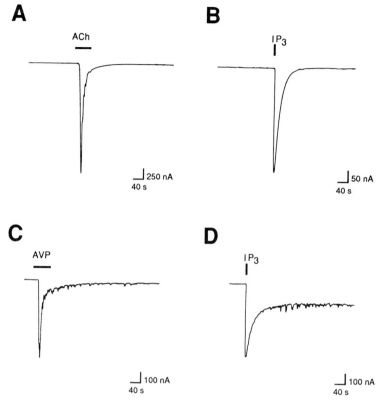

Fig. 5. IP$_3$ injections into native oocytes and RNA-injected oocytes. (A) The native muscarinic response in a collagenase-treated oocyte. ACh is at 1.0 μM. (B) The response of a collagenase-treated oocyte to intracellular injection of IP$_3$ (0.5 pmol). The oocytes in part (A) and (B) are from the same donor frog. Note the lack of the F components (compare to Fig. 3). This is a manifestation of the variability seen between animals. (C) The response to 1.0 μM arginine vasopressin (AVP) in a liver RNA (50 nl of 2.3 mg/ml total RNA solution)-injected cell. (D) The response of an oocyte, from the same group of liver RNA-injected cells, to intracellular injection of IP$_3$ (1.2 pmol). The bars above the responses indicate the duration of transmitter application. The vertical bars indicate the moment of injection of IP$_3$. Modified from Moriarty *et al.* (1989).

Dascal *et al.* (1985a) propose that there are different pools of Ca^{2+} which subserve the two depolarizing components (D$_1$ and D$_2$) of the response. The F component may be due to a Ca^{2+}-dependent Ca^{2+}-release phenomenon (Gillo *et al.*, 1987; Berridge, 1988) similar to that described in other cell types (Fabiato and Fabiato, 1975). Low concentrations of agonist will induce primarily the oscillatory response as will low doses of injected IP$_3$.

Berridge proposes that an IP_3-sensitive pool of Ca^{2+} releases Ca^{2+} in response to IP_3. This increase in cytosolic Ca^{2+} leads to increased uptake of Ca^{2+} by an IP_3-insensitive pool. The overloading of this latter pool is the trigger which causes it to release Ca^{2+} back into the cytosol. Ca^{2+} will again be sequestered and again released until some equilibrium prevails. The complex electrical response of the membrane is produced by Ca^{2+}-sensitive Cl^- channels, passively responding to the changing tides of intracellular Ca^{2+} concentration.

The Ca^{2+}-sensitive Cl^- channels which carry the depolarizing response were first seen separately by Miledi (1982) and Barish (1983). They observed a rapidly inactivating transient outward ("T_{out}") Cl^- current in response to a voltage step from rest to more positive than -20 mV. This current is dependent on the influx of extracellular Ca^{2+} and is sensitive to $[Ca^{2+}]_{out}$. It can be blocked by replacing external Ca^{2+} with either Ba^{2+}, Sr^{2+}, Mg^{2+}, or Mn^{2+}, or by injecting cells with EGTA (Miledi, 1982; Barish, 1983; Miledi and Parker, 1984). The most likely explanation of this transient outward current is a voltage-sensitive Ca^{2+} channel in the plasma membrane which allows the entry of a sufficient quantity of Ca^{2+} into the cytosol to activate the Ca^{2+}-dependent Cl^- channels.

From these observations, the Ca^{2+}-injection studies, and the proposed role of Ca^{2+} in receptor activation of Cl^- current, it was assumed that there existed a Ca^{2+}-sensitive Cl^- channel similar to those seen in other systems (for example Marty et al., 1984 or Owen et al., 1984). The existence of this channel in oocytes was demonstrated directly by Takahashi et al. (1987) with the patch-clamp technique. Using brain-RNA-injected oocytes which expressed serotonin receptors, they looked at serotonin-receptor activation of the IP_3-mediated Cl^- current. In patch clamp, they showed that application of agonist to the extracellular surface of the oocyte activated unitary conductances which were concurrent with the macroscopic current seen in voltage clamp. Analysis of these unitary conductances showed them to have a slope conductance of 3 pS, a reversal potential of -29 mV and a lifetime of 100 ms. Various ion substitution experiments in the patch suggested that these are Cl^--selective channels. In addition, Takahashi et al. (1987) confirmed that the activation of the Cl^- channels was by way of intracellular messengers by demonstrating that iontophoretic application of serotonin to the outside of the oocyte in the cell-attached patch configuration could activate the single channel currents. In this configuration, the agonist could not access the extracellular face of the membrane patch. The role of Ca^{2+} in activating these channels was confirmed in experiments using inside-out patches, in which the inner surface of the plasma membrane was exposed to various concentrations of Ca^{2+}. 10 μM Ca^{2+} activated

3-pS unitary conductances of approximately 100 ms duration which are essentially identical to that of the receptor-activated channels. Reduction of Ca^{2+} to less than 10 nM caused a cessation of the Cl^--channel activity.

Receptor activation of phospholipase C also generates the second messenger diacylglycerol (DAG). There are several examples of DAG effects in the oocyte. Dascal *et al.* (1985b) showed that activation of the native muscarinic response will inhibit cAMP-dependent K^+ currents in oocyte follicles. This inhibition is mimicked by phorbol esters; however, intracellular injections of IP_3 or Ca^{2+} are without effect. They conclude that this inhibition is due to protein kinase C activation.

Another suspected function of DAG in the oocyte involves homologous and heterologous desensitization. Examination of two or more expressed receptors in RNA-injected oocytes has shown that the receptors which couple to the IP_3-mediated Cl^- conductance exhibit marked self- and cross-desensitization (Dascal *et al.*, 1986; Hirono *et al.*, 1987; Nomura *et al.*, 1987). Kato *et al.* (1988) demonstrate that this desensitization is a result of phospholipase C liberating DAG, which activates protein kinase C. They showed that an 8-min pretreatment with nanomolar concentrations of phorbol esters can inhibit the native muscarinic response as well as the expressed ACh and 5-HT responses in cells injected with rat brain RNA. Phorbol ester treatment of oocytes *in vivo* results in a marked increase in phosphorylation of 33 and 45 kDa proteins in native as well as RNA-injected oocytes. Incubation of RNA-injected oocytes with 5-HT, ACh or phorbol esters results in an increase in phosphoproteins as measured in washed membranes *in vitro*. This increase is greatest at ten seconds after addition of transmitter or phorbol ester and is diminished at 5 min after treatment. This pattern of increase in membrane phosphorylation parallels the time course of receptor-activated depolarization. Treatment of oocytes with phorbol esters, however, does not inhibit direct activation of the Cl^- channels by Ca^{2+} or GTPγS (see below). Taken together, these studies suggest that ligand binding to receptors leads to activation of phospholipase C. The lipase liberates IP_3, which initiates the short-term depolarization events, and DAG, which activates protein kinase C and enables a negative feedback inhibition via phosphorylation.

A study by Sigel and Baur (1988) demonstrated multiple effects of DAG on expressed receptors in brain RNA-injected oocytes. Activation of phospholipase C by quisqualate (see Sugiyama *et al.*, 1987) resulted in differential modulation of voltage-gated Na^+ and Ca^{2+} channels and $GABA_A$ receptors. This modulation was mimicked by phorbol esters and 1,2-oleoylacetylglycerol, and was prevented by the protein kinase C inhibitor tamoxifen.

IV. *XENOPUS* OOCYTE AND G PROTEINS

To explore the possible role of G proteins in the receptor-activated IP_3-mediated Cl^- current in *Xenopus* oocyte, investigators have resorted to the "classical" tools of signal transduction research: bacterial toxins and guanine-nucleotide analogs. Early studies on the function of adenylyl cyclase in oocytes used bacterial toxins to identify endogenous G proteins. Sadler *et al.* (1984) showed that pertussis toxin (PTX) ADP-ribosylated a 41-kDa protein in oocyte membranes prepared from manually defolliculated oocytes. They also showed that cholera toxin ADP-ribosylated a number of proteins in the G_s size range. Olate *et al.* (1984) demonstrated PTX-catalyzed ADP-ribosylation of a single 40-kDa protein in membranes from collagenase-treated oocytes which comigrated with purified N_i from human erythrocytes on sodium dodecyl sulfate-polyacrylamide gel electrophoresis (SDS-PAGE). They also showed that cholera toxin labeled a single 42-kDa protein. Both groups demonstrated the likely existence of another G protein, responsible for the guanine nucleotide-sensitive inhibition of adenylyl cyclase by progesterone, which was not PTX-sensitive. Subsequent studies have confirmed the existence of 40-kDa (Dascal *et al.*, 1986), 41-kDa (Kaneko *et al.*, 1987, Moriarty *et al.*, 1988b) and 39-kDa (Kaneko *et al.*, 1987) PTX substrates in native oocytes, and have demonstrated the ability of oocytes to express G proteins after injection of foreign mRNA (Kaneko *et al.*, 1987). It has also been shown, using the specific antiserum U-49, that the β subunit of G proteins is present in native oocyte membranes (Moriarty *et al.*, 1988b) (Fig. 6). Taken together, these studies suggest that the native oocyte contains signal transduction G proteins of G_s, G_o, and G_i sizes.

The potential role of G proteins in coupling receptors to phospholipase C in oocyte is further supported by functional studies utilizing PTX. The native muscarinic response is inhibited approximately 85% by pre-treating oocytes with PTX (2 μg/ml for 26 hr) (Moriarty *et al.*, 1988b). Receptors expressed from exogenous RNA are also uncoupled from phospholipase C by PTX. Dascal *et al.* (1986) showed that the expressed rat brain 5-HT- or ACh receptor-activation of the Cl^- current is inhibited approximately 50% by a 4-hr treatment of cells with 0.5 μg/ml PTX. Others have reported inhibitory effects of PTX on various expressed receptor responses in the oocyte (Table II). The fact that the response to intracellularly injected IP_3 is not inhibited by PTX (Hirono *et al.*, 1987) supports the belief that the PTX effect is localized between the receptor and the formation of IP_3. Cholera toxin apparently has no effect on activation of the Cl^- current (Table II). Taken together, these data strongly suggest that a PTX substrate is coupling some receptors to phospholipase C in the oocyte.

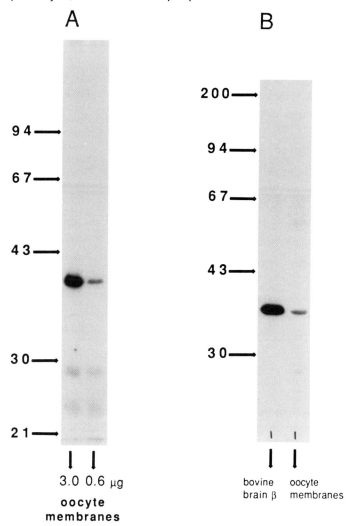

Fig. 6. Demonstration of pertussis toxin substrate and β subunit of G proteins in *Xenopus* oocyte membranes. (A) Pertussis toxin-catalyzed ADP-ribosylation of oocyte membranes. Indicated concentrations of oocyte membranes were incubated with activated pertussis toxin and [^{32}P]NAD$^+$. (B) Immunoblotting of oocyte membranes with the G protein β-subunit-specific antiserum U-49. Modified from Moriarty *et al.* (1988b).

The effects of guanine nucleotides on the receptor-activated Cl$^-$ current have been investigated using the native muscarinic receptor and receptors for ACh and 5-HT from brain RNA expressed in the oocyte (Dascal *et al.*, 1986, Nomura *et al.*, 1987). It has been demonstrated that the Cl$^-$ current

**Table II   Effects of Bacterial Toxins on Receptor-Activated IP$_3$-Mediated Cl$^-$
Conductance in RNA-Injected Oocytes**

Transmitter	Inhibition	Pertussis toxin concentration	Reference
ACh	50	0.5 μg/ml for 4 hr	Dascal et al. (1986)
5-HT	50	0.05 μg/ml for 4 hr	Dascal et al. (1986)
5-HT	50	0.5 μg/ml for 4 hr	Dascal et al. (1986)
5-HT	50	2 μg/ml for 4 hr	Dascal et al. (1986)
ACh	60	4 μg/ml for 48 hr	Sugiyama et al. (1985)
ACh	40	2 μg/ml for 18 hr	Nomura et al. (1987)
5-HT	60	2 μg/ml for 18 hr	Nomura et al. (1987)
glutamate	75	2 μg/ml for 20 hr	Sugiyama et al. (1987)
ACh	60	2 μg/ml for 20 hr	Hirono et al. (1987)
neurotensin	85	2 μg/ml for 20 hr	Hirono et al. (1987)
AVP	85	2 μg/ml for 24 hr	Moriarty et al. (1989)
CCK	0	2 μg/ml for 24 hr	Moriatry et al. (1989)
Bombesin	90	2 μg/ml for 24 hr	Moriatry et al. (1989)
		Cholera toxin concentration	
5-HT	0	0.2 nM for 8 hr	Dascal et al. (1986)
5-HT	0	10 nM for 8 hr	Dascal et al. (1986)
ACh	0	2 μg/ml for 18 hr	Nomura et al. (1987)
5-HT	0	2 μg/ml for 18 hr	Nomura et al. (1987)

can be elicited directly by injecting GTPγS into the cell in the absence of agonist (Dascal *et al.,* 1986; Kaneko *et al.,* 1987). The GTPγS-evoked current is much more persistent, sometimes lasting for tens of minutes, than that seen with physiological activation, which lasts less than five minutes. This persistence may be due to the asynchronous activation of other G proteins over time as the GTPγS diffuses throughout the cell, as well as to the inability of the cell to remove the GTPγS from the activated G protein. Cells injected with GTPγS are unresponsive to subsequent exposure to agonist suggesting that the necessary G proteins are disabled by binding GTPγS. Intracellular injection of GDPβS does not directly evoke a response; however, injection of GDPβS, as well as GDP, does reduce the ability of agonist to evoke the Cl$^-$ current. Cells injected with GTP display a reduced response to low concentrations of agonist. This may be due to a reduced affinity of receptor for agonist, which is known to occur in receptor systems coupled to G proteins (Gilman, 1987).

The mechanism of G protein-coupling of receptors to phospholipase C has been studied using the oocyte and purified subunits of G proteins (Moriarty *et al.,* 1988b). It was shown that intracellular injection of excess βγ subunits can inhibit the native muscarinic response. This inhibition is specific for the $\beta\gamma$ complex, such that boiled $\beta\gamma$ subunits, bovine serum albumin or resolved α subunits do not inhibit the response. The inhibitory effect is dose-dependent with half maximal inhibition at approximately 10 nM. The site of action of the $\beta\gamma$ subunits was determined to be prior to phospholipase C by showing that direct injection of IP$_3$ could evoke the response in the presence of saturating $\beta\gamma$ subunits. These experiments demonstrate that a heterotrimeric G protein is involved in activation of phospholipase C. The mechanism is similar to that of the hormone-activated adenylyl cyclase and the light-sensitive cGMP phosphodiesterase (Gilman, 1987).

V. SUMMARY

In this chapter we have presented some basic physiology of the *Xenopus* oocyte, with emphasis on the receptor-activated IP$_3$-mediated Cl$^-$-current, and have suggested that the oocyte may be a useful tool for investigating the role of G proteins in signal transduction through phospholipase C. It should be possible to take advantage of the intact receptor/phospholipase C/effector pathway of the oocyte, and the availability of purified and recombinant G proteins, to identify the G protein(s) which couples recep-tors to phospholipase C. The ability of the oocyte to express foreign mRNAs, heterologous and clonal, should facilitate the structure–function analysis of the proteins of signal transduction.

REFERENCES

Barish, M. E. (1983). A transient calcium-dependent chloride current in the immature *Xenopus* oocyte. *J. Physiol.* **342,** 309–325.

Berridge, M. J. (1988). Inositol trisphosphate-induced membrane potential oscillations in *Xenopus* oocytes. *J. Physiol.* **403,** 589–599.

Birdsall, N. J. M., Hulme, E. C., and Burgen, A. (1980). The character of the muscarinic receptors in different regions of the rat brain. *Proc. R. Soc. London B* **207,** 1–12.

Brown, C. L., Wiley, H. S., and Dumont, J. N. (1979). Oocyte-follicle cell gap junctions in *Xenopus laevis* and the effects of gonadotropin on their permeability. *Science* **203,** 182–183.

Busa, W. B., Ferguson, J. E., Joseph, S. K., Williamson, J. R., and Nuccitelli, R. (1985). Activation of frog *(Xenopus laevis)* eggs by inositol triphosphate. I. Characterization of Ca^{++} release from intracellular stores. *J. Cell. Biol.* **101,** 677–682.

Dascal, N. (1987). The use of *Xenopus* oocytes for the study of ion channels. *CRC Crit. Rev. Biochem.* **22,** 317–387.

Dascal, N., and Cohen, S. (1987). Further characterization of the slow muscarinic responses in *Xenopus* oocytes. *Pflügers Arch.* **409,** 512–520.

Dascal, N., and Landau, E. M. (1980). Types of muscarinic response in *Xenopus* oocytes. *Life Sci.* **27,** 1423–1428.

Dascal, N., and Landau, E. M. (1982). Cyclic GMP mimics the muscarinic response in *Xenopus* oocytes: Identity of ionic mechanisms. *Proc. Natl. Acad. Sci. U.S.A.* **79,** 3052–3056.

Dascal, N., Landau, E. M., and Lass, Y. (1984). *Xenopus* oocyte resting potential, muscarinic responses and the role of calcium and guanosine 3′,5′-cyclic monophosphate. *J. Physiol.* **352,** 551–574.

Dascal, N., Gillo, B., and Lass, Y. (1985a). Role of calcium mobilization in mediation of acetylcholine-evoked chloride currents in *Xenopus laevis* oocytes. *J. Physiol.* **366,** 299–313.

Dascal, N., Lotan, I., Gillo, B., Lester, H. A., and Lass, Y. (1985b). Acetylcholine and phorbol esters inhibit potassium currents evoked by adenosine and cAMP in *Xenopus* oocytes. *Proc. Natl. Acad. Sci. U.S.A.* **82,** 6001–6005.

Dascal, N., Ifune, C., Hopkins, R., Snutch, T. P., Lübbert, H., Davidson, N., Simon, M. I., and Lester, H. A. (1986). Involvement of a GTP-binding protein in mediation of serotonin and acetylcholine responses in *Xenopus* oocytes injected with rat brain messenger RNA. *Mol. Br. Res.* **1,** 201–209.

Doods, H. N., Mathy, M. J., Davidesko, D., Van Charldorp, K. J., De Jonge, A., and Van Zwieten, P. A. (1987). Selectivity of muscarinic antagonists in radioligand and *in vivo* experiments for the putative M_1, M_2 and M_3 receptors. *J. Pharmacol. Exp. Ther.* **242,** 257–262.

Dumont, J. N. (1972). Oogenesis in *Xenopus laevis* (Daudin). 1. Stages of oocyte development in laboratory maintained animals. *J. Morphol.* **136,** 153–180.

Dumont, J. N., and Brummett, A. R. (1978). Oogenesis in *Xenopus laevis* (Daudin). 5. Relationships between developing oocytes and their investing follicular tissues. *J. Morphol.* **155,** 73–98.

Fabiato, A., and Fabiato, F. (1975). Contractions induced by a calcium triggered release of calcium from the sarcoplasmic reticulum of single skinned cardiac cells. *J. Physiol.* **249,** 469–495.

Gillo, B., Lass, Y., Nadler, E., and Oron, Y. (1987). The involvement of inositol 1,4,5-trisphosphate and calcium in the two-component response to acetylcholine in *Xenopus* oocytes. *J. Physiol.* **392,** 349–361.

Gilman, A. G. (1987). G proteins: Transducers of receptor-generated signals. *Annu. Rev. Biochem.* **56,** 615–649.

Gundersen, C. B., Miledi, R., and Parker, I. (1984). Messenger RNA from human brain induces drug- and voltage-operated channels in *Xenopus* oocytes. *Nature* **308,** 421–424.

Gurdon, J. B., Lane, C. D., Woodland, H. R., and Marbaix, G. (1971). Use of frog eggs and oocytes for the study of messenger RNA and its translation in living cells. *Nature* **233,** 177–182.

Hirono, C., Ito, I., and Sugiyama, H. (1987). Neurotensin and acetylcholine evoke common responses in frog oocytes injected with rat brain messenger ribonucleic acid. *J. Physiol.* **382**, 523–535.

Ito, I., Hirono, C., Yamagishi, S., Nomura, Y., Kaneko, S., and Sugiyama, H. (1988). Roles of protein kinases in neurotransmitter responses in *Xenopus* oocytes injected with rat brain mRNA. *J. Cell. Physiol.* **134**, 155–160.

Julius, D., MacDermott, A. B., Axel, R., and Jessell, T. M. (1988). Molecular characterization of a functional cDNA encoding the serotonin 1c receptor. *Science* **241**, 558–564.

Kaneko, S., Kato, K., Yamagishi, S., Sugiyama, H., and Nomura, Y. (1987). GTP-binding proteins G_i and G_o transplanted onto *Xenopus* oocyte by rat brain messenger RNA. *Mol. Br. Res.* **3**, 11–19.

Kato, K., Kaneko, S., and Nomura, Y. (1988). Phorbol ester inhibition of current responses and simultaneous protein phosphorylation in *Xenopus* oocyte injected with rat brain mRNA. *J. Neurochem.* **50**, 766–773.

Kubo, T., Fukuda, K., Mikami, A., Maeda, A., Takahashi, H., Mishina, M., Haga, T., Haga, K., Ichiyama, A., Kangawa, K., Kojima, M., Matsuo, H., Hirose, T., and Numa, S. (1986). Cloning, sequencing and expression of complementary DNA encoding the muscarinic acetylcholine receptor. *Nature* **323**, 411–416.

Kusano, K., Miledi, R., and Stinnakre, J. (1977). Acetylcholine receptors in the oocyte membrane. *Nature* **270**, 739–741.

Kusano, K., Miledi, R., and Stinnakre, J. (1982). Cholinergic and catecholaminergic receptors in the Xenopus oocyte membrane. *J. Physiol.* **328**, 143–170.

Lane, C. D., Marbaix, G., and Gurdon, J. B. (1971). Rabbit haemoglobin synthesis in frog cells: the translation of reticulocyte 9 s RNA in frog oocytes. *J. Mol. Biol.* **61**, 73–91.

Lester, H. (1988). Heterologous expression of excitability proteins: Route to more specific drugs? *Science* **241**, 1057–1063.

Levitan, E. S. (1988). Cloning of serotonin and substance K receptors by functional expression of frog oocytes. *TINS* **11**, 41–43.

Lotan, I., Dascal, N., Cohen, S., and Lass, Y. (1982). Adenosine-induced slow ionic currents in the *Xenopus* oocyte. *Nature* **298**, 572–574.

Lübbert, H., Hoffman, B. J., Snutch, T. P., van Dyke, T., Levine, A. J., Hartig, P. R., Lester, H. A., and Davidson, N. (1987). cDNA cloning of serotonin $5-HT_{1c}$ receptor by electrophysiological assays of mRNA-injected *Xenopus* oocytes. *Proc. Natl. Acad. Sci. U.S.A.* **84**, 4332–4336.

Maller, J. L., and Krebs, E. G. (1980). Regulation of oocyte maturation. *Curr. Top. Cell. Regul.* **16**, 271–311.

Marty, A., Tan, Y. P., and Trautmann (1984). Three types of calcium-dependent channel in rat lacrimal glands. *J. Physiol.* **357**, 293–325.

Masu, Y., Nakayama, K., Tamake, H., Harada, Y., Kuno, M., and Nakanishi, S. (1987). cDNA cloning of bovine substance-K receptor through oocyte expression system. *Nature* **329**, 836–838.

McIntosh, R. P., and Catt, K. (1987). Coupling of inositol phospholipid hydrolysis to peptide hormone receptors expressed from adrenal and pituitary mRNA in *Xenopus laevis* oocytes. *Proc. Natl. Acad. Sci. U.S.A.* **84,** 9045–9048.

Miledi, R. (1982). A calcium-dependent transient outward current in *Xenopus laevis* oocytes. *Proc. Roy. Soc. London B.* **215,** 491–497.

Miledi, R., and Parker, I. (1984). Chloride current induced by injection of calcium into *Xenopus* oocytes. *J. Physiol.* **357,** 173–183.

Moriarty, T. M., Gillo, B., Sealfon, S., Blitzer, B., Roberts, J. L., and Landau, E. M. (1988a). Functional expression of CCK and BBS receptors in *Xenopus* oocytes. *Mol. Br. Res.* **4,** 75–79.

Moriarty, T. M., Gillo, B., Carty, D. J., Premont, R. T., Landau, E. M., and Iyengar, R. (1988b). $\beta\gamma$-subunits of GTP-binding proteins inhibit muscarinic receptor stimulation of phospholipase C. *Proc. Natl. Acad. Sci. U.S.A.* **85,** 8865–8869.

Moriarty, T. M., Sealfon, S. C., Carty, D. J., Roberts, J. L., Iyengar, R., and Landau, E. M. (1989). Coupling of exogenous receptors to phospholipase C in *Xenopus* oocytes through pertussis toxin sensitive and insensitive pathways: Crosstalk through heterotrimeric G Proteins. *J. Biol. Chem.* **264,** 13524–13530.

Myerhof, W., Morley, S., Schwartz, J., and Richter, D. (1988). Receptors for neuropeptides are induced by exogenous poly(A)$^+$ RNA in oocytes from *Xenopus laevis*. *Proc Natl. Acad. Sci. U.S.A.* **85,** 714–717.

Nadler, E., Gillo, B., Lass, Y., and Oron, Y. (1986). Acetylcholine- and inositol 1,4,5-trisphosphate-induced calcium mobilization in *Xenopus laevis* oocytes. *FEBS Letts.* **199,** 208–212.

Nomura, Y., Kaneko, S., Kato, K., Yamagishi, S., and Sugiyama, H. (1987). Inositol phosphate formation and chloride current responses induced by acetylcholine and serotonin through GTP-binding proteins in *Xenopus* oocyte after injection of rat brain messenger RNA. *Mol. Br. Res.* **2,** 113–123.

Olate, J., Allende, C., Allende, J. E., Sekuar, R. D., and Birnbaumer, L. (1984). Oocyte adenylyl cyclase contains N_i, yet the guanine nucleotide-dependent inhibition by progesterone is not sensitive to pertussis toxin. *FEBS Letts.* **175,** 25–30.

Oron, Y., Dascal, N., Nadler, E., and Lupa, M. (1985). Inositol 1,4,5-trisphosphate mimics muscarinic response in *Xenopus* oocytes. *Nature* **313,** 141–143.

Oron, Y., Straub, R. E., Traktman, P., and Gershengorn, M. (1987). Decreased TRH receptor mRNA activity precedes homologous downregulation: Assay in oocytes. *Science* **238,** 1406–1408.

Oron, Y., Gillo, B., and Gershengorn, M. (1988a). Differences in receptor-evoked membrane electrical responses in native and mRNA-injected *Xenopus* oocytes. *Proc. Natl. Acad. Sci. U.S.A.* **85,** 3820–3824.

Oron, Y., Gillo, B., Straub, R. E., and Gershengorn, M. (1988b). Mechanism of membrane electrical response to thyrotropin-releasing hormone in *Xenopus* oocytes injected with GH_3 pituitary cell messenger ribonucleic acid. *Mol. Endocrinol.* **1,** 918–925.

Owen, D. G., Segal, M., and Barker, J. L. (1984). A Ca-dependent Cl- conductance in cultured mouse spinal neurones. *Nature* **311,** 567–570.

Parker, I., and Miledi, R. (1986). Changes in intracellular calcium and in membrane currents evoked by injection of inositol trisphosphate into *Xenopus* oocytes. *Proc. R. Soc. London B* **228**, 307–315.

Sadler, S. E., Maller, J. L., and Cooper, D. M. F. (1984). Progesterone inhibition of *Xenopus* oocyte adenylate cyclase is not mediated via the *Bordetella pertussis* toxin substrate. *Mol. Pharm.* **26**, 526–531.

Sigel, E., and Baur, R. (1988). Activation of protein kinase C differentially modulates neuronal Na^+, Ca^{2+}, and γ-aminobutyrate type A channels. *Proc. Natl. Acad. Sci. U.S.A.* **85**, 6192–6196.

Snutch, T. P. (1988). The use of Xenopus oocytes to probe synaptic communication. *TIPS* **11**, 250–256.

Snyder, P. M., Krause, K-H., and Welsh, M. J. (1988). Inositol trisphosphate isomers, but not inositol 1,3,4,5-tetrakisphosphate, induce calcium influx in *Xenopus laevia* oocytes. *J. Biol. Chem.* **263**, 11048–11051.

Sugiyama, H., Hisanaga, Y., and Hirono, C. (1985). Induction of muscarinic cholinergic responsiveness in *Xenopus* oocytes by mRNA isolated from rat brain. *Brain Res.* **338**, 346–350.

Sugiyama, H., Ito, I., and Hirono, C. (1987). A new type of gultamate receptor linked to inositol phospholipid metabolism. *Nature* **325**, 531–533.

Takahashi, T., Neher, E., and Sakman, B. (1987). Rat brain serotonin receptors in *Xenopus* oocytes are coupled by intracellular calcium to endogenous channels. *Proc. Natl. Acad. Sci. U.S.A.* **84**, 5063–5067.

VanRhenterghem, C., Penit-Soria, J., and Stinnakre, J. (1984). β-adrenergic induced K^+ current in *Xenopus* oocytes: Involvement of cAMP. *Biochemie* **66**, 135–138.

Van Wezenbeek, L. A. C. M., Tonnaer, J. A. D. M., and Ruigt, G. S. F. (1988). The endogenous muscarinic acetylcholine receptor in *Xenopus* oocytes is of the M3 subtype. *Eur. J. Pharmacol.* **151**, 497–500.

Williams, J. A., McChesney, D. J., Calayag, M. C., Lingappa, V. R., and Logsdon, C. D. (1988). Expression of receptors for cholecystokinin and other Ca^{2+}-mobilizing hormones in *Xenopus* oocytes. *Proc. Natl. Acad. Sci. U.S.A.* **85**, 4939–4943.

CHAPTER 21

Receptor Regulation of Cell Calcium

Itaru Kojima

Cell Biology Research Unit, Fourth Department of Internal Medicine, University of Tokyo School of Medicine, Tokyo 112, Japan

I. INTRODUCTION

Since the role of calcium as a regulator of skeletal muscle contraction was established (Ebashi, 1976), numerous studies have focused on its role as a regulator of cell function in nonmuscle cells. It is now widely accepted that calcium acts as an intracellular messenger of various types of extracellular stimuli and as a critical regulator of cellular functions. Calcium-regulated responses include smooth muscle contraction, cell motility, exocytosis, fertilization, cell growth, and metabolic responses such as glycogenolysis, superoxide production, and steroidogenesis (Rasmussen and Barrett, 1984). In general, when an extracellular stimulus reaches the cell it is recognized by a receptor molecule on the cell surface. Following recognition by the receptor, information is then transferred to the cell interior. GTP-binding proteins (G proteins) act as transducers in this signal transfer system in many cells (Gilman, 1987). Thus, G protein transduces information from the receptor to an effector molecule which eventually generates a second messenger(s) and the intracellular messenger is often the calcium ion.

Our understanding of the calcium messenger system has been greatly enhanced by the elucidation of phosphoinositide metabolism (Berridge, 1987) and by new technological advances in evaluating cell calcium metabolism, such as patch-clamp techniques (Hamil *et al.,* 1981) and techniques to measure cytoplasmic free calcium concentration ($[Ca^{2+}]_c$) (Tsien, 1981; Borle and Snowdone, 1982; Grynkiewicz *et al.,* 1985). Employing these techniques, we can now monitor in more detail a sequence of cellular events in response to an extracellular stimulus. In this chapter, receptor-regulated changes in cell calcium metabolism are discussed. Extracellular agonists which affect cell calcium can be classified into two major categories on the basis of their effects on phosphoinositide turnover. The first group includes agonists which enhance phosphoinositide turnover; receptors linked to phospholipase C are classified in this group. The second group includes agonists which do not affect phosphoinositide metabolism. Receptors for this group of agonists are not coupled to phospholipase C.

II. RECEPTORS LINKED TO PHOSPHOLIPASE C

In a variety of systems, agonists which employ calcium as an intracellular messenger cause breakdown of phosphoinositides. Examples of such agonists are listed in Table I. These agents increase $[Ca^{2+}]_c$ by causing a release

Table I   Agonists Which Induce Phosphoinositide
Turnover

Catecholamine (α_1)	Collagen
Acetylcholine (M)	Platelet-activating factor
Adenosine (P_2)	Thromboxane A_2
Histamine (H_1)	Formyl–Met–Leu–Phe
Vasopressin (V_1)	Opsonized zymosan
Angiotensin II	Leukotriene B_4
TRH	Phytohemagglutinin
GnRH	Concanavalin A
Cholecystokinin	Interleukin 1
5-Hydroxytryptamine	Bombesin
Bradykinin	Platelet-derived growth factor
Thrombin	Fibroblast growth factor

of calcium from an intracellular nonmitochondrial pool (trigger pool). Most of these agents simultaneously stimulate influx of calcium via a specific calcium-gating system in the plasma membrane. The mechanism by which these agonists cause calcium release from the trigger pool appears to be similar. In contrast, the mechanism by which these agonists induce calcium influx may vary depending on the type of agonist and the type of target cell. Thus, they seem to act on different classes of ion channels by different mechanisms.

A. Release of Calcium from Trigger Pool

The agonists listed in Table I elicit an elevation of $[Ca^{2+}]_c$ in their target cells. This involves activation of an enzyme, phospholipase C, specific to a minor component of phospholipids in plasma membranes, namely phosphatidylinositol 4,5-bisphosphate (PIP_2). Hydrolysis of PIP_2 leads to generation of two potentially important second messengers, inositol 1,4,5-trisphosphate (IP_3) and diacylglycerol (DG) (Berridge, 1987).

An initial step in the action of a calcium-mobilizing agonist is binding of the agonist to its specific receptor on the surface of the plasma membrane. Although structures of receptors for the agonists listed in Table I are largely unknown at present, the primary structure of muscarinic acetylcholine receptor has been deduced from analysis of its complementary DNA (Kubo et al., 1986). Hydrophobicity analysis of the muscarinic receptor suggests the presence of seven transmembrane domains. In this regard, the muscarinic acetylcholine receptor resembles β-adrenergic receptor (Dixon et al., 1986) and rhodopsin (Nathans and Hogness, 1983), both of which are coupled to G proteins. Since receptors for this group of agonists appear to be coupled to G proteins, it seems reasonable to speculate that most of the receptors for calcium-mobilizing agonists have similar structures.

After the binding of a ligand to its receptor, the information flows through the plasma membrane to an effector enzyme, phospholipase C. Recent studies have revealed that a G protein is involved in this process. In neutrophils, for example, a chemotactic peptide, formyl-Met–Leu–Phe (fMLP), induces a rapid breakdown of PIP_2 by acting on its specific receptor (Dougherty et al., 1984). Involvement of a G protein in the activation of phospholipase C was first suggested by the observation that pertussis toxin, which ADP-ribosylates the α-subunit of an inhibitory G protein (G_i) in the adenylyl cyclase system (Katada and Ui, 1982), inhibited fMLP-induced activation of phospholipase C (Okajima and Ui, 1984; Ohta et al., 1984; Lad et al., 1985). In addition, the binding of fMLP in a plasma membrane preparation was inhibited by GTP or its stable analog and this inhibition was reversed when the membrane was treated with pertussis toxin (Okajima *et al.,* 1985). Moreover, GTP stimulated hydrolysis of PIP_2 in a ^{32}P-labeled plasma membrane preparation (Smith *et al.,* 1985). These data are consistent with the notion that the fMLP receptor is coupled to a pertussis toxin-sensitive G protein and that binding of GTP to the G protein leads to the activation of phospholipase C. Further support for the idea that a G protein is involved in fMLP-induced activation of phospholipase C was provided by Kikuchi *et al.* (1986). They showed that an addition of either G_i or G_o to a membrane preparation of pertussis toxin-treated neutrophils could restore the ability of fMLP to activate phospholipase C. However, the pertussis toxin-sensitive G protein involved in the action of fMLP differs from G_i or G_o in its antigenic properties (Gierschik *et al.,* 1986). A G protein with a molecular weight of 40K has been recently isolated from DMSO-treated HL-60 cells (Oinuma *et al.,* 1987) and it has been suggested that this may be the G protein involved in the fMLP-induced activation of phospholipase C. Similarly, G protein involvement has been suggested in other systems, including thrombin action in platelets (Baldassare and Fisher, 1986), thyrotropin-releasing hormone (TRH) action in GH_3 cells (Martin *et al.,* 1986; Aub *et al.,* 1986; Straub and Gershengorn, 1986), 5-hydroxytryptamine (5-HT) action in blowfly salivary glands (Litosch and Fain, 1985; Litosch *et al.,* 1985), angiotensin action in hepatocytes (Pobiner *et al.,* 1985; Blackmore *et al.,* 1985; Uhing *et al.,* 1986), and thrombin action in Swiss 3T3 fibroblasts (Murayama and Ui, 1985). Available data suggest that GTP reduces the calcium requirement of phospholipase C so that phospholipase C is activated in the presence of physiological intracellular concentrations of calcium ions (C. D. Smith *et al.,* 1986). A precise mechanism for the regulation of phospholipase C by G protein is discussed in other chapters.

Activation of phospholipase C specific to PIP_2 leads to the production of two products, IP_3 and DG (Berridge, 1987), within seconds of the addition of an agonist (Berridge, 1983). It is now widely accepted that IP_3 acts as an intracellular messenger to mobilize calcium from an intracellular nonmitochondrial pool (Berridge and Irvine, 1984). Using permeabilized cell preparation, it has been demonstrated that IP_3 stimulates calcium release in pancreatic acini (Streb *et al.,* 1983), hepatocytes (Burgess *et al.,* 1983; Joseph *et al.,* 1984), vascular smooth muscle cells (Suematsu *et al.,* 1984), and many other types of cells (Berridge, 1987) suggesting that IP_3 enhances calcium release from an intracellular calcium pool. IP_3 is effective in inducing calcium release at submicromolar concentrations, a range which is expected to occur in stimulated cells (Joseph *et al.,* 1984). When isomers of inositol phosphates were examined for their ability to mobilize calcium in permeabilized cell preparations, IP_3 was the most effective compound, both inositol 2,4,5-trisphosphate and IP_2 were slightly effective, and inositol 1,4-bisphosphate was without effect (Irvine *et al.*, 1984). Therefore, actions of inositol phosphates are stereospecific and *trans*-vicinal phosphates on the 4- and 5-positions are essential for the calcium-mobilizing action of inositol trisphosphate. In addition, phosphate in the 1-position is also important. In agreement with these observations, binding sites for IP_3 were found in permeabilized hepatocytes and neutrophils (Spat *et al.*, 1986a) and in the microsomal fraction of hepatocytes (Spat *et al.*, 1986b). Relative potencies of isomers of inositol phosphates for displacing [^{32}P]IP$_3$ binding correlated well with the potencies for calcium release. These data strongly suggest that IP_3 exerts its action by acting on a specific receptor for IP_3 in the intracellular calcium pool. Moreover, since IP_3-induced calcium release is not affected by a decrease in temperature, IP_3 may regulate gating of calcium through a channel-like molecule (Smith *et al.*, 1985).

Studies have indicated that metabolism of IP_3 may be more complex than originally thought (Batty *et al.*, 1985). IP_3 is further phosphorylated by a calmodulin-dependent kinase to inositol 1,3,4,5-tetrakisphosphate [$I(1,3,4,5)$-P_4] (Yamaguchi *et al.*, 1987); $I(1,3,4,5)$-P_4 is then dephosphorylated to $I(1,3,4)$-P_3 (Downs *et al.*, 1986); and, in some tissue, $I(1,3,4)$-P_3 can be again phosphorylated to $I(1,3,4,6)$-P_4 (Shears *et al.*, 1987). Functions of these isomers of inositol trisphosphate and inositol tetrakisphosphate are not yet clarified, but the time course of production of these inositol phosphates implicates their biological roles. An immediate increase in IP_3 is observed within seconds upon stimulation with an agonist. This is followed after several seconds by gradual increases in both $I(1,3,4,5)$-P_4 and $I(1,3,4)$-P_3 (Hawkins *et al.*, 1986; Merritt *et al.*, 1986; Trimble *et al.*, 1987). The response of IP_3 is rapid, but only transient, whereas $I(1,3,4)$-P_3 remains elevated for a much longer period. After 1 min of stimulation by an agonist, most of the inositol trisphosphate found in the cell is $I(1,3,4)$-P_3. The temporal pattern of the increase in IP_3 is consistent with the hypothesis that IP_3 serves as an intracellular messenger to release calcium from a trigger pool. In addition, it seems reasonable to speculate that $I(1,3,4)$-P_3 may have a functional role, particularly in a late phase. When a calcium-mobilizing agonist stimulates the cell for a period, the trigger pool of calcium appears to remain empty. In exocrine pancreas, for example, addition of carbachol results in a decrease in cell calcium, presumably due to an extrusion by the calcium pump of the cytosolic calcium which is released from a trigger pool. The cell calcium remains decreased in the presence of carbachol but rapidly increases after the blockade of carbachol action by atropine (Schulz, 1980). Likewise, when adrenal glomerulosa cells are stimulated by angiotensin II, the total cell calcium decreases rapidly. It remains decreased in the presence of angiotensin II and recovers quickly after removal of the agonist stimulation (Kojima *et al.*, 1987).

Because $I(1,3,4)$-P_3 is capable of mobilizing calcium from a nonmitochondrial pool at relatively high concentrations (Irvine *et al.*, 1986), and the concentration of $I(1,3,4)$-P_3 in stimulated cells is reasonably high (Daniel *et al.*, 1987), $I(1,3,4)$-P_3 may play a role in the sustained release of calcium from the intracellular pool.

B. Influx of Calcium via Calcium Channel

Agonists listed in Table I increase $[Ca^{2+}]_c$ in their target cells. They not only cause mobilization of calcium from an intracellular pool but most of them simultaneously stimulate an entry of calcium from the extracellular fluid across the plasma membrane. In GH_3 cells, for example, TRH mobilizes calcium from an intracellular pool and stimulates calcium influx (Tan and Tashijian, 1981). In adrenal glomerulosa cells, angiotensin II causes both release of calcium and influx of calcium across the plasma membrane (Kojima *et al.*, 1985). Likewise, vasopressin and angiotensin II increase calcium influx in hepatocytes and vascular smooth muscle cells (Mauger *et al.*, 1984; Wallnofer *et al.*, 1987). Nevertheless, routes through which the increased calcium entry occurs may not be identical. In GH_3 cells, the action of TRH on calcium influx is blocked by a dihydropyridine calcium channel blocker (Tan and Tashijian, 1981). Similarly, angiotensin II-induced calcium influx is also blocked by nitrendipine (Kojima *et al.*, 1985). These agonists appear to induce calcium influx by modulating dihydropyridine-sensitive voltage-dependent calcium channels (L-type channels) in these cells. In contrast, vasopressin-induced calcium influx in smooth muscle cells is insensitive to dihydropyridine calcium channel blockers, suggesting that vasopressin affects calcium channels other than L-type calcium channels (Wallnofer *et al.*, 1987). Noteworthy is the fact that this second class of channels seems to be regulated by second messengers generated in hydrolysis of phosphoinositides.

1. Regulation of Voltage-Dependent Calcium Channel

The mechanism by which an agonist increases entry of calcium via voltage-dependent calcium channels is well-characterized in GH_3 cells. In these cells, TRH induces calcium-dependent action potentials suggesting that these cells are excitable (Hagiwara and Ohmori, 1982). In activating voltage-dependent calcium channels, TRH does not affect the opening of the calcium channel directly. Rather TRH acts indirectly on the voltage-dependent calcium channel by modulating membrane potential (Fig. 1A). Upon stimulation with TRH, an initial hyperpolarization of the membrane is observed, due to an activation of calcium-activated potassium channels by an increase in $[Ca^{2+}]_c$. The initial hyperpolarization is followed by a depolarization of the membrane potential, largely due to an inhibition of a voltage-dependent potassium current (Dubinsky and Oxford, 1985). During the depolarizing phase, the membrane potential elevates to a value higher than the threshold of calcium channels; thus the voltage-dependent calcium channel opens. Hence, a key step in the regulation of the voltage-dependent calcium channel by TRH seems to be a modulation of the voltage-dependent potassium channel. Although the regulation of potassium channels by TRH is not established at present, it should be mentioned that protein kinase C may well be involved in this regulation. In whole-cell patch clamp recording, delayed action of TRH, including activation of voltage-dependent calcium channels, is not observed. Failure of TRH to induce delayed electrical responses may be due to loss of soluble components in whole cell recording, an artifact inherent to this particular technique. When purified protein kinase C is injected, the ability of TRH to induce delayed electrical responses could be restored (Dufy et al., 1987). This observation, together with observations that oleoylacetylglycerol, an activator of protein kinase C, increases calcium influx in GH_3 cells (Albert et al., 1987), suggest a role for protein kinase C in TRH-induced electrical responses in the membrane. In this regard, it is noteworthy that activation of C-kinase results in an inhibition of voltage-dependent potassium channels (M-current) in NG 108-15 cells (Higashida and Brown, 1986). Further studies are clearly needed to clarify these issues. In adrenal cells, angiotensin II stimulates calcium influx by activating the voltage-dependent calcium channel (Kojima et al., 1985; Hescheler et al., 1988). Of interest is the fact that the action of angiotensin II on calcium influx is abolished by pretreatment with pertussis toxin (Kojima et al., 1986; Hescheler et al., 1988). It is possible that a certain type of potassium channel is regulated by a pertussis toxin-dependent mechanism. In addition to the regulation of membrane potential, calcium-mobilizing agonists may regulate voltage-dependent calcium channels by another mechanism (Clapp et al., 1987).

Conversely, it is reported that a certain type of voltage-dependent calcium channel may be regulated negatively by protein kinase C in the central nervous system. This issue is discussed in another chapter.

2. Regulation of Voltage-Independent Cation Channel

Another group of ion channels regulated by agonists have two common properties. First, they are voltage-independent cation channels which are not selective for calcium. Both monovalent and divalent cations permeate through the channels. In fact, Na^+, K^+, Ca^{2+}, and Ba^{2+} pass through these channels. Second, these channels are regulated by second messengers generated in phosphoinositide turnover.

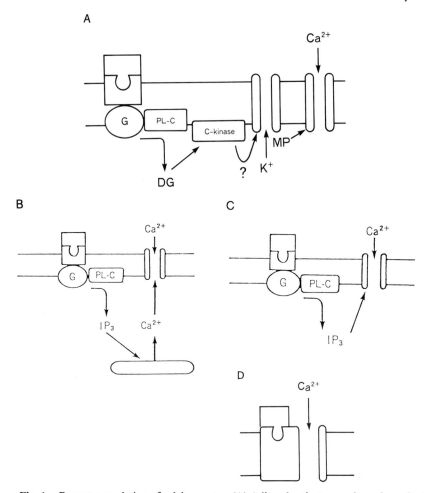

Fig. 1. Receptor regulation of calcium entry. (A) A ligand activates a voltage-dependent calcium channel indirectly by changing membrane potential. In the case of TRH action, TRH causes depolarization by inhibiting a certain class of K^+ channel. (B) A ligand activates a calcium-permeable cation channel by increasing cytoplasmic free calcium concentration. (C) A ligand activates calcium-permeable cation channel by increasing IP_3. (D) A ligand activates a receptor-associated calcium-permeable channel. (E) A ligand activates calcium-permeable channel directly by a G-protein-dependent mechanism. (F) A ligand sensitizes a voltage-dependent calcium channel by increasing cAMP. (G) A ligand regulates a voltage-dependent calcium channel indirectly by changing membrane potential. In adrenal medulla, acetylcholine activates a nicotinic receptor-associated sodium channel. Depolarization of the plasma membrane leads to an activation of the voltage-dependent calcium channel. (H) A ligand regulates a voltage-dependent calcium channel indirectly by modulating membrane potential. In the case of somatostatin action, a potassium channel is activated by a G-protein-dependent mechanism. G, G protein; PL-C, phospholipase C; DG, diacylglycerol; MP, membrane potential; IP_3, inositol (1,4,5)-triphosphate; A–C, adenylyl cyclase.

E

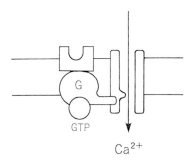

F

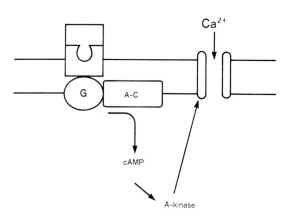

G   H

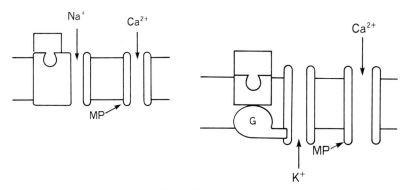

Fig. 1 (Continued)

In neutrophils, a chemotactic peptide, fMLP, stimulates influx of calcium. Using patch clamp techniques, Von Tscharner et al. (1986) demonstrated that fMLP activated a calcium-permeable cation channel which was a nonselective cation channel and was activated by calcium. Thus, when $[Ca^{2+}]_c$ was elevated by the addition of a divalent cation ionophore, this channel became activated and calcium passed through the channel. Furthermore, fMLP did not activate the cation channel when cytoplasmic calcium was chelated by an injection of EGTA into the cytoplasm. Thus, fMLP appears to regulate the influx of calcium into neutrophils by increasing $[Ca^{2+}]_c$. A subsequent study suggested an additional mechanism for fMLP-induced calcium influx (Nasmith and Grinstein, 1987). It is quite possible that the cation channel is regulated by multiple mechanisms. The fact that a calcium-sensitive calcium-permeable cation channel is found in other cells suggests that such a mechanism is not limited to neutrophils but may be more widely present. (Fig. 1B).

A second example for second messenger-operated channels has been described in T lymphocytes. In this system, mitogenic lectins increase $[Ca^{2+}]_c$ by causing both calcium release from an internal pool and calcium influx across the plasma membrane. Kuno et al. (1986) have shown that mitogenic lectins increase the opening probability of a cation channel in T lymphocytes. This channel is a nonselective cation channel and calcium passes through it. Changes in membrane potential do not affect its life-time histogram, indicating that the opening probability of this lectin-regulated cation channel is voltage-independent. An intriguing feature of the cation channel is that it is activated by IP_3. When studied in the excised patch mode, the opening probability of the cation channel is greatly increased by an application of IP_3 to the cytoplasmic site of the plasma membrane (Kuno and Gardner, 1986). These observations, together with the fact that mitogenic lectins stimulate hydrolysis of PIP_2, indicate that mitogenic lectins may modulate the voltage-independent cation channel by increasing IP_3 (Fig. 1C). It is not clear at present whether this is a fairly common mechanism by which calcium-mobilizing agonists increase calcium entry into the cell.

An alternate mechanism for second messenger-induced calcium gating is proposed by Irvine and Moor (1986). They have shown that microinjection of $I(1,3,4,5)-P_4$ into sea urchin eggs leads to raising of the fertilization envelope, which is dependent on extracellular calcium, suggesting that $I(1,3,4,5)-P_4$ augments calcium entry in sea urchin eggs. They have also shown that coexistence of $I(2,4,5)-P_3$ is required for $I(1,3,4,5)-P_4$ to have this effect. The implication of these observations is not totally clear. Moreover, it has not been shown directly whether $I(1,3,4,5)-P_4$ activates calcium gating systems. In this regard, Morris et al. have recently reported that a combination of I(1,3,4,5)-P$_4$ and IP$_3$ evokes a continuous increase in calcium-activated potassium current which is dependent on extracellular calcium in exocrine cells (Morris *et al.*, 1987). I(1,,3,4,5)-P$_4$ is a potentially interesting metabolite which may be involved in the regulation of cellular calcium metabolism at the plasma membrane.

III. RECEPTORS NOT LINKED TO PHOSPHOLIPASE C

Agonists in this category regulate calcium gating either positively or negatively without altering phosphoinositide turnover. There are at least four ways for these agonists to modulate calcium gating. First, calcium-permeable channel is localized in the receptor molecule and binding of a ligand leads to the activation of the channel (Fig. 1D). Second is a regulation of the channel by the receptor, where G protein is involved as a transducer (Fig. 1E). Third is generation of an intracellular messenger, not associated with phosphoinositide turnover, which modulates opening of the channel (Fig. 1F). G protein is involved in the generation of the second messenger. Fourth is an indirect regulation of voltage-dependent calcium channels by changing membrane potential. In this case, the agonist regulates membrane potential by altering fluxes of monovalent cations (Fig. 1G,H).

A. Activation of Receptor-Associated Channel by Agonist

Glutamate is a neurotransmitter which mediates fast excitatory synaptic transmission in the central nervous system. In hippocampal neurons, glutamate increases [Ca^{2+}]$_c$ by stimulating calcium influx. N-methyl-D-aspartate (NMDA), a specific agonist of a subtype of glutamate receptor, stimulates calcium entry by activating a calcium-permeable channel which is considered to be associated with the NMDA receptor. The NMDA receptor thus provides an example of a receptor-associated channel (Fig. 1D). Recent studies have suggested that sodium and potassium also pass through the channel (Jahr and Stevens, 1987; Cull-Candy and Usowicz, 1987).

B. G Protein-Dependent Regulation of Calcium-Permeable Channel by Receptors

Some agonists regulate calcium entry by employing a G protein as a transducer (Fig. 1E). An example of positive regulation is the action of insulin-like growth factors (IGFs). IGFs are potent mitogens which are called progression factors in cell growth. There are two structurally related

IGFs, IGF-I and IGF-II. We have shown recently that IGF-II stimulates calcium influx in BALB/c 3T3 fibroblasts (Nishimoto *et al.*, 1987a). The action of IGF-II is dependent on the cell cycle, and IGF-II elicits its action only in IGF-responsive primed competent cells. To exert its action, IGF-II acts on its receptor, the type-II IGF receptor (Kojima *et al.*, 1988). Of particular interest is the fact that pretreatment with pertussis toxin completely abolishes IGF-II-induced calcium influx (Nishimoto *et al.*, 1987a). A dose–response relationship for the inhibitory action of pertussis toxin correlates well with that for toxin-mediated ADP-ribosylation of a 40kDa protein. Furthermore, binding of [125I]IGF-II to plasma membranes is inhibited by GTPγS. Taken together, these data indicate that IGF-II may stimulate calcium influx by a mechanism involving a pertussis toxin-sensitive G protein. In fact, we have found recently that binding of IGF-II to the type-II IGF receptor leads to dissociation of a 40kDa protein from its $\beta\gamma$ subunit (I. Nishimoto, submitted for publication).

To determine the mechanism by which IGF-II stimulates calcium influx, we analyzed IGF-II-sensitive ion current by employing patch-clamp techniques in BALB/c 3T3 cells. When a tight-seal cell-attached patch was made using a micropipette containing IGF-II and Ba^{2+}, an inward current was detected (Matsunaga *et al.*, 1988). However, an application of IGF-II to outside the patch was without effect, suggesting that the channel is tightly regulated by the receptor. Both divalent and monovalent cations permeated through the IGF-II-sensitive channel and its opening probability was independent of membrane potential. Although this cation channel has similar properties to those of the lectin-sensitive, calcium-permeable cation channel in T lymphocytes (Kuno *et al.*, 1986), the two channels differ in that the IGF-II-sensitive channel in BALB/c 3T3 cells is not regulated by either IP_3 or calcium when studied in an excised patch mode. More important is the fact that IGF-II does not induce phosphoinositide turnover (unpublished observation). Since the IGF-II-sensitive cation channel is activated by GTPγS, we postulate that IGF-II regulates calcium-permeable cation channels directly by employing a G protein as a transducer. Likewise, IGF-I stimulates calcium influx in primed competent BALB/c 3T3 cells (Kojima *et al.*, 1988b). The IGF-I action on calcium influx is slightly different from that of IGF-II in that the action of IGF-I occurs slowly. Since IGF-I action is reproduced by high concentrations of insulin, IGF-I may act mainly on the type-I IGF receptor. Involvement of a G protein is suggested by the observation that pretreatment of the cells with pertussis toxin completely abolishes IGF-I action on calcium influx (Nishimoto *et al.*, 1987b). When a tight-seal cell-attached patch with IGF-I and Ba^{2+} in the pipette solution was applied, an inward current was observed. Analysis of kinetics of IGF-I-sensitive Ba^{2+} currents revealed that both IGF-I and IGF-II act on the same calcium-permeable cation channels. Taken together, it is conceivable that IGFs directly regulate calcium-permeable cation channels by a G-protein-dependent mechanism.

Another example for positive regulation of calcium entry by a G protein is recently reported. In the heart, β-adrenergic stimulants stimulate calcium influx via a voltage-dependent calcium channel. In addition to the well-known regulation of the channel by phosphorylation, Yatani et al. (1987a) have shown an additional mechanism by which G_s directly activates a voltage-dependent calcium channel.

An example of negative regulation of calcium entry by a G protein may be found in neuronal cells. In neuroblastoma–glioma hybrid cells, enkephalin inhibits opening of a voltage-dependent calcium channel (N-type channel). It has been shown that this inhibition is brought about by G_o (Hescheler et al., 1987), a member of the G protein family which is abundant in brain.

C. Regulation of Calcium Channel by Second Messenger

In many systems, calcium and cAMP regulate cell function by interacting with each other. Molecular mechanisms for cAMP-induced changes in cellular calcium metabolism have been well defined in cardiac muscle. In the heart, isoproterenol, a β-adrenergic agonist, increases contractility of cardiac muscle. It is thought that the action of isoproterenol is mediated by an increase in cAMP, and that a cAMP-dependent protein kinase is responsible for the potentiation since addition of the catalytic subunit of the cAMP-dependent protein kinase enhances calcium influx (Osterrieder et al., 1982). When a voltage-dependent calcium channel is phosphorylated by the cAMP-dependent protein kinase, opening periods of the channel are prolonged, and closed periods between openings are shortened (Reutter et al., 1982); this may be a mechanism for the inotropic action of isoproterenol. The significance of cAMP-mediated channel phosphorylation is less clear in nonmuscle cells. In hippocampal neurons, norepinephrine and the β-adrenergic agonist isoproterenol activate voltage-dependent calcium channels (Gray and Johnston, 1987). More indirect observations have been obtained in endocrine cells. In pancreatic β cells, agents which increase cAMP also elevate $[Ca^{2+}]_c$, which is dependent on extracellular calcium. In ACTH-secreting pituitary tumor cells, vasoactive intestinal peptide (VIP) increases $[Ca^{2+}]_c$ (Luini et al., 1985). The VIP-induced increase in $[Ca^{2+}]_c$ is sensitive to dihydropyridine calcium channel blockers and is dependent on extracellular calcium. Since VIP increases cAMP in these cells, it is possible that VIP regulates voltage-dependent calcium channels by cAMP-dependent protein phosphorylation.

D. Regulation of Voltage-Dependent Calcium Channel by Changing Membrane Potential

An alternate way to control calcium gating is to regulate voltage-dependent calcium channels indirectly by modulating membrane potential. In adrenal chromaffin cells, acetylcholine stimulates calcium influx by opening voltage-dependent calcium channels. A resultant increase in $[Ca^{2+}]_c$ leads to a release of catecholamine. In these cells acetylcholine acts primarily on the nicotinic acetylcholine receptor, subunits of which form a sodium channel (Noda *et al.,* 1982). Binding of acetylcholine to the nicotinic receptor results in an opening of the channel gate and sodium enters the cell immediately. Sodium influx depolarizes the plasma membrane, which in turn causes the opening of voltage-dependent calcium channels. Acetylcholine thus regulates calcium entry indirectly by stimulating nicotinic receptor-gated sodium influx (Fig. 1G).

Another way to regulate membrane potential is the modulation of potassium channels (Fig. 1H). In human growth hormone-secreting tumor cells, somatostatin decreases $[Ca^{2+}]_c$ and suppresses secretion of growth hormone. Somatostatin decreases calcium influx by inhibiting the opening of voltage-dependent calcium channels. To regulate voltage-dependent calcium channels, somatostatin activates potassium channels and thereby hyperpolarizes the plasma membrane (Yamashita *et al.,* 1986). These changes in membrane potential are mediated by a pertussis toxin-sensitive G protein, presumably G_i, since the action of somatostatin on potassium channels can be inhibited by pretreatment with pertussis toxin (Yamashita *et al.,* 1987) and since intracellular application of GTPγS reproduces the action of somatostatin (Yamashita and Ogata, 1988). Similar regulation is shown in GH_3 cells (Yatani *et al.,* 1987b). The regulation of potassium channels by somatostatin is therefore analogous to the action of acetylcholine in the heart.

ACKNOWLEDGMENT

The author is indebted to Kiyoshi Kurokawa for critical reading of the manuscript. The author thanks Etsuro Ogata for support and encouragement.

REFERENCES

Albert, P. R., Wolfson, G., and Tashijian, A. H. (1987). *J. Biol. Chem.* **262,** 6577–6581.

Aub, D. L., Frey, E. A., Sekura, R. D., and Cote, T. E. (1986). *J. Biol. Chem.* **261**, 9333–9340.

Baldassare, J. J., and Fisher, G. J. (1986). *J. Biol. Chem.* **261**, 11942–11944.

Batty, I. R., Nahorski, S. R., and Irvine, R. F. (1985). *Biochem. J.* **233**, 211–215.

Berridge, M. J. (1983). *Biochem. J.* **212**, 849–858.

Berridge, M. J. (1987). *Annu. Rev. Biochem.* **56**, 159–193.

Berridge, M. J., and Irvine, R. F. (1984). *Nature* **312**, 315–321.

Blackmore, P. F., Bocckino, S. B., Waynick, L. E., and Exton, J. H. (1985). *J. Biol. Chem.* **260**, 14477–14483.

Borle, A. B., and Snowdone, K. (1982). *Science* **217**, 252–254.

Burgess, G. M., Godfrey, P. P., Mckinney, J. S., Berridge, M. J., Irvine, R. F., and Putney Jr, J. W. (1984). *Nature* **309**, 63–65.

Clapp, L. H., Vivaudou, M. B., Walsh, J. V., and Singer, J. J. (1987). *Proc. Natl. Acad. Sci. U.S.A.* **84**, 2092–2096.

Cull-Candy, S. G., and Usowicz, M. M. (1987). *Nature* **325**, 525–528.

Daniel, J. L., Dangelmaier, C. A., and Smith, J. B. (1987). *Biochem. J.* **246**, 109–114.

Dixon, R. A. F., Kobilka, B. K., Strader, D. J., Benovic, J. L., Dohlman, H. J., Frielle, T., Boranowski, M. A., Benett, C. D., Rands, E., Diehl, R. E., Mumford, R. A., Slater, E. E., Sigal, I. S., Caron, M. G., Lefkovitz, R. J., and Strader, C. D. (1986). *Nature* **321**, 75–79.

Dougherty, R. W., Godfrey, P. P., Hoyle, P. C., Putney, J. W., and Freer, R. J. (1984). *Biochem. J.* **222**, 307–314.

Downs, C. P., Hawkins, P. T., and Irvine, R. F. (1986). *Biochem. J.* **238**, 501–506.

Dubinsky, J. M., and Oxford, G. S. (1985). *Proc. Natl. Acad. Sci. U.S.A.* **82**, 4282–4286.

Dufy, B., Jaken, S., and Barker, J. L. (1987). *Endocrinology* **121**, 793–802.

Ebashi, S. (1976). *Annu. Rev. Physiol.* **38**, 293–309.

Gierschik, P., Fallon, J., Milligan, G., Pine, M., Gallin, J. I., and Spiegal, A. (1986). *J. Biol. Chem.* **261**, 8052–8058.

Gilman, A. G. (1987). *Annu. Rev. Biochem.* **56**, 615–649.

Gray, R., and Johnston, D. (1987). *Nature* **327**, 620–622.

Grynkiewicz, G., Poenie, M., and Tsien, R. Y. (1985). *J. Biol. Chem.* **260**, 3440–3450.

Hagiwara, S., and Ohmori, H. (1982). *J. Physiol.* **331**, 231–252.

Hamil, O. P., Marty, A., Neher, E., Sakmann, B., and Sigworth, F. J. (1981). *Pfluger Arch.* **391**, 85–100.

Hawkins, P. T., Stephens, L., and Downes, C. P. (1986). *Biochem. J.* **238**, 507–516.

Hescheler, J., Rosenthal, W., Trautwein, W., and Schulz, G. (1987). *Nature* **325**, 445–447.

Hescheler, J., Rosenthal, W., Hinsch, K. D., Wulfern, M., Trautwein, W., and Schultz, G. (1988). *EMBO J.* **7**, 619–624.

Higashida, H., and Brown, D. A. (1986). *Nature* **323**, 333–335.

Irvine, R. F., and Moor, R. M. (1986). *Biochem. J.* **240**, 917–920.

Irvine, R. F., Brown, K. D., and Berridge, M. J. (1984). *Biochem. J.* **221**, 269–272.

Irvine, R. F., Letcher, A. J., Lander, D. J., and Berridge, M. J. (1986). *Biochem. J.* **240**, 301–304.

Jahr, C. E., and Stevens, C. F. (1987). *Nature* **325**, 522–525.

Joseph, S. K., Thomas, A. P., Williams, R. J., Irvine, R. F., and Williamson, J. R. (1984). *J. Biol. Chem.* **259**, 3077–3081.

Katada, T., and Ui, M. (1982). *Proc. Natl. Acad. Sci. U.S.A.* **79**, 3129–3133.

Kikuchi, A., Kazawa, O., Kaibuchi, K., Katada, T., Ui, M., and Takai, Y. (1986). *J. Biol. Chem.* **261**, 11555–11562.

Kojima, I., Kojima, K., and Rasmussen, H. (1985). *J. Biol. Chem.* **260**, 9171–9176.

Kojima, I., Shibata, H., and Ogata, E. (1986) *FEBS Lett.* **204**, 347–351.

Kojima, I., Shibata, H., and Ogata, E. (1987). *J. Biol. Chem.* **262**, 4557–4563.

Kojima, I., Nishimoto, I., Iiri, T., Ogata, E., and Rosenfeld, R. (1988a). *Biochem. Biophys. Res. Commun.* **154**, 9–19.

Kojima, I., Matsunaga, H., Kurokawa, K., Ogata, E., and Nishimoto, I. (1988b). *J. Biol. Chem.* **263**, 16561–16567.

Kubo, T., Fukuda, K., Mikami, A., Maeda, A., Takahashi, H., Mishina, M., Haga, T., Haga, K., Ichiyama, A., Kangawa, K., Kojima, M., Matuo, M., Hirose, T., and Numa, S. (1986). *Nature* **323**, 411–416.

Kuno, M., and Gardner, P. (1986). *Nature* **326**, 301–306.

Kuno, M., Goronzy, J., Weyand, C. M., and Gardner, P. (1986). *Nature* **323**, 269–273.

Lad, P. M., Olson, C. V., and Smiley, P. A. (1985). *Proc. Natl. Acad. Sci. U.S.A.* **82**, 869–873.

Litosch, I., and Fain, J. (1985). **260**, 16052–16055.

Litosch, I., Wallis, C., and Fain, J. (1985). **260**, 5464–5471.

Luini, A., Lewis, D., Guild, S., Corda, D., and Axelrod, J. (1985). *Proc. Natl. Acad. Sci. U.S.A.* **82**, 8034–8038.

Martin, T. F. J., Bajjalieh, S. M., Lucas, D. O., and Kowalchyk, J. A. (1986). *J. Biol. Chem.* **261**, 10041–10049.

Matsunaga, H., Nishimoto, I., Kojima, I., Yamashita, N., Kurokawa, K., and Ogata, E. (1988). *Am. J. Physiol.* **255**, C442–C446.

Mauger, J. P., Poggioli, J., Guesdon, F., and Claret, M. (1984). *Biochem. J.* **221**, 121–127.

Merritt, J. E., Taylor, C. W., Rubin, R. P., and Putney J. W. Jr. (1986). *Biochem. J.* **238**, 825–829.

Morris, A. P., Gallacher, D. V., Irvine, R. F., and Petersen, O. H. (1987). *Nature* **330**, 653–655.

Murayama, T., and Ui, M. (1985). *J. Biol. Chem.* **260**, 7226–7233.

Nasmith, P. E., and Grinstein, S. (1987). *FEBS Lett.* **221**, 95–100.

Nathans, J., and Hogness, D. S. (1983). *Cell* **34**, 807–810.

Nishimoto, I., Hata, Y., Ogata, E., and Kojima, I. (1987a). *J. Biol. Chem.* **262**, 12120–12126.

Nishimoto, I., Ogata, E., and Kojima, I. (1987b). *Biochem. Biophys. Res. Commun.* **148**, 403–411.

Noda, M., Takahashi, H., Tanabe, T., Toyosato, M., Furutani, Y., Hirose, T., Asai, M., Inayama, S., Miyata, T., and Numa, S. (1982). *Nature* **299**, 793–797.

Ohta, H., Okajima, F., and Ui, M. (1985). *J. Biol. Chem.* **260**, 15771–15780.

Okajima, F., and Ui, M. (1984). *J. Biol. Chem.* **259**, 13863–13871.

Okajima, F., Katada, T., and Ui, M. (1985). *J. Biol. Chem.* **260**, 6761–6768.

Oinuma, M., Katada, T., and Ui, M. (1987). *J. Biol. Chem.* **262**, 8347–8353.

Osterrieder, W., Brum, G., Hescheler, J., Trautwein, W., and Hofmann, F. (1982). *Nature* **298**, 576–578.

Pobiner, B. F., Hewlett, E. L., and Garrison, J. C. (1985). *J. Biol. Chem.* **260**, 16200–16209.

Rasmussen, H., and Barrett, P. Q. (1984). *Physiol. Rev.* **64**, 938–984.

Reutter, H., Steens, C. F., Tsien, R. W., and Yellen, G. (1982). *Nature* **297**, 501–504.

Schulz, I. (1980). *Am. J. Physiol.* **239**, G335–G347.

Shears, S. B., Parry, J. B., Tang, E. K., Irvine, R. F., Michell, R. H., and Kirk, C. J. (1987). *Biochem. J.* **246**, 139–147.

Smith, C. D., Lane, B. C., Kusaka, I., Verghese, M. W., and Snyderman, R. (1985). *J. Biol. Chem.* **260**, 5875–5878.

Smith, C. D., Cox, C. C., and Snyderman, R. (1986). *Science* **232**, 97–100.

Smith, J. B., Smith, L., and Higgins, B. L. (1985). *J. Biol. Chem.* **260**, 14413–14416.

Spat, A., Bradford, P. G., Mckinney, J. S., Rubin, R. P., and Putney J. W. Jr. (1986a). *Nature* **319**, 514–516.

Spat, A., Fabiato, A., and Rubin, R. P. (1986b). *Biochem. J.* **233**, 929–932.

Straub, R. E., and Gershengorn, M. C. (1986). *J. Biol. Chem.* **261**, 2712–2717.

Streb, H., Irvine, R. F., Berridge, M. J., and Schulz, I. (1983). *Nature* **306**, 67–69.

Suematsu, E., Hirata, M., Hashimoto, T., and Kuriyama, H. (1984). *Biochem. Biophys. Res. Commun.* **120**, 481–485.

Tan, K., and Tashjian, A. H. (1981). *J. Biol. Chem.* **256**, 8994–9002.

Trimble, E. R., Bruzzone, R., Meehan, C. J., and Biden, T. J. (1987). *Biochem. J.* **242**, 289–292.

Tsien, R. Y. (1981). *Nature* **290**, 527–528.

Uhing, R. J., Prpic, B., Jiang, H., and Exton, J. H. (1986). **261**, 2140–2146.

Von Tscharner, V., Prod'hom, B., Baggiolini, M., and Reuter, H. (1986). *Nature* **324**, 369–372.

Wallnofer, A., Caubin, C., and Ruegg, U. (1987). *Biochem. Biophys. Res. Commun.* **148**, 273–278.

Yamaguchi, K., Hirata, M., and Kuriyama, H. (1987). *Biochem. J.* **244**, 787–791.

Yamashita, N., and Ogata, E. (1988). *Proc. Natl. Acad. Sci. U.S.A.* **85**, 4924–4928.

Yamashita, N., Shibuya, N., and Ogata, E. (1986). *Proc. Natl. Acad. Sci. U.S.A.* **83**, 6198–6202.

Yamashita, N., Kojima, I., Shibuya, N., and Ogata, E. (1987). *Am. J. Physiol.* **253**, E28–E32.

Yatani, A., Codina, J., Imoto, Y., Reeves, J. P., Birnbaumer, L., and Brown, A. M. (1978a). *Science* **238**, 1288–1292.

Yatani, A., Codina, J., Sekura, R. D., Birnbaumer, L., and Brown, A. M. (1987b). *Mol. Endocrinol.* **1**, 283–289.

CHAPTER 22

Insulin and Its Interaction with G Proteins

Miles D. Houslay

Molecular Pharmacology Group, Institute of Biochemistry, University of Glasgow, Glasgow G12 8QQ, Scotland, United Kingdom

I. Introduction
II. Structure of Insulin Receptor
   A. Binding Studies
   B. Structure of Insulin Receptor
   C. Insulin Receptor β Subunit Expressing Tyrosyl Kinase Activity
III. Action of Insulin on Cyclic AMP Metabolism
   A. Introduction
   B. Stimulation of Distinct Cyclic AMP Phosphodiesterases
IV. Inhibition of Adenylyl Cyclase
   A. Characterization
   B. G_i Is Not Involved in Transducing Inhibition by Insulin of Adenylyl Cyclase Activity in Liver: Loss of G_i in Diabetes and Action of Hypoglycemic Agent Metformin
   C. Insulin Attenuates Ability of Cholera Toxin to Activate Adenylyl Cyclase
   D. Insulin Attenuates Ability of Cholera Toxin to Cause ADP-ribosylation of 25-kDa Protein
V. Stimulation of Distinct GTPase Activity in Human Platelets by Insulin
VI. Phosphorylation of Defined GTP-Binding Proteins by Human Insulin Receptor
   A. G_i and G_o
   B. Transducin
   C. p21^{ras}

I. INTRODUCTION

Insulin is a key hormone that regulates a wide spectrum of metabolic processes as well as promoting normal cell growth and development. Indeed, it is the very plethora of actions of this hormone which has made it very difficult to conceptualize, let alone discover, how insulin exerts its effects on target tissues.

The fact that 2% of the population of Western societies are diabetic, and that of these some three-quarters exhibit non-insulin-dependent (NIDDM), insulin-resistant diabetes, means that diabetes is a major health problem. In the past diabetes has been perceived as primarily a problem of glucose metabolism but this is undoubtedly a naive outlook and fails to appreciate that the core of the problem in insulin-resistant diabetes must be a lesion within or associated with the signaling apparatus of the insulin receptor itself. Thus, over recent years a great deal of effort has been put into trying to identify the mechanism of action of the insulin receptor and to define the nature of the signal or signals which emanate from it (see e.g. Czech, 1977, 1975; Denton, 1986; Houslay, 1985, 1988).

II. STRUCTURE OF INSULIN RECEPTOR

A. Binding Studies

The availability of monoiodinated ^{125}I-labeled insulin of high specific radioactivity allowed binding studies to be performed on a variety of cells. These demonstrated that even rich sources of receptor, like adipocytes, expressed only some 100,000 copies per cell. In many instances the binding of insulin to target tissues was complex and showed indications of either negative cooperativity or multiple receptor populations. Indeed, this issue has still not been resolved entirely. Certainly, one complication is that in many tissues there are also insulin-like growth factor-1 (IGF-1) receptors, which can bind insulin, albeit with lower affinity than the native insulin receptor. Also, posttranslational modifications to the receptor, in particular glycosylation, have also been demonstrated to modify the binding and functional properties of the receptor and could thus give rise to heterogeneity. Molecular biological studies of the human insulin receptor have, however, been performed by two groups (Ullrich et al., 1985; Ebina et al., 1985), who appear to have isolated two slightly different genes, differing by a small insert on the α subunit. Whether this reflects some artifact of the cloning procedures or true diversity has yet to be resolved. Indeed, these workers also indicated the existence of multiple messages for the insulin receptor, the significance of which has yet to be ascertained. A further contributing factor to the anomalous binding studies may result from the indication that each insulin receptor molecule expresses two distinct binding sites for insulin which exhibit different affinities for the hormone.

However, in contrast to the binding studies, the dose–effect curves for the ability of insulin to stimulate metabolic events show normal saturation curves and are indicative of a single set of high affinity sites. This contrasts with the binding studies in which saturation is achieved over some orders of magnitude. Clearly, there are important issues to resolve as to the nature of the anomalous binding, the relationships between occupancy and effect, and the possibility of receptor multiplicity.

B. Structure of Insulin Receptor

The molecular size of the insulin receptor was determined by covalently cross-linking ^{125}I-labeled insulin to it, either with bifunctional cross-linking reagents or photochemically by incorporating an aryl azide group onto the radiolabeled insulin molecule. The tagged complexes were then analyzed by sodium dodecyl sulfate-polyacrylamide gel electrophoresis (SDS-PAGE) under both reducing and nonreducing conditions. This allowed for the identification of the holomeric receptor as a complex of circa 450 kDa, consisting of two α subunits of 135 kDa and two β subunits of 90 kDa. These subunits were shown to be held together covalently by intersubunit disulfide bridges which could be broken under reducing conditions, leading to the resolution of the individual types of subunits.

The radiolabeled insulin was found to attach itself exclusively to the α subunit, which provides the binding site for insulin on the receptor complex. This subunit is vectorially disposed about the cell surface plasma membrane, being found exposed exclusively at the extracellular surface. The α subunit of the insulin receptor does not appear to have any domain embedded in the core of the bilayer but, rather, is anchored there by virtue of its covalent attachment to the β subunit (Fig. 1). These properties are consistent with those which might be predicted from the gene sequence of the insulin receptor α subunit. However, from the gene sequence a molecular mass of only 82 kDa for the α subunit and 70 kDa for the β subunit would have been predicted. This is readily explainable by the fact that both subunits of the receptor are glycosylated, with the α subunit being particularly heavily so, accounting for the anomalous migration on SDS-PAGE (see for reviews Czech, 1975; Houslay, 1985,1988).

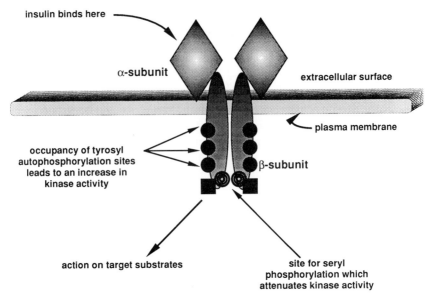

Fig. 1. Structure of the human insulin receptor. The insulin receptor is a heterodimeric molecule consisting of two α and two β subunits. For simplicity, phosphorylation sites and the kinase active site are shown on one β subunit only.

C. Insulin Receptor β Subunit Expressing Tyrosyl Kinase Activity

The β subunit has been shown to be a transmembrane entity, possessing a glycosylated globular domain at the extracellular surface and a globular domain at the cytosol surface of the plasma membrane. These two domains are linked by a single twist of α-helix which spans the bilayer (Fig. 1). Thus one can deduce that the intracellular signal(s) elicited by the insulin receptor emanate from the cytosol-located domain of the β subunit. The vectorial flow of information passes from the insulin binding site on the α subunit to the external domain of the β subunit, across the single transmembrane segment to the cytosol-facing domain of the β subunit.

The purified, solubilized insulin receptor has been shown to exhibit tyrosyl kinase activity. This allows it to phosphorylate artificial substrates and also to undergo autophosphorylation (see for reviews Czech, 1975; Houslay, 1985, 1988). Such properties are shared by the receptors for epidermal growth factor (EGF), platelet-derived growth factor (PDGF) and for the kinases coded for by certain oncogenic retroviruses (Houslay, 1981; Foulkes and Rosner, 1985). The true physiological significance of such an activity has yet to be clearly established. However, studies on receptors from insulin-resistant states, studies using monoclonal antibodies directed against the receptor kinase domain, and site-specific mutagenesis studies have given credence to the idea that such an activity may be integral to at least certain of the actions of insulin (see Czech, 1975; Houslay, 1985, 1988). In one particular study (Ullrich *et al.,* 1985), substitution of Ala for Lys at residue 1018, which is within the tyrosyl kinase domain of the β-subunit, led to the loss of tyrosyl kinase activity and also to the loss of stimulation, by insulin, of a number of processes within the cell in which this mutant receptor was expressed. Although it is tempting to assume that such experiments provide conclusive proof for an involvement of tyrosyl kinase activity in insulin action, there is now considerable evidence which indicates that even very conservative changes in proteins can have dramatic effects on their structure by altering the folding pattern of the nascent protein during its biosynthesis. Thus other domains on the receptor β subunit which may, for example, interact with an effector system or a guanine nucleotide-binding protein (G protein), may also be perturbed, indicating that site-specific mutagenesis studies should be viewed with a little caution.

A key observation (White *et al.,* 1987) that has been made about the insulin receptor is that it undergoes autophosphorylation at three distinct tyrosine residues on its β subunit. This autophosphorylation occurs first at residue 1146, then at 1150 and finally at 1316. Deletion of these sites (Ellis *et al.,* 1986) or blockade with domain-specific monoclonal antibodies appears to block the functioning of the receptor, implying that autophosphorylation itself is a key feature of the receptor signaling system. However, mutagenesis experiments have also been performed to change the tyrosine residue at 960 on the β subunit, which is not a target for autophosphorylation, to a phenylalanine residue. Although, such an alteration neither affected the kinase activity of the receptor nor its ability to undergo autophosphorylation, the ability of insulin to stimulate amino-acid transport was dramatically reduced in cells expressing this mutant receptor compared with that seen in cells which had been transfected with the gene for the normal receptor. In cells expressing this mutant insulin receptor, the ability of insulin to modulate other intracellular processes was also attenuated. Such experiments might be taken to imply that this point mutation altered the affinity of the receptor for its endogenous substrates. However, phosphorylation of the identifiable endogenous substrates in such cells appeared to proceed normally. Thus the possibility remains that this mutation perturbed a domain on the receptor β subunit which interacts with another protein involved in the signal generation system. Could such a species be a G protein?

Certainly this seems to be a particularly important domain of the receptor, because the binding of a monoclonal antibody, made against a synthetic peptide corresponding to 16 amino acids surrounding Tyr^{960}, led to the inhibition of receptor autophosphorylation and action on exogenous substrates (Herrera *et al.*, 1985).

In any event, it would appear that autophosphorylation at the other sites on the receptor would probably be a necessary prelude to any interaction at this domain. Thus it is likely to be the autophosphorylated receptor which provides the interacting species. If this should prove to be the case for a G protein, it is unlikely that the binding of insulin to its receptor would be affected by GTP, as is seen for many, but not all, ligands which transduce actions via G proteins.

III. ACTION OF INSULIN ON CYCLIC AMP METABOLISM

A. Introduction

The discovery of G-protein regulation of signal transduction arose from the pioneering observations of Rodbell, Birnbaumer, and colleagues in their studies on glucagon-stimulated adenylyl cyclase activity in liver plasma membranes. Such studies indicated that the presence of GTP enhanced the stimulatory coupling between glucagon receptors and adenylyl cyclase and also modulated the affinity of glucagon for its receptor (see Birnbaumer, 1973 for review).

Soon after the discovery of cyclic AMP, insulin was shown to cause a marked decrease in the intracellular concentrations of this second messenger in a number of tissues. This led to the suggestion that insulin might exert its actions through such a route. It soon become apparent, however, that such changes in cyclic AMP concentrations could not explain all of the actions that insulin exerted upon target tissues. Nevertheless, it remains an attractive idea that the profound effect on cyclic AMP metabolism, caused by insulin, plays a role in accounting for at least certain of the actions that insulin exerts on target cells. Certainly, it is thought that this extends to the regulation of certain metabolic processes, e.g. antilipolytic effects and enhanced glycogen synthesis (see Czech, 1977 for review).

Irrespective of the physiological significance of the actions of insulin on cyclic AMP metabolism, we have maintained that this well-established effect might provide a useful model system for elucidating the molecular mechanism(s) of signal generation by the insulin receptor itself (see Houslay, 1986).

Control of intracellular cyclic AMP metabolism can, theoretically, be exerted at three points: the site of synthesis provided by adenylyl cyclase, the site of degradation provided by a family of enzymes exhibiting cyclic

AMP phosphodiesterase activity, and also exit of cyclic AMP from the cell itself. To date there is no evidence, from studies performed on either hepatocytes or adipocytes, that insulin can increase the rate of efflux of cyclic AMP. However, there is a substantial body of evidence which demonstrates that insulin can exert inhibitory effects on adenylyl cyclase and stimulatory effects upon specific cyclic AMP phosphodiesterases (see e.g. Houslay, 1986 for review).

Such observations pose two immediate questions. First, does insulin inhibit adenylyl cyclase through G_i? Second, which cyclic AMP phosphodiesterases are the targets for insulin actions and what mechanisms are involved? Indeed, is a G protein involved in regulating cyclic AMP phosphodiesterase activity as has been shown for the ability of rhodopsin to activate a cyclic GMP phosphodiesterase via transducin (T)?

B. Stimulation of Distinct Cyclic AMP Phosphodiesterases

1. Complex Family of Enzymes Hydrolyzing Cyclic AMP Found in All Tissues

Degradation of intracellular cyclic AMP is performed exclusively by a family of cyclic AMP phosphodiesterase enzymes. It is now clear from protein-chemical and immunological studies that such diversity is due to distinct isoenzymes, although apparent multiplicity can arise by virtue of certain isoenzymes being particularly susceptible to proteolysis (Takemoto et al., 1982; Beavo et al., 1982). These enzymes are found associated with various intracellular membrane fractions and also free in the cytosol. The distribution is complex in tissues with, for example, rat hepatocytes containing at least fourteen apparently distinct forms (Houslay, 1986). Their paucity, in terms of protein mass, lability to proteolysis, susceptibility to thermal degradation, and the large number of isoenzymes has meant that relatively few species have been purified to apparent homogeneity. The nomenclature describing such species is thus inadequate; however, it is possible to consider the following identifiable groupings: (1) cyclic GMP-specific phosphodiesterases, of which the retinal enzyme controlled by transducin and rhodopsin is the best example, (2) Ca^{2+}/calmodulin-activated phosphodiesterases, which can hydrolyze both cyclic AMP and cyclic GMP, but in many instances show a preference for the latter, (3) cyclic GMP-activated cyclic nucleotide phosphodiesterases, in which low concentrations of cyclic GMP can bind to a regulatory site which activates the hydrolysis of cyclic AMP by this enzyme, (4) cyclic AMP-specific phosphodiesterases, enzymes which hydrolyze cyclic AMP extremely effectively and cyclic GMP extremely poorly. They also exhibit very low K_m values and often exhibit kinetics indicative of negative cooperativity. It is now clear that a family of enzymes falls into this category. Two distinct members appear to be a cyclic GMP-inhibited species, which may exist in both soluble and particulate forms, and a species which is particularly sensitive to inhibition by the compound Ro-20-1724.

2. Initial Observations on Insulin Activating Intracellular Cyclic AMP Phosphodiesterase Activity

Senft and colleagues (1968) provided the first evidence that insulin could cause the stimulation of cyclic AMP phosphodiesterase activity in mammalian tissues. This initial observation was confirmed rapidly by a number of investigators who extended the original observations (see e.g. Thompson and Strada, 1978; Francis and Kono, 1982; Houslay et al., 1983). The stimulatory effect of insulin was found to be localized to the high-affinity cyclic AMP phosphodiesterase activity exhibited by membrane rather than soluble fractions of both hepatocytes and adipocytes. However, there was an early indication that more than one membrane-bound cyclic AMP phosphodiesterase was activated by insulin. The identity of the activated enzyme(s), their location, and the mechanism of activation remained to be elucidated. However, it was demonstrated that the major insulin-activated cyclic AMP phosphodiesterase in adipocytes was not associated with the plasma membrane (see Francis and Kono, 1982). Also, the activity of this enzyme could be stimulated by glucagon in a cyclic AMP-dependent fashion. This species, we presume, is the so-called "dense-vesicle" enzyme described below (for review see Houslay et al., 1983).

The resolution of membrane-bound cyclic AMP phosphodiesterases undergoing stable activation by insulin awaited the development and utilization of rapid subcellular fractionation techniques.

3. Activation of Peripheral Liver Plasma Membrane Cyclic AMP Phosphodiesterase by Insulin

Detailed analysis of the hormonal regulation of hepatocyte cyclic AMP phosphodiesterase activities not only made the key observation that the activity of more than one cyclic AMP phosphodiesterase isoenzyme was under rapid hormonal control, but also gave the first insight into the possibility that insulin might interact with, or even exert some of its actions through the G protein system (Heyworth et al., 1983).

Fractionation of hepatocyte homogenates on Percoll gradients allows the rapid separation of cytosol and the various cellular membrane components. Such methodology was employed to resolve the cyclic AMP phosphodiesterase activity associated with particular membrane fractions and, importantly, to identify changes which occurred subsequent to the application of hormones to the intact cell. Thus we were able to show (Heyworth *et al.*, 1983) that two distinct species of cyclic AMP phosphodiesterase were activated within 5 min of exposure of hepatocytes to insulin. These enzymes were associated with distinct membrane compartments within the cell (Fig. 2), had very different properties and have subsequently been shown to be separate isoenzymes (Pyne *et al.*, 1987a). One is called the peripheral plasma-membrane cyclic AMP phosphodiesterase (PPM-PDE) and the other the dense-vesicle cyclic AMP phosphodiesterase (DV-PDE).

Insulin appears to activate the PPM-PDE by causing its phosphorylation, in that ATP is required for insulin to activate this enzyme in isolated membranes (Marchmont and Houslay, 1980) and a phosphorylated enzyme has been obtained (Marchmont and Houslay, 1981). In isolated membrane systems this reaction also appeared to require cyclic AMP. This was considered, in the original study, to be required for activation of a cyclic AMP-dependent protein kinase, with insulin playing the role of inducing a conformational change in the PPM-PDE which allowed it to be phosphorylated. Subsequently, however, it has been considered that this

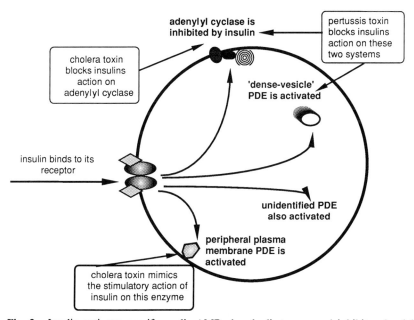

Fig. 2. Insulin activates specific cyclic AMP phosphodiesterases and inhibits adenylyl cyclase activity. The activation of these cyclic AMP phosphodiesterases by insulin has been demonstrated in intact hepatocytes (Heyworth *et al.*, 1983, 1984) as has the inhibition of adenylyl cyclase (Heyworth and Houslay, 1983b; Heyworth *et al.*, 1986).

kinase is not involved in the activation process, but rather that cyclic AMP must bind to the PPM-PDE for activation to ensue. Certainly the purified enzyme is not a substrate for cyclic AMP-dependent protein kinase (A-kinase activity) (Marchmont and Houslay, 1981) and, indeed, our observations imply that insulin-treatment of hepatocytes causes the tyrosyl phosphorylation of this enzyme (M. D. Houslay and N. J. Pyne, unpublished).

In intact hepatocytes, the activity of the PPM-PDE, while being activated by insulin, was not subject to activation by glucagon. However, the intriguing observation was made that treatment of hepatocytes with cholera toxin led to the full activation of this enzyme, with the subsequent addition of insulin having no further effect (Heyworth et al., 1983). Clearly, cholera toxin could not be exerting this action through either G_s activation or by increasing intracellular cyclic AMP, since glucagon, which exerted both such effects, failed to activate this enzyme. Furthermore, elevation of intracellular cyclic AMP concentrations with dibutyryl cyclic AMP activated A-kinase activity but failed to activate the PPM-PDE. This led us to suggest that the control of hormonal regulation of the PPM-PDE might be mediated by a unique G protein, which, like G_s, was a substrate for ribosylation by cholera toxin. We termed this putative G protein "G_{ins}" (Houslay and Heyworth, 1983).

A further interesting aspect of the regulation of the PPM-PDE came from our observations that the prior exposure of hepatocytes to glucagon completely blocked the ability of insulin to activate this enzyme (Heyworth et al., 1983). This at first sight appeared to be somewhat bizarre in that cyclic AMP appeared to be required for the activation of the enzyme in isolated membranes, and also that the addition of dibutyryl cyclic AMP, while not activating this enzyme itself, served to potentiate the stimulatory effect that insulin exerted. It became apparent, however, that the blocking process triggered by glucagon had exactly the characteristics of the process through which glucagon mediated the desensitization of adenylyl cyclase (Fig. 3).

Glucagon desensitization in hepatocytes has been shown to be elicited through an action of glucagon which does not involve cyclic AMP (Heyworth and Houslay, 1983a; Murphy et al., 1987). Indeed, desensitization can be elicited by an analogue of glucagon (TH-glucagon) which cannot elevate intracellular cyclic AMP levels, and by hormones which act to stimulate inositol phospholipid metabolism in hepatocytes and can be mimicked by phorbol esters. It is known that glucagon, as well as activating adenylyl cyclase, can elicit a small stimulation of inositol phospholipid metabolism to produce diacylglycerol and inositol trisphosphate. These actions of glucagon have been suggested to be mediated via two distinct sets of glucagon receptors: GR1 receptors coupled to stimulate inositol

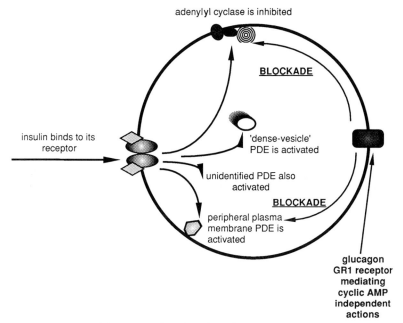

adenylyl cyclase is inhibited

BLOCKADE

insulin binds to its
receptor

'dense-vesicle'
PDE is activated

unidentified PDE also
activated

BLOCKADE

peripheral plasma
membrane PDE is
activated

glucagon
GR1 receptor
mediating
cyclic AMP
independent
actions

Fig. 3. Glucagon elicits a "selective" insulin-resistant state in rat hepatocytes, a cyclic
AMP-independent process. It has been shown (Heyworth *et al.,* 1983) that pretreatment of
hepatocytes with glucagon can block certain, but not all, insulin actions through a cyclic
AMP-independent action. Such a process may be mediated through GR1 glucagon receptors,
which have been postulated to be coupled to the stimulation of polyphosphatidylinositol
metabolism (Wakelam *et al.,* 1986b). This action is believed to cause the desensitization of
GR2 receptors coupled to stimulate adenylyl cyclase (Murphy *et al.,* 1987).

phospholipid metabolism, and GR2 receptors coupled to stimulate adenylyl cyclase (Wakelam *et al.,* 1986a), although the possibility remains that
one receptor could stimulate two G protein-controlled processes. Thus
glucagon desensitization is suggested to be mediated by the activation of
C-kinase, which appears to cause the transient uncoupling of glucagon
receptors from G_s. The site of the lesion is, as yet, unknown, although it has
been suggested to be at the level of G_s because alterations in its kinetics of
activation by non-hydrolyzable GTP analogs have been noted after desensitization (Heyworth and Houslay, 1983a). This desensitization process
also appears to transiently inactivate G_i; thus it is possible that the functioning of a G protein involved in transducing certain insulin actions could
similarly be attenuated, leading to a selective, insulin-resistant state.

One further aspect of these studies was the observation (Wallace *et al.,*
1984) that both adenosine and its nonmetabolizable analog N^6-(phenyliso-

propyl) adenosine were found to be able to prevent both glucagon-mediated desensitization and the glucagon-mediated blockade of activation of the PPM-PDE by insulin. This is particularly intriguing in the light of observations (Klein *et al.,* 1987) that adenosine enhances the tyrosyl kinase activity of the insulin receptor in adipocytes.

The possibility of involvement of a G protein in the activation of this enzyme was further tested in an isolated plasma membrane fraction. Here it was shown that the addition of guanine necleotides could mimic insulin in triggering the activation of the PPM-PDE.

Insulin regulation of the activity of the PPM-PDE remains a potentially very interesting system for elucidating mechanisms because the components all interact in, and are confined to, the plane of the plasma membrane. However, by virtue of the PPM-PDE being a peripheral protein it is easily detached from the plasma membrane by mechanical stress, freeze-thawing, high ionic strength, chelating agents, and the presence of detergents. Thus membranes can easily be denuded of this enzyme. This makes the PPM-PDE a fragile and difficult system to perform studies on, especially those aimed at analyzing the system using reconstitution technologies.

4. Activation of Dense-Vesicle Cyclic AMP Phosphodiesterase By Insulin

A number of investigators have noted that in both intact hepatocytes and adipocytes there is a membrane-bound, high-affinity cyclic AMP-specific phosphodiesterase that can be activated by both insulin and glucagon. The activity of this enzyme toward cyclic AMP can be inhibited by low concentrations of cyclic GMP and inhibited selectively by the compound ICI 118233 (Pyne *et al.,* 1987a,b). We have called this enzyme the dense-vesicle-PDE (DV-PDE). It has been purified to homogeneity from both hepatocytes (Pyne *et al.,* 1987a) and adipocytes (Degerman *et al.,* 1987), and shown to be an isoenzyme which is distinct from the PPM-PDE as regards both its kinetic properties, inhibitor sensitivity, mechanism(s) of hormonal regulation, tissue distribution, immunological cross-reactivity, and by peptide mapping.

Glucagon clearly exerts it stimulatory action on the DV-PDE by increasing intracellular cyclic AMP concentrations, whereupon cyclic AMP-dependent protein kinase elicits the phosphorylation of the DV-PDE. In contrast to the effects seen on the PPM-PDE, treatment of cells with both insulin and glucagon leads to the additive or even synergistic activation of the dense-vesicle-PDE.

The mechanism through which insulin activates this enzyme is unclear. However, the ability of insulin, but not glucagon, to activate the DV-PDE in intact cells can be blocked by treatments which are believed to inhibit membrane recycling phenomena within the cell (see e.g. Wilson *et al.*, 1983; Heyworth *et al.*, 1984). This has been taken to imply that a membrane processing event may be involved in the activation process in an analogous fashion to that seen when insulin stimulates the translocation of glucose carriers and IGF-2 receptors between membrane compartments (see for reviews Houslay, 1985, 1988). Interestingly, treatment of both hepatocytes (Heyworth *et al.*, 1986) and adipocytes (Elks *et al.*, 1983) with pertussis toxin appears to block the ability of insulin to activate the DV-PDE (Fig. 2). Whether this action of pertussis toxin is related to an effect on a G protein or to a nonspecific action remains to be defined. Certainly hepatocytes appear only to express one major substrate for modification by pertussis toxin and that is G_i (Heyworth *et al.*, 1984).

As might be expected, treatment of hepatocytes with cholera toxin leads to the activation of the DV-PDE (Heyworth *et al.*, 1983), presumably by mimicking the action of glucagon in elevating the intracellular concentrations of cyclic AMP.

IV. INHIBITION OF ADENYLYL CYCLASE

A. Characterization

Over the years a number of investigators have attempted to observe the ability of insulin to inhibit adenylyl cyclase activity in isolated plasma membrane fractions (see e.g. Czech, 1977; Heyworth and Houslay, 1983b). Detailed investigation of this matter allowed us to define, using liver plasma membranes, a number of criteria that needed to be satisfied in order to observe this effect (Heyworth and Houslay, 1983b). First, for inhibition to be observed, the presence of GTP is essential. The half-maximal concentration necessary is of the order of 3 μM GTP, which is well above that needed to activate G_s maximally, implying that a distinct G protein is involved. Second, while inhibition by insulin occurred at half maximal levels of 0.1 – 1 nM, such an effect was abolished if concentrations of insulin rose above 10 nM. Indeed, a number of insulin-regulated processes exhibit a similar bell-shaped response curve. Third, the inhibitory effect of insulin could be titrated out by elevating glucagon concentrations, and totally obliterated when glucagon was at a saturating level. Fourth, it would appear that the inhibitory effect of insulin can also be negated in adenylyl cyclase assays employing a 'trap' of unlabeled cyclic AMP, as is common in the α-^{32}P assay. This may be because if the membranes contain A-kinase activity, the exogenous pool of cyclic AMP suffices to stimulate this kinase to inactivate the insulin receptor (see e.g. Tanti *et al.*, 1987; Roth and Beaudoin, 1987).

Thus, in the presence of GTP and either in the absence of a hormone acting through G_s or with such a species at subsaturating levels, insulin can elicit the inhibition of adenylyl cyclase. As with most inhibitory ligands, this effect gave of the order of 25–30% inhibition of activity in either isolated membrane systems (Heyworth and Houslay, 1983b) or in intact hepatocytes (Heyworth *et al.*, 1986). In order to observe this, however, cyclic AMP phosphodiesterase activity must first be blocked with an inhibitor, such as isobutylmethyl xanthine (IBMX), and then the cells must be challenged simultaneously with both insulin (1 nM) and a low concentration (0.1 nM–1 nM) of glucagon (Heyworth and Houslay, 1983b; Heyworth *et al.*, 1986). As with insulin activation of the PPM-PDE in intact hepatocytes, prior challenge of the cells with glucagon causes a 'selective' insulin-resistant state (Heyworth and Houslay, 1983b). This, in this instance, prevents insulin from inhibiting adenylyl cyclase in the intact cell and gives a further indication that the glucagon-desensitization process can trigger a 'selective' insulin-resistant state (Fig. 3).

B. G_i Is Not Involved in Transducing Inhibition by Insulin of Adenylyl Cyclase Activity in Liver: Loss of G_i in Diabetes and Action of Hypoglycemic Agent Metformin

Inhibition of adenylyl cyclase is normally attributed to the functioning of G_i, which can inhibit adenylyl cyclase in two distinct ways: first, by virtue of the direct action of its α_i subunit on the catalytic unit of adenylyl cyclase, and second, by the release of a pool of β subunits. Because the β subunit of G_i is identical to that found in G_s, the free β subunits, released upon the activation and subsequent dissociation of G_i, serve to inhibit the dissociation of G_s. Thus a free α_s, able to activate adenylyl cyclase, is not released (see e.g. Birnbaumer *et al.*, 1985).

Does insulin, then, inhibit adenylyl cyclase by exerting actions through G_i? Certainly the studies employing cholera (Heyworth and Houslay, 1983b) and pertussis toxins (Heyworth *et al.*, 1986) have yielded contradictory answers to this question. Pertussis toxin, which causes the ADP-ribosylation and inactivation of G_i in hepatocytes (Heyworth *et al.*, 1986), does block insulin inhibition of adenylyl cyclase. However, so does treatment of the intact hepatocytes with cholera toxin (Heyworth and Houslay, 1983b), an action which has been shown not to block G_i functioning.

One approach to this problem would be to observe the actions of insulin in a membrane depleted of G_i. We were, somewhat to our surprise, able to achieve a dramatic reduction ($>90\%$) in the concentration of G_i in liver plasma membranes by making rats diabetic, thus providing the first evidence of a G_i lesion in a disease state (Gawler *et al.*, 1987). This was achieved by inducing diabetes with either streptozotocin or alloxan and then measuring the concentration of the α subunit of G_i using a specific anti-peptide antibody. The loss of the α subunit of G_i, as detected immunologically, was paralleled by a similar loss in function as assessed in liver plasma membranes by the ability of low concentrations of Gpp(NH)p to inhibit forskolin-stimulated adenylyl cyclase activity (Fig. 4). Such a condition was readily reversed by insulin therapy, demonstrating that this action was a consequence of the diabetic state rather than the drug used. It remains to be seen, however, whether the levels of G_i are controlled by insulin, glucose, or some other ligand.

In such G_i-depleted membranes from diabetic animals, insulin failed to inhibit adenylyl cyclase activity (Gawler *et al.*, 1988). Such an observation appeared to support the conclusion that could be drawn from using pertussis toxin, namely, that insulin might inhibit adenylyl cyclase via G_i. However, acute insulin-resistant states have been noted in liver and other tissues from animals with chemically-induced diabetes and it is important to identify whether such a state is really attributable to the loss of G_i rather than to another type of defect engendered in the insulin signaling system. In order to approach this question we treated animals with the biguanide drug, metformin (Fig. 5). Biguanide drugs have been used as hypoglycemic agents to treat insulin-resistant states, although the molecular mechanism of action of such agents remains to be elucidated. However, it has been demonstrated that such agents appear to exert postreceptor effects on the insulin signaling system rather than altering the number of receptors themselves. Thus such agents are believed to act by correcting a defect in the insulin–receptor signaling system (see Gawler *et al.*, 1988 for discussion).

Treatment of normal rats with metformin was shown to yield a marked fall in the concentration of G_i in liver plasma membranes. However, this was not accompanied by any diminution whatsoever in the ability of insulin to inhibit adenylyl cyclase activity. Indeed, treatment of streptozotocin-induced diabetic rats with metformin completely restored the ability of insulin to inhibit adenylyl cyclase activity, while G_i concentrations remained at $<10\%$ of those found in control animals (Gawler *et al.*, 1988). Such experiments appear to show quite clearly that G_i cannot be transducing the inhibitory action of insulin upon adenylyl cyclase activity.

The GTP-dependent nature of the inhibitory effect of insulin, its toxin sensitivity, and its abolition by the glucagon-desensitization mechanism suggest that insulin could be exerting this action on adenylyl cyclase either

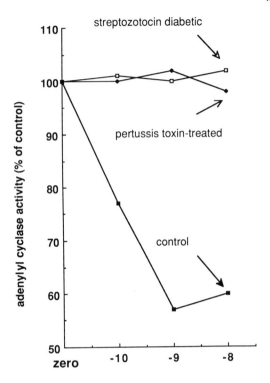

Fig. 4. Chemically induced diabetes results in the loss of G_i from hepatocyte plasma membranes. Treatment of rats with streptozotocin or alloxan induces diabetes with concomitant hypoinsulinemia and hyperglycemia. Functional G_i activity can be assessed by virtue of the ability of low concentrations of Gpp(NH)p to inhibit forskolin-stimulated adenylyl cyclase activity. Here we see that functional G_i activity is not seen in membranes from streptozotocin-treated rats or in membranes from control rats whose hepatocytes have been treated with pertussis toxin. This action of streptozotocin can also be mimicked by using alloxan to induce diabetes, and can be reversed by treating the diabetic animals with insulin. Chemically induced diabetes also has effects upon adenylyl cyclase itself. For further details see Gawler *et al.* (1987).

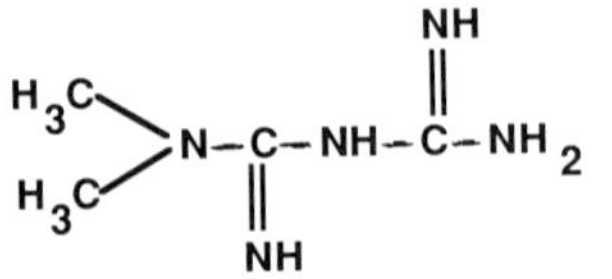

Fig. 5. Structure of metformin (*N',N'*-dimethylbiguanide).

via a novel G protein or by altering the functioning of G_s. Indeed, the ability of glucagon to titrate out the inhibitory action of insulin might imply an action directed at the level of regulation of G_s.

Certainly, a number of G proteins appear to have β subunits identical to those associated with G_s. Thus, such species would be expected to inhibit adenylyl cyclase activity by the release of such β subunits, which would attenuate the dissociation of G_s. If insulin stimulated a novel G protein to elicit such an action then one might expect that any inhibitory effect could be overcome at saturating glucagon concentrations, provided the concentration of G_s was much greater than that of the novel G protein.

The possibility that insulin might interfere directly with G_s, by causing the phosphorylation of one of its components, must also be considered. This is especially true in view of the fact that the GDP-bound holomeric forms of G_i, G_o, and transducin can be phosphorylated by the insulin receptor tyrosyl kinase (see Section VI). Modification of either the α_s subunit or the β subunit, in order to enhance their affinity for each other, appears to be a possible route for stabilizing the GDP-bound, holomeric, and inactive state of G_s. This might be expected to lead to a state of inhibition that also could be overcome at high glucagon (stimulatory) challenge.

C. Insulin Attenuates Ability of Cholera Toxin to Activate Adenylyl Cyclase

Cholera toxin can bind to cells which express G_{M1} gangliosides on their cell surface membrane. After a defined lag period, which is believed to reflect either an endocytotic processing step (Houslay and Elliott, 1979, 1981) or the entry of the active A subunit into the plasma membrane (Vaughan and Moss, 1978), the A subunit of the toxin elicits the constitutive activation of adenylyl cyclase. It does this by causing the NAD^+-dependent ADP-ribosylation of α_s. Such a covalent modification obliterates the inherent GTPase activity of α_s and allows the free GTP-bound α_s subunit to stimulate adenylyl cyclase activity (Vaughan and Moss, 1978).

Cholera toxin can elicit a profound activation of the basal adenylyl cyclase activity of hepatocytes, which it does after a lag period of 10–12 min (Houslay and Elliott, 1979, 1981). However, if insulin is added to the cells together with cholera toxin, the ability of the toxin to activate adenylyl cyclase is markedly attenuated (Irvine and Houslay, 1988). This action occurs in a fashion which is dose-dependent upon insulin and appears to be mediated through high-affinity insulin receptors. A possible explanation for this phenomenon could be that insulin inhibits the ability of cholera toxin to cause the ADP-ribosylation of α_s. However, this does not appear to offer the explanation as in such experiments it was found that the degree of ADP-ribosylation of α_s, by cholera-toxin treatment, was unaffected by the presence of insulin. Thus it would appear that insulin reduces the ability of ADP-ribosylated α_s to activate adenylyl cyclase. This could be because the ADP-ribosylated form of α_s can serve as a substrate for phosphorylation by the insulin receptor tyrosyl kinase, which attenuates its inherent functioning or enhances its affinity for the G-protein β subunit. Alternatively, it may be that the β subunit of G_s can be phosphorylated, leading to an increase in its affinity for the α subunit of G_s.

D. Insulin Attenuates Ability of Cholera Toxin to Cause ADP-Ribosylation of 25-kDa Protein

The interaction of a specific receptor with a particular G protein would lead to a structural or functional change in that G protein, induced as a consequence of hormone (ligand) binding to the receptor. Does insulin elicit such actions? Certainly the preceding section appears to imply the interaction of insulin with the G protein system. One method of identifying an interaction between the insulin receptor and a G protein would be to investigate whether the hormone stimulates a specific high-affinity GTPase activity and that is discussed in Section V. An alternative method applicable to G proteins, which are subject to covalent modification by either pertussis or cholera toxins, would be to assess whether a hormone altered the ability of the toxin to cause the ADP-ribosylation of the G-protein α subunit in an isolated membrane system. The rationale for such experiments came from observations that pertussis toxin only caused the ADP-ribosylation of the α subunit of G_i when this G protein was in its holomeric state. If isolated membranes were treated with a nonhydrolyzable GTP analog, such as Gpp(NH)p or GTPγS, then pertussis toxin failed to ADP-ribosylate the free α_i subunit (Tsai *et al.*, 1984).

From the experiments described in Sections III.B and IV.B it was suggested that insulin might interact with a unique G protein which could act as a substrate for ribosylation and activation by cholera toxin. Interestingly, a number of workers (see Heyworth *et al.*, 1985 for references) had shown autoradiographs of cholera toxin ADP-ribosylated proteins that had been separated by SDS-PAGE, that implied that there were a number of possible substrates for this toxin in plasma membranes of hepatocytes and adipocytes. However, it was unclear, and remains so to date, as to whether such proteins truly represent distinct G proteins, genetically arisen or proteolytically derived variants of G_s, the non-specific ribosylation of species which are not G proteins, or even autoribosylated cholera toxin A subunit.

Experiments were performed (Heyworth *et al.*, 1985) on rat liver plasma membranes to investigate whether insulin altered the ability of cholera toxin to cause the ADP-ribosylation of particular species in an isolated plasma membrane fraction. Such experiments clearly demonstrated that insulin did not alter the ability of cholera toxin to ribosylate G_s. Indeed, both the holomeric and free subunits of G_s can be ribosylated by cholera toxin, with perhaps the free, GTP-bound α_s subunit providing a very slightly better substrate. Insulin did appear to attenuate the ability of cholera toxin to ADP-ribosylate a 25-kDa component. Under some conditions, usually when no membrane substrates are present, cholera toxin can undergo an autoribosylation reaction and its labeled A subunit migrates in approximately this region. However, five pieces of evidence indicated that this did not appear to offer the explanation for the occurrence of the 25-kDa ribosylated species. First, using an azido-GTP derivative we were able to label a GTP-binding protein in liver plasma membranes that was of this size; second, using membranes prepared from hepatocytes which had been pretreated with glucagon to elicit a "selective" insulin-resistant state, we found that insulin was incapable of inhibiting the ability of cholera toxin to label the 25-kDa protein; third, treatment of cells with the phorbol ester TPA prevented cholera toxin from labeling this protein except when insulin was present; fourth, pretreatment of the intact hepatocytes with cholera toxin caused the endogenous ADP-ribosylation of not only α_s but also the 25-kDa species; and fifth, the use of an ADP-ribosylation inhibitor blocked the ability of cholera toxin to ribosylate the 25-kDa species but did not block the autophosphorylation of the cholera toxin A subunit seen when the toxin was incubated with NAD^+ in the absence of membranes.

Such experiments could be taken to imply the existence of a novel G protein, possessing an α subunit of about 25 kDa, with which the insulin receptor could interact. Certainly, the apparent failure of insulin to exert its attenuating effect in membranes from hepatocytes which had been pretreated with glucagon agrees well with the inability of insulin to inhibit adenylyl cyclase (Heyworth and Houslay, 1983b) and activate the PPM-cyclic AMP phosphodiesterase (Heyworth *et al.*, 1983) after such a pretreatment.

The identity of this putative G protein, called G_{ins} (Houslay and Heyworth, 1983), has yet to be established. However, a 25-kDa α subunit of a G protein, of unknown function, has been isolated from brain, placenta, and platelets (Waldo *et al.*, 1987). It remains to be seen whether this is related to the 25-kDa species identified in liver and presumed to interact with the insulin receptor system.

V. STIMULATION OF DISTINCT GTPASE ACTIVITY IN HUMAN PLATELETS BY INSULIN

As has been described elsewhere in this volume, G proteins exhibit a GTPase activity which reflects their functioning. Thus the receptor-mediated stimulation of a G protein will specifically activate the GTPase activity of that particular G protein. In many tissues, however, such ligand-stimulated GTPases are very difficult to identify owing to the low amounts of such G proteins and to the high nonspecific GTPase activities. However, human platelets have been shown to be particularly useful for identifying and characterizing receptor-controlled GTPase activities. In platelet membranes the GTPase activity of G_s can be stimulated by either prostaglandin E_1 (PGE_1) or by a β-adrenoceptor agonist, that of G_i by either α_2-adrenoceptor stimulation or thrombin, and that of the putative "G_p", which, in platelets, is purported to be a pertussis toxin-insensitive G protein mediating the stimulation of inositol phospholipid metabolism and which can be stimulated specifically by vasopressin and by thrombin (see Houslay et al., 1986a,b). However, as well as these high-affinity GTPase activities a further species stimulated by insulin has been identified (Gawler and Houslay, 1987). This species appears to be distinct from G_s, G_i, and G_p, although, like G_s, it would seem to be a substrate for the action of cholera toxin. If such a species can be identified, then certainly the ability of cholera toxin to activate it would account for the stimulatory effect of this toxin on the PPM-cyclic AMP phosphodiesterase. If this putative G protein, which we have called G_{ins}, is also responsible for exerting an inhibitory effect upon adenylyl cyclase, perhaps by the release of β subunits, the cholera-toxin treatment would be expected to obliterate the inhibitory action of insulin applied in isolated membranes as it would lead to the activation of both G_s and G_{ins}.

VI. PHOSPHORYLATION OF DEFINED GTP-BINDING PROTEINS BY HUMAN INSULIN RECEPTOR

A. G_i and G_o

A pure preparation of insulin receptors, solubilized from human placental tissue, has been shown to cause the phosphorylation of the α and β subunits of pure preparations of G_i and G_o from bovine brain (O'Brien et al., 1987a). However, it was found to be crucial that these G proteins were in their GDP-bound state for them to be phosphorylated by the insulin receptor kinase. Thus, activation of these G proteins, using non-hydrolyzable analogs of GTP, obliterated the ability of the insulin receptor to phosphorylate them. In contrast, stabilization of the holomeric, inactive state, using GDPβS, ensured that α and β subunits of G_s and G_i were phosphorylated by the insulin receptor kinase (Fig. 6). We did note, however, that at high Mg^{2+} concentrations the ability of the insulin receptor to phosphorylate the α subunit of G_i was abolished despite the presence of GDPβS. It is known that Mg^{2+} can act to regulate the functioning of G proteins (Birnbaumer *et al.*, 1985) and it is thus possible that the conformational change induced in G_i by this divalent cation acted to prevent the α subunit of G_i from providing a substrate for phosphorylation by the insulin receptor tyrosyl kinase.

Phosphorylation of the α and β subunits of G_s and G_i was alkali-stable and shown to be on tyrosine residues rather than serine. When insulin was used to stimulate the receptor kinase activity, the phosphorylation reaction exhibited a K_m of circa 20 μM for G_s and G_i (O'Brien *et al.*, 1987a). This would thus appear to be a high-affinity reaction when such a value is compared with those found for various exogenous substrates that can be phosphorylated by the pure insulin receptor, e.g., histones (10–100 μM), casein (10–20 μM) and angiotensin (2–4 mM). The rate of phosphorylation of G_s and G_i was found to be comparable to that using casein as a substrate but rather less than that seen with histones which, to date, have provided one of the best exogenous substrates.

The presence of G_s and G_i, however, did not affect the extent or rate of the insulin-stimulated autophosphorylation of the β subunit of the insulin receptor. Thus G_s and G_i neither regulate autophosphorylation or appear to act as competing substrates for the autophosphorylation reaction (O'Brien *et al.*, 1987a).

The stoichiometry and the functional significance of the phosphorylation of G_i subunits have yet to be determined. It may be, however, that the phosphorylation of G_s and G_i acts to alter the stability of the inactive, holomeric state of these species. In view of the experiments describing the interaction of insulin with adenylyl cyclase it is tempting to suggest that tyrosyl phosphorylation of G_s may occur, where such an event might act to stabilize the holomeric, inactive state of this G protein.

It will be of particular interest to determine whether such phosphorylation events can occur *in situ* in the intact cell.

B. Transducin

This G protein is found exclusively in the retina, where distinct forms occur in rods and cones. It plays a key role in the visual transduction system where it acts to couple rhodopsin to the stimulation of a high-affinity cyclic AMP phosphodiesterase.

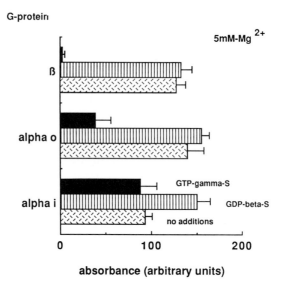

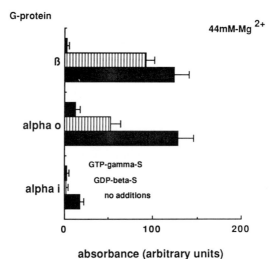

Fig. 6. The insulin receptor phosphorylates the GDP-bound forms of G_i and G_o. Data are adapted from O'Brien *et al.* (1987a). Graph shows the extent of phosphorylation, assessed by densitometric scanning of autoradiographs, of the subunits of G_i and G_o by the insulin receptor tyrosyl kinase. At low [Mg^{2+}], phosphorylation occurs either in the presence of GDPβS or in the absence of added nucleotides, the G proteins being in their GDP-bound holomeric states. GTPγS has a higher affinity for G_o than for G_i, hence the differences in attenuation of association and hence phosphorylation. At high [Mg^{2+}], GTPγS causes dissociation of all species, hence the loss in phosphorylation. At high [Mg^{2+}] the phosphorylation of G_i occurs, despite GDPβS being present, as this divalent cation can presumably effect a conformational change in this G protein.

Purified transducin (T) has also been shown to be a substrate for phosphorylation by the human insulin receptor (Zick *et al.*, 1986). This occurred on tyrosine residues and seemed to occur primarily on the α subunit of transducin (Tα). As with the experiments performed with G_i and G_o, the presence of transducin had no effect on insulin-stimulated receptor autophosphorylation. Transducin, however, appeared to be a good substrate for phosphorylation by the purified insulin receptor, exhibiting a K_m value of circa 1 μM.

As with G_s and G_i, phosphorylation only occurred when the G protein was in its GDP-bound state and did not occur if activating GTP analogs were used. In these experiments, however, it was possible to show that the free α subunit of transducin could be phosphorylated, but that this only occurred if it was in its inactive GDP-bound state. Such a form of the enzyme is unlikely to have any prolonged existence *in situ* in the cell as it would be expected to associate rapidly with $\beta\gamma$ subunits. Thus it would seem from these studies that it is not the holomeric state of the G protein per se which allows it to act as a substrate for the tyrosyl kinase but, rather, that GDP be bound to the α subunit of the G protein. This gives further credence to the notion that phosphorylation may play a regulatory role, perhaps by stabilizing the GDP-bound form of the regulatory protein, presumably in its holomeric state.

The potentially important observation that Tα can be phosphorylated on tyrosyl residues, *in situ* in retinal rod cells, has been made (Zick *et al.*, 1987). Retinal rods express EGF-1 receptors which, like those for insulin, express a tyrosyl kinase activity. Treatment of retinal rod cells with EGF appeared to elicit the phosphorylation of Tα. This was also seen using a solubilized and purified IGF-1 receptor preparation together with pure transducin. Phosphorylation of transducin by the EGF-1 receptor depended upon transducin being in its GDP-bound form, confirming all of the other studies. The functional significance of such a modification has, however, yet to be ascertained, however it has been noted that IGF-1 concentrations in the vitreous of diabetics with severe retinopathy are increased some twofold over those found in control subjects.

C. p21^{ras}

Of all human cancers, 20–30% are characterized by the presence of a mutant *ras* oncogene. This gene codes for a 21-kDa GTP/GDP-binding protein, called p21^{ras}, which is found associated with the cytosol surface of the plasma membrane. Activated oncogenes usually express an attenuated GTPase activity and it has been suggested that the protooncogene product may perform a regulatory role in the cell, where point mutations lead to the constitutive activation of p21^{ras} in much the same way that cholera toxin causes the constitutive activation of adenylyl cyclase by obliterating the GTPase activity of G$_s$. The functional role of the protooncogene and the molecular mechanism of action of its oncogenic variants have yet to be defined (Macara, 1985). However, evidence has been presented which suggests that it may act to couple receptors to stimulate inositol phospholipid metabolism in systems which are insensitive to the action of pertussis toxin (Wakelam et al., 1986a).

1. Attenuation of Receptor Autophosphorylation

Pure preparations of p21^{ras}, from *Escherichia coli* expression systems, are available. These are identical to p21^{ras} found in mammalian systems except that they are not acylated on Cys^{186} and thus do not insert correctly into membranes. Incubation of a pure preparation of p21^{ras} with the pure human insulin receptor failed to elicit the phosphorylation of p21^{ras}, irrespective of whether p21^{ras} was in its inactive, GDPβS, or activated, Gpp(NH)p-bound, state. However, if p21^{ras} was purified by method involving a denaturation/renaturation cycle with urea, then p21^{ras} did act as a substrate for the insulin receptor tyrosyl kinase. Under such circumstances p21^{ras} was phosphorylated on tyrosyl residues (O'Brien et al., 1987b). Of particular note, however, was our observation (O'Brien et al., 1987b) that if p21^{ras} was not purified by the denaturation/renaturation cycle with urea, then, and only then, did it act to inhibit insulin-stimulated receptor autophosphorylation. As no phosphorylation of p21^{ras} occurred, this clearly could not be exerting such an action by providing an alternative substrate. Indeed, the ability of p21^{ras} to inhibit receptor phosphorylation depended upon this G protein being in its GDP-bound form; the addition of GTP analogs prevented such an action. Half-maximal inhibition of receptor autophosphorylation was seen at $10-20$ μM p21^{ras} and at circa 100 μM GDP. The rather high concentration of GDP required certainly requires further explanation if this is to be a physiologically relevant action. Interestingly, both mutant (Val^{12}) forms of Ha-p21^{ras} and mutant forms (Lys^{61}) of N-p21^{ras} were equally effective as their normal protooncogene forms.

Phosphorylation of p21^{ras} purified using the purified insulin receptor has also been noted by other investigators (Korn et al., 1987; Kamata et al., 1987), where it appeared that phosphorylation of p21^{ras} did not affect its GTPase activity. As found also by O'Brien et al. (1987b), receptor autophosphorylation was unaffected by the presence of p21^{ras} under conditions where p21^{ras} was a substrate for phosphorylation by the receptor kinase.

Indeed, under such conditions it also appeared that the binding of insulin to its receptor was unaffected by the presence of p21^{ras}.

Kamata *et al.* (1987) have demonstrated that insulin can stimulate the phosphorylation of v-Harvey-p21^{ras} in both intact HaNRK cells and in membranes prepared from them. However, in contrast to the studies performed above using purified soluble components, in these membrane/cell studies insulin stimulated the phosphorylation of p21^{ras} exclusively on threonine residue(s).

All of these studies indicate that p21^{ras} can interact with the insulin receptor. However, they strongly imply that the conformation of p21^{ras} is crucial in defining the mode of interaction, i.e., whether p21^{ras} can either attenuate receptor autophosphorylation or be subject to phosphorylation on either threonine or tyrosine residues. It may be that such interactions are controlled by the membrane components, e.g. growth factor receptors, with which p21^{ras} interacts and also by the nature of the guanine nucleotide which occupies the binding site on p21^{ras}.

It remains to be seen whether p21^{ras} can exert a tonic inhibitory effect *in situ* in the intact cell. If this should, however, prove to be the case, then one might expect that receptors which could interact with p21^{ras}, activating it by allowing it to bind GTP, would serve to potentiate at least certain of the actions of insulin. Indeed, this might account for the synergizing effect that insulin has with certain growth factors in stimulating cell growth.

2. Action of Anti-p21^{ras} Antibody

The possibility that p21^{ras} may be connected with certain actions of insulin was mooted some time ago (Houslay and Heyworth, 1983). Insulin induces the maturation of *Xenopus* oocytes through a process which appears to involve the tyrosyl kinase activity of the insulin receptor, as maturation can be blocked by the microinjection of an antibody which blocks the functioning of the receptor kinase. However, it has been demonstrated (Birchmeier *et al.*, 1985) that microinjection of the activated, oncogenic form of p21^{ras} into *Xenopus* oocytes also elicits their maturation. Korn *et al.* (1987) have attempted to determine whether there is an interaction in this system between the insulin receptor and p21^{ras}. To do this they microinjected into oocytes a monoclonal antibody directed against a synthetic peptide which corresponded to residues 29–44 of p21^{ras}. Such a treatment was found to block the ability of microinjected v-Ha-p21^{ras} molecules, which were activated by a point mutation (Val) at position 12, to induce maturation. However, *Xenopus* oocytes, as with most cells, contain normal (protooncogenic) forms of p21^{ras}, which are presumed to function as coupling G proteins. Interestingly, microinjection of this specific monoclonal antibody was demonstrated to block the ability of insulin to stimulate oocyte maturation in a dose-dependent fashion. In contrast, this antibody had no effect whatsoever on the ability of progesterone to induce maturation, indicating a specificity in its action.

Such experiments provide tantalizing evidence for an interaction between the insulin receptor and p21^{ras}. Whether p21^{ras} is involved in the signaling mechanism per se or in regulating the functioning of the insulin receptor remains to be seen.

VII. CONCLUDING REMARKS

In 1983 we suggested (Heyworth *et al.,* 1983; Houslay and Heyworth, 1983) that insulin might exert at least certain of its actions by interacting with the G protein system. Since then an increasing amount of evidence has accrued in support of such a contention. What, however, remains unclear is whether insulin exerts actions through a distinct G protein, the putative G_{ins}, or whether it merely acts to modify the functioning of other G proteins of established identity; indeed, it may exert actions through more than one of these routes. Certainly, the technology is now available for investigating the action of insulin on the known G proteins and we can expect to see further fruits of such studies appearing in the near future.

To identify a distinct G protein through which insulin might exert actions is more problematical as it depends upon constructing an appropriate reconstitution assay system along the lines used to identify G_s. Certainly there are now a number of purified or partially resolved membrane-bound GTP-binding proteins which are in search of a function.

The possibility that insulin might interact with a specific G protein begs the question as to what effector/signal generation system might be coupled to it. One possibility would be the control of a serine kinase activity of the nature suggested by Denton *et al.* (1981). Indeed, Walaas *et al.* (1981) noted that the cyclic AMP-independent kinase activity of insulin, found in sarcolemma membranes, was enhanced if micromolar concentrations of GTP were present. They suggested, based upon these observations, that a G protein may act to control this kinase activity.

It has been suggested (Saltiel *et al.,* 1987) that insulin might activate a phosphatidylinositol-specific phospholipase C in order to produce two second messengers. One of these is diacylglycerol, and the other a glycan inositol phosphate (GIP) compound whose precise structure has yet to be elucidated (Saltiel *et al.,* 1987). By analogy with receptor stimulation of (poly)phosphatidylinositol metabolism, in which a specific phospholipase C is activated by a putative G protein, termed G_p (Cockcroft and Gomperts, 1985), it may be that a G protein, specific to the action of insulin, is involved in the process of generating such second messengers. It should be pointed out that the GIP compound(s) has also been suggested to be capable of inhibiting adenylyl cyclase (Saltiel *et al.*, 1987). This requires further investigation as a possible alternative mechanism to the ones proposed above.

A further possible point of interaction of the putative G_{ins} would be at the level of control of glucose transporters as shown by studies (Baker and Carruthers, 1983) in which microinjection of nonhydrolyzable GTP analogs into barnacle muscle cells was found to mimic the action of insulin in stimulating glucose transport. In adipocytes, insulin stimulates glucose transport by a two-step process, the first involving carrier recruitment to the plasma membrane, the second being carrier activation. The regulation of this second process has many analogies with that seen for the regulation of the PPM-PDE and is a possible site for regulation by a G protein (see Houslay, 1985).

G-protein involvement has also recently been suggested (Nishimoto *et al.*, 1987) to play a role in the molecular mechanism of action of the IGF-2 receptor. In this instance, involvement has been inferred from studies using pertussis toxin. It was demonstrated that, in fibroblasts primed by exposure to EGF, IGF-2 could trigger an influx of Ca^{2+} that was blocked when cells were pretreated with pertussis toxin. While blockade of this action by pertussis toxin appeared to show a good relationship with the ribosylation of a 41-kDa protein, it remains to be determined whether this is G_i, G_o, or a related species, and whether direct coupling between the IGF-2 receptor and such a G protein occurs. However, that the nonhydrolyzable GTP analog, GTPγS, appeared to inhibit the binding of IGF-2 to its receptor in a fashion which was blocked by pertussis-toxin treatment, indicates that this receptor might indeed interact with a pertussis toxin-sensitive G protein.

It is apparent from this review that a significant body of information is accumulating which implicates the interaction of the insulin receptor with the G protein system. It remains to be seen whether an insulin-specific species exists, the putative G_{ins}, or whether insulin confines itself to interacting with and modifying the function of established species. Certainly the binding of insulin to its receptor might elicit changes in the functioning of more than one G-protein species; this could lead to a pronounced proliferation of signals emanating from the plasma membrane as a consequence of the binding of insulin to its receptor. Such a response could be interpreted as leading to a multipathway mechanism of action (Fig. 7) for insulin as proposed by us earlier (Houslay and Heyworth, 1983).

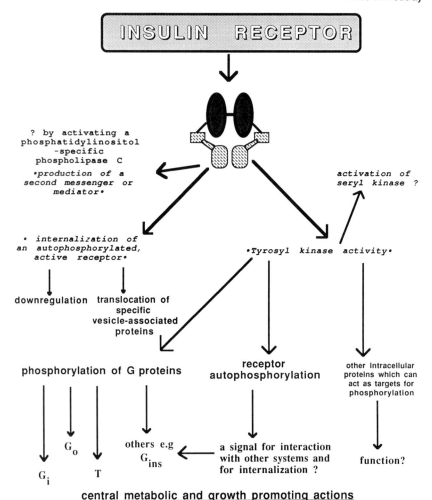

central metabolic and growth promoting actions

Fig. 7. A multipathway mechanism for the action of insulin. Here it is envisaged that information entering the cell, as a consequence of insulin binding to its receptor, diverges rapidly to generate a number of diverse signals. Lesions in certain of these pathways give rise to selective insulin-resistant states.

ACKNOWLEDGMENTS

The work described in this report was supported by the Medical Research Council, United Kingdom, British Diabetic Association, and the Scottish Home and Health Department.

REFERENCES

Baker, P. F., and Carruthers, A. (1983). Insulin regulation of sugar transport in giant muscle fibres of the barnacle. *J. Physiol.* **336,** 397–431.

Beavo, J. A., Hansen, R. S., Harrison, S. A., Hurwitz, R. L., Martins, T. J., and Mumby, M. C. (1982). Multiple forms of cyclic nucleotide phosphodiesterases. *Mol. Cell. Endocrinol.* **28,** 387–410.

Birchmeier, C., Brock, D., and Wigler, M. (1985). RAS proteins can cause meiosis in *Xenopus oocytes. Cell* **43,** 615–621.

Birnbaumer, L. (1973). Hormone-stimulated adenylate cyclase. *Biochim. Biophys. Acta* **300,** 129–158.

Birnbaumer, L., Codina, J., Mattera, R., Cerione, R. A., Hildebrandt, J. D., Sunya, T., Rojas, F. J., Caron, M. J., Lefkowitz, R. J., and Iyengar, R. (1985). Structural basis of adenylate cyclase stimulation and inhibition by distinct guanine nucleotide regulatory proteins. *Mol. Asp. Cellul. Reg.* **4,** 131–182.

Cockcroft, S., and Gomperts, B. D. (1985). Role of a guanine nucleotide binding protein in the activation of polyphosphatidylinositide phosphodiesterase. *Nature (London)* **314,** 534–536.

Czech, M. P. (1975). The nature and regulation of the insulin receptor: structure/function. *Annu. Rev. Physiol.* **47,** 357–381.

Czech, M. P. (1977). Molecular basis of insulin action. *Annu. Rev. Biochem.* **46,** 359–384.

Degerman, E., Belfrage, P., Newman, A. H., Rice, K. C., and Manganiello, V. C. (1987). Purification of the putative hormone-sensitive cyclic AMP phosphodiesterase from rat adipose tissue using a derivative of cilostamide as a novel affinity ligand. *J. Biol. Chem.* **262,** 5957–5807.

Denton, R. M. (1986). Early events in insulin actions. *Adv. Cyclic Nucleotide Prot. Phosph. Res.* **20,** 295–341.

Denton, R. M., Brownsey, R. W., and Belsham, G. J. (1981). A partial view of the mechanism of insulin action. *Diabetologia* **21,** 347–367.

Ebina, Y., Ellis, L., Jarnagan, K., Edery, M., Graf, L., Clauser, E., Ou, J-h., Masiarz, F., Kan, Y. W., Goldfine, I. D., Roth, R. A., and Rutter, W. J. (1985). The human insulin receptor cDNA. *Cell* **40,** 747–758.

Elks, M. L., Watkins, P. A., Manganiello, V. C., Moss, J., Hewlett, E., and Vaughan, M. (1983). Pertussis toxin blocks insulin's activation of adipocyte cyclic AMP phosphodiesterase. *Biochem. Biophys. Res. Commun.* **116,** 593–596.

Ellis, L., Clauser, E., Morgan, D. O., Edery, M., Roth, R. A., and Rutter, W. J. (1986). Replacement of insulin receptor tyrosine residues 1162 and 1163 compromises insulin-stimulated kinase activity and uptake of 2-deoxyglucose. *Cell* **45,** 721–732. (The nomenclature used here is different from that quoted in the text which refers to that cited by Ullrich *et al.* [1985]. Using that 'nomenclature these tyrosyl residues would be referred to as 1150 and 1151.)

Foulkes, J. G., and Rosner, M. R. (1985). Tyrosine-specific protein kinases as mediators of growth control. *Mol. Asp. Cellul. Reg.* **4,** 217–252.

Francis, S. H., and Kono, T. (1982). Hormonal regulation of cyclic AMP phospho-diesterases. *Mol. Cell. Biochem.* **42,** 109–116.

Gawler, D., and Houslay, M. D. (1987). Insulin stimulates a novel GTPase activity in human platelets. *FEBS Lett.* **216,** 94–98.

Gawler, D., Milligan, G., Spiegel, A. M., Unson, C. G., and Houslay, M. D. (1987). Abolition of the expression of inhibitory guanine nucleotide regulatory protein G_i activity in diabetes. *Nature (London)* **327,** 229–232.

Gawler, D. J., Milligan, G., and Houslay, M. D. (1988). Treatment of streptozoto-cin diabetic rats with metformin restores the ability of insulin to inhibit adenylate cyclase activity and demonstrates that insulin does not exert this action through the inhibitory guanine nucleotide regulatory protein G_i. *Bio-chem. J.* **249,** 537–542.

Herrera, R., Petruzzelli, L., Thomas, N., Bramson, H. N., Kaiser, E. T., and Rosen, O. M. (1985). An antipeptide antibody that specifically inhibits insulin recep-tor autophosphorylation and protein kinase activity. *Proc. Natl. Acad. Sci. U.S.A.* **82,** 7899–7903.

Heyworth, C. M., and Houslay, M. D. (1983a). Challenge of hepatocytes by gluca-gon triggers a rapid modulation of adenylate cyclase activity in isolated mem-branes. *Biochem. J.* **214,** 93–98.

Heyworth, C. M., and Houslay, M. D. (1983b). Insulin exerts actions through a distinct species of guanine nucleotide regulatory protein: Inhibition of adeny-late cyclase. *Biochem. J.* **214,** 547–552.

Heyworth, C. M., Wallace, A. V., and Houslay, M. D. (1983). Insulin and glucagon regulate the activation of two distinct membrane-bound cyclic AMP phospho-diesterases in hepatocytes. *Biochem. J.* **214,** 99–110.

Heyworth, C. M., Wallace, A. V., Wilson, S. R., and Houslay, M. D. (1984). An assessment of the ability of insulin-stimulated cyclic AMP phosphodiesterase to decrease hepatocyte intracellular cyclic AMP concentrations. *Biochem. J.* **222,** 183–187.

Heyworth, C. M., Whetton, A. D., Wong, S., Martin, B. R., and Houslay, M. D. (1985). Insulin inhibits the cholera toxin-catalysed ribosylation of a M_r-25,000 protein. *Biochem. J.* **228,** 593–603.

Heyworth, C. M., Grey, A-M., Wilson, S. R., Hanski, E., and Houslay, M. D. (1986). The action of islet activating protein (pertussis toxin) on insulin's ability to inhibit adenylate cyclase and activate cyclic AMP phosphodiesterases in hepatocytes. *Biochem. J.* **235,** 145–149.

Houslay, M. D. (1981). Membrane phosphorylation: An essential role for the action of insulin, EGF and pp60*src? Bioscience Rep.* **1,** 19–34.

Houslay, M. D. (1985). The insulin receptor and signal generation at the plasma membrane. *Mol. Asp. Cellul. Reg.* **4,** 279–333.

Houslay, M. D. (1986). Insulin, glucagon and the receptor-mediated control of cyclic AMP concentrations in liver. *Biochem. Soc. Trans.* **14,** 183–193.

Houslay, M. D. (1988). *In* "Hormones and Their Actions - New Comprehensive Biochemistry" (B. A. Cooke, R. J. B. King, and H. J. Van der Molen, eds.) Ch. 15 pp 321–348. Elsevier Biomedical Press, Amsterdam.

Houslay, M. D., and Elliott, K. R. F. (1979). Cholera toxin-mediated activation of adenylate cyclase in rat hepatocytes. *FEBS Lett.* **104**, 359–363.

Houslay, M. D., and Elliott, K. R. F. (1981). Is the receptor-mediated endocytosis of cholera toxin a pre-requisite for its activation of adenylate cyclase in intact hepatocytes? *FEBS Lett.* **128**, 289–292.

Houslay, M. D., and Heyworth, C. M. (1983). Insulin: In search of a mechanism. *Trends Biochem. Sci.* **8**, 449–457.

Houslay, M. D., Wallace, A. V., Wilson, S. R., Marchmont, R. J., and Heyworth, C. M. (1983). *In* "Hormones and Cell Regulation" J. N. Dumont and J. Nunez eds., Vol. VII, pp 105–120, Elsevier Biomedical Press, Amsterdam.

Houslay, M. D., Bojanic, D., and Wilson, A. (1986a). Platelet activating factor and U44069 stimulate a GTPase activity in human platelets which is distinct from the guanine nucleotide regulatory proteins, N_s and N_i. *Biochem. J.* **234**, 737–740. (The nonmenclature G_s and G_i now replaces the 'old' nomenclature for N_s and N_i, respectively.)

Houslay, M. D., Bojanic, D., Gawler, D., O'Hagan, S., and Wilson, A. (1986b). Thrombin, unlike vasopressin, appears to stimulate two distinct guanine nucleotide regulatory proteins in human platelets. *Biochem. J.* **238**, 109–113.

Irvine, F. J., and Houslay, M. D. (1988). Insulin and glucagon attenuate the ability of cholera toxin to activate adenylate cyclase in intact hepatocytes. *Biochem. J.* **251**, 447–452.

Kamata, T., Kathuria, S., and Fujita-Yamaguchi, Y. (1987). Insulin stimulates the phosphorylation level of v-Ha-*ras* protein in membrane fraction. *Biochem. Biophys. Res. Commun.* **144**, 19–25.

Klein, H. H., Ciaraldi, T. P., Friedenberg, G., and Olefsky, J. M. (1987). Adenosine modulates insulin activation of insulin receptor kinase in intact rat adipocytes. *Endocrinology* **120**, 2339–2345.

Korn, L. J., Siebel, C. W., McCormick, F., and Roth, R. A. (1987). *Ras* p21 as a potential mediator of insulin action in *Xenopus* oocytes. *Science* **236**, 840–843.

Macara, I. (1985). Oncogenes, ions and phospholipids. *Am. J. Physiol.* **248**, C3–C33.

Marchmont, R. J., and Houslay, M. D. (1980). Insulin triggers the cyclic AMP-dependent activation and phosphorylation of a plasma membrane cyclic AMP phosphodiesterase. *Nature (London)* **286**, 904–906.

Marchmont, R. J., and Houslay, M. D. (1981). Characterisation of the phosphorylated form of the insulin-stimulated cyclic AMP phosphodiesterase from rat liver plasma membranes. *Biochem. J.* **195**, 653–660.

Murphy, G. J., Hruby, V. J., Trivedi, D., Wakelam, M. J. O., and Houslay, M. D. (1987). The rapid desensitization of glucagon-stimulated adenylate cyclase is a cyclic AMP-independent process that can be mimicked by hormones which stimulate inositol phospholipid. *Biochem. J.* **243**, 39–46.

Nishimoto, I., Hata, Y., Ogata, E., and Kojima, I. (1987). Insulin-like growth factor 2 stimulates calcium influx in competant BALB/c 3T3 cells primed with epidermal growth factor. *J. Biol. Chem.* **262**, 12120–12126.

O'Brien, R. M., Houslay, M. D., Milligan, G., and Siddle, K. (1987a). The insulin receptor tyrosyl kinase phosphorylates holomeric forms of the guanine nucleotide regulatory proteins G_i and G_o. *FEBS Lett.* **212**, 281–288.

O'Brien, R. M., Siddle, K., Houslay, M. D., and Hall, A. (1987b). Interaction of the human insulin receptor with the *ras* oncogene product p21. *FEBS Lett.* **217**, 253–259.

Pyne, N., Cooper, M., and Houslay, M. D. (1987a). The insulin- and glucagon-stimulated 'dense-vesicle' high affinity cyclic AMP phosphodiesterase from rat liver. *Biochem. J.* **242**, 33–42.

Pyne, N. J., Anderson, N., Lavan, B. E., Milligan, G., Nimmo, H. G., and Houslay, M. D. (1987b). Specific antibodies and the selective inhibitor ICI 118233 demonstrate that the hormonally-stimulated 'dense-vesicle' and peripheral plasma membrane cyclic AMP phosphodiesterases display distinct tissue distribution in the rat. *Biochem. J.* **248**, 897–901.

Roth, R. A., and Beaudoin, J. (1987). Phosphorylation and inhibition of the insulin receptor tyrosyl kinase by cyclic AMP-dependent protein kinase. *Diabetes* **36**, 123–126.

Saltiel, A. R., Sherline, P., and Fox, J. A. (1987). Insulin-stimulated diacylglycerol production results from the hydrolysis of a novel phosphattiidyl inositol glycan. *J. Biol. Chem.* **262**, 1116–1121.

Senft, G., Schultz, G., Munske, K., and Hoffman, M. (1968). Influence of insulin on cAMP-PDE activity in liver, skeletal muscle, adipose tissue and kidney. *Diabetologia* **4**, 322–329.

Takemoto, D. J., Hansen, J., Takemoto, L. J., and Houslay, M. D. (1982). Peptide mapping of the multiple forms of cyclic AMP phosphodiesterase. *J. Biol. Chem.* **257**, 14597–14599.

Tanti, J. F., Gremeaux, T., Rochet, N., Van Obberghen, E., and Le Marchand-Brustel, Y. (1987). Effect of cyclic AMP-dependent protein kinase on insulin receptor protein kinase activity. *Biochem. J.* **245**, 19–26.

Thompson, W. J., and Strada, S. J. (1978). *In* "Receptors and Hormone Action" (L. Birnbaumer and S. J. O'Malley, eds.) Vol.3 pp. 553–575, Academic Press, New York.

Tsai, S. C., Adamik, R., Kanaho, Y., Hewlett, E. L., and Moss, J. (1984). Effects of guanyl nucleotides and rhodopsin on ADP-ribosylation of the inhibitory GTP-binding component of adenylate cyclase by pertussis toxin. *J. Biol. Chem.* **259**, 15320–15323.

Ullrich, A., Bell, J. R., Chen, E. Y., Herrera, P., Petruzzelli, L. M., Dull, T. J., Gray, A., Coussens, L., Liao, Y. C., Tsubokawa, M., Mason, A., Seeburg, P. H., Grunfeld, C., Rosen, O. M., and Ramachandran, J. (1985). The human insulin receptor and its relationship to the tyrosyl kinase family of oncogenes. *Nature (London)* **313**, 756–761.

Vaughan, M., and Moss, J. (1978). Mechanism of action of cholera toxin. *J. Supramolec. Struct.* **8**, 473–488.

Wakelam, M. J. O., Davies, S. A., Houslay, M. D., McKay, I., Marshall, C. J., and Hall, A. (1986a). Normal p21[N-ras] couples bombesin and other growth factor receptors to inositol phosphate production. *Nature (London)* **323**, 173–176.

Wakelam, M. J. O., Murphy, G. J., Hruby, V. J., and Houslay, M. D. (1986b). Activation of two signal transduction systems in hepatocytes by glucagon. *Nature (London)* **323,** 68–71.

Walaas, O., Horn, R. S., Lystad, E., and Adler, L. (1981). ADP-ribosylation of sarcolemma membrane protein sin the presence of cholera toxin and its influence on insulin-stimulated membrane protein kinase activity. *FEBS Lett.* **128,** 133–136.

Waldo, G. L., Evans, T., Fraser, E. D., Northup, J. K., Martin, M. W., and Harden, T. K. (1987). Identification and purification from bovine brain of a guanine nucleotide binding protein distinct from G_s, G_i, and G_o. *Biochem. J.* **246,** 431–439.

Wallace, A. V., Heyworth, C. M., and Houslay, M. D. (1984). N6 (phenylisopropyladenosine) prevents glucagon both blocking insulin's activation of the plasma membrane cyclic AMP phosphodiesterase and uncoupling hormonal stimulation of adenylate cyclase activity in hepatocytes. *Biochem. J.* **222,** 177–182.

White, M. F., Stegmann, E. W., Dull, T. J., Ullrich, and Kahn, C. R. (1987). Characterization of an endogenous substrate of the insulin receptor. *J. Biol. Chem.* **262,** 9769–9777.

Wilson, S. R., Wallace, A. V., and Houslay, M. D. (1983). Insulin activates the plasma membrane and dense-vesicle cyclic AMP phosphodiesterase in hepatocytes by distinct routes. *Biochem. J.* **216,** 245–248.

Zick, Y., Sagi-Eisenberg, R., Pines, M., Gierschik, P., and Spiegel, A. M. (1986). Multisite phosphorylation of the α-subunit of transducin by the insulin receptor kinase and protein kinase C. *Proc. Natl. Acad. Sci. U.S.A.* **83,** 9294–9297.

Zick, Y., Spiegel, A. M., and Sagi-Eisenberg, R. (1987). Insulin-like growth factor 1 receptors in retinal rod outer segments. *J. Biol. Chem.* **262,** 10259–10264.

CHAPTER 23

G Proteins in Growth Factor Action

Jacques Pouysségur

Centre de Biochimie, CNRS, Université de Nice, Parc Valrose, 06034 Nice, France

I. INTRODUCTION

Cell division is a complex biological process, fully active during embryogenesis and development and essential over the entire adult life for hematopoiesis, immunological response, and tissue repair. An extreme variety of growth factors and hormones play a pivotal role in controlling cell proliferation. Therefore, G proteins which mediate the action of some of these hormones (Stryer and Bourne, 1986; Gilman, 1987) appear to represent key signal transducers in processes leading to initiation or inhibition of cell division. Perhaps the first observation supporting this proposal is that growth of cells in culture is often inhibited by bacterial toxins known to specifically ADP-ribosylate G proteins. Pertussis toxin might act by interrupting the growth factor signal (Katada and Ui, 1982), and cholera toxin by stimulating a rise in cAMP, which in many cell systems acts as an inhibitory signal for growth (Pastan *et al.*, 1975). Another strong indication along this line which has aroused the interest of many investigators is that abnormal expression of mutated *ras* gene products (similar to G proteins), suffices to induce uncontrolled cell proliferation (Barbacid, 1987).

In this chapter, we would like to present the nature of the signaling pathways which might be directly involved in the initiation of growth and discuss the possible implication of G proteins. We shall refer mainly to our studies performed with secondary cultures and a diploid cell line (CCL39) of Chinese hamster lung fibroblasts (CHL) and to studies by others with mouse 3T3 fibroblasts.

II. CHINESE HAMSTER LUNG FIBROBLASTS: A MODEL SYSTEM TO ANALYZE GROWTH FACTOR ACTION

Chinese hamster lung (CHL) fibroblasts grow with a generation time of 12 to 15 hr in medium supplemented with 10% fetal calf serum. Removal of serum for 24 hr is sufficient to arrest the entire population in a postmitosis resting state referred to as G_0/G_1. With the sole presence of insulin and transferrin, these quiescent cells remain viable for at least 1 week as judged by their ability to reenter the proliferative state in response to serum. A limited set of pure hormones or growth factors was found to elicit reinitiation of DNA synthesis in these G_0-arrested CHL cells. In decreasing order of potency, we have: α-thrombin (THR), fibroblast growth factor (FGF), epidermal growth factor (EGF), platelet-derived growth factor (PDGF), serotonin (5-HT), and insulin (substituting for insulin-like growth factor-I, IGF-I). For instance, at their optimal concentration, and depending on whether the cell line CCL39 or secondary cultures are used, THR, FGF, or EGF alone triggers DNA replication in respectively 30–40%, 10–25%, or 5–20% of the cells as measured by labeled nuclei. Insulin (IGF-I) has no mitogenic action per se but potentiates the mitogenic action of all other growth factors. 5-HT, a very weak mitogen, potentiates the action of EGF, PDGF, and especially that of FGF. Although weak, this mitogen is of great value because the pharmacology of 5-HT receptors is rather well developed allowing discrimination between different signaling pathways with the use of specific agonists/antagonists (Fozard, 1987).

Another point of interest is that these mitogens are also competent for promoting continuous cell proliferation of CCL39 cells in serum-free medium (Pérez-Rodriguez *et al.*, 1981). The minimal combination, insulin and THR (1 nM) allows an exponential growth with a generation time of 15–18 hrs (Van Obberghen-Schilling and Pouysségur, 1983).

III. MECHANISMS OF GROWTH-FACTOR SIGNAL TRANSDUCTION

A. G Protein-Mediated Transmembrane Signaling Pathways

Receptor-mediated hydrolysis of phosphatidylinositol bisphosphate, (PIP$_2$) which generates two second messenger molecules, inositol trisphosphate (IP$_3$) and diacylglycerol (Berridge, 1984; Nishizuka, 1986), has been strongly implicated in various hormone-activated processes including cell proliferation (Berridge, 1984).

It is of interest to see that THR, the most potent of the mitogens for quiescent CHL fibroblasts, is the most potent effector of inositol lipid hydrolysis (Carney *et al.*, 1985; L'Allemain *et al.*, 1986). IP$_3$ could be detected as early as 5 sec after THR addition, and parallels a transient rise in cytoplasmic Ca^{2+} (Magnaldo *et al.*, 1986). 5-HT (10 μM) is a weak mitogen when added alone on G$_0$-arrested CHL cells, but it significantly induced polyphosphoinositide (PI) hydrolysis. The very low basal rate of PIP$_2$–phospholipase C (PIP$_2$–PLC) of quiescent CHL cells is stimulated ten-fold by 5-HT, a response equivalent to 1 to 10 pM of THR (Seuwen *et al.*, 1988b). Interestingly, ketanserin or ritanserin, two antagonists of 5-HT$_2$ receptors (Fozard, 1987) completely abolished 5-HT-induced PIP$_2$– PLC (Seuwen *et al.*, 1988b).

As reported for other systems (Cockcroft and Gomperts, 1985), hormonal activation of PIP$_2$–PLC in CHL fibroblasts is mediated via a GTP-binding protein (G$_p$). Three observations support this conclusion: (1) both THR- and 5-HT-induced IP release are inhibited (50%) by pretreating the cells with the toxin of *Bordetella pertussis* (Paris and Pouysségur, 1986; Seuwen *et al.*, 1988b); (2) A1F$_4^-$, a general agonist of G proteins (Gilman, 1984; Bigay *et al.*, 1985), activates PLC in intact CCL39 cells (Paris and Pouysségur, 1987), and (3) CCL39 plasma membranes release IP$_3$ in response to THR only in the presence of a GTP analog (Magnaldo *et al.*, 1987).

THR and 5-HT receptors of quiescent CHL cells are also coupled negatively to adenylyl cyclase (AC), an action easily measured on the activated enzyme. Levels of cAMP formed in response to prostaglandin E$_1$, forskolin, or cholera toxin are reduced by about 30% in response to low THR concentration (10 pM) or 5-HT (1 μM) (Magnaldo *et al.*, 1988; Seuwen *et al.*, 1988b). As expected for a G_i-mediated action, this inhibition of adenylyl cyclase is completely released by pretreating the cells with pertussis toxin (Ui, 1984). Another point of interest for the rest of the discussion is that 5-HT-mediated inhibitory action of AC is not affected by the 5-HT$_2$ receptor antagonist (ketanserin) but completely eliminated by a 5-HT$_{1B}$ receptor antagonist (compound 21009; Fozard, 1987). This pharmacological dissection of 5-HT-induced early response strongly supports the existence of at least two distinct classes of 5-HT receptors in CHL cells, one population of receptors coupled to G_p and the other (5-HT$_{1B}$ subclass) coupled to G_i (Fozard, 1987; Seuwen *et al.*, 1988b).

Finally, we must add that THR and 5-HT, two growth factors activating G proteins, are just two members of a larger and increasing family of mitogens including: bombesin, bradykinin, vasopressin, histamine, and substances P and K (Zachary *et al.*, 1986). These hormones and neurotransmitters have been clearly shown to activate inositol lipid breakdown in the cell types in which they stimulate DNA synthesis (Brown *et al.*, 1984; Vicentini and Villereal, 1984; Nilsson *et al.*, 1985).

B. G Protein-Independent Transmembrane Signaling Pathways

The other pure mitogens for quiescent CHL cells, mouse EGF, basic and acidic bovine FGF, human PDGF, and insulin (IGF-I), did not induce a significant release of inositol phosphates when tested in the presence of 20 mM Li$^+$ to amplify the sensitivity of the assay (L'Allemain and Pouysségur, 1986; Magnaldo *et al.*, 1986; Paris *et al.*, 1988). Even a potent mitogenic combination, such as EGF/insulin for G_0-arrested CHL cells, failed to stimulate PIP$_2$–PLC at early, mid, and late G_1 phase or in the logarithmic phase of growth. PLC, however, remained highly activatable by THR at all stages of growth stimulation (L'Allemain and Pouysségur, 1986). This result strongly supports the notion that inositol lipid hydrolysis, if essential for some type of growth factors (see discussion below), is not a prerequisite for reinitiation of DNA synthesis and cell proliferation. Likewise, EGF, PDGF, FGF and insulin failed to significantly decrease basal or hormonally-stimulated cAMP levels in quiescent CHL cells (I. Magnaldo, unpublished results). It is remarkable that all these mitogens, which apparently are not directly coupled to G_p or G_i, specifically activate transmembrane tyrosine kinases (Carpenter, 1987). Most importantly, studies with an engineered point mutation abolishing the tyrosine kinase of insulin-(McClain *et al.*, 1987), EGF- (Chen *et al.*, 1987; Livneh *et al.*, 1987) and PDGF receptor (L. Williams, personal communication) have elegantly demonstrated that transmembrane activation of the tyrosine kinase suffices to elicit all biological effects of the corresponding hormone.

C. Crosstalk between Tyrosine Kinases and G Protein-Mediated Signaling Pathways

We have reported that EGF, PDGF, and FGF do not significantly stimulate inositol lipid hydrolysis in quiescent CHL cells. This result is a matter of high controversy since it appears to differ from one cell type to another; the most difficult case is PDGF (Pouysségur *et al.*, 1988). In the same CHL cells we can show that if we preactivate PIP_2–PLC with either THR or ALF_4^-, then FGF, EGF, PDGF, and to a lesser extent insulin, do potentiate up to twofold the rate of IP release (Paris *et al.*, 1988). This observation, together with the fact that receptors coupled to G proteins form an amazingly conserved family (Dohlman *et al.*, 1987) strikingly distinct from the structure of tyrosine kinase receptors (Ullrich *et al.*, 1986), has led us to propose the following two conclusions: (1) tyrosine kinase receptors for EGF, PDGF, IGF-I, insulin, and FGF(?) cannot directly activate "resting" PIP_2–PLC, a process specifically requiring dissociation of G_p (2) If G_p is partially dissociated by some "leaky or autocrine" mechanism or by addition of an agonist, activated tyrosine kinases do potentiate hydrolysis of polyphosphoinositides. This example of positive crosscommunication between these two transmembrane signaling systems could easily reconcile the divergent reports found in the literature. Alternatively, more extensive studies might demonstrate that some cells could express a different class of EGF or PDGF receptor which might belong to the G-coupled receptor family.

Crosscommunication could also operate in the other direction in that inositol lipid breakdown agonists might activate or potentiate tyrosine kinases. In that context, it is clear that bombesin (Cirillo *et al.*, 1986) and thrombin (Kohno and Pouysségur, 1986) stimulate phosphorylation of proteins at tyrosine residues. Although little is known about the mechanisms eliciting these effects it seems at least that in the case of THR, protein kinase C serves as an intermediate in this tyrosine phosphorylation (Vila and Weber, 1988).

IV. EVIDENCE FOR TWO G PROTEINS INVOLVED IN INITIATION OF GROWTH

Growth factors added to quiescent cells stimulate within seconds a common and complex network of biochemical changes whereas DNA replication never occurs before a lag of 10–15 hr. This delay suggests that growth factors must initiate a cascade of timely programmed events in order to turn on the machinery for DNA replication. Do G proteins play a key role in the initiation of these early and interconnected late events?

A. G_p and Inositol Lipid Hydrolysis

If we postulate that inositol lipid hydrolysis is determinant in THR-induced mitogenicity, two predictions could be made: (1) pertussis toxin, an inhibitor of THR-activated PIP$_2$-PLC, should attenuate THR-induced DNA synthesis, and (2) pertussis toxin should not affect the mitogenic action of growth factors activating tyrosine kinases like EGF, PDGF,

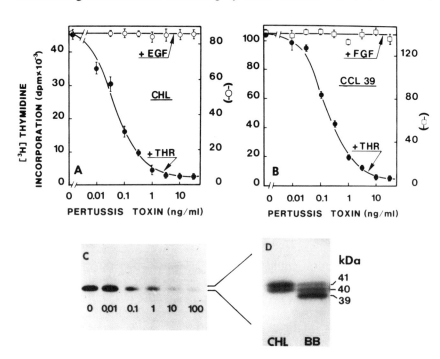

Fig. 1. Parallel effect of pertussis toxin on THR-, EGF- and FGF-induced reinitiation of DNA synthesis and on ADP-ribosylation *in vivo*. G$_0$-arrested secondary cultures of CHL fibroblasts (A) or of CCL39 (B) were preincubated for 4 hr with [³H]thymidine and the indicated concentrations of pertussis toxin. Incubation was continued for 24 hrs in the presence of either 1 nM THR (●), 100 ng/ml basic FGF (□) or 50 ng/ml EGF (○). (C) Dose–response of toxin-catalyzed *in vivo* ADP-ribosylation. (A, B, C were reprinted from Chambard *et al,.* 1987. (D) Pertussis-toxin substrates of CCL39 membranes (CHL) and bovine brain extracts (BB). ADP-ribosylated proteins with [α-³²P]NAD in presence of pertussis toxin were separated by SDS-PAGE (10% acrylamide/0.13% bisacrylamide) as previously reported (Toutant *et al.*, 1987). Note the absence of G$_0$α in CCL39-membranes.

IGF-I, and possibly FGF. The experiments with G_0-arrested CCL39 or CHL cells depicted in Fig. 1A,B entirely satisfied these two predictions (Chambard *et al.*, 1987). THR-induced reinitiation of DNA synthesis in CHL cells is antagonized up to 95% by pertussis toxin at 10 ng/ml. Reinitiation of DNA synthesis by EGF or FGF alone or the synergistic action elicited by IGF-I were not affected by the toxin at concentrations up to 100 ng/ml. The dose–response curve of toxin-induced ADP-ribosylation of a 40–41-kDa protein *in vivo*, paralleled toxin-induced inhibition of DNA synthesis (Fig. 1C). These results clearly indicate: (1) the existence of two mechanistically distinct growth factor signaling pathways capable of operating independently and (2) that at least one G protein, substrate of pertussis toxin, participates in the initiation of growth by THR. We first interpreted these results as a strong indication that inositol lipid hydrolysis plays a crucial role in dictating cells to replicate DNA (Chambard *et al.*, 1987). However, if this conclusion is still valid it needs some reevaluation; in particular, if we consider that THR activates other pertussis toxin-sensitive signaling pathways besides inositol lipid breakdown. Indeed, a more precise analysis of the pertussis toxin substrates in these cells has indicated the presence of two discrete $G\alpha$ subunits of 41 and 40 kDa (Fig. 1D), clearly distinct from α_o (39 kDa) present in brain (B. Rouot, and G. L'Allemain, unpublished results). This result was confirmed by immunoblotting with antibodies to specific peptides (Goldsmith *et al.*, 1987): two α subunits appeared, one specific for α_{i3} (α_{41}) and the other for α_{i2} (α_{40} of neutrophils; P. Gierschik *et al.*, 1987 and unpublished results). This second form of pertussis-toxin substrate, α_{40}, has been postulated to represent the α subunit of G_p; however, direct evidence is still lacking. What is the evidence favoring a role for other pertussis-toxin substrates in growth factor action?

B. G_i and Adenylyl Cyclase and/or Other Effector

If PIP_2–PLC were the only signaling pathway involved in the mitogenic action of THR, pertussis toxin, which attenuates by 50% the rate of IP release at all THR concentrations tested (Paris and Pouysségur, 1986), should shift the THR-induced mitogenic response to the right by at most one order of magnitude. This question, addressed with CCL39 and A71, a clone derivative much more sensitive to growth factors, is depicted in Fig. 2. In both clones, pertussis toxin shifted the dose–response curve to the right by two and three orders of magnitude for CCL39 and A71, respectively. Because A71 and CCL39 display the same dose–response for THR-induced IP release and the same sensitivity to pertussis toxin (50% inhibition), we conclude that besides inositol lipid breakdown, another pathway

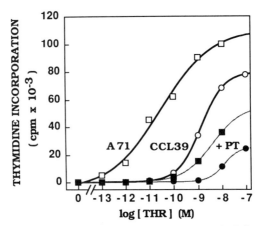

Fig. 2. Concentration dependence of THR-induced DNA synthesis in two clones of CHL cells, showing effects of pertussis toxin. Experiments were conducted with G_0-arrested CCL39 cells (O,●) or a derived clone A71 (□,■) much more sensitive to growth factors (Van Obberghen-Shilling et al., 1983). When present (■,●) pertussis toxin (PT) was added at 20 ng/ml for 5 hr before addition of the indicated concentrations of growth factors. [³H]Thymidine incorporation was measured 24 hrs later.

sensitive to pertussis toxin and operating in the low range of THR concentrations must play a role in mitogenesis.

We confirmed and extended this observation by exploring the mitogenic action of 5-HT. This was a key finding in reason of the rich pharmacology available for 5-HT receptors. 5-HT is not a (or is a weak) mitogen on its own, but significantly enhances the mitogenicity of EGF and especially that of FGF (Fig. 3). Pertussis toxin completely abolished 5-HT-induced mitogenicity, here again leaving intact the FGF response. Ketanserin, a 5-HT$_2$ receptor blocker as shown by complete inhibition of 5-HT-induced IP release, did not prevent 5-HT-induced mitogenicity. In contrast, a specific antagonist of 5-HT$_{1B}$ receptors (compound 21009) which suppressed 5-HT-mediated AC inhibition in CHL cells, suppressed 5-HT-induced mitogenic action (Seuwen et al., 1988b). This finding clearly points out a direct role played by G_i dissociation in the potentiation of DNA synthesis. This G_i-mediated signalling pathway, sensitive to pertussis toxin, is likely to be the same, eliciting DNA synthesis in response to low THR concentration (Fig. 2) or to 5-HT in smooth muscle cells (Kavanaugh et al., 1988), or to serum (Hildebrandt et al., 1986) or bombesin (Letterio et al., 1986; Zachary et al., 1987) in Swiss 3T3 mouse fibroblasts.

A crucial question is whether α_i-mediated inhibition of AC is the triggering event of 5-HT-induced mitogenic response and/or whether α_i is cou-

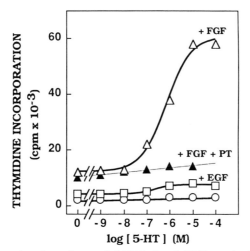

Fig. 3. Concentration dependence of 5-HT-induced DNA synthesis in CCL39 cells, showing effects of FGF and pertussis toxin. Experiments were conducted with G_0-arrested CCL39 cells. When present (▲), pertussis toxin (PT) was added at 20 ng/ml for 5 hr before addition of the indicated concentrations of growth factors: (O) 5-HT alone; 5-HT with 50 ng/ml EGF (□) or with 100 ng/ml FGF (△). [³H]Thymidine incorporation was measured 24 hr later.

pled to some other effector which remains to be discovered. The first hypothesis would imply that a moderate decrease in cAMP levels should potentiate DNA synthesis. This rather old hypothesis (Pastan *et al.*, 1975) has never been directly tested because of technical problems. However, this hypothesis is consistent with the fact that agents which increase cAMP levels in quiescent CHL cells are potent inhibitors of growth factor action. Prostaglandin E_1, cholera toxin, or forskolin alone, by doubling the basal cAMP levels, strongly antagonize (up to 90%) either FGF or THR-induced reinitiation of DNA synthesis (I. Magnaldo, unpublished results). This hypothesis is, however, hardly tenable for Swiss 3T3 cells in which an increase in cAMP potentiates DNA synthesis (Rozengurt *et al.*, 1983). The second hypothesis is that α_i, besides its negative coupling to AC, is coupled to other membrane effectors, the action of which remains to be elucidated. Although some ionic channels have been proposed as possible sites of action (Pfaffinger *et al.*, 1985; Holz *et al.*, 1986), in the context of cell growth it might be worthwhile to investigate a possible coupling with a tyrosine kinase. The membrane-associated tyrosine kinases of the *src* gene family (Hunter and Cooper, 1986) constitute attractive effectors for stimulating DNA synthesis. Their mechanism of activation, which relies upon the interplay of phosphorylation/dephosphorylation at a specific tyrosine residue, would be, according to our hypothesis, controlled by a G_i-mediated action. This hypothesis has the merit of being easily testable.

V. *RAS* AND GROWTH-FACTOR SIGNALING PATHWAYS

Like G-protein α subunits *ras* gene products bind guanine nucleotide, show GTPase activity, and are associated with the cytoplasmic surface of the cell membrane via fatty acid acylation of a cys residue of the carboxyl-terminus. Because the cellular *ras* gene products are involved in regulating cell division and mutated forms of p21, particularly mutations reducing GTPase activity or altering nucleotide bindings cause malignant transformation of cultured fibroblasts (Barbacid, 1987), it was tempting to assign a role of these GTP-binding proteins in growth factor or hormone – receptor coupling.

In several studies with fibroblastic cell lines expressing different levels of normal or activated forms of *ras*, an increase in diacylglycerol content or an increase of PLC – receptor coupling in response to external effectors was reported (Fleischman *et al.*, 1986; Lacal *et al.*, 1987; Wakalam *et al.*, 1986). These results were interpreted as an action of the *ras* gene products at the level of PLC. We analyzed in CHL cells the consequence of moderate and elevated expression of activated Ki- and Ha-*ras* on PI metabolism. None of the predictions expected if activated forms of *ras* were able to play the role of the G-protein-activating PIP_2 – PLC have been observed (Seuwen *et al.*, 1988a). We therefore concluded that uncontrolled growth by mutated *ras* is not explained by constitutive activation of inositol lipid breakdown. Similarly, it has been clearly shown that in contrast to the situation found in *S. cerevisiae, ras* does not activate or inhibit AC in higher eucaryotes (Beckner *et al.*, 1985). Therefore, at the moment, the site of *ras* action remains a real mystery. Instead of searching for alteration of growth factor – receptor coupling it might be worthwhile to investigate a possible role in the dynamics of plasma membrane/cytoskeleton assembly. A profound disruption in the homeostasis of cytoskeleton assembly – disassembly could lead to aberrant gene expression and might be a source of growth autonomy (Puck, 1987).

VI. Conclusions

This chapter was restricted to one facet of growth factor action, namely the capacity to induce in cultured CHL fibroblasts the transition from quiescence to the proliferative state. Our major goal was to delineate, among

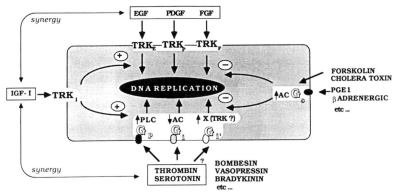

Fig. 4. Schematic representation of growth-factor signal transduction in CHL fibroblasts: TRK, receptor tyrosine kinases; PLC, PIP$_2$–phospholipase C; AC, adenylyl cyclase; X, unknown effector.

transmembrane growth signaling pathways, those involving dissociation of a G protein. This issue is complicated by the existence of crosscommunications between these pathways and the fact that a hormone often acts through different receptors, resulting in the activation of multiple effectors within the same cell. Nevertheless the following conclusions can be drawn (schematically outlined in Fig. 4): (1) Neurohormones, neurotransmitters, vasoactive peptides (thrombin, bombesin, bradykinin, vasopressin, substances P and K, histamine, serotonin) should now be considered as an important new class of growth factors. (2) Their action appears to be mediated through multiple G-protein-coupled effectors. (3) At least two distinct G proteins are involved in signaling growth: G_p, a G protein coupled positively to PLC, and G_i, a G protein coupled negatively to AC and possibly to other unknown effectors. (4) Receptor-tyrosine kinases for EGF, PDGF, IGF-I (FGF?) provide alternate growth signaling pathways independent of G_p- and G_i-mediated actions. (5) Adenylyl cyclase activation, through G_s dissociation, strongly antagonizes all growth signaling pathways.

Hopefully, future work should identify the effectors activated by each subset of isolated α subunits. Presently it is difficult to know whether G_p is a substrate of pertussis toxin or whether the toxin sensitivity of THR-activated PLC, in CHL, is secondary to some other pertussis toxin-sensitive pathways. Many questions remain unanswered. For instance, one would like to know the degree of cross reactivity or of "degeneracy" which might exist between effectors, G protein, and corresponding receptors. Is a receptor capable of interacting with a single or multiple G protein? Similarly, is a single G protein capable of activating one or multiple effectors? Given the possibility existing today of stably (over)expressing in cell lines distinct classes of receptors (Ashkenazi *et al.*, 1987) or Gα subunits (normal or activated forms by appropriate mutations; Masters *et al.*, 1988), these questions should receive a direct answer. Following that approach, the expression in CHL cells of constitutively activated α_p and/or α_i (form 1, 2, or 3) should lead to partial or total growth autonomy if our conclusion, stating that activation of G_p and G_i are key transducers in initiating growth, is indeed valid.

VII. EDITOR'S COMMENTS (BY RAVI IYENGAR)

The conclusions in this chapter have received recent experimental support from the work of Julius *et al.*, who demonstrated that over-expression of the serotonin 1C (5-HT_{1C}) receptor in NIH-3T3 fibroblasts can result in transformation. This transformation is dependent on the continued presence of serotonin 1C receptor agonist, indicating the requirement of an active receptor. Injection of transformed cells into nude mice results in the appearance of tumors. Since it is known that the serotonin 1C receptor acts through G proteins to stimulate phospholipase C, it appears that the observed growth effects are a direct result of activation of this second messenger pathway.

Reference

Julius, D., Livelli, T., Jessel, T. M., and Axel, R. (1989). Ectopic expression of the serotonin 1C receptor and the triggering of malignant transformation. *Science* **244,** 1057–1062.

ACKNOWLEDGMENTS

I am indebted to my colleagues J. C. Chambard, G. L'Allemain, I. Magnaldo, S. Paris, and K. Seuwen for their collaboration and stimulating discussions and to M. Bonacci for preparing the manuscript. This work was supported by grants from the Centre National de la Recherche Scientifique (LP 7300), the Institut National de la Santé et de la Recherche Médicale, the Fondation pour la Recherche Médicale and the Association pour la Recherche contre le Cancer.

REFERENCES

Ashkenazi, A., Winslow, J., Peralta, E., Peterson, G., Schimerlik, M., Capon, D., and Ramachandran, J. (1987). An M2 muscarinic receptor subtype coupled to both adenylyl cyclase and phosphoinositide turnover. *Science* **238**, 672–675.

Barbacid, M. (1987). *ras* genes. *Annu. Rev. Biochem.* **56**, 779–827.

Beckner, S. K., Hattori, S., and Shih, T. Y. (1985). The *ras* oncogene product p21 is not a regulatory component of adenylate cyclase. *Nature* **317**, 71–72.

Berridge, M. J. (1984). Inositol trisphosphate and diacylglycerol as second messengers. *Biochem. J.* **220**, 345–360.

Bigay, J., Deterre, P., Pfister, C., and Chabre, M. (1985). Fluoroaluminate activates transducin GDP by mimicking the γ-phosphate of GTP in its binding site. *FEBS Lett.* **191**, 181–185.

Brown, K. D., Blay, J., Irvine, R. F., Heslop, J. P., and Berridge, M. J. (1984). Reduction of epidermal growth factor receptor affinity by heterologous ligands: Evidence for a mechanism involving the breakdown of phosphoinositides and the inactivation of protein kinase C. *Biochem. Biophys. Res. Commun.* **123**, 377–384.

Carney, D. H., Scott, D. L., Gordon, E. A., and La Belle, E. F. (1985). Phosphoinositide in mitogenesis: Neomycin inhibits thrombin-stimulated phosphoinositide turn-over and initiation of cell proliferation. *Cell* **42**, 479–488.

Carpenter, G. (1987). Receptors for epidermal growth factor and other polypeptide mitogens. *Annu. Rev. Biochem.* **56**, 881–914.

Chambard, J. C., Paris, S., L'Allemain, G., and Pouysségur, J. (1987). Two growth factor signaling pathways in fibroblasts distinguished by pertussis toxin. *Nature* **326**, 800–803.

Chen, W. S., Lazar, C. S., Poenie, M., Tsien, R. Y., Gill, G. N., and Rosenfeld, M. G. (1987). Requirement for intrinsic protein tyrosine kinase in the immediate and late actions of the EGF receptor. *Nature* **328**, 820–823.

Cirillo, D., Guadino, G., Naldini, L., and Comoglio, P. (1986). Receptor for bombesin with associated tyrosine kinase activity. *Mol. Cell. Biol.* **6**, 4641–4649.

Cockcroft, S., and Gomperts, B. D. (1985). Role of guanine nucleotide binding protein in the activation of polyphosphoinositide phosphodiesterase. *Nature* **314**, 534–536.

Dohlman, H. G., Caron, M. G., and Lefkowitz, R. J. (1987). A family of receptors coupled to guanine nucleotide regulatory proteins. *Biochemistry* **26**, 2657–2664.

Fleischman, L. F., Chahwala, S. B., and Cantley, L. (1986). *Ras*-transformed cells: Altered levels of phosphatidylinositol-4,5-bisphosphate and catabolites. *Science* **231**, 407–410.

Fozard, J. (1987). 5-HT: The enigma variations. *Trends Pharmacol. Sci.* **8**, 501–506.

Gierschik, P., Sidiropoulos, D., Spiegel, A., and Jakobs, K., (1987). Purification and immunochemical characterization of the major pertussis-toxin-sensitive guanine nucleotide binding protein of bovine-neutrophils membranes. *Eur. J. Biochem.* **165**, 185–194.

Gilman, A. G. (1984). G proteins and dual control of adenylate cyclase. *Cell* **36**, 577–579.

Gilman, A. G. (1987). G proteins: Transducers of receptor-generated signals. *Annu. Rev. Biochem.* **56**, 615–649.

Goldsmith, P., Gierschik, P., Milligan, G., Unson, C., Vinitsky, R., Malech, H., and Spiegel, A. (1987). Antidobies directed against synthetic peptides distinguish between GTP-binding proteins in neutrophils and brain. *J. Biol. Chem.* **262**, 14683–14688.

Hildebrandt, J. D., Stolzenberg, E., and Graves, J. (1986). Pertussis toxin alters the growth characteristics of Swiss 3T3 cells. *FEBS Lett.* **203**, 87–90.

Holz, G., Rane, S., and Dunlap, K. (1986). GTP-binding proteins mediate transmitter inhibition of voltage-dependent calcium channels. *Nature* **319**, 670–672.

Hunter, T., and Cooper, J. (1986). Viral oncogenes and tyrosine phosphorylation. *In* "The Enzyme" Vol. XVII, Control by phosphorylation, Part A (P. Boyer and E. Krebs, eds.) pp 192–237.

Katada, T., and Ui, M. (1982). Direct modification of the membrane adenylate cyclase system by islet-activating protein due to ADP-ribosylation of a membrane protein. *Proc. Natl. Acad. Sci. U.S.A.* **79**, 3129–3133.

Kavanaugh, W. M., Williams, L. T., Ives, H. E., and Coughlin, S. R. (1988). Serotonin-induced DNA synthesis in vascular smooth muscle cells involves a novel, pertussis toxin-sensitive pathway. *Molec. Endocrin.* **2**, 599–605.

Kohno, M., and Pouysségur, J. (1986). α-thrombin-induced tyrosine phosphorylation of 43,000 and 41,000-M_r proteins is independent of cytoplasmic alkalinization in quiescent fibroblasts. *Biochem. J.* **238**, 451–457.

Lacal, J. C., Moscat, J., and Aaronson, S. A. (1987). Novel source of 1,2-diacylglycerol elevated in cells transformed by Ha-*ras* oncogene. *Nature* **330**, 269–272.

L'Allemain, G., and Pouysségur, J. (1986). EGF and insulin action in fibroblasts: Evidence that phosphoinositide hydrolysis is not an essential mitogenic signaling pathway. *FEBS Lett.* **197**, 344–348.

L'Allemain, G., Paris, S., Magnaldo, I. and Pouysségur, J. (1986). Thrombin-induced inositol phosphate formation in GO-arrested in cycling hamster lung fibroblasts: Evidence for a protein kinase C-mediated desensitization response. *J. Cell. Physiol.* **129**, 167–174.

Letterio, J. J., Coughlin, S. R., and Williams, L. T. (1986). Pertussis toxin-sensitive pathway in the stimulation of c-*myc* expression and DNA synthesis by bombesin. *Science* **234**, 1117–1119.

Livneh, E., Reiss, N., Berent, E., Ullrich, A., and Schlessinger, J. (1987). An insertional mutant of epidermal growth factor receptor allows dissection of diverse receptor functions. *EMBO J.* **9**, 2669–2676.

Magnaldo, I., L'Allemain, G., Chambard, J. C., Moenner, M., Barritault, D. and Pouysségur, J. (1986). The mitogenic-signaling pathway of FGF is not mediated through polyphosphoinositide hydrolysis and protein kinase C activation in hamster fibroblasts. *J. Biol. Chem.* **261**, 16916–16922.

Magnaldo, I., Talwar, H., Anderson, W. B., and Pouysségur, J. (1987). Evidence for a GTP-binding protein coupling thrombin receptor to PIP2-phospholipase C in hamster fibroblasts. *FEBS Lett* **210**, 6-10.

Magnaldo, I., Pouysségur, J., and Paris, S. (1988). Thrombin exerts a dual effect on stimulated adenylate cyclase in hamster fibroblasts: An inhibition via a GTP-binding protein and a potentiation via activation of protein kinase C. *Biochem. J.* **253**, 711-713.

Masters, S. B., Sullivan, K. A., Miller, R. T., Beiderman, B., Lopez, N. G., Ramachandran, J., and Bourne, H. R. (1988). Carboxy terminal domain of Gsα specifies coupling of receptors to stimulation of adenylyl cyclase. (submitted).

McClain, D. A., Mazgawa, H., Lee, H., Dull, T. J., Ullrich, A., and Olefsky, J. M. (1987). A mutant insulin receptor with defective tyrosine kinase displays no biologic activity and does not undergo endocytosis. *J. Biol. Chem.* **262**, 14663-14671.

Nilsson, J., Von Euler, A., and Dalsgaard, C. J. (1985). Stimulation of connective tissue cell growth by substance P and substance K. *Nature* **315**, 61-63.

Nishizuka, Y. (1986). Studies and perspectives of protein kinase C. *Science* **307**, 305-312.

Paris, S., and Pouysségur, J. (1986). Pertussis toxin inhibits thrombin-induced activation of phosphoinositide hydrolysis and Na^+/H^+ exchange in hamster fibroblasts. *EMBO J.* **5**, 55-60.

Paris, S., and Pouysségur, J. (1987). Further evidence for a phospholipase C-coupled G protein in hamster fibroblasts. Induction of inositol phosphate formation by fluoroaluminate and vanadate and inhibition by pertussis toxin. *J. Biol. Chem.* **262**, 1970-1976.

Paris, S., Chambard, J. C., and Pouysségur, J. (1988). Tyrosine kinase-activating growth factors potentiate thrombin- and AlF_4^--induced phosphoinositide breakdown in hamster fibroblasts: Evidence for positive cross-talk between the two mitogenic signaling pathways. *J. Biol. Chem.* **263**, 12893-12900.

Pastan, I., Johnson, G., and Anderson, W. (1975). Role of cyclic nucleotide in growth control. *Annu. Rev. Biochem.* **44**, 491-521.

Pérez-Rodriguez, R., Franchi, A., and Pouysségur, J. (1981). Growth factor requirements of chinese hamster lung fibroblasts in serum free media: High mitogenic action of thrombin. *Biol. Int. Rep.* **5**, 347-357.

Pfaffinger, P., Martin, J., Hunter, D., Nathanson, N., and Hille, B., (1985). GTP-binding proteins couple cardiac muscarinic receptors to a K channel. *Nature* **317**, 536-538.

Pouysségur, J., Chambard, J-C., L'Allemain, G., Magnaldo, I. and Seuwen, K. (1988). Transmembrane signaling pathways initiating cell growth in fibroblasts. *Phil. Trans. R. Soc. London* **320**, 427-436.

Puck, T. (1987). Genetic Regulation of growth control: Role of cAMP and cell cytoskeleton. *Somat. Cell Mol. Genet.* **13**, 451-457.

Rozengurt, E., Collins, M., and Keehan, M. (1983). Mitogenic effect of prostaglandin E_1 in Swiss 3T3 cells: Role of cyclic AMP. *J. Cell. Physiol.* **116**, 379-384.

Seuwen, K., Lagarde, A., and Pouysségur, J. (1988a). Deregulation of hamster fibroblast proliferation by mutated *ras* oncogenes is not mediated by constitutive activation of phosphoinositide-specific phospholipase C. *EMBO J.* **7,** 161–168.

Seuwen, K., Magnaldo, I., and Pouysségur, J. (1988b). Serotonin stimulates DNA synthesis in fibroblasts acting through 5-HT$_{1B}$ receptors coupled to a G$_i$-protein. *Nature* **335,** 254–256.

Stryer, L., and Bourne, H. (1986). G-proteins: A family of signal transducers. *Annu. Rev. Cell Biol.* **2** 391–419.

Toutant, M., Bockaert, J., Homburger, V., and Rouot, B. (1987). G-proteins in Torpedo marmorata electric organ. Differential distribution in pre- and post-synaptic membranes and synaptic vesicles. *FEBS Lett.* **222,** 51–55.

Ui, M. (1984). Islet-activating protein, pertussis toxin: A probe for functions of the inhibitory guanine nucleotide regulatory component of adenylate cyclase. *Trends Pharmacol. Sci.* **5,** 277–279.

Ullrich, A., Gray, A., Tam, A. W., Yang-Feng, T., Tsubokawa, M., Collins, G., Henzel, W., Le Bon, T., Kathuria, S., Chen, E., Jacobs, S., Francke, U., Ramachandran, J., and Fujita-Yamaguchi, Y. (1986). Insulin-like growth factor-I receptor structural determinants that define functional specificity. *EMBO J.* **5,** 2503–2512.

Van Obberghen-Schilling, E., and Pouysségur, J. (1983). Mitogen-potentiating action and binding characteristics of insulin-like growth factors in Chinese hamster fibroblasts. *Exp. Cell. Res.* **147,** 369–378.

Van Obberghen-Schilling, E., Perez-Rodriguez, R., Franchi, A., Chambard, J. C., and Pouysségur, J. (1983). Analysis of growth factor relaxation in Chinese hamster lung fibroblasts required for tumoral expression. *J. Cell Physiol.* **115,** 123–130.

Vicentini, L. M., and Villereal, M. L. (1984). Serum, bradykinin and vasopressin stimulate release of inositol phosphate from human fibroblasts. *Biochem. Biophys. Res. Commun.* **123,** 663–670.

Vila, J., and Weber, M. (1988). Mitogen-stimulated tyrosine phosphorylation of a 42 kDa cellular protein: Evidence for a protein kinase C requirement. *J. Cell. Physiol.* **135,** 285–292.

Wakalam, M. J., Davies, S. A., Houslay, M. D., McKay, I., Marshall, C. J., and Hall, A. (1986). Normal p21^{N-ras} couples bombesin and other growth factor receptors to inositol phosphate production. *Nature* **323,** 173–176.

Zachary, I., Woll, P., and Rozengurt, E. (1986). A role for neuropeptides in the control of cell proliferation. *Dev. Biol.* **124,** 295–308.

Zachary, I., Millar, J., Nanberg, E., Higgins, T., and Rozengurt, E. (1987). Inhibition of bombesin-induced mitogenesis by pertussis toxin: Dissociation from phospholipase C pathway. *Biochem. Biophys. Res. Commun.* **146,** 456–463.

CHAPTER 24

G Proteins in Yeast *Saccharomyces cerevisiae*

Janet Kurjan

Department of Biological Sciences, Columbia University, New York,
New York 10027

I. INTRODUCTION

Members of the G-protein family in vertebrate cells are involved in the transduction of a diverse array of extracellular signals into intracellular responses (reviewed by Stryer and Bourne, 1986; Gilman, 1987). The recent discovery of G proteins in the lower eukaryotes, *Saccharomyces cerevisiae* and *Dictyostelium discoideum*, indicates that these proteins have been conserved over a large evolutionary distance. In addition, the discovery of G proteins in yeast allows the utilization of an extensive genetic analysis not possible in mammalian systems to study the mechanism of action of these proteins.

Two genes that encode homologs to the α subunits of mammalian G proteins have been identified in the yeast, *Saccharomyces cerevisiae* (Dietzel and Kurjan, 1987a; Nakafuku *et al.*, 1987, 1988). The first yeast Gα gene identified has been called alternatively *SCG1* (Dietzel and Kurjan, 1987a) or *GPA1* (Nakafuku *et al.*, 1987); I will refer to it as *SCG1*. Considerable evidence indicates that *SCG1* acts in the pheromone response pathway that is essential for mating (Section IV). Less evidence is available concerning the role of the second yeast Gα homolog, *GPA2*, although there is a suggestion that *GPA2* may be involved in regulating cAMP levels (Nakafuku *et al.*, 1988). It has been shown that two genes involved in pheromone response and mating (*STE4* and *STE18*; Whiteway *et al.*, 1989) probably encode β and γ subunits that act in the pheromone response pathway.

II. PHEROMONE RESPONSE AND MATING IN YEAST

Yeast can grow vegetatively as any of three cell types. Cells of the two haploid mating types, **a** and α, mate with one another to form the third cell type, the **a**/α diploid (reviewed in Sprague *et al.*, 1983). Secretion of and response to peptide pheromones are essential aspects of mating; **a** cells secrete **a**-factor and respond to α-factor, and α cells secrete α-factor and respond to **a**-factor. These responses are mediated by specific receptors; **a** cells express α-factor receptor, encoded by the *STE2* gene, and α cells express **a**-factor receptor, encoded by the *STE3* gene (MacKay and Manney, 1974b; Jenness *et al.*, 1983; Hagen *et al.*, 1986). The two receptors do not show sequence homology (Burkholder and Hartwell, 1985; Nakayama *et al.*, 1985; Hagen *et al.*, 1986), but have structural features similar to one another and to the receptors involved in G-protein-mediated signal transduction systems in mammals (Sibley *et al.*, 1987). These features include seven putative membrane-spanning domains and a hydrophilic carboxy terminus containing many serine and threonine residues. Experiments of Bender and Sprague (1986) and Nakayama *et al.* (1987) suggest that the intracellular signaling pathway is shared by **a** and α cells, i.e. that each of the receptors binds the opposite pheromone and then activates the identical pathway in both haploid mating types.

Exposure to pheromone induces several responses. Some of these responses are essential for mating and others may be involved in modulation of mating and pheromone response; for example, some of the responses are likely to play a role in desensitization to pheromone (Chan and Otte, 1982b; Moore, 1983; Dietzel and Kurjan, 1987b). Cells exposed to the pheromone secreted by cells of the opposite mating type arrest as unbudded cells in the G_1 phase of the cell cycle (Bucking-Throm *et al.*, 1973; Duntze *et al.*, 1970), the point of the cell cycle at which conjugation occurs (Hartwell, 1973). Cells exposed to high levels of pheromone also show morphological changes: the cells become enlarged and elongated (called "shmoo" formation). A number of genes involved in mating and aspects of pheromone response are induced upon exposure to the opposite pheromone (Hagen and Sprague, 1984; Nakayama *et al.*, 1985; Hartig *et al.*, 1986; Dietzel and Kurjan, 1987b; McCaffrey *et al.*, 1987; Trueheart *et al.*, 1987). Mutants that are defective in pheromone production or response are sterile (MacKay and Manney, 1974a; Manney and Woods, 1976; Hartwell, 1980; Kurjan, 1985; Michaelis and Herskowitz, 1988). During response to α-factor, the α-factor receptor is downregulated (Jenness and Spatrick, 1986). Downregulation probably occurs by receptor-mediated endocytosis, and α-factor is internalized and degraded during this process (Chvatchko *et al.*, 1986; Jenness and Spatrick, 1986).

Until recently, no information was available concerning the intracellular signaling pathway. The identification and genetic analysis of homologs of G-protein subunits involved in this pathway indicates that the mechanism of pheromone response is related to the response pathways of vertebrate G-protein-mediated signal transduction systems. This discovery is consistent with the proposed structures of the **a**-factor and α-factor receptors, which are similar to the proposed structures for vertebrate receptors that act through G proteins.

III. IDENTIFICATION OF *SCG1* *(GPA1)*, A Gα HOMOLOG

The *SCG1* gene was initially identified using two different approaches. One approach involved isolation of a clone that showed hybridization to rat α_{i2} and α_o cDNA probes under low stringency conditions (Nakafuku *et al.*, 1987). A second approach involved the analysis of a multicopy plasmid

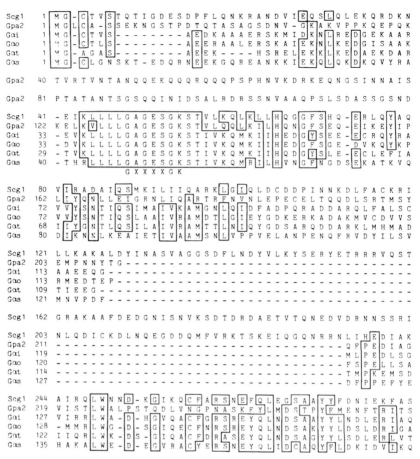

Fig. 1. Comparison of Gα subunits from yeast and vertebrate cells. The amino-acid sequences of yeast Scg1 (Dietzel and Kurjan, 1987a) and Gpa2 (Nakafuku *et al.*, 1988) are shown in the top two lines. These sequences are compared to vertebrate Gα proteins, α_{i2} (Itoh *et al.*, 1986), α_o (Van Meurs *et al.*, 1987), α_{t1} (Tanabe *et al.*, 1985), and α_s (Bray *et al.*, 1986; Robishaw *et al.*, 1986a,b). Amino acids that are identical in at least four of the six G_α subunits are boxed. The consensus sequences (Dever *et al.*, 1987) determined for regions involved in guanine-nucleotide exchange and GTPase activity in the family of guanine nucleotide-binding proteins including the Gα subunits, *ras*, and EF-Tu are shown below the Gα sequences. Percent identities were determined based on the number of identities between pairs of Gα subunits divided by the number of amino acids in the shorter member of the pair. Because the alignments are based on a comparison of all six Gα subunits, alignments between two individual subunits might result in a higher percent identity for some combinations. The percent identities between pairs are: Scg1/Gpa2, 33%; Scg1/α_i, 46%; Scg1/α_o, 42%; Scg1/α_t, 43%; Scg1/α_s, 35%; Gpa2/α_i, 40%; Gpa2/α_o, 38%; Gpa2/α_t, 41%; Gpa2/α_s, 33%.

Scg1 284 P N Y V C T D E D I L K G R I K T T G I T E T E F N I G S - S K F K V L D A G G Q
Gpa2 260 P N Y R P T Q Q D I L R S R Q M T S G I F D T V I D M G S D I K M H I Y D V G G Q
Gαi 166   S D Y I P T Q Q D V L R T R V K T T G I V E T H F T F K D - L H F K M F D V G G Q
Gαo 166   A D Y Q P T E Q D I L R T R V K T T G I V E T H F T F K N - L H F R L F D V G G Q
Gαt 161   P G Y V P T E Q D V L R S R V K T T G I I E T Q F S F K D - L N F R M F D V G G Q
Gαs 174   D D Y V P S D Q D L L R C R V L T S G I F E T K F Q V D K - V Q F H M F D V G G Q
                                                              D X X G

Scg1 324 R S E R K K W I H C F E G I T A V L F V L A M S E Y D Q M L F E D E R V N R M H E
Gpa2 301 R S E R K K W I H C F D N V T L V I F C V S L S E Y D Q T L M E D K N Q N R F Q E
Gαi 206   R S E R K K W I H C F E G V T A I I F C V A L S A Y D L V L A E D E E M N R M H E
Gαo 206   R S E R K K W I H C F E D V T A T I F C V A L S G Y D Q V L H E D E T T N R M H E
Gαt 201   R S E R K K W I H C F E G V T C I I F I A A L S A Y D M V L V E D D E V N R M H E
Gαs 214   R D E R R K W I Q C F N D V T A I I F V V A S S S Y N M V I R E D N Q T N R L Q E

Scg1 365 S I M L F D T L L N S K W F K D T P F I L F L N K I D L F E E K V - - K S M P I R
Gpa2 342 S L V L F D N I V N S R W F A R T S V V L F L N K I D L F A E K L - - R K V P M E
Gαi 247   S M K L F D S I C N N K W F T D T S I I L F L N K K D L F E E K I - - T Q S P L T
Gαo 247   S L M L F D S I C N N K F F I D I S I I L F L N K K D L F G E K I - - K K S P L T
Gαt 242   S L H L F N S I C N H R Y F A T T S I V L F L N K K D V F S E K I - - K K A H L S
Gαs 255   A L N L F K S I W N N R W L R T T S V I L F L N K Q D L L A E K V L A G K S K I E
                                              N K X D

Scg1 404 K Y F P D Y Q G R V G D A E A G - - - - - - - - - - - - - - - - - L K Y F E - K I F
Gpa2 381 N Y F P D Y T G G S D I N K A A - - - - - - - - - - - - - - - - - K Y I L W R - F
Gαi 286   I C F P E Y T G A N K Y D E A G - - - - - - - - - - - - - - - - - S Y I Q S K F E
Gαo 286   I C F P E Y T G S N T Y E D A A - - - - - - - - - - - - - - - - - A Y I Q A Q F E
Gαt 281   I C F P D Y N G P N T Y E D A G - - - - - - - - - - - - - - - - - N Y I K V Q - F
Gαs 296   D Y F P E F A R Y T T P E D A T P E P G E D P R V T R A K Y F I R D E F L - R I S

Scg1 428 L S L N - - K T N K P I Y V K R T C A T D T Q T M K F V L S A V T D L I I Q Q N L
Gpa2 404 V Q L N - R - A N L S I Y F H V T Q A T D T S N I R L V F A A I K E T I L E N T L
Gαi 310   D L N K - R K D T K E I Y T H F T C A T D T K N V Q F V F D A V T D V I I K N N L
Gαo 310   S K N R - - S P N K E I Y C H M T C A T D T N N I Q V V F D A V T D I I I A N N L
Gαt 304   L E L N M R R D V K E L I Y S H M T C A T D T Q N V K F V F D A V T D I I I K E N L
Gαs 336   T A S G - - D G R H Y C Y P H S T C A V D T E N I R R V F N D C R D I I Q R M H L

Scg1 467 K K S G I I
Gpa2 443 K D S G V L Q
Gαi 350   K D C G L F
Gαo 349   R G C G L Y
Gαt 345   K D C G L F
Gαs 375   R Q Y E L L

Fig. 1 (Continued)

from a yeast library that was able to suppress a mutation (*sst2-1*) that results in an alteration in pheromone response (Dietzel and Kurjan, 1987a).[1] *sst2* mutants are supersensitive to pheromone; growth arrest occurs upon exposure to very low levels of pheromone, and *sst2* cells are defective in recovery from pheromone arrest (Chan and Otte, 1982a,b). Supersensitivity to the opposite pheromone is seen in both mating types; **a** *sst2* cells are supersensitive to α-factor, and α *sst2* cells are supersensitive to **a**-factor. The *SCG1* gene, when present on the multicopy plasmid, was able to suppress *sst2* supersensitivity in both **a** and α cells. The *SST2* product probably plays a role in desensitization to pheromone, i.e. it acts to

---

[1] Recessive mutations are indicated by lowercase letters in yeast. The wild-type locus is indicated in uppercase letters. Dominant alleles will be indicated in uppercase letters with a superscript, e.g. *SCG1*Val50.

turn off the pheromone response after exposure to pheromone (Moore, 1983; Dietzel and Kurjan, 1987b). The ability of the *SCG1* gene to partially suppress *sst2* supersensitivity indicated that it was able to inhibit pheromone response when overexpressed on a multicopy plasmid. This phenotype suggested that it might be a component of the pheromone response pathway; however, the ability to inhibit pheromone response might also be an artifact of overexpression of a gene product that normally is not involved in pheromone response. Additional results (Section IV) provide evidence that the *SCG1* product is an important component of the pheromone response pathway.

Sequencing of the *SCG1* gene indicated that it encodes a homolog to the α subunits of vertebrate G proteins (Fig. 1; Nakafuku *et al.*, 1987; Dietzel and Kurjan, 1987a). The presumed *SCG1* product is a 472-amino-acid protein with a predicted molecular weight of 54K; therefore, it is larger than the vertebrate α subunits. The additional amino acids are in a region in which different vertebrate Gα subunits show little similarity. In the regions that have been implicated in guanine-nucleotide binding and GTPase activities, *SCG1* shows a high level of identity to the vertebrate Gα subunits and other guanine nucleotide-binding proteins (Dever *et al.*, 1987). The *SCG1* genes from two different strains show five amino-acid polymorphisms (Dietzel and Kurjan, 1987a; Nakafuku *et al.*, 1987); none of the polymorphisms are in regions implicated in guanine-nucleotide binding, and most of the polymorphisms are in poorly conserved regions of the α subunits.

IV. MUTATIONS IN *SCG1* INDICATING ROLE FOR *SCG1* IN PHEROMONE RESPONSE PATHWAY

A. Null Mutations in *scg1* Resulting in Haploid-Specific Phenotype That Resembles Constitutive Pheromone Response

The identification of a Gα homolog by its ability to inhibit pheromone response was intriguing, especially in conjunction with the similarity in structure of the pheromone receptors and receptors for well-characterized G-protein-mediated pathways. Evidence that *SCG1* is a primary component of the pheromone response pathway has been provided by genetic experiments. Disruption or deletion mutations of *scg1* were constructed [using the gene replacement technique of Rothstein (1983)] in **a**/α diploids to produce **a**/α *SCG1*/*scg1-null* heterozygotes, where the null allele contained a scorable marker (*LEU2*, *URA3*, or *HIS3*). Sporulation and dissection of these diploids resulted in the production in each tetrad of two wild-type (*SCG1*) spores and two null mutant spores with growth and morphological defects. The null phenotype is somewhat variable, depending on the strain used. In some strains the mutant spores do not grow into visible colonies; microscopic examination indicates that the spores undergo several cell divisions and arrest as mostly unbudded cells, some of which show aberrant morphology (Miyajima *et al.*, 1987). In other strains the mutant spores produce barely visible colonies that are composed of mostly unbudded, very large and aberrantly shaped cells (Dietzel and Kurjan, 1987a). The growth and morphological defects of the *scg1-null* mutants seemed reminiscent of the phenotype of temperature-sensitive cell cycle "Start" mutants; the Start mutants arrest at the restrictive temperature at the same point of the cell cycle as pheromone-arrested cells, with a morphology that resembles "shmoos" (Hartwell *et al.*, 1973; Reed, 1980). Diploid cells homozygous for an *scg1-null* mutation (**a**/α *scg1/scg1*) do not show the haploid null phenotype; instead they show wild-type growth and morphology (Dietzel and Kurjan, 1987a; Miyajima *et al.*, 1987). After sporulation and dissection, they produce four spores with the *scg1-null* phenotype. Because **a**/α diploids do not respond to pheromone, the haploid-specificity of the *scg1* phenotype is consistent with the hypothesis that the *scg1* phenotype represents constitutive expression of the pheromone response pathway. Expression of *SCG1* is also haploid-specific as shown by the presence of a 1.7-kb RNA that hybridizes to an *SCG1* probe in **a** and α but not **a**/α cells (Miyajima *et al.*, 1987) and by a similar expression pattern for an *scg1*–β-galactosidase fusion (Dietzel and Kurjan, 1987a).

B. Inactivation of *SCG1* Bypasses Need for Pheromone Receptor

Additional evidence that *SCG1* plays a role in the pheromone response pathway has been obtained by the analysis of strains in which *SCG1* activity or expression is conditional. Temperature-sensitive *scg1* mutants, in which *SCG1* is functional at the permissive temperature and nonfunctional at the restrictive temperature, have been identified (Jahng *et al.*, 1988). These mutants show G1 arrest and morphological changes at the restrictive temperature; i.e. at the restrictive temperature the *scg1-ts* phenotype is similar to the *scg1-null* phenotype. Double mutants containing an *scg1-ts* mutation and a deletion of the α-factor or **a**-factor receptor in **a** and α strains, respectively (**a** *scg1-ts ste2-null* and α *scg1-ts ste3-null*) are able to mate at the restrictive temperature. The ability to mate in the absence of a pheromone receptor suggests that the pheromone response pathway is being turned on at the restrictive temperature independent of exposure to pheromone.

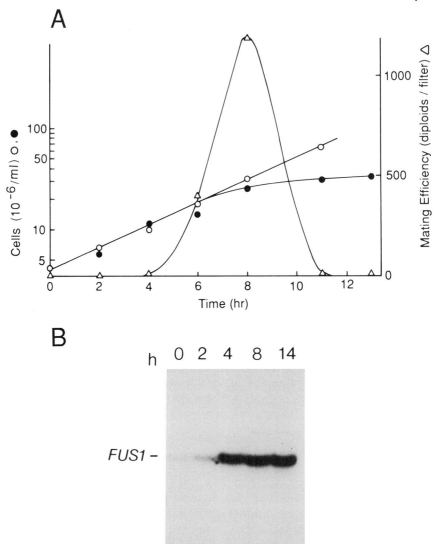

Fig. 2. Induction of division arrest, mating, and *FUS1* RNA by repression of *SCG1* *(GPA1)* expression. **a** *ste2 scg1 (gpa1)* cells containing *SCG1 (GPA1)* expressed under the control of the *GAL1* promoter [p*GAL-SCG1*] were grown in galactose to allow expression of the *SCG1 (GPA1)* gene. At time 0, the cells were shifted to glucose-repressing conditions, which prevent *SCG1(GPA1)* expression. (A) Growth in galactose (open circles), growth in glucose (closed circles), and mating efficiency in glucose (open triangles). Repression of *SCG1* *(GPA1)* results in arrest of cell division and allows transient induction of mating ability. (B) *FUS1* RNA levels. *FUS1* is a gene that is induced by exposure to pheromone and is involved in cell fusion during zygote formation (McCaffrey *et al.*, 1987; Trueheart *et al.*, 1987). *FUS1* RNA levels are induced five- to ten fold by repression of *SCG1 (GPA1)* expression. Results and figure from Matsumoto *et al.* (1988).

Similar results have been obtained using a different approach (Matsumoto *et al.*, 1988). Expression of *SCG1* under control of the *GAL* promoter results in expression of *SCG1* in galactose medium, and therefore allows growth of an *scg1* [p*GAL-SCG1*] strain. When shifted to glucose medium, expression of the *SCG1* gene is repressed, resulting in growth arrest and morphological changes (Fig. 2; Miyajima *et al.*, 1987). In an **a** *ste2-null scg1-null* mutant containing *pGAL-SCG1*, as *SCG1* expression is repressed and cells initiate arrest of cell division, mating ability is seen (Fig. 2; Matsumoto *et al.*, 1988). In addition, *scg1-ts* and *scg1-null pGAL-SCG1* strains show induction (at the restrictive temperature and after glucose repression, respectively) of genes or gene products that are induced in wild-type cells by exposure to pheromone (Fig. 2; Matsumoto *et al.*, 1988; Jahng *et al.*, 1988). These results indicate that repression or inactivation of *SCG1* results in cell division arrest, morphological changes, and induction of pheromone-inducible genes in a manner similar to the response of wild-type cells to pheromone. These responses are occurring in the absence of pheromone receptor. The induction of mating efficiency under these conditions indicates that repression or inactivation of *SCG1* is sufficient to induce all aspects of pheromone response necessary for mating.

C. Site-Directed Mutations in *SCG1* Resulting in Two Opposite Phenotypes: Constitutive Pheromone Response or Sterility

Mutations in *SCG1* constructed by site-directed mutagenesis have provided information concerning the role and mechanism of action of *SCG1* (C. Dietzel and J. Kurjan, in preparation). Alterations in the regions of *SCG1* predicted to be involved in guanine-nucleotide binding and/or GTPase activity have been made based on analogy to mutations in *ras* (McGrath *et al.*, 1984; Seeburg *et al.*, 1984; Sweet *et al.*, 1984; Clanton *et al.*, 1986; Walter *et al.*, 1986). In addition, alterations of the SCG1 carboxy terminus have been made based on the hypothesis that the carboxy termini of Gα subunits are involved in receptor interactions (Masters *et al.*, 1986; Sullivan *et al.*, 1987).

An Asn to Lys mutation (*scg1*[Lys388]) was constructed at a position analogous to amino acid 116 in *ras* (Clanton *et al.*, 1986; Walter *et al.*, 1986; C. Dietzel and J. Kurjan, in preparation). Based on the crystal structure of EF-Tu and *ras*, this region interacts with the guanine ring of GDP (Jurnak, 1985, LaCour *et al.*, 1985; De Vos *et al.*, 1988). Analogous *ras* mutations result in defects in guanine-nucleotide binding and a transforming phenotype (Clanton *et al.*, 1986; Walter *et al.*, 1986). The *scg1*[Lys388] mutant shows growth and morphological changes similar to but less severe than *scg1-null* mutants (C. Dietzel and J. Kurjan, in preparation). This mutation is recessive to the wild type, as are *scg1-null* mutations. The *scg1-null*-like phenotype provides circumstantial evidence that the *SCG1* product is a guanine-nucleotide binding protein and that guanine-nucleotide binding is essential for *SCG1* function. The defect in guanine-nucleotide binding in *ras*[116] mutants reflects an increase in the dissociation rate for guanine nucleotides (Feig *et al.*, 1986; Feig and Cooper, 1988). Under *in vivo* conditions, where GTP is present at higher concentrations than GDP, this increased dissociation rate could result in an increased amount of protein in the GTP-bound form (as discussed by Feig *et al.*, 1986; Walter *et al.*, 1986; Feig and Cooper, 1988). In the models for *SCG1* action (Section VIII), in the absence of pheromone *SCG1*-GDP acts to keep the pheromone response pathway off; guanine-nucleotide exchange eliminates this inhibitory *SCG1* activity and therefore turns on pheromone response. An increase in the level of *SCG1* in the GTP-bound form should result in constitutive activation of the pheromone response pathway and thus result in a phenotype that resembles the *scg1-null* phenotype, as is seen with the *scg1*[Lys388] mutant.

A Gly to Val mutation (*SCG1*[Val50]), analogous to the mammalian *ras*[Val12] and yeast *RAS*[Val19] mutations, has a different effect. In EF-Tu and *ras*, this region is implicated in the interaction with the phosphate groups of GDP (Jurnak, 1985; La Cour *et al.*, 1985, De Vos *et al.*, 1988). The mammalian *ras*[Val12] mutation leads to a GTPase defect and a transforming phenotype (McGrath *et al.*, 1984; Seeburg *et al.*, 1984; Sweet *et al.*, 1984). Yeast *RAS* acts to activate adenylyl cyclase, and the yeast *RAS*[Val19] mutation results in hyperactivation of adenylyl cyclase (Toda *et al.*, 1985). The *SCG1*[Val50] mutation (C. Dietzel and J. Kurjan, in preparation) results in a slight *scg1-null*-like phenotype (somewhat enlarged cells) and a mating defect (a low level of mating in comparison to the wild type). The mating defect is partially dominant to the wild type. This phenotype could be explained by a defect in guanine-nucleotide exchange, which would prevent activation of the pheromone response pathway and therefore result in a mating defect. This phenotype is consistent with the effect of the analogous mutation in α_s, which results in a decrease in the rate of guanine-nucleotide exchange (Bourne *et al.*, 1988).

Mutations of the carboxy terminus also show informative phenotypes (C. Dietzel and J. Kurjan, in preparation). A mutant with a five-amino-acid carboxy-terminal truncation (*SCG1*[5aatrunc]) is defective in pheromone response and mating; this mutation is dominant to the wild type. A Ser to Cys substitution four amino acids from the carboxy terminus (*SCG1*[Cys469]) results in a slight decrease in mating, which is also dominant to the wild type. This mutation gives rise to a carboxy terminal sequence similar to the sequence at the carboxy terminus of *ras* and some $G\alpha$ subunits, which has been implicated in palmitylation and membrane localization of *ras* (Chen *et al.*, 1985; Fujiyama and Tamanoi, 1986). The mating and pheromone-response defects of these carboxy-terminal *SCG1* mutants are consistent with an interaction between the *SCG1* carboxy terminus and the phero-mone receptors. A mutation near the carboxy terminus of α_s has a similar effect, i.e. it uncouples the G protein from the receptor (Sullivan *et al.*, 1987).

A mutant containing a larger carboxy terminal *SCG1* truncation shows a different phenotype (C. Dietzel and J. Kurjan, in preparation). This 22-amino-acid carboxy-terminal truncation (*scg1[22aatrunc]*) results in an *scg1-null* phenotype and is recessive to the wild type. This phenotype is consist-ent with a hypothesis (Bourne *et al.*, 1988; Heideman and Bourne, Chapter 2 of this volume; H. Bourne, personal communication) that the carboxy termini of $G\alpha$ subunits shield the guanine nucleotide-binding site and prevent release of GDP. Interaction with the receptor would result in a conformational change that allows dissociation of GDP. Truncation of $G\alpha$ would eliminate the ability of the carboxy terminus to prevent GDP release. In *SCG1*, such a truncation would allow guanine-nucleotide ex-change in the absence of pheromone, resulting in constitutive activation of the pheromone response pathway, i.e. the *scg1-null* phenotype. The pheno-types of the *SCG1[5aatrunc]* and *scg1[22aatrunc]* mutants suggest that the five-amino-acid truncation eliminates the ability to interact with the receptor, but still allows shielding of the guanine nucleotide-binding site, whereas the 22-amino-acid truncation eliminates the ability of the carboxy termi-nus to shield the guanine nucleotide-binding site.

Together, the mutant studies indicate that there are two opposite pheno-types associated with *SCG1* mutations. One set of mutations results in decreased pheromone response and mating ability (C. Dietzel and J. Kur-jan, in preparation). Such mutations are likely to result from an inability to interact with the receptor or to undergo guanine-nucleotide exchange, both of which should be necessary to turn on the pheromone response pathway. These mutations are dominant (or partially dominant) to the wild-type gene. This dominance suggests that the mutant proteins are able to seques-ter a downstream component of the pathway that is necessary to turn on the response ($\beta\gamma$ according to the current models, Section VIII).

The other set of mutants shows constitutive expression of the phero-mone response pathway in the absence of pheromone, resulting in cell cycle arrest and morphological changes (Dietzel and Kurjan, 1987a; Miya-jima *et al.*, 1987; Jahng *et al.*, 1988; C. Dietzel and J. Kurjan, in prepara-tion). Null mutants are included in this set, indicating that wild-type *SCG1* must play a role in keeping the pathway inhibited in the absence of pheromone. The null mutants are recessive to the wild type, as would be expected for loss-of-function mutations. Some of the temperature-sensitive *scg1* mutations are partially dominant with respect to growth arrest at the restrictive temperature (Jahng *et al.*, 1988). At the permissive temperature, both the wild-type and mutant *SCG1* proteins would act to inhibit the pathway in the absence of pheromone. At the restrictive temperature, the mutant *scg1* protein would be inactivated, allowing some expression of the pathway. In contrast, in a strain heterozygous for the wild type and a null mutation, the only functional *SCG1* present is wild type, and this protein must be present at a level sufficient to efficiently inhibit the pathway.

V. RAT Gα SUBUNITS COMPLEMENTING *scg1* GROWTH DEFECT

The striking similarity of *SCG1* and the vertebrate Gα subunits strongly suggests an evolutionary relationship. Studies have shown that there is functional homology between mammalian *ras* and yeast *RAS* proteins, i.e., expression of mammalian *ras* can partially complement yeast *ras* mutations and expression of yeast *RAS* containing a transforming mutation (and a deletion of sequences not present in mammalian *ras*) results in transformation in mammalian cells (DeFeo-Jones *et al.*, 1985; Kataoka *et al.*, 1985a). Similarly, expression of some mammalian Gα subunits can partially complement *scg1* mutations, indicating functional as well as sequence conservation. Expression of rat α_s, under the control of a yeast promoter, results in complementation of the growth and morphological defects of *scg1-null* mutations (Dietzel and Kurjan, 1987a; Matsumoto *et al.*, 1988). The *scg1-null* [α_s] strains are sterile. Rat α_s, therefore, can act to inhibit the pathway as *SCG1* does in the absence of pheromone, but cannot stimulate the pathway upon exposure to pheromone. This phenotype suggests that α_s, can interact with a downstream component ($\beta\gamma$, see Sections VII and VIII) but cannot interact with the receptor. Rat α_i can weakly complement *scg1-null* mutations, but no complementation has been seen with α_o (Matsumoto *et al.*, 1988; D. Tipper, J. Stadel, and J. Kurjan, in preparation).

Some hybrid genes have also been tested (D. Tipper, J. Stadel, and J. Kurjan, in preparation). Hybrids consisting of approximately the amino-terminal two-thirds of *SCG1* and the carboxy terminal one-third of α_s or α_i efficiently complement the *scg1-null* growth and morphological defects. Again, the resulting strains are sterile. The inability of these hybrids to stimulate the pheromone response pathway is consistent with an interaction between the carboxy terminus of G_α subunits with their respective receptors. A comparison with the $scg1^{22aatrunc}$ mutant suggests that an intact (or almost intact) carboxy terminus is necessary to inhibit the pheromone response pathway, possibly by preventing guanine-nucleotide exchange (Section IV. C). Heterologous G_α carboxy termini (at least of α_s and α_i) can provide the carboxy-terminal function necessary for inhibition of the pheromone response pathway, presumably by preventing GDP release.

VI. IDENTIFICATION OF β AND γ SUBUNITS INVOLVED IN PHEROMONE RESPONSE

Recent results indicate that there are β and γ subunits involved in the pheromone response pathway, further strengthening the analogy to vertebrate G-protein-mediated signal transduction systems (Whiteway et al., 1989). The presumptive β and γ genes (STE4 and STE18, respectively) are members of a group of genes called non-specific steriles; loss-of-function mutations in these genes result in a sterile phenotype in both **a** and α cells (MacKay and Manney, 1974a,b; Hartwell, 1980). Most of these mutants, including the ste4 and ste18 mutants, are unable to respond to pheromone, and therefore may be components of the pheromone response pathway or may regulate or modify components of this pathway.

Cloning and sequencing of the STE4 and STE18 genes revealed 43% identity in comparison to rat β_1 and β_2 subunits and 31% identity in comparison to γ_t, respectively (Fig. 3; Whiteway et al., 1989). In addition, many conservative differences are seen between these sequences. Both the putative yeast β and γ proteins are larger than the corresponding vertebrate proteins, as is also true for the α subunits and RAS. The relatively low level of identity between STE18 and vertebrate γ is not surprising, because different vertebrate γ subunits seem to be less well conserved than the other G protein subunits (Hildebrandt et al., 1985). The degree of similarity between STE4 and vertebrate β subunits suggest that STE4 acts as a β subunit in the pheromone response pathway. Although the level of similarity of STE18 to γ_t is relatively low, genetic results (described below in this section) suggest that STE4 and STE18 act at the same point in the pheromone response pathway, and thus support the hypothesis that STE18 acts as a γ subunit in this pathway. In addition, expression of STE4 and STE18 is haploid-specific (Whiteway et al., 1989), consistent with a role in pheromone response.

The phenotypes of several STE4 and STE18 mutants provide information concerning their role. Recessive ste4 and ste18 loss-of-function mutants are sterile and defective in pheromone response (MacKay and Manney, 1974a,b; Hartwell, 1980; Whiteway et al., 1989), a phenotype similar

A

```
Ste4      1   M A A H Q M D S I T Y S N N V T Q Q Y I Q P Q S L Q D I S A V

Ste4     32   E E E I Q N K I E A A R Q E S K Q L H A Q I N K A K H K I Q D
Beta-1    1           M S E L D Q L R Q E A E Q L K N Q I R D A R K A C A D
Beta-2    1           M S E L E Q L R Q E A E Q L R N Q I R D A R K A C G D

Ste4     63   A S L F Q M A N K V T S L T K N K I N L K P N I V L K G H N N
Beta-1   28   A T L S Q I T N - - N I D P V G R I Q M R T R R T L R G H L A
Beta-2   28   S T L T Q I T A - - G L D P V G R I Q M R T R R T L R G H L A

Ste4     94   K I S D F R W S R D S K R I L - S A S Q D G F M L I W D S A S
Beta-1   57   K I Y A M H W G T D S - R L L V S A S Q D G K L I I W D S Y T
Beta-2   57   K I Y A M H W G T D S - R L L V S A S Q D G K L I I W D S Y T

Ste4    124   G L K Q N A I P L D S Q W V L S C A I S P S S T L V A S A G L
Beta-1   87   T N K V H A I P L R S S W V M T C A Y A P S G N Y Y V A C G G L
Beta-2   87   T N K V H A I P L R S S W V M T C A Y A P S G N F V A C G G L

Ste4    155   N N N C T I Y R V S K E N R V A Q N V A S I F K G H T C Y I S
Beta-1  118   D N I C S I Y - N L K T R E G N V R V S R E L A G H T G Y L S
Beta-2  118   D N I C S I Y - S L K T R E G N V R V S R E L P G H T G Y L S

Ste4    186   D I E F T D N A H I L T A S G D M T C A L W D I P K A K R V R
Beta-1  148   C C R F L D D N Q I V T S S G C T T C A L W D I E T G Q Q T T
Beta-2  148   C C R F L D D N Q I I T S S G D T T C A L W D I E T G Q Q T V

Ste4    217   E Y S D H L G D V L A L A I P E E P N L E N S S N T F A S C G
Beta-1  179   T F T G H T G D V M S L - - - - - - S L A P D T R L F V S G A
Beta-2  179   G F A G H S G D V M S L - - - - - - S L A P D G R T F V S G A

Ste4    248   S D G Y T Y I W D S R S P S A V Q S F Y V N D S D I N A L R F
Beta-1  204   C D A S A K L W D V R E G M C R Q T F I G H E S D I N A I C F
Beta-2  204   C D A S I K L W D V R D S M C R Q T F I G H E S D I N A V A F

Ste4    279   F K D G M S I V A G S D N G A I N M Y D L R S D C S I A T F S
Beta-1  235   F P N G N A F A T G S D - - - - - - - D - - - - - - - A T C R
Beta-2  235   F P N G Y A F T T G S D - - - - - - - D - - - - - - - A T C R

Ste4    310   L F R G Y E E R T P T P T Y M A A N M E Y N T A Q S P Q T L K
Beta-1  252   L F - - - D L R - - - - - - - - A - - - - - - - - - D Q E L -
Beta-2  252   L F - - - D L R - - - - - - - - A - - - - - - - - - D Q E L L

Ste4    341   S T S S Y L D N Q G V V S L D F S A S G R L M Y S C Y T D I
Beta-1  262   M T Y S H D N I I C G I T S V S F S K S G R L L L A G Y D D F
Beta-2  263   M - Y S H D N I I C G I T S V A F S R S G R L L L A G Y D D F

Ste4    372   G C V V W D V L K G E I V G K L E G H G G R V T G V R S S P D
Beta-1  293   N C N V W D A L K A D R A G V L A G H D N R V S C L G V T D D
Beta-2  293   N C N I W D A M K G D R A G V L A G H D N R V S C L G V T D D

Ste4    402   G L A V C T G S W D S T M K I W S P G Y Q
Beta-1  324   G M A V A T G S W D S F L K I W N
Beta-2  324   G M A V A T G S W D S F L K I W N
```

B

```
Ste18      1   M T S V Q N S P R L Q Q P Q E Q Q Q Q Q Q Q L S L K I K Q L K
Gamma-t    1   M P V I - N - - - I E - - - D - - - - - - - L T - E - K D - K

Ste18     32   L K R I N E L N N K L R K E L S R E R I T A S N A C L T I I N
Gamma-t   15   L K - M - E V D Q - L K K E V T L E R M L V S K C C - E - - E

Ste18     63   Y T S N T K D Y T L P E L W G Y P V A G S N H F I E G L K N A
Gamma-t   40   F - - - - R D Y - V E E - - - - - R S G E D P L V K G I - P E

Ste18     94   Q K N S Q M S N S N S V C C T L M
Gamma-t   60   D K N - P F K E L K G G - C V I S
```

to the phenotype of the dominant $SCG1^{5aatrunc}$ mutant (Section IV.C). A dominant $STE4$ mutant ($STE4^{Hpl}$) has been identified that has a phenotype similar to the $SCG1$-$null$ phenotype, i.e., constitutive expression of the pheromone response pathway (Blinder et al., 1989). The similar phenotypes associated with $SCG1$ and $STE4$ mutants suggest that they act in the same pathway.

Novel $ste4$ and $ste18$ mutants have been isolated that are not sterile but show an unusual pattern of pheromone sensitivity: sensitivity to low but not high levels of pheromone (Whiteway et al., 1988, 1989). Although the explanation for this phenotype is not clear, the identification of $ste4$ and $ste18$ alleles with the same unusual phenotype suggests that $STE4$ and $STE18$ are likely to act at the same point in the pheromone response pathway. This result, therefore, is consistent with the suggested correspondence of $STE4$ and $STE18$ to β and γ subunits.

Biochemical results are also consistent with a role for the $STE4$ product in an early step of the pathway (Jenness et al., 1987). Mutations in $ste4$ were shown to have at most a slight effect on levels of α-factor receptor, as determined by α-factor binding assays. The reduction in receptor level was not sufficient to account for the complete loss of pheromone response in $ste4$ mutants. This result suggested that $STE4$ must act (either directly or indirectly) at a step of the response pathway downstream from the receptor. A small decrease in affinity for pheromone was also seen in $ste4$ mutants (whereas the other non-specific steriles tested, $ste5$, $ste7$, $ste11$, and $ste12$, showed no change in affinity for α-factor). This change in affinity suggested that $STE4$ might also control a structural feature of the α-factor receptor. Both of these results can be explained if $STE4$ is a β subunit that acts in the pheromone response pathway. The epistatic studies and models described below (Sections VII and VIII) place $\beta\gamma$ downstream of α in the pathway, consistent with the pheromone response defect seen with close to wild-type levels of α-factor receptor. If the $\alpha\beta\gamma$ heterotrimer interacts with the receptor, a β mutation could result in a change in receptor properties, therefore accounting for the decreased affinity for α-factor in the $ste4$ mutants.

Fig. 3. Comparison of Gβ and Gγ subunits from yeast and vertebrate cells. (A) The presumed Gβ subunit involved in pheromone response in yeast (Ste4) is compared to the vertebrate Gβ_1 and Gβ_2 subunits (Fong et al., 1987). Amino acids that are identical in at least two of the Gβ subunits are boxed. (B) The presumed Gγ subunit involved in pheromone response in yeast (Ste18) is compared to the Gγ_t subunit from rod cells (Yatsunami et al., 1985). Amino-acid identities are boxed. Ignoring gaps, Ste4 shows 42% identity to both β_1 and β_2, and Ste18 shows 31% identity to γ_t. The Ste4 and Ste18 sequences are from Whiteway et al., 1989.

Another nonspecific sterile (*ste5*) shows characteristics that are consistent with the possibility that *STE5* is a component of the pheromone response pathway. The *ste5* mutations, as well as *ste4* mutations, result in only a slight decrease in α-factor receptor levels (Jenness *et al.*, 1987). The severe defect in pheromone response seen in *ste5* mutants, combined with the almost wild-type levels of receptor, indicate that *STE5* is either a downstream component of the pathway or affects the function or expression of a downstream component(s). The characteristics associated with *STE5* suggest that it could encode the effector (or a component of the effector) of the pheromone response pathway, although there is currently no definitive evidence to indicate such a role.

VII. EPISTATIC RELATIONSHIPS

A useful genetic approach for ordering the components of a pathway involves a study of the epistatic relationships between mutations in different genes. This approach involves the construction of double mutants and a comparison of the double-mutant phenotype with the phenotypes of the single mutants. Epistatic relationships can only be tested using double mutants if the individual mutations result in distinguishable phenotypes. If two mutants show similar phenotypes, the double mutant could either show the same phenotype or a more severe phenotype, neither of which provides information concerning the order of the steps controlled by the two genes. If two genes control the same step in a pathway, loss-of-function mutations in both genes should result in the same phenotype, and the double mutant should also show this phenotype. If loss-of-function mutations in two genes give rise to quite different phenotypes, these two gene products cannot control the same step of the pathway.

For the pheromone response pathway, some information is provided by a comparison of the null phenotypes of different components of the pathway. The null phenotype of *scg1* mutants differs from the null phenotype of the sterile mutants, *ste2, ste4,* and *ste18* (*scg1* mutants show growth arrest and morphological changes due to constitutive expression of the pheromone response pathway, and the *ste* mutants show an inability to respond to pheromone resulting in a mating defect). The *SCG1* product and the *STE* products, therefore, must act at different steps in the pathway. The null *ste* mutants all show the same phenotype; therefore, some of these gene products could act at the same step of the pathway.

Epistatic relationships between some of the components of the pheromone response pathway have been tested. If an *scg1-null ste-null* double mutant shows the sterile phenotype, this particular *STE* gene must act downstream of *SCG1*. The rationale for this conclusion is that the pathway is expressed constitutively due to the *scg1* mutation, but some component of the pathway necessary for completion of the pathway, and therefore downstream of *SCG1*, is defective. The *scg1 ste4* and *scg1 ste18* double mutants show such a pattern, i.e. the double mutants show normal growth and morphology, but are sterile (Nakayama *et al.*, 1988; Blinder *et al.*, 1989; Whiteway *et al.*, 1989). This result indicates that *STE4* and *STE18* act downstream of *SCG1*.

If an *scg1-null ste-null* double mutant shows constitutive expression of the pheromone response pathway, the pathway downstream of *SCG1* must be functional; therefore, this particular *STE* product must act upstream of *SCG1*. a *ste2 scg1* double mutants show this pattern (Miyajima *et al.*, 1987; Dietzel and Kurjan, 1987a), indicating that *SCG1* acts downstream of *STE2*, as would be predicted for the relationship between the receptor and a Gα subunit. The results described above (Section IV.B) indicating that inactivation or repression of *SCG1* in a *ste2* strain allows mating also suggests that *SCG1* acts downstream of the receptors (Jahng *et al.*, 1988; Matsumoto *et al.*, 1988).

The identical phenotype in the *ste-null* mutants makes determination of epistatic relationships difficult. The identification of the dominant *STE4^Hpl* mutant (Section VI), which results in growth arrest due to constitutive expression of the pheromone response pathway, made it possible to determine the epistatic relationship between *STE2* and *STE4*. The *ste2 STE4^Hpl* double mutant shows constitutive pheromone response (Blinder *et al.*, 1989), indicating that all components downstream of *STE4* are functional and that *STE2* acts upstream of *STE4* (i.e. *STE2* acts upstream of *SCG1* which acts upstream of *STE4*). The epistatic relationship between *STE4* and *STE18* has not been tested; however, if these genes encode β and γ subunits, their products should act at the same step of the pathway. As described below (Section VIII), the ordering of components of the pathway using epistatic relationships has provided information that has eliminated some of the models that have been proposed for the mechanism involved in the pheromone response pathway.

Epistatic relationships between *ste5* mutants and *scg1-null* and *STE4^Hpl* mutants have also been investigated (Nakayama *et al.*, 1988; Blinder *et al.*, 1989). *scg1 ste5* double mutants are sterile, indicating that *STE5* acts (directly or indirectly) downstream of *SCG1*. A *ste5 STE4^Hpl* double mutant is also sterile, indicating that *STE5* acts either at the same step as *STE4* or downstream of *STE4*. The latter position is consistent with, although again does not provide definitive evidence for, the possibility that *STE5* is the effector (or a component of the effector) of the pheromone response pathway.

VIII. MODELS FOR MECHANISM OF PHEROMONE RESPONSE

Several models have been proposed for the mechanism of action of the G-protein homolog involved in pheromone response (Dietzel and Kurjan, 1987a; Miyajima *et al.*, 1987; Jahng *et al.*, 1988). These models are diagrammed in Figs. 4 and 5 along with the model for vertebrate G-protein action involving G_s stimulation of adenylyl cyclase and G_t stimulation of cGMP phosphodiesterase (Stryer and Bourne, 1986; Gilman, 1987). Each model is shown in two forms: a pictorial diagram and a diagram using a genetic format that illustrates the types of interactions involved, i.e., whether a particular component stimulates or inhibits the activity of the component immediately downstream in the pathway. Each model is also depicted before and after exposure to pheromone. Before pheromone exposure, the G protein exists as an $\alpha\beta\gamma$ heterotrimer and the α subunit has GDP bound. All of the models require that on activation of the receptor, the G protein heterotrimer interacts with the receptor and guanine-nucleotide exchange occurs (this step is not shown). After pheromone (or agonist) exposure, the models are shown after dissociation of the G protein from the receptor and dissociation of α from $\beta\gamma$. The effector(s) in the pheromone response system has not been identified; the diagrams are drawn with a single effector acting to elicit all aspects of pheromone response.

The phenotype of *scg1-null* mutants is inconsistent with a model identical to the vertebrate model (Fig. 4; Dietzel and Kurjan, 1987a). If the mechanism of pheromone response were analogous to the mechanisms involved in these vertebrate systems, an *scg1-null* mutant should be unable to respond to pheromone, leading to sterility. Instead, the opposite phenotype, constitutive expression of the pheromone response pathway in the absence of exposure to pheromone, is seen. The constitutive activation of the pheromone response pathway in *scg1*-null mutants indicates that *SCG1* plays a negative or inhibitory role, acting to keep the pheromone response pathway inactive in the absence of exposure to pheromone (Dietzel and Kurjan, 1987a; Miyajima *et al.*, 1987; Jahng *et al.*, 1988). Various models have been described that propose an inhibitory role (either direct or indirect) for *SCG1*–GDP. In these models, this inhibition is relieved after guanine-nucleotide exchange, allowing activation of the pathway. Two of the proposed models (Fig. 4) are analogous to the vertebrate model in that the α subunit interacts with the effector, although the effect of this interaction is different from the effect in the vertebrate model. Two additional models (Fig. 5) are based on controversial results suggesting that $\beta\gamma$ can act directly on an effector or channel to alter its activity (Jelsema and Axelrod, 1987; Logothetis *et al.*, 1987).

We initially proposed that the GDP-bound form of *SCG1* acts to keep the effector inactive (Fig. 4, Model A; Dietzel and Kurjan, 1987a). After exposure to pheromone and guanine-nucleotide exchange, the GTP-bound form of *SCG1* would be unable to inhibit the effector, and the activated effector would turn on downstream components of the pathway, leading to pheromone response. The role of $\beta\gamma$ would be to promote the G-protein–receptor interaction and guanine-nucleotide exchange. In this model, an α null mutation should be epistatic to $\beta\gamma$ null mutations, i.e. the *scg1 ste4* and *scg1 ste18* strains should show growth arrest due to constitutive expression of the pheromone response pathway. Instead, $\beta\gamma$ mutations are epistatic to α mutations, indicating that $\beta\gamma$ acts downstream of α; therefore this model is inconsistent with the recent genetic results (Nakayama *et al.*, 1988; Blinder *et al.*, 1989; Whiteway *et al.*, 1989). Another model was proposed (Fig. 4, Model B; Miyajima *et al.*, 1987) that is similar to our model in the absence of pheromone response, i.e., *SCG1*-GDP would act to inhibit effector activity. After pheromone exposure and guanine-nucleotide exchange, however, *SCG1*–GTP was proposed to have an additional function necessary for activation of the effector. There is no current evidence that requires this positive function for *SCG1*–GTP in addition to the negative function for *SCG1*–GDP in inhibiting the pathway. In this model, the role of $\beta\gamma$ would also be in the G-protein–receptor interaction and guanine-nucleotide exchange; therefore, the epistatic interactions between the α and $\beta\gamma$ mutations are inconsistent with this model.

Two additional models have been proposed, both of which are consistent with the current evidence. We proposed a model in which $\beta\gamma$ acts to activate the effector (Fig. 5, Model C; Dietzel and Kurjan, 1987a). In the absence of pheromone, *SCG1*–GDP would interact with $\beta\gamma$, thus preventing activation of the effector. After exposure to pheromone and guanine-nucleotide exchange, *SCG1*–GTP would dissociate from $\beta\gamma$, and free $\beta\gamma$ would activate the effector, which would then act to turn on downstream components of the pathway leading to pheromone response. In this model, $\beta\gamma$ acts downstream of α, which is consistent with the genetic results.

Another model has been proposed that also has $\beta\gamma$ acting downstream of α, but the role of the effector is different. Our model proposes that the effector has a positive role, acting to turn on pheromone response after exposure to pheromone (Fig. 5, Model C; Dietzel and Kurjan, 1987a); the alternative model proposes that the effector has a negative role, acting to inhibit pheromone response in the absence of pheromone (Fig. 5, Model D; Jahng *et al.*, 1988). In the latter model, after exposure to pheromone and guanine-nucleotide exchange, $\beta\gamma$ is freed and acts to inhibit the effector. By inhibiting an effector whose role is to inhibit pheromone response, $\beta\gamma$ acts to allow the pheromone response pathway to be activated; in

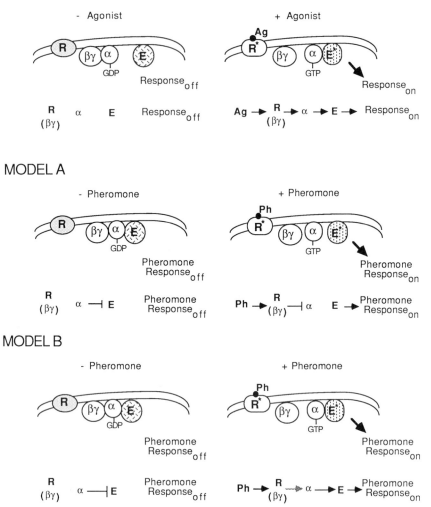

Fig. 4. Models A and B for pheromone response involving a Gα–effector interaction. Two models for the pheromone response pathway that involve a direct *SCG1*–effector interaction (Dietzel and Kurjan, 1987a; Miyajima *et al.*, 1987) are indicated and compared to the mechanism of well-characterized vertebrate G-protein-mediated signal transduction systems (Stryer and Bourne, 1986; Gilman, 1987). Each model is shown in a diagrammatic form on top and a genetic form below. An arrow indicates that one component activates a downstream component, and a line with a terminal bar indicates that a component inhibits a downstream component. The models are shown before and after exposure to pheromone (or agonist). After pheromone (agonist) exposure, the models indicate the state of the system after guanine-nucleotide exchange has occurred. During an intermediate step (not shown) the G protein heterotrimer interacts with the activated receptor, guanine-nucleotide exchange

MODEL C

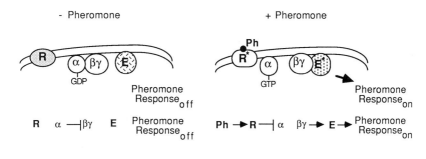

MODEL D

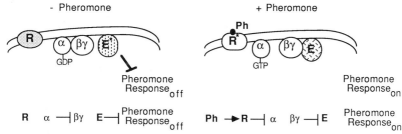

Fig. 5. Models C and D for pheromone response involving a G$\beta\gamma$–effector interaction. Two models for the pheromone response pathway that involve a $\beta\gamma$–effector interaction (Dietzel and Kurjan, 1987a; Jahng *et al.*, 1988) are shown. All details are as indicated in the legend to Fig. 4. These models are consistent with the current results (Nakayama *et al.*, 1988; Blinder *et al.*, 1989; Whiteway *et al.*, 1989) indicating the order of action of the components in the pathway.

occurs, the G protein dissociates from the receptor, and the α subunit dissociates from $\beta\gamma$. In the genetic form of the models $\beta\gamma$ is indicated as acting before α, because the role of $\beta\gamma$ is to promote the receptor interaction and guanine-nucleotide exchange. The steps occurring after the pheromone–receptor interaction are assumed to be the same in **a** and α cells. Further details are described in the text. These models are not consistent with genetic results indicating that the presumptive β and γ subunits act downstream of α. Ag, agonist; Ph, pheromone (**a**-factor or α-factor); R, receptor; R*, activated receptor; α, Gα subunit with bound GDP or GTP; $\beta\gamma$, Gβ and Gγ subunits (in the genetic forms, the parentheses indicate that the role of $\beta\gamma$ is to promote the G-protein–receptor interaction and guanine-nucleotide exchange); E, inactive effector; E*, active effector. In the vertebrate systems, R is rhodopsin or the β-adrenergic receptor, α is α_{t1} or α_s, and E is cGMP phosphodiesterase or adenylyl cyclase. In the yeast pheromone response system, R is the **a**-factor receptor (*STE3*) or the α-factor receptor (*STE2*), α is *SCG1/GPA1*, β is *STE4*, γ is *STE18*, and E is currently unidentified.

genetic terms, $\beta\gamma$ is a negative regulator of a negative regulator, and thus acts as a positive regulator of pheromone response. Because $\beta\gamma$ acts downstream of α in this model, it is also consistent with the genetic results. Further experiments are necessary to distinguish (or disprove) these models and to identify additional components of the pathway, in particular, the effector.

IX. IDENTIFICATION OF SECOND Gα HOMOLOG (GPA2)

A second yeast Gα homolog (*GPA2*) has been identified by low stringency hybridization to α_{i2} and α_o cDNA probes (Nakafuku *et al*, 1988). *GPA2* could encode a protein of 449 amino acids with molecular weight of 50.5K. This protein shows a high level of similarity to *SCG1* and vertebrate α subunits (Fig. 1). Like *SCG1*, the *GPA2* protein is larger than vertebrate Gα proteins. The extra amino acids are at a different position from the extra amino acids in *SCG1*. *GPA2* expression is not cell type-specific as indicated by hybridization to a 1.9-kb RNA in all three cell types (**a**, α, and **a**/α). No phenotype associated with null mutations of *gpa2* (including one mutation that deletes about half of the open reading frame) has been seen.

It has been suggested that *GPA2* may play a role in regulation of cAMP levels (Nakafuku *et al.*, 1988). In yeast, the *RAS* proteins (encoded by *RAS1* and *RAS2*) are involved in regulation of adenylyl cyclase (Broek *et al.*, 1985; Toda *et al.*, 1985). At least one of the *RAS* genes and the *CYR1* (adenylyl cyclase) gene are essential for growth in yeast (Matsumoto *et al.*, 1982, 1984; Kataoka *et al.*, 1984, 1985b; Casperson *et al.*, 1985; Tatchell *et al.*, 1984). The regulation of adenylyl cyclase by G proteins in vertebrate cells suggested the possibility that a yeast G protein might act (in addition to *RAS*) to regulate the cAMP pathway (Nakafuku *et al.*, 1988).

In wild-type cells, addition of glucose to cells arrested in stationary phase leads to a transient stimulation in cAMP levels (Nakafuku *et al.*, 1988). The *gpa2-null* mutant shows a response to glucose similar to the response seen in the wild-type strain. Glucose induction of a wild-type strain containing *GPA2* on a multicopy plasmid (YEp-*GPA2*) results in about a twofold higher level of cAMP stimulation than the wild-type strain lacking the plasmid. In addition, cAMP levels remain high for at least 30 minutes in the YEp-*GPA2*-containing strain, whereas cAMP levels return to the basal level within a few minutes in the wild-type strain that lacks YEp-*GPA2*. The YEp-*GPA2* plasmid is also able to suppress the growth defect of a *ras1-null ras2-ts* mutant at the restrictive temperature, but it is not able to suppress *ras1-null ras2-null* or *cyr1* growth defects. These results have been interpreted as suggesting that *GPA2* plays a role in the regulation of cAMP levels in yeast (Nakafuku *et al.*, 1988). The similarity in glucose induction of cAMP in the wild-type and *gpa2* disruption strains, however, does not support this hypothesis. A speculation that *GPA2* function is redundant (i.e., that there is a second gene that encodes a protein with a similar function) has been made to account for the lack of a phenotype or an effect on cAMP levels in the *gpa2* disruption mutant, although at present there is no evidence to support the existence of a second gene. An alternative possibility is that *GPA2* does not normally play a role in the cAMP pathway, but when overexpressed (by expression on a multicopy plasmid) could act to regulate a component of this pathway. In such a situation, a *gpa2* disruption mutant should not show any differences from the wild-type strain with respect to stimulation of cAMP levels. Further work should help to define the role of *GPA2*.

X. PERSPECTIVES

The discovery of a G protein involved in a signal transduction pathway in yeast indicates the extensive evolutionary conservation of these proteins. The evidence that *SCG1* (α), *STE4* (β), and *STE18* (γ) play roles in pheromone response that are similar to the roles of G-protein subunits in vertebrate signal-transduction systems is convincing. One critical component of the pheromone response pathway, the effector, is currently unidentified. Hopefully, the powerful genetics possible with this system will allow identification of the effector(s) in the near future. As additional components of the pheromone response pathway are identified, and the mechanisms of action of these components are determined, it will be valuable to compare this pathway to the many vertebrate G-protein-mediated pathways that are currently being elucidated. The analogy of this yeast system to vertebrate systems has been quite exciting to those of us working on yeast pheromone response, and we have tried to make use of information provided by the vertebrate systems in the formulation of models and planning of further experiments. It will be equally exciting if information found in the yeast system provides information useful in studies of vertebrate systems.

ACKNOWLEDGMENTS

I thank M. Whiteway, V. MacKay, D. Jenness, and K. Matsumoto for communication of in-press or unpublished results, and H. Bourne for communication of results and ideas. I thank Jeanne Hirsch and Chris Dietzel for discussions and Didi

Robins, Teri Melese, Jeanne Hirsch, and Malcolm Whiteway for comments on the manuscript.

REFERENCES

Bender, A., and Sprague, G. F., Jr. (1986). Yeast peptide pheromones, a-factor and α-factor, activate a common response mechanism in their target cells. *Cell* **47**, 929–937.

Blinder, D., Bouvier, and Jenness, D. (1989). Constitutive elements in the yeast pheromone response: Ordered function of the gene products. *Cell* **56**, 479–486.

Bourne, H. R., Masters, S. B., Miller, R. T., Sullivan, K. A., and Heideman, W. (1988). Mutations probe structure and function of G protein α chains, *Cold Spring Harbor Symp. Quant. Biol.* **53**, 221–228.

Bray, P., Carter, C., Simons, C., Guo, V., Puckett, C., Kamholz, J., Spiegel, A., and Nirenberg, M. (1986). Human cDNA clones for four species of $G_{\alpha s}$ signal transduction protein. *Proc. Natl. Acad. Sci. U.S.A.* **83**, 8893–8897.

Broek, D., Samiy, N., Fasano, O., Fujiyama, A., Tamanoi, F., Northup, J., and Wigler, M. (1985). Differential activation of yeast adenylate cyclase by wild-type and mutant *RAS* proteins. *Cell* **41**, 763–769.

Bucking-Throm, E., Duntze, W., Hartwell, L. H., and Manney, T. R. (1973). Reversible arrest of haploid yeast cells at the initiation of DNA synthesis by a diffusible sex factor. *Exp. Cell Res.* **76**, 99–110.

Burkholder, A. C., and Hartwell, L. H. (1985). The yeast α-factor receptor: Structural properties deduced from the sequence of the *STE2* gene. *Nucleic Acids Res.* **13**, 8463–8475.

Casperson, G. F., Walker, N., and Bourne, H. R. (1985). Isolation of the gene encoding adenylate cyclase in *Saccharomyces cerevisiae*. *Proc. Natl. Acad. Sci. U.S.A.* **82**, 5060–5063.

Chan, R. K., and Otte, C. A. (1982a). Isolation and genetic analysis of *Saccharomyces cerevisiae* mutants supersensitive to G1 arrest by a-factor and α-factor pheromones. *Mol. Cell. Biol.* **2**, 11–20.

Chan, R. K., and Otte, C. A. (1982b). Physiological characterization of *Saccharomyces cerevisiae* mutants supersensitive to G1 arrest by a-factor and α-factor pheromones. *Mol. Cell. Biol.* **2**, 21–29.

Chen, Z. Q., Ulsh, L. S., DuBois, G., and Shih, T. Y. (1985). Posttranslational processsding of p21 ras proteins involves palmitylation of the C-terminal tetrapeptide containing cysteine-186. *J. Virol.* **56**, 607–612.

Chvatchko, Y., Howald, I., and Rizman, H. (1986). Two yeast mutants defective in endocytosis are defective in pheromone response. *Cell* **46**, 355–364.

Clanton, D. J., Hattori, S., and Shih, T. Y. (1986). Mutations of the *ras* gene product p21 that abolish guanine nucleotide binding. *Proc. Natl. Acad. Sci. U.S.A.* **83**, 5076–5080.

DeFeo-Jones, D., Tatchell, K., Robinson, L. C., Sigal, I. S., Vass, W. C., Lowy,

D. R., and Scolnick, E. M. (1985). Mammalian and yeast *ras* gene products: Biological function in their heterologous systems. *Science* **228**, 179–184.

Dever, T. E., Glynias, M. J., and Merrick, W. C. (1987). GTP-binding domain: Three consensus sequence elements with distinct spacing. *Proc. Natl. Acad. Sci. U.S.A.* **84**, 1814–1818.

De Vos, A. M., Tong, L., Milburn, M. V., Matias, P. M., Jancarik, J., Noguchi, S., Nishimura, S., Miura, K., Ihtsuka, E., and Kim, S. -H. (1988). Three-dimensional structure of an oncogene protein: Catalytic domain of human c-H-ras-p21. *Science* **239**, 888–893.

Dietzel, C., and Kurjan, J. (1987a). The yeast *SCG1* gene: a G_α-like protein implicated in the **a**- and α-factor response pathway. *Cell* **50**, 1000–1010.

Dietzel, C., and Kurjan, J. (1987b). Pheromonal regulation and sequence of the *Saccharomyces cerevisiae SST2* gene: a model for desensitization to pheromone. *Mol. Cell. Biol.* **7**, 4169–4177.

Duntze, W., MacKay, V., and Manney, T. R. (1970). *Saccharomyces cerevisiae*; a diffusible sex factor. *Science* **168**, 1472–1473.

Feig, L. A., and Cooper, G. M. (1988). Relationship among guanine nucleotide exchange, GTP hydrolysis, and transforming potential of mutated *ras* proteins. *Mol. Cell. Biol.* **8**, 2472–2478.

Feig, L. A., Pan, B. -T., Roberts, T. M., and Cooper, G. M. (1986). Isolation of *ras* GTP-binding mutants using a *in situ* colony-binding assay. *Proc. Natl. Acad. Sci. U.S.A.* **83**, 4607–4611.

Fong, H. K. W., Amatruda, T. T., III, Birren, B. W., and Simon, M. I. (1987). Distinct forms of the β subunits of GTP-binding regulatory proteins identified by molecular cloning. *Proc. Natl. Acad. Sci. U.S.A.* **84**, 3792–3796.

Fujiyama, A., and Tamanoi, F. (1986). Processing and fatty acid acylation of RAS1 and RAS2 proteins in *Saccharomyces cerevisiae*. *Proc. Natl. Acad. Sci. U.S.A.* **83**, 1266–1270.

Gilman, A. G. (1987). G proteins: Transducers of receptor-generated signals. *Annu. Rev. Biochem.* **56**, 615–649.

Hagen, D. C., and Sprague, G. F., Jr. (1984). Induction of the yeast α-specific *STE3* gene by the peptide pheromone a-factor. *J. Mol. Biol.* **178**, 835–852.

Hagen, D. C., McCaffrey, G., and Sprague, G. F., Jr. (1986). Evidence the yeast *STE*3 gene encodes a receptor for the peptide pheromone **a** factor: Gene sequence and implications for the structure of the presumed receptor. *Proc. Natl. Acad. Sci. U.S.A.* **83**, 1418–1422.

Hartig, A., Holly, J., Saari, G., and MacKay, V. L. (1986). Multiple regulation of *STE2*, a mating-type-specific of *Saccharomyces cerevisiae*. *Mol. Cell. Biol.* **6**, 2106–2114.

Hartwell, L. H. (1973). Synchronization of haploid yeast cell cycles, a prelude to conjugation. *Exp. Cell Res.* **76**, 111–117.

Hartwell, L. H. (1980). Mutants of *Saccharomyces cerevisiae* unresponsive to cell division control by polypeptide mating hormone. *J. Cell Biol.* **85**, 811–822.

Hartwell, L. H., Mortimer, R. K., Culotti, J., and Culotti, M. (1973). Genetic control of the cell division cycle in yeast: V. Genetic analysis of *cdc* mutants. *Genetics* **74**, 267–286.

Heideman, W. R., and Bourne, H. R. (1989). Structure and function of G protein α chains. *In* "G Proteins" (R. Iyengar and L. Birnbaumer, eds.). Academic Press, San Diego (in press; this volume).

Hildebrandt, J. D., Codina, J., Rosenthal, W., Birnbaumer, L., Neer, E. J., Yamazaki, A., and Bitensky, M. W. (1985). Characterization by two-dimensional peptide mapping of the γ subunits of N_s and N_i, the regulatory proteins of adenylyl cyclase, and of transducin, the guanine nucleotide-binding protein of rod outer segments of the eye. *J. Biol. Chem.* **260,** 14867–14872.

Itoh, H., Kozasa, T., Nagata, S., Nakamura, S., Katada, T., Ui, M., Iwai, S., Ohtsuka, E., Kawasaki, H., Suzuki, K., and Kaziro, Y. (1986). Molecular cloning and sequence determination of cDNAs for α subunits of the guanine nucleotide-binding proteins G_s, G_i, and G_o from rat brain. *Proc. Natl. Acad. Sci. U.S.A.* **83,** 3776–3780.

Jahng, K. -Y., Ferguson, J., and Reed, S. I. (1988). Mutations in a gene encoding the α subunit of a *Saccharomyces cerevisiae* G protein indicate a role in mating pheromone signaling. *Mol. Cell. Biol.* **8,** 2484–2493.

Jelsema, C. L., and Azelrod, J. (1987). Stimulation of phospholipase A_2 activity in bovine rod outer segments by the subunits of transducin and its inhibition by the α subunit. *Proc. Natl. Acad. Sci. U.S.A.* **84,** 3623–3627.

Jenness, D. D., and Spatrick, P. (1986). Down regulation of the α-factor pheromone receptor in *Saccharomyces cerevisiae*. *Cell* **46,** 345–353.

Jenness, D. D., Burkholder, A. C., and Hartwell, L. H. (1983). Binding of α-factor pheromone to yeast **a** cells: Chemical and genetic evidence for an α-factor receptor. *Cell* **35,** 521–529.

Jenness, D. D., Goldman, B. S., and Hartwell, L. H. (1987). *Saccharomyces cerevisiae* mutants unresponsive to α-factor pheromone: α-factor binding and extragenic suppression. *Mol. Cell. Biol.* **7,** 1311–1319.

Jurnak, F. (1985). Structure of the GDP domain of EF-Tu and location of the amino acids homologous to ras oncogene proteins. *Science* **230,** 32–36.

Kataoka, T., Powers, S., McGill, C., Fasano, O., Strathern, J., Broach, J., and Wigler, M. (1984). Genetic analysis of yeast *RAS1* and *RAS2* genes. *Cell* **37,** 437–445.

Kataoka, T., Powers, S., Cameron, S., Fasano, O., Goldfarb, M., Broach, J., and Wigler, M. (1985a). Functional homology of mammalian and yeast *RAS* genes. *Cell* **40,** 19–26.

Kataoka, T., Broek, D., and Wigler, M. (1985b). DNA sequence and characterization of the *S. cerevisiae* gene encoding adenylate cyclase. *Cell* **43,** 493–505.

Kurjan, J. (1985). α-Factor structural gene mutations in *Saccharomyces cerevisiae*: Effects on α-factor production and mating. *Mol. Cell. Biol.* **5,** 787–796.

La Cour, T. F. M., Nyborg, J., Thirup, S., and Clark, B. F. C. (1985). Structural details of the binding of guanosine diphosphate to elongation factors Tu from *E. coli* as studied by X-ray crystallography. *EMBO J.* **4,** 2385–2388.

Logothetis, D. E., Kurachi, Y., Galper, J., Neer, E. J., and Clapham, D. E. (1987). The subunits of GTP-binding proteins activate the muscarinic K^+ channel in heart. *Nature* **325,** 321–326.

MacKay, V. L., and Manney, T. R. (1974a). Mutations affecting sexual conjugation and related processes in *Saccharomyces cerevisiae*. I. Isolation and phenotypic characterization of nonmating mutants. *Genetics* 76, 255–271.

MacKay, V., and Manney, T. R. (1974b). Mutations affecting sexual conjugation and related processes in *Saccharomyces cerevisiae*. II. Genetic analysis of nonmating mutants. *Genetics* 76, 273–288.

Manney, T. R., and Woods, V. (1976). Mutants of *Saccharomyces cerevisiae* resistant to the α mating-type factor. *Genetics* 82, 639–644.

Masters, S. B., Stroud, R. M., and Bourne, H. R. (1986). Family of G protein α chains: Amphipathic analysis and predicted structure of functional domains. *Prot. Eng.* 1, 47–54.

Matsumoto, K., Uno, I., Oshima, Y., and Ishikawa, T. (1982). Isolation and characterization of yeast mutants deficient in adenylate cyclase and cAMP-dependent protein kinase, *Proc. Natl. Acad. Sci. U.S.A.* 79, 2355–2359.

Matsumoto, K., Uno, I., and Ishikawa, T. (1984). Identification of the structural gene and nonsense alleles for adenylate cyclase in *Saccharomyces cerevisiae. J. Bacteriol.* 157, 277–282.

Matsumoto, K., Nakafuku, M., Nakayama, N., Miyajima, I., Kaibuchi, K., Miyajima, A., Brenner, C., Arai, K., and Kaziro, Y. (1988). The role of G proteins in yeast signal transduction. *Cold Spring Harbor Symp. Quant. Biol.* 53, 567–575.

McCaffrey, G., Clay, F. J., Kelsay, K, and Sprague, G. F., Jr. (1987). Identification and regulation of a gene required for cell fusion during mating of the yeast *Saccharomyces cerevisiae. Mol. Cell. Biol.* 7, 2680–2690.

McGrath, J. P., Capon, D. J., Goedell, D. V., and Levinson, A. D. (1984). Comparative biochemical properties of normal and activated human ras p21 protein. *Nature* 310, 644–649.

Michaelis, S., and Herskowitz, I. (1988). The a-factor pheromone of *Saccharomyces cerevisiae* is essential for mating. *Mol. Cell. Biol.* 8, 1309–1318.

Miyajima, I., Nakafuku, M., Nakayama, N., Brenner, C., Miyajima, A., Kaibuchi, K., Arai, K. -I., Kaziro, Y., and Matsumoto, K. (1987). *GPA*1, a haploid-specific essential gene, encodes a yeast homolog of mammalian G protein which may be involved in mating factor signal transduction. *Cell* 50, 1011–1019.

Moore, S. A. (1983). Yeast cells recover from mating pheromone α factor-induced division arrest by desensitization in the absence of α factor destruction. *J. Biol. Chem.* 259, 1004–1010.

Nakafuku, M., Itoh, H., Nakamura, S., and Kaziro, Y. (1987). Occurrence in *Saccharomyces cerevisiae* of a gene homologous to the cDNA coding for the α subunit of mammalian G proteins. *Proc. Natl. Acad. Sci. U.S.A.* 84, 2140–2144.

Nakafuku, M., Obara, T., Kaibuchi, K., Miyajima, I., Miyajima, A., Itoh, H., Nakamura, S., Arai, K. -I., Matsumoto, K., and Kaziro, Y. (1988). Isolation of a second yeast *Saccharomyces cerevisiae* gene (*GPA2*) coding for guanine nucleotide-binding regulatory protein: Studies on its structure and possible functions. *Proc. Natl. Acad. Sci. U.S.A.* 85, 1374–1378.

Nakayama, N., Miyajima, A., and Arai, K. (1985). Nucleotide sequences of *STE2* and *STE3*, cell type-specific sterile genes from *Saccharomyces cerevisiae*. *EMBO J.* **4**, 2643–2648.

Nakayama, N., Miyajima, A., and Arai, K. (1987). Common signal transduction system shared by *STE2* and *STE3* in haploid cells in *Saccharomyces cerevisiae*: Autocrine cell-cycle arrest results from forced expression of *STE2*. *EMBO J.* **6**, 249–254.

Nakayama, N., Kaziro, Y., Arai, K. -I., and Matsumoto, K. (1988). Role of *STE* genes in the mating-factor signaling pathway mediated by *GPA1* in *Saccharomyces cerevisiae*. *Mol. Cell. Biol.* **8**, 3777–3783.

Reed, S. I. (1980). The selection of *S. cerevisiae* mutants defective in the start event of cell division. *Genetics* **93**, 561–577.

Robishaw, J. D., Russell, D. W., Harris, B. A., Smigel, M. D., and Gilman, A. G. (1986a). Deduced primary structure of the α subunit of the GTP-binding stimulatory protein of adenylate cyclase. *Proc. Natl. Acad. Sci. U.S.A.* **83**, 1251–1255.

Robishaw, J. D., Smigel, M. D., and Gilman, A. G. (1986b). Molecular basis for two forms of the G protein that stimulates adenylate cyclase. *J. Biol. Chem.* **261**, 9587–9590.

Rothstein, R. J. (1983). One-step gene disruption in yeast. *Methods Enzymol.* **101**, 202–211.

Seeburg, D. H., Colby, W. W., Capon, D. J., Goedell, D. V., and Levinson, A. D. (1984). Biological properties of human c-Ha-*ras*1 genes mutated at codon 12. *Nature* **312**, 71–75.

Sibley, D. R., Benovic, J. L., Caron, M. G., and Lefkowitz, R. J. (1987). Regulation of transmembrane signaling by receptor phosphorylation. *Cell* **48**, 913–922.

Sprague, G. F., Jr., Blair, L. C., and Thorner, J. (1983). Cell interactions and regulation of cell type in the yeast *Saccharomyces cerevisiae*. *Annu. Rev. Microbiol.* **37**, 623–660.

Stryer, L., and Bourne, H. R. (1986). G proteins: A family of signal transducers. *Annu. Rev. Cell. Biol.* **2**, 391–419.

Sullivan, K. A., Miller, R. T., Masters, S. B., Biederman, B., Heidman, W., and Bourne, H. R. (1987). Identification of a receptor contact site involved in receptor-G protein coupling. *Nature* **330**, 758–760.

Sweet, R. W., Yokoyama, S., Kamata, T., Feramisco, J. R., Rosenberg, M., and Gross, M. (1984). The product of *ras* is a GTPase and the T24 oncogenic mutant is deficient in this activity. *Nature* **311**, 273–275.

Tanabe, T., Nukada, T., Nishikawa, Y., Sugimoto, K., Suzuki, H., Takahashi, H., Noda, M., Haga, T., Ichiyama, A., Kangawa, K., Minamino, N., Matsuo, H., and Numa, S. (1985). Primary structure of the α-subunit of transducin and its relationship to ras proteins. *Nature* **315**, 242–245.

Tatchell, K., Chaleff, D. T., DeFeo-Jones, D., and Scholnick, E. M. (1984). Requirement of either of a pair of *ras*-related genes of *Saccharomyces cerevisiae* for spore viability. *Nature* **309**, 523–527.

Toda, T., Uno, I., Ishikawa, T., Powers, S., Kataoka, T., Broek, D., Cameron, S.,

Broach, J., Matsumoto, K., and Wigler, M. (1985). In yeast, *RAS* proteins are controlling elements of adenylate cyclase. *Cell* **40**, 27–36.

Trueheart, J., Boeke, J. D., and Fink, G. R. (1987). Two genes required for cell fusion during yeast conjugation: Evidence for a pheromone-induced surface protein. *Mol. Cell. Biol.* **7**, 2316–2328.

Van Meurs, K. P., Angus, C. W., Lavu, S., Kung, H. -F., Czarnecki, S. K., Moss, J., and Vaughan, M. (1987). Deduced amino acid sequence of bovine retinal $G_{o\alpha}$: Similarities to other guanine nucleotide-binding proteins. *Proc. Natl. Acad. Sci. U.S.A.* **84**, 3107–3111.

Walter, M., Clark, S. G., and Levinson, A. D. (1986). The oncogenic activation of human p21^{ras} by a novel mechanism. *Science* **233**, 649–652.

Whiteway, M. S., Hougan, L., and Thomas, D. Y. (1988). Expression of *MFα1* in *MAT*a cells supersensitive to α-factor leads to self-arrest. *Mol. Gen. Genet.* **214**, 85–88.

Whiteway, M., Hougan, L., Dignard, D., Thomas, D. Y., Bell, L., Saari, G. C., Grant, F. J., O'Hara, P., and MacKay, V. L. (1989). The *STE4* and *STE18* genes of yeast encode potential β and γ subunits of the mating factor receptor-coupled G protein. *Cell* **56**, 467–477.

Yatsunami, K., Pandya, B. V., Oprain, D. D., and Khorana, H. G. (1985). cDNA-derived amino acid sequence of the γ subunit of GTPase from bovine rod outer segments. *Proc. Natl. Acad. Sci. U.S.A.* **82**, 1936–1940.

CHAPTER 25

GTP-Binding Proteins and Exocytotic Secretion

Bastien D. Gomperts
Department of Physiology, University College London
London WC1E 6JJ, United Kingdom

VIII. Conclusion
    A. General Biological Role for G_E
    References

I. Ca^{2+} *SECUNDUS INTER PARES* IN EXOCYTOTIC SECRETION

For most scientists outside the field, and even for many who work within it, the idea of secretion (here we are concerned with exocytotic secretion, the release of preformed materials from storage granules within the cytoplasm) is inextricably bound up with notions concerning calcium. Calcium has been associated with secretion for more than 100 years (Ringer, 1883, 1884; Locke, 1894) and yet it is true to say that we are no nearer understanding what it does to enable the intracellular membrane-fusion processes which occur during exocytosis than when the expression stimulus–secretion coupling, (implying some parallelism with excitation-contraction coupling and a certain role for Ca^{2+} in the activation of myosin ATPase), was first expressed in 1961 (Douglas and Rubin, 1961). In short, Ca^{2+} does everything and nothing; it is at once both the most manipulable and the most visible of the second messengers. Only since the discovery of other intracellular effectors that alone or together with Ca^{2+} cause exocytosis can it be said that we are moving once again toward a solution of this blatant yet cryptic cellular function.

Ca^{2+}-Independent Secretion

That Ca^{2+} is not a universal signal for exocytosis has been known since at least 1974 when it was shown that secretion of amylase from rat parotid glands is mediated by β-adrenergic agonists acting conventionally to elevate the concentration of cyclic AMP (Dormer and Ashcroft, 1974; Leslie *et al.*, 1976; Schramm and Selinger, 1975). Subsequently it has been shown that this process is not necessarily associated with elevation of cytosol Ca^{2+} (Takemura, 1985; McMillian *et al.*, 1988). Secretion of parathyroid hormone, which is actually suppressed by elevation of cytosol Ca^{2+} (Fitzpatrick *et al.* 1986b) may also be stimulated by cyclic AMP (Muff and Fisher, 1986a). Other instances of Ca^{2+}-independent secretion include platelets [stimulated with thrombin and with phorbol ester (Rink and Sanchez, 1984; Rink *et al.*, 1983)], neutrophils [stimulated with phorbol ester (Di Virgilio *et al.*, 1984)], pancreatic islets and insulin-secreting RINm5F cells [stimulated with phorbol ester (Tamagawa *et al.*, 1985; Arkhammer *et al.*, 1986)] and exocrine pancreas [stimulated with caerulein or carbamoylcholine (Bruzzone *et al.*, 1986)] in all of which the release of stored materials can occur with Ca^{2+} at, and even well below, its resting level of approximately $10^{-7}M$.

II. EXOCYTOTIC MECHANISMS

A. Biophysical Aspects

Exocytosis has been described as the controlled interaction of perigranular membranes with the plasma membrane that results in their fusion and subsequent fission to release the granular contents to the exterior (Palade, 1975). This description was based mainly on thin-section and freeze-fracture electron-microscopic evidence of cells undergoing exocytosis. The fusions are absolutely specific, and as far as we know there is no leakage of granule contents into the cytosol. While this description remains valid in a macroscopic sense, and has indeed been well substantiated by more recent developments (Lawson et al., 1977; Orci et al., 1977; Orci and Perrelet, 1978; Ornberg and Reese, 1981; Chandler and Heuser, 1979), it also conceals a number of problems which must be faced if we are ever to understand the molecular mechanisms involved.

We should consider some problems of temporal and spatial resolution. Membranes are regarded as being in close apposition, and hence in a preexocytotic state when the fixed and stained thin section reveals a pentalaminar structure. It is at this point, as the hydrated surfaces come within about 20 Å of each other (i.e., close to the limit of resolution in electron microscopy), that they become subject to an overwhelming force of repulsion (the hydration force) which increases according to a high-order exponent as the distance of separation diminishes (Parsegian et al. 1984; Parsegian and Rau, 1984; Le Neveu et al., 1976). Yet, in order to fuse, the membranes must necessarily come within molecular feeling distance of one another.

Next, it must be realized that the phospholipid bilayer of which membranes are composed is a stable structure in the biological environment in which bulk water is normally present (i.e. at 55 M). It is hydration which maintains phospholipids in this arrangement (Bangham, 1968), and the polar surfaces of the bilayers have a high affinity for water (Le Neveu et al., 1976). In order to disorganize the bilayer organization to induce fusion between cells it is common practice to dehydrate the environment, though this in itself is insufficient. Thus, a high concentration of polyethylene glycol, sufficient to remove all traces of free water, is commonly used as a fusogen in the formation of cell hybrids, but the propensity to fuse cells varies greatly between polyethylene glycols from different sources (Honda *et al.*, 1981; Smith *et al.*, 1982) depending on the presence of destabilizing impurities in the commercial material (Wojcieszyn *et al.*, 1983). It must be obvious that biological fusions occur by mechanisms other than those which involve the removal of bulk water.

B. Biochemical Aspects

In the light of experience with artificial systems it becomes axiomatic that the controlled fusions which occur in biology are catalytically mediated and proceed on a molecule-by-molecule basis. In seeking to understand the mechanism of exocytosis we are searching for a fusogenic catalyst; in effect, what we discover is a sequence of control steps, the last of which, so far unidentified, being that which regulates the activity of the fusogen.

1. Gaining Access to Cytosol

Until recently, the main problem confronting investigations into the mechanism of exocytosis involved the plasma membrane, which acts as a barrier preventing access to the secretory granules, but which is at the same time an integral component involved in exocytotic fusions and also in the transduction of the receptor-mediated signals which control secretion. With the development of permeabilized cell preparations, in which the exocytotic process is seemingly well preserved, receptor functions can be obviated and we now have direct access to the cytosol and the intracellular sites at which exocytosis occurs (Knight and Scrutton, 1986a; Gomperts and Fernandez, 1985). Table I lists a number of procedures and reagents which have been used to gain access to the cytosol of cells in the investigation of secretion. It should be stressed at the outset that all these reagents may have effects on the target cells in addition to permeabilization of the plasma membrane; some of the differences which have been recorded when similar experiments have been carried out in different laboratories might possibly be ascribed to such effects. Some of the reagents (e.g., Sendai virus) might be expected to have properties of a receptor-directed agonist. More than this, the filtration properties and the lifetimes of the lesions generated vary greatly, depending on the reagents, the cells, and other conditions which may not be apparent or controllable. The use of the patch pipette requires that the cells be adherent to a surface with all that this implies in terms of cytoskeletal organization. When reading literature on exocytosis it is therefore worth bearing in mind the methods used to permeabilize cells. Of the methods listed, only the use of ATP^{4-} (Gomperts, 1983) and the patch pipette (Almers and Neher, 1985) provide the possibility of precisely timed resealing.

Table I Methods of Cell Permeabilization

System	Effective filtration diameter[a]	Method of assessment	Reference[b]
Sendai virus	Approx. 1 nm	Exclusion of fluorescent peptides	Impraim et al. (1980)
Staphylococcal α-toxin	2–3 nm	Electron microscopy	Füssle et al. (1981)
High-voltage discharge	2–4 nm	Rate of uptake and efflux of various markers in chromaffin cells	Knight and Baker (1982)
ATP^{4-}	Variable dimensions (α[ATP^{4-}])	Rate of ^{32}P-metabolite efflux and [^{57}Co]HEDTA uptake	Cockcroft and Gomperts, (1979) Bennett et al. (1981)
Lysolecithin	Variable dimensions	RNase (M_r 14K) uptake and LDH efflux	Miller et al. (1978)
Plant glycosides	Macromolecular dimensions	Slow leakage of LDH (M_r 142K)	Dunn & Holz, (1983)
Streptolysin O	Greater than 13 nm	Efflux of urease (M_r 483K) from sheep red cell membrane vesicles	Buckingham and Duncan (1983)
Patch pipette	Micron dimensions	Measurement of tip resistance	Hamill et al. (1981), Marty and Neher (1983)

[a]The filtration dimensions are given only as a very rough guide. For any reagent or method there will be wide variation depending on membrane composition and other conditions. Furthermore, the different methods used to assess the filtration properties of membrane lesions must necessarily give different results.

III. G PROTEINS AND STIMULUS–SECRETION COUPLING

The idea that GTP might be an intracellular mediator of secretion was born of frustrations concerning earlier proposals for a mechanism of Ca^{2+} mobilization involving hydrolysis of phosphatidylinositol (PI) (Gomperts, 1983). While the association between PI-linked events and Ca^{2+} mobilization had been steadily strengthened since it was first proposed (Michell, 1975), by 1983 no mechanism was evident and it was only clear that the original idea of phospholipase C activation to yield inositol 1-phosphate and diglyceride (DG) was in need of modification (Cockcroft et al., 1980, 1981; Cockcroft, 1982). At about this time it was shown that in liver cells hydrolysis of the polyphosphoinositides precedes the decline in PI in response to stimulation by vasopressin and other Ca^{2+}-mobilizing hormones (Creba et al., 1983). This was shortly followed by the well-known demonstration that the hydrolytic product of phosphatidylinositol 4,5-bisphosphate (PIP$_2$), inositol 1,4,5-trisphosphate (IP$_3$), when introduced into the cytosol of permeabilized pancreatic cells, would release Ca^{2+} from the same intracellular stores as those that are involved following hormonal stimulation (Streb *et al.*, 1983, 1985). These stores are now thought to represent a subset of the light membrane microsomal fraction, probably distinct from the endoplasmic reticulum (Volpe *et al.*, 1988; Henne *et al.*, 1987), and have been termed *calciosomes*.

A. Discovery of G_p

When membranes are treated with GTP (or one of its analogs) the affinity of agonist binding at a number of calcium-mobilizing receptors declines [e.g., muscarinic cholinergic receptors (Berrie *et al.*, 1979), α_1-adrenergic receptors (Goodhardt *et al.*, 1982), angiotensin II receptors (Glossman *et al.*, 1974), and the neutrophil receptor for formylmethionyl peptides (Koo *et al.*, 1983)]. In view of this we realized that the ATP^{4-}-permeabilized mast cell preparation could provide the means to test whether GTP, and hence a G protein, might be involved in the mechanism of cellular activation for secretion. By introducing GTP analogs into the permeabilized cells and then trapping them (on resealing the cells by conversion of ATP^{4-} to its inactive Mg^{2+} salt), we obtained a cell which secreted simply by addition of Ca^{2+} to the extracellular phase (in the millimolar range of concentrations), without involvement of receptors (Gomperts, 1983).

It was a further two years before we (Cockcroft and Gomperts, 1985) and others (Smith *et al.*, 1985) showed that one mode of action of GTP in the sequence of events leading to exocytosis might be in the activation of polyphosphoinositide specific phosphodiesterase (PPI-PDE, a phospholipase C). We have called this class of G proteins (i.e., both those that are sensitive and those that are insensitive to pertussis toxin) G_p.[1]

Having recognized that there is a clearly defined function for GTP, it would be quite wrong to assume that GTP, and hence the G proteins, play no further role in secretion or other cell activation processes. The purpose of this essay is to show that GTP acts again at a late stage in the stimulus–secretion sequence in some (but certainly not all) cell types, and that it might do so by controlling, either directly or indirectly, a protein dephosphorylation which is an enabling reaction for exocytotic membrane fusion to occur.

[1]There is confusion in the nomenclature concerning G_p. In our terminology the subscript P is indicative of the functional relationship with polyphosphoinositides. Others have isolated a new class of GTP-binding proteins of unknown function from platelets, placenta, and polymorphonuclear leukocytes which they call G_p out of deference to their original cellular sources (Evans *et al.*, 1986). Besides, P follows O in the alphabet and at the time of the discovery of these new GTP-binding proteins, G_o was the last to be discovered. The logic here is irreproachable, but futile. The best hope is that their G_p and our G_p will turn out to be one and the same entity.

B. GTP: The Second Site

The experiments which led up to the discovery of G_P were initiated using the permeabilized resealed mast cell preparation which, as with intact cells, obscures the exocytotic site due to the reestablishment of an impermeant plasma membrane. In order to study the final stages of the exocytotic process it is necessary to work in an open cell configuration, and all the experiments to be described have been carried out in this manner. This is exemplified by those presented in Fig. 1 (panels A – D) which illustrates the effect of guanine nucleotides on the Ca^{2+}-induced secretory responses of four different cell types.

IV. ROLE OF G PROTEINS IN CONTROL OF SECRETION

A. Secretion from Various Cells and Tissues

1. Platelets

In Fig. 1A we see the effect of providing two analogs of GTP along with Ca^{2+} in the stimulation of serotonin secretion from permeabilized platelets (Haslam and Davidson, 1984b; Knight and Scrutton, 1986b). The effect of the guanine nucleotides is mainly to shift the dependence of secretion on Ca^{2+} to lower concentrations but it always remains absolutely dependent on Ca^{2+}; the effect on the maximal extent of secretion at high concentrations of Ca^{2+} is marginal. Since provision of thrombin or agonistic phorbol ester (Haslam and Davidson, 1984a; Knight et al., 1984) produces a very similar effect on the Ca^{2+} dependence of amine secretion, it has been concluded that the effect of the GTP analogs in this system is exerted through G_P with consequent generation of diglyceride (Knight and Scrutton, 1986b). The enhancement in the effective affinity for Ca^{2+} in the secretory process is then due to protein phosphorylation catalyzed by protein kinase C.

2. Neutrophils

A very different picture is presented when a similar experiment is carried out on permeabilized rabbit neutrophils (Barrowman et al., 1986). Here, secretion of β-glucuronidase from the lysosomal (or azurophilic) granules can be stimulated either by Ca^{2+} or by analogs of GTP. When low concentrations of GTPγS are used together with Ca^{2+} in the range pCa 7 – 5, their effects are roughly additive. Unlike the platelets, the effect of GTP in neutrophils is to increase the extent of secretion at all concentrations of Ca^{2+}; there is very little (if any) alteration in the effective affinity for Ca^{2+}. Since stimulation of secretion by GTPγS is not only maintained, but

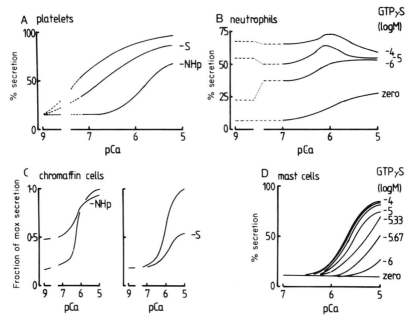

Fig. 1. Evidence regarding the involvement of a G protein (G_E) at a late stage in the exocytotic pathway of various cells. (A) Platelets, permeabilized by high-voltage discharge, secrete serotonin in response to Ca^{2+} in the range pCa 6.5–5 (in the presence of MgATP). When GTPγS (S) or Gpp(NH)p (NHp) are provided, secretion is elicited at concentrations of Ca^{2+} below pCa 7. (Data from Knight and Scrutton, 1986b). (B) Rabbit neutrophils, treated with cytochalasin B and permeabilized with Sendai virus, secrete lysosomal enzymes in the presence of Ca^{2+} in the range pCa 6.5–5 (ATP is required). When GTP analogs are provided, the extent of secretion is enhanced but there is no shift in the dependence on Ca^{2+}. In the absence of Ca^{2+} (even at concentrations below $10^{-10}\ M$), GTPγS still elicits secretion. (From Barrowman *et al.*, 1986.) (C) Permeabilized adrenal chromaffin cells release catecholamines in response to Ca^{2+} in the presence of ATP, and this can be modulated by analogs of GTP. In the figure, the data are presented as the fraction of maximal release, which approximates 20%. Left panel: Cells in primary culture, permeabilized with digitonin. Here, Gpp(NH)p (NHp) induces Ca^{2+}-independent secretion, but unlike neutrophils [see panel (B)] this is nonadditive with the secretion due to Ca^{2+}. Right panel: Freshly disaggregated cells permeabilized by high-voltage discharge. Here, GTPγS (S) inhibits Ca^{2+}-induced secretion without affecting the concentration dependence. (Data from Bittner *et al.*, 1986 and Knight and Baker, 1985.) (D) Rat mast cells, treated with antimycin A and deoxyglucose and then permeabilized by streptolysin O in the absence of ATP, secrete histamine in the presence of a dual effector system comprising Ca^{2+} *plus* GTPγS. (From Howell *et al.*, 1987.)

actually enhanced when the cells are initially permeabilized in the presence of very high concentrations of chelating agents (EGTA or BAPTA[2] at 30 mM) for 2 min prior to supplying the GTP analogue, it can be said that this is truly an example of Ca^{2+}-independent secretion. More than this, since generation of IP_3 and DG do not occur when Ca^{2+} is suppressed below the normal cytosol resting level of $\sim 10^{-7}M$ (Cockcroft, 1986), it is unlikely that activation of protein kinase C is involved. We can also say that GTP is likely to be involved in Ca^{2+}-mediated secretion, since GDP (and its analogs) suppress secretion elicited by a Ca^{2+}-only stimulus (Barrowman *et al.*, 1986).

It was from these experiments that we first learned that GTP can act at a late stage in the stimulus secretion sequence, quite independent of any reactions mediated by G_p. We have called this second site of action of GTP G_E, the E indicating its relationship to exocytosis (Gomperts, 1986; Gomperts *et al.*, 1986).

The question of whether GTP and G_E are involved in exocytosis from intact neutrophils following stimulation with receptor-directed agonists has not been resolved. What is certain is that there must be other signaling systems acting in concert with Ca^{2+} since the concentration of Ca^{2+} required in a Ca^{2+}-only stimulus far exceeds that which is achieved in the normal physiological situation (Lew *et al.*, 1986; Barrowman *et al.*, 1987; Cockcroft and Gomperts, 1988).

3. Insulin-Secreting Cells

GTP-induced, Ca^{2+}-independent secretion of insulin has been demonstrated in RINm5F (insulinoma) cells and islet cells permeabilized by high-voltage discharge (Vallar *et al.*, 1987; Wollheim *et al.*, 1987). Under the conditions of low Ca^{2+} (<pCa9) there was no release of IP_3 from the RIN5mF cells so it is likely that GTP acts directly to cause exocytosis through reactions not involving activation of protein kinase C. Unlike the neutrophils, the effects of Ca^{2+} and GTP (analogs) on insulin secretion are not additive and this suggests that they might act at a common late stage in the exocytotic control pathway. Similar to the neutrophils, however, GDP (and its analogs) are inhibitory to Ca^{2+}-stimulated secretion.

There is evidence that GTP and a G protein (G_{Ei}) also mediate the inhibition of insulin secretion at a late stage following activation of inhibi-

---

[2]Bis (*o*-aminophenoxy)ethane-*N,N,N',N'*-tetraacetic acid (Tsien, 1980), a chelator sharing with EGTA the property of great $Ca^{2+}:Mg^{2+}$ selectivity, but having the virtue of complete proton dissociation throughout the physiological pH range. This means that unlike EGTA, the affinity of BAPTA for Ca^{2+} remains almost constant in the face of alteration of pH. BAPTA is the parent compound for the synthesis of the fluorescent Ca^{2+}-indicator, quin2.

tory α_2-adrenergic receptors. Although these receptors remain coupled to adenylyl cyclase in permeabilized pancreatic islets (Jones *et al.*, 1987a) and RINm5F cells (Ullrich and Wollheim, 1988) (conventionally by suppression of cyclic AMP production, through G_i), there is no correlation between the amount of cyclic AMP generated and the rate of secretion. When GTP is provided to permeabilized RINm5F cells, adrenaline inhibits both Ca^{2+}- and diglyceride-induced secretion, an effect which is abolished by pretreatment with pertussis toxin (Ullrich and Wollheim, 1988). The non-hydrolyzable analogs of GTP do not support inhibition, presumably because they support the activation response.

4. Parathyroid

Unlike most secretory systems, secretion from the parathyroid is associated with a reduction in the concentration of extracellular Ca^{2+}. If this is reduced to less than 1.25 mM then secretion of parathyroid hormone occurs, achieving a maximal rate at 1 mM (Fitzpatrick *et al.*, 1986a); it has been suggested that there is a specific Ca^{2+} receptor on the external surface of these cells (Nemeth and Scarpa, 1986). Concurrent with reduction of extracellular Ca^{2+}, the concentration of intracellular Ca^{2+} also declines from a high resting level of about 0.5 μM (Muff and Fischer, 1986b). However, it is unlikely that the secretory response is elicited simply as a consequence of reducing $[Ca^{2+}]_i$ since dopamine, noradrenaline, and Li^+ are all capable of inducing secretion, and do so without any alteration in the concentration of cytosolic Ca^{2+} (Nemeth *et al.*, 1986).

There is evidence for the involvement of GTP binding proteins at two distinct stages in the stimulus–secretion sequence of these cells. In pertussis-treated parathyroid cells, exocytosis becomes insensitive to the inhibitory effects of high extracellular Ca^{2+} (in this system it is common to regard high Ca^{2+} as an inhibitor, rather than low Ca^{2+} as an activator of secretion) and appears to proceed constitutively. Since the application of Ca^{2+}-ionophore inhibits release regardless of whether cells have been treated with the toxin, the pertussis substrate is probably mediating events at the receptor level (Fitzpatrick *et al.*, 1986a), possibly associated with production of inositol phosphates (Brown *et al.*, 1987).

Surprisingly, it was found that exocytosis from permeabilized cells is associated not with depletion but with an elevation of cytosol Ca^{2+}, although the extent of Ca^{2+}-induced secretion is rather small (Oetting *et al.*, 1987). This occurred in the same range of concentrations ($10^{-7} - 10^{-5} M$) that causes activation of most other secretory systems, and from this viewpoint the control mechanism for exocytosis appeared initially to be perfectly conventional. Nonmetabolizable analogs of GTP (i.e., GTPγS and Gpp(NH)p) are also capable of inducing exocytosis in the permeabilized cells and this occurs to an extent that is much greater than that due to elevation of Ca^{2+}. This effect can be regarded as Ca^{2+}-independent since it is fully expressed at pCa 9. When loaded with Gpp(NH)p (10^{-5} M) exocytosis remains maximal as the level of Ca^{2+} is raised up to 200 nM but above this it declines steeply to a nil response at pCa 6 (Oetting et $al.$, 1986) exactly in line with the physiological response of intact cells.

5. Melanotrophs

Secretion of α-melanocyte stimulating hormone (α-MSH) from cells of rat pituitary gland intermediate lobe, which is Ca^{2+}-dependent, can be enhanced by agonists acting through β-adrenergic receptors acting conventionally to elevate cyclic AMP (Tsuruta et $al.$, 1982). However, in cells permeabilized by high voltage discharge, cyclic AMP is without effect on Ca^{2+}-induced secretion unless GTP (or one of its analogs) is also provided. (Yamamoto et $al.$, 1987). Mg^{2+} is also a requirement for the potentiation, but not for the exocytotic reaction. The finding that neomycin, an inhibitor of inositide-specific phospholipase C, is without effect on the enhancement due to GTP suggests that this is another example of involvement of a G protein at a late stage in the control pathway for exocytosis. Confirmation of this latter point will require an actual demonstration of maintained α-MSH secretion in the face of full suppression of IP_3 release.

6. Neurosecretory Cells

A role for GTP in catecholamine secretion from adrenal chromaffin cells is not quite as apparent as it is in the neutrophils and the other cells discussed above, and here its involvement might be that of an inhibitor. Experiments from two laboratories (Bittner et $al.$, 1986; Knight and Baker, 1985) are presented in Fig. 1C. Using digitonin-permeabilized cells in primary culture, Holz and colleagues find that GTP analogs induce Ca^{2+}-independent secretion but, unlike the neutrophils, this is in no way additive to that due to Ca^{2+}, and at high concentrations of Ca^{2+} (i.e., around pCa 5) the guanine nucleotide may even become inhibitory. In the experiments of Knight and Baker (1985) using freshly disaggregated bovine cells permeabilized by high voltage discharge, GTP analogs are inhibitory to Ca^{2+}-induced secretion.

In the bovine cells, normal stimulus–secretion coupling is mediated by nicotinic cholinergic receptors which allow Ca^{2+} to enter the cells through receptor-controlled channels, and following membrane depolarization through voltage-sensitive Ca^{2+} channels (Boarder et $al.$, 1987). Muscarinic receptors (which could be coupled to G_p) are very sparse and so it is unlikely that effects of GTP are exerted at an early stage in the stimulus–secretion sequence (i.e., at G_p with consequent generation of diglyceride). However, as with the neutrophils already mentioned, the levels of Ca^{2+} normally achieved following stimulation of nicotinic receptors would be insufficient to cause secretion unless supported by another activated signaling system (Knight and Baker, 1982; Burgoyne, 1984; Cockcroft and Gomperts, 1988, but see Kao and Schneider, 1986).

It should not be presumed that GTP is a universal effector or modulator of neurosecretory systems; it has a very small effect (and that only when applied at concentrations approaching 10^{-3} M, but synthetic analogs were not tested) on secretion of vasopressin and oxytocin when applied to (digitonin-) permeabilized neurohypophyseal nerve endings (Cazalis et al., 1987).

Suggestive evidence for involvement of a G protein in the control of catecholamine secretion comes from the finding of inhibition of Ca^{2+}-induced secretion from cells poisoned with botulinum neurotoxins (Knight et al., 1985; Knight, 1986). This has been shown to catalyze ADP-ribosylation of 21-kDa substrates [but not ras protein (Adam-Vizi et al., 1987)] in a number of cell types including the bovine adrenal medullary cells, and G-protein substrates have been inferred (Ohashi and Narumiya, 1987; Adam-Vizi and Knight, 1987). This interpretation is by no means certain since other evidence indicates that the inhibitory effects of botulinum neurotoxin can be separated from the ADP-ribosylating enzyme (Rosener et al., 1987; Ashton et al., 1988). The question of whether botulinum toxin acts at a stage close to the exocytotic site is also not fully resolved. In cerebral cortical synaptosomes it is suggested that the toxin blocks the entry of Ca^{2+} into an intracellular hydrophobic milieu since the blockade of Ca^{2+}-dependent glutamate secretion can be reversed by the Ca^{2+}-ionophore ionomycin (Sanchez-Prieto et al., 1987).

7. Mast Cells

The pattern of Ca^{2+}/GTP interaction in the control of secretion from permeabilized mast cells is again different (Howell et al., 1987; Neher, 1988). Here, exocytosis (measured as release of histamine or hexosaminidase) requires the presence of both Ca^{2+} and a guanine nucleotide. Unlike the other cell types so far discussed, secretion from these cells shows no requirement for ATP and, indeed, in the experiments to be discussed, the cells were pretreated with metabolic inhibitors to the point of nonresponsiveness to receptor-directed agonists before permeabilization.[3] Other non-

[3]In part, the requirement for the presence of ATP in certain systems could be an artifact of procedure, particularly when high-voltage electric discharge is used to generate permeability

metabolizable analogs of GTP (Gpp(NH)p and Gpp(CH$_2$)p) may be used in place of GTPγS but higher concentrations are required. Some nucleoside triphosphates (other than ATP) are also supportive of Ca^{2+}-dependent secretion; these include ITP, XTP, and GTP, and the rank order of their effectiveness, which resembles that which operates to support (G$_s$-mediated) pigeon erythrocyte adenylyl cyclase (Bilezikian and Aurbach, 1974), gives another clear indication of involvement of a G protein in the exocytotic mechanism.

8. HL-60 cells

The exocytotic reaction of HL-60 cells, like the mast cells, requires a pair of effectors, but here any two of Ca^{2+}, GTP (or an analog), or phorbol ester are supportive (Stutchfield and Cockcroft, 1988). Fig. 2A illustrates the dependence on concentration of a range of different nucleotides in support of Ca^{2+}-dependent lysosomal enzyme secretion following permeabilization with streptolysin 0 (SLO). As with the mast cells there is no requirement for ATP though the extent of secretion is enhanced when it is provided. Fig. 2B illustrates the effect of the same nucleotides on inositol 1,4,5-trisphosphate production in the same experiment. While GTPγS, followed by Gpp(NH)p and Gpp(CH$_2$)p, is the most effective in causing both secretion and inositide hydrolysis, other nucleotides (XTP and ITP) have negligible effects in the phospholipase reaction while causing substantial Ca^{2+}-dependent secretion. From this it would appear that the G protein involved in exocytosis is distinct from that which activates phospholipase C.

B. Constitutive Secretion

All the above examples of secretory processes refer to situations in which exocytosis is under the control of external stimuli. In fact, by far the greatest amount of protein secretory activity occurs as an ongoing process at the rate at which the protein is synthesized. The primary example of this must be the plasma proteins (adult humans secrete approx 1 g of albumin per day) which are secreted by the liver. There is now evidence that G proteins are also involved in such constitutive secretory processes in both

---

lesions. In this case it is normal to permeabilize all the cells in bulk, and then to distribute them in tubes in the presence of Ca^{2+} and other effectors. Were such a procedure applied to mast cells, then they too would require a supply of ATP since they become refractory to stimulation within a few minutes of permeabilization (Howell and Gomperts, 1987; Howell et al., 1988). Responsiveness is considerably prolonged when ATP is provided to the permeabilized cells.

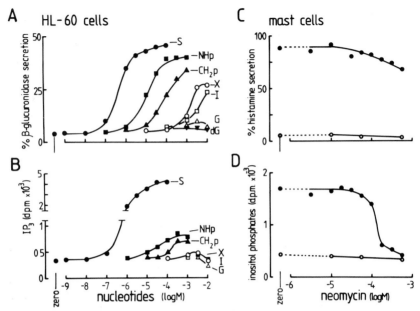

Fig. 2. G_P and G_E have different specificities towards nucleotides, and mediate separate reactions. (A) Concentration dependence on a selection of nucleotides for β-glucuronidase secretion from HL-60 cells permeabilized with streptolysin O. These were stimulated in the presence of MgATP at pCa 5. S, GTPγS; NHp, Gpp(NH)p; CH$_2$p, Gpp(CH$_2$)p; X, XTP; I, ITP; G, GTP; dG, dGTP. (B) Concentration dependence on a selection of nucleotides for generation of IP$_3$ in HL-60 cells: conditions as for panel (A). (C) Concentration dependence on neomycin for Ca^{2+}- *plus* GTPγS-induced secretion of hexosaminidase from mast cells permeabilized with streptolysin O. (D) Concentration dependence on neomycin for inhibition of Ca^{2+}- *plus* GTPγS-induced generation of inositol phosphates in mast cells: conditions as for panel C. [Panels A and B, data of Stutchfield and Cockcroft, 1988; panels C and D, data of Cockcroft *et al.*, 1987]

mammalian cells (Melançon *et al.*, 1987; Imamura *et al.*, 1988) and in yeast (Salminen and Novick, 1987).

1. Monocytes

Secretion of tumor necrosis factor from human blood monocytes stimulated with bacterial lipopolysaccharide is inhibited when the cells are cultured in the presence of botulinum toxin D (Imamura *et al.*, 1988). There is no effect on secretion induced by treatment with phorbol ester. The toxin causes ADP-ribosylation of a protein (M_r 21K) and since the reaction is inhibited in membranes treated with GTPγS the substrate is likely to be a G protein (but see the strictures above concerning ADP-ribosyltransferase activity of botulinum neurotoxin (Section IV. A. 6). As with the mast cells, it has been possible to introduce and trap GTPγS in these cells by permeabilization with ATP^{4-}. This accelerates secretion in a manner subject to partial inhibition by toxin pretreatment, and partial inhibition by cointroduction of neomycin. These findings suggest that there are two control pathways for secretion in monocytes, one controlled by activation of protein kinase C, and the other directly controlled by a G protein, which is not G_p.

2. Inhibition by GTP Analogs of Transport Through Golgi Stack

Enveloped viruses have been used to investigate the transport of proteins from their point of synthesis to the plasma membrane. Their pathway through the Golgi is shared with the constitutively secreted proteins (Marsh and Quinn, 1988). It has been shown that GTP analogs and AlF^{4-} inhibit transport of vesicular stomatitis virus (VSV) coat glycoprotein between Golgi cisternae (Melançon et al., 1987). The assay is an in vitro system measuring transport of newly synthesized protein from the Golgi of mutant cells lacking N-acetylglucosaminyltransferase to wild-type membranes. Addition of N-[^{3}H] acetylglucosamine to the viral spike glycoprotein by the wild-type membranes is an indicator of nonleaky transfer between the Golgi cisternae and as such represents an early stage in the pathway of constitutive secretion. The effect of the guanine nucleotide is expressed at the acceptor membranes since preincubation of donor membranes (i.e. the mutant containing the virus) with GTPγS fails to cause inhibition; likewise, preincubation of the cytosol with GTPγS is without effect. By contrast, preincubation of acceptor membranes with GTPγS causes irreversible inhibition and so it is likely that the inhibitory G protein concerned is an integral component of the Golgi membranes. Inhibition also requires the participation of soluble components of the cytosol and is exerted at a step corresponding to the processing of Golgi vesicles prior to fusion.

3. Involvement of ras-like Protein in Secretion from Yeast

A protein having close structural homology with human ras proteins appears to be involved at a late stage in the secretory pathway of Saccharomyces cerevisiae. There are a large number of temperature-sensitive mutants in which exposure of surface proteins and secretion [e.g. of invertase (β-fructofuranosidase)] is blocked as the temperature is raised from 25° to 37°. As secretion halts there is a corresponding accumulation of secretory material within the cells and, depending on the mutant in question, morphological distortion of one of the organelles of the secretory pathway. The gene product of one of these mutants is a protein having 32% overall homology with human H-*ras*, but those regions which bind and catalyze hydrolysis of GTP show 73% identity between these two proteins. The product of *SEC4*, like *YPT1* (with which it shares 47.5% homology) is essential for viability unlike the yeast *RAS* genes (*RAS1* and *RAS2*) which control adenylyl cyclase (Toda et al. 1985) and which are interchangeable (Katoaka *et al.*, 1984). In the temperature-sensitive mutant, the blockade is expressed with a rapidity (80% in 5 min, 100% in 15 min) which implies a direct role for the gene product at a late stage of the secretory pathway. The question has been raised as to why a protein of the class generally accepted to act in signal-transduction chains should be involved in the constitutive secretory process of yeast. Possibly, as suggested, its role is that of a spatial and not a temporal regulator, restricting secretion to the growing region of the cells.

V. G_p AND G_E ACT IN SERIES TO CONTROL EXOCYTOSIS IN MAST CELLS

We are confronted with the problem of deciding whether the action of GTP and its analogs in support of secretion from mast cells (and HL-60 cells) is mediated through G_p and hydrolysis of PIP_2, or as we have concluded from experiments with neutrophils, that GTP might interact at a late stage in the exocytotic mechanism. Of the two hydrolytic products of PIP_2, any effects of the water-soluble IP_3 can be readily dismissed. In these experiments in which cells are permeabilized with SLO to the point of leakage of lactate dehydrogenase (LDH, MW 150K) there is a rapid loss of IP_3, besides which its effects on Ca^{2+} release are fully compensated by the provision of Ca^{2+} buffers. We should, however, be concerned with any possible effects of diglyceride as this is likely to be retained in the plasma membrane and could activate protein kinase C.

Two findings indicate that the obligatory requirement for GTP is unlikely to be in the activation of protein kinase C, and point once again to a direct role in exocytosis. As already mentioned, there is no requirement for ATP in the exocytotic mechanism and this in itself strongly suggests that phosphorylation reactions such as might be catalyzed by protein kinase C are inoperative. More direct evidence against the idea of a necessary role of G_p involvement comes from the use of neomycin as an inhibitor of PPI-PDE (Cockcroft *et al.*, 1987); this compound, an aminoglycoside antibiotic, is known to bind to the headgroups of the polyphosphoinositides (but not their water-soluble hydrolytic products) (Swann and Whitaker, 1986). In these experiments mast cells were labeled by overnight culture in the presence of *myo*[^{3}H]inositol. They were permeabilized at pCa 5 in the presence and absence of GTPγS to demonstrate GTP activation of both PPI-PDE (measured as release of IP$_3$) and exocytosis (release of histamine). As shown in Fig. 2 (panels C and D), inclusion of neomycin at concentrations above 10^{-4} M inhibits generation of IP$_3$ but leaves the exocytotic mechanism relatively unscathed. Neomycin does become inhibitory to secretion at higher concentrations, but under the conditions of this experiment we have a clear dissociation of GTPγS-mediated activation of PPI-PDE through G$_P$ and its effects on exocytosis. As with secretion from HL-60 cells (Fig. 2A and B) and neutrophils, we would say that GTP activates exocytosis through its action as a ligand at G$_E$. In the case of the permeabilized mast cells the requirement for GTP as an activating ligand for G$_E$ is obligatory.

As already remarked, such differences between cell types could be real, but could also occur as artifacts of the permeabilization methods used. In the neutrophil experiments cited earlier, the permeabilization agent was a 3-day egg-grown culture of hemolytic Sendai virus (Wyke et al., 1980). This preparation is known to contain diglyceride which could cause activation of protein kinase C. In comparison with SLO used to permeabilize mast cells in recent work, Sendai virus generates very small lesions in the plasma membrane (Impraim et al., 1980), allowing dialysis, but certainly not a full control and exchange of the soluble constituents of the cytosol. Furthermore, the virus binds to cell surface sialoglycoproteins and sialoglycolipids (Haywood, 1974; Haywood and Boyer, 1984) and these may have properties of receptors in their own right.

1. Evidence for Second G Protein in Stimulus–Secretion Pathway of Intact Mast Cells

In respect of the experiments with permeabilized mast cells it is certainly worth asking whether intact cells stimulated with receptor-directed agonists also require the presence of GTP in order to undergo exocytotic secretion. Evidence from experiments with RBL-2H3 cells (a rat basophilic cell line, which like mast cells secretes histamine in response to antigens that crosslink identical receptors for IgE) indicates that there is a parallel, but essential, second messenger system which acts in synergy with Ca^{2+} (Beaven et al., 1987). In these cells (as in adrenal chromaffin cells and neutrophils) no significant secretion occurs in response to concentrations of Ca^{2+}-ionophore (A23187), which cause elevation in cytosol Ca^{2+} comparable to that commensurate with a maximal stimulus to secretion through the IgE receptor system. The possibility that the synergizing signal might involve GTP is currently raised only through exclusion of other possibilities. In particular, synergy is unlikely to arise through activation of protein kinase C since treatment of the cells with phorbol ester has the effect of inhibiting the antigen-induced inositide and Ca^{2+} responses, while having little effect on secretion (Sagi-Eisenberg *et al.*, 1985, 1987).

The idea that GTP could provide the synergistic signal comes from experiments involving prolonged culture of mast cells with the antiviral agent ribavirin, a guanosine analog known to depress guanine nucleotide levels rather selectively [i.e., ATP is maintained (Johnson and Mukku, 1979)]. This has the effect of inhibiting secretion due not only to activation through the IgE system, but also to the Ca^{2+}-ionophore A23187 (Marquardt *et al.*, 1987; Wilson *et al.*, 1988). While one would expect receptor-mediated secretion to be suppressed following depletion of GTP due to inability to activate G_p and polyphosphoinositide hydrolysis, failure of the Ca^{2+}-only stimulus points to the requirement for GTP even under conditions in which the receptors have been bypassed.

VI. MODULATION OF EXOCYTOSIS BY ATP

Although there is no absolute requirement for ATP in mast cell exocytosis (see page 612 and Figure 1D) this is not to say that ATP is without effect in this reaction. ATP can best be understood as a modulator of exocytosis and it probably serves this role by acting in its conventional manner as a phosphoryl donor in protein phosphorylation reactions catalyzed by membrane-bound or membrane-adherent kinases.

A. Prolongation of Responsiveness

After mast cells are permeabilized, the ability to respond to subsequent provision of Ca^{2+} plus GTPγS rapidly declines such that no secretion is elicited if Ca^{2+} (pCa 5) and GTPγS (20 μM) are provided after 5 min (Howell *et al.*, 1988). If ATP is provided, the duration of responsiveness can be extended in a practical sense up to 20 min (Fig. 3). During this time the exocytotic reaction is characterized by a progressive shift in the requirement for Ca^{2+} to higher and higher concentrations. If ATP is provided 5 min postpermeabilization (by which time there is no response at pCa 5), responsiveness can be fully restored (Howell *et al.*, 1988). Since nearly all of the releasable LDH leaks from the cells within the first 5 min of permeabilization (Howell and Gomperts, 1987), and the exocytotic mechanism is demonstrably tolerant of this procedure, we should ask why exocytotic responsiveness then declines in the longer term, and why responsiveness can be prolonged in the presence of ATP. Clearly, the readily soluble proteins of the cytosol are unlikely to be involved in exocytosis, but one possibility is that sticky or membrane-adherent proteins required in the

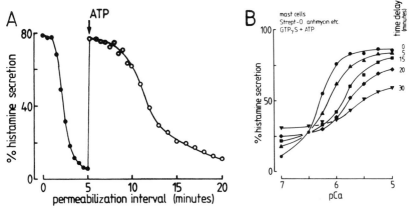

Fig. 3.   Effect of ATP on the duration of responsiveness of permeabilized mast cells. (A) Mast cells were permeabilized initially at pCa 8 in the absence of ATP. Ca^{2+} (pCa 5) and GTPγS (20 μM) were provided at times indicated. After 5 min the cells became refractory to stimulation but responsiveness could be fully restored by addition of ATP. (B) Mast cells were permeabilized at pCa 8 in the presence of ATP and transferred to GTPγS (20 μM) and a range of [Ca^{2+}] at times indicated. The decline in responsiveness is characterized by a time-dependent shift in the effective affinity for Ca^{2+}. [Panel (A), from Howell et al., 1988; panel (B), from Gomperts et al., 1988]

membrane fusion process detach and leak slowly through the SLO-induced lesions. Possibly such leaching is retarded when ATP is provided.

B. ATP Controls Affinity for Ca^{2+} and GTPγS

ATP, and therefore in all probability a protein phosphorylation, is also involved in controlling the affinity of both Ca^{2+} and GTP at their respective binding sites (Howell et al., 1987). When the Ca^{2+} dependence of exocytosis is measured in the presence of an optimal concentration of GTPγS, addition of ATP shifts the midpoint [Ca^{2+}] from pCa 5.9 to 6.35, a change of approximately threefold (Fig. 4A). Conversely, with Ca^{2+} held at pCa 5, the addition of ATP shifts the midpoint [GTPγS] approximately tenfold so that exocytosis now commences at concentrations of GTPγS below $10^{-8}M$ (Fig. 3B). These effects of ATP are negated when the cells are permeabilized in the presence of neomycin at concentrations which inhibit PPI-PDE (Cockcroft et al., 1987), and pretreatment of the cells with phorbol ester (Fig. 5) enhances the affinity for Ca^{2+} even when the cells are subsequently permeabilized in the absence of ATP (Fig. 5) (Howell et al., 1988). On this basis we have deduced that the phosphorylations controlling the affinity of the two obligatory effectors are catalyzed by protein kinase

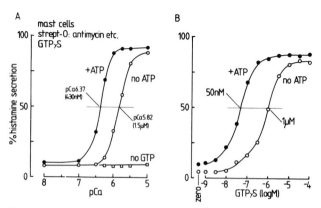

Fig. 4. Effect of ATP on the effective affinity for Ca^{2+} and GTPγS in the exocytotic reaction of mast cells. (A) Mast cells permeabilized with streptolysin O were triggered to secrete by addition of 20 μM GTPγS and Ca^{2+} as indicated, in the presence or absence of ATP. (B) Mast cells permeabilized with streptolysin O were triggered to secrete by addition of Ca^{2+} (pCa 5) and GTPγS as indicated, in the presence or absence of ATP (Gomperts *et al.*, 1987).

C. If this is the case, then protein kinase C would have to be one of the proteins retained in a membrane-bound form, in the face of the rapid leakage of LDH from SLO-permeabilized cells. Following permeabilization the effective affinity for Ca^{2+} declines, albeit more slowly if ATP is present, and it is the failure to sense normal activating concentrations of

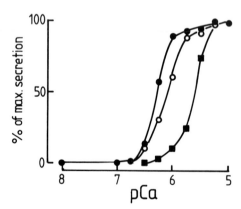

Fig. 5. Phorbol ester pretreatment enhances effective affinity for Ca^{2+}: a priming effect of protein phosphorylation. Mast cells were treated for 5 min with phorbol ester. They were then permeabilized in the presence or absence of ATP, GTPγS (20 μM), and Ca^{2+} at concentrations indicated. Controls (■), no pretreatment, no ATP; (○), phorbol ester pretreatment, no ATP; (●), no pretreatment, plus ATP (Howell *et al.*, 1988).

Ca^{2+} and GTPγS that characterizes the loss of exocytotic responsiveness. One possibility is that all this is dictated by the leakage of protein kinase C (or another phosphorylating enzyme) from the cells (Cockcroft *et al.*, 1987; Howell *et al.*, 1988).

There is another important conclusion to be derived from these experiments. While a phosphorylation reaction does not comprise an essential step in the triggering of exocytosis, a certain state of phosphorylation is a *sine qua non* for progress into the late stages of the pathway. It is only in this primed condition that the exocytotic mechanism is able to recognize Ca^{2+} and GTP.

C. ATP: An Inhibitor of Exocytosis

In nearly every investigation into the mechanism of secretion (from intact cells) or exocytosis (from permeabilized cells), conclusions have been reached by relating the extent of release to the concentration of some agonist or effector. Nearly all the experiments so far described in this essay have been of this general type; they set the scene, and give information about which effectors or second messengers are important under various sets of conditions. They give very little information about the mechanism of exocytosis. With respect to the mast cells, all we have learned can be summarized by saying that there is a dual requirement for two effectors (Ca^{2+} and GTP) and that ATP, for which there is no absolute requirement, acts as a modulator. Kinetic experiments now indicate that ATP also acts as an inhibitor of exocytosis (Tatham and Gomperts, 1989; Gomperts and Tatham, 1989).

1. Kinetics of Intracellular Reaction Sequence

In the experiments to be described, mast cells were always pretreated with metabolic inhibitors as in the experiments described above, to the point of nonresponsiveness to application of receptor-directed ligands.[4] The cells were then permeabilized and loaded with one of the two obligatory effectors; after a further two minutes of incubation the exocytotic reaction was

---

[4]The reason why intact cells must be metabolically competent should by now be fairly obvious. Apart from cellular housekeeping duties (e.g. Ca^{2+} homeostasis), ATP is needed at three points in the complete stimulus–secretion sequence. (1) ATP is needed as a phosphoryl donor in the reactions catalyzed by the phosphoinositide kinases to maintain PIP_2; (2) ATP is needed as the phosphoryl donor in the reactions catalyzed by protein kinase C, by which affinity for Ca^{2+} and GTP are maintained; and (3) ATP is needed in the transphosphorylation reaction to maintain GTP. None of these constraints apply in the permeabilized cells.

initiated by addition of the other effector, and timed samples were then withdrawn and quenched.

2. G_E Regulates Onset of Exocytosis

Figure 6A illustrates the time course of exocytosis from cells loaded with GTPγS, and subsequently triggered by addition of Ca^{2+} (pCa 5). In the absence of ATP, exocytosis commences abruptly (within 3 sec) on completion of the effector combination. In contrast, when ATP is provided (Fig. 6B), there is a delay at all concentrations of GTPγS before exocytosis commences. The duration of this delay is dependent on the concentration of GTPγS such that it becomes shorter as the concentration of GTPγS is increased. The primary effect of varying GTPγS concentration is certainly on the duration of the delays and not on the maximal rates of exocytosis which are affected to a much smaller degree. A tenfold reduction in the delay requires that the concentration of GTPγS is increased by a factor of about 100. The effect of varying the concentration of Ca^{2+} is very similar.

D. Inhibition by ATP: A Triggering Role for Protein Dephosphorylation

The effect of ATP on the time course of exocytosis is that of an inhibitor since the process occurs much more rapidly in its absence. The pattern of delays followed by a relatively constant rate of secretion over a wide range

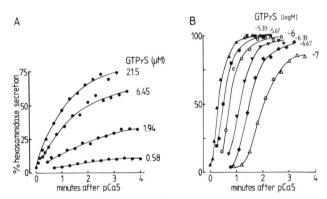

Fig. 6. Kinetics of GTPγS-dependent, Ca^{2+}-triggered exocytosis from permeabilized mast cells. (A) Mast cells, pretreated with metabolic inhibitors to the point of total refractoriness towards receptor-directed agonists, were permeabilized (using streptolysin O) in the presence of CaEGTA (to maintain pCa 8) and GTPγS at concentrations indicated. They were triggered to secrete 2 min later by transfer to pCa 5. (B) Conditions as for (A) except that the cells were permeabilized in the presence of 1 mM MgATP.

of effector concentrations is also characteristic of coupled enzyme reactions in which the delay, or transition time, is the sum of the transition times of each intermediate state (Easterby, 1981). The lower the concentration of the effector the longer the time to set up steady state conditions. The relationship between effector concentration and delay time is systematic and displays an unusual square-root relationship such that the delay is proportional to $\{[Ca^{2+}]\,[GTP\gamma S]\}^{-1/2}$. Once the exocytotic reaction is underway, its rate is rather insensitive to the concentration of GTPγS and hence the activation state of the G protein (G_E).

As in all reactions involving ATP, first thoughts must turn to its function as a phosphoryl donor in a reaction catalyzed by a protein kinase. The phosphorylation state of this protein controls the rate at which this steady-state condition is achieved. The effect of GTPγS and Ca^{2+} is to increase this rate and so raises the possibility that the target enzyme controlled by G_E is a protein phosphatase.

1. Protein Dephosphorylation in Mechanism of Exocytosis

From these experiments we have drawn two conclusions, one relating to the mechanism of exocytosis, the other concerning the target enzyme of G_E. We have seen that maintenance of a phosphorylating environment is inhibitory to the onset of secretion; from this we can understand that there could exist a protein, phosphorylated in the resting situation, which must be dephosphorylated in order to allow exocytosis to proceed. Such a mechanism certainly appears to obtain in the control of trichocyst discharge in *Paramecium tetraurelia*. Here it has been found that a unique 65-kDa phosphoprotein (pp65) undergoes dephosphorylation within 1 sec of stimulation. Mutants lacking this protein are unable to undergo trichocyst discharge suggesting that the dephosphorylated pp65 plays a positive role in the control sequence (Gilligan and Satir, 1982; Zieseniss & Plattner, 1985). Discharge can be inhibited by injecting the organism with antibodies raised against calcineurin, a protein phosphatase of neural origin, and conversely, discharge can be induced by injection of alkaline phosphatase (Momayezi et al., 1987).

In many mammalian endocrine and exocrine cells, dephosphorylation (as well as phosphorylation) of proteins occurs following stimulation, but these experiments do not reveal whether such reactions are intrinsic to the exocytotic mechanism (Wrenn, 1984; Burnham and Williams, 1982; Burnham, 1985; Jones et al., 1987b; Cote et al., 1986; Knight et al., 1984). Better evidence for dephosphorylation comes from the observation that in permeabilized adrenal chromaffin cells, ATPγS inhibits Ca^{2+}-induced secretion. This is probably due to irreversible protein thiophosphorylation (Brooks et al., 1984).

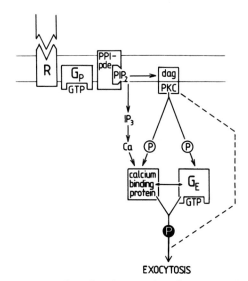

Fig. 7. Schematic representation of the interrelationships between G_P, G_E, and hypothetical phosphorylation/dephosphorylation reactions in the stimulus-secretion pathway of the rat mast cell. R, receptor; dag, diacylglycerol; PKC, protein kinase C.

3. Protein Phosphatase as Target for G_E

In the light of the observation that preexocytotic delays are only observed when ATP is provided, and that their duration is controlled by the concentrations of GTPγS and Ca^{2+} applied to the permeabilized cells, we would suggest that the target enzyme of the exocytotic G protein G_E is likely to be a protein phosphatase. The relationships between the two G proteins, G_P and G_E, and the hypothetical phosphorylation and dephosphorylation reactions of the stimulus–secretion pathway are indicated schematically in Fig. 7.

VII. G-PROTEIN REGULATION OF DEGRANULATION IN SINGLE CELLS

The advantage of the sampling technique that was used in the above kinetic experiments is that it can be coupled with analytical measurements related to the metabolic status of the cells and the various control pathways. The disadvantage of sampling is that we can learn nothing about short-lived intermediate states; for this we need a method having the temporal and spatial resolution capable of revealing the pre-steady-state condition of unitary fusion events. To do this it is necessary to follow the kinetics of degranulation at the level of the single cell.

A. Measurement of Membrane Capacitance

Recent developments exploiting the use of the patch pipette have made it possible to monitor changes in the electrical capacitance (proportional to membrane area) of single cells undergoing exocytotic degranulation. The first such investigation concerned the adrenal chromaffin cell (Marty and Neher, 1983). Here it was shown that membrane expansion could occur simply in response to depolarizing pulses under voltage clamp, or to infusion of Ca^{2+} (buffered with EGTA to concentrations in the micromolar range) through the measuring pipette, which is used at the same time to control the cytosol composition of the cell. The secretory granules of the chromaffin cell are small, and although evidence for individual granule membrane–plasma membrane fusions was obtained, these were close to the limit of resolution. The mast cell appeared to provide more fertile territory for further developments in this area.

Attempts to monitor exocytosis on mast cells were initially held up since it was assumed that, like the chromaffin cell, they would degranulate as soon as intracellular Ca^{2+} was elevated into the micromolar range. As is now well recognized, GTP is an absolute requirement in mast cell exocytosis and there is even some doubt, especially among those working at the single cell level, concerning the necessity for Ca^{2+} (Neher and Almers, 1986; Penner et al., 1987). With mast cells coupled to patch pipettes containing GTPγS, exocytosis can be seen under Nomarski optics, and the capacitance of the plasma membrane can be monitored simultaneously (Fernandez et al., 1984). A fourfold expansion in membrane area (equivalent to an increase in capacitance from 5 to 20 picofarads) is typical. The onset of degranulation following entry of GTPγS into the cytosol ensues after a delay, and just as with the secretion experiments described earlier, the duration of the delay is inversely related to the concentration of the nucleotide (Fernandez et al., 1987; Lindau and Nüsse, 1987). The possible contribution of ATP in retarding degranulation has not so far been examined in this work.

B. Unit Exocytotic Events

1. Capacitance Flicker

With scale expansion it becomes possible to identify discrete steps in the capacitance trace which are equivalent to unit exocytotic events. This opens the way to kinetic analysis of exocytosis, especially as there was an early indication of "flicker," an apparent instability in the capacitance traces, which could be related to intermediate events in the fusion process.[5]

2. Beige Mouse Mast Cells

Subsequent advances have exploited the use of mast cells from beige (bg^j/bg^j) mice (Zimmerberg, 1987; Zimmerberg *et al.*, 1987; Breckenridge and Almers, 1987). The mast cells and granulocytes of these animals (which suffer from Chediak–Higashi syndrome, characterized by poor resistance to bacterial infection) contain a very few but very enlarged secretory granules. These are typically three microns in diameter. In other respects, their propensity to secrete in response to receptor-directed agonists, or internal perfusion with solutions containing GTP, is similar to wild-type mast cells. The beige mouse mast cells offer uniquely favorable conditions for studying the kinetics of exocytosis at the level of individual fusions between granules and the plasma membrane.

On-line capacitance measurements, coupled with measurements of membrane conductance, secretion (monitored by fluorescence measurement of release of previously loaded quinacrine), and measurement of granule expansion indicate that the stepwise increase in plasma membrane area is the preliminary to the final release process. Capacitance flicker is much more readily monitored in these mutant cells and it clearly registers an intermediate stage, at which reversal to a fully enclosed unfused vesicle can still occur. The development of the fusion pore can be followed as long as its dimensions prevent the maximal rate of transfer of charge needed to polarize the granule membrane. Thus, measurements of the rise time of the flicker events provide an indication of the conductivity, and hence the diameter, of the pore structure. If this is assumed to contain bulk water and to be a cylinder 100 Å long (i.e., sufficient to span two membrane thicknesses), then the fusion pore has diameter between 18 and 52 Å (Zimmerberg, 1987). Such channels may correlate with the bridging structures that have been observed in mast cells (Chandler and Heuser, 1980), neutrophils (Chandler *et al.*, 1983), and in *Limulus* amebocytes (Ornberg and Reese, 1981) undergoing exocytosis and is in agreement with the observation that little loaded quinacrine is released during the intermediate capacitance flicker state (Breckenridge and Almers, 1987). Only after the final commitment to fusion does the vesicle matrix swell and rapid exteriorization of its

---

[5]In passing it should be noted that in well-degranulated cells, downward steps in the capacitance trace, likely to be related to endocytotic events, were also recorded (Fernandez *et al.*, 1984). Such steps are of a scale similar to those which occur during exocytosis and are larger by far than those which would occur during endocytosis of coated pits.

contents proceed, with the fusion pore being replaced by a wide opening and the separate membranes merging into one.

VIII. CONCLUSION

A. General Biological Role for G_E

As this account has moved from a general description of exocytosis and focused particularly on this process as it occurs in the mast cell, one central control point linking all events in the stimulation–secretion sequence stands out. This is the (as yet unidentified) G protein G_E. In the mast cell G_E is pivotal since there appears to be no circumstance under which its involvement can be obviated. In other cell types in which GTP can trigger Ca^{2+}-independent exocytosis it is also possible to trigger exocytosis in a manner independent of GTP. This is why the mast cell should provide the most direct clues concerning the possible target of action of G_E. The indications provided by the kinetic experiments suggest that this might be a protein phosphatase. If this proves to be a correct diagnosis of the situation then G_E will assume a more general significance in the regulation of cellular processes.

1. A Challenge and an Invitation

It was through the use of permeabilized secretory cells that we were led to the discovery of two new sites of action of GTP in the stimulus–secretion sequence. We named these G_P and G_E to indicate their functional significance in this process, and in so doing we assumed that when isolated and characterized, these proteins would prove to be members of the family of G-protein heterotrimers. At the present time there are many more G proteins than there are cellular functions that can be ascribed to them, and several of these have been derived from the secretory cells and tissues discussed in this essay (Evans et al., 1986; Falloon et al., 1986; Gierchik et al., 1986; Iyengar et al., 1987; Bokoch and Parkos, 1988; Lo and Hughes, 1987). Sequences encoding many more G proteins have been identified on the basis of genomic analysis. The marriage of structure and function represents a considerable challenge. Simple reconstitution experiments, involving direct microinjection or introduction into permeabilized cells, are possible, but this strategy is likely to produce ambiguous results. Already it has been shown that mast cells injected with Ha-$ras^{Val,12}$ will undergo exocytosis (Bar-Sagi and Gomperts, 1988) but this should not fool anyone into thinking that it follows that the ras protein and G_E are synonymous. A more fruitful approach is likely to derive from the introduction of specific antibodies, but even this approach is far from trivial. This chapter constitutes an invitation to those in the possession of G proteins of unknown function to collaborate with us in the testing of such immunological and other specific probes for the identification of G_E.

ACKNOWLEDGMENTS

Work carried out in our laboratory has been financed by the Wellcome Trust, the Arthritis and Rheumatism Council for Research, the Vandervell Foundation, and the Medical Research Council.

REFERENCES

Adam-Vizi, V., and Knight, D. E. (1987). Does botulinum toxin type D inhibit exocytosis by ADP-ribosylation? *J. Physiol.* **394,** 96P.

Adam-Vizi, V., Knight, D. E., and Hall, A. (1987). The *ras* protein is not associated with exocytosis. *Nature* **328,** 581.

Almers, W., and Neher, E. (1985). The Ca-signal from fura-2 loaded mast cells depends strongly on the method of dye loading. *FEBS Lett.* **192,** 13–18.

Arkhammar, P., Nilsson, T., and Berggren, P. -O. (1986). Stimulation of insulin release by the phorbol ester 12-0-tetradecanoylphorbol 13-acetate in the clonal cell line RINm5F despite a lowering of the free cytoplasmic Ca^{2+} concentration. *Biochim. Biophys. Acta* **887;** 236–241.

Ashton, A. C., Edwards, K., and Dolly, O. (1988). Lack of detectable ADP-ribosylation in synaptosomes associated with inhibition of transmitter release by botulinum neurotoxins A and B. *Biochem. Soc. Trans.* **16,** 883–884.

Bangham, A. (1968). Membrane models with phospholipids. *Prog. Biophys. Mol. Biol.* **18,** 30–95.

Bar-Sagi, D., and Gomperts, B. D. (1988). Stimulation of exocytotic degranulation by microinjection of the *ras* oncogenic protein into rat mast cells. *Oncogene* **3,** 463–469.

Barrowman, M. M., Cockcroft, S., and Gomperts, B. D. (1986). Two roles for guanine nucleotides in stimulus secretion sequence of neutrophils. *Nature* **319,** 504–507.

Barrowman, M. M., Cockcroft, S., and Gomperts, B. D. (1987). Differential control of azurophilic and specific granule exocytosis in Sendai virus permeabilized rabbit neutrophils. *J. Physiol.* **383,** 115–124.

Beaven, M. A., Guthrie, D. F., Moore, J. P., Smith, G. A., Hesketh, T. R., and Metcalfe, J. C. (1987). Synergistic signals in the mechanism of antigen-induced exocytosis in 2H3 cells: Evidence for an unidentified signal required for histamine release. *J.Cell Biol.* **105,** 1129–1136.

Bennett, J. P., Cockcroft, S., and Gomperts, B. D. (1981). Rat mast cells permeabilized with ATP secrete histamine in response to calcium ions buffered in the micromolar range. *J. Physiol. London* **317,** 335–345.

Berrie, C. P., Birdsall, N. J. M., Burgen, A. S. V., and Hume, E. C. (1979). Guanine nucleotides modulate muscarinic receptor binding in the heart. *Biochem. Biophys. Res. Commun.* **87**, 1000–1005.

Bilezikian, J. P., and Aurbach, G. D. (1974). The effects of nucleotides on the expression of *β*-adrenergic adenylate cyclase activity in membranes from turkey erythrocytes. *J. Biol. Chem.* **249**, 157–161.

Bittner, M. A., Holz, R. W., and Neubig, R. R. (1986). Guanine nucleotide effects on catecholamine secretion from digitonin-permeabilized adrenal chromaffin cells. *J. Biol. Chem.* **261**, 10182–10188.

Boarder, M. R., Marriott, D., and Adams, M. (1987). Stimulus-secretion coupling in cultured chromaffin cells. Dependency on external sodium and on dihydropyridine-sensitive calcium channels. *Biochem. Pharmacol.* **36**, 163–167.

Bokoch, G. M., and Parkos, C. A. (1988). Identification of novel GTP-binding proteins in the human neutrophil. *FEBS Lett.* **227**, 66–70.

Breckenridge, L. J., and Almers, W. (1987). Final steps in exocytosis observed in a cell with giant secretory granules. *Proc. Nat. Acad. Sci. (U.S.A.)* **84**, 1945–1949.

Brooks, J. C., Treml, S., and Brooks, M. (1984). Thiophosphorylation prevents secretion by chemically skinned chromaffin cells. *Life Sci.* **35**, 569–574.

Brown, E., Enyedi, P., LeBoff, M., Rotberg, J., Preston, J., and Chen, C. (1987). High extracellular Ca^{2+} and Mg^{2+} stimulate accumulation of inositol phosphates in bovine parathyroid cells. *FEBS Lett.* **218**, 113–118.

Bruzzone, R., Pozzan, T., and Wollheim, C. B. (1986). Caerulein and carbamoylcholine stimulate pancreatic amylase release at resting free Ca^{2+}. *Biochem. J.* **235**, 139–143.

Buckingham, L., and Duncan, J. (1983). Approximate dimensions of membrane lesions produced by streptoylsin S and streptolysin O. *Biochim. Biophys. Acta.* **729**, 115–122.

Burgoyne, R. D. (1984). The relationship between secretion and intracellular free calcium in bovine adrenal chromaffin cells. *Biosci. Rep.* **4**, 605–611.

Burnham, D. B. (1985). Characterization of Ca^{2+}-activated protein phosphatase in exocrine pancreas. *Biochem. J.* **231**, 335–341.

Burnham, D. B., and Williams, J. A. (1982). Effects of carbachol, cholecystokinin and insulin on protein phosphorylation in isolated pancreatic acini. *J. Biol. Chem.* **257**, 10523–10528.

Cazalis, M., Dayanithi, G., and Nordmann, J. J. (1987). Requirements for hormone release from permeabilized nerve endings isolated from the rat neurohypophysis. *J. Physiol.* **390**, 71–91.

Chandler, D. E., and Heuser, J. E. (1979). Membrane fusion during secretion: Cortical granule exocytosis in sea urchin eggs as studied by quick freezing and freeze-fracture. *J. Cell Biol.* **83**, 91–108.

Chandler, D. E., and Heuser, J. E. (1980). Arrest of membrane fusion events in mast cells by quick-freezing. *J. Cell Biol.* **86**, 666–674.

Chandler, D. E., Bennett, J., and Gomperts, B. (1983). Freeze fracture studies of chemotactic peptide induced exocytosis in neutrophils: Evidence for two patterns of secretory granule fusion. *J. Ultrastructur. Res.* **82**, 221–232.

Cockcroft, S. (1982). Phosphatidylinositol metabolism in mast cells and neutrophils. *Cell Calcium* **3**, 337–349.

Cockcroft, S. (1986). The dependence on Ca^{2+} of the guanine nucleotide-activated polyphosphoinositide phosphodiesterase in neutrophil plasma membranes. *Biochem. J.* **240**, 503–507.

Cockcroft, S., and Gomperts, B. D. (1979). ATP induces nucleotide permeability in rat mast cells. *Nature* **279**, 541–542.

Cockcroft, S., and Gomperts, B. D. (1985). Role of guanine nucleotide binding protein in the activation of polyphosphoinositide phosphodiesterase. *Nature* **314**, 534–536.

Cockcroft, S., and Gomperts, B. D. (1988). Some new questions concerning the role of calcium in exocytosis. *In* "Calcium and Drug Action" (P. F. Baker, ed.). pp 305–338. Springer-Verlag, Heidelberg.

Cockcroft, S., Bennett, J. P., and Gomperts, B. D. (1980). Stimulus-secretion coupling in rabbit neutrophils is not mediated by phosphatidylinositol breakdown. *Nature* **288**, 275–277.

Cockcroft, S., Bennett, J. P., and Gomperts, B. D. (1981). The dependence on Ca^{2+} of phosphatidylinositol breakdown and enzyme secretion in rabbit neutrophils stimulated by formylmethionylleucylphenylalanine or ionomycin. *Biochem. J.* **200**, 501–508.

Cockcroft, S., Howell, T. W., and Gomperts, B. D. (1987). Two G-proteins act in series in series to control stimulus-secretion coupling in mast cells: Use of neomycin to distinguish between G-proteins controlling polyphosphoinositide phosphodiesterase and exocytosis. *J. Cell Biol.* **105**, 2745–2750.

Cote, A., Doucet, J. P., and Trifaro, J. M. (1986). Phosphorylation and dephosphorylation of chromaffin cell proteins in response to stimulation. *Neuroscience* **19**, 629–645.

Creba, J., Downes, C., Hawkins, P., Brewster, G., Michell, R., and Kirk, C. (1983). Rapid breakdown of phosphatidylinositol 4-phosphate and phosphatidylinositol 4,5-bisphosphate in rat hepatocytes stimulated by vasopressin and other Ca^{2+}-mobilizing hormones. *Biochem. J.* **212**, 733–747.

Di Virgilio, F., Lew, D. P., and Pozzan, T. (1984). Protein kinase C activation of physiological processes in human neutrophils at vanishingly small cytosolic Ca^{2+} levels. *Nature* **310**, 691–693.

Dormer, R. L., and Ashcroft, S. J. H. (1974). Studies on the role of calcium ions in the stimulation by adrenaline of amylase release from rat parotid. *Biochem. J.* **144**, 543–550.

Douglas, W. W., and Rubin, R. P. (1961). The role of calcium in the secretory response of the adrenal medulla to acetylcholine. *J. Physiol.* **159**, 40–57.

Dunn, L. A., and Holz, R. W. (1983). Catecholamine secretion from digitonin-treated adrenal medullary cells. *J. Biol. Chem.* **258**, 4989–4993.

Easterby, J. S. (1981). A generalised theory of the transition time for sequential enzyme reactions. *Biochem. J.* **199**, 155–161.

Evans, T., Brown, M. L., Fraser, E. D., and Northup, J. K. (1986). Purification of the major GTP-binding proteins from human placental membranes. *J. Biol. Chem.* **261**, 7052–7059.

Falloon, J., Malech, H., Milligan, G., Unson, C., Kahn, R., Goldsmith, P. and Spiegel, A. (1986). Detection of the major pertussis toxin substrate of human leukocytes with antisera raised against synthetic peptides. *FEBS Lett.* **209,** 352–356.

Fernandez, J. M., Neher, E., and Gomperts, B. D. (1984). Capacitance measurements reveal stepwise fusion events in degranulating mast cells. *Nature* **312,** 453–455.

Fernandez, J. M., Lindau, M., and Eckstein, F. (1987). Intracellular stimulation of mast cells with guanine nucleotides mimic antigenic stimulation. *FEBS Lett.* **216,** 89–93.

Fitzpatrick, L. A., Brandi, M. -L., and Aurbach, G. D. (1986a). Calcium-controlled secretion is effected through a guanine nucleotide regulatory protein in parathyroid cells. *Endocrinology* **119,** 2700–2703.

Fitzpatrick, L. A., Brandi, M. L., and Aurbach, G. D. (1986b). Control of PTH secretion is mediated through calcium channels and is blocked by pertussis toxin treatment of parathyroid cells. *Biochem. Biophys. Res. Commun.* **138,** 960–965.

Füssle, R., Bhakdi, S., Sziegoleit, A., Tranum-Jensen, J., Kranz, T., and Wellensiek, H. J. (1981). On the mechanism of membrane damage by *Staphylococcus aureus* alpha-toxin. *J. Biol. Chem.* **91,** 83–94.

Gierchik, P., Falloon, J., Milligan, G., Pines, M., Gallin, J. I., and Spiegel, A. (1986). Immunochemical evidence for a novel pertussis toxin substrate in human neutrophils. *J. Biol. Chem.* **261,** 8058–8062.

Gilligan, D. M., and Satir, B. H. (1982). Protein phosphorylation/dephosphorylation and stimulus-secretion coupling in wild type and mutant Paramecium. *J. Biol. Chem.* **257,** 13903–13906.

Glossman, H., Baukal, A., and Catt, K. J. (1974). Angiotensin II receptors in bovine adrenal cortex. Modification of angiotensin II by guanyl nucleotides. *J. Biol. Chem.* **249,** 664–666.

Gomperts, B. D. (1983). Involvement of guanine nucleotide-binding protein in the gating of Ca^{2+} by receptors. *Nature* **306,** 64–66.

Gomperts, B. D. (1986). Calcium shares the limelight in stimulus-secretion coupling. *Trends in Biochem. Sci.* **11,** 290–292.

Gomperts, B. D., and Fernandez, J. M. (1985). Techniques for membrane permeabilisation. *Trends in Biochem. Sci.* **10,** 414–417.

Gomperts, B. D., and Tatham, P. E. R. (1989). GTP-binding proteins in the control of exocytosis. Molecular Biology of Signal Transduction, Cold Spring Harbor Symposium of Quantitative Biology, 53 (M. Wigler, J. R. Feramisco, and J. D. Watson, eds). pp. 983–992.

Gomperts, B. D., Barrowman, M. M., and Cockcroft, S. (1986). Dual role for guanine nucleotides in stimulus-secretion coupling: An investigation of mast cells and neutrophils. *Fed. Proc. Fed. Am. Soc. Exp. Biol.* **45,** 2156–2161.

Gomperts, B. D., Cockcroft, S., Howell, T. W., Nüsse, O., and Tatham P. E. R. (1987). The dual effector system for exocytosis in mast cells: Obligatory requirement for both Ca^{2+} and GTP. *Biosci. Rep.* **7,** 360–381.

Gomperts, B. D., Cockcroft, S., Howell, T. W., and Tatham, P. E. R. (1988).

Intracellular Ca^{2+}, GTP and ATP as effectors and modulators of exocytotic secretion from rat mast cells. *In* "Molecular Mechanisms in Secretion" (N. A. Thorn, M. Treiman, O. H. Petersen, and J. H. Thaysen, eds). pp. 248–258. Munksgaard, Copenhagen.

Goodhardt, M., Ferry, N., Geynet, P., and Hanoune, J. (1982). Hepatic alpha-1 adrenergic receptors show agonist-specific regulation by guanine nucleotides. Loss of nucleotide effect after adrenalectomy. *J. Biol. Chem.* **257**, 11577–11583.

Hamill, O., Marty, A., Neher, E., Sakman, B., and Sigworth, F. (1981). Improved patch-clamp techniques for high-resolution current recording from cells and cell-free membrane patches. *Eur. J. Physiol.* **391**, 85–100.

Haslam, R. J., and Davidson, M. M. L. (1984a). Potentiation by thrombin of the secretion of serotonin from permeabilised platelets equilibrated with Ca^{2+} buffers. *Biochem. J.* **222**, 351–361.

Haslam, R. J., and Davidson, M. M. L. (1984b). Guanine nucleotides decrease the free $[Ca^{2+}]$ required for secretion of serotonin from permeabilized blood platelets: Evidence of a role for a GTP-binding-protein in platelet activation. *FEBS Lett.* **174**, 90–95.

Haywood, A. M. (1974). Characteristics of Sendai virus receptors in a model membrane. *J. Mol. Biol.* **83**, 427–436.

Haywood, A., and Boyer, B. (1984). Effect of lipid composition upon fusion of liposomes with Sendai virus membranes. *Biochemistry* **23**, 4161–4166.

Henne, V., Piiper, A., and Söling, H. -D. (1987). Inositol 1,4,5-trisphosphate and 5′-GTP induce calcium release from different intracellular pools. *FEBS Lett.* **218**, 153–158.

Honda, K., Maeda, Y., Sasakawa, S., Ohno, H., and Tsuchida, E. (1981). The components contained in polyethylene glycol of commercial grade (PEG-6,000) as cell fusogen. *Biochem. Biophys. Res. Commun.* **101**, 165–171.

Howell, T. W., and Gomperts, B. D. (1987). Rat mast cells permeabilized with streptolysin-O secrete histamine in response to Ca^{2+} at concentrations buffered in the micromolar range. *Biochim. Biophys. Acta* **927**, 177–183.

Howell, T. W., Cockcroft, S., and Gomperts, B. D. (1987). Essential synergy between Ca^{2+} and guanine nucleotides in exocytotic secretion from permeabilised mast cells. *J. Cell Biol.* **105**, 191–197.

Howell, T. W., Kramer, I., and Gomperts, B. D. (1988). Protein phosphorylation and the dependence on Ca^{2+} for GTPγS stimulated exocytosis from permeabilised mast cells. *Cell. Signalling* **1**, 157–163.

Imamura, K., Ohno, T., Spriggs, D., and Kufe, D. (1989). Effects of botulinum toxin type D on secretion of tumor necrosis factor from human monocytes. *Mol. Cell Biol.* **9**, 2239–2243.

Impraim, C. C., Foster, K. A., Micklem, K. J., and Pasternak, C. A. (1980). Nature of virally mediated changes in membrane permeability to small molecules. *Biochem. J.* **186**, 847–860.

Iyengar, R., Rich, K. A., Herberg, J. T., Grenet, D., Mumby, S., and Codina, J. (1987). Identification of a new GTP-binding protein. A $M_r = 43,000$ substrate for pertussis toxin. *J. Biol. Chem.* **262**, 9239–9245.

Johnson, G. S., and Mukku, V. R. (1979). Evidence in intact cells for an involvement of GTP in the activation of adenylate cyclase. *J. Biol. Chem.* **254**, 95–100.

Jones, P. M., Fyles, J. M., Persaud, S. J., and Howell, S. L. (1987a). Catecholamine inhibition of Ca^{2+}-induced insulin secretion from electrically permeabilised islets of Langerhans. *FEBS Lett.* **219**, 139–144.

Jones, P., Salmon, D. M. W., and Howell, S. L. (1987b). Effects of noradrenaline on protein phosphorylation in electrically permeabilised islets of Langerhans. *Diabetologia* **30**, 536A.

Kao, L. -S., and Schneider, A. S. (1986). Calcium mobilisation and catecholamine secretion in adrenal chromaffin cells: A quin-2 fluorescence study. *J. Biol. Chem.* **261**, 4881–4888.

Katoaka, T., Powers, S., McGill, C., Fasano, O., Strathern, J., Broach, J., and Wigler, M. (1984). Genetic analysis of yeast *RAS1* and *RAS2* genes. *Cell* **37**, 437–445.

Knight, D. E. (1986). Botulinum toxin types A, B and D inhibit catecholamine secretion from bovine adrenal medullary cells. *FEBS Lett.* **207**, 222–226.

Knight, D. E., and Baker, P. F. (1982). Calcium-dependence of catecholamine release from bovine adrenal medullary cells after exposure to intense electric fields. *J. Membrane Biol.* **68**, 107–140.

Knight, D. E., and Baker, P. F. (1985). Guanine nucleotides and Ca-dependent exocytosis. *FEBS Lett.* **189**, 345–349.

Knight, D. E., and Scrutton, M.C. (1986a). Gaining access to the cytosol: The technique and some applications of electropermeabilisation. *Biochem. J.* **234**, 497–506.

Knight, D. E., and Scrutton, M. C. (1986b). Effects of guanine nucleotides on the properties of 5-hydroxytryptamine secretion from electro-permeabilised human platelets. *Eur. J. Biochem.* **160**, 183–190.

Knight, D. E., Niggli, V., and Scrutton, M. C. (1984). Thrombin and activators of protein kinase C modulate secretory responses of permeabilized human platelets induced by Ca^{2+}. *Eur J. Biochem.* **143**, 437–446.

Koo, C., Lefkowitz, R., and Snyderman, R. (1983). Guanine nucleotides modulate the binding affinity of the oligopeptide chemoattractant receptor on human polymorphonuclear leukocytes. *J. Clin. Invest.* **72**, 748–753.

Lawson, D., Raff, M., Fewtrell, C., Gomperts, B., and Gilula, N. (1977). Molecular events during membrane fusion: A study of exocytosis in rat peritoneal mast cells. *J.Cell Biol.* **72**, 242–259.

LeNeveu, D. M., Rand, P. M., and Parsegian, V. A. (1976). Measurement of forces between lecithin bilayers. *Nature* **259**, 601–603.

Leslie, B. A., Putney, J. W., and Sherman, J. M. (1976). α-Adrenergic, β-adrenergic and cholinergic mechanisms for amylase secretion by rat parotid gland *in vitro*. *J. Physiol.* **260**, 351–370.

Lew, P. D., Monod, A., Waldwogel, F. A., DeWald, B., Baggiolini, M., and Pozzan, T. (1986). Quantitative analysis of the cytosolic free calcium dependency of exocytosis from three subcellular compartments in intact human neutrophils. *J. Cell Biol.* **102**, 2197–2204.

Lindau, M., and Nüsse, O. (1987). Pertussis toxin does not affect the time course of exocytosis in mast cells stimulated by intracellular application of GTPγS. *FEBS Lett.* **222**, 317–321.

Lo, W. W. Y., and Hughes, J. (1987). A novel cholera toxin-sensitive G-protein (G_c) regulating receptor-mediated phosphoinositide signalling in human pituitary clonal cells. *FEBS Lett.* **220**, 327–331.

Locke, F. S. (1894). Notiz über den Einflüss, physiologischer Kochsalzläsung auf die Errebarkeit von Muskel und Nerv. *Zentralblat. Physiol.* **8**, 166–167.

Marquardt, D. L., Gruber, H. E., and Walker, L. L. (1987). Ribavirin inhibits mast cell mediator release. *J. Pharmacol. Exp. Ther.* **240**, 145–149.

Marsh, M., and Quinn, P. (1988). Membrane cycling through the endocytic and exocytic pathways. *In* "Cellular Membrane Fusion" (J. Wilschut and D. Hoekstra eds.) Marcell Dekker, New York.

Marty, A., and Neher, E. (1983). Tight seal whole-cell recording. *In* "Single Channel Recording" (B. Sakmann and E. Neher, eds.) pp 107–121. (Plenum, New York).

Melançon, P., Glick, B. S., Malhotra, V., Weidman, P., Serafini, J. T., Gleason, M. L., Orci, L., and Rothman, J. E. (1987). Involvement of GTP-binding "G" proteins in transport through the Golgi stack. *Cell* **51**, 1053–1062.

Michell, R. H. (1975). Inositol phospholipids in cell surface receptor function. *Biochim. Biophys. Acta* **415**, 81–147.

Miller, M. R., Castellot, J. J., and Pardee, A. B. (1978). A permeable animal cell preparation for studying macro molecular synthesis, DNA synthesis and the role of deoxy ribonucleotides in S phase initiation. *Biochemistry* **17**, 1073–1080.

Momayezi, M., Lumpert, C. J., Kerksen, H., Gras, U., Plattner, H., Krinks, M. H., and Klee, C. B. (1987). Exocytosis induction in paramecium tetraurelia cells by exogenous phosphoprotein phosphatase *in vivo* and *in vitro*: Possible involvement of calcineurin in exocytotic membrane fusion. *J. Cell Biol.* **105**, 181–189.

Muff, R., and Fischer, J. A. (1986a). Parathyroid hormone secretion does not respond to changes of free calcium in electropermeabilised bovine parathyroid cells, but is stimulated with phorbol ester and cyclic AMP. *Biochem. Biophys. Res. Commun.* **139**, 1233–1238.

Muff, R., and Fischer, J. A. (1986b). Stimulation of parathyroid hormone secretion by phorbol esters is associated with a decrease of cytosolic calcium. *FEBS Lett.* **194**, 215–218.

Neher, E. (1988). The influence of intracellular calcium concentration on degranulation of dialysed mast cells from rat peritoneum. *J. Physiol.* **395**, 193–214.

Neher, E., and Almers, W. (1986). Fast calcium transients in rat peritoneal mast cells are not sufficient to trigger exocytosis. *EMBO J.* **5**, 51–53.

Neher, E., and Marty, A. (1982). Discrete changes of cell membrane capacitance observed under conditions of enhanced secretion in bovine adrenal chromaffin cells. *Proc. Nat. Acad. Sci. U.S.A.* **79**, 6712–6716.

Nemeth, E. F., and Scarpa, A. (1986). Cytosolic Ca^{2+} and the regulation of secretion in parathyroid cells. *FEBS Lett.* **203**, 15–19.

Nemeth, E. F., Wallace, J., and Scarpa, A. (1986). Stimulus-secretion coupling in bovine parathyroid cells: Dissociation between secretion and net changes in cytosolic Ca²⁺. *J. Biol. Chem.* **261**, 2668–2674.

Oetting, M., LeBoff, M., Swiston, L., Preston, J., and Brown, E. (1986). Guanine nucleotides are potent secretagogues in permeabilised parathyroid cells. *FEBS Lett.* **208**, 99–104.

Oetting, M., Leboff, M. S., Levy, S., Swiston, L., Preston, J., Chen, C., and Brown, E. M. (1987). Permeabilization reveals classical stimulus-secretion coupling in bovine parathyroid cells. *Endocrinology* **121**, 1571–1576.

Ohashi, Y., and Narumiya, S. (1987). ADP-ribosylation of a M_r21,000 membrane protein by type D Botulinum toxin. *J.Biol. Chem.* **262**, 1430–1433.

Orci, L., and Perrelet, A. (1978). Ultrastructural aspects of exocytotic membrane fusion. *Cell Surf. Rev.* **5**, 629–656.

Orci, L., Perrelet, A., and Friend, D. (1977). Freeze-fracture of membrane fusions during exocytosis in pancreatic B-cells. *J. Cell Biol.* **75**, 23–30.

Ornberg, R. L., and Reese, T. S. (1981). Beginning of exocytosis captured by rapid freezing of *Limulus* amebocytes. *J. Cell Biol.* **90**, 40–54.

Palade, G. (1975). Intracellular aspects of the process of protein synthesis. *Science* **189**, 347–358.

Parsegian, V. A., and Rau, D. C. (1984). Water near intracellular surfaces. *J. Cell Biol.* **99**, 196s–200s.

Parsegian, V. A., Rand, R. P., and Gingell, D. (1984). Lessons for the study of membrane fusion from membrane interactions in phospholipid systems. *Ciba Found. Symp.* **103**, 9–27.

Penner, R., Pusch, M., and Neher, E. (1987). Washout phenomena in dialyzed mast cells allow discrimination of different steps in stimulus-secretion coupling. *Biosci Rep.* **7**, 313–321.

Ringer, S. (1883). Regarding the action of the hydrate of soda, hydrate of ammonia, and hydrate of potash on the ventricle of the frog's heart. *J. Physiol.* **3**, 195–202.

Ringer, S. (1884). A further contribution regarding the influence of the different constituents of the blood on the contraction of the heart. *J. Physiol.* **4**, 29–42.

Rink, T., and Sanchez, A. (1984). Effects of prostaglandin I₂ and forskolin on the secretion from platelets evoked at basal concentrations of cytoplasmic free calcium by thrombin, collagen, phorbol ester and exogenous diacylglycerol. *Biochem. J.* **222**, 833–836.

Rink, T. J., Sanchez, A., and Hallam, T. J. (1983). Diacylglycerol and phorbol ester stimulate secretion without raising cytoplasmic free calcium in human platelets. *Nature* **305**, 317–319.

Rosener, S., Chhatwal, G. S., and Aktories, K. (1987). Botulinum ADP-ribosyltransferase C3 but not botulinum neurotoxins C1 and D ADP-ribosylates low molecular mass GTP-binding proteins. *FEBS Lett.* **224**, 38–42.

Sagi-Eisenberg, R., Lieman, H., and Pecht, I. (1985). Protein kinase C regulation of the receptor-coupled calcium signal in histamine-secreting rat basophilic leukaemia cells. *Nature* **313**, 59–60.

Sagi-Eisenberg, R., Foreman, J. C., Raval, P. J., and Cockcroft, S. (1987). Protein and diacylglycerol phosphorylation in the stimulus-secretion coupling of rat mast cells. *Immunology* **61**, 203–206.

Salminen, A., and Novick, P. J. (1987). A *ras*-like protein is required for a post-Golgi event in yeast secretion. *Cell* **49**, 527–538.

Sanchez-Prieto, J., Sihra, T., Evans, D., Ashton, A., Dolly, J. O., and Nicholls, D. G. (1987). Botulinum toxin A blocks glutamate exocytosis from guinea-pig cerebral cortical synaptosomes. *Eur. J. Biochem.* **165**, 675–681.

Schramm, M., and Selinger, Z. (1975). The functions of cyclic AMP and calcium as alternative second messengers in parotid gland and pancreas. *J. Cyclic Nucleotide Res.* **1**, 181–192.

Smith, C. D., Lane, B. C., Kusaka, I., Verghese, M. W., and Snyderman, R. (1985). Chemoattractant receptor induced hydrolysis of phosphatidylinositol 4,5-bisphosphate in human polymorphonuclear leukocytes membranes: Requirement for a guanine nucleotide regulatory protein. *J. Biol. Chem.* **260**, 5875–5878.

Smith, C. L., Ahkong, Q. F., Fisher, D., and Lucy, J. A. (1982). Is purified poly(ethyleneglycol) able to induce cell fusion? *Biochim. Biophys. Acta* **682**, 109–114.

Streb, H., Irvine, R., Berridge, M., and Schultz, I. (1983). Release of Ca^{2+} from a non-mitochondrial intracellular store in pancreatic acinar cells by inositol-1,4,5-trisphosphate. *Nature* **306**, 67–69.

Streb, H., Heslop, J., Irvine, R., Schulz, I., and Berridge, M. (1985). Relationship between secretogogue-induced Ca^{2+} release and inositol polyphosphate production in permeabilized pancreatic acinar cells. *J. Biol. Chem.* **260**, 7309–7315.

Stutchfield, J., and Cockcroft, S. (1988). Guanine nucleotides stimulate polyphosphoinositide phosphodiesterase and exocytotic secretion from HL-60 cells permeabilised with streptolysin O. *Biochem. J.* **250**, 375–382.

Swann, K., and Whitaker, M. (1986). The part played by inositol trisphosphate and calcium in the propagation of the fertilization wave in sea urchin eggs. *J. Cell Biol.* **103**, 2333–2342.

Takemura, H. (1985). Changes in free cytosolic calcium concentration in isolated rat parotid cells by cholinergic and β-adrenergic agonists. *Biochem. Biophys. Res. Commun.* **131**, 1048–1055.

Tamagawa, T., Niki, H., and Niki, A. (1985). Insulin release independent of a rise in cytosolic free Ca^{2+} by forskolin and phorbol ester. *FEBS Lett.* **183**, 430–432.

Tatham, P. E. R., and Gomperts, B. D. (1989). ATP inhibits onset of exocytosis in permeabilized mast cells. *Biosci. Reports* **9**, 99–109.

Toda, T., Uno, I., Ishikawa, T., Powers, S., Kataoka, T., Broek, D., Cameron, S., Broach, J., Matsumoto, K., and Wigler, M. (1985). In yeast, Ras proteins are controlling elements of adenylate cyclase. *Cell* **40**, 27–36.

Tsien, R. Y. (1980). New calcium indicators and buffers with selectivity against magnesium and protons: Design, synthesis and properties of prototype structures. *Biochemistry* **19**, 2396–2404.

Tsuruta, K., Grewe, C. W., Cote, T. E., Eskay, R. L., and Kebabian, J. W. (1982).

Coordinated action of calcium ion and adenosine 3', 5' -monophosphate upon the release of α-melanocyte-stimulating hormone from the intermediate lobe of the rat pituitary gland. *Endocrinology* **110**, 1133–1140.

Ullrich, S., and Wollheim, C. B. (1988). GTP-Dependent inhibition of insulin secretion by epinephrine in permeabilized RINm5F cells: Lack of correlation between insulin secretion and cyclic AMP levels. *J. Biol. Chem.* (in press).

Vallar, L., Biden, T. J., and Wollheim, C. B. (1987). Guanine nucleotides induce Ca^{2+} independent secretion from permeabilized RINm5F cells. *J. Biol. Chem.* **262**, 5049–5056.

Volpe, P., Krause, K. -H., Hashimoto, S., Zorzato, F., Pozzan, T., Meldolesi, J., and Lew, D. P. (1988). "Calciosome," a cytoplasmic organelle: The inositol 1,4,5-trisphosphate-sensitive Ca^{2+} store of non-muscle cells? *Proc. Nat. Acad. Sci. (U.S.A.)* **85**, 1091–1095.

Wilson, B., Deanin, G., Stump, R., and Oliver, J. (1988). Depletion of guanine nucleotides suppresses IgE-mediated degranulation in rat basophilic leukemia cells. *FASEB J.* **2**, A1236.

Wojcieszyn, J. W., Schlegel, R. A., Lumley-Sapanski, K., and Jacobson, K. A. (1983). Studies on the mechanism of polyethylene glycol-mediated cell fusion using fluorescent membrane and cytoplasmic probes. *J. Cell Biol.* **96**, 151–159.

Wollheim, C. B., Ullrich, S., Meda, P., and Vallar, L. (1987). Regulation of exocytosis in electrically permeabilized insulin-secreting cells: Evidence for Ca^{2+} dependent and independent secretion. *Biosci. Rep.* **7**, 443–454.

Wrenn, R. W. (1984). Phosphorylation of a pancreatic zymogen membrane protein by endogenous calcium/phospholipid-dependent protein kinase. *Biochim. Biophys. Acta* **775**, 1–6.

Wyke, A., Impraim, C., Knutton, S., and Pasternak, C. (1980). Components involved in virally mediated membrane fusion and permeability changes. *Biochem. J.* **190**, 625–638.

Yamamoto, T., Furuki, Y., Kebabian, J. W., and Spatz, M. (1987). α-Melanocyte-stimulating hormone secretion from permeabilized intermediate lobe cells of rat pituitary gland. *FEBS Lett.* **219**, 326–330.

Zieseniss, E., and Plattner, H. (1985). Synchronous exocytosis in *Paramecium* cells involves very rapid (≤ 1 s), reversible dephosphorylation of a 65-kD phosphoprotein in exocytosis-competent strains. *J. Cell Biol.* **101**, 2028–2035.

Zimmerberg, J. (1987). Molecular mechanisms of membrane fusion: Steps during phospholipid and exocytotic membrane fusion. *Biosci. Rep.* **7**, 251–268.

Zimmerberg, J., Curran, M., Cohen, F. S., and Brodwick, M. (1987). Simultaneous electrical and optical measurements show that membrane fusion precedes secretory granule swelling during exocytosis of beige mouse mast cells. *Proc. Nat. Acad. Sci. U.S.A.* **84**, 1585–1589.

INDEX